D1129665

INORGANIC BIOCHEMISTRY

INORGANIC BIOCHEMISTRY

VOLUME 2

Edited by

GUNTHER L. EICHHORN

National Institutes of Health, Gerontology Research Center,
Baltimore City Hospitals, Baltimore, U.S.A.

ELSEVIER SCIENTIFIC PUBLISHING COMPANY
AMSTERDAM — LONDON — NEW YORK
1973

ELSEVIER SCIENTIFIC PUBLISHING COMPANY
335 JAN VAN GALENSTRAAT
P.O. BOX 1270, AMSTERDAM, THE NETHERLANDS

AMERICAN ELSEVIER PUBLISHING COMPANY, INC.
52 VANDERBILT AVENUE
NEW YORK, NEW YORK 10017

LIBRARY OF CONGRESS CARD NUMBER: 72-135493

ISBN 0-444-41022-8

VOLUME 1: 114 ILLUSTRATIONS AND 81 TABLES

VOLUME 2: 200 ILLUSTRATIONS AND 58 TABLES

PRINTED IN THE NETHERLANDS

CONTENTS OF VOLUME 2

Part V. Oxidation—Reduction

Chapter 19. Oxidation—Reduction in Coordination Compounds . . 611

by N. SUTIN

Introduction 611
Theoretical considerations 612
Potential energy surfaces, 612; The resonance description, 615; The perturbation treatment, 616; Tunneling calculations, 617; The interaction energy, 618; Elementary steps, 619; Formation of the precursor complex, 622; Inner shell reorganization, 623; Solvent reorganization, 624
Comparisons with experimental data 626
Electron exchange reactions, 626; Cross reactions, 633; One-electron and two-electron transfers, 637
Bridging groups 639
Hydroxide and other inorganic anions, 639; Organic bridging groups, 641
Conformation- and substitution-controlled oxidation—reduction reactions . . . 643
Oxidative addition 645
Some further thoughts on biological systems 645
Notes added in proof 649
References 651

Chapter 20. Metal Ion Catalysis of Reactions of Molecular Oxygen . 654

by A.E. MARTELL and M.M. TAQUI KHAN

Introduction 654
Energetics of oxygen reduction, 656
I. Insertion reactions 656
A. Total insertion with O—O bond cleavage, 656
II. Partial insertion reactions 659
A. Peroxide as an oxidant, 659; B. Hydroxylation by direct reduction of molecular oxygen, 662
III. Non-insertion reactions in which oxygen is reduced to peroxide 669
A. Synthetic and natural oxygen carriers, 669; B. Model systems, 676; C. Enzymic systems, 679
IV. Disproportionation of hydrogen peroxide 684
References 686

Chapter 21. The Copper-containing Oxidases 689

by R. MALKIN

I. Introduction 689

II. The forms of copper in copper-containing oxidases 691
A. The "blue" Cu^{2+} ion, 691; B. The "non-blue" Cu^{2+} ion, 695; C. The "e.p.r. non-detectable" copper, 700
III. The catalytic activity of the copper-containing oxidases 703
A. Kinetic studies of laccase-catalyzed oxidations, 703; B. Possible roles of the different copper atoms in the oxidase reaction, 704
IV. Conclusion 706
Acknowledgements 706
Note added in proof 706
References 707

Chapter 22. Ferredoxins and Other Iron—Sulfur Proteins 710

by W.H. ORME-JOHNSON

I. Introduction 710
II. Nomenclature 711
III. 1-Fe—S and 2-Fe—S proteins 715
IV. 2-Fe—S* proteins 719
V. 4-Fe—S* proteins 731
VI. 8-Fe—S* proteins 734
VII. Concluding remarks 739
Acknowledgements 740
References 741

Chapter 23. Bioinorganic Chemistry of Dinitrogen Fixation 745

by R.W.F. HARDY, R.C. BURNS and G.W. PARSHALL

I. Introduction 745
II. Chemistry of dinitrogen 746
III. Biological dinitrogen fixation 748
A. Chemical and physical characteristics of N_2 ase, 749; B. Nitrogenase reaction, 754; C. Postulated mechanisms for biological N_2 fixation, 768; D. Ancillary roles of metals in biological N_2 fixation, 775
IV. Abiological dinitrogen fixation 776
A. Haber-Bosch process, 777; B. Homogeneous catalysis, 777; C. Transition metal complexes of N_2, 781; D. Non-catalytic N_2 fixation reactions, 786; E. Summary, 786
References 787

Part VI. Porphyrin Compounds and Related Ligands

Chapter 24. Iron Porphyrins — Hemes and Hemins 797

by W.S. CAUGHEY

Introduction 797

Structures of porphyrins of natural hemes 798
Derivatives of natural hemes 801
Heme *a* derivatives, 802; Derivatives of heme *b* and heme *s*, 803; Heme *c* derivatives, 804; Synthetic porphyrins, 804
Effects of porphyrin structure on properties 805
Hemins with one axial ligand, 818; Hemins with two *trans* axial ligands, 824; μ-Oxo-bis-hemins, 826; Iron(II) porphyrins, 827
References 829

Chapter 25. Hemoglobin and Myoglobin 832
by J.M. RIFKIND

Introduction 832
Coordination chemistry of hemoglobin and myoglobin 837
Geometry of the complex, 837; The spin state, 839; The requirement for iron(II), 841; Role of the porphyrin, 843; Role of the fifth ligand, 846
The iron—oxygen bond 848
Orientation of oxygen relative to the heme, 849; Electron transfer from iron to oxygen, 850
Reversible oxygenation 851
Environment of the heme, 851; Role of the protein in preventing autooxidation, 851
The binding of ligands to myoglobin 853
Role of the protein residues in the vicinity of the heme, 853; Involvement of the total protein conformation, 854
The binding of ligands to hemoglobin 858
Conformational changes produced by the binding of ligands to hemoglobin, 860; Functionally significant conformational changes, 871; Equilibria and kinetics of the binding of ligands to hemoglobin, 874; Interaction between subunits (heme—heme interaction), 883
Acknowledgement 891
References 891

Chapter 26. Cytochromes *b* and *c* 902
by H.A. HARBURY and R.H.L. MARKS

Cytochromes *c* 903
"Mammalian-type" cytochromes *c*, 904; Other cytochromes *c* of eukaryotes, 925; Cytochromes *c* of prokaryotes, 926
Cytochromes *b* 933
Cytochromes *b*, 933; Cytochromes b_5, 934; Cytochrome *b*-555, 937; Cytochrome *b*-562, 937; Cytochrome b_2, 939; Cytochromes P-450, 942
Note added in proof 946
Acknowledgement 946
References 947

Chapter 27. Cytochrome Oxidase 955

by D.C. WHARTON

Introduction 955
Purified preparations 957
Methods, 957; Molecular weight, 957; Composition, 959
Spectral properties 959
Absorption spectra, 959; Circular dichroism and optical rotatory dispersion, 962
Prosthetic groups 963
Heme *a*, 963; Properties of iron, 968; Properties of copper, 971; E.p.r. studies on copper, 974; Oxidation—reduction potentials, 976
Reaction of cytochrome oxidase 977
Reaction with cytochrome *c*, 978; Reaction with oxygen, 979; Reaction with copper, 981; Reactions with inhibitors, 981
Conclusions 982
Acknowledgements 983
References 983

Chapter 28. Peroxidases and Catalases 988

by B.C. SAUNDERS

Introduction 988
Peroxidase 988
Historical, 988
Occurrence 989
Isozymes, 990; Commission on Enzymes of the International Union of Biochemistry, 990
Properties of horseradish peroxidase (HRP) 990
Compounds of horseradish peroxidase (HRP) and hydrogen peroxide 992
Peroxidase oxidation of amines 993
Aniline, 993; *p*-Toluidine, 995; Mesidine, 996; 4-Methoxy-2,6-dimethylaniline, 997; Dimethylaniline, 997; *p*-Anisidine, 999; *p*-Chloroaniline, 999; Enzymic rupture of a C—F bond, 1000
Transiodination and related processes 1001
Transalkylation reactions 1004
Investigations into the benzidine blood test 1005
Peroxidic oxidation of mixtures 1008
Oxidation of phenols 1009
Enzymic oxidation of —CH_3 to —CHO, 1009; Durenol, 1009; Guaiacol, 1010
Hydroxyl radicals and the peroxidase—hydrogen peroxide system 1011
Mechanism of peroxidase action 1012
Peroxidase oxidations involving other inorganic materials 1014
Manganese, 1014; Chelated iron, 1015; Further recent applications of peroxidase, 1015
Catalases 1015
Historical, 1015; Mechanism of catalase action, 1016; Some inorganic catalases, 1018; A recently recorded joint action of catalase and peroxidase, 1018

Conclusion 1018
Acknowledgement 1019
References 1019

Chapter 29. Chlorophyll 1022

by J.J. KATZ

1.0 Introduction 1022
1.1 General, 1022; 1.2 Literature, 1024; 1.3 Properties of chlorophyll, 1024; 1.4 Coordination properties of magnesium in chlorophyll, 1025; 1.5 Spectroscopy of chlorophyll, 1026; 1.6 Chlorophyll as electron donor—acceptor, 1027
2.0 Chlorophyll—chlorophyll (endogamous) interactions 1028
2.1 Infra-red spectra in the 1800—1600 cm^{-1} (carbonyl) region, 1028; 2.2 Nuclear magnetic resonance spectra, 1031; 2.3 Chlorophyll—chlorophyll oligomers, 1037
3.0 Chlorophyll—ligand (exogamous) interactions 1042
3.1 Chlorophyll—ligand interactions from infra-red spectroscopy, 1043; 3.2 Chlorophyll—ligand interactions from n.m.r. spectroscopy, 1043; 3.3 Chlorophyll interactions with bifunctional ligands, 1045; 3.4 Chlorophyll—water interactions, 1046; 3.5 Chlorophyll—pheophytin interaction, 1049; 3.6 Equilibria between chlorophyll species, 1050
4.0 Long wavelength forms of chlorophyll 1054
4.1 Deconvolution of electronic transition spectra, 1054; 4.2 Electronic transition spectra of chlorophyll *a* dimers, 1055; 4.3 Electronic transition spectra of chlorophyll *a* oligomers, 1055; 4.4 Electronic transition spectra of chlorophyll—bifunctional ligand adducts, 1057; 4.5 Electronic transition spectra of *Tribonema aequalis*, 1058
5.0 Photo-activity of chlorophyll species 1058
5.1 E.s.r. of monomeric chlorophyll, 1059; 5.2 E.s.r. of chlorophyll dimer and oligomer, 1059; 5.3 E.s.r. of $(Chl{\cdot}H_2O)_n$, 1060; 5.4 E.s.r. of photosynthetic organisms, 1060; 5.5 E.s.r. linewidths, 1061; 5.6 Origin of signal I, 1062; 5.7 Model of the photosynthetic unit, 1062
Acknowledgement 1063
References 1063

Chapter 30. Corrinoids 1067

by H.A.O. HILL

Introduction 1067
Nomenclature 1068
General chemical features 1069
Comparative chemistry of cobalt(III) corrinoids 1071
Structural features, 1072; Some properties associated with cobalt and axial ligands, 1076; *Cis-* and *trans*-effects, 1079; Electronic absorption spectra, 1082; Cobalt(II) corrinoids, 1085; Cobalt(I) corrinoids, 1090
Reactions of cobalt corrinoids 1091
Reactions with sulfur-containing ligands, 1091; Reaction with carbon monoxide, 1093

Reactions of organocobalt complexes 1094
Photolysis, 1094; Thermolysis, 1095; Reductive cleavage, 1096; Nucleophilic displacement of carbon ligands, 1096
Reaction with electrophiles .1098
Reaction with iodine, 1098; Reaction with metal ions, 1099
Corrinoid-dependent enzymatic reactions 1099
Biosynthesis of 5′-deoxyadenosyl coenzymes, 1103; Methylmalonyl-CoA mutase, 1103; Glutamate mutase and α-methyleneglutarate, 1104; Dioldehydrase, 1105; Ethanolamine deaminase, 1108; L-β-Lysine mutase, 1109; Ribonucleotide reductase, 1110; Methonine synthetase, 1112; Methane synthetase and methyl transferase, 1114; Acetate synthetase, 1115
Mechanism of reactions requiring coenzyme B_{12} 1116
Acknowledgements 1120
Appendix 1120
References 1122

Part VII. Metal Interactions with other Prosthetic Groups

Chapter 31. Vitamin B_6 Complexes 1137

by R.H. HOLM

I. Introduction 1137
II. Types and structures of complexes 1140
III. Mechanism of formation 1144
IV. Reactions in model systems 1147
A. Reactions resulting from labilization of α-hydrogen, 1150;
B. Reactions resulting from labilization of the carboxyl group, 1159;
C. Reactions resulting from labilization of an R group, 1160;
D. Oxidative deamination, 1162
Abbreviations 1164
References 1165

Chapter 32. The Structure and Reactivity of Flavin—Metal Complexes 1168

by P. HEMMERICH and J. LAUTERWEIN

Introduction 1168
Charge-transfer chelates 1171
Flavosemiquinone metal chelates 1180
Conclusions about the biochemical function of flavins 1185
Acknowledgements 1189
References 1189

Chapter 33. Complexes of Nucleosides and Nucleotides 1191

by G.L. EICHHORN

Introduction 1191
Potential metal binding sites 1192
Stabilities of metal complexes of bases, nucleosides and nucleotides 1193
Structures of metal complexes of nucleosides and nucleotides 1195
Complexes of the bases, 1195; Adenine nucleoside and nucleotide complexes, 1197; Metal complexes of other nucleosides and nucleotides, 1203
Acknowledgement .1206
Note added in proof .1207
References 1207

Chapter 34. Complexes of Polynucleotides and Nucleic Acids . . 1210

by G.L. EICHHORN

Introduction 1210
Metal ions and DNA replication 1212
Metal ions and transcription 1214
Metal ions and translation 1216
Stability of metal complexes of polynucleotides 1217
Binding sites of metals on polynucleotides 1218
Effect of metal ions on DNA 1219
Stabilization, 1219; Unwinding and rewinding, 1220; Reaction of Hg(II) and Ag(I) with DNA, 1226
Effect of metal ions on RNA 1228
Stabilization, 1228; Stabilization and destabilization of different conformations by different metal ions, 1230; Destabilization of ordered structure of polyribonucleotides by metal ions, 1231; Depolymerization of RNA and polyribonucleotides by metal ions, 1232
Effect of metal ions on the enzymatic degradation of RNA and DNA 1235
Metal ions and nucleoproteins 1237
Acknowledgement .1239
Note added in proof .1239
References 1240

Index 1245

The following chapters are included in Volume 1.

Chapter 1. Structure and stereochemistry of coordination compounds
by D.A. BUCKINGHAM
Chapter 2. Stability of coordination compounds
by R.J. ANGELICI
Chapter 3. Electronic structures of iron complexes
by H.B. GRAY and H.J. SCHUGAR
Chapter 4. Metal complexes of amino acids and peptides
by H.C. FREEMAN
Chapter 5. Microbial iron transport compounds (siderochromes)
by J.B. NEILANDS
Chapter 6. Alkali metal chelators — the ionophores
by B.C. PRESSMAN
Chapter 7. Metal—protein complexes
by ESTHER BRESLOW
Chapter 8. Ferritin
by PAULINE M. HARRISON and T.G. HOY
Chapter 9. The transferrins (siderophilins)
by P. AISEN
Chapter 10. Ceruloplasmin
by I.H. SCHEINBERG and A.G. MORELL
Chapter 11. Hemerythrin
by M.Y. OKAMURA and I.M. KLOTZ
Chapter 12. Hemocyanin
by R. LONTIE and R. WITTERS
Chapter 13. Metal induced ligand reactions involving small molecules
by M.M. JONES and J.E. HIX, Jr.
Chapter 14. Metal enzymes
by M.C. SCRUTTON
Chapter 15. Carboxypeptidase A and other peptidases
by MARTHA L. LUDWIG and W.N. LIPSCOMB
Chapter 16. Carbonic anhydrase
by J.E. COLEMAN
Chapter 17. Phosphate transfer and its activation by metal ions; alkaline phosphatase
by T.G. SPIRO
Chapter 18. Kinases
by W.J. O'SULLIVAN

PART V

OXIDATION—REDUCTION

Chapter 19

OXIDATION–REDUCTION IN COORDINATION COMPOUNDS

N. SUTIN

Chemistry Department, Brookhaven National Laboratory, Upton, N.Y. 11973, U.S.A.

INTRODUCTION

This chapter is concerned with reactions in which one or more electrons are transferred between two metal centers. This class of reactions, called oxidation–reduction or redox reactions, has attracted a great deal of interest in recent years. Work in this area has led to the recognition of two types of oxidation–reduction reactions, called inner-sphere and outer-sphere reactions. The two metal centers are connected by a common bridging group in inner-sphere reactions, while the coordination shells of the metal centers remain intact during outer-sphere reactions. An example of an inner-sphere, chloride-bridged reaction is

$$(NH_3)_5CoCl^{2+} + Cr^{2+} \rightarrow Co^{2+} + CrCl^{2+} + 5NH_3 \qquad (1)$$

while the electron exchange* between the tris-phenanthroline complexes of iron(II) and iron(III) is an example of an outer-sphere reaction**.

$$Fe(phen)_3{}^{2+} + Fe(phen)_3{}^{3+} \rightleftharpoons Fe(phen)_3{}^{3+} + Fe(phen)_3{}^{2+} \qquad (2)$$

The major questions dealt with in this chapter are: What is the detailed nature of the individual steps leading from reactants to products? How do the natures of the metal centers, and the structures of the bridging groups affect the rate of the oxidation–reduction reaction? To what extent can the observations be interpreted in terms of simple theoretical models? No attempt has been made in this chapter to review the entire field of oxidation–reduction reactions involving coordination complexes. Instead the author has tried to select those aspects of inorganic oxidation–reduction reactions which may have biochemical significance. More comprehensive reviews of other aspects of inorganic oxidation–reductions are available[1–9].

---

* Oxidation–reduction reactions which involve the transfer of one or more electrons between two oxidation states of a single element are called electron exchange reactions. Since the reactants and products of electron exchange reactions are identical, the overall free energy change accompanying the oxidation–reduction reaction is zero.

** The following abbreviations are used in this article: phen (1,10-phenanthroline); en (ethylenediamine); cyt-c^{II} (ferrocytochrome *c*); cyt-c^{III} (ferricytochrome *c*).

References pp. 651–653

THEORETICAL CONSIDERATIONS

Consider the reaction:

$$A + B \rightarrow C + D \tag{3}$$

in which an electron is transferred from A to B. The progress of the reaction may be described in terms of the motion of a point representing the system on a potential energy surface. Such a potential energy surface may be constructed by solving the Schrödinger wave equation for the system. In order to solve this equation, the Born–Oppenheimer approximation is usually used. This approximation states that the nuclear motions in ordinary molecular vibrations are so slow that they do not affect the electronic states of molecules. This means that the potential energy of the system may be calculated for a set of fixed nuclear coordinates and then plotted as a function of these nuclear coordinates to give the potential energy surface for the reaction.

Potential energy surfaces

If ψ_a and ψ_b are the electronic wave functions of the reactants when they are infinitely far apart, then $\psi_i = \psi_a\psi_b$, $H_a\psi_a = E_a\psi_a$, $H_b\psi_b = E_b\psi_b$, and $(H_a + H_b)\psi_i = (E_a + E_b)\psi_i$, where ψ_i is the wave function for the system in its initial state, and E_a and E_b are the energies of the separated reactants. The zero-order potential energy of the state having the electronic configuration of the reactants is:

$$\int \psi_i H \psi_i d\tau = H_{ii} \tag{4}$$

where H is the total Hamiltonian operator for the system. In other words, $H = (H_a + H_b + H_{ab})$ where H_{ab} denotes the interaction terms and, unlike H_a and H_b, is a function of r_{ab}. The zero-order potential energy surface for the reactants may now be constructed by plotting H_{ii} as a function of the nuclear configuration of the reactants. This surface will have valleys corresponding to the more stable nuclear configurations of the reactants. Similarly, if ψ_c and ψ_d are the electronic wave functions of the products when they are infinitely far apart, then $\psi_f = \psi_c\psi_d$, $H_c\psi_c = E_c\psi_c$, $H_d\psi_d = E_d\psi_d$, and $(H_c + H_d)\psi_f = (E_c + E_d)\psi_f$, where ψ_f is the wave function for the system in its final state, and E_c and E_d are the energies of the separated products. The zero-order energy of the products is:

$$\int \psi_f H \psi_f d\tau = H_{ff} \tag{5}$$

where $H = (H_c + H_d + H_{cd}) = (H_a + H_b + H_{ab})$. The zero-order potential energy surface of the products may be constructed by plotting H_{ff} as a function of the nuclear configuration of the products. This surface will have its own valleys, corresponding to the more stable nuclear configurations of the products. Note that ψ_i and ψ_f are not true eigenfunctions of the system (unless H_{ab} and H_{cd} are equal to zero). A profile of the zero-order energy surfaces is shown in Fig. 1.

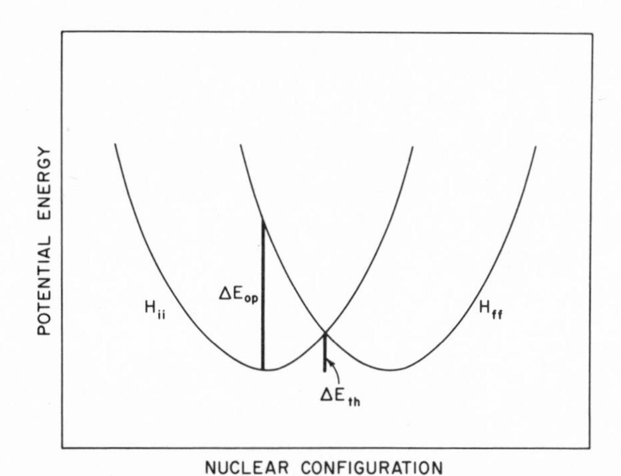

Fig. 1. Profile of the potential energy as a function of the nuclear configuration. ΔE_{op} is the energy for the photochemical electron transfer and ΔE_{th} is the activation energy for the thermal electron transfer.

The zero-order surfaces will intersect on a surface in the nuclear configuration space on which the reactants and products have the same nuclear configurations and the same energies. The degeneracy at the intersection will, in general, be removed by interaction between the initial and final states of the system, and two new surfaces will be formed. This is shown in Fig. 2. In the intersection region the system can no longer be described by the separate wave functions ψ_i and ψ_f. Instead it is necessary to form linear combinations of these wave functions. The true eigenfunctions for the system are*:

$$\psi_+ = c_1\psi_i + c_2\psi_f \tag{6}$$

and

$$\psi_- = c_2\psi_i - c_1\psi_f \tag{7}$$

where the coefficients c_1 and c_2 are functions of the nuclear coordinates.

* **Any contribution from excited states of the system is ignored.**

Also

$$H\psi_+ = E_+\psi_+, \qquad H\psi_- = E_-\psi_-$$

$$E_+ = \frac{(H_{ii} + H_{ff})}{2} + \frac{[(H_{ii} - H_{ff})^2 + 4H_{if}^2]^{\frac{1}{2}}}{2} \tag{8a}$$

and

$$E_- = \frac{(H_{ii} + H_{ff})}{2} - \frac{[(H_{ii} - H_{ff})^2 + 4H_{if}^2]^{\frac{1}{2}}}{2} \tag{8b}$$

Since ψ_i and ψ_f are eigenfunctions of the Hamiltonian operator for the unperturbed system, the only contribution to H_{if} arises from the interaction term in the total Hamiltonian operator.

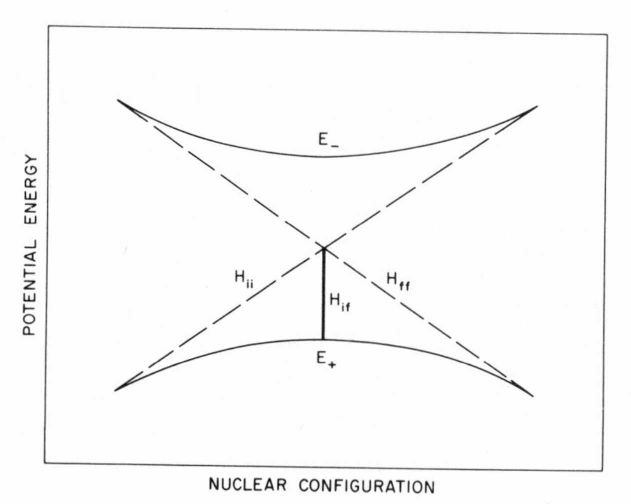

Fig. 2. Splitting at the intersection of the zero-order potential energy surfaces.

The intersection region (transition state) can be reached as a result of a suitable fluctuation in the nuclear coordinates of the reactants. Depending on the behavior of the system on reaching the intersection region, it is possible to distinguish two classes of reactions, called adiabatic and non-adiabatic reactions. In adiabatic reactions p, the probability of the reactants being converted into products in the transition state, is unity, and all the systems remain on the lower potential energy surface on passing through the intersection region. Thus in adiabatic reactions the potential energy of the system changes in a continuous manner from the energy of the reactants to that of the products, and the wave function changes continuously from ψ_i to ψ_f. In non-adiabatic reactions, on the other hand, p is less than unity and some of the systems jump to the upper potential energy surface on passing through the intersection region. Two approaches which have been used to describe the behavior of the system in the intersection region will be considered next.

The resonance description

In the intersection region the system can exist in two states. These states have the same nuclear configurations and the same energies, but one state has the electron configuration of the reactants while the other has the electron configuration of the products. Consequently the electron exchange in the intersection region can be described in terms of resonance between these two states. On the intersection surface $c_1 = c_2$ and $H_{ii} = H_{ff}$, therefore

$$\psi_+ = \frac{1}{(2)^{\frac{1}{2}}}(\psi_i + \psi_f), \qquad \psi_- = \frac{1}{(2)^{\frac{1}{2}}}(\psi_i - \psi_f) \tag{9}$$

and

$$E_+ = (H_{ii} + H_{if}), \qquad E_- = (H_{ii} - H_{if}) \tag{10}$$

The stationary state solutions to the wave equation including the time:

$$H\psi(x,t) = \frac{i\hbar\partial\psi(x,t)}{\partial t} \tag{11}$$

are

$$\psi_+(x,t) = \psi_+ \exp(-iE_+t/\hbar) \tag{12}$$

and

$$\psi_-(x,t) = \psi_- \exp(-iE_-t/\hbar) \tag{13}$$

The linear combination of these solutions:

$$\psi(x,t) = a_+\psi_+(x,t) + a_-\psi_-(x,t) \tag{14}$$

is also a solution to the wave equation including the time. However, this particular solution does not describe a stationary state of the system because in this case $\psi(x,t)\psi^*(x,t)$ is not independent of time. Since the system initially has the electron distribution appropriate to the reactants, a_+ and a_- are each equal to $1/(2)^{\frac{1}{2}}$. Therefore

$$\psi(x,t) = \frac{(\psi_i + \psi_f)}{2}\exp(-iE_+t/\hbar) + \frac{(\psi_i - \psi_f)}{2}\exp(-iE_-t/\hbar) \tag{15}$$

The probability distribution of the electron is given by:

$$\psi(x,t)\psi^*(x,t) = \psi_i^2\cos^2(E_- - E_+)t/2\hbar + \psi_f^2\sin^2(E_- - E_+)t/2\hbar \tag{16}$$

and thus

$$p = \sin^2(E_- - E_+)t/2\hbar \tag{17}$$

and the electron configuration changes from the configuration of the reactants to that of the products with a frequency given by:

$$\nu_e = \frac{(E_- - E_+)}{h} = \frac{2|H_{if}|}{h} \tag{18}$$

Evidently $\nu_e = 10^{13}\ \text{sec}^{-1}$ for interaction energies of only 0.5 kcal mole^{-1}. Since Δt, the time spent in the intersection region, is about 10^{-13} sec, it is apparent that $p \approx 1$ if the interaction energy is larger than about 500 cal. Moreover, $p = (\pi\nu_e\Delta t)^2$ for $\nu_e\Delta t$ small compared to $1/\pi$, in other words, in terms of this model the probability of electron transfer increases with the square of the time rather than linearly with the time, as might be more reasonable from a physical point of view. More important, perhaps, the above treatment focuses attention on the intersection region and glosses over the perturbation which brings the system to this region. This perturbation is a suitable fluctuation in the positions of the ligand and solvent nuclei.

The perturbation treatment

A better treatment of the intersection region is provided by first-order time-dependent theory. The total wave function for the system may be written as:

$$\psi(x,t) = c_i(t)\psi_i \exp(-iH_{ii}t/\hbar) + c_f(t)\psi_f \exp(-iH_{ff}t/\hbar) \tag{19}$$

At any time t_o there are $c_i^*(t_o)c_i(t_o)$ systems in state i and $c_f^*(t_o)c_f(t_o)$ systems in state f. The probability of electron transfer in the intersection region is given by the value of $c_f^*c_f$ at $t = \infty$, subject to the condition that at time $t = -\infty$, $c_i^*(-\infty)c_i(-\infty) = 1$ and $c_f^*(-\infty)c_f(-\infty) = 0$. If ψ_i and ψ_f are normalized and orthogonal, then

$$H\psi_i = H_{ii}\psi_i + H_{if}\psi_f \tag{20}$$

and

$$H\psi_f = H_{ff}\psi_f + H_{if}\psi_i \tag{21}$$

Substitution of these equations into the time-dependent Schrödinger wave equation:

$$\left(H - i\hbar\frac{\partial}{\partial t}\right)(c_i(t)\psi_i \exp(-iH_{ii}t/\hbar) + c_f(t)\psi_f \exp(-iH_{ff}t/\hbar)) = 0 \tag{22}$$

gives

$$i\hbar \frac{\partial c_i}{\partial t} = H_{if} \exp(-i(H_{ii} - H_{ff})t/\hbar)c_f \tag{23}$$

and

$$i\hbar \frac{\partial c_f}{\partial t} = H_{if} \exp(i(H_{ii} - H_{ff})t/\hbar)c_i \tag{24}$$

These two equations may be solved subject to the boundary condition given above with the aid of a few simplifying assumptions to give:

$$p = 1 - \exp\left[\frac{-4\pi^2 H_{if}{}^2}{hv|s_i - s_f|}\right] \tag{25}$$

a result first derived by Landau[10], and by Zener[11]. Evidently p is equal to:

$$\frac{4\pi^2 H_{if}{}^2}{hv|s_i - s_f|} \tag{26}$$

if the electron transfer probability is small. In the above expression s_i and s_f are the slopes of the zero-order surfaces at the intersection, and v is the velocity with which the point representing the system moves through the intersection region. It is noteworthy that if $v \sim 3 \times 10^4$ cm sec^{-1} [estimated from $v = (kT/m)^{\frac{1}{2}}$] and $|s_i| = |s_f| = 50 \times 10^8$ kcal mole^{-1} cm^{-1} (which is the slope of the barrier in a typical electron transfer reaction) then $p \approx 1$ provided $|H_{if}|$ is more than a few hundred calories. This estimate of the interaction energy required for an adiabatic reaction is consistent with the value obtained in the resonance treatment.

The limitations of the Landau–Zener expression have been discussed by Coulson and Zalewski[12]. An informative treatment of this problem has been given by Kauzmann[13].

Tunneling calculations

Tunneling formulae have also been used to calculate κ_e, the probability of the redox electron tunneling through the appropriate energy barrier. If ν_e' is the number of times that the electron strikes the barrier per second (about 10^{16} sec^{-1}), then the product electron configuration is formed from the reactant configuration $\nu_e'\kappa_e$ times per second. This frequency may be equated with the resonance frequency $2|H_{if}|/h$, and for this reason it has been suggested[14] that electron tunneling calculations actually serve as a crude method for estimating the interaction energy, which may then be substituted in the Landau–Zener formulation.

The interaction energy

It is evident from the previous discussion that the probability of electron transfer in the intersection region increases with the interaction energy. The interaction energy will be small when the separation between the reactants is large, when the electron transfer is accompanied by a change in spin multiplicity, or when the orbitals involved in the electron transfer do not have appropriate symmetries. Large overlap of the orbitals involved in the electron transfer will lead to large interaction energies and therefore to adiabatic reactions. If the ligands have filled or empty π orbitals ($p\pi$, $d\pi$, or π systems), these orbitals will interact with the *d* orbitals of the metal ion having t_{2g} symmetry. Ligands having empty π orbitals of higher energy than the metal t_{2g} orbitals may accept electrons from the metal ($M \rightarrow L$ π- or back-bonding), whereas ligands having filled π orbitals of lower energy than the metal t_{2g} orbitals may donate electrons to the metal ($L \rightarrow M$ π-bonding). The *d* orbitals of the central metal having e_g symmetry will overlap with the ligand orbitals having σ symmetry. Evidently electron transfer between t_{2g} orbitals (as occurs, for example, between low-spin iron(II) and iron(III) complexes, and between low-spin ruthenium(II) and ruthenium(III) complexes) will be favored by π-bridging orbitals, while electron transfer between e_g orbitals (as occurs between high-spin chromium(II) and chromium(III) complexes) will be favored by σ-bridging ligands. This is shown in Fig. 3. Detailed calculations of the interaction energy in these systems have not yet been published. Halpern and Orgel[15] have examined some aspects of electron transfer through conjugated bridging systems. They showed that ν_e, the frequency of electron transfer in the activated complex, is proportional to the mobile bond order of the conjugated system. Manning, Jarnagin, and Silver[16] have extended these calculations to include the effects of electrostatic factors on the lifetime of the activated complex.

In many oxidation–reduction systems of biological interest the overall reaction involves the transfer of more than one electron. There is theoretical and experimental evidence that the simultaneous transfer of two electrons in the reactions[17]:

$$Ne^{2+} + Ne \rightarrow Ne + Ne^{2+}$$

$$Ar^{2+} + Ar \rightarrow Ar + Ar^{2+}$$

is only about two to four times less probable than the corresponding one-electron transfers:

$$Ne^{+} + Ne \rightarrow Ne + Ne^{+}$$

$$Ar^{+} + Ar \rightarrow Ar + Ar^{+}$$

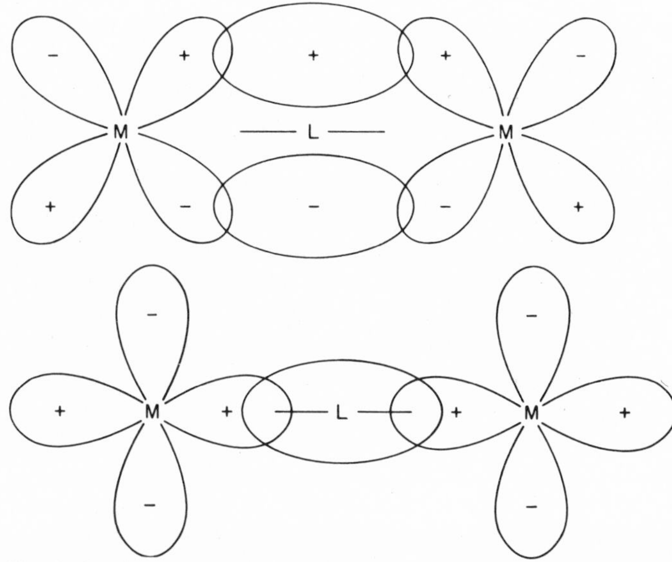

Fig. 3. Overlap of orbitals of different symmetry. Above, overlap of metal t_{2g} orbitals with ligand orbitals having π symmetry; below, overlap of metal e_g orbitals with ligand orbitals having σ symmetry.

However, as will be seen, additional factors are involved in comparing the rates of one- and two-electron transfers in more complex systems.

Elementary steps

The previous sections have focused attention on the frequency or probability of electron transfer in the intersection region, or transition state, for the reaction. We turn next to a consideration of how the system reaches the intersection region[14,18-20]. In general, the system reaches the intersection region in one or more steps. In outer-sphere reactions the steps leading from reactants to products are:

$$A + B \rightleftharpoons A\|B$$

$$A\|B \rightleftharpoons [A\|B]^*$$

$$[A\|B]^* \rightleftharpoons \overset{-}{A}\|\overset{+}{B}$$

$$\overset{-}{A}\|\overset{+}{B} \rightleftharpoons A^- + B^+$$

where || denotes that no chemical bonds have been made or broken. In inner-sphere reactions these steps are:

$$AX + B(H_2O) \rightleftharpoons AXB + H_2O$$

$$AXB \rightleftharpoons [AXB]^*$$

$$[AXB]^* \rightleftharpoons \overset{-}{A}X\overset{+}{B}$$

$$\overset{-}{A}X\overset{+}{B} + H_2O \rightleftharpoons A(H_2O)^- + BX^+$$

References pp. 651–653

The first step involves the formation of a precursor complex from the separated reactants. The distance between the centers of the reactants in the precursor complex is approximately the same as in the activated complex. No chemical bonds are made or broken in forming the precursor complex in outer-sphere reactions, whereas substitution (or addition) is involved in forming the precursor complex in inner-sphere reactions. Reorganization of the precursor complex to form the activated complex occurs in the second step. In this step the inner coordination shells of the reactants, as well as the polarization of the surrounding medium, adjust to the configuration appropriate to the activated complex. The electron transfer may, and usually does, take place during the latter stages of the reorganization of the precursor complex. The third step involves the deactivation of the activated complex to form the successor complex (or to re-form the precursor complex if no electron transfer had occurred in the activated complex). The electron distribution in the precursor complex corresponds to that of the reactants, and the electron distribution in the successor complex corresponds to that of the products of the reaction. Dissociation of the successor complex to form the separated products occurs in the fourth step. Although transfer of the bridging group usually accompanies the dissociation of the successor complex in an inner-sphere mechanism, such transfer need not necessarily occur and is not an essential feature of an inner-sphere reaction.

A profile of the reaction is shown in Fig. 4. States i and f are the initial and final states of the system, while p and s are the precursor and successor states, respectively[20]. State t is the transition state for the reaction. The upper curve in Fig. 4 corresponds to the "normal" reaction described above, namely a reaction in which the reorganization of the precursor complex is rate determining. The lower curve, by contrast, describes a reaction in which the formation of the precursor complex is rate determining.

In terms of the above scheme, the rate constant for an oxidation–reduction reaction (in which the formation of the precursor complex is not rate-determining) is given by:

$$k = K_o k_p \tag{27}$$

$$= K_o \nu_p \exp(-\Delta G^{**}/RT) \tag{28}$$

where K_o is the equilibrium constant for the formation of the precursor complex, k_p is the first-order rate constant for the conversion of the precursor to the successor complex, ν_p is a frequency factor to be discussed later, and $\Delta G^{**} = (\Delta G_i^* + \Delta G_o^*)$ is the free energy required to reorganize the coordination shells of the reactants prior to the electron transfer. ΔG_i^* is the energy required to reorganize the inner coordination shells of the reactants before the electron transfer, and ΔG_o^* is the energy required to reorganize the surrounding medium.

We have seen that when the interaction energy is relatively small the system in the intersection region can exist in two states. These two states

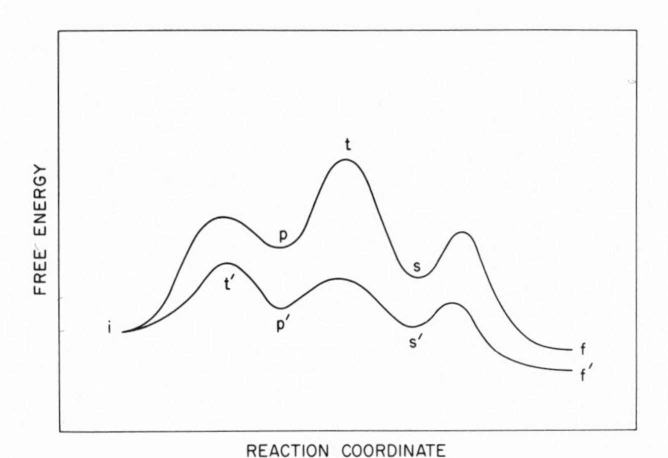

Fig. 4. Profile of the free energy as a function of the reaction coordinate. States i and f are the initial and final states of the reaction, and states p and s are the precursor and successor states, respectively. State t is the transition state for the reaction. The upper curve describes a "normal" electron transfer reaction, while the lower curve describes an oxidation–reduction reaction in which the formation of the precursor complex is rate-determining.

have the same nuclear configurations, but one state has the electron configuration of the reactants, p^*, and the other the electron configuration of the products, s^*. Following Marcus[18], the two states p^* and s^* may be regarded as two activated complexes for the reaction, and the conversion of the precursor complex to the successor complex can be considered to occur in the following sequence:

$$p \underset{\nu_n}{\overset{k_2}{\rightleftharpoons}} p^*$$

$$p^* \underset{\nu_e}{\overset{\nu_e}{\rightleftharpoons}} s^*$$

$$s^* \overset{\nu_n'}{\rightleftharpoons} s$$

The steady state assumption for the concentrations of p^* and s^* gives

$$k_p = \frac{k_2}{[1 + \nu_n/\nu_e(1 + \nu_e/\nu_n')]} \tag{29}$$

where ν_n and ν_n' are typical frequencies for nuclear motion ($\sim 10^{13}$ sec^{-1}). Since k_2 is given by

$$k_2 = \nu_n'' \exp(-\Delta G^{**}/RT)$$

it follows that

$$k_p = \frac{\nu_e \nu_n}{\nu_n + 2\nu_e} \exp(-\Delta G^{**}/RT) \tag{30}$$

provided $\nu_n'' \approx \nu_n' \approx \nu_n$. Two limiting cases may be distinguished.

(a) When $\nu_n \gg \nu_e$, in other words, when the probability of electron transfer in the activated complex is very low, then

$$k_p = \nu_e \exp(-\Delta G^{**}/RT) \tag{31}$$

and $\nu_p = \nu_e$. Evidently this is the case for interaction energies of less than about 0.2 kcal mole^{-1}. Under these conditions the reaction is nonadiabatic.

(b) When $\nu_n \ll \nu_e$, in other words, when the frequency of electron transfer in the activated complex is very high, then

$$k_p = \frac{\nu_n}{2} \exp(-\Delta G^{**}/RT) \tag{32}$$

Under these conditions $\nu_p = \nu_n/2$, and a description of the system in terms of two activated complexes is no longer appropriate. The oxidation–reduction reaction now resembles rather closely an ordinary reaction in which the reaction is consummated by a nuclear motion rather than by the transfer of the redox electron. Evidently the rate constant for the reaction no longer depends on the electron transfer frequency, and the reaction is adiabatic.

A somewhat different expression for the rate constant for an oxidation–reduction reaction has been derived by Marcus[14]. This expression is:

$$k = pZ \exp(-U(r)/RT) \exp(-\Delta G^{**}/RT) \tag{33}$$

where Z is the collision frequency between two uncharged reactants in solution ($\sim 10^{11}$ M^{-1} sec^{-1}), $U(r)$ is the work required to bring the reactants from an infinite distance apart to their separation in the activated complex, and p, as we have seen, is equal to unity in adiabatic reactions and is less than unity is nonadiabatic reactions.

Formation of the precursor complex

Usually only electrostatic factors are considered in the formation of the precursor complex in outer-sphere reactions. Approximate values of the equilibrium constant for the formation of an outer-sphere precursor complex may be calculated from the following expressions:

$$K_o = \frac{4\pi N r^3}{3000} \exp(-U(r)/RT) \tag{34}$$

$$U(r) = \frac{q_1 q_2}{D_s r(1 + \kappa r)} \tag{35}$$

where r is the distance between the centers of the reactants, $U(r)$ is the Debye–Hückel interaction potential, q_1 and q_2 are the charges on the reactants, D_s is the (static) dielectric constant of the medium, and κ is the Debye–Hückel ionic strength parameter.

The value of the equilibrium constant for the formation of the precursor complex in an inner-sphere reaction is probably best estimated by assuming that it is equal to the equilibrium constant for the formation of an analogous binuclear complex in a substitution reaction which is not followed by an electron transfer step.

Inner shell reorganization

The calculation of the inner shell reorganization energy will be illustrated by considering the electron exchange between $Fe(H_2O)_6^{2+}$ and $Fe(H_2O)_6^{3+}$ ions.

$$\underset{1}{Fe(H_2O)_6^{2+}} + \underset{2}{Fe(H_2O)_6^{3+}} \rightleftharpoons \underset{1'}{Fe(H_2O)_6^{3+}} + \underset{2'}{Fe(H_2O)_6^{2+}}$$

The following symbols will be used: $r_1 = r_2'$ = equilibrium Fe^{2+}–O distance, $r_2 = r_1'$ = equilibrium Fe^{3+}–O distance, r_1^* = the Fe^{2+}–O distance in the transition state, r_2^* = the Fe^{3+}–O distance in the transition state, $\Delta r^\circ = (r_2' - r_2) = (r_1 - r_1')$, $\Delta r_2^* = (r_2^* - r_2)$, $\Delta r_1^* = (r_1 - r_1^*)$. If it is assumed that the vibrations of the hydration shells of the ions are harmonic, then the potential energy of the reorganized reactants before the electron transfer is:

$$\Delta E^*_{(\text{before})} = \frac{f_1}{2}(\Delta r_1^*)^2 + \frac{f_2}{2}(\Delta r_2^*)^2$$

where f_1 and f_2 are the force constants of the Fe^{2+}–O and Fe^{3+}–O bonds respectively. After the electron transfer the potential energy of the products is:

$$\Delta E^*_{(\text{after})} = \frac{f_1}{2}(\Delta r^\circ - \Delta r_2^*)^2 + \frac{f_2}{2}(\Delta r^\circ - \Delta r_1^*)^2$$

Since the vibrational energy is conserved during the transition, $E_{\text{(before)}} = E_{\text{(after)}}$, and therefore

$$\Delta r^\circ = \Delta r_1^* + \Delta r_2^*$$

or $\quad r_1^* = r_2^*$

Evidently the radii of the $Fe(H_2O)_6{}^{2+}$ and $Fe(H_2O)_6{}^{3+}$ ions adjust to the same value prior to the electron transfer[21]. If $Q_{\text{vib}}^* = Q_{\text{vib}}$, where Q_{vib}^* and Q_{vib} are the vibrational partition functions of the activated complex and of the reactants, respectively, then $\Delta G_i^* = \Delta E^*$. In other words, the free energy required to reorganize the inner coordination shells of the reactants is given by:

$$\Delta G_i^* = \frac{f_1}{2} (\Delta r^\circ - \Delta r_2^*)^2 + \frac{f_2}{2} (\Delta r_2^*)^2$$

For minimum reorganization energy, $(\partial \Delta G_i^*)/(\partial \Delta r_2^*) = 0$, and therefore

$$\Delta r_2^* = \frac{f_1}{(f_1 + f_2)} \Delta r^\circ \tag{36}$$

$$\Delta G_i^* = \frac{f_1 f_2}{2(f_1 + f_2)} (\Delta r^\circ)^2 \tag{37}$$

Evidently the inner sphere reorganization energy can be calculated if the appropriate force constants and bond distances are known. A different expression for the potential energy of the inner coordination shells can be used, but this leads to a more complicated expression for the reorganization energy[7].

Solvent reorganization

In order to derive an expression for the solvent reorganization energy, the medium outside the inner shells of the reactants is treated as a dielectric continuum having electronic, atomic, and orientation polarization[19]. The total polarization of the medium is equal to the sum of these contributions. The work required to charge a conducting sphere of radius a in such a medium is equal to $q^2/2aD_s$, where q is the charge on the sphere. In the case of two ions separated by a distance r, the free energy of the system is:

$$G_{12} = \frac{q_1{}^2}{2a_1 D_s} + \frac{q_2{}^2}{2a_2 D_s} + \frac{q_1 q_2}{r D_s} \tag{38}$$

where q_1 and q_2 are the charges on the ions, and a_1 and a_2 are the ionic radii. The first two terms are the solvation energies of the separated ions, and the third term is the interaction energy. Similarly the work required to charge a conducting sphere in a medium which responds only through electronic polarization is equal to $q^2/2aD_{op}$, where D_{op} is the dielectric constant of the medium at optical frequencies. D_{op} measures the electronic polarizability of the medium and is equal to the square of the refractive index. The energy of a charged sphere due to the (equilibrium) atomic and orientation polarization of the medium is equal to the difference between these two work terms. In other words, this energy is given by:

$$G = \frac{q^2}{2a}\left(\frac{1}{D_{op}} - \frac{1}{D_s}\right) \tag{39}$$

Consider next the electron exchange reaction. In the initial state the polarization (electronic, atomic, and orientation) of the medium is in equilibrium with the charges q_1 and q_2 on the reactants. Since the solvent electrons move very rapidly, the electronic polarization of the medium can change in phase with the transferring electron. Consequently the electronic polarization of the medium need not adjust to some non-equilibrium value prior to the electron transfer. In the zero interaction approximation, the charge distribution in the transition state is the same as that existing on the reactants. Consequently, within this approximation the electronic polarization of the medium (in the transition state as well is in the initial state) is in equilibrium with the charges on the reactants. On the other hand, the atomic and orientation polarization relaxes too slowly to change in phase with the transferring electron. Energy conservation in the electron transfer reaction therefore requires that the atomic and orientation polarization of the medium change to a suitable, non-equilibrium value prior to the electron transfer. By using arguments analogous to those employed in calculating the inner shell reorganization, it can be shown that in the case of exchange reactions this non-equilibrium atomic and orientation polarization is in equilibrium with the hypothetical charges $(q_1 + n/2)$ and $(q_2 - n/2)$, where $n = (q_2 - q_1)$. In the transition state for an exchange reaction the atomic and orientation polarization of the medium is thus appropriate to a system in which a charge $n/2$ has been transferred from q_2 to q_1. It can be shown that the energy required to form the non-equilibrium atomic and orientation polarization appropriate to the transition state is simply equal to the work required to place charges $+n/2$ and $-n/2$ on two initially uncharged spheres of radius a_1 and a_2 a distance r apart in a medium which responds solely by atomic and orientation polarization. As before, this work is equal to the Born charging energy of the separated spheres:

$$\frac{(n/2)^2}{2a_1}\left(\frac{1}{D_{op}} - \frac{1}{D_s}\right) + \frac{(-n/2)^2}{2a_2}\left(\frac{1}{D_{op}} - \frac{1}{D_s}\right)$$

plus the energy required to bring the two charged spheres to a separation distance r in this medium:

$$\frac{(n/2)(-n/2)}{r}\left(\frac{1}{D_{op}}-\frac{1}{D_s}\right)$$

The value of ΔG_o^* is given by the sum of these terms:

$$\Delta G_o^* = \frac{n^2}{4}\left(\frac{1}{2a_1}+\frac{1}{2a_2}-\frac{1}{r}\right)\left(\frac{1}{D_{op}}-\frac{1}{D_s}\right) \tag{40}$$

Evidently the energy required for solvent reorganization is about four times as large for a two-electron transfer reaction as for a reaction in which one electron is transferred, provided all other factors are equal. If $r = (a_1 + a_2)$ and a_1 and a_2 are not too different, then the above equation reduces to:

$$\Delta G_o^* = \frac{n^2}{4r}\left(\frac{1}{D_{op}}-\frac{1}{D_s}\right) \tag{41}$$

It should be remembered that the expressions derived so far for ΔG_i^* (eqn. 37) and ΔG_o^* (eqn. 40) are valid only for exchange reactions ($\Delta G^\circ = 0$) which are outer-sphere, and in which the interaction between the redox orbitals of the reactants is so small (< 0.5 kcal mole^{-1}) that it can be neglected in calculating the free energy of activation for the reaction.

COMPARISONS WITH EXPERIMENTAL DATA

The previous discussion has indicated that four factors are important in determining the rates of electron exchange reactions. These are the interaction energy, the stability of the precursor complex, the inner shell reorganization energy, and the solvent reorganization energy. An attempt will next be made to interpret some measured rate constants in terms of these concepts.

Electron exchange reactions

Second-order rate constants for a number of electron exchange reactions are presented in Table I. It is apparent from this Table that replacing the water molecules coordinated to iron(II) and iron(III) by phenanthroline groups greatly increases the rate of electron exchange between iron(II) and iron(III). On the other hand, a similar coordination shell replacement in the cobalt(II)–cobalt(III) system barely affects the rate of electron exchange between cobalt(II) and cobalt(III). However, a

TABLE I

SECOND ORDER RATE CONSTANTS FOR SOME ELECTRON EXCHANGE REACTIONS

Reaction	T (°C)	k (M^{-1} sec^{-1})	Ref.
$Fe(H_2O)_6^{2+}$ + $Fe(H_2O)_6^{2+}$ $(t_{2g})^4(e_g)^2$ $(t_{2g})^3(e_g)^2$	25	4.0	22
Cyt-c^{II} + Cyt-c^{III} $(t_{2g})^6$ $(t_{2g})^5$	?	5×10^4	23
$Fe(phen)_3^{2+}$ + $Fe(phen)_3^{3+}$ $(t_{2g})^6$ $(t_{2g})^5$	25	$\geqslant 3 \times 10^7$	24
$Ru(NH_3)_6^{2+}$ + $Ru(ND_3)_6^{3+}$ $(t_{2g})^6$ $(t_{2g})^5$	25	8.2×10^2	25
$Ru(phen)_3^{2+}$ + $Ru(phen)_3^{3+}$ $(t_{2g})^6$ $(t_{2g})^5$	25	$\geqslant 10^7$	25[a]
$Co(H_2O)_6^{2+}$ + $Co(H_2O)_6^{3+}$ $(t_{2g})^5(e_g)^2$ $(t_{2g})^6$	25	~5	26
$Co(NH_3)_6^{2+}$ + $Co(NH_3)_6^{3+}$ $(t_{2g})^5(e_g)^2$ $(t_{2g})^6$	64.5	$\leqslant 10^{-9}$	27
$Co(en)_3^{2+}$ + $Co(en)_3^{3+}$ $(t_{2g})^5(e_g)^2$ $(t_{2g})^6$	50	1.4×10^{-4}	28
$Co(phen)_3^{2+}$ + $Co(phen)_3^{3+}$ $(t_{2g})^5(e_g)^2$ $(t_{2g})^6$	0	1.1	29

[a] The rate constant for the $Ru(phen)_3^{2+}$ + $Ru(phen)_3^{3+}$ exchange is an estimated value based on the expectation that the $Ru(phen)_3^{2+}$ + $Ru(phen)_3^{3+}$ and $Fe(phen)_3^{2+}$ + $Fe(phen)_3^{3+}$ reactions will have similar rate constants. See also footnote 35 of ref. 25.

large decrease in the electron exchange rate is found in the cobalt(II)–cobalt(III) system when the coordinated water molecules are replaced by ethylenediamine groups. Note also that the electron exchange between $Ru(NH_3)_6^{2+}$ and $Ru(NH_3)_6^{3+}$ is some 10^{12} times faster than the exchange between $Co(NH_3)_6^{2+}$ and $Co(NH_3)_6^{3+}$. These effects are illustrated in Fig. 5.

The iron(II)–iron(III) and ruthenium(II)–ruthenium(III) reactions will be considered first. The redox electron in these systems is in a t_{2g} orbital, which is essentially nonbonding (see Fig. 6). The observed and calculated rate constants for the $Fe(H_2O)_6^{2+}$–$Fe(H_2O)_6^{3+}$ and $Fe(phen)_3^{2+}$–$Fe(phen)_3^{3+}$ reactions are compared in Table II. The inner shell reorganization energy for the hydrated ions was calculated from the expression given by Hush[7,20] since the appropriate force constants are not known. The inner shell reorganization energy for the $Fe(phen)_3^{2+}$–$Fe(phen)_3^{3+}$ reaction

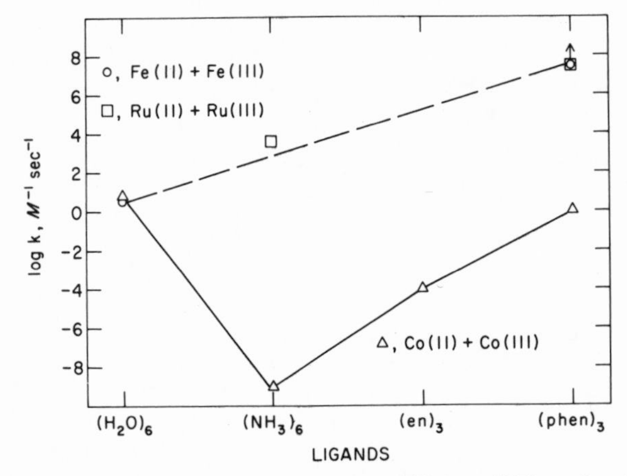

Fig. 5. Reactivity pattern for iron(II)–iron(III), ruthenium(II)–ruthenium(III) and cobalt(II)–cobalt(III) exchange reactions.

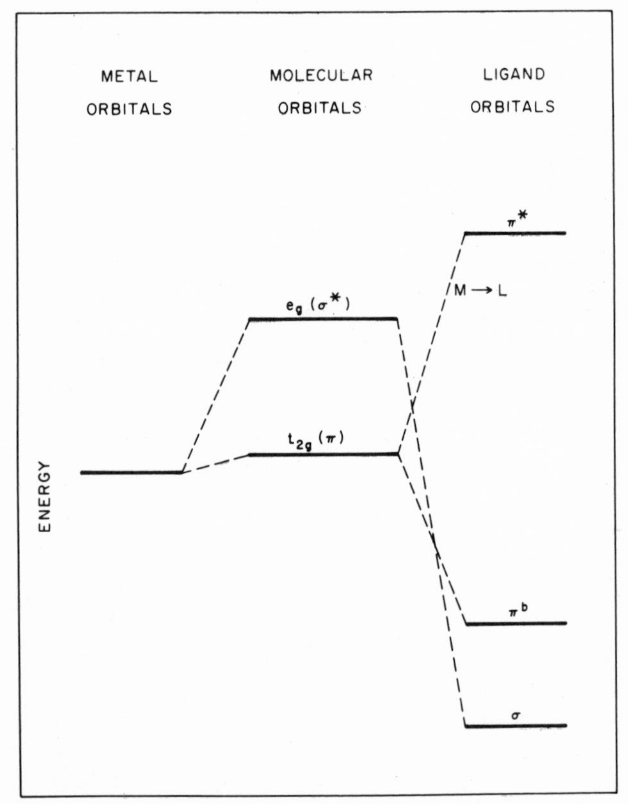

Fig. 6. Part of a molecular orbital energy level diagram for octahedral metal complexes containing ligands which have π^b and relatively stable π^* orbitals.

TABLE II

COMPARISON OF OBSERVED AND CALCULATED RATE CONSTANTS OF EXCHANGE REACTIONS

Parameter	$Fe(H_2O)_6^{2+} + Fe(H_2O)_6^{3+}$	$Fe(phen)_3^{2+} + Fe(phen)_3^{3+}$
$\nu_p (sec^{-1})$	10^{13}	10^{13}
r (Å)	7	14
$K^\circ (M^{-1})$	1.9×10^{-3}	3.3×10^{-1}
ΔG_o^* (kcal mole^{-1})	6.4	3.2
ΔG_i^* (kcal mole^{-1})	6.5	3.25
$k_{calc} (M^{-1} sec^{-1})$	6	6×10^7
$k_{obs} (M^{-1} sec^{-1})$	4.0	$\geqslant 3 \times 10^7$

has been rather arbitrarily assumed to be equal to half the inner shell reorganization energy for the $Fe(H_2O)_6^{2+}$–$Fe(H_2O)_6^{3+}$ exchange because the reorganization energy for the former reaction will be reduced by $M \rightarrow L$ π-bonding. Since the added electron in $Fe(phen)_3^{2+}$ will be distributed over the empty π^* orbitals of the ligand, the configurations of $Fe(phen)_3^{2+}$ and $Fe(phen)_3^{3+}$ will tend to become similar. It is apparent from Table II that the agreement between the observed and calculated rates is quite good. The very rapid rate of the $Fe(phen)_3^{2+}$–$Fe(phen)_3^{3+}$ exchange is seen to be due, in part, to the large size of the phenanthroline groups, which leads to the formation of a relatively stable precursor complex and to a relatively low solvent reorganization energy, and, in part, to the π-accepting ability of the phenanthroline groups, which has been assumed to lead to a low inner shell reorganization energy. If the inner shell reorganization energy for the $Fe(phen)_3^{2+}$–$Fe(phen)_3^{3+}$ exchange turns out to be higher than has been assumed above, then it would indicate that the π-orbital system of the phenanthroline groups provides such a strong coupling between the iron atoms that the electronic interaction substantially lowers the activation energy★.

It is apparent from Table I that the $Ru(NH_3)_6^{2+}$–$Ru(NH_3)_6^{3+}$ exchange proceeds somewhat faster than the $Fe(H_2O)_6^{2+}$–$Fe(H_2O)_6^{3+}$ reaction. The increased rate of the ruthenium reaction is not unexpected in view of the larger size of the reactants (the radii of Ru^{3+}, Fe^{3+}, H_2O, and NH_3 are 0.70, 0.64, 1.38 and 1.40 Å, respectively). Since both H_2O and NH_3 are non-π-bonding ligands, the inner shell reorganization energy should be very similar for the $Ru(NH_3)_6^{2+}$–$Ru(NH_3)_6^{3+}$ and $Fe(H_2O)_6^{2+}$–$Fe(H_2O)_6^{3+}$ reactions. Similarly, the $Ru(phen)_3^{2+}$–$Ru(phen)_3^{3+}$ reaction is likely to

★ The activation energy is lowered as a consequence of the large splitting at the intersection.

be rapid for the same reasons that the $Fe(phen)_3^{2+}$–$Fe(phen)_3^{3+}$ reaction is rapid.

At first sight the cytochrome-c^{II}–cytochrome-c^{III} * reaction seems somewhat slower than might be expected on the basis of the π-bonding ability of the porphyrin system. Recent crystallographic data[30] confirm that the heme group in cytochrome-*c* from horse heart is situated in a crevice and indicate that the fifth coordination site on the iron is occupied by histidine-18 while the remaining coordination site is occupied by methionine-80 (Chapter 26). Because of their locations and relatively small sizes the two heme groups will not be brought close together on every collision of the two reactants. This will introduce a steric factor *S*. The crystallographic data show that the cytochrome-*c* molecule is a prolate spheroid approximately 25 x 25 x 37 Å. The surface area of cytochrome-*c* calculated from these dimensions is 2.6×10^3 Å^2. The effective area of the heme group is about 30 to 100 Å^2 so that the active site of cytochrome-*c* occupies only about 3% of the total surface area of the molecule. Consequently the value of *S* for the cytochrome-c^{II}–cytochrome-c^{III} exchange reaction is estimated to be about 10^{-3}. Multiplication of the rate constant for the $Fe(phen)_3^{2+}$–$Fe(phen)_3^{3+}$ exchange reaction ($\sim 10^8\ M^{-1}\ sec^{-1}$) by this estimate of *S* gives $1 \times 10^5\ M^{-1}\ sec^{-1}$ in excellent agreement with the observed rate constant for the cytochrome-c^{II}–cytochrome-c^{III} exchange reaction**. It should be noted that the location of the heme group could also lead to relatively poor overlap of the redox orbitals of the two reactants, and this factor together with the poor "electron conductivity" of the polypeptide chain could make the cytochrome-c^{II}–cytochrome-c^{III} exchange reaction nonadiabatic. Nonadiabaticity may also be responsible for the extremely slow electron exchange rates observed in certain synthetic hemoprotein systems[31].

This may be a suitable place to consider whether it is appropriate to assume that a molecule such as cytochrome-*c* can be regarded simply as a large coordination complex. One might also ask whether the theoretical expressions which were derived for relatively simple systems are applicable

---

* **Editor's note.**—This nomenclature follows inorganic practice. In other chapters (26 and 27) the more usual biochemical names, ferrocytochrome *c* and ferricytochrome *c* are used.

** Although this "calculation" is done somewhat "tongue-in-cheek", the agreement of the observed and calculated rate constants is encouraging. The calculation assumes, of course, that the steric factor is mainly responsible for the difference in the rates of the $Fe(phen)_3^{2+}$–$Fe(phen)_3^{3+}$ and cytochrome-c^{II}-cytochrome-c^{III} exchange reactions and that other effects tend to cancel. Electrostatic effects (see, for example, eqn. 34) are almost certainly different for the two reactions. Similarly, the collision frequency (eqn. 33) will be about four times larger for the cytochrome-c^{II}-cytochrome-c^{III} exchange than for the $Fe(phen)_3^{2+}$–$Fe(phen)_3^{3+}$ reaction. This factor of 4 arises from the difference in the sizes of the two reactants and is equal to the square of the ratio of the radii of cytochrome-*c* and iron tris-phenanthroline. The above calculation also assumes, of course, that the electron transfer takes place through the edge of the porphyrin ring system and that no electron transfer takes place through the protein.

without modification to the cytochrome-c^{II}-cytochrome-c^{III} exchange reaction. We have already mentioned the problem of steric effects and the question of the detailed pathway for the electron transfer in cytochrome-*c*. Although these problems complicate the calculations of K_oS and the interaction energy, they do not introduce any radically new features. The location of the heme group will introduce some specificity in the electron transfer reactions of cytochrome-*c*, and this problem will be explored later. Another problem which arises is the calculation of the inner- and outer-shell reorganization energies in the cytochrome-*c* system. In some respects it might be convenient to regard the protein (excluding those groups that are directly bonded to the heme) as part of the "surrounding medium" and to treat the polarization of the medium (in the limiting case) as an additive function of the polarizations of the protein and of the solvent. In any event it will be shown that one can go a long way towards correlating the rates of electron transfer reactions without being able to perform an *ab initio* calculation of the electron transfer rate in a particular system.

We turn next to the cobalt(II)-cobalt(III) exchanges. These reactions involve the transfer of an e_g electron. The e_g orbital is an antibonding orbital (see Fig. 6) and when an electron (actually two electrons, see below) is added to this orbital there is a decrease in the strength of the metal-ligand bond. This will tend to lengthen the metal-ligand bond in cobalt(II), thereby increasing the difference between the configurations of the cobalt(II) and cobalt(III) complexes. The situation is complicated by the fact that the cobalt(II) complexes presented in Table I are high spin, while the cobalt(III) complexes are low spin. The exchange reactions in these systems involve not only the transfer of an electron, but also the rearrangement of the other *d* electrons of the two reactants. The electron transfer is partly spin-forbidden and should proceed very slowly. Despite this restriction it is apparent from Table I that the $Co(H_2O)_6^{2+}$–$Co(H_2O)_6^{3+}$ exchange is as rapid as the $Fe(H_2O)_6^{2+}$–$Fe(H_2O)_6^{3+}$ reaction. This result may be rationalized by postulating that only a small amount of energy is necessary to excite the low spin $Co(H_2O)_6^{3+}$ to the high spin state[29]. More striking is the relatively slow rate of the $Co(phen)_3^{2+}$–$Co(phen)_3^{3+}$ exchange. Since the added electron in $Co(phen)_3^{2+}$ is in an e_g orbital, the increased electron density on the cobalt is no longer removed by M → L π-bonding as was the case in $Fe(phen)_3^{2+}$. This results in a relatively large inner shell reorganization energy for the $Co(phen)_3^{2+}$–$Co(phen)_3^{3+}$ reaction. However, it is most unlikely that the inner shell reorganization energy for this reaction is so large that it can account for the factor of more than 10^7 between the rates of the $Fe(phen)_3^{2+}$–$Fe(phen)_3^{3+}$ and $Co(phen)_3^{2+}$–$Co(phen)_3^{3+}$ reactions. An additional factor contributing to the slow rate of the latter reaction may be some nonadiabaticity brought about by the poor overlap of the e_g orbital on the cobalt with the π-orbital system of the phenanthroline. If it is assumed that the inner shell reorganization energy for the $Co(phen)_3^{2+}$–$Co(phen)_3^{3+}$ exchange is three times larger than that for the $Fe(phen)_3^{2+}$–$Fe(phen)_3^{3+}$ reaction (probably an overestimate), then ν_e is calculated to be 10^{10} sec^{-1} for the former reaction (compared with 10^{13} sec^{-1} assumed for the $Fe(phen)_3^{2+}$–$Fe(phen)_3^{3+}$ exchange). The calculated value of ν_e will decrease if a lower value of the inner shell reorganization energy is assumed.

It has long been believed that the very slow rate of the $Co(NH_3)_6^{2+}$–$Co(NH_3)_6^{3+}$ reaction is due to a large value of $(r_2 - r_3)$. However, recent measurements[33] have shown that $(r_2 - r_3) = 0.18$ Å in this system (compare $(r_2 - r_3) = 0.15$ Å for the $Fe(H_2O)_6^{2+}$–$Fe(H_2O)_6^{3+}$ reaction) and consequently the very slow rate of the $Co(NH_3)_6^{2+}$–$Co(NH_3)_6^{3+}$ reaction cannot be ascribed to a large inner shell reorganization energy provided the two assumptions mentioned above are valid. The results suggest, instead, that ν_e for this reaction is very small, and that a somewhat larger ν_e obtains in the $Co(en)_3^{2+}$–$Co(en)_3^{3+}$ exchange reaction.

The above discussion shows that the application of the theoretical concepts is restricted by the absence of quantitative information concerning the interaction and inner shell reorganization energies. Fortunately some information about the inner shell reorganization energy can be obtained from a somewhat different approach, namely the position of the absorption band in intervalence compounds[32]. Thus it is apparent from Fig. 1 that $\Delta E_{op} \approx 4\Delta E_{th}$ if the potential energy curves are harmonic and the splitting at the intersection is relatively small. The value of ΔE_{op} can be calculated from the position of the intervalence absorption band. Creutz and Taube[34] have used the position of this band (1570 nm) to calculate the value of ΔE_{op} for the electron transfer within complex I.

$$\left[(NH_3)_5Ru^{II}N\bigcirc N\,Ru^{III}(NH_3)_5\right]^{5+}$$

(I)

ΔE_{op} turns out to be 18.22 kcal $mole^{-1}$ and consequently $(\Delta G_i^* + \Delta G_o^*)$ for the thermal electron transfer within this complex is $\approx$ 4.55 kcal $mole^{-1}$, provided the two assumptions mentioned above are valid. The frequency of the thermal intramolecular electron transfer, calculated from $10^{13} \exp(-\Delta E_{th}/RT)$, is 5×10^9 sec^{-1}. The magnitude of the thermal reorganization energy seems reasonable since the π-orbital system of the pyrazine group should provide relatively strong coupling between the two ruthenium atoms, thereby tending to decrease the difference between the configurations of their coordination shells. The data in Table I indicate that the energy required to make the configurations of the ruthenium(II) and ruthenium(III) equivalent decreases when NH_3 is replaced by back-bonding ligands such as phenanthroline[25]. On this basis Creutz and Taube speculate that the replacement of the NH_3 molecules by ligands such as phenanthroline or bipyridine should tend to make the two ruthenium atoms in the pyrazine-bridged complex equivalent.

Cross-reactions

The expressions derived earlier for the inner- and outer-shell reorganization energies for exchange reactions lead to a relatively simple relation between the rate constant for an oxidation–reduction reaction accompanied by a net chemical change, for example, between the rate constant for the reaction:

$$\underset{\text{A}}{Co^{3+}} + \underset{\text{B}}{Fe^{2+}} \xrightarrow{k_{12}} \underset{\text{A}'}{Co^{2+}} + \underset{\text{B}'}{Fe^{3+}}$$

and the rate constants for the related exchange reactions:

$$\underset{\text{A}}{Co^{3+}} + \underset{\text{A}'}{Co^{2+}} \xrightarrow{k_1} \underset{\text{A}'}{Co^{2+}} + \underset{\text{A}}{Co^{3+}}$$

and

$$\underset{\text{B}'}{Fe^{3+}} + \underset{\text{B}}{Fe^{2+}} \xrightarrow{k_2} \underset{\text{B}}{Fe^{2+}} + \underset{\text{B}'}{Fe^{3+}}$$

Consider the reorganization of only the inner coordination shells of the reactants. The following notation is used: $\Delta r_A^\circ = (r_A' - r_A)$, $\Delta r_B' = (r_B - r_B')$, $\Delta r_A^* = (r_A^* - r_A)$, $\Delta r_B^* = (r_B - r_B^*)$, $F_A = F_A' = 2f_A f_A'/(f_A + f_A')$, $F_B = F_B' = 2f_B f_B'/(f_B + f_B')$. Evidently F_i, the breathing force constant of species i, is assumed to have the same value regardless of whether species i is in the reduced or oxidized state. It follows from eqn. (37) that

$$\Delta G_A^* = \frac{F_A(\Delta r_A^\circ)^2}{4} \text{ and } \Delta G_B^* = \frac{F_B(\Delta r_B^\circ)^2}{4}$$

The standard free energy change for the cross-reaction is equal to the free energy of activation for the forward reaction minus the free energy of activation for the reverse reaction. Therefore:

$$\Delta G^\circ = \Delta G_{AB}^* - \Delta G_{BA}^* \tag{42}$$

$$= \frac{F_A(\Delta r_B^*)^2}{2} + \frac{F_B(\Delta r_B^*)^2}{2} - \frac{F_A(\Delta r_A^\circ - \Delta r_A^*)^2}{2} - \frac{F_B(\Delta r_B^\circ - r_B^*)^2}{2}$$

$$\therefore \quad \Delta r_B^* = \frac{1}{F_B \Delta r_B^\circ}\left[\alpha - F_A(\Delta r_A^\circ)(\Delta r_A^*)\right] \tag{43}$$

References pp. 651–653

where

$$\alpha = \Delta G^\circ + \frac{F_A(\Delta r_A^\circ)^2}{2} + \frac{F_B(\Delta r_B^\circ)^2}{2}$$

Substitution for Δr_B^* in the expression for ΔG_{AB}^* gives:

$$\Delta G_{AB}^* = \frac{F_A(\Delta r_A^*)^2}{2} + \frac{1}{2F_B(\Delta r_B^\circ)^2}[\alpha - F_A(\Delta r_A^\circ)(\Delta r_A^*)]^2 \tag{44}$$

Since $(\Delta G_{AB}^*)/(\Delta r_A^*) = 0$, it follows that:

$$\Delta r_A^* = \frac{(\Delta r_A^\circ)\alpha}{F_A(\Delta r_A^\circ)^2 + F_B(\Delta r_B^\circ)^2} \tag{45}$$

Substitution for Δr_A^* in eqn. (44) gives:

$$\Delta G_{AB}^* = \frac{\alpha^2}{2[F_A(\Delta r_A^\circ)^2 + F_B(\Delta r_B^\circ)^2]} \tag{46}$$

and finally, substitution for α gives the expressions:

$$\Delta r_A^* = \frac{\Delta r_A^\circ}{2}\left[1 + \frac{\Delta G^\circ}{2(\Delta G_A^* + \Delta G_B^*)}\right]$$

$$\Delta r_B^* = \frac{\Delta r_B^\circ}{2}\left[1 + \frac{\Delta G^\circ}{2(\Delta G_A^* + \Delta G_B^*)}\right]$$

$$\Delta G_{AB}^* = \frac{(\Delta G_A^* + \Delta G_B^*)}{2} + \frac{\Delta G^\circ}{2} + \frac{(\Delta G^\circ)^2}{8(\Delta G_A^* + \Delta G_B^*)} \tag{47}$$

Equation (47) can easily be generalized to include the outer shell reorganization energy to give the relation:

$$\Delta G_{12}^{**} = \frac{(\Delta G_1^{**} + \Delta G_2^{**})}{2} + \frac{\Delta G_r^\circ}{2} + \frac{(\Delta G_r^\circ)^2}{8(\Delta G_1^{**} + \Delta G_2^{**})} \tag{48}$$

where ΔG_{12}^{**} refers to the cross-reaction, ΔG_1^{**} and ΔG_2^{**} refer to the appropriate exchange reactions, $\Delta G^{**} = (\Delta G_i^* + \Delta G_o^*)$, and ΔG_r° is the free energy change for the cross-reaction when the two reactants are a distance r apart★. Equation (48), which was first derived by Marcus[14,19], also follows from relatively simple geometrical considerations[7]. Some of

---

★ The quadratic character of eqn. (48) is a consequence of two approximations: (a) the vibrational potential energy of the inner coordination shells of the reactants is assumed to be a quadratic function of fluctuations in the vibrational coordinates, and (b) the free energy of the system is assumed to be a quadratic function of fluctuations in the dielectric polarization of the medium.

the approximations and assumptions made in deriving eqn. (48) have been examined by Newton[35]. Equation (48) can be related to the appropriate rate constants by means of eqns. (28) or (33). Provided $\nu_e \approx 10^{-13}\,\text{sec}^{-1}$ for all three of the reactions being considered, and all three of the precursor complexes as well as all three of the successor complexes have similar stabilities, then

$$k_{12} = (k_1 k_2 K_{12} f)^{\frac{1}{2}} \tag{49}$$

where $\log f = (\log K_{12})^2/[4\log(k_1k_2/Z^2)]$ and K_{12} is the equilibrium constant for the cross-reaction. Evidently $f \to 1$ as $K_{12} \to 1$.

The rate constants for a number of cross-reactions are presented in Table III. The agreement of the observed and calculated rates is, on the

TABLE III

COMPARISON OF OBSERVED AND CALCULATED RATE CONSTANTS

Reaction	k_{obs} (M^{-1} sec^{-1})	k_{calc} (M^{-1} sec^{-1})	Ref.
$W(CN)_8^{4-} + IrCl_6^{2-}$	6.1×10^7	8.1×10^7	36
$Fe(CN)_6^{4-} + IrCl_6^{2-}$	3.8×10^5	5.7×10^5	36
$Mo(CN)_8^{4-} + IrCl_6^{2-}$	1.9×10^6	1.0×10^6	36
$W(CN)_8^{4-} + Mo(CN)_8^{3-}$	5.0×10^6	1.7×10^7	36
$Fe(CN)_6^{4-} + Mo(CN)_8^{3-}$	3.0×10^4	2.7×10^4	36
$W(CN)_8^{4-} + Fe(CN)_6^{3-}$	4.3×10^4	5.1×10^4	36
Cyt-c^{II} + $Fe(CN)_6^{3-}$	8.4×10^6	1×10^6	59
$V^{2+} + Ru(NH_3)_6^{3+}$	7.3×10^2	8.1×10^2	43
$Cr^{2+} + Ru(NH_3)_6^{3+}$	2×10^2	$\leqslant 1.5 \times 10^3$	43
$V^{2+} + Co(NH_3)_6^{3+}$	3.7×10^{-3}	$\leqslant 1.3 \times 10^{-3}$	44
$Cr^{2+} + Co(NH_3)_6^{3+}$	8.8×10^{-5}	$\leqslant 2.9 \times 10^{-3}$	44

whole, satisfactory. The rate constants for the reactions of Cr^{2+} and V^{2+} with $Ru(NH_3)_6^{3+}$ and $Co(NH_3)_6^{3+}$ illustrate several interesting features[42,43]. The oxidation of V^{2+} by either $Ru(NH_3)_6^{3+}$ or $Co(NH_3)_6^{3+}$ proceeds more rapidly than the oxidation of Cr^{2+}. Evidently the more favorable standard free energy changes for the Cr^{2+} reactions are counterbalanced by the larger inner shell reorganizations associated with the transfer of an e_g electron from Cr^{2+} as compared to the transfer of a t_{2g} electron from V^{2+} (the rate constants for the V^{2+}–V^{3+} and Cr^{2+}–Cr^{3+} exchanges at 25°C are 1.0×10^{-2} and $\leqslant 2 \times 10^{-5}\,M^{-1}\,\text{sec}^{-1}$, respectively). It is also apparent from Table III that $Ru(NH_3)_6^{3+}$ reacts by a factor of 10^6 faster than $Co(NH_3)_6^{3+}$ with both Cr^{2+} and V^{2+}. This factor is predicted by eqn. (49) for it is equal to the square root of the ratio of the rate constants for the $Ru(NH_3)_6{}^{2+}$–$Ru(NH_3)_6{}^{3+}$ and $Co(NH_3)_6{}^{2+}$–$Co(NH_3)_6{}^{3+}$ exchange reactions. However, while this agreement is gratifying,★ most reactions of cobalt(III) which have been studied do not satisfy eqn. (49). These exceptions could be due to spin multiplicity restrictions or other non-adiabatic factors as well as to differences in the stabilities of the precursor complexes involved in the comparisons[38]. Evidence for the formation of relatively stable precursor complexes in certain oxidations by cobalt(III) is provided by the activation enthalpies for these reactions. The general expression for the enthalpy of activation for an electron transfer reaction is:

$$\Delta H_{\text{obs}} = \Delta H_o + \Delta H_p^* \tag{50}$$

where ΔH_o is the enthalpy of formation of the precursor complex, and ΔH_p^* is the enthalpy of activation for the subsequent electron transfer step. It is apparent from eqn. (50) that ΔH_{obs} will be small or even negative if ΔH_o is sufficiently negative. Such negative activation enthalpies have been found for the reactions of some cobalt(III) complexes with chromium(II)[9,39]★★.

According to eqn. (48):

$$\Delta G_{12}^{**} = \frac{(\Delta G_1^{**} + \Delta G_2^{**})}{2} + \frac{\Delta G_r^\circ}{2} \tag{51}$$

provided $(\Delta G_r^\circ)^2 \ll 8(\Delta G_1^{**} + \Delta G_2^{**})$ (in other words, provided $f = 1$). Under these conditions a linear relationship with a slope of 0.50 is predicted between the free energies of activation and the standard free energy changes for a series of reactions for which $(\Delta G_1^{**} + \Delta G_2^{**})$ is a constant. Such a linear dependence has been observed[36]. It should be remembered that the linear dependence predicted by eqn. (51) obtains only between the free energy of activation and the standard free energy change for the actual electron transfer step, and does not necessarily obtain between the free energy of activation for the overall reaction ($\log k_{12}$) and the overall free energy change ($\log K_{12}$)[40,41].

To summarize, the barrier to electron transfer is composed of two parts. There is an intrinsic contribution $(\Delta G_1^{**} + \Delta G_2^{**})$ and a thermodynamic contribution (ΔG_r°). The intrinsic contribution is a sum of two terms, one term depends only on the properties of species 1 in its initial and final states, and the other depends only on the properties of species 2 in its initial and final states. The intrinsic contribution may be calculated from the appropriate exchange rate constants. Differences in reaction rates

---

★ Note though that only a lower limit has been established for the rate constant for the $Co(NH_3)_6{}^{2+}$–$Co(NH_3)_6{}^{3+}$ reaction.

★★ See notes added in proof (2), p. 650.

may be due to differences in either or both of these contributions. In many instances these contributions oppose one another. As we have seen, this occurs in the reactions of Cr^{2+} and V^{2+} with $Ru(NH_3)_6{}^{3+}$ and $Co(NH_3)_6{}^{3+}$. Another example is the oxidation of Fe^{2+} and $Fe(phen)_3{}^{2+}$ by cerium(IV) sulfate solutions. In these reactions the relatively low intrinsic barrier for the oxidation of $Fe(phen)_3{}^{2+}$ (see Table II) is offset by a less favorable free energy change for its oxidation ($E°$ for the Fe^{2+}/Fe^{3+} and $Fe(phen)_3{}^{2+}/Fe(phen)_3{}^{3+}$ couples are 0.68 and 1.08 V, respectively). The net result of these two opposing factors is that the two oxidations proceed at comparable rates[45].

One-electron and two-electron transfers

The model discussed above may be used to determine whether two one-electron transfers or one two-electron transfer is favored in reactions in which two equivalents need to be transferred. For example, the stable oxidation states of thallium differ by two equivalents. Exchange between these two oxidation states can be brought about either by the "simultaneous" transfer of two electrons:

$$Tl^{3+} + Tl^{+} \rightleftharpoons Tl^{+} + Tl^{3+} \quad (52)$$

or by a one electron transfer to generate Tl^{2+} as an intermediate

$$Tl^{3+} + Tl^{+} \rightleftharpoons Tl^{2+} + Tl^{2+} \quad (53)$$

The latter reaction is favored by a lower ΔG_o^* (see eqn. 40). On the other hand, the former reaction is accompanied by a more favorable free energy change ($\Delta G° = 0$ for reaction 52 and $\gg 0$ for reaction 53). The net result of these opposing factors appears to be that the one electron transfer is favored[20]. Ashurst and Higginson[46] have shown that the reaction between Fe^{2+} and Tl^{3+} is inhibited by Fe^{3+} ions. This observation establishes that one equivalent is transferred in this non-complementary reaction★:

$$Tl^{3+} + Fe^{2+} \rightleftharpoons Tl^{2+} + Fe^{3+} \quad (54)$$

$$Tl^{2+} + Fe^{2+} \rightarrow Tl^{+} + Fe^{3+} \quad (55)$$

The above reaction scheme provides a mechanism for the catalysis of the Tl^{+}–Tl^{3+} exchange by Fe^{2+} ions★★. Many other electron transfer reactions

★ A non-complementary reaction occurs between a one-equivalent couple and a two-equivalent couple, while a complementary reaction occurs between two one-equivalent couples, or between two two-equivalent couples.
★★ This catalysis has been observed by R. W. Dobson and B. Warnqvist, *Inorg. Chem.*, 10 (1971) 2624.

are catalyzed by transition metal ions[1,47]. Thus the oxidation of Fe^{2+} by Co^{3+} is catalyzed by Ag^+ ions according to the following scheme[48]

$$Co^{3+} + Ag^+ \rightleftharpoons Co^{2+} + Ag(II)$$

$$Ag(II) + Fe^{2+} \rightarrow Ag^+ + Fe^{3+}$$

A similar scheme accounts for the catalysis of the reaction between V^{3+} and Fe^{3+} by Cu^{2+} ions[49,50]:

$$V^{3+} + Cu^{2+} \rightleftharpoons V(IV) + Cu^+$$

$$Fe^{3+} + Cu^+ \rightarrow Fe^{2+} + Cu^{2+}$$

The Ag^+/Ag^{2+} and Cu^+/Cu^{2+} couples find wide application as electron mediators. The ability of these couples to function as mediators is probably due to a number of factors including the relatively low charges on the ions, the stable d^{10} configurations of the lower oxidation states, as well as to the Jahn–Teller distortions of the upper oxidation states.

A binuclear complex could act as a two-equivalent oxidizing or reducing agent if electron transfer within the complex were sufficiently rapid. For example, we have seen that the frequency of intramolecular electron transfer within complex I is about 5×10^9 sec^{-1}. The Ru^{II}–Ru^{II} analog of this complex (perhaps with phenanthroline replacing the coordinated NH_3 groups) could conceivably accept two electrons "simultaneously" from a suitable two-electron donor. This type of mechanism could obtain in the reactions of metal complexes with hydrogen peroxide or oxygen. Thus Wang has synthesized a copolymer of ferroheme, 4,4′-dipyridine, and poly(L-lysine)[51]. Part of the active center of this copolymer is shown in structure II with the Fe forming part of a

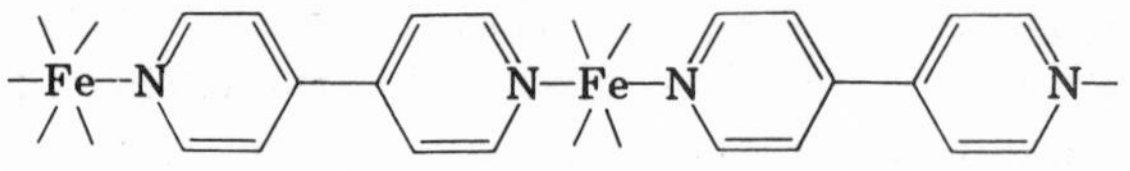

(II)

heme group. A tunneling calculation (which uses an interesting blend of quantum mechanics and solid geometry, and which neglects reorganization energies) shows that the frequency of electron transfer between two neighboring Fe^{III} heme units through the bridging dipyridine group is $\geqslant 6 \times 10^9$ sec^{-1}*. Since this frequency is much larger than the rate of oxidation of the Fe^{II} heme copolymer by oxygen, the Fe^{II} heme units of the copolymer could cooperate in the reduction of O_2[51].

* Note the similarity of this frequency with the value calculated for complex I using the position of the intervalence absorption bond.

BRIDGING GROUPS

The free energy equations discussed in the previous section were derived on the assumptions that the reactions were outer sphere and that the splitting at the intersection region was small enough so that it could be neglected in calculating the free energy of activation, but large enough so that the value of p was close to unity. More recently these restrictions have been removed[19]. Changing the bridging group in an inner-sphere reaction may thus alter the rate of electron transfer by changing the interaction energy, the stability of the precursor complex, the reorganization energy, as well as the standard free energy change for the reaction. It is not possible to assess the relative importance of all these factors in a bridged system. For this reason the following discussion of bridging groups will be largely qualitative in nature, and emphasis will be placed upon the experimental observations.

Hydroxide and other inorganic anions

The rates of reduction of complexes in which one or more of the ligands are water molecules generally increase with decreasing acidity. These rate increases are ascribed to the displacement of the hydrolysis equilibrium

$$MOH_2^{n+} \rightleftharpoons MOH^{(n-1)+} + H^+ \qquad (58)$$

with $MOH^{(n-1)+}$ undergoing more rapid reduction than MOH_2^{n+} [22]. The evidence indicates that hydroxide is acting as a bridging ligand in those reactions in which it exerts a marked catalytic effect[6,44,52-54]. Thus there is very little difference between the effects of H_2O and OH^- on the rates of outer-sphere reactions.★ The large discrimination between H_2O and OH^- observed in inner-sphere reactions may be related to the greater stability of the hydroxide-bridged precursor complex III compared to the water-bridged precursor complex IV[55,56].

★ Since $M^{III}OH_2^{3+}$ complexes are generally much more acidic than $M^{II}OH_2^{2+}$ complexes, the driving force for the outer-sphere reaction:

$$M^{III}OH^{2+} + M^{2+} \rightarrow M^{II}OH^{+} + M^{3+}$$

will be very unfavorable. On the other hand, the corresponding inner-sphere exchange reaction:

$$M^{III}OH^{2+} + M^{2+} \rightarrow M^{2+} + M^{III}OH^{2+}$$

is thermoneutral.

References pp. 651–653

$$\left[\begin{matrix} & H & \\ & | & \\ M- & O & -M \end{matrix}\right]^{(n-1)+} \qquad \left[\begin{matrix} & H & \\ & | & \\ M- & O & -M \\ & | & \\ & H & \end{matrix}\right]^{n+}$$

(III) (IV)

Direct evidence for the formation of oxygen- or hydroxide-bridged intermediates has been obtained from ^{18}O labeling experiments[57★], from the rate law for the reaction[52,56,58], and from the detection of the successor complexes in suitable systems[6].

An interesting pH dependence has been observed in the reaction of cytochrome-c^{II} with ferricyanide ions[59]. The ratio of the forward to reverse rate constants for the reaction:

$$\text{cyt-}c^{II} + Fe(CN)_6{}^{3-} \rightleftharpoons \text{cyt-}c^{III} + Fe(CN)_6{}^{4-}$$

determined by the temperature-jump technique is different from the equilibrium constant determined spectrophotometrically. The difference between the kinetic and spectrophotometric equilibrium constants is ascribed to a relatively slow pH-dependent change in the conformation of cytochrome-c^{III} [59]. This conformation change, which can be studied by the stopped flow technique, is too slow to be detected by the faster relaxation method.

Halide ions can also function as efficient bridging ligands. A normal order of bridging efficiency ($I^- > Br^- > Cl^- > F^-$) is observed in certain reactions, and an inverted order ($F^- > Cl^- > Br^- > I^-$) in others. For example, $(NH_3)_5CoX^{2+}$ complexes react about 10^{11} times faster with Cr^{2+} than $Co(NH_3)_6{}^{3+}$ does, and the relative rates Cl:Br:I are 1:2:5[61]. On the other hand, $(NH_3)_5RuX^{2+}$ complexes do not react very much faster with Cr^{2+} than $Ru(NH_3)_6{}^{3+}$ does, and the relative rates Cl:Br:I are 10^2:10:1[42]. Although these bridging orders have been much discussed in recent years[6,42,54,60,61], the reactivity pattern for a particular system can not always be predicted with confidence.

Interesting effects have been observed with unsymmetrical bridging ligands. The transfer of an unsymmetrical group in an inner-sphere reaction may yield a product in which the "wrong end" of the bridging ligand is attached to a metal center:

$$FeNCS^{2+} + Cr^{2+} \rightarrow Fe^{2+} + CrSCN^{2+} \tag{59}$$

A number of "wrong-bonded" complexes, or linkage isomers, have been prepared in this manner[62-66]. The difference in rates observed with symmetrical and unsymmetrical bridging ligands forms a basis for determining whether a particular reaction proceeds by an inner-sphere or an outer-sphere mechanism. Comparisons of the effects of azide and thiocyanate on electron

★ The pathway in the reaction between Cr^{2+} and $(NH_3)_5CoOH_2{}^{3+}$ previously ascribed to water bridging has been shown to be a medium effect[108].

transfer rates have been widely used in this connection[67,68]. If the metal centers of the oxidizing and reducing agents are hard, and if transfer of the bridging group occurs during the reaction, then an inner-sphere reaction should proceed faster when the bridging group is azide than when it is isothiocyanate. This is principally due to the increased stability of a metal–nitrogen over a metal–sulfur bond. Similar free energy considerations predict that an inner-sphere reaction should proceed faster when the bridging group is thiocyanate than when it is either azide or isothiocyanate. This is found to be the case in practice[40].

Organic bridging groups

Studies of the reactions of chromium(II) with more than 100 carboxylatopentaamminecobalt(III) complexes have shown that electron transfer is slow unless the bridging ligand contains a conjugated bond system as well as a group which is capable of binding the chromium(II). The latter condition is a requirement for the formation of a relatively stable precursor complex.

In some systems the formation of the precursor complex is inhibited by steric effects[69]. Thus the rate of reaction of Cr^{2+} with carboxylatopentaamminecobalt(III) complexes decreases in the order $HCOO^- >$ $CH_3COO^- > (CH_3)_3CCOO^-$. This is the order expected if the group R in the binuclear complex V repels the chromium(II). [There is good evidence

$$(NH_3)_5Co^{III}{-}O{-}\overset{\overset{\displaystyle R}{|}}{C}{=}O{-}Cr^{II}(H_2O)_5$$

(V)

that the entering Cr^{2+} ion attacks the carboxylate oxygen atom which is not attached to the cobalt(III)[42,69,73].] On the other hand, the formation of a relatively stable precursor complex is favored if the Cr^{2+} ion can be chelated to the bridging group as occurs, for example, in the reactions of Cr^{2+} with pentaamminecobalt(III) complexes of α-hydroxy acids[70].

No evidence has been obtained for electron transfer through extended saturated chains[69]. However, electron transfer can take place through suitable unsaturated systems. Thus the reaction of Cr^{2+} with VI (isonicotinamidepentaamminecobalt(III)) yields VII[71].

$$\left[(NH_3)_5CoN\langle\bigcirc\rangle{-}CONH_2\right]^{3+} \longrightarrow \left[HN\langle\bigcirc\rangle{-}\underset{\underset{\displaystyle NH_2}{|}}{C}{=}OCr(H_2O)_5\right]^{4+}$$

(VI) (VII)

The site of attachment of the chromium(III) in the product establishes that the chromium(II) attacks the remote end of the isonicotinamide group. The available evidence indicates that this reaction as well as other cobalt(III) reactions that take place by remote attack, proceed by a radical-ion intermediate mechanism in which the electron transfer is initially to a low-lying, unoccupied π orbital of the bridging ligand rather than to the cobalt center[69]. For this type of mechanism to obtain it is necessary that the bridging ligand be reducible.

The studies with organic bridging ligands provide many examples of the importance of the orbital symmetry in determining the rate and mechanism of the electron transfer reaction. Thus the reaction of Cr^{2+} with the ruthenium complex VIII proceeds about 3×10^4 times as rapidly

$$\left[(NH_3)_5Ru\text{N}\langle\bigcirc\rangle\text{—CONH}_2\right]^{3+}$$

(VIII)

as the reaction of Cr^{2+} with the analogous cobalt complex (VI)[72]. Remote attack occurs also with the ruthenium(III) complex, but in this system the favorable overlap of the acceptor t_{2g} orbital on the ruthenium with the π-orbital system of the bridging ligand facilitates the transfer of the redox electron to the metal center and a relatively long-lived radical ion intermediate is not formed[69,72]. As discussed above, such an intermediate is formed in the reaction of Cr^{2+} with the corresponding cobalt(III) complex, where the poor overlap of the acceptor e_g orbital on the cobalt with the π-orbital system of the bridging ligand will tend to stabilize the intermediate. Of course, the lifetime of the intermediate in the cobalt(III) system will also be prolonged by the relatively large amount of energy required to reorganize the inner coordination shell of the cobalt(III) before the electron transfer from the bridging ligand to the cobalt center can occur.

Another illustration of the importance of orbital symmetries is provided by the observation that the three-atom carboxylate group and halide ions are comparable bridging groups in the reactions of Cr^{2+} with $(NH_3)_5RuX^{2+}$. This result is not unexpected since the t_{2g} acceptor orbital on the ruthenium(III) can overlap efficiently with the carboxylate π system. On the other hand, the carboxylate group is less effective than halide ions in the reactions of Cr^{2+} with $(NH_3)_5CoX^{2+}$ by a factor of about 10^7. This result, too, seems reasonable in view of the poor overlap of the cobalt e_g acceptor orbital with the carboxylate π-system[42].

Almost all of the organic bridging groups which have been studied so far are attached to the oxidizing metal centers through either oxygen or nitrogen atoms. Some preliminary results with sulfur as the "lead-in" atom have been reported recently[74]. It was found that Cr^{2+} reacts more than 10^3 times more rapidly with $(en)_2Co(SCH_2COO)^+$ than with $(en)_2Co(OCH_2COO)^+$. Steric repulsion of the entering Cr^{2+} ion by the methylene group should be smaller for the sulfur than for oxygen complex and this should favor the formation of a more stable precursor complex and lead to more rapid reaction rates for the sulfur compound, as is indeed found[74]. Another important factor may be the availability of the *d* electrons on the sulfur. The rapid electron transfer through sulfur is of interest in view of the suggestion[21] that in certain cases the electron transfer to cytochrome-*c* might proceed through the thioether linkage of cysteines 14 and 17. The mode of attachment of the heme group to the polypeptide chain is shown in **IX**.

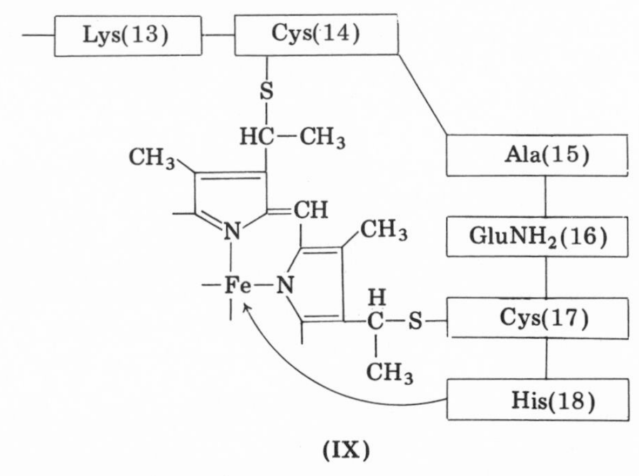

(IX)

The electron transfer may be aided by the loss of a proton from the carbon atom adjacent to the sulfur since this would result in the formation of a favorable π system from the sulfur to the iron[80]. Thus there is some evidence that the methylene group adjacent to the sulfur atom is acidic.

CONFORMATION- AND SUBSTITUTION-CONTROLLED OXIDATION–REDUCTION REACTIONS

There is an important group of oxidation–reduction reactions in which the rate-determining step is not the conversion of the precursor to the successor complex. Instead the rate-determining step may be a slow hydrolysis reaction, dissociation of a dimer, change in the conformation of a protein, an isomerization, or the formation of a reactant. The common feature of this class of reactions is that under suitable conditions the rate of the oxidation–reduction reaction becomes independent of the concentration of the reducing (or oxidizing) agent. Under these conditions the rate of electron transfer in the system is equal to the rate of a slow step preceding the formation of the precursor complex. This type of mechanism occurs, for example, in the oxidation of chromium(II) by iron(III) in the presence of chloride ions[75]. Under certain conditions the rate of formation of chromium(III) is independent of the chromium(II) concentration of the solution. When this happens the slow step in the oxidation–reduction reaction is the formation of the monochloroiron(III) complex which is very rapidly reduced by chromium(II) in a subsequent inner-sphere reaction.

$$Fe^{3+} + Cl^- \xrightarrow{\text{slow}} FeCl^{2+} \tag{60}$$

$$FeCl^{2+} + Cr^{2+} \xrightarrow{\text{rapid}} Fe^{2+} + CrCl^{2+} \tag{61}$$

Evidently reactions of this type provide information not only about the composition of the activated complex but also about its structure.

There is good evidence that the formation of the precursor complex is rate determining in certain reactions involving vanadium(II) and chromium(II) (see Fig. 4). These substitution-controlled reactions show reactivity patterns which are different from those found for "ordinary" oxidation–reduction reactions[6,76]. Instead of varying as the bridging group is changed, the rate constants for the substitution-controlled reactions are insensitive to the nature of the bridging group and of the oxidizing agent used, and, as expected, are related to the rate constants for water replacement on the reducing agent[65,77].

The successor complexes formed in the reaction of a number of ruthenium complexes with chromium(II) are relatively stable, and the first-order dissociations of these complexes have been studied[42,78,79]. These successor complexes owe their stabilities to the substitution inertness of the d^3 and low spin d^6 electron configurations of chromium(III) and ruthenium(II), respectively. Oxygen-bridged and hydroxide-bridged successor complexes have also been observed in a number of systems[6]. An attempt has been made to use chromium(II) as a probe of the site of electron transfer in cytochrome-*c* by taking advantage of the stability of the chromium(III)–cytochrome-c^{III} complex formed in this system[80]. Although this system is rather complicated and the interpretation of the results is not free of ambiguities, this approach holds much promise for use with simpler systems. Finally, as discussed earlier, cytochrome-c^{III} formed in the reaction between cytochrome-c^{II} and $Fe(CN)_6^{3-}$ undergoes an acid-dependent conformation change after the electron transfer step[59]. Since this oxidation–reduction reaction is reversible, the opposite conformation change must precede the reaction of cytochrome-c^{III} with $Fe(CN)_6^{4-}$*.

* The interpretation of the results reported in ref. 59 is not free of ambiguities. Thus the slow reaction ascribed there entirely to a conformation change could be complicated by the pH-dependent formation of relatively stable complexes between the reactants or products of the reaction. The formation of such complexes would also cause the thermodynamics of the electron transfer step to be different from the thermodynamics of the overall reaction.

Evidently protein conformation changes can occur prior to the formation of a precursor complex, in the precursor complex, in the successor complex, or after the dissociation of the successor complex.

OXIDATIVE ADDITION

This section is concerned with oxidation–reduction reactions in which the formation of the precursor complex involves an addition rather than a substitution step. An example of this type of reaction is the oxidation of $Co(CN)_5^{3-}$ by $(NH_3)_5CoX^{2+}$ complexes[81]:

$$(NH_3)_5CoX^{2+} + Co(CN)_5^{3-} \rightarrow Co^{2+} + Co(CN)_5X^{3-} + 5NH_3$$

The common feature of these reactions is that the reducing agent is coordinatively unsaturated, that is, all the potential coordination sites on the reducing agent are not occupied. Coordinatively unsaturated complexes find wide application as catalysts because of their ability to add a variety of ligands. The following reactions illustrate the oxidative addition of XY to various coordinatively unsaturated low-spin complexes, where XY = HCl, Br_2, H_2, HO-OH, CH_3-I, *etc.*[82,83]:

(a) five-coordinate d^7 complexes → six-coordinate d^6 complexes:

$$2Co^{II}(CN)_5^{3-} + XY \rightarrow Co^{III}(CN)_5X^{3-} + Co^{III}(CN)_5Y^{3-} \tag{62}$$

(b) four-coordinate d^8 complexes → six-coordinate d^6 complexes:

$$Rh^{I}Cl(PPh_3)_3 + XY \rightarrow Rh^{III}Cl(PPh_3)_3XY \tag{63}$$

(c) two-coordinate d^{10} complexes → six-coordinate d^6 complexes

$$Pt^{\circ}(PPh_3)_2 + 2XY \rightarrow Pt^{II}(PPh_3)_2XY + XY \rightarrow Pt^{IV}(PPh_3)_2X_2Y_2 \tag{64}$$

The driving force for the above reactions is the relatively stable closed-shell configuration (of 18 valence electrons) of the six-coordinate d^6 complexes produced[82]. Since molecular oxygen adds to a variety of d^8 and d^{10} complexes[83], these systems may serve as simple models for more complex biological systems.

SOME FURTHER THOUGHTS ON BIOLOGICAL SYSTEMS

Although it is not yet possible to perform an *ab initio* calculation of the electron exchange rate in even the simplest biological system, it is possible to rationalize the rates of cross-reactions in terms of their standard

References pp. 651–653

free energy changes and the rate constants for the component exchange reactions. Some further applications to biological systems will be given in this section.

A detailed study has been made of the rate of reaction of cytochrome-c^{II} with the peroxide complex of cytochrome-c peroxidase[84,85]. The data are consistent with the following scheme:

$$E + H_2O_2 \rightleftharpoons E(H_2O_2) \quad (65)$$

$$E(H_2O_2) \rightleftharpoons (ES)'' \quad (66)$$

$$(ES)'' + \text{cyt-}c^{II} \rightarrow (ES)' + \text{cyt-}c^{III} \quad (67)$$

$$ES' + \text{cyt-}c^{II} \rightarrow E + \text{cyt-}c^{III} \quad (68)$$

where E represents cytochrome-c peroxidase. The first step in the reaction is the formation of a complex between E and hydrogen peroxide with a rate constant of about $1.4 \times 10^8\ M^{-1}\ \text{sec}^{-1}$. This complex then undergoes an intramolecular electron transfer with a rate constant of about $10^4\ \text{sec}^{-1}$ into a two-equivalent oxidant ES″. One equivalent in ES″ is believed to be on the iron and the second on an aromatic amino acid (histidine, tryptophan, methionine, or tyrosine) near the heme group. The rate constant for the reaction of ES″ with cyt-c^{II} is about $5.9 \times 10^8\ M^{-1}\ \text{sec}^{-1}$ and the rate of reaction of ES′ with a second molecule of cyt-c^{II} is $> 5.9 \times 10^8\ M^{-1}\ \text{sec}^{-1}$. These rapid electron transfer rates may reflect the favorable free energy change for the reaction as well as a favorable overlap of the redox orbitals in the (ES)″–cyt-c^{II} complex.

The rate constants for the reaction of ferricyanide with cytochrome-c^{II} ($k_{21} = 1.6 \times 10^7\ M^{-1}\ \text{sec}^{-1}$)[37] and for the reaction of the peroxide complex of cytochrome-c peroxidase with cytochrome-c^{II} ($k_{31} = 5.0 \times 10^8\ M^{-1}$ sec^{-1})[84] ⋆ can be used to predict the rate constant k_{32} for the oxidation of ferrocyanide by the peroxide complex of cytochrome-c peroxidase. It follows from equation 49 (taking $f = 1$)⋆⋆ that:

$$k_{21} = (k_2 k_1 K_{21})^{\frac{1}{2}} \quad (69)$$

$$k_{31} = (k_3 k_1 K_{31})^{\frac{1}{2}} \quad (70)$$

where k_1 is the rate constant for the cytochrome-c^{II}–cytochrome-c^{III} exchange reaction. Therefore:

$$k_{32} = (k_3 k_2 K_{32})^{\frac{1}{2}} \quad (71)$$

$$= k_{31} k_2 / k_{21} \quad (72)$$

⋆ The peroxide compound used in these studies is C_2H_5OOH.

⋆⋆ This assumption is not strictly-speaking necessary since the f values tend to cancel in equation (72).

where k_2 is the rate constant for the ferrocyanide–ferricyanide exchange reaction ($3 \times 10^2\ M^{-1}\ sec^{-1}$). This calculation gives $k_{32} = 9 \times 10^3\ M^{-1}\ sec^{-1}$, in good agreement with the experimental value of $3 \times 10^3\ M^{-1}\ sec^{-1}$ [86].

Another system which has been thoroughly studied is the reaction between oxygen and the reduced form of cytochrome-*c* oxidase[87]. Cytochrome-*c* oxidase contains two heme groups (heme-*a* and heme-a_3)★ and two atoms of copper. The first step in the reaction of oxygen with cytochrome-*c* oxidase is the formation of a precursor complex with a rate constant of $1 \times 10^8\ M^{-1}\ sec^{-1}$. This step is followed by the three-step transfer to oxygen of four equivalents, one from heme-a_3^{II} ($k = 3 \times 10^4\ sec^{-1}$), two from the two Cu(I) atoms ($k = 7 \times 10^3\ sec^{-1}$), and one from heme-$a^{II}$ ($k = 7 \times 10^2\ sec^{-1}$), resulting in the formation of two molecules of H_2O[87]. An unusual feature of these studies is the evidence obtained for the transfer of a single equivalent to oxygen in the first oxidation step. This is a most surprising result in view of the unfavorable driving force for the one-equivalent reduction of molecular oxygen[89,90]. Contrary to an earlier report[88], the recent studies do not provide evidence for a rapid intramolecular transfer of an electron from heme-a^{II} to heme-a_3^{III} [87]. It would be of interest to determine whether the first equivalent transferred to the oxygen is an electron from the iron or a hydrogen atom from an aromatic amino acid located near the heme group[91,92]. More important, perhaps, it may be that such an aromatic amino acid cooperates with heme-a_3^{II} to transfer two equivalents "simultaneously" to the oxygen molecule. In terms of this model the oxidized aromatic amino acid (free radical) is then reduced in the second oxidation step by one of the copper atoms in the enzyme. Alternatively a "simultaneous" two-equivalent transfer to oxygen might occur from heme-a_3^{II} and one of the copper atoms, without involving a free radical intermediate. Clearly the role of the copper in these reactions needs further elucidation.

Heme-a^{III} of cytochrome-*c* oxidase reacts with cytochrome-c^{II} with a rate constant of $4 \times 10^7\ M^{-1}\ sec^{-1}$ [93]. Heme-a_3^{III} oxidizes cytochrome-c^{II} much more slowly (Chapter 27, page 978). Studies of the cytochrome-*c* oxidase catalysis of the oxidation of cytochrome-c^{II} by oxygen:

$$\text{cyt-}c^{II} + \tfrac{1}{4}O_2 + H^+ \rightarrow \text{cyt-}c^{III} + \tfrac{1}{2}H_2O \qquad (73)$$

have shown that this reaction is inhibited by cytochrome-c^{III} [94]. This inhibition is ascribed to the formation of a complex between the enzyme and cytochrome-c^{III} at the cytochrome-c^{II} binding site of the enzyme. The oxidation of cytochrome-c^{II} by oxygen proceeds very slowly in the absence

★ Although there is still some confusion on this point, the available evidence seems to indicate that the environments of the two heme groups in cytochrome-*c* oxidase are different, and that the structures of the heme groups are similar but not necessarily identical.

References pp. 651–653

of cytochrome-*c* oxidase, presumably because cytochrome-c^{II} does not readily accept more than one equivalent. The overall scheme for the oxidation of cytochrome-c^{II} by oxygen in the presence of cytochrome-*c* oxidase may thus be summarized as follows:

$$O_2 + 4H^+ + a_3^{II}, a^{II} \xrightarrow{2Cu?} 2H_2O + a_3^{III}, a^{III} \quad (74)$$

followed by:

$$a^{III} + \text{cyt-}c^{II} \rightarrow a^{II} + \text{cyt-}c^{III} \quad (75)$$

Reaction (75) may proceed by direct electron transfer between the heme groups on the two reactants. Alternatively, a protruding aromatic amino acid on the cytochrome-*c* oxidase (which is connected to heme-a^{III} by a conjugated bond system) may be inserted into the crevice containing the heme group in cytochrome-c^{II} (or *vice versa*), thereby providing an indirect but still favorable path for the electron transfer[91,92].

In addition to the heme crevice, cytochrome-c^{III} possesses two other interesting structural features, which have been called the "right" and "left channels"[30,95]. These channels lead from the surface of the protein to the sides of the heme group: the right channel leads to the histidine-18 side and the left channel to the methionine-80 side. Dickerson *et al.*[95] speculate that the right channel may provide access to the heme group for small molecules or side chains, and they propose that this channel and the heme crevice are involved in electron transfer from cytochrome-c^{II} to oxidases, perhaps along the lines discussed above. However, following a suggestion made earlier by Winfield[92], they favor a different pathway for electron transfer to cytochrome-c^{III}. They suggest that this pathway involves the left channel and they propose the detailed electron transfer sequence: reductase → tyrosine-74 → tyrosine-67 → methionine-80 sulfur → heme iron(III). This mechanism involves motion of the tyrosine-67 side chain — however, it has the advantage that separate paths are involved for electron transfer to and from the heme. This latter feature is attractive because it removes the requirement that the cytochrome-*c* molecule in the mitochondrial membrane rotate so as to expose the same electron transfer site successively to its reductase and oxidase, as has been proposed, for example, by Chance *et al.*[96]. As we have seen, sulfur is a good "lead-in" atom and, moreover, the lone pair of electrons on the methionine sulfur can be used for hydrogen-bonding of the hydroxyl proton of tyrosine-67. Thus the electron transfer from tyrosine-67 to the iron(III) could occur in a proton-bridged Tyr–O–H–S–Fe intermediate. Whether net transfer of the bridging proton would occur in this process (as Dickerson *et al.* propose) would depend on the relative basicities of the oxygen of the tyrosyl radical cation and of the sulfur coordinated to the iron(II), and is difficult to predict.

An alternative mechanism for the reduction of heme iron(III) could involve motion of methionine-80 rather than of tyrosine-67. In this case the electron transfer path would be: reductase → methionine-80 sulfur → heme iron(III). Thus in the presence of a reductase the heme iron(III) might remove an electron from the methionine sulfur, and the polypeptide chain 79–83 might then rotate so as to bring the methionine-80 close to the reductase. A model of cytochrome-c^{III} indicates that this type of rotation of the polypeptide chain 79–83 is possible and that this rotation would bring the ϵ-amino nitrogen of lysine-79 into a position to occupy the sixth coordination site of the heme iron[95]. The methionine-80 could return to a position near the iron(II) after accepting an electron from the reductase.

These are intriguing conjectures (see also Chapter 26, p. 924) and additional information is necessary before it is possible to rule out one (or both) of the above mechanisms.

NOTES ADDED IN PROOF

(1) There have been several important developments in the area of oxidation–reduction reactions since the manuscript for the above chapter was submitted to the Editor. A subjective and incomplete selection from some recent papers is presented here.

Stynes and Ibers[97] have determined the molecular structures of hexaammineruthenium(II) and hexaammineruthenium(III). They find that the ruthenium(II)–NH_3 and ruthenium(III)–NH_3 bond distances differ by 0.040 Å, which is considerably less than the 0.178 Å difference between the cobalt(II)–NH_3 and cobalt(III)–NH_3 distances (page 632). They conclude that the slowness of the $Co(NH_3)_6{}^{2+}$–$Co(NH_3)_6{}^{3+}$ exchange reaction is due to differences in energy between the spin states of the two reactants and is not due to the inner shell reorganization energy. Stynes and Ibers propose the following reaction scheme

$$\underset{(t_{2g})^5(e_g)^2,\ \text{high spin}}{Co(NH_3)_6{}^{2+}} \rightleftharpoons \underset{(t_{2g})^6(e_g)^1,\ \text{low spin}}{Co(NH_3)_6{}^{2+}}$$

$$\underset{(t_{2g})^6(e_g)^1 + (t_{2g})^6}{Co(NH_3)_6{}^{2+} + Co(NH_3)_6{}^{3+}} \rightarrow \underset{(t_{2g})^6 + (t_{2g})^6(e_g)^1}{Co(NH_3)_6{}^{3+} + Co(NH_3)_6{}^{2+}}$$

This scheme involves the prior electronic excitation of the high-spin cobalt(II) to a low-spin form. The actual electron transfer takes place between the two low-spin forms of cobalt and is presumed to be rapid.

Theoretical treatments of the electronic structure of the pyrazine-bridged, binuclear complex I (page 632) have been published[98,99]. An attempt to prepare binuclear complexes containing ruthenium(II) and ruthenium(III) and 4,4′-bipyridine or 1,2-bis(4-pyridyl)ethylene as bridging groups was not successful[100]. The stability constants of the precursor complexes formed in a number of inner-sphere[101,102] and outer-sphere reactions[103] (page 623) have been determined. It should be noted that one reaction's precursor complex is the reverse reaction's successor complex.

The 2.8 Å resolution map of cytochrome-c^{III} (page 912) has been published[104] and the rate constant for the opening of the heme crevice and/or for the rupture of the bond between the heme iron and the methionine-80 sulfur has been determined[105]. The possibility that some electron transfer reactions of cytochrome-c^{III} involve the rupture of the iron–sulfur bond has been examined[105,106]. The reactivity pattern of the anion-catalyzed chromium(II) reduction of cytochrome-c^{III} is consistent with a mechanism involving electron transfer through the edge of the porphyrin ring system[106,107]. However, evidence was also obtained for an electron-transfer pathway involving opening of the heme crevice and/or rupture of the iron–sulfur bond. Thus, depending upon the conditions used, the electron transfer to cytochrome-c^{III} may occur either directly to the iron (probably via a suitable bridging group) or through the exposed edge of the porphyrin ring system. Other pathways may obtain under other conditions.

(2) The following more general form of eqn. (49) may prove useful in rationalizing some of the reactions involving the cobalt(II)–cobalt(III) couple

$$k_{12} = p_{12}\left[\frac{k_1 k_2 K_{12} f}{p_1 p_2}\right]^{1/2} \tag{76}$$

where f now contains the p factors. Equation (76) allows for the possibility that all of the reactions involved in the comparison may not be adiabatic; it reduces to eqn. (49) when $p_{12} = p_1 = p_2 = 1$ (that is, when the reactions are adiabatic) and also when $p_{12} = (p_1 p_2)^{1/2}$ (and $f = 1$), a condition that has been previously discussed[6]. Note that if $p_1 = p_2 = 1$ but $p_{12} \ll 1$, eqn. (49) will still give the correct free energy dependence, but the k_{12} values calculated from eqn. (49) will be higher than the observed values. This type of behavior has been seen, for example, in the oxidation of a series of $Fe(phen)_3^{2+}$ complexes by aquo cobalt(III)[36]. On the other hand, if p_1 or $p_2 \ll 1$ the f factors calculated assuming $p_1 = p_2 = 1$ will be too high, and the free energy dependence calculated using these f factors will be too large.

Since small p factors will tend to show up as relatively negative entropies of activation, some information about the magnitudes of the p factors in real systems may be obtained from the temperature dependence of the electron transfer rates. If the reactions involved in the rate comparisons have very different temperature-independent factors, then it may be possible to correct for these differences by using eqn. (76) or a similar equation. This procedure assumes, of course, that the breakdown of eqn. (49) is due to non-adiabaticity; additional considerations will have to be introduced if, for example, as proposed above, prior electronic excitation of one of the reactants is required.

REFERENCES

1 A. G. Sykes, *Adv. Inorg. Chem. Radiochem.*, 10 (1967) 153.
2 F. Basolo and R. G. Pearson, *Mechanisms of Inorganic Reactions*, Wiley, New York, 1967, p. 454.
3 W. L. Reynolds and R. W. Lumry, *Mechanisms of Electron Transfer*, The Ronald Press, New York, 1966.
4 N. Sutin, *A. Rev. Phys. Chem.*, 17 (1966) 119.
5 H. Taube and E. S. Gould, *Accounts Chem. Res.*, 2 (1969) 321.
6 N. Sutin, *Accounts Chem. Res.*, 1 (1968) 225.
7 N. Sutin, *A. Rev. Nucl. Sci.*, 12 (1962) 285.
8 J. Halpern, *Quart. Rev.*, 15 (1961) 217.
9 H. Taube, *Adv. Inorg. Chem. Radiochem.*, 1 (1959) 1.
10 L. Landau, *Phys. Z. Sowjet.*, 2 (1932) 46.
11 C. Zener, *Proc. R. Soc.*, A137 (1932) 696; A140 (1933) 660.
12 C. A. Coulson and K. Zalewski, *Proc. R. Soc.*, A268 (1962) 437.
13 W. Kauzmann, *Quantum Chemistry*, Academic Press, New York, 1957, p. 535.
14 R. A. Marcus, *A. Rev. Phys. Chem.*, 15 (1964) 155.
15 J. Halpern and L. E. Orgel, *Discuss. Faraday Soc.*, 29 (1960) 32.
16 P. V. Manning, R. C. Jarnagin and M. Silver, *J. Phys. Chem.*, 68 (1964) 265.
17 E. F. Gurnee and J. L. Magee, *J. Chem. Phys.*, 26 (1957) 1237.
18 R. A. Marcus, *J. Chem. Phys.*, 24 (1956) 966; 44 (1965) 679.
19 R. A. Marcus, *J. Phys. Chem.*, 72 (1968) 891.
20 N. S. Hush, *Trans. Faraday Soc.*, 57 (1961) 557.
21 P. George and J. Griffith, in P. Boyer, H. Lardy and K. Myrback, *The Enzymes*, Vol. 1, Academic Press, New York, 1959, p. 347.
22 J. Silverman and R. W. Dodson, *J. Phys. Chem.*, 56 (1952) 846.
23 A. Kowalsky, *Biochemistry*, 4 (1965) 2382.
24 D. W. Larsen and A. C. Wahl, *J. Chem. Phys.*, 43 (1965) 3765.
25 T. J. Meyer and H. Taube, *Inorg. Chem.*, 7 (1968) 2369.
26 N. A. Bonner and J. P. Hunt, *J. Am. Chem. Soc.*, 82 (1960) 3826.
27 D. R. Stranks, *Discuss. Faraday Soc.*, 29 (1960) 73.
28 F. P. Dwyer and A. G. Sargeson, *J. Phys. Chem.*, 65 (1961) 1892.
29 B. R. Baker, F. Basolo and H. M. Neuman, *J. Phys. Chem.*, 63 (1959) 371.
30 R. E. Dickerson, L. M. Kopka, J. Weinzierl, J. Varnum, D. Eisenberg and E. Margoliash, *J. Biol. Chem.*, 242 (1967) 3015.
31 H. R. Gygax and J. Jordan, *Discuss. Faraday Soc.*, 45 (1968) 227.
32 N. S. Hush, *Progr. Inorg. Chem.*, 8 (1967) 391.
33 M. T. Barnet, B. M. Craven, H. C. Freeman, N. E. Kime and J. A. Ibers, *Chem. Commun.*, (1966) 307; N. E. Kime and J. A. Ibers, *Acta Cryst.*, B25 (1969) 168.
34 C. Creutz and H. Taube, *J. Am. Chem. Soc.*, 91 (1969) 3988.
35 T. W. Newton, *J. Chem. Educ.*, 45 (1968) 571.
36 R. J. Campion, N. Purdie and N. Sutin, *Inorg. Chem.*, 3 (1964) 1091.
37 N. Sutin and D. R. Christman, *J. Am. Chem. Soc.*, 83 (1961) 1773.
38 D. P. Fay and N. Sutin, work in progress.
39 R. C. Patel, R. E. Ball, J. F. Endicott and R. G. Hughes, *Inorg. Chem.*, 9 (1970) 23.
40 D. P. Fay and N. Sutin, *Inorg. Chem.*, 9 (1970) 1291.
41 G. Davies, N. Sutin and K. O. Watkins, *J. Am. Chem. Soc.*, 92 (1970) 1892.
42 J. A. Stritar and H. Taube, *Inorg. Chem.*, 8 (1969) 2281.
43 J. F. Endicott and H. Taube, *J. Am. Chem. Soc.*, 86 (1964) 1686.
44 A. Zwickel and H. Taube, *J. Am. Chem. Soc.*, 83 (1961) 793.

45 G. Dulz and N. Sutin, *Inorg. Chem.*, 2 (1963) 917.
46 K. G. Ashurst and W. C. E. Higginson, *J. Chem. Soc.*, (1954) 587.
47 W. C. E. Higginson, D. R. Rosseinsky, J. B. Stead and A. G. Sykes, *Discuss. Faraday Soc.*, 29 (1960) 49.
48 D. H. Huchital, N. Sutin and B. Warnqvist, *Inorg. Chem.*, 6 (1967) 838.
49 W. C. E. Higginson and A. G. Sykes, *J. Chem. Soc.*, (1962) 2841.
50 O. J. Parker and J. H. Espenson, *Inorg. Chem.*, 8 (1969) 1523.
51 J. H. Wang, *Accounts Chem. Res.*, 3 (1970) 90.
52 J. H. Espenson, *Inorg. Chem.*, 4 (1965) 1025.
53 B. Baker, M. Orhanovic and N. Sutin, *J. Am. Chem. Soc.*, 89 (1967) 722.
54 R. C. Patel and J. F. Endicott, *J. Am. Chem. Soc.*, 90 (1968) 6364.
55 R. D. Cannon and J. E. Earley, *J. Am. Chem. Soc.*, 88 (1966) 1872.
56 M. P. Litelpo and J. F. Endicott, *J. Am. Chem. Soc.*, 91 (1969) 3982.
57 W. K. Kruse and H. Taube, *J. Am. Chem. Soc.*, 82 (1960) 526.
58 A. Haim, *Inorg. Chem.*, 5 (1966) 2081.
59 K. G. Brandt, P. C. Parks, G. H. Czerlinski and G. P. Hess, *J. Biol. Chem.*, 241 (1966) 4180.
60 A. Haim, *Inorg. Chem.*, 7 (1968) 1475.
61 J. P. Candlin and J. Halpern, *Inorg. Chem.*, 4 (1965) 766.
62 J. Halpern and S. Nakamura, *J. Am. Chem. Soc.*, 87 (1965) 3002.
63 J. H. Espenson and J. P. Birk, *J. Am. Chem. Soc.*, 87 (1965) 3280.
64 A. Haim and N. Sutin, *J. Am. Chem. Soc.*, 88 (1966) 434.
65 M. Orhanovic and N. Sutin, *J. Am. Chem. Soc.*, 90 (1968) 4286.
66 J. P. Birk and J. H. Espenson, *J. Am. Chem. Soc.*, 90 (1968) 1153.
67 D. L. Ball and E. L. King, *J. Am. Chem. Soc.*, 80 (1958) 1091.
68 J. H. Espenson, *Inorg. Chem.*, 4 (1965) 121.
69 H. Taube and E. S. Gould, *Accounts Chem. Res.*, 2 (1969) 321.
70 R. D. Butler and H. Taube, *J. Am. Chem. Soc.*, 87 (1965) 5597.
71 F. Nordmeyer and H. Taube, *J. Am. Chem. Soc.*, 90 (1968) 1162.
72 R. G. Gaunder and H. Taube, quoted in ref. 69.
73 R. J. Balahura and R. B. Jordan, *J. Am. Chem. Soc.*, 92 (1970) 1553.
74 R. H. Lane and L. E. Bennett, *J. Am. Chem. Soc.*, 92 (1970) 1089.
75 G. Dulz and N. Sutin, *J. Am. Chem. Soc.*, 86 (1964) 829.
76 H. J. Price and H. Taube, *Inorg. Chem.*, 7 (1968) 1.
77 M. V. Olson, Y. Kanazawa and H. Taube, *J. Chem. Phys.*, 51 (1969) 289.
78 W. G. Movius and R. G. Linck, *J. Am. Chem. Soc.*, 91 (1969) 5394.
79 D. Seewald, N. Sutin and K. O. Watkins, *J. Am. Chem. Soc.*, 91 (1969) 7307.
80 A. Kowalsky, *J. Biol. Chem.*, 244 (1969) 6619.
81 J. P. Candlin, J. Halpern and S. Nakamura, *J. Am. Chem. Soc.*, 85 (1963) 2517.
82 J. Halpern, *Pure Appl. Chem.*, 20 (1969) 59.
83 J. P. Collman, *Accounts Chem. Res.*, 1 (1968) 136.
84 T. Yonetani and G. S. Ray, *J. Biol. Chem.*, 241 (1966) 700.
85 B. Chance, D. DeVault, V. Legallais, L. Mela and T. Yonetani, in *Fast Reactions and Primary Processes in Chemical Kinetics*, Interscience Publishers, New York, 1967, p. 437.
86 T. Yonetani and H. Schleyer, *J. Biol. Chem.*, 242 (1967) 1974.
87 C. Greenwood and Q. H. Gibson, *J. Biol. Chem.*, 242 (1967) 1782.
88 Q. H. Gibson and C. Greenwood, *Biochem. J.*, 86 (1963) 541.
89 P. George, in *Oxidases and Related Redox Systems*, Vol. 1, Wiley, New York, 1965, p. 3.
90 N. Sutin, in *Oxidases and Related Redox Systems*, Vol. 1, Wiley, New York, 1965, p. 37.
91 M. E. Winfield, in *Oxidases and Related Redox Systems*, Vol. 1, Wiley, New York, 1965, p. 115.

92 M. E. Winfield, *J. Mol. Biol.*, 12 (1965) 600.
93 Q. H. Gibson, C. Greenwood, D. C. Wharton and G. Palmer, *J. Biol. Chem.*, 240 (1965) 888.
94 T. Yonetani and G. S. Ray, *J. Biol. Chem.*, 240 (1965) 3392.
95 R. E. Dickerson, T. Takano, O. B. Kallai and L. Samson, *Proceedings of the Wenner-Gren Symposium*, 1970.
96 B. Chance, C-P. Lee, L. Mela and D. DeVault, in *Structure and Function of Cytochromes*, University of Tokyo Press, Tokyo, 1968, p. 475.
97 H. C. Stynes and J. A. Ibers, *Inorg. Chem.*, 10 (1971) 2304.
98 J. H. Elias and R. S. Drago, *Inorg. Chem.*, 11 (1972) 415.
99 B. Mayoh and P. Day, *J. Am. Chem. Soc.*, 94 (1972) 2885.
100 E. B. Fleischer and D. K. Lavallee, *J. Am. Chem. Soc.*, 94 (1972) 2599.
101 K. M. Davies and J. H. Espenson, *J. Am. Chem. Soc.*, 91 (1969) 3093.
102 R. D. Cannon and J. Gardiner, *J. Am. Chem. Soc.*, 92 (1970) 3800.
103 D. Gaswick and A. Haim, *J. Am. Chem. Soc.*, 93 (1971) 7248.
104 R. E. Dickerson, T. Takano, D. Eisenberg, O. B. Kallai, L. Samson, A. Cooper and E. Margoliash, *J. Biol. Chem.*, 246 (1971) 1511.
105 N. Sutin and J. K. Yandell, *J. Biol. Chem.*, 247 (1972) 6932.
106 J. K. Yandell, D. P. Fay and N. Sutin, *J. Am. Chem. Soc.*, submitted for publication.
107 N. Sutin, *Chem. Brit.*, 8 (1972) 148.
108 D. L. Toppen and R. G. Linck, *Inorg. Chem.*, 10 (1971) 2635.

Chapter 20

METAL ION CATALYSIS OF REACTIONS OF MOLECULA OXYGEN*

A. E. MARTELL AND M. M. TAQUI KHAN
Department of Chemistry, Texas A & M University, College Station, Texas, U.S.A.

INTRODUCTION

This chapter is a conceptual review of metal ion-catalyzed reactions of molecular oxygen, and of its partial reduction product, hydrogen peroxide. Since these reactions are very numerous and varied in nature, this short chapter cannot include an exhaustive review of the literature; on the other hand an attempt has been made to include all of the important reaction types found in chemical and biological systems, and to classify them in a logical manner. Another purpose of this review is to examine and compare the mechanisms of metal ion catalysis of chemical oxidation reactions, and possibly to apply these mechanisms toward a better understanding of corresponding reactions in biological systems.

As indicated in Table I, reactions of oxygen may be classified in two general ways: those in which oxygen itself combines with and enters the substrate (called insertion reactions), and those in which the oxygen merely serves as an oxidizing agent, and becomes reduced to hydrogen peroxide or water, depending on the nature of the substrate and the reaction conditions. The latter are called non-insertion reactions. Non-insertion reactions may be further classified on the basis of whether the oxidant, oxygen, undergoes a four-electron reduction to water, or a two-electron reduction to hydrogen peroxide. The insertion reactions, on the other hand, can be classified on the basis of whether one or both oxygen atoms enters the substrate.

Because of the importance of peroxides as intermediates in the reduction of oxygen to water, pertinent examples of metal-catalyzed reactions of peroxides are included in this review. Also included is the disproportionation of hydrogen peroxide to oxygen and water (catalase action) because of its close relationship to other reactions of hydrogen peroxide and oxygen, particularly those that occur in biological systems.

---

*Presented in part at the Symposium on Inorganic Biochemistry, 151st Meeting of the American Chemical Society, Pittsburgh, Pa., April, 1966.

TABLE I

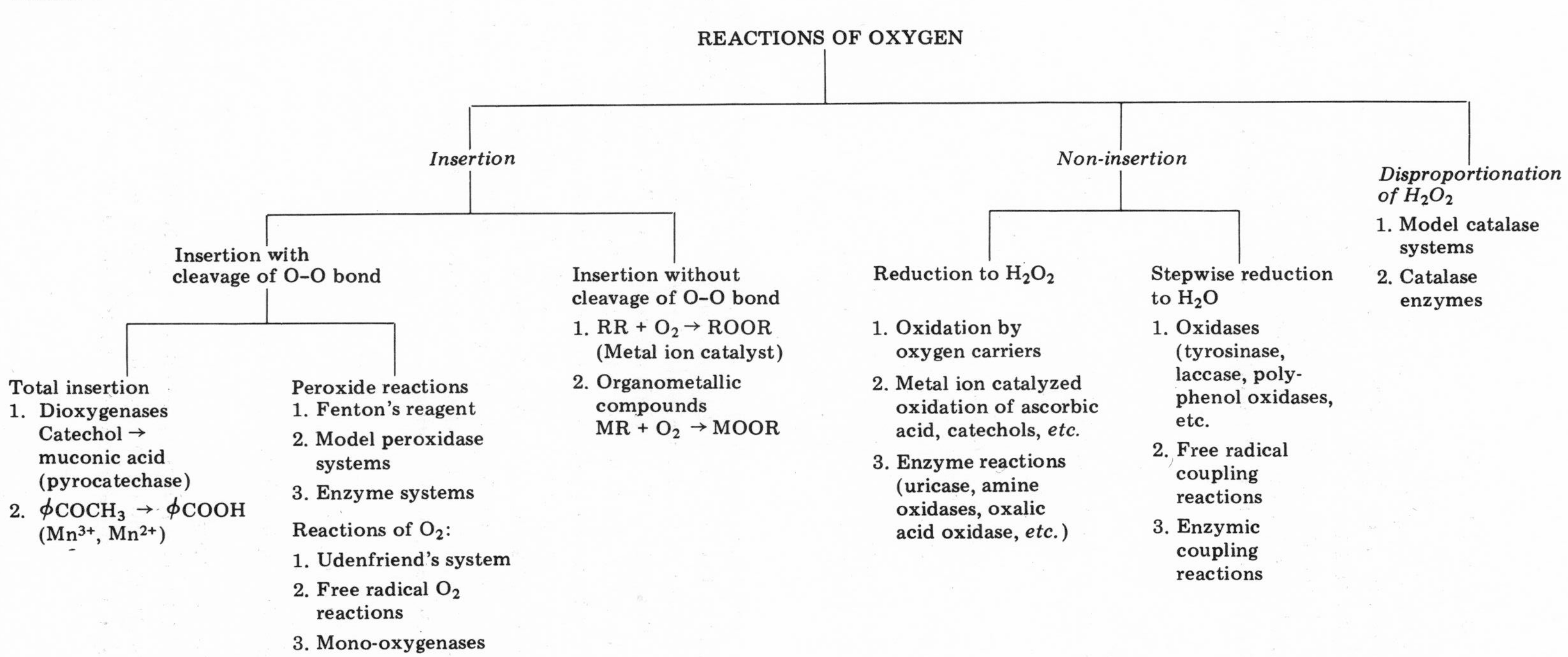
REACTIONS OF OXYGEN
Insertion
Non-insertion
Disproportionation of H_2O_2
1. Model catalase systems
2. Catalase enzymes
Insertion with cleavage of O–O bond
Insertion without cleavage of O–O bond
1. $RR + O_2 \rightarrow ROOR$ (Metal ion catalyst)
2. Organometallic compounds $MR + O_2 \rightarrow MOOR$
Reduction to H_2O_2
1. Oxidation by oxygen carriers
2. Metal ion catalyzed oxidation of ascorbic acid, catechols, *etc.*
3. Enzyme reactions (uricase, amine oxidases, oxalic acid oxidase, *etc.*)
Stepwise reduction to H_2O
1. Oxidases (tyrosinase, laccase, polyphenol oxidases, etc.
2. Free radical coupling reactions
3. Enzymic coupling reactions
Total insertion
1. Dioxygenases Catechol → muconic acid (pyrocatechase)
2. $\phi COCH_3 \rightarrow \phi COOH$ (Mn^{3+}, Mn^{2+})
Peroxide reactions
1. Fenton's reagent
2. Model peroxidase systems
3. Enzyme systems
Reactions of O_2:
1. Udenfriend's system
2. Free radical O_2 reactions
3. Mono-oxygenases

References pp. 686–688

Energetics of oxygen reduction

Before considering details in the mechanisms of oxygen reactions, it is of interest to consider the nature of the oxidation states that are possible between oxygen and water. Dissociation energies of the more common valence states of oxygen are presented in Table II. The data show the expected correlation between bond dissociation energy and bond distance, the O-O distance increasing as electrons are successively added to the O_2 molecule.

TABLE II

PROPERTIES OF OXYGEN AND ITS REDUCTION PRODUCTS

	Oxidation state	Bond distance (Å)	Dissociation energies	
			Reaction	ΔH (kcal/mole)
O_2	0	1.20	$O_2 \rightarrow 2O$	117[1]
O_2^*	0	—	$O_2^* \rightarrow 2O$	95[2]
O_2^-	$-\frac{1}{2}$	1.27	$O_2^- \rightarrow O + O^-$	88[2]
O_2^{2-}	-1	1.35	$H_2O_2(aq) \rightarrow 2OH(aq)$	~46[3]
			$H_2O_2(aq) \rightarrow H^+(aq) + O(aq) + OH^-(aq)$	~46[3]

From the energies involved, it is apparent that reactions involving cleavage of the O-O bond probably proceed through the energetically more favorable peroxide intermediate which may be hydrogen peroxide itself, or a complex in which the two oxygens are held together through single O-O bonds. On the other hand, it is possible to break the double bond either homolytically or heterolytically in a single step when oxygen can form more stable bonds with other elements in the course of concerted electron shifts. Examples of such mechanisms, which do not involve the formation of hydrogen peroxide or peroxy complexes as stable intermediates, will be suggested for many of the types of oxygenation and oxidation reactions that will be described in this review.

I. INSERTION REACTIONS

A. *Total insertion, with O-O bond cleavage*

1. *Enzymic systems*

Reactions in which both oxygen atoms become attached to the substrate may occur by one of two paths: (1) direct association of the oxygen with the substrate, followed by rearrangement to stable molecular groupings;

or (2) the breaking up of the oxygen to reactive fragments, followed by combination with the substrate. The latter type usually results in partial insertion because of the many side reactions that free radicals undergo. Let us consider first the direct insertion of the oxygen molecule, followed by rearrangement.

In biological systems this type of reaction is exemplified by the dioxygenases, such as pyrocatechase, metapyrocatechase, and tryptophan-pyrrolase[4-7]. Suggestions concerning the Fe(II) ion-catalyzed action of pyrocatechase were made by Hayaishi and Hashimoto[8,9] on the oxidation of catechol to *cis–cis* muconic acid. The reaction sequence proposed involves the formation of a hypothetical peroxy intermediate, indicated by the following:

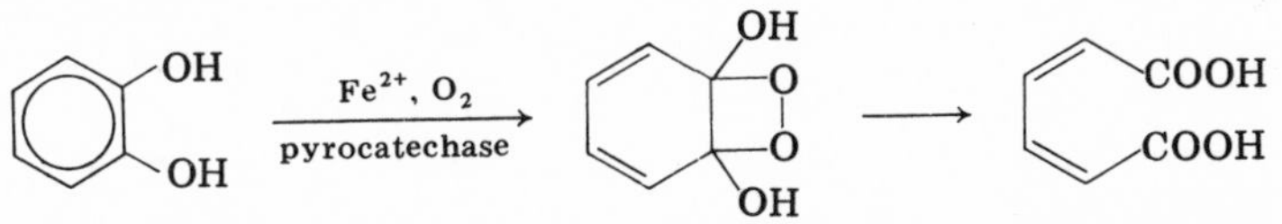

Fe^{2+}, O_2 / pyrocatechase

This mechanism is in conformity with the experimental observations that *o*-quinone and hydrogen peroxide could not be detected as reaction intermediates, and that all of the oxygen is incorporated into the reaction product, as indicated by ^{18}O tracer studies. A more detailed mechanism, involving an intermediate iron(II)–oxygen complex, was proposed by Mason[10].

The peroxo-bridged intermediate proposed by Hayaishi has been criticized by Hamilton[11] because of the energy required in its formation, resulting from the loss of aromatic ring resonance and the strain of the four-membered ring. The mechanism illustrated in Fig. 1, which is analogous to a mechanism recently suggested by Hamilton[11] for tryptophan oxidation, is now proposed as a reasonable alternative.

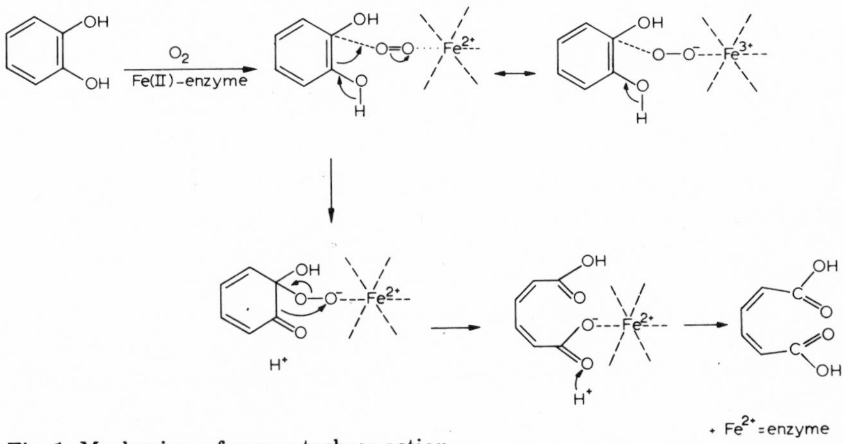

Fig. 1. Mechanism of pyrocatechase action.

References pp. 686–688

This mechanism is in accord with ionic interpretations of metal ion catalysis in enzyme models[11,12] and enzymic systems[11], and seems to offer a better explanation of the general experimental observations on the behavior of these oxygenases. Thus the metal ion is ascribed a more significant role in the reaction mechanism. Also, the only substrates that show activity with the dioxygenase enzymes are those which can dissociate a proton in the initial step, as indicated above. The rearrangement of the organoperoxo intermediate has precedents in organic chemistry[13]. Similar mechanisms may be set up for the oxidation of other substrates catalyzed by dioxygenases, with total insertion of molecular oxygen.

2. Model systems

A non-enzymatic reaction which seems to involve total insertion of oxygen with cleavage of the O–O bond is the Mn^{III}/Mn^{II}-catalyzed oxidation of acetophenone to benzoic acid[14]. The rate of oxidation was found to be independent of the concentration of the metal ion. Enolization of the ketone is proposed as the rate-determining step, followed by rapid free radical reactions initiated and propagated by the oxidized and reduced forms of the metal ion:

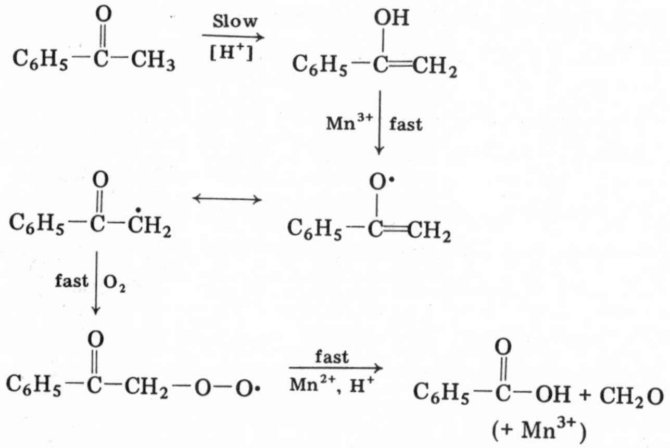

$$C_6H_5-C(=O)-CH_3 \xrightarrow[\text{[H}^+\text{]}]{\text{Slow}} C_6H_5-C(OH)=CH_2$$

$$C_6H_5-C(OH)=CH_2 \xrightarrow[\text{fast}]{Mn^{3+}} C_6H_5-C(O\cdot)=CH_2 \longleftrightarrow C_6H_5-C(=O)-\dot{C}H_2$$

$$C_6H_5-C(=O)-\dot{C}H_2 \xrightarrow[O_2]{\text{fast}} C_6H_5-C(=O)-CH_2-O-O\cdot$$

$$C_6H_5-C(=O)-CH_2-O-O\cdot \xrightarrow[Mn^{2+},\ H^+]{\text{fast}} C_6H_5-C(=O)-OH + CH_2O \quad (+\ Mn^{3+})$$

Comparison of the above reaction mechanism with those suggested for the total insertion enzymic reactions indicates that metal-catalyzed oxidation of acetophenone is not a satisfactory model for the dioxygenases. The main points of dissimilarity are the free-radical nature of the mechanism, and the fact that the oxygen atoms derived from the reacting oxygen molecule are found in two different reaction products. Model metal ion-catalyzed reactions in which both atoms of molecular oxygen enter the substrate molecule and are found in a single oxidation product apparently have not yet been reported.

II. PARTIAL INSERTION REACTIONS

A. *Peroxide as an oxidant*

1. *Hydroxylation reactions*

Reactions in which the O–O bond is ruptured and the oxygen enters the substrate to form hydroxyl groups are generally termed hydroxylation reactions. This means that the O–O bond must be broken in such reactions, and one or both of the fragments may then enter the substrate, depending on the reaction mechanism. Since the steps leading to such hydroxyl mechanism. Since the steps leading to such hydroxyl insertion reactions may or may not go through hydrogen peroxide, it is of interest to include hydroxylation reactions in which hydrogen peroxide is itself a reactant. Such reactions are strongly catalyzed by the Fe(II) ion. Thus hydroxylation may occur by reaction with Fe(II) ion and hydrogen peroxide (Fenton's reagent), or by a model peroxidase system, such as Fe(II) ion and hydrogen peroxide in the presence of ascorbic acid.

2. *Hydroxylation by Fenton's reagent*

Barb *et al.*[15] have suggested the following mechanism for the oxidation of an organic substrate (H_2A) by Fenton's reagent.

$$Fe^{2+} + H_2O_2 \rightarrow Fe^{3+} + \cdot OH + OH^-$$
$$Fe^{2+} + \cdot OH \rightarrow Fe^{3+} + OH^-$$
$$HA + \cdot OH \rightarrow A\cdot + H_2O$$
$$A\cdot + Fe^{3+} \rightarrow A^+ + Fe^{2+}$$
$$A^+ + OH^- \rightarrow AOH$$

Examples:

phenol → dihydroxybenzenes

anisole → hydroxyanisoles

cyclohexane → cyclohexanol

Cleavage of a HO–OH bond by Fe(II) produces $\cdot OH$ and OH^-. The hydroxyl radical then reacts either with Fe(II) or the substrate to yield Fe(III) or the substrate radical, respectively. The substrate radical then undergoes oxidation to AOH. Continuation of the free radical substitution reactions results in higher hydroxylation products. Kolthoff and Medalia[16] have studied the reaction of Fe(II) with H_2O_2 in the absence of oxygen and substrate. On the basis of their kinetic evidence, the first step shown above was suggested as the rate-determining step in the reaction.

3. Model peroxidase systems

Model peroxidase systems consist of Fe(II) ion, hydrogen peroxide, and a two electron donor, such as ascorbic acid, isoascorbic acid, dihydroxyfumaric acid, ninhydrin, alloxan or 2,4,5-triamino-6-hydroxypyrimidine. The presence of a complexing agent such as EDTA enhances the catalytic activity of the metal ion, possibly by increasing its oxidation potential. The hydroxylation reaction requires equimolar amounts of the two-electron donor and hydrogen peroxide. The free radical mechanism proposed by Breslow[17] and Grinstead[18] for the hydroxylation of salicylic acid is:

$$Fe^{II}\text{-EDTA} + H_2O_2 \longrightarrow Fe^{III}\text{-EDTA} + \cdot OH + OH^-$$

$$\text{(salicylate: HO, } COO^-\text{)} \xrightarrow{\cdot OH} \text{(radical intermediate: O, H}^{\bullet}\text{, } COO^-\text{)} \xrightarrow{\cdot OH} \text{(dihydroxybenzoate: HO, HO, } COO^-\text{)}$$

$(+ H_2O)$

Regeneration of Fe(II):

$$Fe^{III}\text{-EDTA} + OH^- + H_2A \rightarrow HA\cdot + Fe^{II}\text{-EDTA} + H_2O$$
$$Fe^{III}\text{-EDTA} + OH^- + HA\cdot \rightarrow A + Fe^{II}\text{-EDTA} + H_2O$$

The first step involves cleavage of the HO–OH bond by one electron reduction of hydrogen peroxide. This is followed by reaction of the OH radical with the substrate to give an organic radical similar to that proposed by Barb *et al.*[15] for hydroxylation by Fenton's reagent. A second attack of the substrate by the hydroxyl radical yields the hydroxylated product. In the last two steps, reduction of Fe(III)–EDTA to the original catalyst, Fe(II)–EDTA, is accomplished by reaction with ascorbic acid or some other suitable two-electron donor.

The above mechanism, if correct, indicates that such systems are not appropriate models for peroxidase action, in view of the fact that metal ions do not seem to be vitally involved with the rate-determining steps of the reaction. Also, it has been shown[19] that intermediates in enzymic systems appear to have higher oxidation states of the iron–porphyrin complex of the enzyme. These intermediates are not iron–peroxide complexes because they can be formed by a variety of oxidizing agents in the absence of peroxide.

Recently, Hamilton *et al.*[20-22] investigated the oxidation of aromatic compounds and aliphatic alcohols by hydrogen peroxide with the Fe(III)–catechol complex as a catalyst, and found that the formation of a reactive intermediate is the rate-determining step in the reaction, that both alcohols and aromatic compounds are oxidized at the same rate, and that the observed rate does not depend on the concentrations of these substrates. The mechanism illustrated in Fig. 2 was suggested.

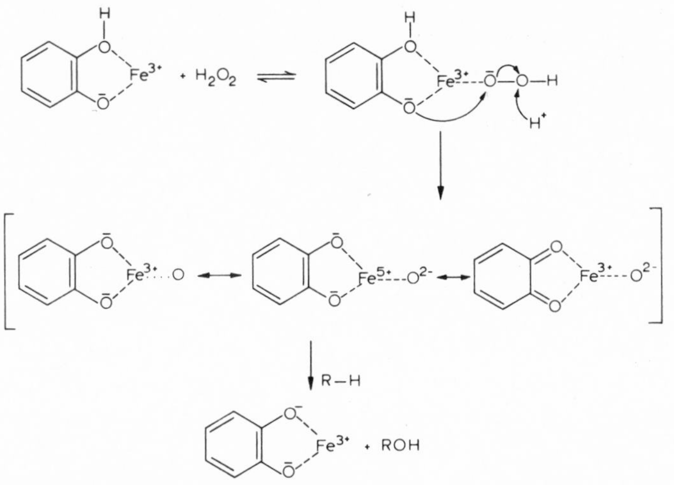

Fig. 2. Model peroxidase system.

This mechanism involves the formation of a mixed ligand complex containing an oxygen atom. This complex exists in, and is stabilized by, several delocalized electronic configurations, three of which are shown. This mixed ligand complex has an oxidation state two units higher than that of the Fe(III) ion. A description of the oxidation level applies to the complex as a whole, since it is not possible to assign specific oxidation states to the metal ion, ligand, and oxygen atom separately.

4. Peroxidase enzyme

It has been suggested by Hamilton[11] that an analogous structure may be assigned to the oxidized intermediate in reaction systems in which peroxidase (Chapter 28) is acting as the catalyst. Thus the porphyrin ring coordinated to Fe(III) in the active site of catalase, indicated by formula I below, may be oxidized to a quinoid structure (II) two oxidation units higher than the parent compound, through an electron shift, without the need to transfer or rearrange any atoms or groups.

Present knowledge of the active site of peroxidases is incomplete, and the nature of the reactions taking place, as well as the most favored electronic configuration around the coordination sphere will depend on the nature of X, Y, and substituents on the porphine ring. On the other hand, it is clear that structures such as **II** can accept electrons or donate an oxygen atom in its reaction with substrates. Thus, as recently suggested by Hamilton[11], structure **II** offers a reasonable interpretation in the light of present experimental information for the oxidized intermediate (compound I) in the reactions catalyzed by peroxidase[19]. Alternatively, it is clear that conjugated systems such as **I** and **II** can exist in several electronic states, including a number of radical (unpaired electronic) structures, that offer many possible mechanistic alternatives for the electron transfer reactions of peroxidase enzymes.

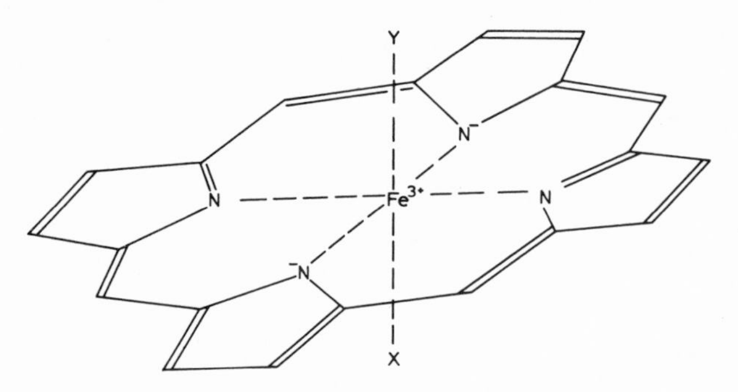

(**I**) Active site in peroxidase enzymes.

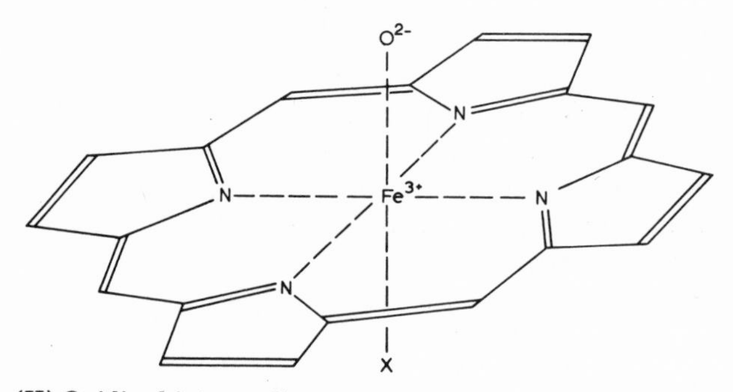

(**II**) Oxidised intermediate.

B. Hydroxylation by direct reduction of molecular oxygen

1. Model systems

The system Fe(II)–EDTA–ascorbic acid–molecular oxygen was first employed by Udenfriend and coworkers[23-25] for the hydroxylation of an organic substrate. Udenfriend's system is similar to the model peroxidase systems described above, with oxygen substituting for hydrogen peroxide. As with model peroxidase systems, Udenfriend's system involves a two electron donor and a suitably complexed transition metal ion catalyst, such as Cu(II), Co(II), Mn(II), Sn(II)[26], and Fe(II)[20,27]. The most widely used complexing agent is EDTA, though pyrophosphate, amino acids, F^-, N_3^- and CN^- have also been employed with varying success. The presence of a complexing agent is not essential, although the reaction seems to be considerably enhanced in its presence. Stoichiometric quantities of oxygen and ascorbic acid (or other electron donor) are consumed in the reaction. The nature of the hydroxylating species in Udenfriend's system has not been determined conclusively. Grinstead[18] suggested that the hydroperoxide free radical, $HO_2^{\cdot}$ is the reactive species and that hydrogen peroxide is an intermediate in the reaction.

Recent work on isomer distribution of reaction products by Norman and Rodda[28] and by others[29–31] has revealed that hydrogen peroxide is not an intermediate in the reaction, and that the hydroxylating agent is not a free radical. More recently Hamilton and coworkers[27] demonstrated that the Udenfriend system would not only hydroxylate aromatic compounds, but would also convert saturated hydrocarbons to alcohols, and olefins to epoxides. The relationship of these reactions to those of the mono-oxygenases, described below, was demonstrated by the fact that tetrahydropteridines[32] (which are reducing agents for the oxygenases) and related pyrimidine derivatives[27] may replace ascorbic acid in the Udenfriend system.

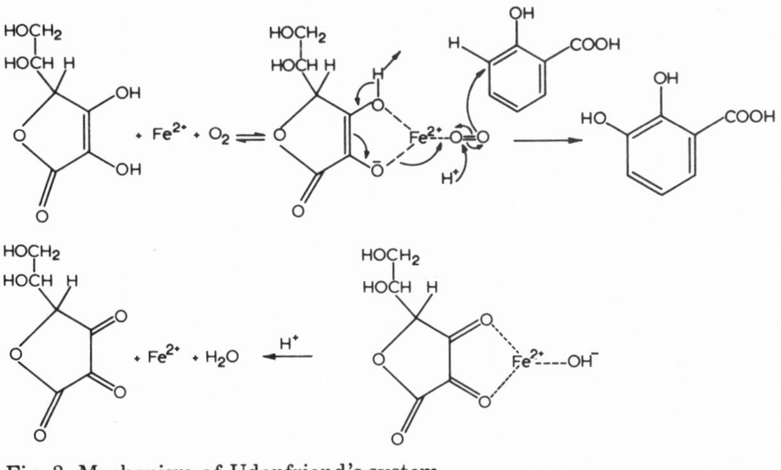

Fig. 3. Mechanism of Udenfriend's system.

The reaction sequence in Fig. 3 illustrates the mechanism suggested by Hamilton[11]. The concerted electron transfer reactions occurring in this intermediate involve the oxidation of the ascorbic acid moiety to dehydroascorbic acid, electron transfer through the metal ion to the oxygen atom which is not transferred, and transfer of the other oxygen as a neutral electrophilic species to the substrate. The energy required to break the oxygen–oxygen bond is regained by the concerted formation of more stable oxygen bonds with other elements.

The metal ion, in addition to holding together the activated complex and providing the pathway for the electron transfer from the reductant to molecular oxygen, may also have the function of providing some stability for the oxygen complex prior to reaction with the substrate. Thus the Fe(II) ion may combine with oxygen in a prior equilibrium to give a complex which is stabilized somewhat by the resonance forms indicated below:

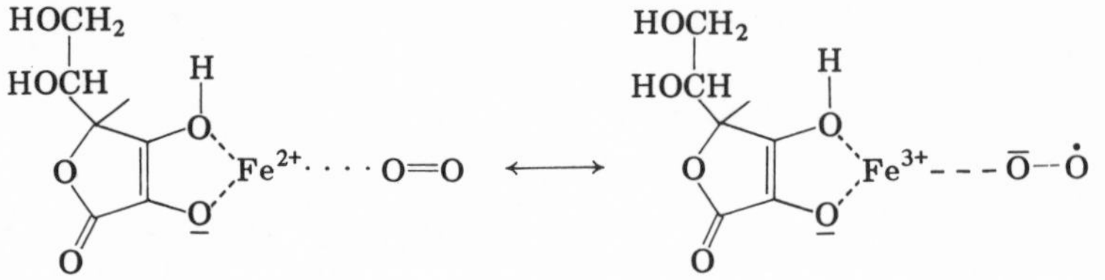

$HOCH_2$ $HOCH_2$
HOCH H HOCH H
$Fe^{2+}\cdots O{=}O \longleftrightarrow Fe^{3+}\text{---}\bar{O}\text{–}\dot{O}$

2. *Mono-oxygenase or mixed function oxidase (MFO) enzymes*

The mono-oxygenases, described in reviews by Mason, Hayaishi and coworkers[5–7] in most cases require a transition metal ion and a reducing agent, and catalyze many partial oxygen insertion reactions such as hydroxylation of aromatic and aliphatic hydrocarbons, the epoxidation of olefins, oxidative decarboxylation, and oxidative dehydroxy-methylation. The reducing agents are usually tetrahydropteridines, reduced flavins, or ferridoxins, but may also be pairs of reduced metal ions at active enzymic sites.

It is of interest to consider possible mechanistic pathways for the mono-oxygenase enzymes. It is clear that while O_2 is a four electron oxidant, only two of the oxidizing equivalents (*i.e.*, one oxygen atom) are used to oxidize the substrate. It is also known that substrate peroxides are not involved in the mechanisms of reaction of these enzymes. Therefore O–O bond fission must occur prior to combination of one oxygen atom with the substrate, or there must be a concerted reaction for the three processes: O–O fission, insertion of one oxygen atom, and reduction of the other oxygen atom to H_2O. Further, the reduced enzyme is now known[5-7] to be the form that reacts with the oxygen and (for a concerted mechanism) with the substrate.

Possible mechanisms for both the concerted and consecutive processes are provided by the enzyme models described above. The concerted mechanism could be similar to that suggested for the Udenfriend system, whereby the oxygen is bonded simultaneously to the reduced metal ion (Fe^{2+}) and to the substrate, at least in the transition state, allowing a concerted electron shift which results in O–O bond fission. Alternatively it is possible that an oxidized intermediate such as that indicated above for the peroxidase model system could be formed, and that this in turn could donate an oxygen atom to the substrate. The two possible mechanisms can therefore be represented by the following reaction sequences:

Concerted reaction

$$EH_2 + S + O_2 \rightleftharpoons EH_2 \cdot S \cdot O_2 \rightarrow E \cdot SO \cdot H_2O \rightleftharpoons E + SO + H_2O$$

Oxidation of enzyme, followed by oxygen insertion

$$EH_2 + S + O_2 \rightleftharpoons EH_2 \cdot S \cdot O_2 \rightarrow EO \cdot S \cdot H_2O \rightleftharpoons EO \cdot S + H_2O$$

$$EO \cdot S \overset{\bullet}{\rightarrow} E \cdot SO \rightleftharpoons E + SO$$

where: EH_2 = reduced enzyme or enzyme + reduced coenzyme,
S = substrate.
SO = initial product after oxygen insertion.
$EH_2 \cdot S \cdot O_2$ = ternary enzyme-substrate-oxygen complex (activated (complex).

Careful kinetic studies of these enzyme systems should make it possible to distinguish between these two general mechanisms, as well as between the several sequential alternatives possible in the second mechanism. Apparently such experiments have not yet been carried out.

3. Nature of the reducing agent in MFO enzyme systems

A concerted mechanism similar to that suggested above for Udenfriend's system has been proposed[11,27,32–34] for mixed function oxidases requiring Fe(II) and a tetrahydropteridine, which undergoes a two-electron oxidation in the reaction. Figure 4 is a tentative reaction sequence analogous to the Udenfriend mechanism.

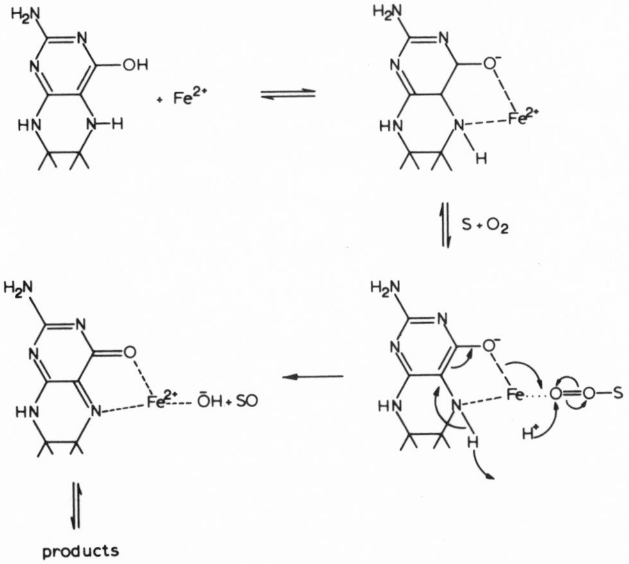

Fig. 4. Tetrahydropteridine as reductant in Udenfriend's system.

The active site of the enzyme, which contains Fe^{2+}, serves to bring the reactants together perhaps through initial formation of a resonance-stabilized Fe(II)–oxygen complex before or after combination with the reducing agent. The latter is itself a coordinating ligand and may combine initially with metal ion. As suggested by Hamilton[11], the enzyme also probably assists the proton transfer pictured above through acid–base catalysis mediated by proton donors and acceptors at the active site.

Many other types of reducing agents could serve the function of the tetrahydropteridine indicated in the above mechanism, the essential requirement being that it be capable of transferring two electrons to one of the oxygen atoms while the other undergoes insertion into the substrate. Thus when the metal ion at the active site is a single electron transfer agent, such as the copper ions in dopamine-β-hydroxylase, the metal ions are believed to react in electronically linked pairs so that they can transfer two electrons to oxygen. In this case it has been suggested that each pair of copper atoms, originally in the +1 oxidation state, combine with one oxygen molecule to form a complex analogous to the one described above for the Fe(II)-enzyme:

$$Cu^{1+}\text{---}Cu^{1+}\cdots O{=}O\cdots S \rightleftharpoons Cu^{2+}\text{---}Cu^{2+}\text{---}OH^- + SO$$

$$H^+$$

4. Mechanism of insertion step

Several mechanisms are possible for the transfer of the oxygen atom to the substrate in mixed function oxidases. The transition states illustrated above may exist in several resonance forms, so that the oxygen moiety being inserted may enter the substrate as a positive, neutral, or even negative species. Whatever the detailed mechanism that applies for a particular enzyme, it is expected to be strongly influenced by the type of ligand coordinated to the metal ion. Thus it has been suggested[11] that the insertion reaction is electrophilic in the epoxidation of olefins, by analogy with known chemical and model reactions. Comparisons with chemical systems for aromatic hydroxylation suggest either an electrophilic or radical addition of oxygen. In either case, the hydrogen shifts detected in these reactions[35] are consistent with the formation of a benzene epoxide intermediate:

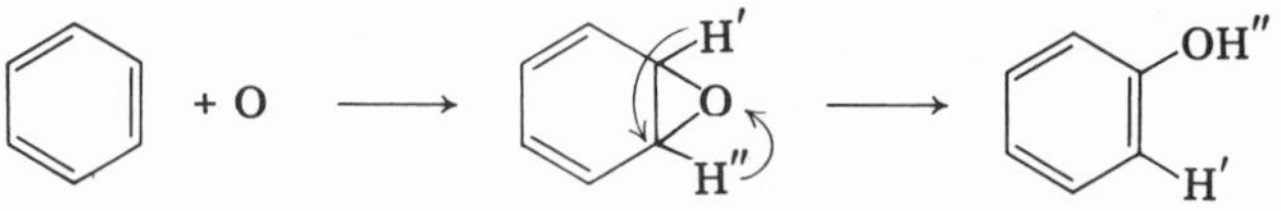

$$\text{C}_6\text{H}_6 + O \longrightarrow \text{(epoxide; } H', O, H''\text{)} \longrightarrow \text{C}_6\text{H}_4(OH'')(H')$$

Another MFO-catalyzed reaction, which is readily rationalized by assuming a benzene epoxide-type intermediate, is the oxidative dehydroxymethylation of aromatic compounds[36] as follows:

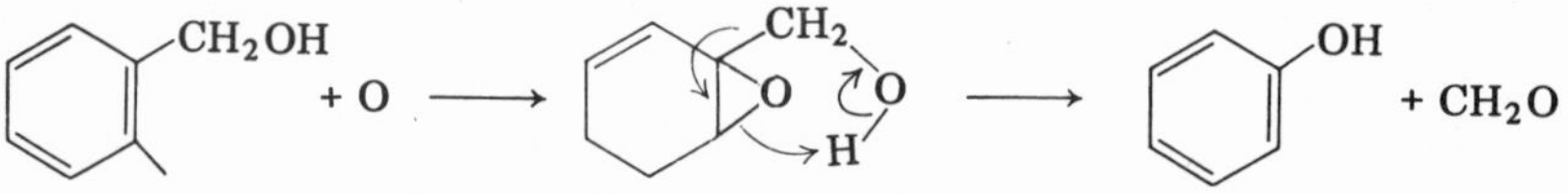

$$\text{Ar}{-}CH_2OH + O \longrightarrow \text{(epoxide; } CH_2, O, O, H\text{)} \longrightarrow \text{Ar}{-}OH + CH_2O$$

The resonance stabilization of the aromatic ring can be regenerated if the proton is transferred to the electron pair in the oxygen-bridge, thus opening the three-membered ring, as indicated above.

5. Partial insertion through free radical addition of oxygen

Partial insertion reactions may result from free radical addition of oxygen to form peroxides, which may then decompose to more stable products. Decomposition of the peroxide is often accomplished by the catalytic action of metal ions. An example of this type of reaction is the oxidation of cumene to acetophenone in the presence of Co(III) as a catalyst[37]. The rate of the catalytic oxidation of cumene is proportional to the first power of the catalyst, the substrate and of molecular oxygen concentrations; in accordance with the following mechanism:

$$C_6H_5CH(CH_3)_2 + X \longrightarrow C_6H_5\dot{C}(CH_3)_2 + XH$$

(X is a free radical initiator)

$$C_6H_5\dot{C}(CH_3)_2 \xrightarrow{O_2} C_6H_5C(CH_3)_2{-}O{-}O\cdot$$

$$C_6H_5C(CH_3)_2{-}O{-}O\cdot + C_6H_5CH(CH_3)_2 \longrightarrow C_6H_5C(CH_3)_2OOH + C_6H_5\dot{C}(CH_3)_2$$

$$C_6H_5C(CH_3)_2OOH + Co^{3+} \longrightarrow C_6H_5C(CH_3)_2OO\cdot + Co^{2+} + H^+$$

$$C_6H_5C(CH_3)_2OOH + Co^{2+} \longrightarrow C_6H_5C(CH_3)_2O\cdot + Co^{3+} + OH^-$$

$$C_6H_5C(CH_3)_2O\cdot \longrightarrow C_6H_5C(=O)CH_3 + \cdot CH_3$$

Cumene hydroperoxide is produced from the free radical by reaction with oxygen followed by free radical exchange with cumene. In the presence of Co(II), cumene hydroperoxide decomposes rapidly to acetophenone.

Thus far natural, enzyme-catalyzed reactions involving reactive free radical intermediates analogous to those described above have not been reported. Apparently nature prefers to carry out reactions smoothly and efficiently, without the formation of free radicals that are highly reactive and lead to many side reactions that are difficult to control. The ionic mechanisms described above for partial insertion of oxygen provide more reasonable kinetic pathways for biological systems.

References pp. 686–688

6. Insertion without O–O bond cleavage

The combination of oxygen with an organic or organometallic compound sometimes occurs without cleavage of the O–O bond. The product of such a reaction must therefore be a peroxide.

Formation of organoperoxides. Many organic compounds undergo free radical reactions with molecular oxygen to produce organic peroxides. Free radical initiation may be readily accomplished with a metal ion, resulting in a chain reaction involving direct combination of oxygen with an organo free radical.

An example of this type of reaction is the formation of a bistetralin peroxide from tetralin and molecular oxygen in the presence of decanoates of Mn(II), Cu(II), Ni(II), and Fe(II) as catalysts[33]. Initiation of the reaction chain takes place by reaction of a small amount of tetralin hydroperoxide with the metal ion. The reaction proceeds with a small steady state concentration of tetralin hydroperoxide:

Initiation:

$$C_{10}H_{11}OOH \xrightarrow{M^{n+}} C_{10}H_{11}OO\cdot + M^{(n-1)+} + H^+$$

Propagation:

$$C_{10}H_{12} + C_{10}H_{11}OO\cdot \rightarrow C_{10}H_{11}\cdot + C_{10}H_{11}OOH$$

$$C_{10}H_{11}\cdot + O_2 \rightarrow C_{10}H_{11}OO\cdot$$

Termination:

$$2C_{10}H_{11}OO\cdot \rightarrow C_{10}H_{11}OOC_{10}H_{11} + O_2$$

A similar reaction is the Cu(II)-phthalocyanine catalyzed oxidation of cumene to cumene hydroperoxide by molecular oxygen. The following mechanism has been proposed for this reaction by Kropf[38]:

$$\text{Cu(II)-Pc} + O_2 \rightleftharpoons (\text{Cu(II)-Pc})^{\delta+}\, O_2^{\delta-}$$

$$(\text{Cu(II)-Pc})^{\delta+}\, O_2^{\delta-} + RH + ROOH \longrightarrow \text{Cu(II)-Pc} + ROO\cdot + R\cdot + H_2O_2$$

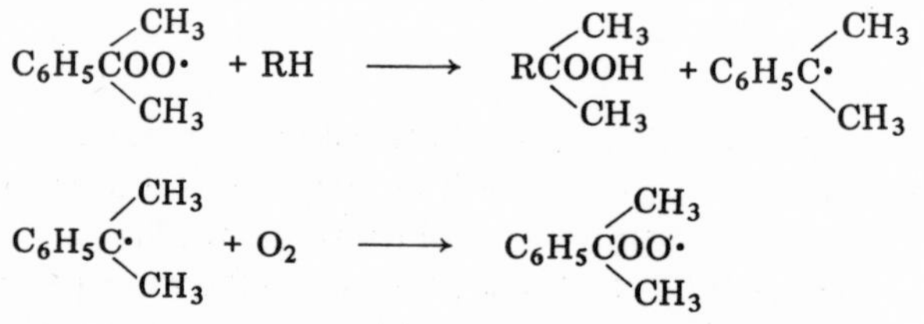

$$C_6H_5C(CH_3)_2OO\cdot + RH \longrightarrow RC(CH_3)_2OOH + C_6H_5C(CH_3)_2\cdot$$

$$C_6H_5C(CH_3)_2\cdot + O_2 \longrightarrow C_6H_5C(CH_3)_2OO\cdot$$

Apparently the function of the copper phthalocyanine oxygen complex is simply the generation of free radicals, with the formation of hydrogen peroxide. The main coversion to the peroxide is carried out by a chain reaction involving molecular oxygen.

Metal organoperoxy compounds. Organometallic compounds of B, Al, Zn, Cd, Mg and Li react with molecular oxygen in ether solution to yield organoperoxy metallic compounds. The mechanism suggested[39] for the insertion of molecular oxygen in the organometallic compound is an initial nucleophilic attack of molecular oxygen on the metal ion, followed or accompanied by a 1,3-rearrangement of the alkyl group from the metal ion to oxygen in the manner depicted below:

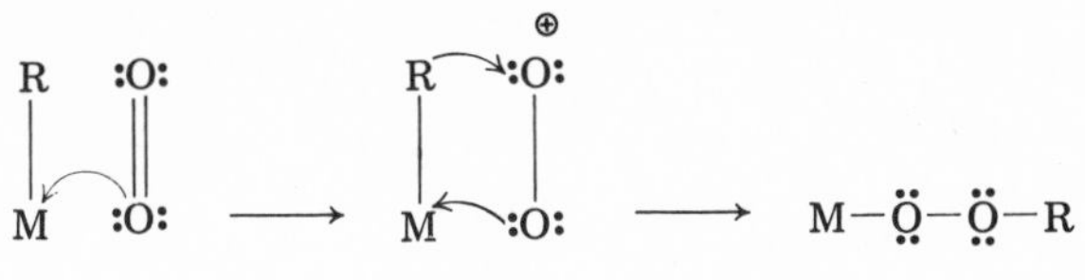

$$\begin{matrix} R & :O: \\ | & \| \\ M & :O: \end{matrix} \longrightarrow \begin{matrix} R & :\overset{\oplus}{O}: \\ | & | \\ M & :O: \end{matrix} \longrightarrow M{-}\ddot{\underset{..}{O}}{-}\ddot{\underset{..}{O}}{-}R$$

III. NON-INSERTION REACTIONS IN WHICH OXYGEN IS REDUCED TO PEROXIDE

A. *Synthetic and natural oxygen carriers*

Oxygen-carrying compounds are generally metal complexes that combine reversibly with molecular oxygen. As will be shown below, chemical and structural evidence indicate that these compounds may be considered formally to be compounds in which molecular oxygen is reduced to a peroxo or superoxo ligand, and the metal ion undergoes a corresponding oxidation. The reversibility of the formation of the oxygen complex and decomposition to the original molecular oxygen and oxygen-free complex requires facile electron shifts within the complex and near-equivalence in stability of the two states.

1. *Formation and properties of synthetic oxygen carriers*

Although a few examples of the formation of oxygen carriers in solution have been known for some time[40] only during the past few years has sufficient data been accumulated to indicate the large number of complexes that may be formed, and the generality of the oxygenation–coordination reaction with synthetic chelate compounds in solution. Fallab[41,42] and Miller and Wilkins[43,44] have described the composition and kinetics of formation of some oxygenated complexes of cobalt(II)–polyamine chelates at high pH.

The equilibria between oxygen and cobalt(II) complexes have been measured quantitatively by Nakon[45,46]. These studies have revealed new acid–base stoichiometry for the oxygen complexes in solution, and relatively high stability constants for their formation from gaseous oxygen, the metal ion, and the ligand. The stoichiometries of oxygenation are indicated by Figs. 5 and 6. Potentiometric data showed the existence of the hydroxo

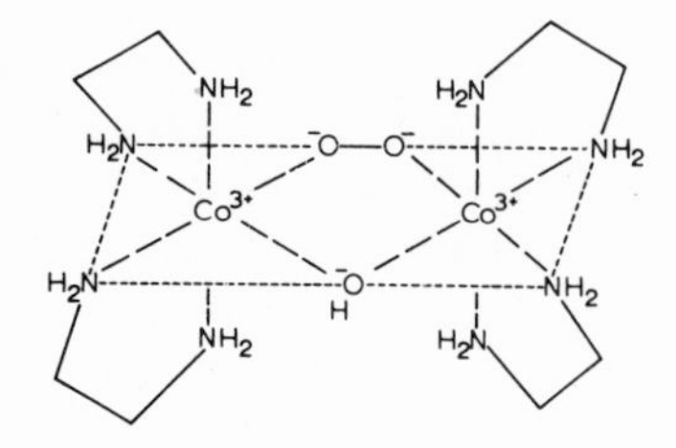

Tetrakisethylenediamine-μ-hydroxo-μ-peroxo-dicobalt ion, $Co_2L_4O_2OH^{3+}$ (ethylenediamide = L)

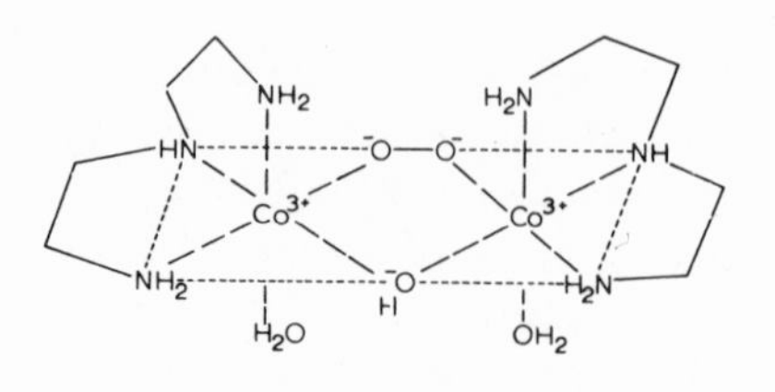

Bisdiethylenetriamine-μ-hydroxo-μ-peroxo-dicobalt(III) ion, $Co_2L_2O_2OH^{3+}$ (diethylenetriamine = L)

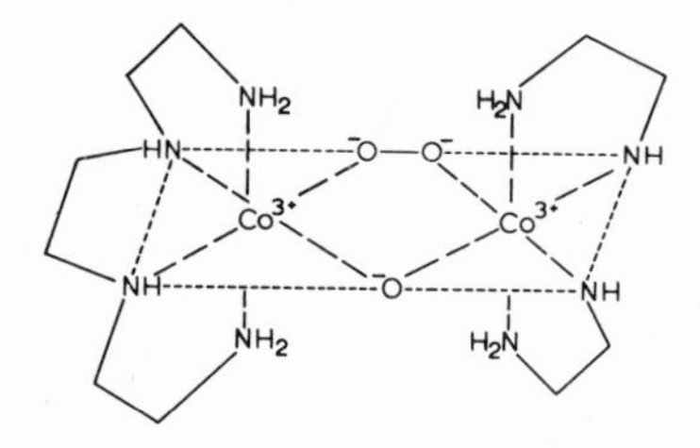

Bistriethylenetetramine-μ-hydroxo-μ-peroxo-dicobalt (III) ion, $Co_2L_2O_2OH^{3+}$ (triethylenetetramine = L)

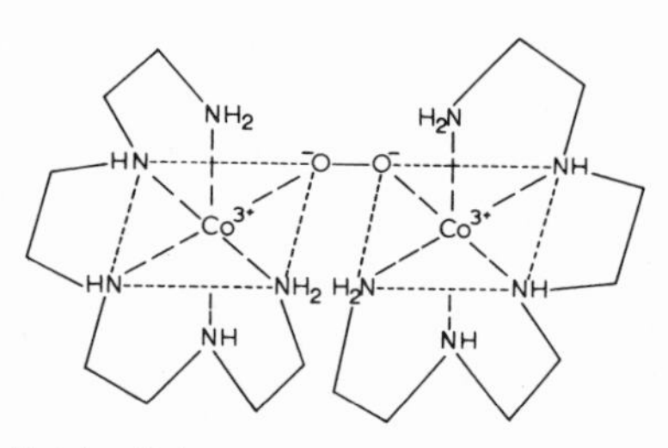

Bistetraethylenepentamine-μ-peroxo-dicobalt (III) ion, $Co_2L_2O_2^{4+}$ (tetraethylenepentamine = L)

Fig. 5. Oxygen complexes of Co(II) and polyamines.

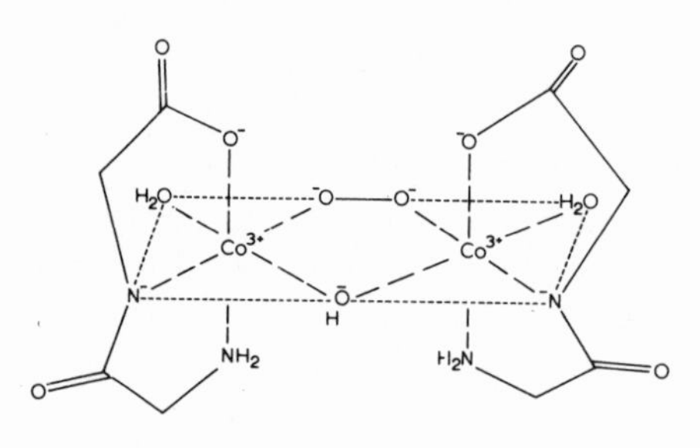

Bisglycylglycinato-μ-hydroxo-μ-peroxodicobalt (III) ion, $Co_2(H_{-1}L)_2O_2(OH)^-$ (glycylglycine = HL)

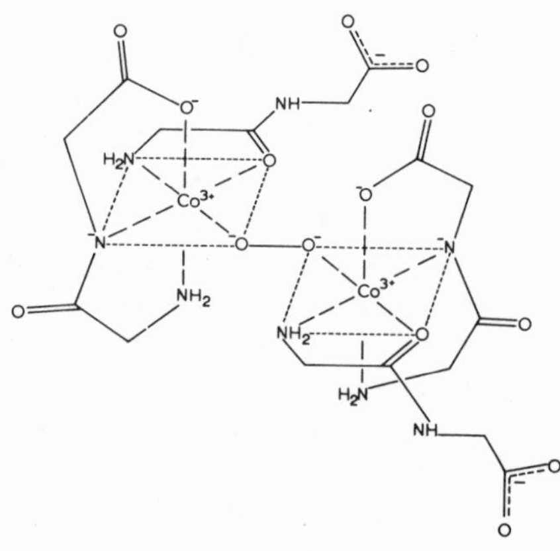

Tetrakisglycylglycinato-μ-peroxodicobalt (III) ion, $Co_2(H_{-1}L)_2L_2O_2$

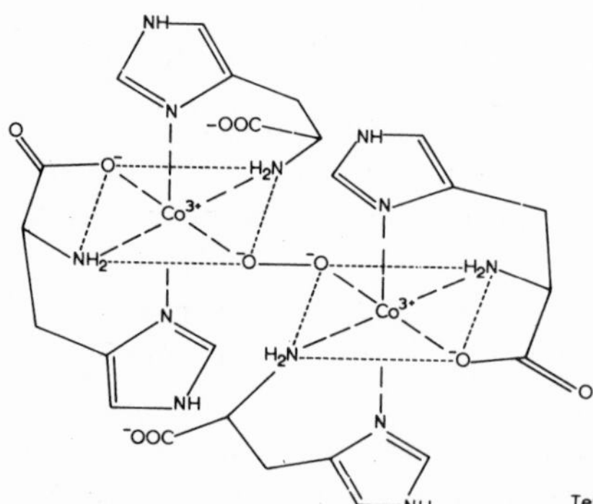

Tetrakishistidinato-μ-peroxodicobalt (II) ion, $Co_2O_2L_4$

Fig. 6. Oxygen complexes of glycylglycine and histidine.

bridging group in the binuclear oxygen complexes of ethylenediamine, diethylenetriamine, triethylenetetramine, and the 1:1 glycylglycine–cobalt chelate. For histidine and glycylglycine, a 2 : 1 or higher molar ratio of ligand to metal ion gave binuclear oxygen complexes without hydroxo bridges, presumably because all of the available coordination positions of the cobalt atoms are occupied by the ligand donor groups and the oxygen molecule. This interpretation was supported by the fact that the 1 : 1 tetraethylenepentamine cobalt complex combines with oxygen to give a binuclear oxygen-bridged complex without a hydroxo bridge, whereas hydroxo bridges formed in all cobalt oxygen complexes containing lower polyamines. In the latter the binuclear oxygen complexes would have cobalt atoms with one or more coordination sites occupied by water molecules, which would be very reactive toward hydroxo group coordination, since such groups are formed by dissociation of a proton from a coordinated water molecule.

The cobalt atom in all of the oxygen complexes shown below may be considered to be in the +3 oxidation state, with the coordinated oxygen reduced to a binegative peroxide anion. Evidence pointing to this type of electronic structure include diamagnetism of the complex, the unusually high stability of the oxygen complex, and the fact that coordination with oxygen greatly increases the interaction of the metal ion with the ligand, as measured potentiometrically.

2. Structures of some oxygen carriers

Most of the synthetic oxygen carriers are polynuclear complexes, and frequently have structures in which two atoms of a metal ion are coordinated to a molecule of oxygen. A compound that has recently been thoroughly investigated is the decammine-μ-peroxodicobalt(II) cation, that exists in both the reduced form, with an ionic charge of +4, and in an oxidized form, with a charge of +5. The species with +4 charge is brown and diamagnetic, while that with +5 charge is green and paramagnetic. Vleck[47] suggested a perpendicular bonding of oxygen to the *d* orbitals of both the cobalt ions of the complex; similar to that found in olefin–metal complexes. X-ray analysis of nitrate salt of the oxidized form of the decammine–cobalt oxygen carrier by Brosett and Vannerberg[48,49] seemed to support the theoretical conclusion of Vleck. Recently Schaefer and Marsh[50] have determined the X-ray structure of a crystalline sample of the paramagnetic form, $[(NH_3)_5Co–O_2–Co(NH_3)_5](SO_4)_2(HSO_4) \cdot 3H_2O$, and concluded that the coordination of two cobalt ions in the paramagnetic cation is nearly exactly octahedral, with an average deviation of the ligand–cobalt–ligand angle from 90° of less than 2° (Co–N distance is 1.95 ± 0.02 Å). The bridging peroxo group was found to be not perpendicular to the Co–Co axis and not bonded to the cobalt ions by $d\pi$ bonds. As indicated in Fig. 7, each oxygen atom of the peroxy group is σ-bonded to each of the cobalt atoms resulting in a staggered arrangement of the Co–O–O–Co bonds, so that the two horizontal planes of the coordination spheres are nearly coplanar. The O–O distance in the paramagnetic complex was found to be 1.31 Å, considerably shorter than that reported by Brosett and Vannerberg (1.45 Å), for the paramagnetic $[(NH_3)_5Co\text{-}O_2\text{-}Co(NH_3)_5](NO_3)_5$ and slightly longer than the O–O distance in the superoxide anion (1.28 Å). On the basis of the O–O bond distance in $[(NH_3)_5Co\text{-}O_2\text{-}Co(NH_3)_5]^{3+}$, it is concluded[50] that the oxygen in this complex is a superoxide anion, $O_2^{\cdot-}$.

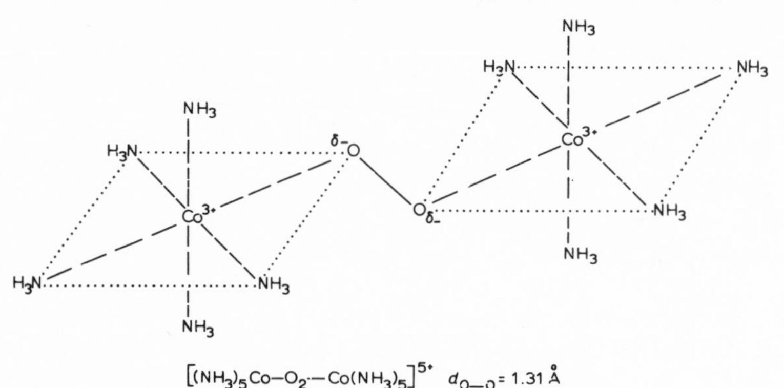

Fig. 7. Decammine-μ-peroxodicobalt(III).

It may be added here that the X-ray analysis of Schaeffer and Marsh[50] indicates that oxygen in the pseudo oxygen carrier, the decaammine-μ-peroxo-dicobalt complex, is a superoxide anion. The suggestion of σ-bonding of oxygen to cobalt is based on the nearly perfect octahedral symmetry of each cobalt ion. Partial π-bonding to the $Co\text{-}O_2$ bonds, because of some overlap of properly oriented *d* orbitals of cobalt with partly vacant anti-bonding orbitals of $O_2^{\cdot-}$, can not be completely ruled out. The electron spin resonance (e.s.r.) spectra[51] of the decaammine-μ-peroxo-dicobalt cation indicate that the unpaired electron interacts equally with both cobalt nuclei and resides for most of the time on the oxygen molecule.

Vaska[52] has synthesized a mononuclear Ir(I) oxygen carrier, formulated as $IrClCO[(C_6H_5)_3P]_2 \cdot O_2$ (see also Chapter 13, p. 361). Bis(triphenylphosphine)chlorocarbonyl-Ir(I) in benzene solution takes up one molecule of oxygen per metal atom to yield a photosensitive diamagnetic oxygen complex, that has a dipole moment of 5.9 D in benzene. The structure of this Ir–oxygen carrier, determined by Ibers and LaPlaca[53] by X-ray analysis, has a planar arrangement of Cl^-, CO, Ir and O_2, with the triphenylphosphines above and below the plane, as indicated in Fig. 8. This structure may be described as trigonal pyramid, if oxygen is considered as a single ligand, or a distorted octahedron if two oxygen atoms are considered to be separately bonded to iridium(I). The two oxygen atoms are however, equidistant from iridium, with an Ir–O distance of 2.06 ± 0.02 Å, and an O–O bond distance of 1.28 Å. The O–O bond distance is thus characteristic of a superoxide radical, $O_2^{\cdot -}$. The observed diamagnetism of the complex may be due to spin coupling of the unpaired electron in Ir(II) and $O_2^{\cdot -}$.

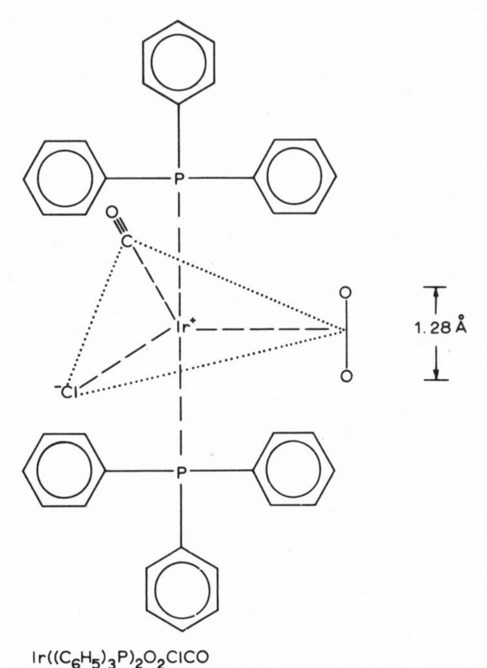

Fig. 8. "Vaska" complex.

An alternative arrangement of bonds would involve sigma bonding to Ir(I) by the π-electrons of the oxygen, with back donation of electrons from the appropriate *d* orbitals of the metal of the antibonding π orbitals of the oxygen. These exchange phenomena account for the coordination of the oxygen. Also, the removal of electrons from the bonding orbitals of the oxygen, and the filling of its antibonding orbitals, could easily account for the lengthening of the O=O bond of molecular oxygen to the value observed. In a sense this picture is similar to the alternate one described above, with the possible exception that it gives a better picture of the coordination of the oxygen molecule.

For the natural oxygen carrier haemoglobin, Weiss[53a] has suggested an oxidation state of +3 for iron and an unpaired electron on the coordinated oxygen, to give the superoxide anion, O_2^-. This suggestion is based on the observed diamagnetism, the dissociation constant, and the absorption spectrum of the oxygen carrier. In the strong field of the porphyrin ligand, Fe(III) ion is considered to have one unpaired electron, which couples with the unpaired electron on $\cdot O_2^-$ to give a diagmagnetic complex.

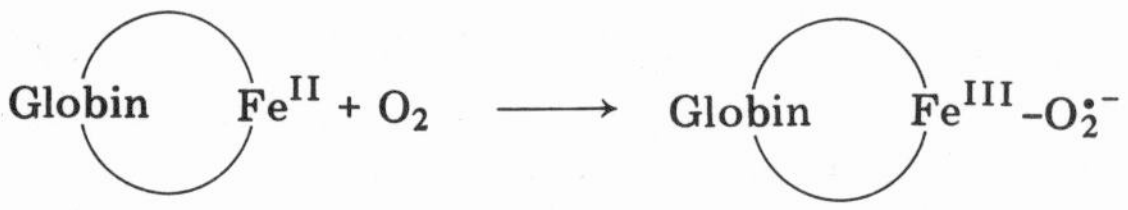

$$\text{Globin}\ \text{Fe}^{II} + O_2 \longrightarrow \text{Globin}\ \text{Fe}^{III}\text{-}O_2^{\cdot -}$$

3. Reactions of oxygen carriers

Oxygen carriers may participate in redox reactions in two ways: (1) the reaction may take place directly with the coordinated oxygen, which, as we have seen above, may have properties that vary from molecular oxygen through superoxide to peroxide, depending on the metal ion, substrate, and reaction conditions; (2) the metal ion itself may act as an electron acceptor, by virtue of its tendency to be oxidized by the coordinated oxygen. The possibility of the metal ion serving as the reactive site would be enhanced by the presence of one or more free coordination sites about the metal ion. The following are a few examples of the reactions of oxygen carriers.

The oxidation of Fe(II) has been reported by Sykes[54,55] by utilizing $[(NH_3)_5Co\text{-}O_2\text{-}Co(NH_3)_5]^{5+}$ as oxidant. The rate-determining step in the reaction of this cation is the formation of the normal diamagnetic oxygen carrier from the paramagnetic species:

$$[(NH_3)_5Co\text{-}O_2\text{-}Co(NH_3)_5]^{5+} + Fe^{2+} \xrightarrow{\text{slow}} Fe^{3+} + [(NH_3)_5Co\text{-}O_2\text{-}Co(NH_3)_5$$

$$[(NH_3)_5Co\text{-}O_2\text{-}Co(NH_3)_5]^{4+} \xrightarrow{\text{fast}} 2[Co(NH_3)_5]^{2+} + O_2$$

Beck and Gorog[56] have reported the concerted oxidation of the Co(II)-glycylglycine (GG) complex and ascorbic acid by molecular oxygen. The formation of the oxygen carrier $[GG\text{-}Co\text{-}O_2\text{-}Co\text{-}GG]$ increases in the presence of ascorbic acid. At the end of the reaction Co(II)–glycylglycine is irreversibly oxidized to the corresponding Co(III) complex. According to Beck, molecular oxygen is initially activated by ascorbic acid, resulting in more complete formation of the Co(II)-glycylglycine–oxygen complex. The oxygen carrier then oxidizes ascorbic acid more effectively than does molecular oxygen.

Further studies in the authors' laboratory[45,46] revealed that only the glycylglycine–cobalt–oxygen system with excess glycylglycine, and the incompletely coordinated cobalt oxygen complex in the polyamine systems (1 : 1 diethylenetriamine) accelerate the rate of ascorbic acid oxidation. In the 1 : 1 diethylenetriamine oxygen complex, it is seen that two coordination sites are available on each cobalt atom for coordination with the substrate. In the glycylglycine complex, it is noted that one glycylglycine is strongly bound with peptide proton dissociation, while the other is weakly bound in a bidentate fashion, and may be readily displaced by the substrate.

These observations led to the proposal of a new mechanism for the oxidation of ascorbic acid by oxygen carriers, indicated in Fig. 9 which assumes direct coordination of ascorbic acid to the cobalt, followed by electron transfer to convert the latter to Co(II) and the oxygen to peroxide.

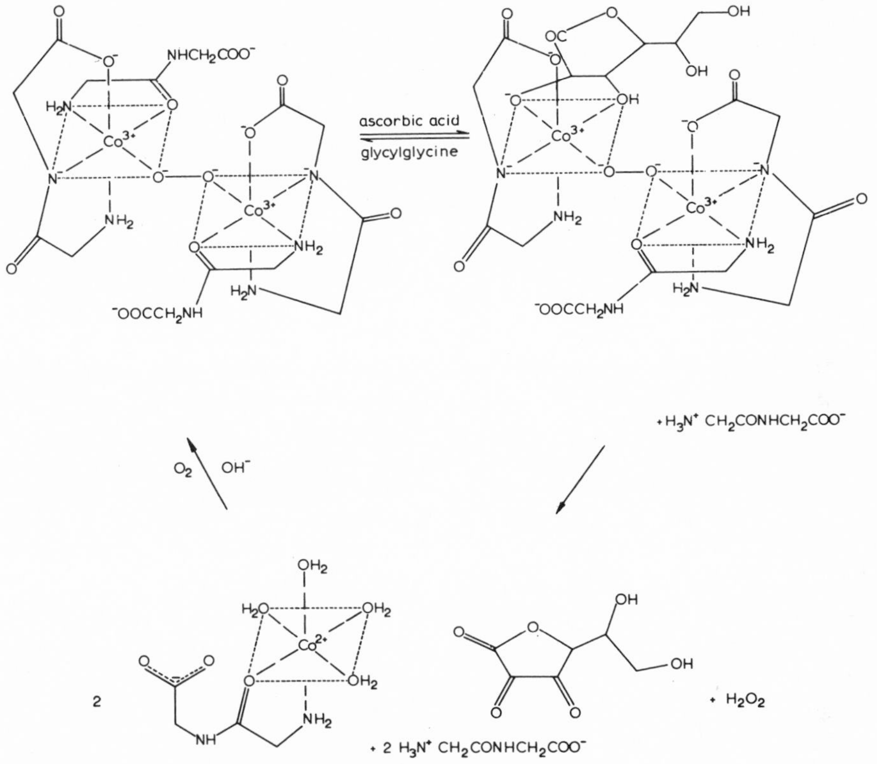

Fig. 9. Ascorbate oxidation by oxygen complex.

While the oxygen complex itself may be considered to be in a form in which the oxygen ligand and the metal ions do not have any particular oxidation state, the evidence cited above indicates that the "average" or "hybrid" electronic structure lies very close to a state in which the metal ion is Co(III) and the oxygen bridge is O_2^{2-}. The many examples cited above of the ionic mechanism for electron transfer from reductant to oxidant lends confidence in the proposed mechanism. Thus electrons are transferred from the ascorbate anion to the metal which as Co(II) can then complete the remaining shift (which may be slight) of electrons to the coordinated oxygen converting it to O_2^{2-}. Transfer of protons through the solvent completes the conversion of O_2 to hydrogen peroxide without the formation of intermediate free radicals.

References pp. 686–688

B. Model systems

1. Metal ion and metal chelate catalyzed oxidation of ascorbic acid

Metal ion catalysis of the oxidation of ascorbic acid, which has been known for a long time, is an interesting example of a non-insertion reaction in which the oxygen is reduced to peroxide or water, and the metal ion serves as an electron transfer agent between the substrate and the oxidant. This reaction has recently been given detailed study by the authors of this review and will be described in some detail. The Cu(II)-ion catalyzed oxidation of ascorbic acid by molecular oxygen was reported by Weissberger *et al.*[57], and the Fe(III)-EDTA catalyzed oxidation of ascorbic acid by hydrogen peroxide was described by Grinstead[18]. In the case of the Cu(II) and Fe(III)-ion catalyzed oxidation of ascorbic acid, the rate of the reaction has been found[58,59] to be first order with respect to ascorbate anion, metal ion catalyst, oxygen concentration, and the reciprocal of the hydrogen ion concentration. The kinetic observations support the mechanism outlined in Fig. 10 for Cu(II) and Fe(III) ion catalyzed oxidation of ascorbic acid.

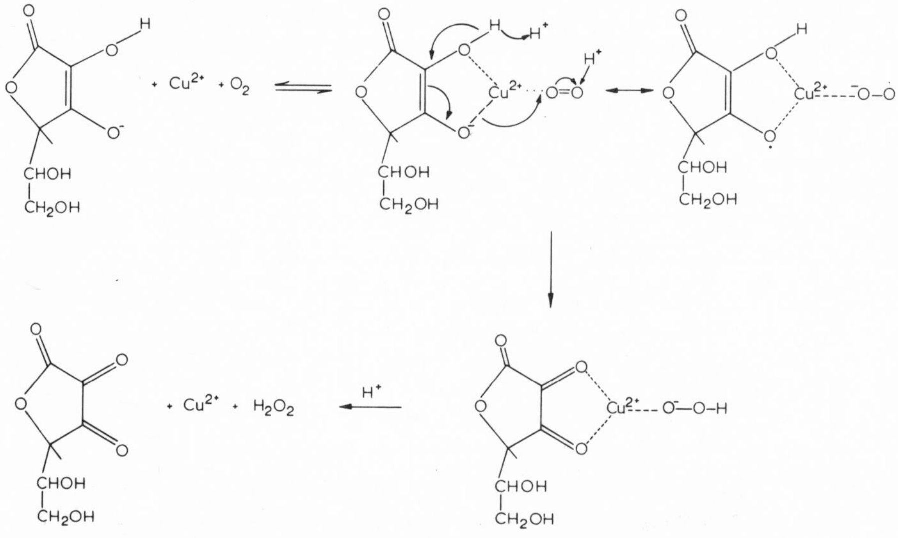

Fig. 10. Mechanism of copper(II)-catalyzed oxidation of ascorbic acid.

The free radical intermediate that is formed as the result of the first electron shift (Fig. 10) is merely a resonance form of the initial oxygen–ascorbate mixed ligand copper(II) complex, since no nuclei have actually moved. Therefore the first and probably rate-determining step is the one in which a proton is shifted to form the dehydroascorbic acid–hydroperoxide mixed ligand complex. Since the now electron-deficient dehydroascorbic acid is a weak donor, the latter complex rapidly dissociates to regenerate the catalyst.

In the presence of Cu(II) chelates[60], the reaction mechanism appears to be different, since in this case the reaction rate is not dependent on the concentration of oxygen employed, and the oxygen is reduced to water, whereas H_2O_2 was produced in the absence of chelating agents. Here the rate determining step, as indicated in Fig. 11, appears to be the reduction of Cu(II) to Cu(I) and the formation of a semiquinone-like radical, involving a single electron transfer. Rapid reoxidation of the Cu(I) chelate by oxygen in the solution makes possible a relatively rapid second electron transfer step to produce the dehydroascorbic acid.

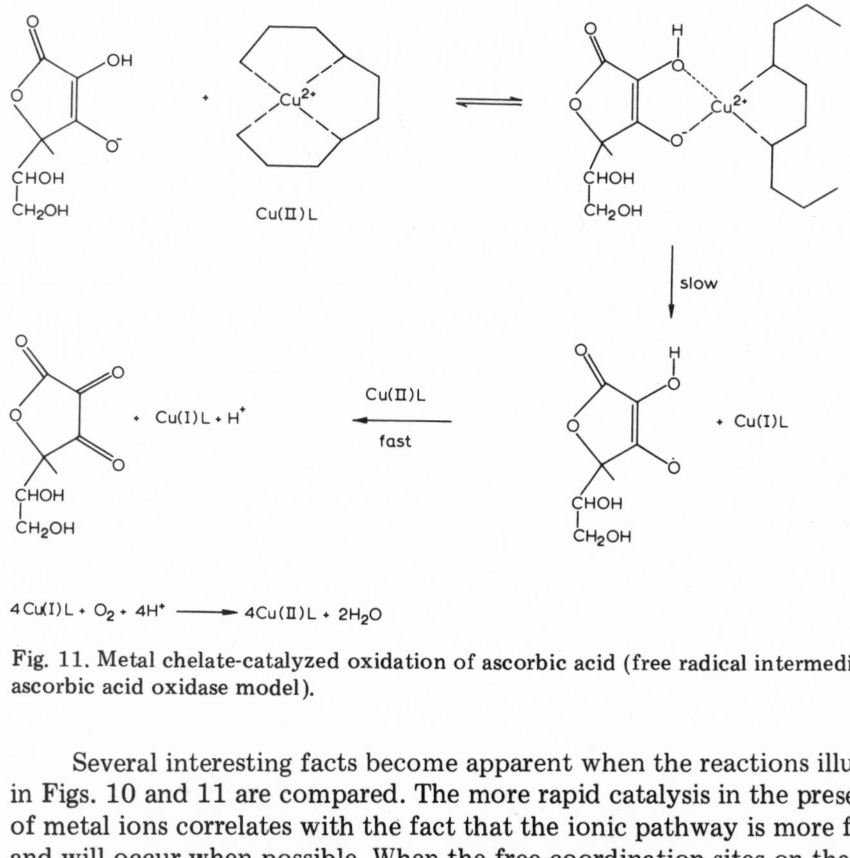

$$4Cu(I)L + O_2 + 4H^+ \longrightarrow 4Cu(II)L + 2H_2O$$

Fig. 11. Metal chelate-catalyzed oxidation of ascorbic acid (free radical intermediate-ascorbic acid oxidase model).

Several interesting facts become apparent when the reactions illustrated in Figs. 10 and 11 are compared. The more rapid catalysis in the presence of metal ions correlates with the fact that the ionic pathway is more facile and will occur when possible. When the free coordination sites on the metal ion are blocked by coordination with a chelating ligand, it is now difficult to form a complex containing both the oxygen and the substrate, and the reaction occurs through the alternative but more difficult free radical pathway.

A radical mechanism such as is indicated in Fig. 11 would not be expected to occur in cases where the formation of a free radical intermediate is not stabilized by resonance to the extent possible in the ascorbate radical.

2. *Oxidation of catechols*

The oxidation of catechols to the corresponding quinones in enzymic and model systems is very similar to the oxidation of ascorbic acid, described above.

Metal ion catalysis of the non-enzymic oxidation of 3,5-ditertiary-butylcatechol by molecular oxygen has been reported by Grinstead[61]. A two-step free radical mechanism, in which the metal ion is reversibly oxidized by oxygen and reduced by the substrate, was proposed for the conversion of the catechol derivative to the corresponding quinone. The catalytic activities of metal ions were reported to vary in the order:

$$Mn(II) > Co(II) > Fe(II) > Cu(II) > Zn(II) > Ni(II)$$

A recently-completed study of the oxidation of pyrocatechol and ditertiarybutylpyrocatechol by O_2 in the presence of various divalent metal chelates showed the formation of variable amounts of H_2O_2 (depending on the metal ion employed) with little or no formation of free radical intermediates[62]. In this case the carrier ligand is always bidentate so that additional coordination positions are available around the metal ion for combination with molecular oxygen. The reaction mechanism which we now propose for the first time for this reaction, illustrated in Fig. 12, involves the formation of a substrate-oxygen carrier ligand complex of the metal ion, in which transfer of two electrons can occur through the metal ion from substrate to oxygen via an ionic mechanism.

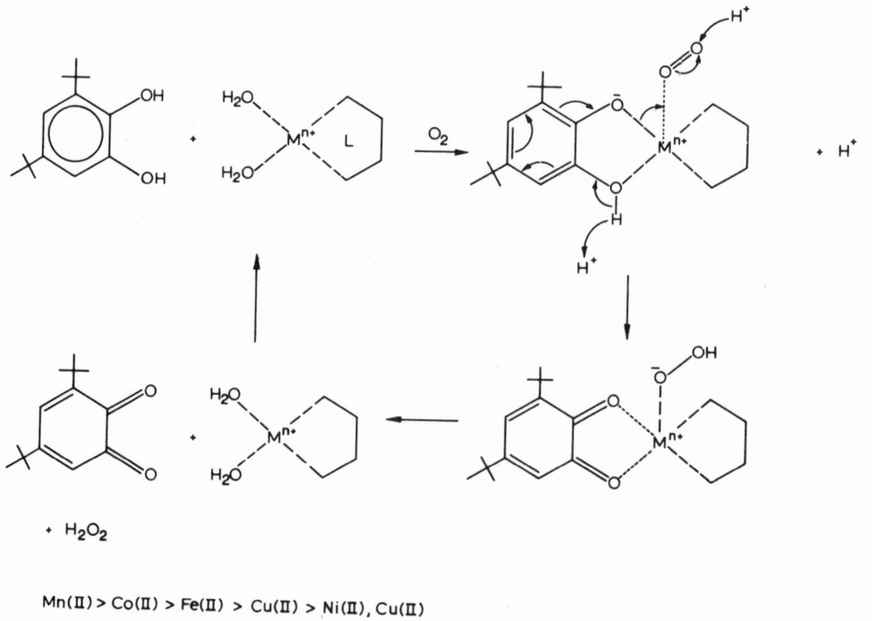

+ H_2O_2

Mn(II) > Co(II) > Fe(II) > Cu(II) > Ni(II), Cu(II)

Fig. 12. Metal chelate catalyzed oxidation of di-t-butyl catechol.

3. *Amine oxidase models*

A model system for the oxidation of α-amino acids to α-keto acids has recently been studied by Hamilton and Revesz[63], by Hill and Mann[64], and in the authors' own laboratory[46]. Many amino acids and amino acid derivatives are oxidized to α-keto acids and ammonia by molecular oxygen in the presence of pyridoxal and transition metal ions at room temperature and somewhat elevated pH (pH ~9). The mechanism of this reaction is related to those of other pyridoxal-catalyzed reactions[65,66] in that a Schiff base chelate is formed as an intermediate (Chapter 31). However, the reaction seems to occur without change of valence of the metal ion, and without the formation of free radical intermediates. Moreover pyridoxamine is not produced in the reaction so that it does not proceed via the familiar transamination route to the keto acid.

In view of this experimental information, an ionic mechanism outlined in Fig. 13 is proposed. According to this mechanism, an electron shift from the substrate Schiff base through the metal ion to loosely coordinated oxygen produces a quinonoid structure in the organic ligand while converting the oxygen ligand to hydroperoxide. Metal-catalyzed hydrolysis of the Schiff base and subsequent hydrolytic removal of ammonia regenerates pyridoxal and the original catalyst.

C. *Enzymic systems*

The enzyme ascorbic acid oxidase seems to oxidize ascorbic acid by a radical mechanism similar to that indicated in Fig. 11[67]. An oxidized form of this copper enzyme apparently oxidizes the ascorbic acid to dehydroascorbic acid in two successive one-electron steps, with the intermediate formation of a free radical, which has been detected by e.s.r. spectroscopy[68]. In the course of the reaction the copper is reversibly reduced and oxidized, the oxidized form being regenerated by reaction with O_2.

An ionic mechanism similar to that illustrated in Fig. 13 for model systems may apply to the amine and amino acid oxidase enzymes that require copper(II) and pyridoxal phosphate for activity, and in fact are the only pyridoxal phosphate-containing enzymes that require O_2 as a reactant. Also it is known from e.s.r. measurements that the Cu(II) does not change its valence during enzymic catalysis[69-71].

There are many more enzymes that catalyze dehydrogenation reactions in which the oxidant, molecular oxygen, is reduced to hydrogen peroxide. Many of these systems are believed to react through the ionic mechanisms similar to those described above for the metal ion-catalyzed oxidation of ascorbic acid, catechols and amino acids. Some of these enzymes, and experimental observations pointing toward their mechanisms of reaction, are listed in Table III.

References pp. 686–688

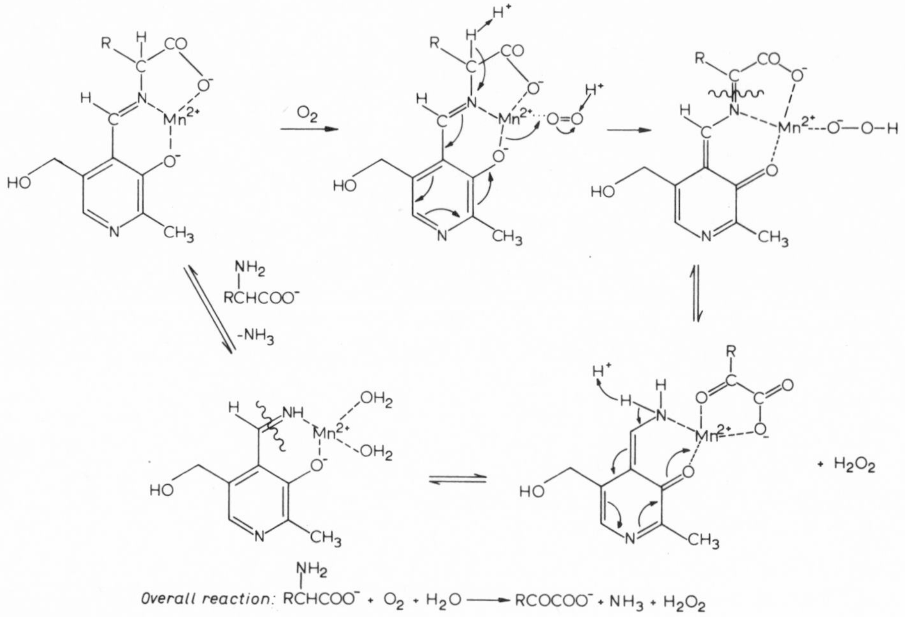

Overall reaction: $RCH(NH_2)COO^- + O_2 + H_2O \longrightarrow RCOCOO^- + NH_3 + H_2O_2$

Fig. 13. Amine and amino acid oxidase models.

TABLE III

EXAMPLES OF DEHYDROGENATION REACTIONS CATALYZED BY OXIDASE ENZYMES

Enzyme	Redox reactions	Mechanistic information	Ref.
Ascorbic acid oxidase	Ascorbic acid → dehydroascorbic acid	Semiquinone intermediates	67
	$O_2 \rightarrow H_2O_2$	Copper reversibly oxidized and reduced	68
Amino acid oxidases	Amino acid → keto acid + NH_3	Cu(II) and pyridoxal phosphate required	69–71
	$O_2 \rightarrow H_2O_2$	Cu(II) does not change valence	
Uricase	Uric acid → intermediates → allantoin	One active site per molecule containing Cu II)	72
	$O_2 \rightarrow H_2O_2$	Cu(II) does not undergo redox reactions	
		No free radical intermediates	
Galactose oxidase	Galactose → corresponding aldehyde	Cu(II) at active site does not change valence	73
	$O_2 \rightarrow H_2O_2$		
		No free radicals detected	
Impure enzyme preparation	Side chain of cholesterol cleaved to two compounds	Metal ion not known	74
		No free radical intermediates	
	$O_2 \rightarrow H_2O_2$ (NADPH required) (H_2O_2 reacts with aldehyde)		
Inositol oxygenase	Myoinositol → dialdehyde → D-glucuronic acid	One atom of iron per mole	75–77
	$O_2 \rightarrow H_2O_2$ (H_2O adds to aldehyde)		
Oxalic acid oxidase	Oxalic acid → CO_2	Metal ion possible but not detected	78
	$O_2 \rightarrow H_2O_2$		

1. Non-insertion reactions involving complete reduction of O_2 to H_2O

Enzymic oxidation of diphenols to quinones. The oxidation of diphenols to quinones, with the corresponding reduction of oxygen to water, occurs in biological systems by both ionic and free radical reactions. These reactions generally occur through a dehydrogenation mechanism, but free radicals (semiquinones) may or may not be formed depending on the enzymes and substrates involved. Thus the catecholase action of tyrosinase seems not to involve the formation of semiquinone intermediates and the metal ion (Cu(I)) does not seem to change valence[79]. On the other hand laccase, which also catalyzes the dehydrogenation of diphenols to quinones through the reduction of oxygen to water, is known[80,81] to involve semiquinone radicals as intermediates, and the metal ion is found to be reversibly oxidized and reduced. An ionic mechanism suggested for catechol oxidation is illustrated in Fig. 14.

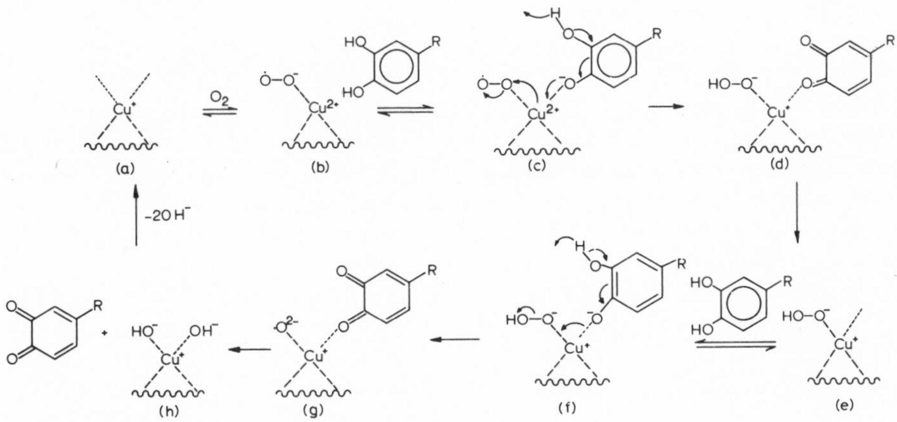

Fig. 14. Mechanism of tyrosinase action.

Thus the catecholase activity of tyrosinase can be explained by a series of electronic shifts through the metal ion from the catechol derivative to the oxygen, with redistribution of protons through the solvent to neutralize the negative charges that are transferred. It is seen that two successive two-electron catechol oxidations are necessary to reduce one oxygen molecule to the elements of water. There is no valence change in the metal species shown, it being remembered that the combination of Cu^{2+} with a superoxide anion (b) is electronically equivalent to the coordination of Cu^{+} with molecular oxygen.

The mechanism of laccase action (Chapter 21) may be considered somewhat similar to that given above for catecholase with the insertion of intermediate steps involving the formation of semiquinone structures between (c) and (d), and between (g) and (h). In order that these free radical intermediates and metal redox reactions be detectable, it is necessary

References pp. 686–688

that they have considerable lifetimes and appreciable concentrations, and be at least partly dissociated from the metal site. Such considerations suggest that the semiquinone may be generated at a different metal center from that to which the oxygen is attached, and that the relatively stable semiquinone intermediate may migrate from one site to the other to complete the redox process.

Apparently with substrates that can readily form stable free radical intermediates such as ascorbic acid and catechols, both the radical and ionic mechanisms may occur. The question of whether free radicals may or may not be formed depend on the nature of the coordination sphere of the metal ion at the one or more active sites of the enzyme, where the semiquinone may be generated and released or further reacted.

2. Non-Insertion Free Radical Reactions

Metal-catalyzed non-insertion free radical oxidation reactions with molecular oxygen. There are many examples of metal catalysis of free radical reactions of organic substances in which oxygen serves to regenerate the oxidized form of the metal ion. In these reactions the metal ion acts as the free radical initiator, and thus serves as the intermediate in the transfer of electrons from the substrate to the oxygen.

Oxidation of benzoic acid to phenol. A typical free radical reaction is the Cu(II) catalyzed oxidation of benzoic acid to phenol in the presence of molecular oxygen[82]. The stoichiometry of the reaction is indicated by the following equation:

$$C_6H_5COOH + 1/2\ O_2 \longrightarrow C_6H_5OH + CO_2$$

Cu(II)-benzoate is fused in benzoic acid and a mixture of air and steam is bubbled into the melt at a temperature of 200–240°C. (Benzoic acid is a desirable solvent for the reaction because of its high boiling point which permits the reaction to be conducted at atmospheric pressure.) The following overall reaction sequences have been proposed for the production of the phenol from benzoic acid[82]:

$$2Cu^{2+}(^{-}OOCC_6H_5)_2 \longrightarrow C_6H_5COO\text{-}C_6H_4\text{-}COOH\ (ortho) + 2Cu^{+}\text{--}^{-}OOCC_6H_5$$

$$2CuOOCC_6H_5 + 2C_6H_5COOH + 1/2\ O_2 \longrightarrow 2Cu(OOCC_6H_5)_2 + H_2O$$

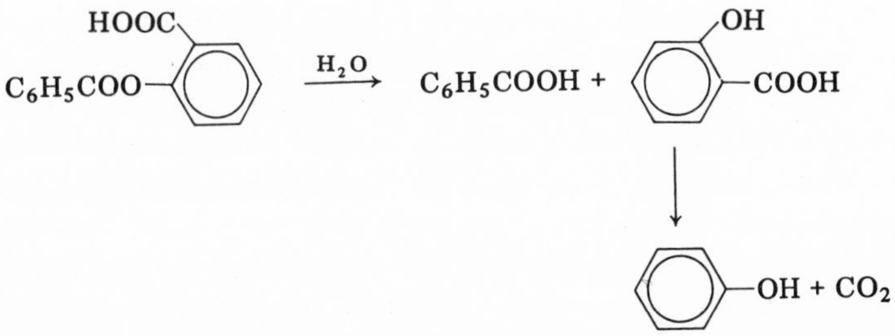

$$C_6H_5COO\text{–}C_6H_4\text{–}COOH \xrightarrow{H_2O} C_6H_5COOH + HO\text{–}C_6H_4\text{–}COOH$$

$$HO\text{–}C_6H_4\text{–}COOH \longrightarrow C_6H_5OH + CO_2$$

The important step in this reaction is the Cu(II)-catalyzed reaction of benzoate to give salicyl benzoate. The reaction probably goes through two successive electron transfers to Cu(II) to form Cu(I) as indicated. The first step probably produces the benzoic radical ϕCOO which then attacks an adjacent benzoate ion in the *ortho* position. Completion of the reaction is accomplished by a second electron transfer to copper(II). The copper (II) is regenerated by direct oxidation, and the phenol is formed by hydrolysis of the salicyl benzoate and subsequent decarboxylation.

Coupling reactions of organic compounds. Metal ions may generate organo free radicals that then undergo coupling reactions. The oxygen serves to reoxidize the metal ion, which is the reactive intermediate in the formation of the organic free radical.

An example of this type of reaction, Fig. 15, is the conversion of 2,6-ditertiarybutylphenol to 3,3′,5,5′-tetra(tertiarybutyl)diphenoquinone in the presence of Cu(I) and oxygen, described by Ochiai[83]. Amines such as pyridine, ethylenediamine, diethylenetriamine, or triethylenetetramine, were used as coordinating ligands for the copper(I)/copper(II) system. The mechanism in Fig. 15 seems to apply to this reaction.

$$2CuL_n^+ + 1/2\,O_2 + 2(m-n)L + 4H^+ \longrightarrow 2CuL_m^{2+} + H_2O$$

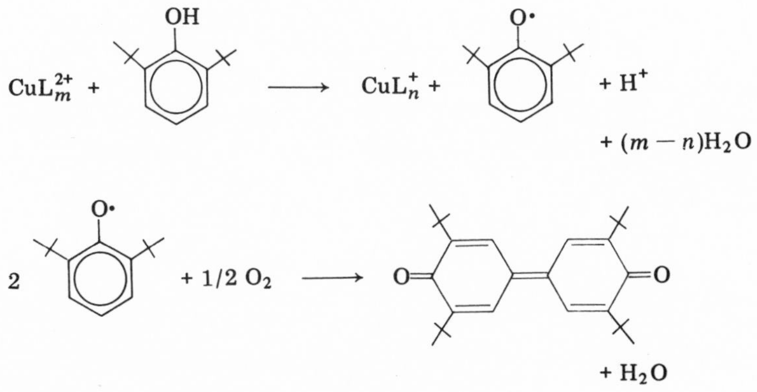

$$CuL_m^{2+} + \text{(2,6-di-}t\text{-butylphenol, ArOH)} \longrightarrow CuL_n^+ + \text{ArO}\cdot + H^+ + (m-n)H_2O$$

$$2\,\text{ArO}\cdot + 1/2\,O_2 \longrightarrow O\text{=}(\text{diphenoquinone})\text{=}O + H_2O$$

Fig. 15. Metal ion-catalyzed coupling reaction.

References pp. 686–688

Enzyme-catalyzed coupling reactions. It is now recognized that oxidative coupling reactions are the basis for the biosynthesis of many natural products[84]. The mechanisms for such reactions are probably quite similar to the free-radical process described above for the non-enzymic system, although Hamilton[11] has suggested an ionic mechanism for biological coupling reactions. An observation that favors the free-radical process is that both enzymic and non-enzymic reactions give very low yields of coupled reaction products. Thus far no enzymes have been isolated that are specific for individual coupling reactions, but it is expected that such enzymes will eventually be isolated and identified.

IV. DISPROPORTIONATION OF HYDROGEN PEROXIDE

1. *Catalase models*

Cu(II)–polyamine complexes

In an interesting review of model catalase and peroxidase reaction, Sigel[85] has compared the catalase activities of a number of Cu(II) chelates, some of which are indicated in Fig. 16. It is seen that for the polyamine–Cu(II) chelates two free coordination sites are required for full catalytic activity. The fully coordinated Cu(II) complexes are inactive.

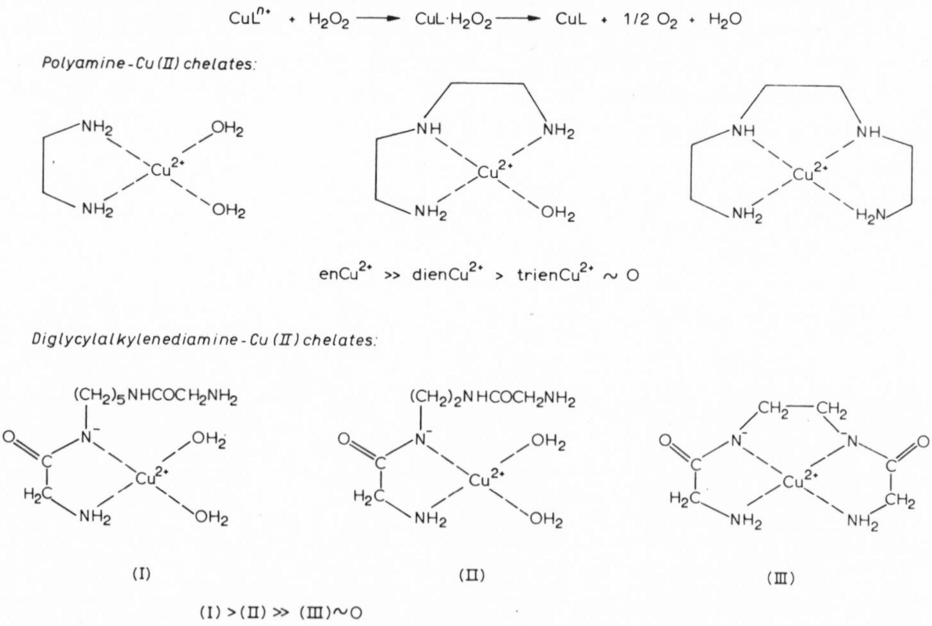

$$CuL^{n+} + H_2O_2 \longrightarrow CuL \cdot H_2O_2 \longrightarrow CuL + 1/2\ O_2 + H_2O$$

Polyamine-Cu(II) chelates:

$$enCu^{2+} \gg dienCu^{2+} > trienCu^{2+} \sim 0$$

Diglycylalkylenediamine-Cu(II) chelates:

(I) (II) (III)

$$(I) > (II) \gg (III) \sim 0$$

Fig. 16. Copper(II) chelates as models of catalase action.

Similarly, for the diglycyl derivatives of diamines that coordinate Cu(II) with the displacement of protons from the amide nitrogens, the most active species are those in which the ligand occupies two coordination sites on the metal ion, and in which the probability of forming more completely coordinated species is minimal. Here again the highly stable complex in which the ligand is tetradentate is inactive as a catalyst.

The reaction mechanism therefore seems to involve the combination of the metal ion with two hydroxide moieties, followed by an intramolecular electron shift and a corresponding transfer of protons through the solvent, in a manner suggested by the ionic redox mechanism indicated below in Fig. 17.

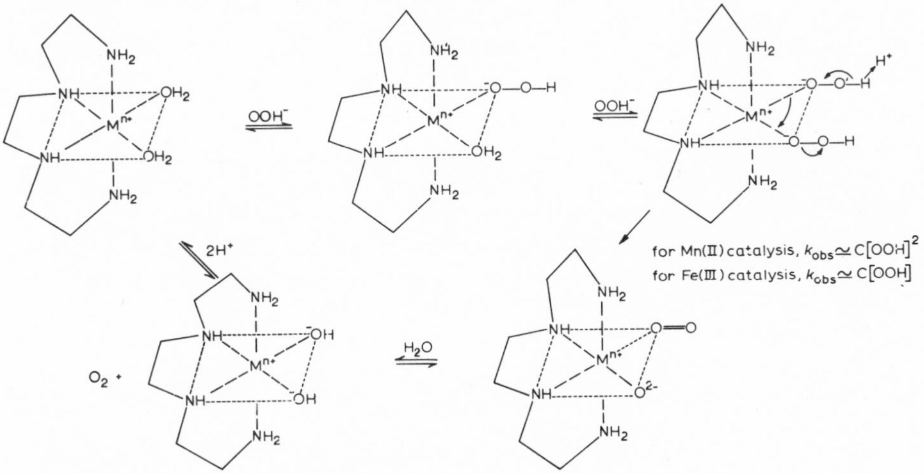

for Mn(II) catalysis, $k_{obs} \simeq C[OOH]^2$
for Fe(III) catalysis, $k_{obs} \simeq C[OOH]$

Fig. 17. Triethylenetetramine chelates as models of catalase action.

Mn(II) and Fe(III)-trien systems

The concept of the requirement of two free coordination positions about the metal ion is strengthened by the observation that for Mn(II) and Fe(III), which have coordination numbers of six, the diethylenetriamine chelate is again an active catalyst. The catalase activities of these metal chelate systems were first described by Wang[86,87] who proposed that the reaction takes place through metal coordination with one peroxide anion in an bidentate fashion so as to occupy two *cis* positions of the metal ion. On the basis of the fact that the reaction is second order in peroxide concentration when Mn^{2+} is the catalytic metal ion, Hamilton[11] has proposed the mechanism described in Fig. 17, in which two hydroperoxide groups are simultaneously coordinated to the two available sites on the metal ion. The electron transfer can then occur smoothly through the metal ion, to give oxygen and water, the protons being transferred through the aqueous solvent. The first order dependence of the Fe(III)-trien reaction on peroxide

References pp. 686–688

concentration was considered due to the fact that the higher coordinating ability of the Fe(III) ion resulted in one of the peroxide anions being bound to the Fe(III) ion at the beginning of the reaction. In view of all the facts available this reaction mechanism certainly seems to be a reasonable one, and also applies to the observations made on the Cu(II) catalase model systems.

2. *Catalase enzymes*

The catalases[88] (Chapter 28) which have iron(III)-porphyrin prosthetic groups at the active site, are very efficient catalysts for the disproportionation of hydrogen peroxide to water and oxygen, known as the catalatic reaction. Catalases are very closely related to the peroxidases[89,90], and the two types of enzymes having certain mechanistic steps in common. An oxidized form of the enzyme, compound (I), forms when the free enzyme reacts with hydrogen peroxide or an organic hydroperoxide. The oxidized intermediate, described on p. 661, then reacts directly with hydrogen peroxide to give O_2. As suggested by George[19], compound (I) probably does not contain coordinated peroxide, but consists of an oxidized form of the iron-porphyrin system as indicated by the formula on page 662.

REFERENCES

1 K. S. Pitzer, *J. Am. Chem. Soc.*, 70 (1948) 2140.
2 A. D. Walsh, *J. Chem. Soc.*, (1948) 331.
3 K. B. Yatsimirski, *Izvest. Vysshikh Ucheb, Zavedenii Chim., Khim. Teknol.*, 2 (1959) 480; *Chem. Abstr.*, 54 (1960) 6289h.
4 O. Hayaishi, *Proc. 6th Int. Congr. Biochem.*, Plenary Sessions, New York, IUB Vol. 33, 31 (1964).
5 H. Mason, *A. Rev. Biochem.*, 34 (1965) 595.
6 T. E. King, N. S. Mason and M. Morrison, (Eds.) *Oxidases and Related Redox Systems*, Vols. 1 and 2, John Wiley, New York, 1965.
7 K. Block and O. Hayaishi, (Eds.) *Biological and Chemical Aspects of Oxygenases*, Maruzen, Tokyo, 1966.
8 O. Hayaishi and K. Hashimoto, *J. Biochem. (Japan)*, 37 (1950) 371.
9 O. Hayaishi, M. Katagiri and S. Rothberg, *J. Biol. Chem.*, 229 (1957) 905.
10 H. S. Mason, *Adv. Enzymol.*, 19 (1957) 79.
11 G. A. Hamilton, *Adv. Enzymol.*, 33 (1969) 55.
12 A. E. Martell, *Proc. 3rd Int. Conf. Coordination Chem.*, Plenary Sessions, Debrecen, Hungary, 1970.
13 E. S. Gould, *Mechanism and Structure in Organic Chemistry*, Henry Holt & Co., New York, 1959, p. 633.
14 O. Hayaishi, M. Katagiri and S. Rothberg, *J. Am. Chem. Soc.*, 77 (1955) 5450.
15 W. G. Barb, J. H. Baxendale, P. George and K. R. Hargrave, *Trans. Faraday Soc.*, 47 (1951) 462.
16 I. M. Kolthoff and A. I. Medalia, *J. Am. Chem. Soc.*, 71 (1949) 3777.
17 R. Breslow and L. N. Lukens, *J. Biol. Chem.*, 235 (1960) 292.

18 R. R. Grinstead, *J. Am. Chem. Soc.*, 82 (1960) 3472.
19 P. George in D. E. Green, *Currents in Biochemical Research*, Interscience Publishers, New York, 1956, p. 338.
20 G. A. Hamilton and J. P. Friedman, *J. Am. Chem. Soc.*, 85 (1963) 1008.
21 G. A. Hamilton, J. P. Friedman and P. M. Campbell, *J. Am. Chem. Soc.*, 88 (1966) 5266.
22 G. A. Hamilton, J. W. Hanifin, Jr. and J. P. Friedman, *J. Am. Chem. Soc.*, 88 (1966) 5269.
23 J. Axelrod, S. Udenfriend and B. Brodie, *J. Pharmacol. Exp. Therap.*, 111 (1954) 176.
24 S. Udenfriend, C. T. Clark, J. Axelrod and B. B. Brodie, *Federation Proc.*, 31 (1952) 301.
25 S. Udenfriend, C. T. Clar, J. Axelrod and B. B. Brodie, *J. Biol. Chem.*, 208 (1954) 731, 741.
26 C. Nofre, A. Cier and A. Lefier, *Bull. Soc. Chem. (France)*, (1961) 530.
27 G. A. Hamilton, R. J. Workman and L. Woo, *J. Am. Chem. Soc.*, 86 (1964) 3390.
28 R. O. C. Norman and G. K. Rodda, *Proc. Chem. Soc.*, (1962) 130.
29 V. Ullrich and H. Standinger, in K. Bloch and O. Hayaishi, *Biological and Chemical Aspects of Oxygenases*, Maruzen, Tokyo, 1966, p. 235.
30 H. Standinger, B. Kerekjarto, V. Ullrich and Z. Zubrzychi, in T. E. King, H. S. Mason and M. Morrison, *Oxidases and Related Redox Systems*, John Wiley, New York, 1965, p. 815.
31 R. O. C. Norman and J. R. L. Smith, in T. E. King, H. S. Mason and M. Morrison, *Oxidases and Related Redox Systems*, John Wiley, New York, 1965, p. 131.
32 A. Bobst and M. Viscontini, *Helv. Chim. Acta*, 49 (1966) 884.
33 Y. Kamiya and K. U. Ingold, *Can. J. Chem.*, 42 (1964) 1027.
34 R. van Heldon, A. Bickel and E. Kooyman, *Rec. Trav. Chim.*, 80 (1961) 1237, 1257.
35 G. Guroff, J. W. Daly, D. M. Jerina, J. Renson, B. Witkop and S. Udenfriend, *Science*, 157 (1967) 1524.
36 N. H. Sloane and K. G. Untch, *Biochemistry*, 3 (1964) 1160.
37 R. van Helden and E. C. Kooyman, *Rec. Trav. Chim.*, 80 (1961) 57.
38 H. Kropf, *Ann.*, 637 (1960) 73.
39 A. G. Davies, D. G. Hare and R. F. H. White, *J. Chem. Soc.*, (1961) 341.
40 A. E. Martell and M. Calbin, *Chemistry of the Metal Chelate Compounds*, Prentice Hall, Englewood Cliffs, 1952.
41 S. Fallab, *Chimia*, 21 (1967) 538.
42 S. Fallab, *Chimia*, 23 (1969) 177.
43 F. Miller, J. Simplicio and R. G. Wilkins, *J. Am. Chem. Soc.*, 91 (1969) 1962.
44 F. Miller and R. G. Wilkins, *J. Am. Chem. Soc.*, 92 (1970) 2687.
45 R. Nakon, *Ph. D. Dissertation*, Texas A & M University, 1971.
46 R. Nakon and A. E. Martell, submitted for publication.
47 A. Vleck, *Trans. Faraday Soc.*, 56 (1960) 1137.
48 C. Brosset and N. G. Vannerberg, *Nature*, 190 (1961) 714.
49 N. G. Vannerberg and C. Brosset, *Acta Cryst.*, 16 (1963) 247.
50 W. P. Schaefer and R. E. Marsh, *J. Am. Chem. Soc.*, 88 (1966) 178.
51 E. Ebsworth and J. Weil, *J. Phys. Chem.*, 63 (1959) 1890.
52 L. Vaska, *Science*, 140 (1963) 809.
53 I. A. Ibers and S. J. LaPlaca, in V. Gutman, *Proc. VIIIth Int. Conf. Coordination Chem.*, Springer-Verlag, Vienna, 1964, 10 AI.
53a J. Weiss, *Nature*, 202 (1964) 83.
54 A. G. Sykes, *Trans. Faraday Soc.*, 59 (1963) 1325.
55 A. G. Sykes, *Trans. Faraday Soc.*, 59 (1963) 1334.

56 M. T. Beck and S. Gorog, *Acta Chem. Acad. Sci. Hung.*, 29 (1961) 401.
57 A. Weissberger, J. E. LuValle and O. S. Thomas, *J. Am. Chem. Soc.*, 65 (1943) 1934.
58 M. M. Taqui Khan and A. E. Martell, *J. Am. Chem. Soc.*, 89 (1967) 4176.
59 M. M. Taqui Khan and A. E. Martell, *J. Am. Chem. Soc.*, 90 (1968) 6011.
60 M. M. Taqui Khan and A. E. Martell, *J. Am. Chem. Soc.*, 89 (1967) 7014.
61 R. R. Grinstead, *Biochemistry*, 3 (1964) 1308.
62 C. A. Tyson and A. E. Martell, *J. Am. Chem. Soc.*, submitted for publication.
63 G. A. Hamilton and A. Revesz, *J. Am. Chem. Soc.*, 88 (1966) 2069.
64 J. M. Hill and P. J. G. Mann, *Biochem. J.*, 99 (1966) 454.
65 A. E. Braunstein, in P. Boyer, H. Lardy and K. Myrback, *The Enzymes*, Vol. 2, Academic Press, New York, 1966, p. 113.
66 F. Bufoni, in E. E. Snell, *Pyridoxal Catalysis*, Interscience Publishers, New York, 1968.
67 C. R. Dawson, in J. Peisach, P. Aisen and W. E. Blumberg, *The Biochemistry of Copper*, Academic Press, New York, 1966, p. 305.
68 I. Yamasaki and L. H. Piette, *Biochim. Biophys. Acta*, 50 (1961) 62.
69 H. Yamada, K. Yasonobu, Y. Yamano and H. S. Mason, *Nature*, 198 (1963) 1092.
70 A. Van Heuvelen, *Nature*, 208 (1965) 888.
71 E. V. Goryachenkova, L. J. Stcherbatiuk and C. I. Zamaraev, in E. E. Snell, *Pyridoxal Catalysis*, Interscience Publishers, New York, 1968, p. 391.
72 H. R. Mahler, in P. Boyer, H. Lardy and K. Myrback, *The Enzymes*, Vol. 8, Academic Press, New York, 1963, p. 285.
73 W. E. Blumberg, B. L. Horecker, F. Kelly-Falcoz and J. Peisach, *Biochim. Biophys. Acta*, 96 (1965) 336.
74 G. Constantopoulos, A. Caroebter, P. A. Satch and T. T. Chen, *Biochemistry*, 5 (1966) 1650.
75 F. C. Charalampous, *J. Biol. Chem.*, 234 (1959) 220.
76 F. C. Charalampous, *J. Biol. Chem.*, 235 (1960) 1286.
77 D. I. Crandall, in T. E. King, H. S. Mason and M. Morrison, *Oxidases and Related Redox Systems*, John Wiley, New York, 1965, p. 263.
78 J. Chiriboga, *Arch. Biochem. Biophys.*, 116 (1966) 516.
79 H. S. Hason, E. Spencer and I. Yamazaki, *Biochem. Biophys. Res. Commun.*, 4 (1961) 236.
80 W. G. Levine, in J. Peisach, P. Aisen and W. E. Blumberg, *The Biochemistry of Copper*, Academic Press, New York, 1966, p. 371.
81 T. Nakamura and Y. Ogura, in J. Peisach, P. Aisen and W. E. Blumberg, *The Biochemistry of Copper*, Academic Press, New York, 1966, p. 389.
82 W. W. Kaeding, R. O. Lindblom and R. G. Temple, *Ind. Eng. Chem.*, 53 (1961) 805.
83 E. Ochiai, *Tetrahedron*, 20 (1964) 1831.
84 W. I. Taylor and A. R. Battersby, *Oxidative Coupling of Phenols*, Marcel Dekker, New York, 1967.
85 H. Sigel, *Angew. Chem. (Int. Edn.)*, 8 (1969) 167.
86 J. H. Wang, *J. Am. Chem. Soc.*, 77 (1955) 4715.
87 R. C. Jarnigan and J. H. Wang, *J. Am. Chem. Soc.*, 80 (1958) 6477.
88 P. Nichols and G. R. Schonbaum, in B. Boyer, H. Lardy and K. Myrback, *The Enzymes*, Vol. 8, Academic Press, New York, 1963, p. 147.
89 K. G. Paul, in P. Boyer, H. Lardy and K. Myrback, *The Enzymes*, Vol. 8, Academic Press, New York, 1963, p. 227.
90 B. C. Saunders, A. G. Holmes-Siedle and B. P. Stark *Peroxidases*, Butterworths, London, 1964.

Chapter 21

THE COPPER-CONTAINING OXIDASES

R. MALKIN

Department of Cell Physiology, University of California, Berkeley, California, U.S.A.

I. INTRODUCTION

Copper-containing proteins have been found to be widely distributed in both plants and animals and have been related to such metabolic processes as hydroxylation, oxygen transport, electron transfer and oxidative catalysis. In this chapter, however, only one group of copper-containing proteins will be considered — the oxidases and the related electron-transfer proteins. A recent review has discussed the relationship between the structure and function of copper in proteins in greater detail[1], and several other review articles have dealt with related topics[2-4]. In addition, a symposium on the biochemistry of copper summarizes much of the earlier work in this area[5].

As shown in Table I, copper-containing oxidases may be classified on the basis of their catalytic activity or function. Thus, the enzymes are divided into three groups: the "blue" copper-containing oxidases, all of which contain at least four gram-atoms of copper per mole and catalyze the reduction of O_2 to H_2O; the "non-blue" copper-containing oxidases which have only one or two gram-atoms of copper per mole and catalyze the reduction of O_2 to H_2O_2; and the oxygenases. The "blue" electron-transfer proteins, which have one or two gram-atoms of copper per mole, are apparently non-autooxidizable[30] and have no known catalytic functions. The oxygenases or mixed-function oxidases will, however, not be considered in any great detail since a current volume has summarized work on these proteins[31].

In this chapter, a description of the different forms of copper found in proteins will be given, and this will be the background for a discussion of the relationship of these forms to the catalytic activity of the enzymes. Particular attention will be paid to the forms of copper in the "blue" copper-containing oxidases since they have been most extensively characterized with regard to these different forms and some preliminary attempts have been made towards describing their overall mechanism in detail.

References pp. 707–709

TABLE I

MAIN PROPERTIES OF COPPER-CONTAINING OXIDASES AND RELATED PROTEINS

Protein and source	Molecular weight	Copper content (gram-atoms/mole)	Activity or function	Ref.
Azurin (*Pseudomonas, Bordetella*)	16,000	1	"Blue" electron carriers: non-autooxidizable	6, 7
Stellacyanin (lacquer tree)	20,000	1		8, 9
Plastocyanin (chloroplasts)	21,000	2		10, 11
Tyrosinase (*Neurospora*)	33,000	1	Oxygenases	12
Dopamine-β-hydroxylase (adrenal glands)	290,000	4–7		13,14
Laccase (lacquer tree, fungi)	{64,000, 110,000}	4	"Blue" oxidases: catalyze the reduction of O_2 to H_2O	15–18
Ascorbate oxidase (cucumber, squash)	130,000	8		19, 20
Ceruloplasmin (animal serum)[a]	160,000	8		21, 22
Benzylamine oxidase (pig plasma)	190,000	2	"Non-blue" oxidases: catalyze the reduction of O_2 to H_2O_2	23
Diamine oxidase (pig kidney)	190,000	2		24, 25
Galactose oxidase (fungi)	43,000	1		26, 27
Uricase (liver)	120,000	1		28
Cytochrome *c* oxidase (mitochondria)[b]	~100,000 (monomer)	1	Terminal oxidase	29

[a]Has weak oxidase activity but unknown physiological function.
[b]Contains 1 mole of heme iron per 100,000 as well as copper.

II. THE FORMS OF COPPER IN COPPER-CONTAINING OXIDASES

Three different forms of copper have been identified in oxidases: a Cu^{2+} ion (identified as Type 1 in some of the current literature[32,33]) which is responsible for the intense blue color of some of the proteins; a Cu^{2+} ion (identified as Type 2[32,33]) which is considered as "non-blue" because of its low visible extinction relative to the "blue" Cu^{2+}; and copper in an electron paramagnetic resonance (e.p.r.)-non-detectable state. To summarize the presence of these forms in the different proteins, it has been found that the "blue" copper-containing oxidases contain all three forms of copper while other copper proteins have only one form[1].

A. The "blue" Cu^{2+} ion

1. Spectroscopic properties

The visible and near ultra-violet absorption spectra of two "blue" copper proteins are shown in Fig. 1. The spectrum of stellacyanin, a "blue"

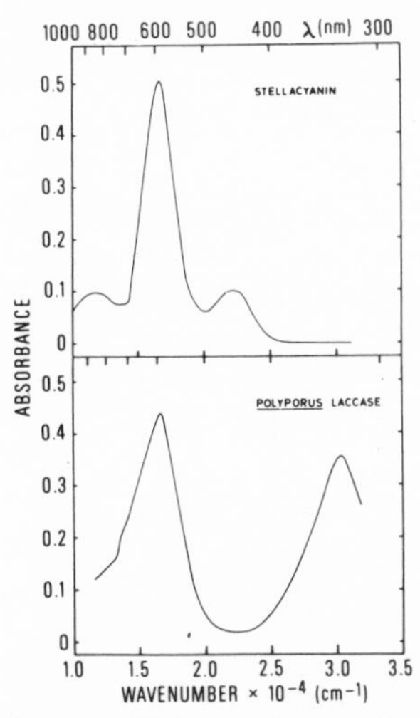

Fig. 1. Visible and near ultra-violet absorption spectra of stellacyanin and laccase. The spectra are presented as the difference between the oxidized and reduced proteins with the latter being obtained by the addition of ascorbate.

protein having only one Cu^{2+} ion per mole, is resolved into three absorption bands while that of laccase shows two major bands and a shoulder in the near infra-red region. By comparison of the optical spectra of the "blue" electron-transfer proteins with those of the "blue" copper-containing oxidases, it is possible to conclude that the major bands which are present in

References pp. 707–709

the spectra of the former group are associated with the "blue" Cu^{2+} and that in the multi-copper enzymes with more than two gram-atoms of copper per mole, these three bands are also present but that an additional band at 330 nm is present[1,30]. This latter band, however, is not associated with the "blue" Cu^{2+} but is related to a different form of copper found in the oxidases[34].

One of the more unusual features of the "blue" copper-containing proteins is the intensity of the blue color. The extinction coefficients on a protein basis for the band in the 600 nm region vary from 3500 M^{-1} cm^{-1} in azurin[35] to 11,300 M^{-1} cm^{-1} for ceruloplasmin[36]. However, ceruloplasmin contains two "blue" Cu^{2+} ions per mole[37] and thus the extinction per gram-atom of "blue" Cu^{2+} is 5600 M^{-1} cm^{-1}. A fact which is commonly overlooked in considerations of the "blue" copper proteins is that these extinctions are at least one order of magnitude greater than those found in low molecular weight cupric complexes. Extinctions due to *d–d* transitions in tetragonal Cu^{2+} complexes are generally less than 100 M^{-1} cm^{-1} [38] while the protein extinctions are of the order of 4000 M^{-1} cm^{-1}. In cases where higher extinctions have been found in the low molecular weight complexes, charge-transfer transitions have been assumed[4] although when the cupric ion is in a distorted symmetry, this can also lead to higher extinctions[39].

Optical rotatory dispersion (o.r.d.) studies of the "blue" electron-transfer proteins have shown that the Cu^{2+} ion is in an optically active environment[9,35,40,41]. Circular dichroism (c.d.) measurements on azurin and *Polyporus* laccase have extended these earlier o.r.d. observations[42]. In the case of azurin, it has been found that the c.d. curve could be resolved into six Gaussian components instead of only three which would be expected on the basis of three *d–d* transitions observed in the visible spectrum. Tang *et al.*[42] have suggested that some of these additional bands in the c.d. spectrum of azurin arise from charge-transfer transitions from the metal to the ligands. Besides the multiplicity of the bands observed, the magnitudes of the c.d. bands in azurin were much larger than those found in low molecular weight cupric complexes and this was taken as evidence of a greater degree of asymmetry for the Cu^{2+} site in the protein relative to the other complexes. C.d. and o.r.d. studies on laccase are open to more complex interpretations because of the different forms of copper in this enzyme[42,43] although the "blue" Cu^{2+} also is in an optically active site in this protein.

One of the most useful spectroscopic tools applied in the study of copper-containing proteins is electron paramagnetic resonance. The method is applicable to systems having unpaired electrons and is based on observing transitions between energy levels produced by an applied magnetic field. Since Cu^{2+} is paramagnetic but Cu^{1+} has no unpaired electrons, it is only the former which can be studied by e.p.r. A detailed description of e.p.r. is beyond the scope of this chapter although some of the relevant features of the e.p.r. of cupric complexes will be presented. The reader is referred to other sources which discuss the general aspects of e.p.r.[44] and which specifically discuss the e.p.r. of oxidative enzymes and copper proteins[45–47].

Three experimentally obtainable parameters of importance in the e.p.r. work on Cu^{2+} proteins are the g value, the A value (the hyperfine splitting constant), and the total area under the absorption curve.

The g value is obtained from the equation which describes the basic resonance condition of e.p.r.:

$$h\nu = g\beta B \qquad (1)$$

where h is Planck's constant, ν is the frequency, g is a dimensionless quantity, β is the Bohr magneton, and B is the strength of the applied magnetic field. In this equation, g, ν and B are variables. Since one usually records an e.p.r. spectrum at constant frequency and B is varied until absorption occurs, one can calculate the g value from the spectrum. In the case of tetragonal Cu^{2+} complexes, two g values are obtained because of the orientation relative to the applied magnetic field, *i.e.*, Cu^{2+} complexes show anisotropy. The value of $g_{\perp}$ is obtained when the applied field is perpendicular to the z-axis and $g_{\parallel}$ when the field is parallel to the z-axis. Typical g values for Cu^{2+} complexes are $g_{\parallel} \approx 2.3$ and $g_{\perp} \approx 2.05$. In cases of lower symmetry, three g values may be obtained.

Because the copper nucleus has a nuclear spin of $\frac{3}{2}$, the e.p.r. absorption is split into four equally spaced peaks (the hyperfine structure) for each Cu^{2+} site. The separation between the absorption peaks is the hyperfine splitting constant, A. The g values and the A values of a cupric complex are related to both the symmetry of the site and the covalency of the cupric–ligand bonds. A direct demonstration of covalent bonding in a complex can be obtained if superhyperfine structure is observed in the e.p.r. spectrum. Under these conditions, the unpaired electron of the cupric ion is influenced by the nuclear moment of the ligand atom, resulting in additional splittings.

The area under the absorption curve (obtained by double integration since e.p.r. spectra are presented as first-derivative curves) is proportional to the amount of unpaired electrons and can be converted to the actual concentration of Cu^{2+} in the sample by calibrating the spectrometer with a suitable standard. In the case of studies with proteins, quantitative e.p.r. measurements allow one to calculate the Cu^{2+} content of the protein by a non-destructive physical measurement thus avoiding the problems involved in the attempted chemical determination of the valence state of the metal in the protein.

The e.p.r. spectrum of spinach plastocyanin is shown in Fig. 2 as one example of the e.p.r. of the "blue" Cu^{2+} ion. The lower portion of the spectrum is recorded at a higher sensitivity and the four copper hyperfine lines in the parallel direction are clearly resolved. Although plastocyanin contains two gram-atoms of copper per mole, the e.p.r. spectrum indicates

References pp. 707–709

that these ions must be bound in essentially identical environments because only one set of e.p.r. parameters is resolved. This is to be contrasted with our later discussion of the "blue" copper-containing oxidases where the e.p.r. spectra have been interpreted as indicating non-equivalently bound cupric ions.

The hyperfine splitting constant, $A_{\parallel}$, for plastocyanin has been calculated to be 0.005 cm^{-1}. As first pointed out by Malmström and Vänngård in studies with *Polyporus* laccase and ceruloplasmin[48], the $A_{\parallel}$ values of the

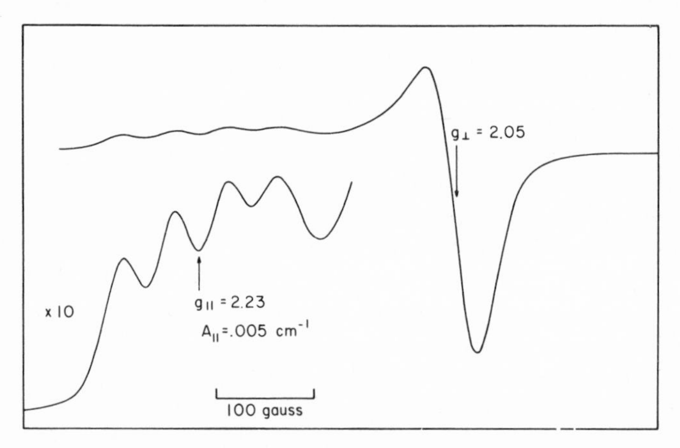

Fig. 2. The e.p.r. spectrum of spinach plastocyanin at 9 GHz. The low-field part of the spectrum is shown at a higher gain in the lower portion of the figure.

"blue" Cu^{2+} proteins are unusually small when compared with those observed in other Cu^{2+} complexes. The values of $A_{\parallel}$ extend from 0.003 cm^{-1} for stellacyanin to 0.009 cm^{-1} for *Polyporus* laccase. In contrast, Cu^{2+}–EDTA has an $A_{\parallel}$ value of 0.015 cm^{-1} [48] and values of 0.020 cm^{-1} are common in cupric complexes.

The first hypothesis which attempted to explain the small $A_{\parallel}$ values of the e.p.r. spectra of "blue" copper proteins was based on the covalency of the complex and assumed a tetragonal symmetry for the calculations[48]. More recent e.p.r. studies at several frequencies have, however, shown that three different *g* values are necessary to describe the spectra of both *Polyporus* laccase[32] and stellacyanin[49], indicating that the Cu^{2+} site in these proteins is one of lower symmetry than tetragonal. Although these findings do not exclude the possibility that a high degree of covalency is present in the copper binding, no direct demonstration of covalency as evidenced by superhyperine structure has been observed.

A recent electron-nuclear double resonance study of stellacyanin[50] has given the first direct indication that nitrogen atoms are coordinated to the Cu^{2+} in the protein. The study also established that the cupric ion is in a hydrophobic environment in the protein and is inaccessible to the solvent. This latter finding agrees with those obtained by chemical studies on the protein[9].

2. Models for the "blue" Cu^{2+} ion

The unusual spectroscopic properties of the "blue" Cu^{2+} have led to several considerations of the type of bonding responsible for these parameters. It is clear that interactions between copper ions, as initially proposed in the case of ceruloplasmin[51,52], need not be considered since two of the "blue" electron-transfer proteins have only one gram-atom of copper per mole and yet exhibit the same type of properties as observed with the "blue" multi-copper enzymes.

The presently accepted view of the nature of the "blue" Cu^{2+} is related to the symmetry of the cupric site. Two groups have presented detailed calculations which have attempted to explain the experimentally observed spectroscopic properties[53,54]. In both calculations, the cupric bonding was assumed to be entirely ionic and charge-transfer transitions were not considered. From both models, it was concluded that the "blue" cupric ion must be in a site which is distorted from the normal Cu^{2+} square planar geometry towards a more tetrahedral configuration. Studies of low molecular weight model compounds with small $A_{\parallel}$ values have also supported the conclusion that the Cu^{2+} is in a tetrahedral or near-tetrahedral environment in these complexes[55,56].

The hypothesis that the site of the "blue" Cu^{2+} is approximately tetrahedral would agree with the recent e.p.r. work on stellacyanin[49] and laccase[32] previously referred to where three g values have been observed in the e.p.r. spectra, indicating a site of low symmetry. The model would also agree with the observed optical activity of the proteins and the high oxidation–reduction potentials which have been found for some of these proteins[57] since these properties are associated with a Cu^{2+} ion in a distorted geometry. However, it must still be stressed that the exact geometry of the "blue" Cu^{2+} site is not yet known with any precision. Preliminary X-ray investigations which have been reported on azurin[58] and ceruloplasmin[59] lead one to hope that the three-dimensional structure of a "blue" Cu^{2+} protein may soon be elucidated and that this will allow an unequivocal statement to be made concerning the environment of the cupric ion in these proteins.

B. The "non-blue" Cu^{2+} ion

The term "non-blue" cupric ion has been adopted to describe a second form of copper found in oxidases. This type of copper is found in all the "blue" oxidases which have been carefully analyzed, in all the "non-blue" oxidases, and in one of the oxygenases (dopamine-β-hydroxylase). However, this form of copper is not present in the "blue" electron-transfer proteins. It is important to stress that the classification of this form of copper in these groups of enzymes does not imply the "non-blue" Cu^{2+} has identical properties or an identical function in the catalytic reactions of the different enzymes. This point will be discussed in greater detail in the consideration of the function of this form of copper.

1. Spectroscopic properties

The spectroscopic properties of the "non-blue" cupric ion, where determined in any detail, are more comparable to those of low molecular weight cupric complexes and have led to the view that this form of copper does not have an unusual type of bonding in proteins. It has been difficult to estimate accurately the positions and intensities of any visible absorption bands associated with this Cu^{2+} because the visible absorption region is complicated by the presence of other chromophoric groups in most of the proteins: pyridoxal phosphate in the amine oxidases[23,60,61], and the "blue" cupric ion in the "blue" oxidases. In those proteins which have no other chromophoric groups, it can be stated with certainty that if any absorption is present which is associated with the "non-blue" Cu^{2+}, it is of low intensity since this absorption has not been observed at the usual concentrations of material obtainable in work with purified proteins[13,26].

The e.p.r. properties of the "non-blue" cupric ion have been studied in more detail than the optical properties and most of the e.p.r. parameters are similar to those found in other cupric complexes. Thus, $A_{\parallel}$ values are in the range of 0.014–0.020 cm^{-1} (see ref. 1 for a complete compilation of the e.p.r. parameters of the "non-blue" Cu^{2+}).

2. The requirement for the "non-blue" Cu^{2+} in the catalytic activity of oxidases

With most of the "non-blue" oxidases (galactose oxidase[26,27], diamine oxidase[24] and dopamine-β-hydroxylase[13]), it has been possible to demonstrate the requirement of the "non-blue" Cu^{2+} for enzymic activity in a rather straightforward manner by the removal of the copper, resulting in an inactive apoprotein, followed by subsequent reactivation upon the readdition of copper. Although in some cases complete reactivation has not been obtained, this result does indicate that the Cu^{2+} is required for the enzymic activity of these proteins.

The study of the role of the "non-blue" cupric ion in "blue" oxidases has been more complicated although recent observations indicate this form of copper is required for the catalytic activities of these enzymes[33,37,62]. In the case of the "blue" oxidases, early studies were mainly concerned with the "blue" Cu^{2+} since this copper was most easily studied by its optical absorption, and there was no reason to suspect any other form of copper was present in the protein. The identification of a form of copper which was distinguishable from the "blue" Cu^{2+} was first accomplished by e.p.r. for ceruloplasmin[63]. This "non-blue" Cu^{2+} had been previously observed and was considered as Cu^{2+} in denatured protein molecules[64,65]; it was only

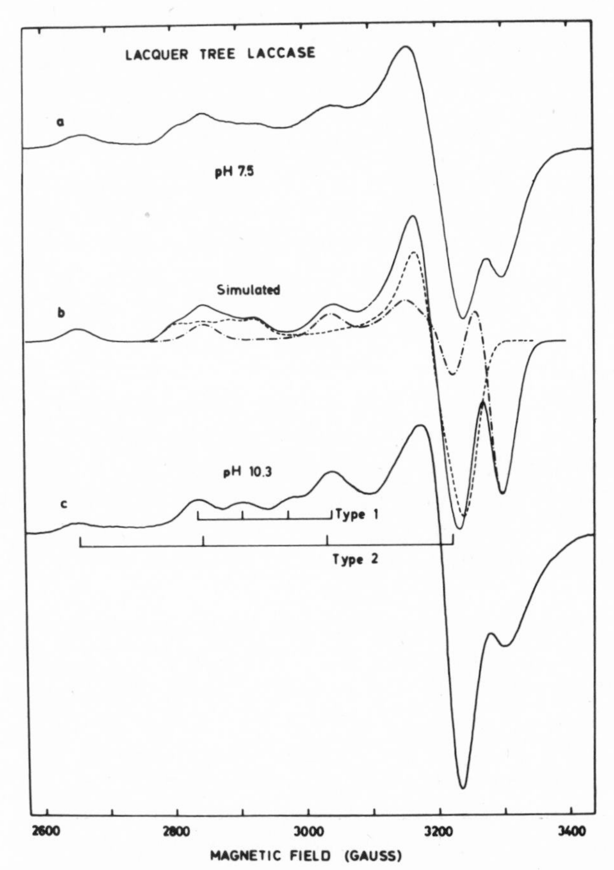

Fig. 3. Experimental and computer-simulated e.p.r. spectra (at 9 GHz) of *Rhus* laccase at (a) pH 7.5 and (c) 10.3. The simulated spectrum (b) was computed on the assumption that two different forms of Cu^{2+} were present (- - - - Type 1; – · – · – · Type 2) in equal amounts. The full line in the simulated spectrum represents the sum of these two components. From ref. 49.

detailed studies on the laccases from *Polyporus*[32] and the lacquer tree[18], followed by a subsequent reinvestigation with ceruloplasmin[37], which led to the acceptance of the view that two different forms of Cu^{2+} are present in these proteins, one form being a "non-blue" Cu^{2+}.

The e.p.r. spectra of lacquer tree laccase at two pH values are shown in Fig. 3(a) and (c) to illustrate the presence of the "non-blue" Cu^{2+} in the "blue" copper-containing oxidases. In these spectra, the four hyperfine lines due to the "blue" Cu^{2+} are unresolved because of the small $A_{\parallel}$ value and are centered at approximately 2900 gauss. However, two additional lines are present in the $g_{\parallel}$ region, one at a lower field (approximately 2650 gauss) and one at a higher field (approximately 3000 gauss). One explanation of these spectra is that two Cu^{2+} centers are present in the molecule in different sites, each having a different $g_{\parallel}$ value (compare with Fig. 2 in which only one Cu^{2+} site is present in the protein). It is possible to test this interpretation by computer-simulation of the e.p.r. spectra on the basis of two components being present in equal amounts but having different e.p.r. parameters. The simulated spectrum in Fig. 3(b) shows a narrowly spaced component (labeled Type 1) and a broadly spaced component (labeled Type 2), and the solid line represents their sum. The agreement between the spectra in Fig. 3(a) and (b) essentially confirms this interpretation. By integration of the spectra according to standard procedures[45], it is possible to show that each form of Cu^{2+} in the protein corresponds to 50% of the total e.p.r. intensity and to one gram-atom of copper per mole. Thus, the protein contains two non-equivalent Cu^{2+} ions, only one of which has a small $A_{\parallel}$ value, and the second form has a more normal $A_{\parallel}$ value. This type of study has also been carried out with *Polyporus* laccase[37] and ceruloplasmin[57], and the results have demonstrated non-equivalent cupric ions in these proteins.

The identification of a form of "non-blue" Cu^{2+} in these "blue" copper-containing oxidases has led to a series of studies which have proven that this form of copper does not exist in denatured molecules but is required for the catalytic activity of these enzymes. In the case of *Polyporus* laccase, it is possible to remove specifically the "non-blue" Cu^{2+} from the protein, resulting in a sample which has no catalytic activity but still contains three gram-atoms of copper per mole[33]. The original copper content and enzymic activity can be restored by the addition of copper to this sample.

Another approach to the question of the involvement of the "non-blue" Cu^{2+} in the catalytic reaction of the "blue" copper-containing oxidases has utilized reactions of these enzymes with anions. Experiments with ceruloplasmin and laccase have shown that anions such as azide, cyanide and fluoride inhibit oxidase activity at low concentrations (10^{-5}–10^{-6} M)[37,62,66-70]. By studying the e.p.r. and optical properties of the respective protein in the presence of some of these inhibitors, it has been possible to show that they bind specifically to the "non-blue" Cu^{2+} [37,62]. One example of the interaction of an anion with the "non-blue" Cu^{2+} is the reaction of *Polyporus* laccase with fluoride as studied by e.p.r. that is shown in Fig. 4. The addition of one equivalent of fluoride causes the hyperfine lines in the g_z region of the "non-blue" Cu^{2+} signal (labelled Type 2 in the Figure) to shift and to split into doublets (Fig. 4b). This would be expected if the unpaired electron from the cupric ion interacted with the fluoride nucleus, the latter having a nuclear spin of $\frac{1}{2}$. No change is observed in the e.p.r. signal of the "blue" Cu^{2+} (labeled Type 1), indicating the reaction is specific for the "non-blue" Cu^{2+}. Upon the addition of more fluoride (Fig. 4c), each of the hyperfine lines of the "non-blue" Cu^{2+} is split into three components with an approximate intensity ratio of 1 : 2 : 1, as

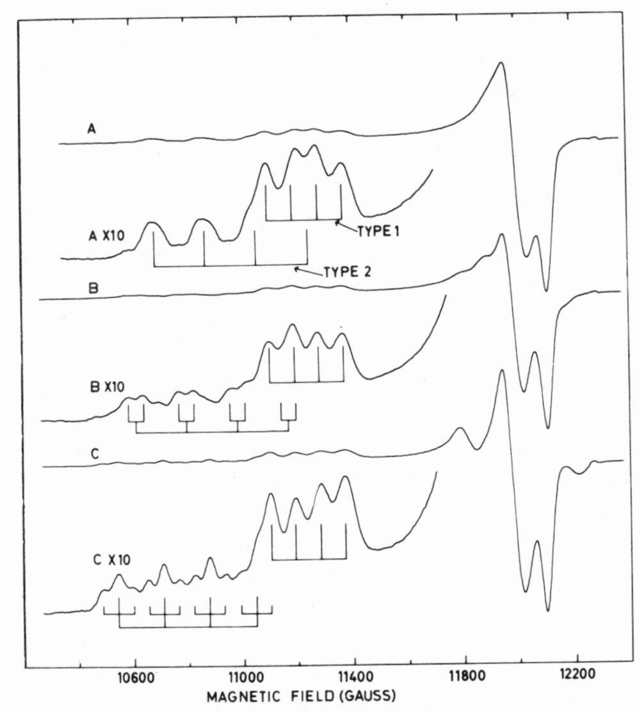

Fig. 4. The reaction of *Polyporus* laccase with fluoride as studied by e.p.r. at 35 GHz (from ref. 62). (a) native laccase; (b) laccase and one equivalent of NaF; (c) laccase and approximately 15 equivalents of NaF. The splittings due to the two forms of Cu^{2+} and the superhyperfine splitting due to F^- are indicated.

expected for two equivalent fluoride nuclei interacting with this Cu^{2+}. With some of the anions, it is also possible to remove the inhibitor from the inactivated enzyme and activity is regained[32]. These types of experiments are therefore a direct demonstration that modification of the "non-blue" cupric ion in these proteins leads to a loss of catalytic activity.

In summary, although the spectroscopic properties of the "non-blue" Cu^{2+} are not as unique as those of the "blue" Cu^{2+}, this form of copper has been demonstrated to be a component of the "blue" copper-containing oxidases and to be required for the catalytic activity. In the case of the "non-blue" oxidases, some of which only contain "non-blue" Cu^{2+} as a

References pp. 707–709

chromophoric center, this demonstration for the activity of the enzymes has been more direct. The possible role of this form of copper in the overall catalytic reaction of these enzymes will be considered in more detail in the final section.

C. The "e.p.r.-non-detectable" copper

The third form of copper which is found in the copper-containing oxidases is less clearly defined than the previously described forms and has only an operational definition: a form of copper which is present in the protein but gives no e.p.r. signal. "e.p.r.-non-detectable" copper can only be detected by a comparison of the total copper content with the total amount of e.p.r.-detectable copper. In the "blue" copper-containing oxidases, it has been found that approximately 50% of the copper is "e.p.r.-non-detectable" while in *Neurospora* tyrosinase, all the copper is in an "e.p.r.-non-detectable" state.

Various states of copper can result in an "e.p.r.-non-detectable" form. The most common form is Cu^{1+} which is diamagnetic and gives no e.p.r. signal. The "e.p.r.-non-detectable" copper in tyrosinase is considered to be cuprous since this protein has only one gram-atom of copper per mole[12]. It is also possible that Cu^{2+} centers which are paramagnetic but are weakly interacting can have no e.p.r. signal due to extensive line broadening under certain conditions, but such copper can be detected by susceptibility measurements. In the case of most of the "blue" copper-containing oxidases, the diamagnetic state of the "e.p.r.-non-detectable" copper has been confirmed by susceptibility data[71,72]. A third state which can lead to "e.p.r.-non-detectable" copper involves strong interactions between cupric ions and experimental evidence has led to the suggestion that this type of interaction is present in the "blue" copper-containing oxidases and is responsible for the "e.p.r.-non-detectable" copper in these proteins.

Anaerobic titrations of *Polyporus* laccase indicate that the diamagnetic copper ions in the protein exist as a spin-paired Cu^{2+}–Cu^{2+} couple in the oxidized protein[73]. This copper pair acts as an electron-accepting unit in the enzyme and is also associated with the near-ultra-violet absorption band at 330 nm which was previously described for laccase[34].

In experiments with various reductants, it was found that the "blue" Cu^{2+} in laccase was reduced linearly with the addition of approximately 3.5 electron-equivalents of reductant[73]. In conjunction with e.p.r., it was demonstrated that the protein contains two additional reducible sites other than the "blue" and "non-blue" Cu^{2+} ions. The use of high potential reductants has allowed an estimate of the oxidation–reduction potential of these sites to be made and has indicated that they must have a potential of at least 0.5 V[73].

The relationship between the additional electron-accepting sites in laccase and the absorption bands at 610 and 330 nm has been more clearly defined in further titration studies[34]. As shown in Fig. 5, titration of native laccase with ascorbate causes a concomitant decrease at both wavelengths over the entire course of the titration. This indicates that the component associated with the 330 nm band is related to an electron-accepting site in the protein but does not give any indication of which of the possible sites.

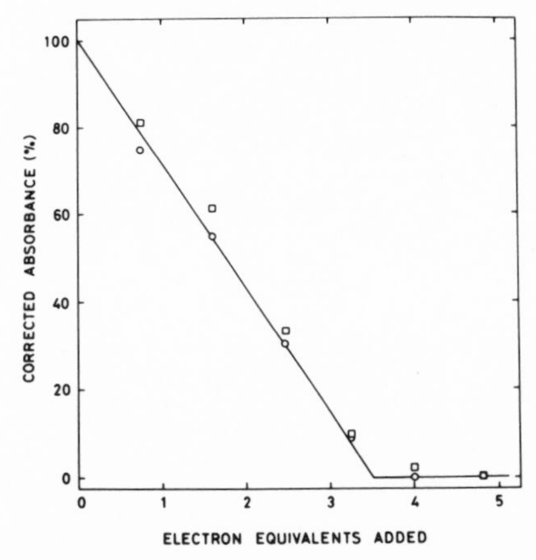

Fig. 5. Anaerobic titration of *Polyporus* laccase with ascorbate. ○—○, absorbance at 610 nm; □—□, absorbance at 330 nm. All absorbance values are corrected for the contribution of the reduced protein. From ref. 34.

By carrying out a titration in the presence of fluoride, it has been possible to differentiate the electron-accepting sites in laccase. As shown in Fig. 6, the optical titration shows two separate phases: the first electron added decreases the absorbance at 610 nm while the absorbance at 330 nm is essentially unchanged; the addition of the next two electrons causes a linear decrease of the absorbance at 330 nm. No new e.p.r. signals were observed during the titration and, in addition, the e.p.r. data showed that the "non-blue" Cu^{2+} was not reduced to any significant extent by the addition of the first three electrons. This experiment therefore demonstrates that the 330 nm band is associated with a two electron-accepting unit which only accepts electrons in pairs and which is distinct from the "blue" and "non-blue" cupric ions in the molecule[34].

The nature of the additional electron-accepting sites in laccase has been considered in some detail[73] and will only be briefly reviewed here. The sites must have the following properties: (1) a redox potential greater than 0.5 V; (2) diamagnetism in the oxidized protein at room temperature; (3) no e.p.r. signal in the oxidized or reduced protein at 77°K; and (4) association with a near ultra-violet absorption band. It has been proposed that

References pp. 707–709

this site in laccase is associated with the two remaining copper ions in the molecule and that these exist as a Cu^{2+}–Cu^{2+} couple having total spin-pairing[73]. The existence of cupric–cupric pairs in low molecular weight complexes is known[74] and is best exemplified by cupric acetate. Some of the properties of the unit proposed for laccase, such as spin-pairing at room temperature, have been observed in other low molecular weight complexes[74].

In addition to the experiments on *Polyporus* laccase, studies on the lacquer tree laccase also show that a two electron-accepting unit is present

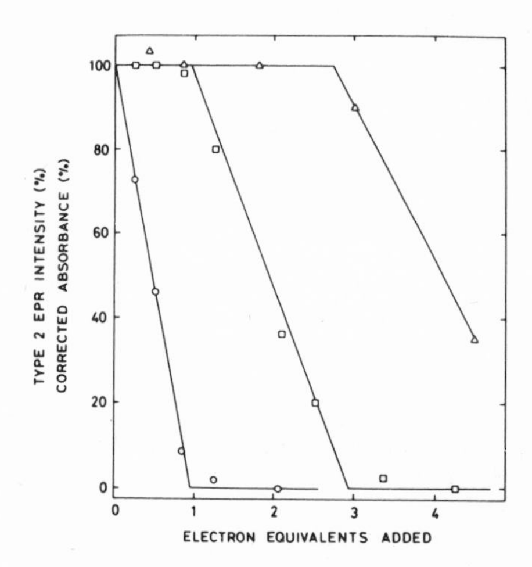

Fig. 6. Anaerobic titration of *Polyporus* laccase with ascorbate in the presence of 3×10^{-3} *M* NaF. ○—○, absorbance at 610 nm; □—□, absorbance at 330 nm and △—△, percentage of initial "non-blue" Cu^{2+} e.p.r. intensity. From ref. 34.

in this enzyme and is associated with a 330 nm absorption band[75]. It is tempting to suggest that in the other "blue" copper-containing oxidases, ceruloplasmin and ascorbate oxidase, a similar unit is present since these enzymes also have an absorption band at 330 nm[20,51].

To summarize the recent findings on the "blue" copper-containing oxidases, experiments on the laccases from *Polyporus* and lacquer tree have given the first experimental indication of the nature of the "e.p.r.-non-detectable" copper in these enzymes. Titration data, in conjunction with physical measurements, are consistent with a structure in which the copper ions are present in the oxidized enzyme as a cupric–cupric pair capable of acting as a reducible two electron-accepting site in the molecule. The involvement and function of this unit in the catalytic reaction of these enzymes will be considered in the next section.

III. THE CATALYTIC ACTIVITY OF THE COPPER-CONTAINING OXIDASES

Our knowledge of the relationship between the forms of copper and their role in the catalytic reaction of the copper-containing oxidases is most clearly defined in the case of the *Polyporus* laccase. As has been discussed in the previous section, the nature of the different forms of copper in this enzyme has been elucidated by a combination of physical and chemical techniques, but most of the understanding which has been gained into the mechanism of laccase-catalyzed reactions has been obtained with the use of kinetic techniques. In this section, we will attempt to describe the overall catalytic activity of the oxidases and to relate the different forms of copper in the enzymes to this activity.

In the most simple hypothesis to describe the catalytic activity of the copper-containing oxidases, the substrate is oxidized by a Cu^{2+} ion in the protein and the Cu^{1+} ion formed is reoxidized by molecular oxygen. Early work on these enzymes supported this general idea since the proteins were found to contain Cu^{2+} which was reducible by substrate[76] and the first substrate product was identified as a free radical[77-79], an indication that oxidation had occurred in a one-electron step. The difficulty in this scheme, however, has involved the problem of reoxidation since the formation of water by the "blue" copper-containing oxidases requires four electrons while the reduction of a cupric ion is a one-electron reaction. Thus, there is a difficulty in the coupling of the two partial reactions catalyzed by the "blue" copper-containing oxidases. To overcome this problem, several hypotheses have been put forward in which each oxidase molecule is pictured as containing a unit of four cupric ions, each of which accepts one electron from the substrate, and all the cuprous ions formed donate their electrons to oxygen in some type of cooperative process[80,81]. However, the discussion of the previous section has stressed the recent findings which show that the copper ions in these enzymes are not equivalent, as required in these models, but exist in different physical states in the protein. This has led to a more recent proposal that the "blue" copper-containing oxidases are more asymmetric than previously pictured[82]. These considerations must also lead ultimately to an understanding of the specific functions of the different forms of copper in the catalytic reaction before the mechanism of these reactions will be understood in any more detail.

A. *Kinetic studies of laccase-catalyzed oxidations*

Rapid kinetic measurements with *Polyporus* laccase have shown that the "blue" Cu^{2+} in the molecule undergoes both reduction and reoxidation at rates which are comparable with the overall rate of the reaction[83]. In addition, it was found that approximately two additional sites in the protein are reduced more slowly than the "blue" cupric ion but still rapidly enough to function in the catalytic sequence. This type of experiment has been performed by the stopped-flow technique using hexacyanoferrate(II) ion as the substrate since it is possible to follow the formation of the oxidized reaction product with this substrate under identical conditions as are used to follow absorbance changes of the protein.

Besides demonstrating that there are three sites which undergo reduction in the protein, kinetic measurements on the reoxidation of the "blue" Cu^{2+} in laccase were extremely informative[83]. It was observed that the reoxidation of a fully reduced protein, *i.e.* a protein which had been reduced with approximately four electron-equivalents of substrate per mole of protein, is very rapid but that after reduction with low substrate concentrations (less than four electron-equivalents per mole), reoxidation was slow. These findings are consistent with a scheme in which rapid reoxidation of the "blue" Cu^{2+} only occurs when the additional electron-accepting sites in the protein are also reduced and there is some type of cooperative interaction between all these sites in the reduction of oxygen.

The evidence with *Polyporus* laccase that the additional reducible sites in the protein are involved in the reduction of oxygen would agree with results obtained with other "blue" Cu^{2+} proteins. It has been found by Nakamura and Ogura[30] that rapid reduction of the "blue" Cu^{2+} occurred in all "blue" Cu^{2+} proteins studied (stellacyanin, plastocyanin, lacquer tree laccase, ceruloplasmin and ascorbate oxidase). However, rapid reoxidation of the reduced proteins only occurred with the multi-copper enzymes (ceruloplasmin, laccase and ascorbate oxidase). The "blue" Cu^{2+} proteins, which have no additional reducible sites other than the "blue" Cu^{2+} ion, were essentially unreactive towards oxygen after reduction. These findings suggest that these additional reducible sites in the "blue" copper-containing oxidases are necessary for the rapid reoxidation of the reduced protein.

B. Possible roles of the different copper atoms in the oxidase reaction

Although the experiments described have indicated that the reducible sites of *Polyporus* laccase undergo rates of reduction and reoxidation which are rapid enough to be involved in the catalytic reaction, little discussion of any specific roles of the different copper atoms in the catalytic sequence was presented. Any discussion of this point must at this time be primarily speculative, however, since detailed experiments are only beginning to be performed on these questions.

The kinetic studies on *Polyporus* laccase have indicated that the "blue" Cu^{2+} is the most rapidly reduced site in the protein[83], and it is tempting to attribute substrate oxidation to this site on the basis of this observation. The details regarding the nature of the substrate oxidation, however, are not known. It appears that direct electron transfer between the substrate and the "blue" Cu^{2+} in laccase or other "blue" cupric proteins is unlikely since proton relaxation measurements have shown that the "blue" cupric ion is not directly accessible to the solvent[32,40] and thus would not be expected to be directly accessible to the substrate. It is possible that the mode of interaction between the substrate and the enzymes can be clarified by studying the reaction of the reductant with the "blue" Cu^{2+} electron-transfer proteins as models since the latter do not undergo any significant reaction with oxygen after reduction.

On the basis of the reoxidation studies, the proposed Cu^{2+}–Cu^{2+} pair in the enzyme can be considered as the site of interaction with oxygen. This site could function both as a reducible center and also as the oxygen-binding site. The interaction of this unit with oxygen would explain the lack of any significant reaction of the "blue" electron-transfer proteins with oxygen since they lack this unit.

Frieden *et al.*[4] first pointed out that only the multi-copper enzymes, such as laccase, which have at least four gram-atoms of copper per mole, can reduce oxygen to water. This was mentioned in relation to Table I in this chapter where it was shown that the enzymes which have only one or two gram-atoms of copper per mole all produce hydrogen peroxide as the product of oxygen reduction. This correlation would also suggest that the group of enzymes which are capable of reducing oxygen to water utilize a mechanism (and a form of copper) which is absent in the other proteins.

From a mechanistic standpoint, a two electron-accepting unit in the copper-containing oxidases would be highly advantageous. On the basis of thermodynamic considerations[84], oxygen is a poor one-electron oxidant and any single electron pathway from oxygen to water would encounter an initial energetic barrier in the first step of the pathway. The presence of the reducible cupric–cupric pair suggests a mechanism of oxygen reduction in which direct double electron transfer to oxygen can occur as the first step in the overall reduction to water. In this type of model, the cupric–cupric unit would not directly interact with the oxidizable substrate but electron transfer to this site would occur from the initial acceptor, the "blue" Cu^{2+}. This would agree with the finding that the "blue" Cu^{2+} is reduced more rapidly than the other two sites in the protein.

The "non-blue" Cu^{2+} in laccase does not appear to undergo reduction at rates compatible with the overall reaction rate[32] and has therefore not been considered as a valence-changing cupric ion. The "non-blue" Cu^{2+} in the "non-blue" copper-containing oxidases was also considered as a non-valence changing cupric ion but recent experiments on diamine oxidase have shown that part of the Cu^{2+} in this enzyme can function as an electron-accepting site[85]. This finding may lead to a reinvestigation of this question with the "blue" copper-containing oxidases.

Another possible function for the "non-blue" cupric ion in the "blue" copper-containing oxidases could, however, be related to the reduction of

References pp. 707–709

oxygen to water. If, as we have pictured the molecule, electrons enter at the "blue" Cu^{2+} site and are supplied to oxygen at the Cu^{2+}–Cu^{2+} site, it would appear that there is still a requirement for the stabilization of some type of oxygen intermediate, such as peroxide, in order to allow two more electrons to enter the oxygen moiety before it is released from the protein as water. The "non-blue" Cu^{2+} could function in this stabilization process and would not therefore be required to undergo reduction. Preliminary data with laccase[86] has shown that a H_2O_2-protein complex does form and that the "non-blue" Cu^{2+} is the site of bonding of the peroxide. In addition, preliminary kinetic measurements obtained with laccase and ceruloplasmin in the presence of anion inhibitors indicate that the rate of reduction of the "blue" Cu^{2+} is unaffected by the presence of the anions, but that the inhibitors, which react with the "non-blue" Cu^{2+}, reduce the activity of the enzymes by inhibiting some later step in the reaction sequence[37,83].

To summarize, kinetic studies have shown that several electron-accepting sites are involved in the catalytic reaction of laccase and that a cooperation between these sites seems to be operative in the reduction of oxygen to water. The detailed nature of the mechanism of the reaction, however, is still unknown and no intermediates in the reaction with oxygen have been positively identified. Thus, the exact manner in which oxygen is reduced to water is not known.

IV. CONCLUSION

In this chapter we have reviewed the various forms of copper which are found in the copper-containing oxidases. The physical and chemical properties of these forms have been described in detail and the relationship between the state of the copper and the catalytic activity of the enzymes was considered.

ACKNOWLEDGEMENTS

The author would like to thank Dr. B. G. Malmström for his help in the preparation of this chapter and Drs. J. A. Fee, B. Reinhammar, and T. Vänngård for many discussions which have led to several of the topics discussed in this chapter.

NOTE ADDED IN PROOF

Recent investigations on lacquer tree laccase and ceruloplasmin have extended the findings on *Polyporus* laccase discussed in detail in this chapter. Anaerobic redox titrations have shown that one electron per copper ion is required to fully reduce these proteins[75,87]. In addition, the near ultra-violet absorbance band at 330 nm has been found to be associated with a two-electron-accepting unit. For lacquer tree laccase the oxidation–reduction potentials of the electron-accepting sites have been determined (at pH 7.5): "blue" Cu^{2+}, E_m = 420 mV; "non-blue" Cu^{2+}, E_m = 390 mV; and 330-nm chromophore, E_m = 460 mV[75]. Rapid kinetic measurements with ceruloplasmin have confirmed that the 610-nm chromophore and the 330-nm chromophore are due to different electron-accepting sites; the mechanism of oxidase action appears to involve the transfer of electrons from the different electron-accepting sites in the protein[88].

REFERENCES

1 R. Malkin and B. G. Malmström, *Adv. Enzymol.*, 33 (1970) 177.
2 B. G. Malmström and L. Rydén, in T. P. Singer, *Biological Oxidations*, Interscience, New York, 1968, p. 415.
3 A. S. Brill, R. B. Martin and R. J. P. Williams, in B. Pullman, *Electronic Aspects of Biochemistry*, Academic Press, New York, 1964, p. 519.
4 E. Frieden, S. Osaki and H. Kobayashi, *J. Gen. Physiol.*, 49 (1965) 213.
5 J. Peisach, P. Aisen and W. E. Blumberg (Eds.), *The Biochemistry of Copper*, Academic Press, New York, 1966.
6 T. Horio, *J. Biochem. (Tokyo)*, 45 (1958) 195.
7 I. W. Sutherland and J. F. Wilkinson, *J. Gen. Microbiol.*, 30 (1963) 105.
8 T. Omura, *J. Biochem. (Tokyo)*, 50 (1961) 394.
9 J. Peisach, W. G. Levine and W. E. Blumberg, *J. Biol. Chem.*, 242 (1967) 2847.
10 S. Katoh, *Nature*, 186 (1960) 533.
11 S. Katoh, I. Shiratori and A. Takamiya, *J. Biochem. (Tokyo)*, 51 (1962) 32.
12 M. Fling, N. H. Horowitz and S. F. Heinemann, *J. Biol. Chem.*, 238 (1963) 2045.
13 S. Friedman and S. Kaufman, *J. Biol. Chem.*, 240 (1965) 4763.
14 M. Goldstein, E. Lauber and M. R. McKereghan, *J. Biol. Chem.*, 240 (1965) 2066.
15 W. G. Levine, in J. Peisach, P. Aisen and W. E. Blumberg, *The Biochemistry of Copper*, Academic Press, New York, 1966, p. 371.
16 R. Mosbach, *Biochim. Biophys. Acta*, 73 (1963) 204.
17 T. Nakamura, *Biochim. Biophys. Acta*, 30 (1958) 44.
18 B. Reinhammar, *Biochim. Biophys. Acta*, 205 (1970) 35.
19 C. Dawson, in J. Peisach, P. Aisen and W. E. Blumberg, *The Biochemistry of Copper*, Academic Press, New York, 1966, p. 305.
20 T. Nakamura, N. Makino and Y. Ogura, *J. Biochem. (Tokyo)*, 64 (1968) 189.
21 C. G. Holmberg and C.-B. Laurell, *Acta Chem. Scand.*, 2 (1948) 550.
22 I. H. Scheinberg, in J. Peisach, P. Aisen and W. E. Blumberg, *The Biochemistry of Copper*, Academic Press, New York, 1966, p. 513.
23 H. Blaschko and F. Buffoni, *Proc. R. Soc.*, B163 (1965) 45.
24 B. Mondovi, G. Rotilio, M. T. Costa *et al.*, *J. Biol. Chem.*, 242 (1967) 1160.
25 H. Yamada, H. Kumagai, H. Kawasaki, H. Matsui and K. Ogata, *Biochem. Biophys. Res. Commun.*, 29 (1967) 723.
26 D. Amaral, L. Bernstein, D. Morse and B. Horecker, *J. Biol. Chem.*, 238 (1963) 2281.
27 F. Kelly-Falcoz, H. Greenberg and B. Horecker, *J. Biol. Chem.*, 240 (1965) 2966.
28 H. Mahler, in P. D. Boyer, H. Lardy and K. Myrbäck, *The Enzymes*, Vol. 8, Academic Press, New York, 1963, p. 213.

29 H. Beinert, in J. Peisach, P. Aisen and W. E. Blumberg, *The Biochemistry of Copper*, Academic Press, New York, 1963, p. 285.
30 T. Nakamura and Y. Ogura, *J. Biochem. (Tokyo)*, 64 (1968) 267.
31 K. Bloch and O. Hayaishi (Eds.), *Biological and Chemical Aspects of Oxygenases*, Maruzen, Tokyo, 1966.
32 B. G. Malmström, B. Reinhammar and T. Vänngård, *Biochim. Biophys. Acta*, 156 (1968) 67.
33 R. Malkin, B. G. Malmström and T. Vänngård, *Eur. J. Biochem.*, 7 (1969) 253.
34 R. Malkin, B. G. Malmström and T. Vänngård, *Eur. J. Biochem.*, 10 (1969) 324.
35 A. S. Brill, G. F. Bryce and H. Maria, *Biochim. Biophys. Acta*, 154 (1968) 342.
36 H. F. Deutsch, in A. C. Maehly, *Biochemical Preparations*, Vol. 11, Wiley, New York, 1966, p. 10.
37 L.-E. Andréasson and T. Vänngård, *Biochim. Biophys. Acta*, 200 (1970) 247.
38 F. A. Cotton and G. Wilkinson, *Advanced Inorganic Chemistry*, Interscience, London, 1966, p. 906.
39 R. L. Belford and W. A. Yeranos, *Mol. Phys.*, 6 (1963) 121.
40 W. E. Blumberg and J. Peisach, *Biochim. Biophys. Acta*, 126 (1966) 269.
41 S.-P. W. Tang and J. E. Coleman, *Biochem. Biophys. Res. Commun.*, 27 (1967) 281.
42 S.-P. W. Tang, J. E. Coleman and Y. P. Myer, *J. Biol. Chem.*, 243 (1968) 4286.
43 F. Bossa, G. Rotilio, P. Fasella and B. G. Malmström, *Eur. J. Biochem.*, 10 (1969) 395.
44 J. W. Orton, *Electron Paramagnetic Resonance*, Iliffe, London, 1968.
45 T. Vänngård, in H. M. Swartz, J. Bolton and D. Borg, *Biological Applications of EPR*, in preparation.
46 H. Beinert and G. Palmer, *Adv. Enzymol.*, 27 (1965) 105.
47 D. C. Gould and A. Ehrenberg, in F. Ghiretti, *Physiology and Biochemistry of Haemocyanins*, Academic Press, New York, 1968, p. 95.
48 B. G. Malmström and T. Vänngård, *J. Mol. Biol.*, 2 (1960) 118.
49 B. G. Malmström, B. Reinhammar and T. Vänngård, *Biochim. Biophys. Acta*, 205 (1970) 48.
50 G. H. Rist, J. S. Hyde and T. Vänngård, *Proc. Natn. Acad. Sci. U.S.*, 67 (1970) 79.
51 W. E. Blumberg, J. Eisinger, P. Aisen, A. G. Morell and I. H. Scheinberg, *J. Biol. Chem.*, 238 (1962) 1675.
52 H. Beinert, D. E. Griffiths, D. C. Wharton and R. H. Sands, *J. Biol. Chem.*, 237 (1962) 2337.
53 W. E. Blumberg, in J. Peisach, P. Aisen and W. E. Blumberg, *The Biochemistry of Copper*, Academic Press, New York, 1966, p. 49.
54 A. S. Brill and G. F. Bryce, *J. Chem. Phys.*, 48 (1968) 4398.
55 D. C. Gould and A. Ehrenberg, *Eur. J. Biochem.*, 5 (1968) 451.
56 D. Forster and V. W. Weiss, *J. Phys. Chem.*, 72 (1968) 2669.
57 J. A. Fee and B. G. Malmström, *Biochim. Biophys. Acta*, 153 (1968) 299.
58 G. Strahs, *Science*, 165 (1969) 60.
59 B. Magdoff-Fairchild, F. M. Lovell and B. W. Low, *J. Biol. Chem.*, 244 (1969) 3497.
60 H. Yamada and K. T. Yasunobu, *J. Biol. Chem.*, 238 (1963) 2669.
61 B. Mondovi, M. T. Costa, A. Finazzi Agrò and G. Rotilio, *Arch. Biochem. Biophys.*, 119 (1967) 373.
62 R. Malkin, B. G. Malmström and T. Vänngård, *FEBS Lett.*, 1 (1968) 50.
63 T. Vänngård, in A. Ehrenberg, B. G. Malmström and T. Vänngård, *Magnetic Resonance in Biological Systems*, Wenner-Gren Center International Symposium Series, Vol. 9, Pergamon Press, Oxford, 1967, p. 213.
64 L. Broman, B. G. Malmström, R. Aasa and T. Vänngård, *J. Mol. Biol.*, 5 (1962) 301.
65 C. B. Kasper, H. F. Deutsch and H. Beinert, *J. Biol. Chem.*, 238 (1963) 2338.
66 G. Curzon, *Biochem. J.*, 77 (1960) 66.

67 G. Curzon, *Biochem. J.*, 100 (1966) 295.
68 G. Curzon and B. E. Speyer, *Biochem. J.*, 105 (1967) 243.
69 G. Curzon and B. E. Speyer, *Biochem. J.*, 109 (1968) 25.
70 B. E. Speyer and G. Curzon, *Biochem. J.*, 106 (1968) 905.
71 A. Ehrenberg, B. G. Malmström, L. Broman and R. Mosbach, *J. Mol. Biol.*, 5 (1962) 450.
72 P. Aisen, S. H. Koenig and H. R. Lilienthal, *J. Mol. Biol.*, 28 (1967) 225.
73 J. A. Fee, R. Malkin, B. G. Malmström and T. Vänngård, *J. Biol. Chem.*, 244 (1969) 4200.
74 M. Kato, H. B. Jonassen and J. C. Fanning, *Chem. Rev.*, 64 (1964) 99.
75 B. Reinhammar and T. Vänngård, *Eur. J. Biochem.*, 18 (1971) 463.
76 B. G. Malmström, R. Mosbach and T. Vänngård, *Nature*, 183 (1959) 321.
77 L. Broman, B. G. Malmström, R. Aasa and T. Vänngård, *Biochim. Biophys. Acta*, 75 (1963) 365.
78 T. Nakamura, in M. S. Blois, Jr., H. W. Brown, R. M. Lemmon, R. O. Lindblom and M. Weissbluth, *Free Radicals in Biological Systems*, Academic Press, New York, 1961, p. 169.
79 I. Yamazaki and L. H. Piette, *Biochim. Biophys. Acta*, 50 (1961) 62.
80 G. Curzon and J. N. Cummings, in J. Peisach, P. Aisen and W. E. Blumberg, *The Biochemistry of Copper*, Academic Press, New York, 1966, p. 545.
81 W. E. Blumberg, in J. Peisach, P. Aisen and W. E. Blumberg, *The Biochemistry of Copper*, Academic Press, New York, 1966, p. 578.
82 B. G. Malmström in A. Engström and B. Strandberg, *Symmetry and Function of Biological Systems at the Macromolecular Level*, Nobel Symposium No. 11, Almqvist and Wiksell, Uppsala, 1969, p. 513.
83 B. G. Malmström, A. Finazzi Agrò and E. Antonini, *Eur. J. Biochem.*, 9 (1969) 383.
84 P. George, in T. E. King, H. S. Mason and M. Morrison, *Oxidases and Related Redox Systems*, Vol. 1, Wiley, New York, 1965, p. 3.
85 B. Mondovi, G. Rotilio, A. Finazzi Agrò, *et al.*, *FEBS Lett.*, 2 (1969) 182.
86 R. Brändén, B. G. Malmström and T. Vänngård, *Eur. J. Biochem.*, 18 (1971) 238.
87 R. J. Carrico, B. G. Malmström and T. Vänngård, *Eur. J. Biochem.*, 20 (1971) 518.
88 R. J. Carrico, B. G. Malmström and T. Vänngård, *Eur. J. Biochem.*, 22 (1971) 127.

Chapter 22

FERREDOXINS AND OTHER IRON–SULFUR PROTEINS

W. H. ORME-JOHNSON

Department of Biochemistry and Institute for Enzyme Research, University of Wisconsin, Madison, Wisconsin 53706, U.S.A.

I. INTRODUCTION

The iron–sulfur proteins are those in which iron atoms are bonded to sulfur-containing ligands. This definition encompasses a large number of proteins, including some of those discussed in Chapter 23. This discussion will be mainly devoted to iron–sulfur proteins containing iron *not* bound to porphyrins (which is the sense in which the name "iron–sulfur proteins" is commonly used), as well as containing sulfur ligands of two chemically distinct classes, *viz.* cysteine sulfhydryl groups and an incompletely understood species — "labile sulfur" — that appears as H_2S, HS^-, or S^{2-} either upon acidification of protein solutions or during oxidation and other denaturation processes at neutral and alkaline pH. Also mentioned will be biochemically related classes such as the rubredoxins, which contain one or two atoms, cysteine ligands, and no labile sulfur, and in passing, certain more complex enzymes and enzyme systems which contain iron and labile sulfur in addition to flavin (DPNH, dihydroorotate, and succinate dehydrogenase), flavin and heme groups (sulfite reductase), flavin and molybdenum (xanthine and aldehyde oxidases) and cytochromes (ubiquinone–cytochrome *c* reductase).

The intensity of interest in this subject among biochemists, physicists, and inorganic chemists may be gauged from five reviews on the topic[1-5] that have appeared since the publication of a symposium volume[6] summarizing progress in the field up to 1965. A second book, devoted to the biology, chemistry and physics of the iron–sulfur proteins, is in preparation[7]. This chapter will briefly summarize those facets of research to date which seem to most strongly bear on the bonding and resultant chemical properties of this class of proteins.

The reasons that a wide range of talents have been attracted to the study of iron–sulfur proteins include: (a) these proteins participate in oxidation–reduction processes in all forms of life thus far tested; (b) they are involved in electron transfer during photosynthesis, nitrogen fixation, and mitochondrial respiration; (c) they possess chemical structures in which the peptide chain by suitable folding forms the basis of a complete ligand arrangement, in contrast to heme proteins which utilize a preformed porphyrin group; (d) they transfer electrons at a redox potential in the vicinity of the $H_2 : 2H^+$ couple; (e) at least until quite recently there were no known model complexes incorporating the features of low redox potential, non-heme iron, and sulfhydryl and labile sulfur ligands; (f) they undergo reversible changes in susceptibility, optical, n.m.r., Mössbauer, and e.p.r. properties during oxidation–reduction. One of their most interesting features is that these proteins transfer either one or in some cases two electrons and exhibit corresponding quantitative changes in magnetic properties, while the iron–sulfur clusters responsible for this property contain one to eight iron atoms.

II. NOMENCLATURE

The name ferredoxin was originally applied by Mortenson *et al.* to the 8-iron-8-labile sulfur protein which transfers electrons among pyruvate dehydrogenase, hydrogenase, and nitrogenase in *Clostridium pasteurianum*[8]. This name was then extended by Tagawa and Arnon[9] to include the 2-iron-2-labile-sulfur protein derived from chloroplasts, which had previously been called "methemoglobin reducing factor" and "photosynthetic pyridine nucleotide reductase." One difficulty with extending the ferredoxin nomenclature is that the etymological origins of the word (fer = iron; redoxin = protein transferring electrons) have been ignored in forming combinations such as adrenodoxin, testodoxin, and even flavodoxin (a protein containing no iron). A second problem has been that expressions such as "plant ferredoxins", meaning 2-iron-2-labile sulfur proteins, have been applied to proteins of plant, animal, and bacterial origin. A committee of the IUB–IUPAC is at present debating official solutions to these problems, but a widely-adopted suggestion of Beinert[10] that these proteins be collectively referred to as "iron–sulfur proteins" offers one interim solution that will be followed in this chapter. They will also be referred to as *n*-Fe–S(*) proteins, where *n* is the number of iron atoms present in each molecule. The presence of a superscript star indicates that *labile* sulfur atoms, generally in amounts equal to the iron content, are present. Thus rubredoxin is a 1-Fe–S protein, spinach ferredoxin a 2-Fe–S* protein, and so forth. The source of the protein and other necessary modifiers will be used as prefixes as needed. A system similar to this has been used by Tsibris and Woody[4] in their review of progress in the field to 1969. The chief drawback to this notation is that it is unhandy for oral communication. These proteins will also be referred to as "plant-type ferredoxins" *etc.*, where this usage is unambiguous. The present state of flux in the nomenclature of iron–sulfur proteins makes it necessary that the reader be acquainted with at least the systems having usage and/or logic as recommendations, and Table I summarizes the types and some of the characteristics of proteins dealt with here.

References pp. 741–744

TABLE I

CHARACTERISTICS[a] OF REPRESENTATIVE IRON–SULFUR PROTEINS

Protein type	Names	Sources	Molecular weight (g/mole)	$E^{\circ\prime}$ (mV)	No. of e^- transferred	Physiological functions
1-Fe–S	Rubredoxin[b]	*Clostridium pasteurianum*, other anaerobes	6380	-57	1	not known, but will replace 8-Fe–S* ferredoxins in some reactions
2-Fe–S	Rubredoxin[b]	*Pseudomonas oleovorans*	~20,000	—	2	electron transfer in hydrocarbon oxidation
2-Fe–S*	1. Chloroplast[c] ferredoxin, photosynthetic pyridine nucleotide reductase, methemoglobin reducing factor	plants, *e.g.* spinach	10,650	-420	1	electron transfer in photosynthesis
	2. Adrenodoxin[c]	adrenal cortex, *e.g.* beef adrenals	13,090	-370	1	electron transfer in steroid hydroxylation
	3. Testodoxin[c]	testes	Presumably the same characteristics as adrenodoxin			
	4. Putidaredoxin[c]	*Pseudomonas putida*	~ 12,000	-240	1	electron transfer in camphor hydroxylation
	5. Paramagnetic[c] protein	*Clostridium pasteurianum*	~ 24,000	—	1	not known
	6. Azotobacter[c] type I protein	*Azotobacter vinelandii*	~ 24,000	—	1	not known, spectroscopically resembles *C. pasteurianum* 2-Fe–S*
	7. Azotobacter[c] type II protein	*Azotobacter vinelandii*	~ 21,000	—	1	not known, found in partly purified nitrogenase, spectroscopically resembles chloroplast 2-Fe–S*
	8. Non-heme[c] iron protein	CoQ-cytochrome *c*. reductase from mitochondria	~ 26,000	+220	1	transfer of electrons in mitochondrial oxidative phosphorylation

References pp. 741–744

TABLE I—*continued*

Protein type	Names	Sources	Molecular weight (g/mole)	$E^{\circ\prime}$ (mV)	No. of e^- transferred	Physiological functions
4-Fe–S*	1. High potential[b] iron protein	*Chromatium*	~ 10,000	+350	1	not known
Conjugated 4-Fe–S*	1. Dihydroorotate[c,d] dehydrogenase	*Zymobacterium oroticum*	~ 120,000	—	—	oxidation of dihydroorotate to orotate
8-Fe–S*	1. Ferredoxin[e]	*Clostridium pasteurianum*, other clostridia and micrococci	6200	-395	2	reductant in N_2 assimilation and H_2 evolution, oxidant in phosphoroclastic pyruvate oxidation
	2. Azotobacter type III protein	*Azotobacter vinelandii*	~ 17,000	—	—	electron donor in N_2 assimilation
	3. Ferredoxin	*Chromatium*	9630	-490	2	reductant in assimilation of CO_2, TPN reduction
	4. Ferredoxin	*Chlorobium thiosulfatophilum*	~ 6000	—	—	reductant in assimilation of CO_2, TPN reduction
Conjugated 8-Fe–S*	1. Xanthine[f] oxidase	Milk	~ 250,000	—	—	oxidation of xanthine and other purines, and aldehydes by oxygen
	2. Xanthine[f] dehydrogenase	*Clostridium acidi-urici*, other anaerobes	—	—	—	reduction of uric acid by 8-Fe–S* protein
	3. Aldehyde[f] oxidase	Liver	~ 280,000	—	—	oxidation of aromatic and aliphatic aldehydes
	4. Succinate[g] dehydrogenase	Mitochondria	~ 200,000	—	—	oxidation of succinate in oxidative phosphorylation
15-Fe–S*	Sulfite reductase[h]	*E. coli*	~ 760,000	—	—	reduction of sulfite (to sulfide) by TPNH
18-Fe–S*	DPNH dehydrogenase[i]	Mitochondria	~ 550,000	—	—	oxidation of DPNH in oxidative phosphorylation

TABLE I—*continued*

[a] The data in this Table should be viewed with varying degrees of skepticism. The *numbers of iron atoms* reported to be in each of the proteins has varied considerably over the past few years, particularly in the bacterial 8-Fe-S* cases where the indicated assignments are based on the careful work of Hong and Rabinowitz with the *Clostridium acidi-urici* protein[87], recent work on the *Chromatium* protein[5], as well as on the "principle of minimum astonishment". The assignments for the bottom three items in the Table are approximate only. The *molecular weights* determined on sequencing the proteins have often been 10% or more different from the hydrodynamic value (cf. refs. 2 and 37, concerning the adrenal protein). The *oxidation reduction potentials* at pH 7 ($E^{o\prime}$), on a scale on which the hydrogen electrode (1 atm.) has a potential of -420 mV, have tended to become more negative as more extensive measurements are made. For a general caveat on this subject, see Clark[118], and for a specific comment on the difficulties involved see Wilson[50]. The *number of electrons transferred* presumably can only be an integer but even here a considerable amount of effort has to be expended. General agreement possibly exists only for the 1-Fe–S, 2-Fe–S*, and the first 4-Fe–S* and 8-Fe–S* cases. The list of *physiological functions* can only be expected to expand, of course. Further discussion of these problems can be found in the text. Two proteins, the components of nitrogenase, have been omitted from this Table, not because there is any reasonable doubt that they are iron–sulfur proteins, but because there does not seem to be good agreement in the literature as to their composition. This is understandable since they are extremely oxygen labile in addition to being sensitive to the same denaturants affecting other iron–sulfur proteins. One component seems to contain 2 molybdenum atoms and 15–35 iron and labile sulfur atoms; the other component may be a 2-Fe–S* protein but does not yield a g = 1.94 e.p.r. signal as far as has been reported. An extensive discussion of these matters will be found in Chapter 23.

[b] These proteins possess an e.p.r. signal in the oxidized state and possess no signal (at least, below 4.2K) when reduced. The 1-Fe–S and 2-Fe–S proteins yield signals near g = 4.3, while the 4-Fe–S* protein has a signal near g = 2. This behavior should be contrasted to that of the other species in this Table.

[c] On reduction, these proteins yield an e.p.r. signal near g = 1.94.

[d] This protein has two flavin groups per molecule, from which semiquinone (g = 2.00) e.p.r. signals can be obtained[57].

[e] On reduction, these proteins yield a pair of overlapping e.p.r. signals, between g = 2.06 and g = 1.94[104,111].

[f] On reduction these proteins yield e.p.r. signals, attributed to iron centers, near g = 1.94, and in addition a signal with a g value near 2.11. Signals can also be elicited from the Mo and flavin moieties (2 of each per molecule)[123].

[g] This protein can be prepared in modifications containing 2–8 Fe atoms per molecule[119]. It contains a single flavin group which yields an e.p.r. signal on reduction to the semiquinone.

[h] This enzyme has 8 flavin and two heme groups, from which e.p.r. signals can be elicited. No signal attributable to Fe–S* clusters has been observed[120].

[i] This protein has been prepared with a range of iron contents[121] E.p.r. signals can be obtained from the single flavin group as well as from several iron–sulfur centers[122].

III. 1-Fe–S AND 2-Fe–S PROTEINS

The first example of this class, the *Clostridium pasteurianum* 1-Fe–S protein rubredoxin, was isolated and studied by Lovenberg and Sobel[11]. A unique physiological function for the protein has not yet been found, though it will substitute for the 8-Fe–S* protein from the same organism, in reactions such as TPN reduction by hydrogen in the presence of ferredoxin-free extracts from *C. pasteurianum.* Since the 1-Fe–S protein has an oxidation–reduction potential about 350 mV more positive than the *ferredoxin*, catalyses by these two proteins can have substantially different kinetic patterns, because the concentration ratio of oxidized to reduced carrier in the steady state may differ for the two proteins.

The 1-Fe–S proteins from five anaerobic bacteria have been analyzed for amino acid content[12] and the following generalizations emerge from these analyses. The proteins contain 51 to 60 amino acids with resulting molecular weights of about 6000–7000. They lack arginine and histidine residues, they have large amounts of carboxylic and aromatic residues, and each protein has four cysteine groups. The iron atom can be released from the protein by acidification *but no* H_2S *is evolved, i.e.*, for this class alone, no labile sulfur is present. The iron chromophore is also destroyed, albeit slowly, upon treatment with organic mercurials. Addition of thiols to aporubredoxin in the presence of iron salts will cause reincorporation of the iron atom into the chromophore[13]. Further studies, where aporubredoxin was treated to modify amino, thioether, indole, and phenol side chains, followed by attempts to incorporate iron into the modified apoprotein, led to the proposal that tryptophan and tyrosine residues, as well as the four cysteine thiols, are ligands of the iron atom[14].

The covalent sequences of apoproteins derived from two of these 1-Fe–S proteins have been determined. These are presented in Table II. The presence of two cys-*x*-*x*-cys pairs in each of these proteins is evident, and Bachmayer *et al.*[15] proposed that the protein doubles back on itself to present the two cysteine pairs to the iron atom.

The absorption and circular dichroism spectra of the 1-Fe–S proteins have been recorded by a number of authors. Data for the *P. elsdenii* protein, typical of the class, are shown in Fig. 1, taken from Garbett, *et al.*[16]. Noteworthy is the almost complete disappearance of absorption and optical activity in the visible range, on passing to the reduced protein. Tsibris and Woody[4] have discussed the optical data for the visible region with the following conclusions: the low Kuhn's anisotropy factor for the observed bands make it unlikely that magnetically allowed $d \rightarrow d$ transitions occur. They do find agreement between observation and theory for a ligand $\rightarrow$ metal charge-transfer model of an iron atom with four sulfur ligands. More recently, using the reduced protein, Eaton and Lovenberg have sought and

TABLE II

SEQUENCES[a] OF 1-Fe–S PROTEINS

Organism	
(1) *P. elsdenii*	met-asp-lys-tyr-glu-cys-ser-ile-cys-gly-tyr-ile-tyr-asp-glu-ald-glu-gly-
(2) *M. aerogenes*	met-gln-lys-phe-glu-cys-thr-leu-cys-gly-tyr-ile-tyr-asp-pro-ala-leu-val-
	5 10 15
(1)	asp-asp-gly-asn-val-ala-ala-gly-thr-lys-phe-ala-asp-leu-pro-ala-asp-trp-
(2)	gly-pro-asp-thr-pro-asp-gln-asp-gly-ala-phe-glu-asp-val-ser-glu-asn-trp-
	20 25 30 35
(1)	val-cys-pro-thr-cys-gly-ala-asp-lys-asp-ala-phe -- val-lys-met-asp
(2)	val-cys-pro-leu-cys-gly-ala-gly-lys-glu-asp-phe-glu-val-tyr-glu-asp
	40 45 50

[a] Data from ref 15.

found weak transitions in the *near infra-red*[17]. At 1630 nm, $\Delta\epsilon/\epsilon = 0.05$, suggesting a magnetically allowed $d \rightarrow d$ character for this band. They conclude from the low energy of this transition that the coordination of the iron atom is tetrahedral.

As will better be appreciated later, the finding of no labile sulfur was not the only significant difference between the 1-Fe–S protein and the other systems. The protein in the oxidized state yields an electron paramagnetic resonance (e.p.r.) signal around $g = 4.3$, (at low temperatures, additional signals, *e.g.* $g = 9$, are also observed) while there was no signal seen from the reduced protein. It is known that this type of signal is characteristic of high-spin iron(III) ($S = \frac{5}{2}$) in a rhombic environment[18]. This characterization

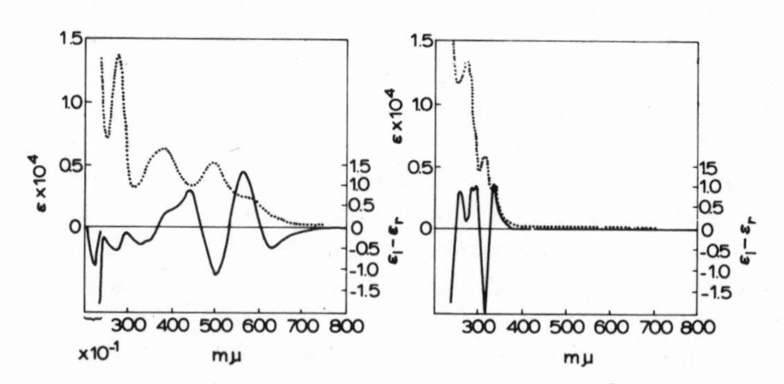

Fig. 1. Spectral properties of *P. elsdenii* rubredoxin (adapted from ref. 16). Left ordinates refer to absorption (..........) and right ordinates to circular dichroism (————). The spectra on the left are of the oxidized protein; those on the right were obtained with reduced protein.

of the oxidized species of the protein has been confirmed by Phillips *et al.* using a nuclear magnetic resonance (n.m.r.) method to measure the magnetic susceptibility near room temperature[19]. These workers find an effective magnetic moment of 5.85 Bohr magnetons (Bm) for the oxidized protein, and 5.05 Bm for the reduced species. This latter finding suggests that the reduced species is a high spin iron(II) ($S = \frac{4}{2}$) case, which was in turn confirmed by observation of a diagnostic large quadrupole splitting (3.24 mm/sec) in the Mössbauer spectrum. A compilation of physical data on the *C. pasteurianum* 1-Fe–S protein is given in Table III.

TABLE III

PHYSICAL PARAMETERS[a] FOR *CLOSTRIDIUM PASTEURIANUM* RUBREDOXIN

	Oxidized		Reduced	
E.p.r. signal, (77K)	$g = 4.3$		absent	
Magnetic moment, μ_{eff} (300K)	5.85 ± 0.2 Bm		5.05 ± 0.2 Bm	
Mössbauer effect[b]	Isomer shift	Quadrupole splitting	Isomer shift	Quadrupole splitting
300K	0.150	0.739	0.36	3.24
77K	0.144	0.779	0.37	3.38
4.2K	0.125	[c]	0.37	3.36

[a] Taken from Bachmayer *et al.*[14] and Phillips *et al.*[19].
[b] Isomer shifts are mm/sec relative to ^{57}Fe in a Cu matrix; add 0.483 mm/sec to the given values to compare to a sodium nitroprusside standard. The errors were 0.02 mm/sec in all cases.
[c] Hyperfine splitting was observed.

This protein has recently yielded to X-ray structure determination. Herriott *et al.*[20] reported a study of the oxidized protein to 2.5 Å resolution, in which it was found that the protein has an irregular folded structure, with no α-helix but with some anti-parallel pleated sheet conformation. The tracing out of the peptide chain can be seen in Fig. 2, and Table IV gives the angles and bond lengths in the iron–cysteine complex. Since the sequence of this 1-Fe–S protein has not been reported, assignments of positions to amino acids other than the cysteines is uncertain, though it is thought that the ends labeled A and B are the amino and carboxyl termini, respectively. Clearly, the sulfur ligands are arranged in a tetrahedral array and the presence of other ligands, suggested by the chemical data, is not evident at this resolution. It may be that the modification of tyrosinyl and tryptophyl residues prevent reformation of the holoprotein by interfering

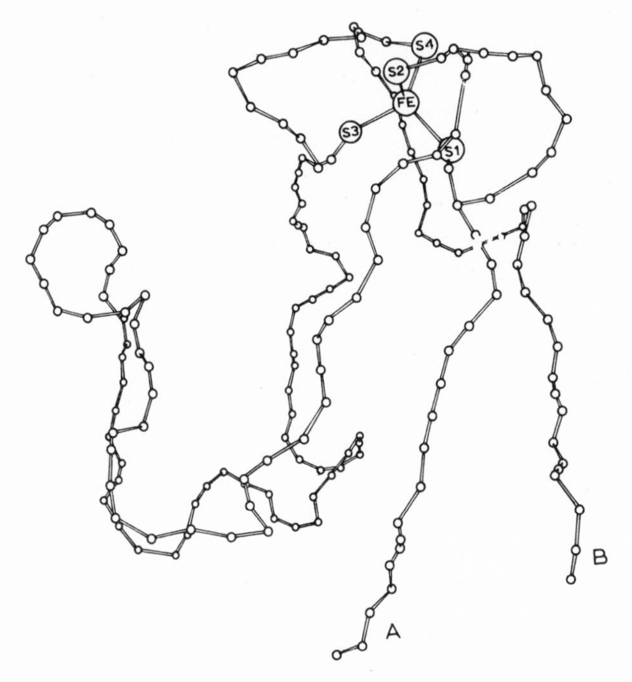

Fig. 2. Outline of the peptide chain of *C. pasteurianum* rubredoxin, as determined by X-ray crystallography to 2.5 Å resolution[20]. The ends of the chain labelled A and B are thought to correspond to the *N*- and *C*-termini, respectively. The small open circles are not to be construed as atomic sites; the numbering cf the sulfur atoms refers to the data of Table IV.

TABLE IV

BOND ANGLES AND LENGTHS[a] IN THE IRON-SULFUR COMPLEX OF *CLOSTRIDIUM PASTEURIANUM* 1-Fe-S PROTEIN

Bond angle (degrees)		Bond length (Å)	
S(1)-Fe-S(2)	110	S(1)-Fe	2.21
S(1)-Fe-S(3)	108	S(2)-Fe	2.42
S(1-)-Fe-S(4)	121	S(3)-Fe	2.37
S(2)-Fe-S(3)	97	S(4)-Fe	2.22
S(2)-Fe-S(4)	96		
S(3)-Fe-S(4)	120		

[a] The data are from Herriott *et al.*[20]; the numbering of atoms corresponds to that in Fig. 2. The authors estimated that the r.m.s. errors are ± 8° for the angles and ± 0.3 Å for the lengths.

with chain folding rather than by preventing these residues from entering the coordination sphere of the iron atom.

Phillips and his co-workers, anticipating that the paramagnetism of the iron–sulfur cluster would cause contact-shifts of n.m.r. resonances of β protons on the cysteine residues, have searched for these at 220 megacycles[19]. The reduced [high-spin iron(II)] form of the protein gave weakly shifted resonances while none were observed with the oxidized protein. The authors interpret these findings, which they contrast to studies on the 8-Fe–S* cases (see below), as indicating that the truly *contact*-shifted resonances are so broadened by dipolar interactions with the strongly paramagnetic center as to be unobservable, and that the observed shifts are of the *pseudocontact* variety.

In summary then, the following picture of 1-Fe–S proteins emerges: they are low molecular weight, acidic proteins transferring a single electron with a potential of about 0 mV. The single iron atom is tetrahedrally coordinated by four sulfur ligands in both oxidation states, and a number of charge-transfer transitions are possible, and both oxidized and reduced proteins are high-spin complexes.

Of the 2-Fe–S class of proteins, only one example, from *Pseudomonas oleovorans*, is at present known. Based on optical and e.p.r. spectra and chemical studies[21,22], one may expect that each molecule of protein will be found to contain two iron–sulfur clusters of the type found in the 1-Fe–S cases. The e.p.r. signal is complex in that resonances at g = 4.3 and 9.4 can be seen[18]. A function of the 2-Fe–S protein is known: It transfers electrons during aliphatic chain hydroxylation[23]. The 2-Fe–S protein has been shown to be cleavable into two fragments of 6000 and 13,000 molecular weight; each fragment contains one of the iron-binding centers[21]. The intact apoprotein can be converted to forms containing one or two iron atoms; both forms are functional in the hydroxylase system and exhibit similar amounts of visible absorbance and e.p.r. signals per iron atom. The primary sequence of the protein is under investigation[22].

IV. 2-Fe–S* PROTEINS

Proteins in this category have been isolated from plants, animals, and micro-organisms. Buchanan and Arnon[5] have recently reviewed the chemistry and functions of the ferredoxin from chloroplasts, which transfers electrons during photoreduction of NADP, oxygen evolution, and the photophosphorylation of ADP to form ATP. The photoreduction of ferredoxin in intact chloroplasts has recently been demonstrated by e.p.r. spectroscopy[24]. Reduced ferredoxin also activates the fructose-1,6-diphosphatase of chloroplasts, thus exhibiting a control effect as well as a direct oxidation–reduction function. This effect may be reductive in nature since dithiothreitol, a reducing agent, will also activate the enzyme. Proteins called "adrenodoxin" and "testodoxin", from mammalian adrenals and testes, respectively, transfer electrons to the oxygenases which catalyze hydrolyxation and side-chain cleavage during steroid hormone biosynthesis[2,25-27]. A similar protein, "putidaredoxin", is part of the camphor hydroxylase complex of *Pseudomonas putida*[28]. Some nitrogen-fixing bacteria contain iron–sulfur proteins of this class, though it is not yet clear what relationship, if any, they have to nitrogen fixation. Thus *Clostridium pasteurianum* (an anaerobe) and *Azotobacter vinelandii* (an aerobe) contain similar plant-type ferredoxins[29,30]. Finally, a protein of this type has been separated (as a succinylated derivative) from the cytochrome b–c_1 segment of the electron-transfer apparatus in heart mitochondria[31].

The wide occurrence and considerable importance of the chloroplast ferredoxins have led to intensive study of the chemistry of these proteins. The amino acid sequences of five of these are given in Table V[32-36]. As can be seen, several regions of the proteins have evidently been resistant to pressures for evolutionary variation. Which of these regions are associated with specific interprotein contact and which contribute folding and liganding properties needed for the iron–sulfur cluster are not yet known, but for reasons elaborated on below, the presence of five invariant cysteine residues is especially to be noted. The sequence of the bovine adrenal iron–sulfur protein has been reported[37] and that of the *P. putida* protein is near completion[38]. Though of similar size to the chloroplast proteins, these last two do not share many sequence features with the plant ferredoxins, but again five or more cysteine residues are present. The above data refer to the apoproteins. The corresponding holoproteins have the following general features: they are acidic proteins (with the exception of the "*Azotobacter* protein II", which is a neutral protein). They contain two iron atoms and two atoms of sulfur which are released upon acidification or treatment with mercurials. The known biological functions of these proteins depend on the integrity of the iron–sulfur complex. These proteins also suffer from degradation on storage which can in many cases be prevented by excluding oxygen and heavy metals, and by keeping the proteins in high ionic strength solutions at neutral pH. A recent report describing these properties also indicates that the source of oxygen denaturation is the oxidation of labile sulfur to zero-valent sulfur[39]. This, and the observation that the labile sulfur is released as sulfide upon acidification suggests the presence of S^{2-} or a protonated form in the intact proteins. The apoproteins can be reconstituted to the native materials by procedures analogous to those developed for the 8-Fe–S* proteins[40,41]. Addition of iron(III) or iron(II) salts, sulfide, and a mercaptan to solutions of the apoprotein at neutral pH results in formation of holoprotein. Substitution of selenide for sulfide leads to a holoprotein containing labile selenium. In the cases of the *adrenal* and *P. putida* proteins the selenoprotein is equally biologically active to the native ferredoxin[10,42].

TABLE V

SEQUENCES[a] OF 2-Fe–S* PROTEINS

Source	1									10										20
(1) Taro	ala-	thr-	tyr-	lys-	val-	lys-	leu-	val-	thr-	pro-	ser-	gly-	gln-	gln-	glu-	phe-	gln-	cys-	pro-	asp-
(2) Spinach		ala	tyr			thr		val			thr		asn	val	glu	phe	gln			
(3) Alfalfa (lucerne)		ser	tyr			lys		val			glu		thr	gln	glu	phe	glu			
(4) *L. glauca*	-	ala	phe			lys		leu			asp		pro	lys	glu	phe	glu			
(5) *Scenedesmus*		thy	tyr			thr		lys			ser		asp	gln	thr	ile	glu			

	21									30										40
1.	asp-	val-	tyr-	ile-	leu-	asp-	gln-	ala-	glu-	glu-	val-	gly-	ile-	asp-	leu-	pro-	tyr-	ser-	cys-	arg-
2.		val					ala				glu		ile	asp						
3.		val					his				glu		ile	val						
4.		val					gln				leu		ile	asp						
5.		thr					ala				ala		leu	asp						

	41									50										60
1.	ala-	gly-	ser-	cys-	ser-	ser-	cys-	ala-	gly-	lys-	val-	lys-	val-	gly-	asp-	val-	asp-	gln-	ser-	asp
2.			ser								leu	lys	thr		ser	leu	asn		asp	
3.			ser								val	ala	ala		glu	val	asn		ser	
4.			ser								leu	val	glu		asp	leu	asp		ser	
5.			ala								val	glu	ala		thr	val	asp		ser	

	61									70										80
1.	gly-	ser-	phe-	leu-	asp-	asp-	glu-	gln-	ile-	gly-	glu-	gly-	trp-	val-	leu-	thr-	cys-	val-	ala-	tyr-
2.	gln						asp		ile	asp	glu		trp					ala		
3.	gly						asp		ile	glu	glu		trp					val		
4.	gln						glu		ile	glu	glu		trp					ala		
5.	gln						ser		met	asp	gly		phe					val		

	81									90										
1.	pro-	val-	ser-	asp-	gly-	thr-	ile-	glu-	thr-	his-	lys-	glu-	glu-	glu-	leu-	thr-	ala			
2.	pro	val			val	thr		glu						glu		thr				
3.	ala	lys			val	thr		glu						glu		thr				
4.	pro	arg			val	val		glu						glu		thr				
5.	pro	thr			cys	thr		ala						asp		phe				

[a] As compiled in ref. 36.

The proteins have been found to undergo a single-electron oxidation–reduction step. Much attention has been given to this property. Though by the criteria of e.p.r. signal integrations, titrations followed by changes in optical and e.p.r. spectra, and low temperature magnetic susceptometry, only one electron is transferred[43-49], accurate potentials for the process remain to be determined. Problems associated with such determinations include the autooxidizability of the proteins, their relative instability, and their slow reaction with electrodes[50]. Available evidence, obtained with enzymatic systems such as hydrogenase, and with redox dyes, suggests potentials in the region –300 to –450 mV at pH 7, though a report of +220 mV obtained with a mitochondrial protein, has appeared[31]. It should not be overlooked that the more negative potentials are not consistent with maximum efficiency for all the functions which these ferredoxins respectively perform, *i.e.*, where a sequence of electron transfer such as pyridine nucleotide → flavoprotein → ferredoxin → electron acceptor is suspected, as in the steroid hydroxylase system, the fact that the potential of the pyridine nucleotide couple is -320 mV means that a considerably more negative potential for the ferredoxin involved may result in an inefficient catalytic system unless the [reduced pyridine nucleotide]/[oxidized pyridine nucleotide] ratio is kept high. Clearly it will be desirable to determine the *effective* potentials in such complex systems as well as the potentials for the isolated components.

The plant-type ferredoxins have optical properties which distinguish them from other iron–sulfur proteins. Their absorption spectra show bands near 320, 420 and 460 nm, with a shoulder near 550 nm. Low temperature (77K) spectra of the adrenal and spinach iron–sulfur proteins are sufficiently sharpened to reveal an additional transition near 515 nm, in the oxidized state[51]. The absorptivity in the 420–460 nm region is about 5000 $M^{-1}\,cm^{-1}$ per iron atom (Table VI), and this absorbance declines about 50% on

TABLE VI

EPR AND VISIBLE ABSORPTION DATA[a] FOR REPRESENTATIVE 2-Fe–S* PROTEINS

Protein	e.p.r.[b] (reduced proteins)			Absorption maxima[c] (oxidized proteins) (nm)		
	g_x	g_y	g_z			
(1) Putidaredoxin	1.93	1.935	2.02	325 (7.5)	415 (5.0)	455 (4.8)
(2) Adrenal protein	1.93	1.935	2.02	320 (7.3)	415 (5.0)	455 (4.8)
(3) *Azotobacter* protein I	1.93	1.94	2.01	331 (7.8)	419 (4.7)	460 (4.9)
(4) *C. pasteurianum* paramagnetic protein	1.92	1.94	2.01	333 (8.3)	425 (4.6)	463 (5.0)
(5) Spinach ferredoxin	1.89	1.96	2.05	325 (6.4)	420 (4.8)	465 (4.9)
(6) *Azotobacter* protein II	1.91	1.96	2.04	344 (8.3)	418 (7.0)	460 (5.3)

[a] From refs. 2, 3, 29, 30 and 61.
[b] The *g* values of proteins (1), (2) and (5) were those derived from spectral synthesis[61]; the other *g* values were estimated from the positions of extrema in first-derivative e.p.r. spectra.
[c] The values in parentheses are m*M* absorptivities per iron atom.

reduction. The spectral changes observed during the reductive titration of the adrenal iron–sulfur protein are shown in Fig. 3. Generally, a broad maximum near 550 nm is found to be the only pronounced feature of spectra of reduced preparations. The circular dichroism spectra of the spinach and adrenal proteins have been compared[51], the interest in doing this being that the former shows a rhombic e.p.r. spectrum in the reduced state, while the latter yields a nearly axial spectrum. Nonetheless, the proteins show a very similar and complex set of Cotton effects, revealing an underlying similarity of their chromophores. The transitions have Kuhn's asymmetry factors of 10^{-3} to 10^{-4}; consequently they are electric-dipole allowed. Whether there are transitions in the near infrared of the magnetic-dipole allowed variety, as seems to be the case with rubredoxin (see II), has not been reported.

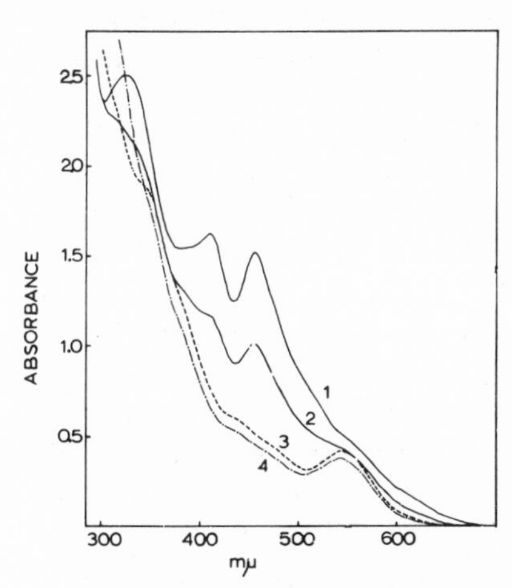

Fig. 3. Absorbance spectra of the adrenal iron-sulfur protein, at various stages of reduction by sodium dithionite (cf. refs. 48, 49). (1) No additions — the molar absorptivity of the peak near 415 nm is 10^4 1 mole^{-1} cm^{-1}; (2) 0.5 reducing equivalent per molecule; (3) 1 reducing equivalent per molecule; (4) excess dithionite present.

The 2-Fe–S* proteins also exhibit a distinctive e.p.r. signal in the reduced form, when specimens are examined in the frozen state. The relaxation properties, and thus the extent to which the temperature must be lowered for convenient observation of the signal, varies for the different cases. The adrenal, *P. putida*, *C. pasteurianum*, and the first *Azotobacter* protein (cf. Table I) can all be easily studied at 100K, whereas the second *Azotobacter* protein and the chloroplast ferredoxins are most easily observed at temperatures well below 77K. The discovery of this latter fact by Palmer and Sands[44] did much to emphasize essential similarities of the 2-Fe–S* class. Prior to their observations it was thought that the plant proteins, which had been studied at the warmer temperatures, gave no e.p.r. signal and were therefore somehow different from the proteins from *Azotobacter* and in mitochondria. No signal is observed in oxidized samples, and magnetic susceptibility observations at low temperatures[47,52,53] confirm the e.p.r. result that the oxidized proteins are diamagnetic and the reduced proteins contain a single unpaired electron, at these temperatures. Table VI

References pp. 741–744

lists e.p.r. and optical absorption parameters for selected examples of these proteins. The resonance near $g = 1.94$ has proved to be a hallmark of this class of proteins, and it is suspected, on the basis of the occurrence of this signal as well as the presence of non-heme-bound iron and labile sulfur, that similar iron–sulfur clusters exist in the mitochondrial electron-transfer system, in xanthine and aldehyde oxidases, and in dihydroorotic dehydrogenase[54-57]. In these more complex cases, the optical properties of the iron–sulfur centers may be obscured by absorptions due to heme and flavin, and the e.p.r. signal, though observable only in the frozen state, often provides the most reliable way of assessing the oxidation–reduction status of this chromophore.

An example of the utility of electron paramagnetic resonance in these latter cases may be seen in a recent report[122], where differences in temperature sensitivity and microwave power saturation properties allowed the dissection of four overlapping iron–sulfur protein signals in mitochondrial NADH dehydrogenase.

Attempts to discern whether one or both iron atoms are present in the electron accepting center, by looking for nuclear hyperfine structure (nhfs) in signals from proteins enriched in ^{57}Fe ($I = \frac{1}{2}$), have met with fair success. Azotobacter iron–sulfur protein I, from organisms grown on media enriched by ^{57}Fe, exhibited broadened e.p.r. signals[58] as did spinach ferredoxin enriched in the isotope by chemical exchange[40], but no *resolved* nhfs was observed. Experiments with the *P. putida* and adrenal proteins were more successful[41]. Figure 4 shows the signal obtained from *P. putida* ferre-

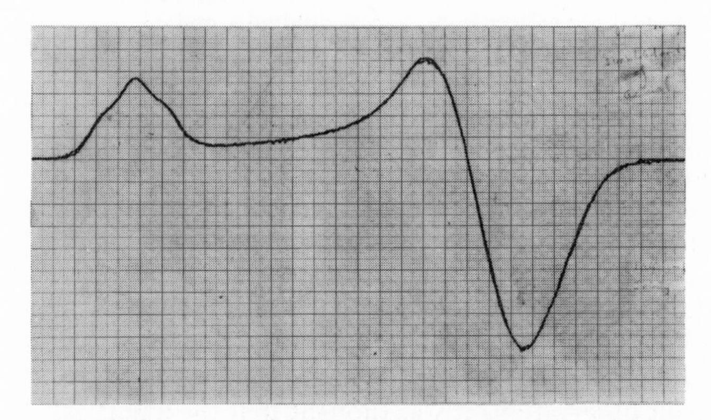

Fig. 4. Superposition of e.p.r. spectral of reduced putidaredoxin, one from a sample in which the ^{57}Fe enrichment was raised above 90% by chemical exchange, and the other calculated on the assumption that both Fe nuclei contribute 14 gauss hyperfine splitting and that the ^{57}Fe content was 94% of the analytical iron content. For the latter calculation the hyperfine interaction was assumed to be isotropic and a spectrum of ^{56}Fe-containing putidaredoxin was taken as the unbroadened line-shape. Spectra were obtained at X-band and at 107K (from ref. 41).

doxin in which the ^{57}Fe content had been raised to 94%. A trace computed on the assumption that *both* iron atoms were present in the center and that both contributed 14 gauss isotropic nhfs is superimposed on the experimental spectrum. No calculated trace assuming *one* iron atom to be present in the center would match the observed spectrum. A similar experiment with the adrenal protein gave similar results, the match between experimental and calculated spectra being excellent near g = 2.02, and only approximate near g = 1.94. It appears likely that the presence of a binuclear iron complex which can accept a single electron is a common feature of plant-type ferredoxins.

Similar efforts, aimed at finding out if the cysteine and labile sulfur moieties participate in the paramagnetic cluster, have been hampered by the low (~ 50%) enrichment of available ^{33}S, the high multiplicity ($I = \frac{3}{2}$) of

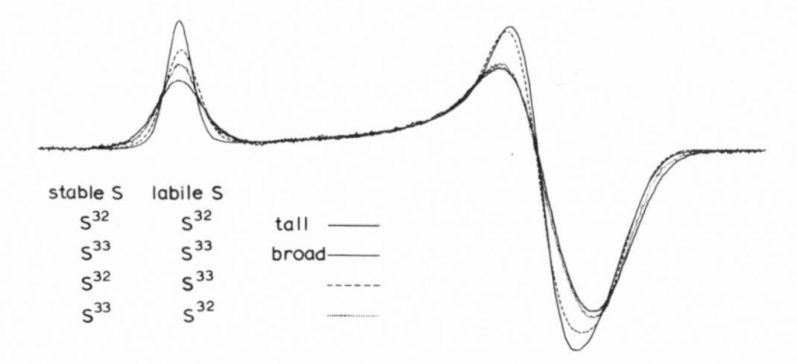

Fig. 5. E.p.r. spectra of samples of reduced *P. putida* iron–sulfur protein, prepared as follows: tall solid line spectrum, organism grown on ^{32}S; broad solid line spectrum, organism grown on 48% ^{33}S–52% ^{32}S; tall dashed spectrum, organism grown on ^{32}S and the acid labile sulfur alone exchanged chemically for 48% ^{33}S; dotted spectrum, organism grown on 48% ^{33}S and the acid labile sulfur alone exchanged chemically for ^{32}S. Spectra were taken at X-band and at 80K, and are normalized so that each spectrum represents an equal number of spins according to double integration of the signal.

the isotope, and, as it turns out, its relatively small nhfs in these substances. Initial attempts to observe broadening in proteins from *A. vinelandii* and *P. putida* grown on enriched media showed that some type or types of sulfur ligands are present[59]. Figure 5 shows the results of a more complex experiment, where signals from samples of *P. putida* ferredoxin, prepared by combinations of growth and chemical exchange techniques, are compared. Broadening occurs when either the labile sulfur or the protein bound (cysteine or methionine) sulfur atoms are ^{33}S, and both are therefore present in the center. A more oblique approach, based on the chemical exchange of selenide for labile sulfide, gave the results depicted (for the adrenal protein) in Fig. 6[10]. Both ^{80}Se (I = 0) and ^{77}Se ($I = \frac{1}{2}$) were available in high (~ 90%) enrichment, and a comparison of signals from proteins incorporating these isotopes, indicates that both ^{77}Se atoms contribute to the observed nhfs at g = 2.03, though the nhfs is markedly anisotropic (no broadening at g = 1.95). This suggests by inference that *both* labile sulfur atoms are ligands in the iron–sulfur cluster in the native protein. An attempt to detect nitrogenous ligands in an *Azotobacter* ferredoxin by similar methods was not successful[60]. The smallness of the nhfs with respect to the line width of the signal in unenriched proteins has precluded quantitative studies with other plant-type ferredoxins.

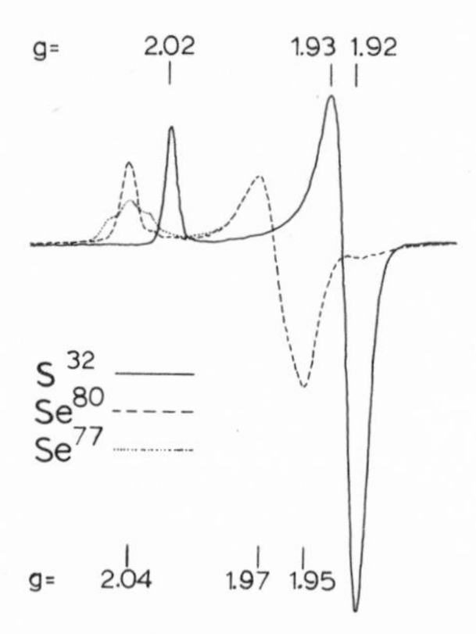

Fig. 6. Spectra of the adrenal iron–sulfur protein, with normal abundance ^{32}S and after chemical exchange of the acid labile sulfur alone for either ^{77}Se or ^{80}Se[10]. Spectra were obtained at X-band and at 80K, and are normalized to represent equal numbers of spins.

The foregoing e.p.r. studies with the adrenal and *P. putida* ferredoxins were satisfying to the extent that they indicated that both iron atoms, both labile sulfur atoms, (as well as amino acid sulfur atoms in the *Pseudomonas* protein) participate in a one-electron accepting cluster. Although one might suspect, based on the success of calculations of the ^{57}Fe nhfs in the cases of the adrenal and *P. putida* proteins, that the iron atoms are equivalent, this is *not* necessarily the case as was pointed out by early workers in the field[41]. It is conceivable that the apparent hyperfine contributions from an iron(III) and iron(II) atom would be equal[69], and of course whether this is so cannot be discerned from e.p.r. data alone. In addition to this, a further complication might be present, namely that the principal axes of the nuclear hyperfine interaction tensors may not be colinear with the axes of the g value tensors. The work of Sands and his colleagues, on the electron-nuclear double resonance (e.n.d.o.r.) spectroscopy of this same group of proteins[61] provides answers to some of these questions. The e.n.d.o.r. experiment involves the observation of the e.p.r. signal intensity at some value of applied magnetic field, while the sample is irradiated with a second radio-frequency field which sweeps through frequencies for which the resonance condition of magnetic nuclei in the sample may be satisfied. If the e.p.r. transition is saturated by a sufficiently large microwave field intensity, passage of the second radiofrequency through the resonance value of a nucleus coupled to the unpaired electron may observably perturb the e.p.r. signal by altering the relaxation rate of the unpaired electron. Thus the e.p.r. signal intensity is used to monitor n.m.r. transitions of just those magnetic nuclei which are involved in the paramagnetic center. Comparisons of e.n.d.o.r. spectra from materials of high and low ^{57}Fe enrichment, for example, allow unequivocal identification of the ^{57}Fe e.n.d.o.r. signals. These are pairs of lines centered at $A/2h$, where A is the hyperfine coupling constant, and having a splitting determined by the nuclear g value and the value of the imposed magnetic field ($\Delta\nu = g_n\beta B_0/h$).The values of the hyperfine coupling constants were thus determined for four of these iron–sulfur proteins. In each case at least one coupling constant for each of two types of iron atoms (identified by the authors as "iron(III)" and "iron(II)" sites, although they recognized that such descriptions of atoms in this type of cluster may be inaccurate) was identified; thus both iron atoms in the reduced protein are present in the paramagnetic centers of all four examples, but the iron atoms are *not* equivalent. Second, the principal values of the coupling tensor are not equal, that is, the hyperfine interaction is *not* isotropic. A third deduction made by the authors was that the principal axes of the hyperfine tensor for the "iron(III)" site are not strictly colinear with the g value tensor except in the putidaredoxin case. The earlier conclusion that both iron atoms are present in the paramagnetic complex is completely verified by these results, but the details of the magnetic properties of the iron atoms are quite complex.

Mössbauer spectroscopy of the ^{57}Fe-enriched proteins has also proven feasible[62-65]. The general features of Mössbauer spectra of samples of these proteins are quite similar, and spectral parameters for two examples are given in Table VII. The oxidized samples exhibit at low temperature a single quadrupole split pair of lines. Since the oxidized proteins are diamagnetic at these temperatures, the Mössbauer spectra are compatible with low-spin iron(II) atoms, or antiferromagnetically coupled iron(III) atoms. Mössbauer spectra of *reduced* plant-type ferredoxins show a more complex behavior (Fig. 7). At low temperatures, the relaxation rate of the unpaired electron is slow, compared to the lifetime of the nuclear event, giving rise to a non-zero net magnetic field component at the nucleus. This produces spectral broadening due to Zeeman splitting of the nuclear energy levels. At higher temperatures, the electron spin relaxation rate is increased. The net electron magnetic field component averages to zero, leaving two quadrupole split pairs of lines visible. The more widely split pair of lines have a quadrupole splitting characteristic of high-spin Fe(II) iron, while the narrow pair could be due to high or low spin Fe(III) iron. Since the reduced protein has an effective magnetic moment equivalent to an $S = \frac{1}{2}$ case, a reasonable

References pp. 741–744

TABLE VII

MÖSSBAUER PARAMETERS[a] FOR TWO IRON–SULFUR PROTEINS

		Protein[b]	
(A) Oxidized proteins		Spinach ferredoxin (4.2K)	Adrenal protein (4.2K)
I.S.		$-0.08\ ^{+0.06}_{-0.03}$	$-0.08\ ^{+0.06}_{-0.03}$
Q.S.		$0.65\ ^{+0.09}_{-0.07}$	$0.61\ ^{+0.09}_{-0.07}$
η		0.5 ± 0.3	0.5 ± 0.3
(B) Reduced proteins		(above 50K)	(above 100K)
[Iron(III) atom]	I.S.	-0.06	-0.07
	Q.S.	0.64	0.81
[Iron(II) atom]	I.S.	+0.21	+0.19
	Q.S.	2.63	2.72
		(4.2K)	(4.2K)
[Iron(III) atom]	I.S.	-0.1	—
	Q.S.	+0.64	—
	η	0.6 ± 0.3	—
[Iron(II) atom]	I.S.	0.19	—
	Q.S.	-3.00	—
	η	0 ± 0.2	—

[a] Data from ref. 65. Isomer shifts are relative to ^{57}Co diffused into platinum; both isomer shifts (I.S.) and quadrupole splittings (Q.S.) are in mm sec^{-1}; η is the asymmetry parameter.
[b] The two examples given seem to be representatives of two different types of spectroscopic behavior; ref. 65 should be consulted for further detail on these and other 2-Fe–S* proteins.

interpretation of the paramagnetic center, based on these data, is that a high-spin iron(II) ($S = \frac{4}{2}$) and a high-spin iron(III) ($S = \frac{5}{2}$) atom are antiferromagnetically coupled to give a net electron spin of $\frac{1}{2}$ in the ground state. Attempts to compute theoretical Mössbauer spectra, based on this interpretation and on the hyperfine coupling constants provided by the e.n.d.o.r. data, have been reasonably successful[65], lending further support to the above picture.

The foregoing studies have all been conducted on materials in the frozen state, but high-resolution proton magnetic resonance (p.m.r.) studies by Poe and his collaborators[66] suggest that at least some features of the picture previously developed apply at biologically relevant temperatures. Chloroplast ferredoxins from parsley and spinach were examined. Contact shifted resonances, thought to be due to the β-CH_2 protons of four cysteine residues, were detected in both oxidized and reduced samples. The

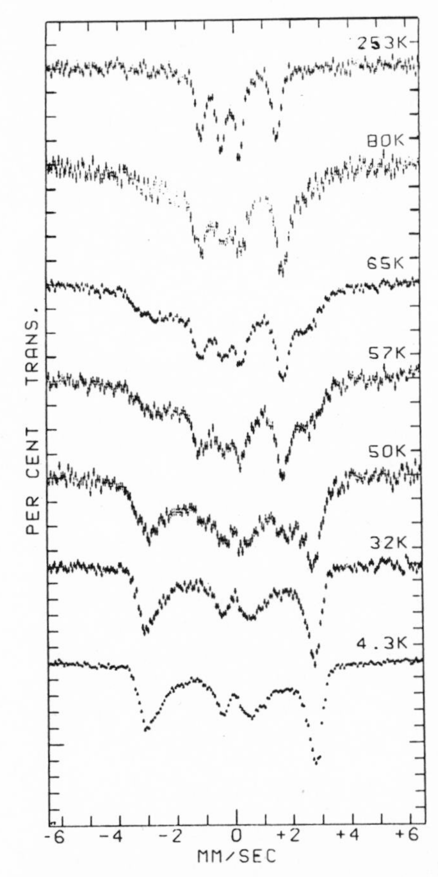

Fig. 7. Mössbauer spectra of reduced spinach iron–sulfur protein at temperatures between 4.3 and 253K[65]. The ^{57}Co source was in a platinum matrix and a field of 580 gauss was applied parallel to the gamma-ray direction. The protein had been enriched to about 90% in ^{57}Fe by chemical exchange.

temperature-dependences of these resonances suggest the presence of antiferromagnetism in both oxidation states. Evidently, upper magnetic states of the complex are populated at room temperature, so that the oxidized protein is no longer strictly diamagnetic as was observed near 4K[47]. Furthermore, the contact shifted resonances fall in two groups of four protons each, with different temperature dependencies. This observation was interpreted as arising from inequivalence in the two iron atoms, to which cysteine residues were thought to be bonded. The arresting feature is that the observed antiferromagnetic behavior would be expected of a binuclear complex with bridging sulfur ligands.

This last conclusion again supports a picture of the paramagnetic complex of plant-type ferredoxins, that was in fact put forward before most of the physical studies described here were undertaken. Brintzinger *et al.*[67] and Gibson *et al.*[68,69] suggested models of the iron–sulfur cluster, using reasonable crystal field parameters to obtain calculated optical and e.p.r.

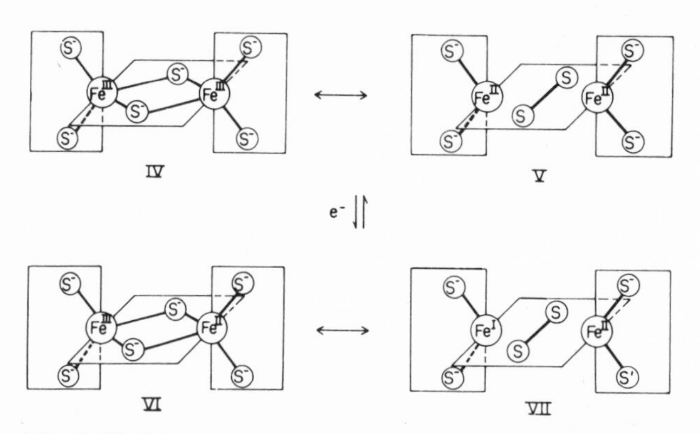

Fig. 8. Model structures proposed by Brintzinger *et al.* for the iron–sulfur cluster in the 2-Fe–S* proteins[67]. The upper structures are possible forms of the oxidized state while the lower structures refer to the one-electron reduced state. Modification of the metal coordination geometry to octahedral would not change the essential point of this model; namely that while the details of the localization of electrons in the iron–sulfur cluster is not certain, the reduced state is proposed to contain iron atoms in two different valence states. The Roman numerals in the figure refer to the original text[67]. The antiferromagnetic coupling mechanism[66] could also operate in structures such as are shown on the left side of this rather comprehensive figure.

absorption parameters in agreement with observation. Since the suspected covalent nature of these cluster systems as well as lack of knowledge of ligand geometry makes these proposals rather tentative, the ligand-field features of these models will not be pursued here. The important point in both these theoretical efforts was that an *iron–sulfur* cluster could reasonably be expected to produce an average electron *g* value below 2, in accord with the early nuclear hyperfine interaction experiments[58,59]. It should be noted that Gibson *et al.* did postulate a model in which pairs of iron atoms are antiferromagnetically coupled through sulfur ligands. Both the p.m.r. data as well as a room temperature magnetic susceptibility measurement by Ehrenberg[70], and also the e.p.r., e.n.d.o.r., and Mössbauer data, suggest that this model is correct in concept. Very recent magnetic susceptibility measurements on spinach ferredoxin show a departure from Curie-law behavior above 77K, for both oxidized and reduced samples, in accord with the proposed antiferromagnetic coupling of high-spin iron atom pairs[71]. It will be interesting to see, when X-ray diffraction studies on these systems are completed, if the iron–sulfur cluster in the proteins also has one of the structures put forward by Brintzinger *et al.* wherein two tetrahedrally coordinated iron atoms share a pair of sulfide ligands and individually have two cysteine ligands each, as in Fig. 8. It will also be of interest if correlations can be drawn between the relatively small variations in the physical parameters of these proteins, and the chemical differences in the environments of the magnetic center, brought about by differences in amino acid sequences and folding properties of the molecules.

V. 4-Fe–S* PROTEINS

Two proteins of this type have been isolated from photosynthetic bacteria[72]. The protein from *Chromatium* (strain D) has a molecular weight of about 9600, a redox potential of +350 mV, and an isoelectric point of 3.68 in the reduced form, while the protein from *Rhodopseudomonas gelatinosa* has a molecular weight of 10,100, a redox potential of +330 mV, and an isoelectric point of 9.50. The proteins can undergo a single electron change but their biochemical function has not been reported. Both proteins have four iron atoms, four labile sulfur atoms, and four cysteine residues, and have similar *N*-terminal sequences. They are often referred to as "high potential iron-proteins", abbreviated "Hipip", and are not generally regarded as "ferredoxins". Most physical and chemical studies reported to data have been concenred with the *Chromatium* protein, and in what follows this iron–sulfur protein will be the one referred to.

E.p.r. and optical spectral parameters for the protein are given in Figs. 9 and 10. On passing from the oxidized to the reduced state, about 20% of the absorbancy near 380 nm is lost, and an overlapping triple-shouldered structure gives way to a single prominent absorption at 388 nm. Unlike the 2-Fe–S* and 8-Fe–S* cases, the oxidized protein has paramagnetism equivalent to one unpaired electron per molecule, and the reduced protein is diamagnetic, in the low temperature range (< 77K) where both e.p.r. and susceptibility measurements have been made[47]. The e.p.r. signal is temperature-sensitive and is increasingly difficult to observe above 30K.

In contrast to the behavior of other iron–sulfur proteins, the iron remains bound to the protein even after acidification has displaced the labile sulfur and destroyed the chromophore[47]. No reports of reconstitution of the protein from apoprotein and inorganic constituents has appeared, nor have studies of nuclear hyperfine interactions of the type utilized with the plant-type ferredoxins. Mössbauer spectroscopy on either the oxidized or reduced forms shows a quadrupole-split single pair of lines, which in the

References pp. 741–744

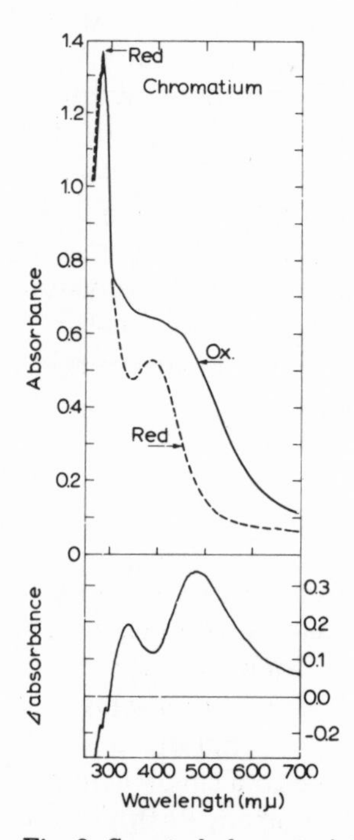

Fig. 9. Spectral characteristics of *Chromatium* high-potential iron protein (from ref. 72). The proteins were at a concentration of 3.52×10^{-5} *M* in 5×10^{-2} *M* phosphate buffer at pH 7.0. The upper spectra are of the oxidized and reduced samples, while the lower curve is the difference, oxidized *minus* reduced, between the upper spectra.

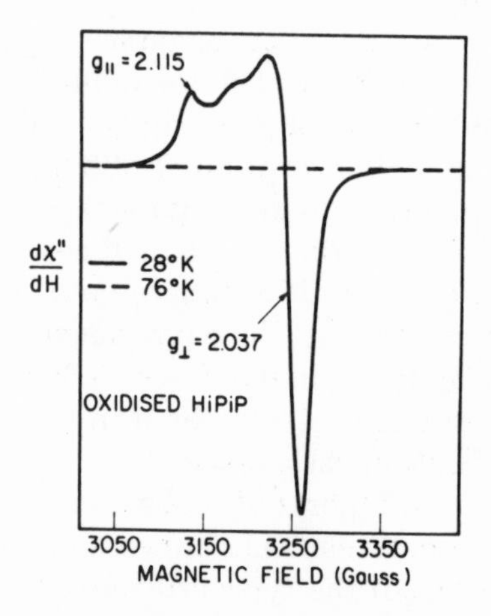

Fig. 10. E.p.r. spectrum of *oxidized chromatium* high potential iron protein, at 28 and 76K. Reproduced from ref. 47.

case of the paramagnetic oxidized form broadens at liquid helium temperatures[73]. As in the 2-Fe–S* case, this is attributed to a decrease in the relaxation rate of the unpaired electron. In the presence of a few gauss of applied magnetic field, the spectrum is resolved into numerous lines[74]. The Mössbauer evidence has been interpreted as indicating that the four iron atoms are in two inequivalent sites. Recent p.m.r. studies[75] have shown that in the oxidized protein two kinds of contact-shifted proton resonances can be distinguished by their temperature behavior. One set, attributed to two β-CH_2 groups of cysteine, have a Curie-law dependence of their contact shift, consistent with coupling to the paramagnetic ($S = \frac{1}{2}$) center. The other set, also attributed to two β-CH_2 groups on cysteine, have a temperature dependence as if they are coupled to an antiferromagnetic system. In the reduced, formally diamagnetic (at low temperatures) protein, the room-temperature contact shifts of resonances, attributed to *all four* β-CH_2 groups of cysteine, increase with temperature. This is interpreted as meaning that the cysteine residues are bonded to an iron–sulfur cluster in which all iron atoms are antiferromagnetically exchanged coupled with a diamagnetic ground state and thermally accessible (at room temperatures) upper magnetic states. Based on this evidence, as well as as on the report by Strahs and Kraut that the X-ray diffraction of this protein as 2.5 Å resolution is compatible with a single cluster of four iron atoms[76], Phillips and his co-workers suggest that the iron–sulfur cluster in this protein consists of iron and sulfur atoms at alternate corners of a distorted cube, with cysteine sulfur atoms additionally bonded to the iron atoms to give a tetrahedral ligand shell to each metal site[75]. They suggest that the electron transfer takes place at one end of the cube only, and that electron exchange between ends of the complex is slower than 10^4 sec^{-1}. This last interpretation depends on the finding of two kinds of magnetic environments for the contact-shifted resonances observed with the oxidized protein. This is consistent with recent Mössbauer evidence[74], since in the presence of a small magnetic field the spectrum can be resolved into two sets of six lines each, though the resolution is not completely unequivocal. If the localization of the unpaired electron is different on the two ends of the iron–sulfur cluster, it is easy to qualitatively imagine how the Mössbauer effect of all four iron atoms would be broadened into a complex, overlapped spectrum by the electron magnetic field at 4K. Indications of this can be seen in an earlier Mössbauer study, where an applied magnetic field was not employed[73]. Finally, pending the appearance of a high-resolution X-ray diffraction study, it might be well to recall the study of Dus *et al.*[72]. They proposed, on the basis of optical absorbance (in the u.v. region) which depends on the binding of iron atoms but not on the presence of the intact iron–sulfur chromophore, that binding to carboxyl groups is also present in the intact protein. Clearly this protein, like all the others discussed in this chapter, still presents many fascinating structural and functional problems, particularly as to how the iron–sulfur cluster is

References pp. 741–744

modulated to produce a quite positive oxidation–reduction potential as well as an odd-electron *oxidized* state and even–electron *reduced* state.

It was earlier thought that some proteins with optical and e.p.r. spectra similar to the 8-Fe–S* proteins (see below) did in fact contain ~ 4 iron and ~ 4 labile sulfur atoms. Careful analytical work has established that one of these, the ferredoxin of *Chromatium*, is an 8-Fe–S* protein[5]. Based on an assumed molar absorptivity of 3 x 10^4 at 390 nm, a similar conclusion can be drawn for ferredoxin from *Chlorobium thiosulfatophilum*[77,78] However, Shethna *et al.* have recently isolated a protein (from *Bacillus polymyxa*) which has 3–4 atoms of iron, a molecular weight of ~ 9000, and like the 8-Fe–S* cases, a very temperature sensitive e.p.r. signal in the reduced state[79]. In contrast to the 8–Fe–S* proteins, this signal seems to amount to about one spin/molecule, and it has a rhombic form reminiscent of spinach ferredoxin. This protein may therefore represent an intermediate case between the 2-Fe–S* and 8-Fe–S* ferredoxins, where a single cluster of *four* iron and sulfur atoms accepts a single electron. This would be a 4-Fe–S* *ferredoxin* in contrast to the 4-Fe–S* *high potential iron proteins* discussed above.

VI. 8-Fe–S* PROTEINS

As mentioned above (II) the term "ferredoxin" was originally applied to the protein of this class from *Clostridium pasteurianum*[8]. The protein appears to shuttle electrons between oxidative and reductive metabolic pathways in a number of anaerobic and photosynthetic bacteria. Thus the discovery of the *Clostridium pasteurianum* protein depended on the fact that this ferredoxin links the oxidative decarboxylation of pyruvate, catalyzed by pyruvate dehydrogenase, to the reduction of nitrogen to ammonia, catalyzed by nitrogenase[80]. Other oxidations in which ferredoxin is an electron acceptor include that of acetaldehyde to acetate, xanthine to urate, α-ketoglutarate to succinate and carbon dioxide, and formate to carbon dioxide. The protein is photoreduced in the presence of chlorophyll-containing particles from *Chlorobium thiosulfatophilum.* In addition to supplying electrons for nitrogen fixation, reduced ferredoxin participates in the following reductions: protons to hydrogen, hydroxylamine to ammonia, sulfite to thiosulfate, pyridine nucleotide reduction, and carbon dioxide assimilations such as acetyl coenzyme A to pyruvate and succinyl coenzyme A to α-ketoglutarate in photosynthetic bacteria. A more complete summary of these matters may be found in ref. 5.

Although these iron–sulfur proteins have so far been found only in lower organisms, interest in them has been intense, partly because of their role in nitrogen fixation (see Chapter 23). The amino acid sequences of five of these proteins are given in Table VIII[81-85]. As in the 2-Fe–S* proteins,

TABLE VIII

SEQUENCES[a] OF 8-Fe–S* PROTEINS

Organism	1–20	21–40	41–60
(1) *M. aerogenes*	ala-tyr-val-ile-asn-asp-ser-cys-ile-ala-cys-gly-ala-cys-lys-pro-glu-cys-pro-val-	asn -- ile-gln-gln-gly -- ser-ile-tyr-ala-ile-asp-ala-asp-ala-cys-ile-asp-cys-	gly-ser-cys-ala-ser-val-cys-pro-val-gly-ala-pro-asn-pro-glu-asp
(2) *Clostridium acidi-urici*	ala-tyr-val-ile-asn-glu-ala-cys-ile-ser-cys-gly-ala-cys-asp-pro-glu-cys-pro-val-	asp-ala-ile-ser-gln-gly-asp-ser-arg-tyr-val-ile-asp-ala-asp-thr-cys-ile-asp-cys-	gly-ala-cys-ala-gly-val-cys-pro-val-asp-ala-pro-val-gln-ala
(3) *C. butyricum*	ala-phe-val-ile-asn-asp-ser-cys-val-ser-cys-gly-ala-cys-ala-gly-glu-cys-pro-val-	ser-ala-ile-thr-gln-gly-asp-thr-gln-phe-val-ile-asp-ala-asp-thr-cys-ile-asp-cys-	gly-asn-cys-ala-asn-val-cys-pro-val-gly-ala-pro-asn-gln-glu
(4) *C. pasteurianum*	ala-val-lys-ile-ala-asp-ser-cys-val-ser-cys-gly-ala-cys-ala-ser-glu-cys-pro-val-	asn-ala-ile-ser-gln-gly-asp-ser-ile-phe-val-ile-asp-ala-asp-thr-cys-ile-asp-cys-	gly-asn-cys-ala-asn-val-cys-pro-val-gly-ala-pro-val-gln-glu
(5) *Chromatium*	ala-leu-met-ile-thr-asp-gln-cys-ile-asn-cys-asn-val-cys-gln-pro-glu-cys-pro-asn-	gly-ala-ile-ger-gln-gly-asp-glu-thr-tyr-val-ile-glu-pro-ser-leu-cys-thr-glu-cys-	val asp-cys-val-glu-val-cys-pro-ile-lys-asp-pro-ser-his-glu-glu-thr-glu-asp-glu- gly-his-tyr-glu-thr-ser-gln-cys-val
(5)			leu-arg-ala-lys-tyr-glu-arg-ile-thr-gly-glu-gly

[a] From refs. 81–85.

extensive regions of the sequences have been resistant to change during speciation. The ferredoxins from non-photosynthetic anaerobes have molecular weights around 6000, so that they are among the smallest of proteins, and it is therefore doubly remarkable that they appear to have arisen from a gene duplication[86]. They contain two clusters of four cysteine residues each, and careful studies by Hong and Rabinowitz indicate that the *Clostridium acidi-urici* ferredoxin contains in addition eight iron atoms and eight labile sulfur atoms[87]. The same stoichiometry is found for the inorganic components of ferredoxins from *Clostridium pasteurianum*[88] and from the photosynthetic bacteria *Chromatium* and *Chlorobium thiosulfatophilum*[10,77].

The holoproteins are strongly acidic, as would be predicted from the amino acid composition of the apoproteins. The resultant adsorption of the proteins from dilute salt solutions onto anion exchange materials is often used during the purification of these proteins[89–92]. They show a characteristic optical absorption spectrum with broad maxima near 280, 320, and 390 nm. The molar absorptivity at 390 nm is about 3×10^4 M^{-1} cm^{-1}. This absorption declines about 35% during the two-electron reduction of the protein[49,93]. Essentially all the absorption from 320 nm to longer wavelengths is abolished by the addition of either mercurials or acid to the protein solution, and for the case of *Clostridium acidi-urici* ferredoxin (Fig. 11), 85% of the absorption at 280 nm is due to the iron–sulfur chromophore which is destroyed by the above treatments[87]. As Malkin, Hong and Rabinowitz showed, the iron and sulfide thus released can be separated from the apoprotein, and the holoprotein can be regenerated from the apoprotein by treatment with thiol compounds, sulfide, and iron salts, at pH 7[94,95]. The intact chromophore is required for the electron-transferring activity of ferredoxin[90]. Denaturants such as urea or guanidine hydrochloride in high concentrations cause degradation of the iron–sulfur center,

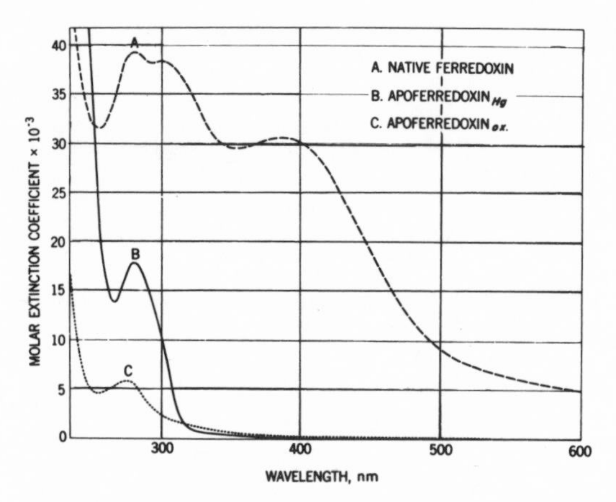

Fig. 11. Absorption spectra of *Clostridium acidi-urici* ferredoxin (A) and the apoproteins resulting from bleaching the protein by exposure to (B) sodium mersalyl or (C) trichloroacetic acid[87]. The spectra were measured with solutions of the proteins in 0.15 *M* Tris–chloride buffer, pH 7.4.

particularly if oxygen, iron-chelating agents, or sulfhydryl reagents are present[96]. The oxygen sensitivity may inhere in the labile-sulfur moiety as was found for the plant-type proteins. The origin of the labile sulfur was repeatedly claimed to be covalently bound cysteine[97,98], but these claims have not withstood further scrutiny[99,100]. Whatever the nature of the labile sulfur when bound to the protein, these atoms belong to a separate class from the sulfur-containing amino acids since the labile sulfur may be exchanged for radioactive sulfide without labeling of the cysteine taking place.

Hong and Rabinowitz have prepared a series of derivatives of *Clostridium acidi-urici* ferredoxin, modified at the NH_2– and COOH termini[101]. Acylated derivatives with neutral or positive charges at the NH_2 terminal end could be reconstituted to the corresponding holoprotein, though of lesser stability and lower biological activity. A succinylated derivative as well as an apoprotein where the tyrosine residue was iodinated did not undergo the reconstitution reaction. A derivative from which the COOH terminal alanine and glutamine residues were removed by carboxypeptidase A, leaving a COOH- terminal valine residue, also was reconstitutable to a less stable ferredoxin. It is apparent from this that both the NH_2- and COOH- terminal residues participate in stabilizing the iron–sulfur cluster system. These residues may be tucked into the interior of the protein, since the holoprotein is not acetylatable nor is it attacked by carboxypeptidase A. The total synthesis of the amino acid sequence of *Clostridium pasteurianum* protein has been reported[102], but the authors were unable to reconstitute significant amounts of the apoprotein to holoprotein. The origin of this difficulty is obscure, since reconstitution from the natural holoprotein can be made to proceed in 60–90% yield[94]. The multiplicity of sequences now known shows that about half the sequence of a given protein is invariant which may mean that some subtle failure of the synthesis has occurred in a critical region of the protein *e.g.* the isomerization of an aspartate linkage from a 1-carboxyl to a 4-carboxyl.

It was early recognized that the redox potential of the bacterial ferredoxin is quite low, near the hydrogen electrode potential at pH 7[45], in keeping with the function of the protein in such processes as hydrogen evolution and carbon dioxide and nitrogen fixation. Though a report that a single electron is transferred has appeared[45], studies on the stoichiometry of dye and pyridine nucleotide reduction by reduced ferredoxin indicate that two electrons are transferred[46,93]. Dithionite titrations, monitored by optical absorbance changes, give the same result[49]. Eisenstein and Wang, analyzing their data on the equilibrium between the methyl viologen and *Clostridium pasteurianum* ferredoxin couples, conclude that the two electrons are transferred with midpoint potentials of –367 mV and –398 mV[103]. An independent approach to this problem has been made using reductive titrations monitored by e.p.r. spectroscopy[104]. During such titrations with dithionite two signals are seen to arise sequentially, as depicted in Fig. 12. These evidently correspond to two paramagnetic centers, and an integration of the e.p.r. signal of the fully reduced protein showed that two spins were present per molecule. The same result is found if hydrogenase and varying pressures of hydrogen are used to elicit the intermediate states of reduction. At this writing, no nuclear hyperfine interaction studies implicating the iron or sulfur components have appeared, though all of the indirect chemical evidence points to the iron–sulfur complex as the site of the paramagnetism and electron acceptance.

In comparison to the 2-Fe–S* cases, Mössbauer spectroscopy of the bacterial proteins has been relatively unrewarding. In both oxidized and reduced proteins, broad quadrupole split doublets are seen, with little change after transfer of the electrons[105,106]. It may be that magnetic hyperfine structure cannot be observed in this case even at 4K, because of the rapid relaxation of the electron spin, a feature that also makes very low

References pp. 741–744

temperatures (< 30K) necessary for observation of the e.p.r. of these substances[107].

Recently, Poe and his collaborators have reported n.m.r. studies of *Clostridium pasteurianum* and *Clostridium acidi-urici* ferredoxins[108,109]. For the latter protein, the oxidized form gives a magnetic susceptibility of 1.2 Bm/iron atom. Sixteen contact-shifted protons are observed and are attributed to the sixteen β-protons on the cysteine residues. The contact shifts

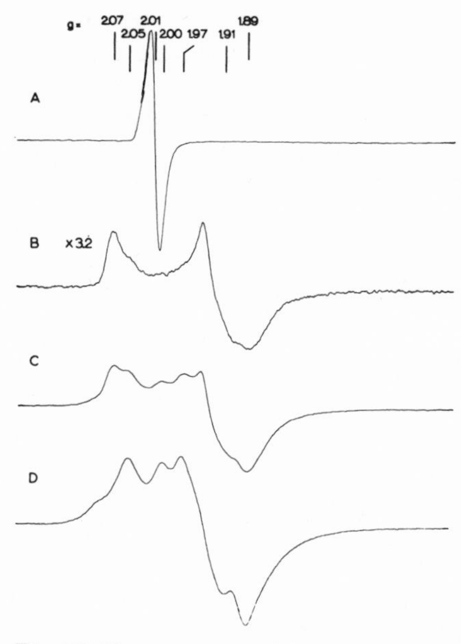

Fig. 12. E.p.r. spectra of 8.2 x 10^{-4} *M* solutions of *Clostridium acidi-urici* ferredoxin in 0.1 *M* Tris–chloride, pH 7.5, reduced with the indicated equivalents of sodium dithionite, frozen, and observed at X-band at 4.2K. (A) No dithionite, (B) 0.44 electrons per molecule, (C) 1.44 electrons per molecule, (D) 2 electrons per molecule. Spectra (A), (C) and (D) were taken at unit instrument gain, while for (B) the gain was increased to 3.2. From ref. 111.

increase with temperature, contrary to the Curie law. Since the magnetic susceptibility is less than that which would be observed for Fe(III) iron atoms, since it increases with temperature, and since the contact shifts behave in a parallel manner, it is reasonable to suppose that the iron–sulfur clusters, in which the cysteine residues are bonded *via* their sulfur atoms, are antiferromagnetically coupled arrays. When the protein is reduced, taking up two electrons, again sixteen contact-shifted resonances are observed, but eight obey the Curie law and eight do not. The authors interpret this as meaning that each unpaired electron is mostly localized on about half its cluster, so that half the cysteines experience the added paramagnetism, while the other half do not.

X-ray studies have been initiated on a number of these ferredoxins. Sieker and Jensen found with *M. aerogenes* ferredoxin that at low resolution they could not place the iron atoms in a single linear array[110] as had earlier been suggested in a model for the *Clostridium pasteurianum* protein[105]. More recently, Sieker *et al.*[111] have obtained the structure of *M. aerogenes* ferredoxin, using 2.5 Å data and heavy-metal derivatives. The molecule contains two tetrameric (4 Fe–S*) clusters, separated by about 12 Å, each of which is approximately cubic in shape. Iron and sulfide atoms share alternate corners of the cube and each iron atom is further complexed to a cysteine sulfur. Each of the two tyrosine side chains lies close to one of the clusters. Cysteines 8, 11, 14 and 45 coordinate to one cube and cysteines 18, 35, 38 and 42 coordinate to the other. (*c.f.* Table VIII).

As was the case with the other iron–sulfur proteins described in this chapter, the story of the bacterial ferredoxins falls naturally into three parts, *viz.* the discovery and appreciation of the wide range of biological functions of the proteins, the study of the chemistry of the proteins and attempts to understand the unique iron–sulfur chromophores involved, along with remarkably rapid advances in knowledge of the amino acid sequences, and finally the beginnings of physical studies of the electron transfer process itself. Again, the fullest understanding of the structure-function relationships, if it is to come, the *chemistry* of the system as modern chemists understand simpler substances, awaits solution of the X-ray structures and perfection of the other physical measurements.

VII. CONCLUDING REMARKS

The "simple" iron–sulfur proteins that have been dealt with in the preceding pages contain one to eight iron atoms bonded to cysteine residues, in peptide chains that are among the shorter examples known. The evident invariance of the positions of these residues within various classes of these proteins is convincing proof that the spatial disposition of the cysteines is an important determinant of the distinctive properties of the four classes. In addition, the proteins with two or more iron atoms evidently contain *clusters* of iron, sulfur (sulfide?) and cysteine thiols (thiolates?) so that a polynuclear complex deals with a single electron. Model systems incorporating some of these features have recently become available.

Connelly and Dahl[112] have synthesized and determined the structure of $[(C_5H_5)Fe(CO)(SCH_3)]_2^+$, which is a sulfur-bridged paramagnetic iron dimer. They show by comparison to an analogous reduced species that there is a single-electron iron–iron bond present, and suggest that a similar interaction may be present in the iron–sulfur proteins. This proposal has the

References pp. 741–744

attractive feature of being testable, at least in principle, since the iron–iron distance shrinks from 3.39 Å in the (non-bonded) reduced to 2.92 Å in the (bonded) oxidized state. Similar sulfur-bridged iron dimers, using dithiolene ligands, have been synthesized[113], as have comparable complexes with carbonyl and sulfide[114], and thioxanthate[115] ligands. Though the ligands used are not those found in proteins, the thioxanthate system in particular may prove a useful analog of the protein cases. For the tetrameric (4-Fe–S*, and, apparently, 8-Fe–S*) proteins, the models $[C_5H_5FeS]_4$ and $[C_5H_5FeS]_4^+$ are available[116,117] and they show structural alterations similar to the dimeric case cited above[112] on passing from the reduced to the oxidized state. Again, whether iron–iron bonds are involved should be discernible when precise X-ray data become available for the oxidized and reduced proteins.

Based mainly on e.p.r. and optical absorbance evidence, the plant-type ferredoxin cluster seems to have been singled out as the principal iron–sulfur cluster type in higher organisms, since it is found in the steroid hydroxylating system, in liver aldehyde oxidase, milk xanthine oxidase, and in mitochondria. This is not to deny the importance of the other types, particularly the bacterial ferredoxins, which may be the prototype of the iron–sulfur proteins in nitrogenase and other enzymes in anaerobic metabolism. Aside from the many remaining problems in the structure and function of the "simple" iron–sulfur proteins, the possibility of their being just such models for electron-transfer centers in the less easily approachable complex proteins makes their continued study worthwhile.

ACKNOWLEDGEMENTS

Work reported here was supported by the Institute of General Medical Sciences, U.S. Public Health Service, Research Grant GM 12394 and Research Career Development Award (5-KO3-GM-10,236); and by the Graduate Research Committee, Wisconsin Alumni Research Fund, The University of Wisconsin.

The author would like to thank Drs. A. J. Bearden, B. B. Buchanan, R. H. Burris, L. F. Dahl, D. V. Der Vartanian, R. Dunham, J. Fritz, I. C. Gunsalus, S. Kerestez-Nagy, T. Kimura, C. C. McDonald, R. Malkin, E. Margoliash, L. E. Mortenson, T. Moss, W. Lovenberg, G. Palmer, W. D. Phillips, M. Poe, J. C. Rabinowitz, I. Salmeen, R. H. Sands, Y. I. Shethna, T. P. Singer, R. Tsai, J. C. M. Tsibris, O. Wallasek, and C. A. Yu, for discussions of these matters, as well as for permission to cite work in advance of publication. The author also would like to record his appreciation of the enthusiasm, support, and occasional forebearance of his colleagues in Madison, Dr. H. Beinert and Mr. R. E. Hansen, in our long-standing collaboration on the studies discussed herein.

REFERENCES

1 R. Malkin and J. C. Rabinowitz, *A. Rev. Biochem.*, 36 (1967) 113.
2 T. Kimura, *Structure and Bonding*, 5 (1968) 1.
3 D. O. Hall and M. C. W. Evans, *Nature*, 223 (1969) 1342.
4 J. C. M. Tsibris and R. W. Woody, *Coordination Chem. Rev.*, 5 (1970) 417.
5 B. B. Buchanan and D. I. Arnon, *Adv. Enzymol.*, 33 (1970) 119.
6 A. San Pietro, (Ed.), *Non-heme Iron Proteins*, Antioch Press, Yellow Springs, Ohio, 1965.
7 W. Lovenberg, (Ed.), *Iron–Sulfur Proteins*, in preparation.
8 L. E. Mortenson, R. C. Valentine and J. E. Carnahan, *Biochem. Biophys. Res. Commun.*, 7 (1962) 448.
9 K. Tagawa and D. I. Arnon, *Nature*, 195 (1962) 537.
10 W. H. Orme-Johnson, R. E. Hansen, H. Beinert, *et al.*, *Proc. Natn. Acad. Sci. U.S.*, 60 (1968) 368.
11 W. Lovenberg and B. E. Sobel, *Proc. Natn. Acad. Sci. U.S.*, 54 (1965) 193.
12 D. J. Newman and J. R. Postgate, *Eur. J. Biochem.*, 7 (1968) 45.
13 W. Lovenberg and W. M. Williams, *Biochemistry*, 8 (1) (1969) 141.
14 H. Bachmayer, K. T. Yasunobu and H. R. Whiteley, *Proc. Natn. Acad. Sci. U.S.*, 59 (1968) 1273.
15 H. Bachmayer, K. T. Yasunobu, J. L. Peel and S. J. Mayhew, *J. Biol. Chem.*, 243 (1968) 1022.
16 N. M. Atherton, K. Garbett, R. D. Gillard, *et al.*, *Nature*, 212 (1966) 590.
17 W. A. Eaton and W. Lovenberg, *J. Am. Chem. Soc.*, 92 (1970) 7195.
18 W. E. Blumberg, in A. Ehrenberg, B. G. Malmström and T. Vänngård, *Magnetic Resonance in Biological Systems*, Pergamon Press, London, 1967, p. 119; R. Aasa, *J. Chem. Phys.*, 52 (1970) 3919.
19 W. D. Phillips, M. Poe, J. F. Weiher and C. C. McDonald, *Nature*, 227 (1970) 574.
20 J. R. Herriott, L. C. Sieker and L. H. Jensen, *J. Mol. Biol.*, 50 (1970) 391.
21 E. T. Lode and M. J. Coon, *Federation Proc.*, 29 (1970) 3724.
22 A. M. Benson, M. Haniu, E. T. Lode, M. J. Coon and K. T. Yasunobu, *Federation Proc.*, 29 (1970) 3478.
23 E. J. McKenna and M. J. Coon, *J. Biol. Chem.*, 245 (1970) 3882.
24 R. Malkin and A. J. Bearden, *Proc. Natn. Acad. Sci. U.S.*, 68 (1971) 16.
25 T. Omura, E. Sanders, D. Y. Cooper and R. W. Estabrook, in S. P. Colowick and N. O. Kaplan, *Methods in Enzymology*, Vol. X, Academic Press, New York, 1967.
26 K. Suzuki and T. Kimura, *Biochem. Biophys. Res. Commun.*, 19 (1965) 340.
27 H. Ohno, K. Suzuki and T. Kimura, *Biochem. Biophys. Res. Commun.*, 26 (1967) 651.
28 D. W. Cushman, R. L. Tsai and I. C. Gunsalus, *Biochem. Biophys. Res. Commun.*, 26 (1967) 577.
29 R. W. F. Hardy, E. Knight, Jr., C. C. McDonald and A. J. d'Eustachio, in ref. 6, p. 275.
30 Y. I. Shethna, D. V. DerVartanian and H. Beinert, *Biochem. Biophys. Res. Commun.*, 31 (1968) 862; D. V. DerVartanian, Y. I. Shethna and H. Beinert, *Biochim. Biophys. Acta*, 194 (1969) 548.
31 J. S. Rieske, D. H. MacLennan and R. Coleman, *Biochem. Biophys. Res. Commun.*, 15 (1964) 338.
32 H. Matsubara and R. M. Sasaki, *J. Biol. Chem.*, 243 (1968) 1732.
33 K. Sugeno and H. Matsubara, *Biochem. Biophys. Res. Commun.*, 32 (1968) 951.
34 A. M. Benson and K. T. Yasunobu, *J. Biol. Chem.*, 244 (1969) 955.

35 S. Keresztes-Nagy, F. Perini and E. Margoliash, *J. Biol. Chem.*, 244 (1969) 981.
36 K. K. Rao and H. Matsubara, *Biochem. Biophys. Res. Commun.*, 38 (1970) 500.
37 M. Tanaka, M. Haniu and K. T. Yasunobu, *Biochem. Biophys. Res. Commun.*, 39 (1970) 1182.
38 K. T. Yasunobu, in W. Lovenberg, *Iron–Sulfur Proteins*, in preparation.
39 D. U. Petering, J. A. Fee and G. Palmer, *J. Biol. Chem.*, 246 (1971) 643.
40 G. Palmer, *Biochem. Biophys. Res. Commun.*, 27 (1967) 315.
41 J. C. M. Tsibris, R. L. Tsai, I. C. Gunsalus, W. H. Orme-Johnson, R. E. Hansen and H. Beinert, *Proc. Natn. Acad. Sci. U.S.*, 59 (1968) 959.
42 J. C. M. Tsibris, M. J. Namtvedt and I. C. Gunsalus, *Biochem. Biophys. Res. Commun.*, 30 (1968) 323.
43 K. T. Fry, R. A. Lazzarini and A. San Pietro, *Proc. Natn. Acad. Sci. U.S.*, 50 (1963) 652.
44 G. Palmer and R. H. Sands, *J. Biol. Chem.*, 241 (1966) 253.
45 K. Tagawa and D. I. Arnon, *Biochim. Biophys. Acta*, 153 (1968) 602.
46 M. C. W. Evans, D. O. Hall, H. Bothe and F. R. Whatley, *Biochem. J.*, 110 (1968) 485.
47 T. H. Moss, D. Petering and G. Palmer, *J. Biol. Chem.*, 244 (1969) 2275; G. Palmer, H. Brintzinger, R. W. Estabrook and R. H. Sands, in A. Ehrenberg, B. G. Malmström and T. Vänngård, *Magnetic Resonance in Biological Systems, Proc. 2nd Int. Conf., Stockholm*, 1966, Pergamon Press, London, 1967, p. 169.
48 W. H. Orme-Johnson and H. Beinert, *J. Biol. Chem.*, 244 (1969) 6143.
49 S. G. Mayhew, D. Petering, G. Palmer and G. P. Foust, *J. Biol. Chem.*, 244 (1969) 2830.
50 G. S. Wilson, *Univ. Illinois Bull.* 67 (1969) 20.
51 G. Palmer, H. Brintzinger and R. W. Estabrook, *Biochemistry*, 6 (1967) 1658.
52 T. Kimura, A. Tasaki and H. Watari, *J. Biol. Chem.*, 245 (1970) 4450.
53 C. Moleski, T. H. Moss, W. H. Orme-Johnson and J. C. M. Tsibris, *Biochim. Biophys. Acta*, 214 (1970) 548.
54 H. Beinert and R. H. Sands, *Biochem. Biophys. Res. Commun.*, 3 (1960) 41.
55 J. F. Gibson and R. C. Bray, *Biochim. Biophys. Acta*, 153 (1968) 721.
56 K. V. Rajagopalan, P. Handler, G. Palmer and H. Beinert, *J. Biol. Chem.*, 243 (1968) 3784.
57 H. Beinert and G. Palmer, *Adv. Enzymol.*, 27 (1965) 105.
58 Y. I. Shethna, P. W. Wilson, R. E. Hansen and H. Beinert, *Proc. Natn. Acad. Sci. U.S.*, 52 (1964) 1263.
59 D. V. DerVartanian, W. H. Orme-Johnson, R. E. Hansen, H. Beinert, *et al.*, *Biochem. Biophys. Res. Commun.*, 26 (1967) 569.
60 T. C. Hollocher and J. K. Luechauer, *Biochem. Biophys. Res. Commun.*, 31 (1968) 417.
61 J. Fritz, R. Anderson, J. Fee, *et al.*, *Biochim. Biophys. Acta*, 253 (1971) 110.
62 R. Cooke, J. C. M. Tsibris, P. G. Debrunner, R. Tsai, I. C. Gunsalus and H. Frauenfelder, *Proc. Natn. Acad. Sci. U.S.*, 59 (1968) 1045.
63 C. E. Johnson, E. Elstner, J. F. Gibson, G. Benfield, M. C. W. Evans and D. O. Hall, *Nature*, 220 (1968) 1291.
64 C. E. Johnson, R. C. Bray, R. Cammack and D. O. Hall, *Proc. Natn. Acad. Sci. U.S.*, 63 (1969) 1234.
65 R. Dunham, A. Bearden, D. Salmeen, *et al.*, *Biochim. Biophys. Acta*, 253 (1971) 134.
66 M. Poe, W. D. Phillips, J. D. Glickson, C. C. McDonald and A. San Pietro, *Proc. Natn. Acad. Sci. U.S.*, 68 (1971) 68.
67 H. Brintzinger, G. Palmer and R. H. Sands, *Biochemistry*, 55 (1965) 397.

68 J. F. Gibson, D. O. Hall, J. H. M. Thornley and F. R. Whatley, *Proc. Natn. Acad. Sci. U.S.*, 56 (1966) 987.
69 J. H. M. Thornley, J. F. Gibson, F. R. Whatley and D. O. Hall, *Biochem. Biophys. Res. Commun.*, 24 (1966) 877.
70 A. Ehrenberg, quoted in ref. 69.
71 G. Palmer, personal communication.
72 K. Dus, H. De Klerk, K. Sletten and R. G. Bartsch, *Biochim. Biophys. Acta*, 140 (1967) 291.
73 T. H. Moss, A. J. Bearden, R. G. Bartsch, M. A. Cusanovitch and A. San Pietro, *Biochemistry*, 7 (1968) 1591.
74 M. C. W. Evans, D. O. Hall and C. E. Johnson, *Biochem. J.*, 119 (1970) 289.
75 W. D. Phillips, M. Poe, C. C. McDonald and R. Bartsch, *Proc. Natn. Acad. Sci. U.S.*, 67 (1970) 382.
76 G. Strahs and J. Kraut, *J. Mol. Biol.*, 35 (1968) 503.
77 B. B. Buchanan, personal communication.
78 B. B. Buchanan, H. Matsubara and M. C. W. Evans, *Biochim. Biophys. Acta*, 189 (1969) 46.
79 Y. I. Shethna, N. Stombaugh and R. H. Burris, *Biochem. Biophys. Res. Commun.*, 42 (1971) 1108.
80 L. E. Mortenson, *A. Rev. Microbiol.*, 7 (1963) 115.
81 M. Tanaka, T. Nakashima, A. Benson, H. Mower and K. T. Yasunobu, *Biochemistry*, 5 (1966) 1666.
82 A. Benson, H. Mower and K. Yasunobu, *Arch. Biochem. Biophys.*, 121 (1967) 563.
83 J. Tsunoda and K. Yasunobu, *J. Biol. Chem.*, 243 (1968) 6262.
84 H. Matsubara, R. Sasaki, D. Tsuchiya and M. C. W. Evans, *J. Biol. Chem.*, 245 (1970) 2121.
85 S. C. Rall, R. E. Bolinger and R. D. Cole, *Biochemistry*, 8 (1969) 2486.
86 R. V. Eck and M. O. Dayhoff, *Science*, 152 (1966) 366.
87 J. S. Hong and J. C. Rabinowitz, *J. Biol. Chem.*, 245 (1970) 4982.
88 W. Lovenberg, personal communication.
89 L. E. Mortenson, *Biochim. Biophys. Acta*, 81 (1964) 71.
90 W. Lovenberg, B. B. Buchanan and J. C. Rabinowitz, *J. Biol. Chem.*, 238 (1963) 3899.
91 R. Bachofen and D. I. Arnon, *Biochim. Biophys. Acta*, 120 (1966) 259.
92 T. Devanathan, J. M. Akagi, R. T. Hersh and R. H. Himes, *J. Biol. Chem.*, 244 (1969) 2846.
93 B. E. Sobel and W. Lovenberg, *Biochemistry*, 5 (1966) 6.
94 R. Malkin and J. C. Rabinowitz, *Biochem. Biophys. Res. Commun.*, 23 (1966) 822.
95 J.-S. Hong and J. C. Rabinowitz, *Biochem. Biophys. Res. Commun.*, 29 (1967) 246.
96 R. Malkin and J. C. Rabinowitz, *Biochemistry*, 6 (1967) 3880.
97 E. Bayer, W. Parr and B. Katzmaier, *Arch. Pharm.*, 298 (1965) 196.
98 K. Gersonde and W. Druskeit, *Eur. J. Biochem.*, 4 (1968) 391.
99 R. Malkin and J. C. Rabinowitz, *Biochemistry*, 5 (1966) 1262.
100 J.-S. Hong, A. B. Champion and J. C. Rabinowitz, *Eur. J. Biochem.*, 8 (1969) 307.
101 J.-S. Hong and J. C. Rabinowitz, *J. Biol. Chem.*, 245 (1970) 4988.
102 E. Bayer, *3rd Int. Conf. Magnetic Resonance in Biology, Warrenton, Va.*, October 1968.
103 K. K. Eisenstein and J. H. Wang, *J. Biol. Chem.*, 244 (1969) 1720.

104 W. H. Orme-Johnson and H. Beinert, *Biochem. Biophys. Res. Commun.*, 36 (1969) 337.
105 D. C. Blomstrom, E. Knight, Jr., W. D. Phillips and J. F. Weiher, *Proc. Natn. Acad. Sci. U.S.*, 51 (1964) 1085.
106 A. J. Bearden and W. H. Orme-Johnson, unpublished results.
107 G. Palmer, R. H. Sands and L. E. Mortenson, *Biochem. Biophys. Res. Commun.*, 23 (1966) 357.
108 M. Poe, W. D. Phillips, C. C. McDonald and W. Lovenberg, *Proc. Natn. Acad. Sci. U.S.*, 65 (1970) 797.
109 M. Poe, W. D. Phillips, C. C. McDonald and W. H. Orme-Johnson, *Biochem. Biophys. Res. Commun.*, 42 (1971) 705.
110 L. W. Sieker and L. H. Jensen, *Biochem. Biophys. Res. Commun.*, 20 (1965) 33.
111 L. C. Sieker, E. Adman and L. H. Jensen, *Nature*, 235 (1970) 40; *Abstr. Am. Cryst. Assoc., Winter Meeting, April 3-7 1972*, p. 66.
112 N. G. Connelly and L. F. Dahl, *J. Am. Chem. Soc.*, 92 (1970) 7472.
113 A.G. Balch, I. G. Donce and R. H. Holm, *J. Am. Chem. Soc.*, 90 (1968) 1135.
114 C. H. Wei and L. F. Dahl, *Inorg. Chem.*, 4 (1965) 1.
115 D. Coucouvanis, S. J. Lippard and J. A. Zubieta, *J. Am. Chem. Soc.*, 91 (1969) 761.
116 C. H. Wei, G. R. Wilkes, P. M. Treichel and L. F. Dahl, *Inorg. Chem.*, 5 (1966) 900.
117 T. Toan, W. P. Fehlhammer, L. F. Dahl, submitted for publication.
118 W. M. Clark, *Oxidation–Reduction Potentials of Organic Systems*, Williams and Wilkins, Baltimore, 1960.
119 D. V. DerVartanian, in T. E. King and M. Klingenberg, *Treatise on Electron and Coupled Energy in Biological Systems*, Vol. II, in preparation.
120 L. M. Siegel and H. Kamin, in K. Yagi, *Flavins and Flavoproteins*, Univ. Park Press, Baltimore, 1968, p. 15.
121 T. P. Singer, in *Biological Oxidations*, Interscience, New York, 1968, p. 339.
122 N. R. Orme-Johnson, W. H. Orme-Johnson, R. E. Hansen, H. Beinert and Y. Hatefi, *Biochem. Biophys. Res. Commun.*, 44 (1971) 446.

Chapter 23

BIOINORGANIC CHEMISTRY OF DINITROGEN* FIXATION

R. W. F. HARDY, R. C. BURNS AND G. W. PARSHALL

Central Research Department, Experimental Station, E. I. du Pont de Nemours & Co., Inc., Wilmington Delaware 19898, U.S.A.

I. INTRODUCTION

Dinitrogen fixation is any reaction of N_2 resulting in nitrogen covalently bonded to any other atom. There appears to be only a single mechanism involved in biological fixation, while several methods of abiological N_2 fixation are known; essentially all N_2-fixing systems contain metals. In biological N_2 fixation a molybdenum- and iron-containing enzyme, nitrogenase, reduces N_2 to NH_3 in the only known facile reaction of this type under ambient conditions. The commercial abiological systems use heterogeneous catalysts for reduction of N_2 to NH_3 under conditions of high pressure and temperature. Abiological systems being actively explored, but not yet of practical utility, involve the formation of transi-

---

* Dinitrogen, the systematic name for N_2, clearly differentiates N_2 from N and will be used in accord with its increasing acceptance; nitrogenyl, by analogy with carbonyl, has been used for $M \cdot N_2$ complexes, Diazene will be used for diimide. Dihydrogen will be used for H_2 and dioxygen for O_2.

Nitrogenase, the enzyme that catalyzes N_2 reduction, will be abbreviated as N_2ase. It is a complex of two proteins designated here as Mo–Fe protein or Fe protein (Table II). N_2ase isolated from cells grown on V in place of Mo is designated V–N_2ase.

The authors have proposed a revised nomenclature, specifically the terms azofermo and azofer, for the Mo–Fe and Fe–proteins, respectively. The suggested designations identify the proteins as derived from N_2ase, yet retain the metallo roots which reflect their composition; other nomenclature currently in the literature is less descriptive, *e.g.*, fraction 1 and fraction 2, or is confusing, *e.g.*, molybdoferredoxin and azoferredoxin, which can be erroneously interpreted to imply ferredoxin-type electron transfer functions for N_2ase proteins.

Other abbreviations used: Fd for ferredoxin; Fld for flavodoxin (or azotoflavin); CoASH for coenzyme A; NADPH for reduced nicotinamide adenine dinucleotide phosphate; LHb for leghemoglobin; ATP, CTP, UTP, GTP, and XTP for the 5′-triphosphate salts of adenosine, cytidine, uridine, guanosine and any of these, respectively; ADP and XTP for the 5′-diphosphate salts adenosine and of any of the above nucleosides, respectively; and XMP for the corresponding 5′-monophosphate salts; Cp for cyclopentadiene; DEPE for *P,P,P,P′*-tetraethylethylenediphosphine; and Ph_3P for triphenyl phosphine.

References pp. 787–793

tion metal complexes of dinitrogen and reductions of N_2 to NH_3 by homogeneous catalysts under mild conditions.

The objective of this chapter is to provide a comprehensive integrated presentation of biological and abiological N_2 fixation, an area with almost unparalleled opportunity for bioinorganic interaction. The scope will include the chemistry of N_2, characteristics, reactions and mechanisms of biological fixation and abiological fixation reactions. Developments of the past decade of research in the biochemistry[1-13] and inorganic chemistry[14-24] of N_2 fixation will be emphasized. This progress occurred for the most part in strictly inter-disciplinary studies so that bioinorganic correlations emerged mainly as the result of chance rather than direction. It is hoped that interdisciplinary reviews will promote convergence of the two approaches and produce important bioinorganic synergisms in the future.

II. CHEMISTRY OF DINITROGEN

Stable nitrogen compounds exist in formal oxidation states from +5 to -3[25]:

+5	+4	+3	+2	+1	0	-1	-2	-3
HNO_3	NO_2	HNO_2	NO	N_2O	N_2	N_2H_2	N_2H_4	NH_3

Most reductions and oxidations occur with ease, and N in valence states from +2 to -3 interacts with the biological N_2 fixation system: N_2 is the substrate and NH_3 the immediate product of N_2 fixation; N_2H_2 and N_2H_4 are proposed intermediates; N_2O is an alternate substrate, and NO is an inhibitor.

The chemistry of dinitrogen is dominated by the extraordinary kinetic and thermodynamic stability of the N_2 molecule[25]. The energy required to cleave $N{\equiv}N$ to a pair of N atoms is 226 kcal/mole. Hence, any reaction requiring initial dissociation of the dinitrogen molecule has an extremely high activation energy, generally requiring temperatures of 1000°C or greater. Reduction to ammonia, the only significant natural or commerical fixation process, however, is subject to catalysis and occurs under much milder conditions. The biological process occurs at biospheric temperatures ranging from polar to hot springs but mainly in the range of 15-40°C, while the commerical Haber–Bosch process occurs at about 450°C. The reductive fixation processes, in addition to being kinetically more attractive than oxidation, are favored thermodynamically[26]. For example, at 25°C,

$$N_2 + 3H_2 \rightleftharpoons 2NH_3\ (g) \quad \Delta F^\circ = -7.95\ \text{kcal/mole}$$

$$N_2 + O_2 \rightleftharpoons 2NO\ (g) \quad \Delta F^\circ = +41.44\ \text{kcal/mole}$$

The extraordinary stability of N_2 may be explained by a simple molecular orbital treatment that puts six valence electrons in a σ_g and a pair of π_u orbitals (illustrated in Fig. 1). The four additional valence

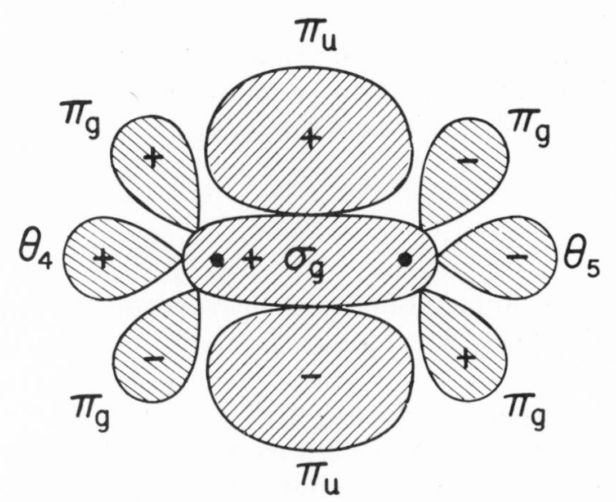

Fig. 1. A cross-section of the N_2 molecule showing the nuclei (•), bonding [hatched] and antibonding [hatched] orbitals. The π_u and π_g orbitals shown are duplicated in a plane perpendicular to the plane of the paper. The θ_4 and θ_5 localized orbitals are mathematical hybrids of the conventional $3\sigma_g$ and $2\sigma_u$ molecular orbitals.

electrons reside in *sigma* orbitals that may be mathematically treated[27] as being localized on the two N atoms (θ_4, θ_5 set) and are not exposed to the extent that they interact with protonic acids. However, this filled set

TABLE I

REDUCTION ENERGIES OF NITROGENASE SUBSTRATES
(All energies expressed as kcal/mole for gaseous reactants at 25° C)

Substrate	Bond dissociation energy[28]	Reduction products	ΔF^a reduction
N≡N	226	$2NH_3$	-7.95 ($2NH_3$)
HC≡N	224	CH_4, NH_3	-44.8
CH_3C≡N	~230	C_2H_6, NH_3	-37.0
CH_3N≡C	—	CH_3NH_2, CH_4	-45.5
HC≡CH	230	C_2H_4	-33.7
N≡N=O	"N≡N" 115 "N=O" 40	N_2, H_2O	-79.40
C≡O	257	CH_4, H_2O^b	-33.97

[a] Calculated from data in ref. 26.
[b] Hypothetical products; CO is not reduced by N_2ase.

References pp. 787-793

together with the unoccupied antibonding π_g orbitals play a significant role in stabilizing N_2 complexes of the transition metals (Section IV C3).

It is interesting to compare N_2 with a variety of other small molecules that are reduced by N_2ase (Section III B3). Bond dissociation energies[28] and reduction energies[26] of these other substrates are compared with those of N_2 in Table I. These substrates have the common characteristic of possessing an actual or potential triple bond and the chemistry of these molecules may be as relevant to N_2ase catalysis as is that of N_2.

III. BIOLOGICAL DINITROGEN FIXATION

Biological dinitrogen fixation is, in fact, biological N_2 reduction, specifically the conversion of N_2 to $2NH_3$. The biological catalyst was named nitrogenase in 1934[29]. Evolution of N_2ase can be conceived as a biological response to a need for biospheric recycling of N_2. Presumably the massive primeval ammonia resources of the earth were depleted by conversion of this readily assimilated nitrogen compound to unusable N_2 by nitrification and denitrification reactions. Nitrogenase may have evolved as a counter reaction. In spite of the extreme physiological and metabolic range found among known N_2-fixing organisms — aerobic to anaerobic, heterotrophic to photosynthetic and free-living to symbiotic — it appears that only a single successful biological N_2-fixing process has developed.

Current understanding of N_2ases from various sources permits the following general definition: N_2ases are complexes of Mo–Fe and Fe proteins whose syntheses are repressed by fixed nitrogen and whose activity couples ATP hydrolysis to electron transfer for the reduction of N_2 to $2NH_3$, N_3^- to $N_2 + NH_3$, N_2O to $N_2 + H_2O$, RCCH to $RCHCH_2$, RCN to $RCH_3 + NH_3$, RNC to RNH_2 + alkanes and alkenes, and/or $2H_3O^+$ to $2H_2O + H_2$, with H_2 a competitive inhibitor of N_2 reduction and CO an inhibitor of all reductions except that of H_3O^+ (Fig. 2).

Of the wide range of N_2-fixing agents in nature two have been most intensively investigated at the molecular level: *Clostridium pasteurianum*[30–68] and *Azotobacter*[45,51,53,56,57,69–113]. While both are free-living bacteria they represent anaerobic and aerobic extremes and thus provide a pool of N_2ase information more revealing than if obtained from two closely related species. Supporting investigations have been made with other aerobes (*Mycobacterium flavum*[114]), symbionts (soybean nodules[57,115–128]), facultative anaerobes (*Klebsiella pneumoniae*[56,94,129–132], *Bacillus polymyxa*[56,133–135]), photosynthetic bacteria (*Chromatium*[136–137], *Rhodospirillum rubrum*[138–139]), and most recently with blue-green algae (*Anabaena cylindrica*[140–142]). In the following discussion of biological N_2 fixation information on the best defined systems will be emphasized and

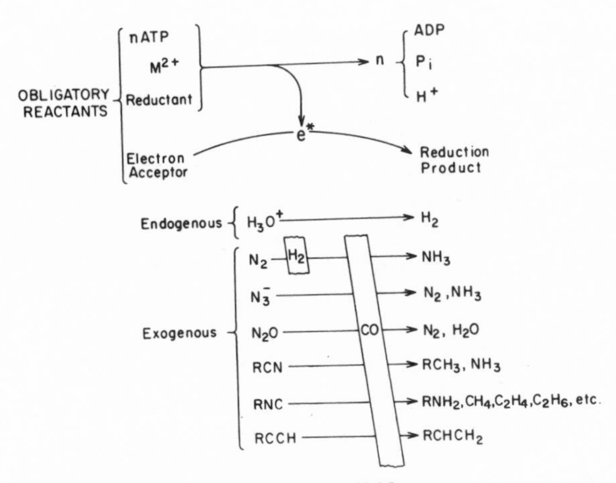

Fig. 2. Reactions common to all N_2ases.

in the majority of cases *Clostridium* and/or *Azotobacter* N_2ases will dominate.

A. *Chemical and physical characteristics of N_2ase*

(1) Nitrogenase

Nitrogenase is a complex of Mo–Fe protein and Fe protein (Table II) suggested to be in dynamic equilibrium with its components as follows[57,63,65,106]:

$$\text{Mo–Fe protein} + 2\ \text{Fe protein} \rightleftharpoons N_2\text{ase}$$

or in terms of proposed nomenclature[8,9]:

$$\text{Azofermo} + 2\ \text{Azofer} \rightleftharpoons N_2\text{ase}$$

TABLE II

PROPOSED, CURRENT AND PRIOR DESIGNATIONS OF NITROGENASE COMPONENTS

Designations		Usage
Azofermo	Azofer	Proposed
Mo–Fe Protein	Fe Protein	Current
Fraction 1	Fraction 2	Current
Molybdoferredoxin	Azoferredoxin	Limited current
0.25 *M* NaCl eluate	0.35 *M* NaCl eluate	No longer used
Enzyme I	Enzyme II	No longer used
Component I	Component II	No longer used

Interspecies as well as intraspecies recombinations of isolated Mo–Fe protein and Fe protein yield a reconstituted N_2ase; however, activity is related to the physiological similarity of the sources[56,57,94]. Reports of three protein types in N_2ase[62,86] and the requirement of an additional protein for N_2 fixation vs. C_2H_2 reduction appear to be erroneous[60]. The enzyme complex has never been isolated in homogeneous form and, therefore, can perhaps best be described on the basis of summation of the characteristics of its defined components according to the above probable stoichiometry. *Azotobacter* N_2ase[105,107,110,111] on this basis, is calculated to have a complex weight of about 200,000 daltons, a Mo, Fe, S^{2-} and CySH content of about 1, 20, 20 and 20+, and a specific activity of ~ 225 nmoles N_2 reduced/min · mg protein; its electron paramagnetic resonance (e.p.r.) spectra are identical to those of the isolated Mo–Fe protein. The participation of iron in the activation of N_2 is suggested by changes in the electron spin resonance (e.s.r.) absorption at $g = 1.94$ and 2.01 produced by specific reactants of N_2ase, such as CO[92]. For comparison, clostridial N_2ase on the same basis can be calculated to have a complex weight of about 225,000–300,000 daltons[61,65–67], a Mo and Fe content of about 1–2 and 20–24, and a specific activity of 175–215 nmoles N_2 reduced/min · mg protein. *Klebsiella* N_2ase[132], reconstituted from its ^{57}Fe-enriched components, gives a Mössbauer spectrum containing a signal characterized by $\delta = 0.90$ mm/sec*, $\Delta E = 3.0$ mm/sec and dependent on ATP, $Na_2S_2O_4$ and both proteins.

Azotobacter N_2ase preparations examined by electron microscopy[113] were shown to consist of a variety of particles including hexagons, rectangular columns of subunits and tetrads, as well as monomeric structures. Dimensions of the hexagons were 150 x 165 Å. Cellular localization of *Azotobacter* N_2ase on inner[101,102] or outer membranes has been proposed. N_2ase has been purified from *Azotobacter* grown on media supplemented with vanadium in place of Mo[112,112a]. These preparations contain V and only traces of Mo and are similar to Mo–N_2ase in some but not all characteristics, suggesting replacement of Mo by V.

Neither Mo–Fe protein nor Fe protein possess known biological activity alone; the contribution each protein makes to the N_2ase complex activity is calculated of necessity from a N_2ase reaction in which one protein is limiting and the other saturating, and specific activity is expressed as activity of N_2ase complex/min · mg limiting protein.

(2) Mo–Fe protein

Crystalline *Azotobacter* Mo–Fe protein[105,110] has been most investigated and its characteristics will be described; where available those of

* To facilitate comparisons of Mössbauer spectra all isomer shifts are given relative to sodium nitroprusside standard; literature values based on other Fe references have been re-calculated to conform to this convention.

clostridial[65,67] or *Klebsiella*[132] Mo–Fe protein will be included. Specific activities of *Azotobacter* and clostridial Mo–Fe protein are similar, ~350 nmoles N_2 reduced/min · mg protein, indicating a turnover number of ~50 moles of N_2 reduced/min · mole Mo. *Azotobacter* Mo–Fe protein molecular weight is 270,000 daltons by ultracentrifugation and is suggested to be dimeric on the basis of Mo content; two types of similar subunits of ~40,000 daltons are produced by various denaturing agents[111]. Electron microscopy[113] reveals rectangular or square structures of 80 to 90 Å from which a molecular weight of ~275,000 is calculated. Purified but not crystallized clostridial Mo–Fe protein has an indicated molecular weight of 170,000–190,000 daltons by gel filtration analysis and is dissociated by sodium dodecyl sulfate to two types of subunits of molecular weights 60,000 and 51,000, with twice as many large subunits as small. Alkali also dissociates *Azotobacter*[85] and clostridial[143] Mo–Fe protein into about six smaller pieces. Purified *Klebsiella* Mo–Fe protein has a suggested molecular weight of about 216,000 daltons. *Azotobacter* Mo–Fe protein has 2 Mo, 32 Fe, ~25 S^{2-} and 41 cysteine residues per 270,000 daltons; no other[86] metals at a level of 1 atom per molecule have been verifiable[105]. Clostridial Mo–Fe protein has 1–2 Mo, 12–15 Fe and 10–15 S^{2-} per 160–200,000 daltons. *Klebsiella* has 1 Mo and 10 Fe per 216,000 daltons.

Mo–Fe proteins are acidic; *Azotobacter* Mo–Fe protein[105,110,111] contains almost twice as many acidic as basic amino acids and has approximately equivalent amounts of iron, S^{2-}, and cysteine, all features common among other iron–sulfur proteins. The u.v.–visible absorption spectra are similar to ferredoxins which contain four or more Fe atoms; an absorption maximum occurs at 280 nm and a shoulder at 412 nm in the native protein, and on reduction with $Na_2S_2O_4$ weak maxima develop at 525 nm and 557 nm and the shoulder at 412 shifts to ~420 nm. The native protein shows electron spin resonance (e.s.r.) at g values of 2.01, 3.67 and 4.30; on reduction with $Na_2S_2O_4$, the intensity of resonances at 4.30 and 3.67 increases and a new resonance appears at $g = 1.94$. The resonance at $g = 1.94$ is common among iron–sulfur proteins, but those at $g = 3.67$ and 4.30 are unique to Mo–Fe protein and may represent an iron environment that is critical to N_2 fixation. The resonances at $g = 3.67$ and 4.30 have been variously attributed to high spin Fe(II) and to intermediate spin Fe(III). The origin of the resonance at $g = 2.01$ is obscure

Mössbauer spectra of *Azotobacter* Mo–Fe protein[110,111] show a doublet with an isomer shift of 0.56 mm/sec and a quadrupole splitting of 0.84 mm/sec indicative of high spin Fe(III); a small contribution from high spin Fe(I) or Fe(II) is suggested by a small absorption peak on the high field side of the main doublet. Reduction with hydrosulfite at pH 7.2 converts ~50% of the iron to high spin Fe(I) or Fe(II) and more complete reduction occurs at pH 4[87]. Magnetic susceptibility is 3.0 ± 0.3 Bohr magnetons per iron. Mössbauer spectra of *Klebsiella* Mo–Fe protein[132] in dilute $Na_2S_2O_4$ show a doublet with isomer shift of 0.62 mm/sec and quadrupole splitting of 0.75 mm/sec; intensity diminishes both on increasing and decreasing $Na_2S_2O_4$ concentration. Other spectral components are less well resolved, but some influence on their intensities by N_2ase substrates and inhibitors is reported.

Azotobacter Mo–Fe protein[105] is crystallized as needles about $2.4\mu m \times 40\text{–}60\mu m$ by taking advantage of the insolubility of the pure protein at ionic strength of < 0.075. The heat of solution is calculated at –5.4 kcal/mole. The protein is dark brown in solution, but the crystalline form is white or slightly yellow, suggesting that the environment of the chromophore is altered on crystallization such as has been observed for transition metal complexes that undergo ligand dissociation on solubilization[144], *e.g.*,

$$\underset{\text{white}}{[(o\text{-Tol O})_3P]_4Ni} \rightleftharpoons \underset{\text{red}}{[(o\text{-Tol O})_3P]_3Ni} + (o\text{-Tol O})_3P$$

Mo–Fe proteins are irreversibly sensitive to O_2. Metal has been removed chemically[143,146], or by dissociation[85] or during fractionation[145] in the case of *Azotobacter* V–N_2ase. A possible apoprotein of Mo–Fe protein has been prepared in attempts to isolate a V–Fe protein from V–N_2ase and may represent the least modified apo Mo–Fe protein[145]. The apparent lability of V in the V–Fe protein suggests an imperfect fit of the smaller (ionic radius 0.54 Å)[269] V^{5+} ion in a site arranged to accommodate Mo^{6+} (radius 0.60 Å)[269]. Treatments with alkali (pH 8.5–9.5) or up to 6 *M* urea are reported to selectively release Mo from an *Azotobacter* fraction, presumably the Mo–Fe protein[85]. Mersalyl treatment of clostridial Mo–Fe protein released all of the Mo[143] and about 50–60% of the Fe[146], while anaerobic dialysis with α,α-dipyridyl removed the iron in two steps, indicative of two types of iron.

(3) Fe protein

Clostridial Fe protein[61,65,66] has been most investigated and its characteristics will be described with those of *Azotobacter*[107] and *Klebsiella*[132] included for comparative purposes. The specific activity of the clostridial and *Azotobacter* Fe proteins are 460 and 530 nmoles N_2 reduced/min · mg protein, respectively. The clostridial protein has been purified to homogeneity as determined by the absence of tryptophan. A molecular weight of 55,000 daltons based on gel electrophoresis, has recently been reported with 4 Fe and 4 S^{2-} per mole. Two identical subunits of 27,500 daltons are produced by sodium dodecyl sulfate treatment. The native protein is yellow-brown with broad absorption from 300 to 550 nm; like ferredoxins, the protein is bleached by mersalyl.

Klebsiella Fe protein is suggested to have 2 Fe atoms per molecular weight of 60,000 daltons. The predominant species in preparations of purified *Azotobacter* Fe protein examined by electron microscopy[113] consisted of a basic repeating subunit with a 40 Å diameter and an estimated molecular weight of 37,000 daltons, a value consistent with earlier ultracentrifuge data indicating a molecular weight in the 35,000–40,000 range[73].

Fe protein is, in general, sensitive to cold with[73,123,147] almost complete loss of ability to reconstitute effective N_2ase after 24 h at 0°C. The protein is highly sensitive to O_2 and loses activity after 5 min exposure to air. This treatment also causes development of an e.s.r. signal at $g \cong 2$[148]; native Fe protein has been reported to show no e.s.r. absorption. Mössbauer spectra of *Azotobacter* Fe protein show a single dipole with $\delta = 0.64$ mm/sec and $\Delta E = 0.84$ mm/sec in the native state with complete conversion to a dipole of $\delta = 1.58$ mm/sec and $\Delta E = 2.95$ mm/sec on reduction with $Na_2S_2O_4$[149]. The Mössbauer spectrum of *Klebsiella* Fe protein[132] in dilute $Na_2S_2O_4$ solution showed some absorption with $\delta = 1.65$ mm/sec and $\Delta E = 2.8$ mm/sec and a narrow doublet characterized by $\delta = 0.60$ mm/sec and $\Delta E = 1.1$ mm/sec.

(4) Models

Both N_2ase proteins are iron–sulfur proteins in which there is approximate Fe: S^{2-}:CySH equivalence. Presumably much of the iron is in an environment similar to that of ferredoxin Fe, although no e.s.r. signal at a *g* value of about 1.94 has been observed with reduced Fe protein. Examples of models for the ferredoxin-type of iron include: Roussin's salts[150], $[Fe(NO)_2SC_2H_5]_2$, $[Fe(MeCS \cdot CH \cdot CSMe)_2Cl_2]^{2+}$, $[Fe(NO)_2(CN)_5]^{4-}$ [151,152] and an iron–mercaptoethanol–sulfide complex[153]; others have been discussed in Chapter 22. In addition N_2ases possess Fe with spectral characteristics uniquely different from those of conventional iron–sulfur or other iron proteins, and no models yet duplicate these species.

A number of complexes of Mo with cysteine and other amino acids and organic sulfur compounds have been synthesized as models for Mo-containing enzymes[155–157]. Of particular current interest are molybdothiol complexes (hereafter designated as molybdothiol-reductant systems) which are being extensively examined as models of N_2ase[158,159]. Molybdenum compounds that have produced catalytically active systems with thiols include Na_2MoO_4, MoO_3, $MoCl_3$, $MoCl_3O$ and polyheteromolybdates. Effective thiols include dithioerythritol, 1-thioglycerol, 2-mercaptoethanol and cysteine. The binuclear Mo–cysteine complex is considerably more active than the Mo complex of histidine. The thiol component could not be replaced by thioethers. It is suggested that binuclear complexes may be inactive and may be hydrolyzed to active mononuclear species, *e.g.*,

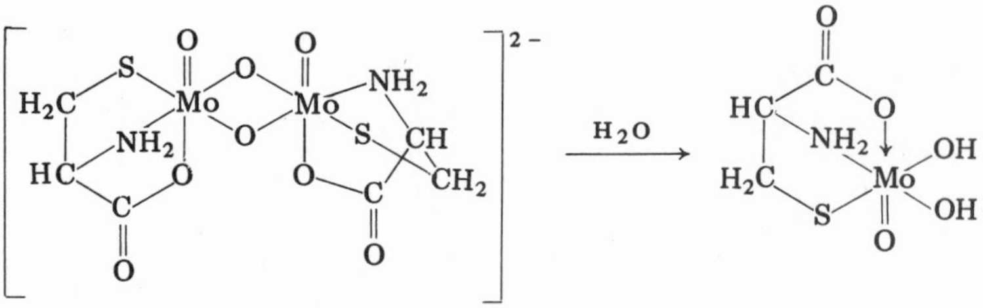

Structural models which contain both Mo and Fe with S ligands are also relevant, and a specific example is $Cp_2Mo(SR)_2FeCl_2$. Attempts to reduce N_2 with these complexes have not yet been successful[160].

B. *Nitrogenase reaction*

(1) Electron source and transfer agents

All the reactions catalyzed by N_2ase are reductive; consequently, a physiological source of electrons as well as a compatible electron transfer agent which interacts with, but is not part of, N_2ase are required for activity. Natural agents are electron transfer proteins of the ferredoxin-[34,36,84,124,125,128] and flavodoxin-types[45-47,83,128]; artificial agents are viologen[36,119] or hydrosulfite[70]. All are strongly reducing with an $\Sigma'_0 \leqq -0.2$ V, but contain diverse redox groups.

Ferredoxins (Chapter 22) are acidic iron–sulfur proteins first isolated as a by-product of N_2 fixation research[10,161]. They possess the lowest redox potential of any defined biological reductant. Flavodoxins[45–47], also isolated as a by-product of N_2 fixation research are acidic proteins with flavin mononucleotide prosthetic groups and the lowest redox potential of any known flavoproteins. The clostridial Fds and Flds are functionally interchangeable *in vitro* and *in vivo* and their biosynthesis is regulated by iron concentration in the bacterial growth medium, possibly the only example of protein synthesis controlled by a metal ion.

Clostridial Fd, $\Sigma'_0 = -0.42$ V[162], or Fld, $\Sigma'_0 = -0.17$ and -0.38 V[163], transfer electrons to N_2ase from pyruvate, the physiological electron source[34-36] via the[45-47] phosphoroclastic reaction, *e.g.*,

$$\text{pyruvate} + \text{CoASH} \longleftrightarrow \text{acetylSCoA} \longleftrightarrow \text{acetyl} \sim \text{P} \longleftrightarrow \text{ATP}$$

$$\downarrow$$

$$\text{Fld or Fd} \longrightarrow N_2\text{ase}$$

It is only in this system that the natural transfer agent and the natural source of electrons for N_2 fixation have been defined. Turnover numbers are 32 and 10 electron moles per min · mole of Fd and Fld, respectively.

In an artificial system clostridial Fd and Fld can transfer electrons to various N_2ases from H_2 via hydrogenase,[69,128] *e.g.*,

$$H_2 \xleftrightarrow{\text{Hydrogenase}} \begin{matrix}\text{Fld or}\\ \text{Fd}\end{matrix} \rightarrow N_2\text{ase}$$

Azotobacter[84] and soybean nodule Fds[124,125,128] and *Azotobacter*[83] Fld (azotoflavin) can couple electrons with a relatively low turnover number to their N_2ases from illuminated modified chloroplasts[137], an artifical source. The physiological source of electrons in both organisms is still undefined; definite but low activities have been obtained with a variety of sources including NADH and NADPH tested in multi-component systems[99,128]. The polymer of β-hydroxybutyrate has been considered as a possible natural source since it constitutes up to 40% of the dry weight of bacteroids[119,128], the pleiomorphic form of the N_2-fixing bacteria in nodules; however, recent results clearly indicate that the stored polymer is not used to support *in vivo* N_2 fixation[164].

For almost all *in vitro* assays, hydrosulfite is used since substrate amounts can be tolerated, thus eliminating the need for a regeneration system. The K_m of hydrosulfite is 1–10 x 10^{-3} M[96,117]. Confirmation of results produced by hydrosulfite with natural electron sources and transferring agents is necessary to avoid attributing artificial reactions to N_2ase. This has been confirmed for the reduction of most classes of substrates, but the ATP : 2e ratio discussed in the next section has not been examined with natural systems.

Some of the important unanswered questions in electron transfer to N_2ase involve the nature of coupling between electron transfer agent and N_2ase, the number of electrons transferred at each step, the N_2ase protein and prosthetic group (metal) to which electrons are transferred, the sequence of the interaction of electrons and ATP with N_2ase, and the physiological electron source in organisms other than *Clostridium*.

(2) Energy

Under standard conditions, all N_2ase-catalyzed reductions are thermodynamically favorable (Table I). However, energy in the specific form of ATP is required for all reactions[33–37,68,117,129,134,136,137,138,140]; the other tested nucleoside triphosphates, UTP, CTP and GTP, are essentially ineffective[52,57]. The utilization of ATP by N_2ase is dependent on electron transfer[34,37,39,41,74].

$$x\text{H}_2\text{O} + x\text{ATP} + 2e \xrightarrow[\text{M}^{2+}]{\text{N}_2\text{ase}} x\text{ADP} + x\text{Po}_4{}^{2-} + x\text{H}^+ + 2e^*$$

and is consequently referred to as a reductant-dependent ATP utilization to distinguish it from other, non-reductant-dependent ATPase activities.

The $2e^*$ are considered to be activated electrons and are experimentally detected only through identification of reduction reactions, which in the simplest case is $2H_3O^+ \rightarrow H_2 + 2H_2O$ (Fig. 2). The x reflects existing uncertainty in the ATP/$2e$ ratio; see (9) below.

Reductant-dependent ATP utilization by N_2ase is fundamental to all N_2ase reactions, and it occurs concomitantly with the more complex reactions described in the next section. Although this part of the N_2ase reaction is poorly understood, it is emphasized since inorganic models will have to reproduce many of its characteristics if they are in fact to mimic N_2ase.

Characteristics of the ATP reaction are:

(1) The K_m for ATP is $1\text{–}3 \times 10^{-4}$ M[96,131].

(2) A divalent cation is required with a M^{2+}:ATP ratio of about 0.5 for optimum activity, and a decreasing order of effectiveness of Mg^{2+}, Mn^{2+}, Co^{2+}, Fe^{2+} and Ni^{2+}, while Ca^{2+}, Cu^{2+} and Zn^{2+} are ineffective or inhibitory[52,74,96].

(3) Inorganic phosphate and ADP are products of the reaction[74,165].

(4) ADP but not AMP is an inhibitor[52,70,106] and in *in vitro* experiments a phosphatidokinase ATP-generating system is commonly used to prevent ADP accumulation[34,35]. This inhibition by ADP may be of physiological significance, and a negative modifier role has been proposed for ADP in regulation of *in vivo* N_2ase activity[52].

(5) A bimolecular ATP reaction has been indicated for the limiting reaction of clostridial N_2ase[52] and an intensive study of *Azotobacter* N_2ase[106] supports this conclusion. However, these interpretations may be equivocal in view of recent evidence that indicates the kinetics of N_2ase to be more complex than those on which the above conclusions were based[126].

(6) Both Mo–Fe and Fe protein are required. Interaction between ATP and Fe protein was inferred from reported specific binding of ^{14}C-ATP and ^{14}C-ADP to clostridial Fe protein but not Mo–Fe protein[59]; however, non-specific binding of ATP to *Klebsiella* Fe and Mo–Fe proteins, as well as other proteins does not support this inference[131].

(7) The reaction rate is not altered by substrates, products or inhibitors of substrate reduction such as N_2, H_2, NH_3 and CO[57,74,106]. However, the proportion of electrons used for N_2 *vs.* H_3O^+ reduction varies directly with ATP concentration[106].

(8) The pH for maximum activity varies among N_2ases, but it is generally between 6 and 8.

(9) The reported ratios of ATP utilized per $2e$ transferred vary from one to five for N_2ase *in vitro*[58,73,74,89,106,165], while a value as low as two has been calculated from *in vivo* measurements[55,91]. Recent *in vitro* measurements appear to support the original report of a ratio of two[74],

and higher values are attributed to ATP hydrolysis independent of reductant[63,64,106]. The ratio is reported to be increased by increasing temperature[89] and decreasing pH[58]. A substantial reductant-independent as well as a reductant-dependent ATP utilization has been found in reconstituted clostridial N_2ase[63,64]. Variations in the ATP : 2*e* ratio observed with reconstituted N_2ase indicate that artificial combinations may occur on recomplexation; moreover, it has been suggested that the optimum Mo–Fe : Fe protein ratio is greater for ATP hydrolysis than for substrate reduction[63,64,93]. These considerations emphasize the desirability of using unfractionated N_2ase in reaction analyses.

(10) The activation energy curve is uniquely biphasic with a value of 14.6 kcal/mole above 20°C and 35–50 kcal/mole below 20°C[57,96]. A value of 20 kcal/mole has also been reported, but this may be in error due to the failure to recognize the biphasicity of the curve[88]. The reductant-dependent ATP utilization reaction is the limiting reaction of all N_2ase-catalyzed reactions. Vanadium–N_2ase shows a similar biphasic activation energy with values somewhat lower than Mo–N_2ase: 30 kcal/mole below 20°C and 10 kcal/mole above[112].

(11) ATP utilization is not altered by the usual levels of a variety of inhibitors of electron transport and uncouplers of oxidative and photosynthetic phosphorylation, indicating the uniqueness of the reaction of ATP with N_2ase relative to its roles in photosynthesis and oxidative phosphorylation[74].

(12) Inhibition of *Azotobacter* N_2ase by *p*-chloromercuribenzene-sulfonate, *o*-phenanthroline and α,α-dipyridyl, suggest involvement of thiols and metals in the overall reaction but do not indicate whether Mo and/or Fe are involved[72,74].

(13) Neither an exchange reaction nor reversibility have been demonstrated.

(14) The sequence of ATP and electron interaction with N_2ase is not defined.

(15) The reductant activity of the molybdothiolborohydride model of N_2ase is stimulated by ATP but not absolutely dependent on it[159].

Proposed functions for ATP include (i) electron activation[34,41], (ii) activation of ferredoxin[37], (iii) conformational change of dinuclear N_2ase site to accommodate the required elongation of the N–N bond during reduction[72], (iv) dehydration of a N_2ase site involved in a H_2O sensitive reaction such as nitriding, (v) formation of solvated electrons and[63], (vi) site-specific proton source associated with reductions in a non-aqueous environment, *e.g.*, N_2 to NH_3[64].

The acceptance of the electron-activation hypothesis for the major role of ATP is indicated by the usual designation of this reaction as the electron-activating reaction. Requirement for ATP in non-ferredoxin as well as ferredoxin systems does not support the proposed ATP function

References pp. 787–793

in ferredoxin activation. The direct involvement of ATP utilization in electron transport and not substrate reduction, argues against conformational change based on a separation of a dinuclear site during substrate reduction, although other mechanisms involving conformational change[126] are possible. The energy requirement for the formation of solvated electrons with a standard redox potential of about −2.7 V is in excess of the energy used by the biological N_2 fixation system; moreover, solvated electrons fail to reduce N_2 and H_2 solutions to NH_3[166]. In general, it can be concluded that the mechanism of ATP interaction with N_2ase is the least understood activity of N_2ase, and with the exception of the recent observations with the molybdothiolborohydride system, there are no relevant models[159].

(3) Nitrogenase-catalyzed reductions

The variety of substrates reduced by N_2ase is extraordinary, although Pt metal, a heterogeneous catalyst, may be more versatile. The N_2ase reductions provide a most useful approach for exploration of N_2ase function and mechanism. In addition they reveal novel chemistry, and several reactions or species first found or indicated with N_2ase have only recently been shown with synthetic homogeneous systems. Examples include transition metal N_2 and N_2O complexes, reduction of N_2O, and isonitrile insertion reactions. Others, such as the facile catalysis of the reduction of N_2 to NH_3 and of nitriles to hydrocarbons and ammonia have not yet been shown with synthetic homogeneous systems.

Reactions catalyzed by N_2ase are grouped below according to the type of reduction (Fig. 3). Characteristics of the enzymic reactions and related inorganic chemistry are included. These reactions all have requirements identical to reductant-dependent ATP utilization.

N_2ase REDUCTIONS

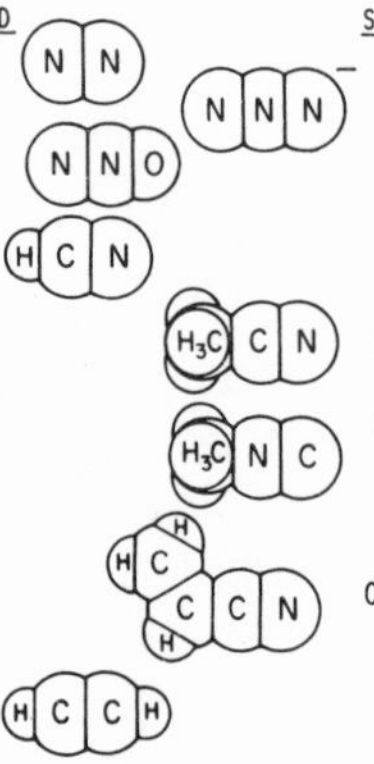

BOND REDUCED		SUBSTRATE	PRODUCTS	ELECTRONS	K_m mM	RATE	RATE X ELECTRON
NN	N N	N_2	$2NH_3$	6	0.03-0.1	1	6
	N N N	N_3^-	N_2, NH_3	2	0.02-1.0	3	6
NO	N N O	N_2O	N_2, H_2O	2	1.0	3	6
CN	H C N	HCN	CH_4, NH_3 CH_3, NH_2	6 4	0.4	0.6	3
	H_3C C N	CH_3CN	C_2H_6, NH_3	6	ca 500	0.02	0.1
	H_3C N C	CH_3NC	CH_3NH_2, CH_4 C_2H_6, C_2H_4 C_3H_8, C_3H_6	6 8, 10 12, 14	0.2-1.0	0.8	5
	H C H C C N	CH_2CHCN	C_3H_6, NH_3 C_3H_8	6 8	10-25	0.25	1.5
CC	H C C H	C_2H_2	C_2H_4	2	0.1-0.3	4	8
		$2H^+$	H_2	2		4	8

Fig. 3. Examples of the classes of substrates reduced by N_2ase arranged according to type of bond reduced[8].

(i) ATP-dependent H_2 evolution.[39,41,71,117,134,138] Dihydrogen is the exclusive reduction product of N_2ase in the absence of added acceptors, *i.e.*,

$$2H_3O^+ \xrightarrow{2e^*} H_2 + 2H_2O.$$

This energy-dependent and irreversible H_2 evolution of N_2ase is called an ATP-dependent H_2 evolution to distinguish it from reversible ATP-independent reactions of H_2 catalyzed by hydrogenases[10]. Hydronium ions are the ultimate source of H_2, based on the ratios of H_2:HD:D_2 evolved from H_2O–D_2O mixtures[80]. The absence of an isotope effect on rates indicates that H_3O^+ is not rate limiting.

Inorganic systems which evolve H_2 and are of possible relevance include acidic solutions of vitamin B_{12s} [167], which form H_2 presumably via a hydride intermediate, and alkaline solutions of the molybdothiolborohydride model of N_2ase[159]. The latter system is stimulated up to 7-fold by XTP or XDP but not by XMP or $Na_5P_3O_{10}$. Like N_2ase, the H_2 evolution by this system is inhibited by C_2H_2 and unaffected by CO.

In the presence of added acceptors N_2ase still evolves some H_2 as H_3O^+ competes with acceptor for electrons, but the electron transfer rate remains constant[57,73,74]. With saturating concentrations of N_2, about 25% and 75% of the electrons are used for H_3O^+ and N_2 reduction[106,110], respectively, while with saturating concentrations of C_2H_2 almost all the electrons are used for C_2H_2 reduction, at least with *Azotobacter* N_2ase[110]. A key role for Mo in terminal electron transfer is indicated by the less effective allocation of electrons to added acceptors *vs.* H_3O^+ by V–N_2ase than by Mo–N_2ase[112].

H_2-evolution *in vitro* is a non-productive reaction, and it is difficult to rationalize its physiological role. One suggestion is that it may be coupled to a conventional hydrogenase to scavenge O_2 ($2H_2 + O_2 \rightarrow 2H_2O$) and thereby maintain an anaerobic environment for N_2ase, at least in aerobic organisms[7,91].

(ii) *Cleavage of NN triple or potential triple bonds.* The two known examples are[50,53,120]:

$$N_2 \rightarrow 2NH_3$$

$$N_3^- \rightarrow N_2 + NH_3$$

The reduction of N_2 to NH_3 is the physiological reaction of N_2ase. The K_m for N_2 is 0.02–0.05 atm. *in vivo* and is somewhat higher, 0.05–0.20 atm., *in vitro*[7,10]. Aerial concentrations of N_2 should saturate N_2ases. All evidence is consistent with NH_3 as the first enzyme-free product of N_2 fixation by N_2ase. Intermediates such as enzyme-free diazene or hydrazine have not been found[40,43], and added hydrazine or diazene generated *in situ* from azodicarboxylate are not reduced by N_2ase[40,50]. Ammonia shows no apparent affinity for N_2ase and is not a specific inhibitor of N_2 fixation[57]. Dinitrogen exchange

$$^{14}N_2 + {}^{15}N_2 \rightarrow 2\ {}^{14}N^{15}N$$

a characteristic reaction of Haber–Bosch catalysts, is not catalyzed by N_2ase, indicating no initial reversible cleavage of the N≡N bond by N_2ase. Also, no reversal of the N_2ase-catalyzed N_2 fixation reaction has been demonstrated[10].

Related inorganic chemistry includes: (1) the numerous stable N_2 complexes of the following three types, $M(N_2)$, $M(N_2)M$, and $M(N_2)_2$; (2) reduction of N_2 to NH_3 under mild but anhydrous conditions by the combination of Ti or Fe systems with strong reducing agents such as RMgX and naphthalide ion in which unstable N_2 complexes appear to be intermediates; and (3) a number of Mo complexes in conjunction with less drastic reducing agents which are reported to reduce N_2 to NH_3 in aqueous media. The rates in the Mo systems are much lower than that of the N_2ase-catalyzed reaction and the reproducibility of several has been questioned. Information on these systems is presented in Section IV.

The potential triple bond of azide is reduced by two electrons to equal amounts of NH_3 and N_2 [9,50,53]. The K_m of N_3^- is about 1 m*M*. Product N_2 is only further reduced to NH_3 by subsequent reaction with N_2ase. The rate of azide reduction is about three times that of N_2; the rates of electron consumption are similar.

An inorganic reaction of interest is the oxidative decomposition of an azidoruthenium(III) complex to form dinitrogen complexes[168]:

$$[(NH_3)_5RuN_3]^{2+} \rightarrow [(NH_3)_5Ru(N_2)]^{2+} + [(NH_3)_5Ru\text{–}N{\equiv}N\text{–}Ru(NH_3)_5]^{4+} + N_2$$

via a proposed reactive intermediate in which a nitrene is coordinated to Ru:

$$[(NH_3)_5RuN_3]^{2+} \xrightarrow{+H} [(NH_3)_5RuNH]^{3+} + N_2$$

A similar reaction of azide coordinated to N_2ase followed by reduction of nitrene to ammonia would provide both the products and stoichiometry of N_2ase-catalyzed azide reduction. The molybdothiol-borohydride system also is reported to reduce N_3^- to N_2 and NH_3 [159].

(iii) Cleavage of NO bond. The sole example is[48,169,170]:

$$N_2O \xrightarrow{2e^*} N_2 + H_2O$$

The stoichiometry and rate of N_2O reduction are similar to those of N_3^-; product N_2 is also not reduced without subsequent reaction with N_2ase.

The K_m of N_2O is 0.05 atm.[53] and the reported K_i for N_2O inhibition of N_2 fixation is also 0.05 atm.[42], suggesting complexation of N_2O to a site involved in N_2 reduction.

Recent inorganic reactions, including the first transition metal catalyzed reductions of N_2O, may be related to steps in N_2ase-catalyzed N_2O reduction. Nitrous oxide reacts rapidly in solution at room temperature and atmospheric pressure with a variety of Co complexes such as vitamin B_{12} according to the stoichiometry[171]:

$$2Co^{I} + N_2O \rightarrow 2Co^{II} + N_2.$$

A slow reduction of N_2O to N_2 and H_2O has been observed with the molybdothiol–borohydride system[159]. An N_2O complex[172,173], $[(NH_3)_5Ru(N_2O)]^{2+}$ has been isolated from $[(NH_3)_5RuH_2O]^{2+}$ and N_2O; an N_2 complex $[(NH_3)_5Ru(N_2)]^{2+}$ has been formed from the same reactants through reduction by Cr^{2+} or amalgamated zinc[172,174]. Surprisingly, the strength of the NO bond is decreased in the only known N_2O complex, $[(NH_3)_5Ru(N_2O)]^{2+}$ [172]. Nitrous oxide has been reduced to N_2 by $(Ph_3P)_3CoH_3$ coupled with the oxidation of Ph_3P to Ph_3PO[175].

Nitric oxide is a potent inhibitor of N_2 fixation by at least clostridial N_2ase[42]. Additional studies are needed to define this activity[132]. The reported K_i of 4.3×10^{-7} M is similar to that of CO. Attempts to reduce NO with N_2ase have been inconclusive, since NO is non-enzymically reduced by $Na_2S_2O_4$ [53]. The formation of nitrosyl complexes such as those of Ru, analogous to the N_2 complexes, and the high affinity of NO relative to N_2 are similar to the activity of NO as an inhibitor of N_2 fixation[176]. The apparent $4e$ reduction of NO to NH_2OH by Ir complexes treated with HCl[177] is a possible N_2ase reaction but is not yet substantiated by enzyme studies.

(iv) Cleavage of NC triple bonds. A variety of cyanide and isonitrile NC bonds are reductively cleaved to hydrocarbons and amines by N_2ase. A tabulation of cyanides[4,53,75,77,81,103,104,114,120,126] and their products follows; multiple products are listed in decreasing order of importance; ammonia in insufficient quantities to detect is indicated as (NH_3).

$$CN^- \text{ or } HCN \rightarrow CH_4 + NH_3 + CH_3NH_2$$

$$CH_3CN \rightarrow C_2H_6 + (NH_3)$$

$$C_2H_5CN \rightarrow C_3H_8 + (NH_3)$$

$$CH_2{=}CHCN \rightarrow C_3H_6, C_3H_8 + NH_3$$

$$cis\text{-}CH_3CH{=}CHCN \rightarrow \text{1-butene, } cis\text{-2-butene, n-butane, } trans\text{-2-butene} + (NH_3)$$

$$CH_2{=}CHCH_2CN \rightarrow \text{1-butene} + (NH_3)$$

trans-$CH_3CH{=}CHCN$ → *trans*-2-butene, 1-butene, n-butane + (NH_3)

$CH_2{=}C(CH_3)CN$ → isobutylene + (NH_3)

n-C_3H_7CN → n-butane + (NH_3)

$(CH_3)_2CHCN$ → no product

Cyanide[9,53,81,132], the parent compound of the nitrile series, is an effective substrate with a reduction rate about 50% that of N_2. It is reported to have a K_m of 0.2–1.0 m*M* based on HCN as the assumed substrate. The actual substrate has not been established since HCN and CN^- are in equilibrium at physiological pH. Cyanide is reduced by 6 electrons, as is N_2, and forms equivalent amounts of CH_4 and NH_3 in the only confirmed homogeneously catalyzed reductive cleavage of a nitrile. In addition, about 10% as much product has been tentatively identified as CH_3NH_2, the sole four electron-addition product found in a N_2ase-catalyzed reduction. Substrate CH_3NH_2 is not reduced to CH_4. Traces of C_2H_6 and C_2H_4 have been reported and may arise from an insertion reaction as proposed for isonitrile (see below). Binding studies of $^{14}CN^-$ with either N_2ase protein have been equivocal[59,131], but a recent report shows that $^{14}CN^-$-binding to reconstituted N_2ase was increased by ATP in an atmosphere of Ar but not CO, indicating the requirement of a functional N_2ase for cyanide binding[132].

Saturated[9,75], unbranched nitriles up to and including butyronitrile are reduced to the corresponding alkanes and presumably ammonia, while the branched nitrile, isobutyronitrile, is not reduced. This failure indicates the need for a sterically unhindered nitrile carbon. The reduction rates of the saturated nitriles are slow — 0.2 to 0.5% of that of N_2 for CH_3CN and C_2H_5CN and 0.02% for C_3H_7CN — and the K_m values are high (500 to 2000 times that of HCN). The reduction rates of all substituted nitriles are increased 2- to 5-fold by D_2O[103,104]. This increase is due to greater allocation of electrons to nitrile *vs.* H_3O^+ reduction. The absence of a similar isotope effect on other substrates, especially CH_3NC, is suggested to reflect the facility of CN reduction in isonitriles *vs.* nitriles.

The extensively studied N_2ase-catalyzed acrylonitrile reduction[75,103,104] occurs at up to 50% of the rate of N_2 reduction and may provide one of the best guides for design of an effective N_2ase model. Acrylonitrile is reduced by 6 or 8 electrons to ammonia and either C_3H_6, the major hydrocarbon product, or C_3H_8, the minor. No intermediate imines or amines representative of 2 and 4 electron addition products were found. The K_m is 10–50 m*M*, indicating about 20-fold greater affinity of N_2ase for unsaturated *vs.* saturated nitriles. Similar increases in affinity are produced by unsaturation with 4 and 5C nitriles. The improved affinity of unsaturated nitriles, especially in the absence of conjugation, *e.g.*, 3-butenenitrile, suggests both olefinic and nitrilic complexation with N_2ase. About 20% or less of the olefinic bonds of acrylonitrile and of other conjugated nitriles are reduced. In most cases, the conjugated nitriles form the corresponding alkene products via double bond shifts. The product C_3H_6:C_3H_8 ratio from acrylonitrile is about 6. Vanadium-N_2ase reduces acrylonitrile and other nitriles giving a C_3H_6:C_3H_8 ratio of about 3[112]. The alteration in product ratio of V-N_2ase *vs.* Mo-N_2ase implies that Mo plays an intimate role in product formation, including reduction and release, as well as in substrate complexation.

Steric effects on N_2ase-catalyzed reductions are indicated by comparison of the rates of reduction of methyl substituted acrylonitriles[9,75]. The reduction rates of *cis*-crotononitrile, *trans*-crotononitrile and methacrylonitrile are about 0.7, 0.07 and 0.03%, respectively, of that of N_2. Steric retardation of the reduction of the two latter substrates further indicates the need for an exposed nitrile carbon.

Transition metal-nitrile complexes are well known. Recently nitrile analogs of the dinitrogen complexes of Co[178] and Ru[179,180] have been prepared. These nitrile complexes all show an abnormal decrease in C≡N stretching frequency which is attributed to metal-to-nitrile π-backbonding in a linear metal-nitrogen-carbon structure. The Co-acetonitrile[178] complex shows reversible exchange between N_2 and acetonitrile, *i.e.*:

$$(Ph_3P)_3CoH(RCN) + N_2 \rightleftarrows (Ph_3P)_3CoH(N_2) + RCN$$

Hydrogen cyanide and acrylonitrile[180] complexes have been prepared with a Ru complex. In the acrylonitrile complex, $[(NH_3)_5Ru(acrylonitrile)]^{2+}$, the nitrile is N bonded; no evidence was found for olefin bonding in this mononuclear complex. Possibly the first report of abiological homogeneous catalysis of nitrile reduction is the conversion of acetonitrile to ethane with the molybdothiol-borohydride system[159].

Methyl, ethyl and vinyl isocyanides are reduced by N_2ase, while phenyl isocyanide is not[75,77,81,93,94,95,114].

$$CH_3NC \rightarrow CH_4, C_2H_4, C_2H_6, C_3H_6, C_3H_8 + CH_3NH_2$$

$$C_2H_5NC \rightarrow CH_4, C_2H_4, C_2H_6 + C_2H_5NH_2$$

$$CH_2{=}CHNC \rightarrow CH_4, C_2H_4, C_2H_6 + [amine(s)?]$$

The K_m values of methyl and ethyl isocyanide are 0.2–1.0 m*M* and 10–25 m*M*, which are about 0.1 and 2% of those of the corresponding nitriles[9,81,95]. These differences indicate superior end-on bonding of the isonitrile carbon *vs.* the nitrile nitrogen. One-, two- and three-carbon alkanes and alkenes in addition to CH_3NH_2 are synthesized from CH_3NC by N_2ase, while CH_4 and two-carbon alkanes and alkenes and $C_2H_5NH_2$ are formed from C_2H_5NC. These products involve the addition of 6, 8, 10, 12 and 14 electrons;

References pp. 787–793

fourteen is the largest number of electrons reported for N_2ase-catalyzed reduction. Formation of perdeuterated methane, ethane and ethylene from CH_3NC reduction in D_2O establishes the isonitrile carbon as the source of one- and two-carbon hydrocarbons. An alternating reduction and insertion mechanism involving the formation of alkyl · N_2ase intermediates which release either alkanes by proton attack or alkenes by hydride elimination is supported by kinetic measurements under a variety of conditions so that:

$$dC_2H_6/dt = K \cdot dCH_4/dt \cdot [CH_3NC]$$

$$dC_3H_6/dt = K' \cdot dC_2H_6/dt \cdot [CH_3NC]$$

with $K = 6$ and $K' = 2$ for *A. vinelandii* N_2ase in H_2O at pH 7.0. Decreased proton concentration increases rate constants; other inserting agents such as CO and CH_2O alter rate constants. The isonitrile reduction reactions provide strong indirect evidence for a metal-complexing and/or reducing site in N_2ase. The reactions indicate the ability of N_2ase to accommodate much larger substrates than N_2, but also show its limitations as reflected in the product balance from C_2H_5NC reduction.

The ratio of C_2H_6/C_2H_4 synthesized from CH_3NC is characteristic of the N_2ase source[131]. Cross-combination experiments suggest that the Mo–Fe protein contains the complexation and reduction site for at least isonitrile reductions.

Transition metal isonitrile complexes are well known, and insertion reactions into metal–carbon bonds are now documented[181-183]. For example, isonitrile insertions have been reported for organometallic derivatives of Ni^{2+}, Pd^{2+} and Mo^{2+}. The molybdothiol–borohydride model system slowly reduces alkyl isonitriles to CH_4, C_2H_6, C_2H_4 and a C_3 product[159].

The insertion reaction with $CH_3Pd\,I(PPh_2Me)_2$ is particularly interesting because multiple insertions occur to give

$$Pd{-}\underset{\displaystyle NR}{\underset{\displaystyle \|}{C}}{-}CH_3 \text{ and } Pd{-}\underset{\displaystyle RN}{\underset{\displaystyle \|}{C}}{-}\underset{\displaystyle NR}{\underset{\displaystyle \|}{C}}{-}CH_3$$ [182].

Reduction of such double insertion products could provide an alternate route to the C_3 products of N_2ase-catalyzed isonitrile reductions.

(v) Reduction of CC triple bonds (refs. 9,49,51,53,57,75,81,110,112, 114,117,126,131,140,141,184,185). Examples include:

$$C_2H_2 \rightarrow C_2H_4$$

$$CH_3C{\equiv}CH \rightarrow C_3H_6$$

$$C_2H_5C{\equiv}CH \rightarrow C_4H_8$$

Acetylene reduction, because of its utility as a universal assay for N_2 fixation[8,57,185] and use for model studies of N_2 fixation[158,159], has been investigated more than any other alternate N_2ase substrate. The K_m of C_2H_2 is 0.004–0.01 atm. for N_2ase in a wide variety of *in vitro* and *in vivo* systems. Acetylene is reduced by both Mo- and V-N_2ase but the K_m of V-N_2ase is about four times that of Mo-N_2ase, implicating Mo in complexation[112]. The rate of C_2H_2 reduction by N_2ase *in vitro* is four times that of N_2 and the resultant rate of electron consumption is about 30% greater than that of any other substrate[110]. Acetylene is thus the most effective substrate in competing with H_3O^+ for electrons, and saturating concentrations of C_2H_2 almost eliminate H_2 evolution. Alkyne substrates are limited to those possessing at least one acetylenic hydrogen, presumably due to steric effects[9]. Allene is also reduced to propylene but it may be isomerized to methyl-acetylene prior to reduction. Single or conjugated olefinic bonds are not reduced and, based on the absence of an inhibitory effect of C_2H_4 on N_2 fixation[49], it is suggested that C_2H_4 shows very weak, if any, affinity for the active site. However, by analogy with nitriles, olefinic bonds in conjugation specifically with acetylenic bonds may be reduced. Two electrons are used for all CC triple or potential triple bond reductions. The extremely specific reductions produce only the corresponding alkenes[9,49,57,75]. The reduction is stereospecific with *cis*-1,2-dideuteroethylene as the major product of C_2H_2 reduction in D_2O[49,57,94].

Transition metal–acetylene complexes are well-known. A series of Ir^I-acetylene complexes has recently been used to exemplify types of mononuclear acetylene complexes including "π-bonded", "doubly σ-bonded" and terminally bound[187]. The orientation of acetylene in complexation with N_2ase is unknown; however, the failure to lose acetylenic protons does not support acetylide-type bonding ($M-C\equiv CR$).

Some transition metal systems show specific reductions of alkynes to alkanes. Complexes of $(PPh_3)_2Pt$(alkyne) undergo oxidative addition with strong acids (HX) to give $(PPh_3)_2PtX_2$ with the hydrogen transferring to the alkyne to form alkene but not alkane[188]. However, the stereochemistry is different from N_2ase with the *trans*-isomer as the major product. Another system, the molybdothiol–borohydride model[158,159], has been extensively investigated with respect to C_2H_2 reduction and shows remarkable similarities to N_2ase-catalyzed acetylene reduction (Table III). An aqueous dinitrogen-fixing system based on Ti^{3+} as reductant and a Mo catalyst is also reported to reduce C_2H_2 specifically to C_2H_4[189].

Molybdenum[158,159] with thiol ligands such as cysteine or thioglycerol with an optimum Mo:thiol ratio of two and borohydride or hydrosulfite as reductant converts C_2H_2 to C_2H_4. In ligand-free systems, the catalytic activity is usually much lower. Other metals with the exception of Ir, *e.g.*, Fe, Nb, Ru, Rh, Pd, Ta, W, Re, Os, and Pt, are $<$ 5% as active as Mo; Ti, V, Cr, Mn, Co, Ni, Cu, Zn, Y, Zr, Ag, Au, and Hg are inactive. An initial

References pp. 787–793

TABLE III

COMPARISON OF CHARACTERISTICS OF N_2ASE AND MOLYBDOTHIOL-CATALYZED REDUCTION OF ACETYLENE

	N_2ase[49,51,57]	Molybdothiol[158,159]
Substrate	C_2H_2	C_2H_2
Products	C_2H_4 (only)	C_2H_4 (major)
		C_2H_6 (minor)
Activation energy (kcal/mole)	14.5	13
K_m (mM)	0.4–1.0	0.3
Requirements	N_2ase	Mo, thiol
	ATP	XTP stimulates
	$S_2O_4^{2-}$	BH_4^- or $S_2O_4^{2-}$
Metal specificity	Fe + Mo or V	Mo or Ir
Optimum pH	6–8	9+
CO	strong inhibitor	weak inhibitor
H_2	no effect	no effect
Turnover number (moles/min · mole Mo)	200	0.05
H_2 evolution	ATP-dependent	XTP-stimulated

lag in rate of C_2H_2 reduction by binuclear molybdenum cysteine complexes is attributed to the need for formation of an active mononuclear species (see Section A4). The maximum C_2H_4:C_2H_6 product ratio approaches 20 and the reaction is first order with respect to C_2H_2. The stereochemistry of reduction is predominantly *cis*, as in N_2ase. 2-Butyne is reduced. Nucleoside tri- and di- but not mono-phosphates stimulate C_2H_4 formation 7-fold and C_2H_6, 45-fold. Acetylene reduction is inhibited by RNC > O_2 > CN^- > CO. Iron, although much less active than Mo in unimetal systems, stimulates C_2H_4 formation by the Mo system, possibly functioning as an electron transfer agent.

(4) Dihydrogen reactions

Dihydrogen reactivity and inhibition provided major approaches for the study of the biochemistry of N_2 fixation during 1940–1960 prior to the advent of *in vitro* N_2ase[190]. The interpretations in each area have been refined by studies with purified N_2ases and these will be presented.

Four reactions of H_2 which will be considered: (i) reversible activation of H_2 by hydrogenase; (ii) ATP-dependent H_2 evolution; (iii) inhibition of N_2 fixation by H_2; and (iv) HD formation from either D_2 and H_2O or H_2 and D_2O. The last three are activities of N_2ase, while the first is not.

Co-occurrence of hydrogenase and N_2ase activities in almost all N_2-fixing organisms led to the concept during 1940–1960 that N_2ase may be a special hydrogenase[191] and at least one N_2 fixation mechanism (see Section C) was proposed on this basis. However, the two activities have been conclusively shown to be associated with separate enzymes.

ATP-dependent H_2 evolution, as discussed already, is a specialized function of N_2ase in which H_3O^+ ions are reduced.

Dihydrogen is a competitive inhibitor of N_2 reduction with a K_i of about 0.2 atm.[42,78,80,81,117]; no isotope effect has been found. The H_2 inhibition of N_2 fixation by N_2ase:

$$N_2ase \cdot N_2 + H_2 \rightleftarrows N_2ase \cdot H_2 + N_2$$

is similar to the competitive reaction between H_2 and N_2 with a Co complex[192]:

$$(Ph_3P)_3Co(N_2)H + H_2 \rightleftarrows (Ph_3P)_3CoH_3 + N_2$$

This N_2 and H_2 complexing site is different from that of all of the other reducible substrates, since H_2 does not inhibit the reduction of any other substrates[79]. Inconsistent reports of H_2 inhibition of the reduction of substrates other than N_2 may be due to an impurity, possibly CO, in commercial H_2.

The failure, except in one unverifiable report[193], to find a major exchange between D_2 and H_2O in the absence of N_2 indicates that $N_2ase \cdot D_2$ does not freely exchange with protons of the media. Homolytic splitting or oxidative addition, but not a heterolytic splitting of H_2, is suggested for the interaction of H_2 and N_2ase. *Ortho–para* conversion of hydrogen may be a useful technique for studies of the N_2 complexation site of N_2ase.

Nitrogenase catalyzes an N_2-dependent formation of HD from D_2 and H_2O or H_2 and D_2O[80,122,169]. Other substrates of N_2ase do not support HD formation and, as indicated, their reductions are not inhibited by H_2. Requirements for HD formation are identical to those for N_2 fixation. Formation of HD as well as N_2 fixation is inhibited by CO and increased by N_2 and the ratio of HD:NH_3 increases with pD_2[80]. Formation of HD has been attributed to exchange between D_2 bonded to a transition metal(s) of N_2ase and protons of intermediates of N_2 fixation bound via a transition metal(s) of N_2ase, *e.g.*, $N_2H^+ \cdot N_2ase$, $N_2H_3^+ \cdot N_2ase$, $NH_3^+ \cdot N_2ase$. A similar exchange occurs between D_2 and protons of both aryldiazene- and arylhydrazine–Pt complexes. The HD exchange reaction of N_2ase is catalytic with HD:NH_3 ratios exceeding 3; the HD exchange reaction of the Pt complexes is also catalytic in protonic media.

(5) *Inhibitors*

Carbon monoxide is not reduced by N_2ase, but is a potent inhibitor of all N_2ase-catalyzed reductions except ATP-dependent H_2 evolution. The inhibition has generally been described as competitive, but recent results suggest that it may not always be competitive[79,112] and multiple sites of CO complexation are a distinct possibility. Reported K_i values are 1–3 x 10^{-4} atm.[42,57,81,121] for Mo–N_2ase and 4–6 x 10^{-4} atm. for V–N_2ase[112], indicating that CO interacts with Mo.

Carbonyl complexes of Mo and V are well known especially for the lower valence states. However, reversible interaction of N_2 and CO has not been demonstrated with synthetic N_2 complexes; in all systems studied to date CO irreversibly displaces N_2.

Inhibitory effects of H_2 and NO have been discussed (Sections B,3,iii and B,4).

Dioxygen inactivates N_2ase. All purified N_2ases, like most N_2 complexes isolated to date, are very sensitive to O_2 and are consequently maintained under strict anaerobic conditions. Nitrogenase is probably the most O_2-sensitive iron–sulfur protein. Aerobic N_2-fixing organisms have developed systems to protect N_2ase from O_2; these possibly include hydrogenase activity and exaggerated respiration rates[7,13,91]. Most aerobic organisms fix N_2 more effectively at pO_2 values lower than atmospheric, and recent dramatic increases in algal fixation have been produced by very low pO_2[140,142]. If it could be tested, reduction of O_2 to H_2O_2 or H_2O would probably be shown by N_2ase.

Mutual inhibitions by various reducible substrates have been examined to further define the similarities or differences of complexing sites for the various classes of substrates. The following groupings were suggested: (i) N_2, H_2; (ii) HCN, NH_3, CH_3NC; (iii) C_2H_2; (iv) CO[4,79].

C. Postulated mechanisms for biological N_2 fixation

Various mechanisms have been proposed for N_2ase-catalyzed N_2 fixation. All proposals involve N_2ase metals in key roles, although as summarized below only indirect evidence for the most part exists for metal involvement. Evidence supporting an involvement of metals in N_2 fixation includes: (i) nutritional requirement for additional Fe and Mo for growth on N_2 *vs.* fixed N[6,10]; (ii) occurrence of metals in both N_2ase proteins. Removal of Fe and Mo correlated with loss of activity; (iii) spectral changes in Fe produced by reductants effective in N_2 fixation; (iv) alteration of substrate affinities, reduction rate, product mixture and electron allocation in V–N_2ase *vs.* Mo–N_2ase. (v) nature of substrates: N_2, N_3^-, N_2O, RCN, RNC, and RCCH; and nature of inhibitors: CO, NO and H_2; and similarity of their affinities for N_2ase and synthetic transition metal complexes; and (vi) similarity of N_2ase-catalyzed reactions to transition metal catalyzed reactions, *e.g.*, N_2 reduction, N_3^- reduction, N_2O reduction, RCN reduction, CH_3NC insertion reaction and C_2H_2 reduction.

A few non-metal chemical systems have been reported to fix N_2[194,195], but these systems do not appear to be reproducible, and, in fact, almost all chemical systems for N_2 fixation involve metals.

Oxidative, as well as reductive routes have been suggested for biological N_2 fixation, but currently only two basic mechanisms, both reductive, are considered, in accord with the demonstrated reductive nature of the N_2ase reaction. One route involves the reduction of N_2 to NH_3 via a nitride intermediate, while the other involves diazene and/or hydrazine intermediates. Specific proposals for both routes will be presented.

(1) Nitride

The simplest proposed mechanisms for N_2 fixation assume metal nitrides as intermediates. In one variant N_2 is coordinated to a single metal atom and six electrons are introduced stepwise:

$$M + N_2 \longrightarrow M{-}N{\equiv}N \xrightarrow[3H^+]{3e} NH_3 + M{\equiv}N \xrightarrow{2H_2O} NH_3 + M(=O){-}OH$$

(cycle: $M(=O){-}OH \xrightarrow{3e + 3H^+} M + N_2$)

In another proposal, six electrons are supplied simultaneously from two metal atoms:

$$2M + N_2 \longrightarrow M{-}N{\equiv}N{-}M \longrightarrow 2M{\equiv}N \xrightarrow{4H_2O} 2NH_3 + 2M(=O){-}OH$$

(cycle: $2M(=O){-}OH \xrightarrow{6e + 6H^+} 2M + N_2$)

These nitriding systems find support in Li-, Ti-, V-, Cr-, and Fe-based inorganic N_2 fixation systems which use RMgX, Al, Mg or naphthalide as reductants (see Section IV). Several catalytic cycles have been developed with Ti, and at least one case has been proposed as a crude model for the biological N_2-fixation process[196]:

$$Ti(OR)_2 + N_2 \longrightarrow [Ti(OR)_2(N_2)]_n \xrightarrow{4e} [\text{nitride}] \xrightarrow{ROH} 2NH_3 + Ti(OR)_4$$

(cycle: $Ti(OR)_4 \xrightarrow{2e} Ti(OR)_2$)

These nitride proposals for biological N_2 fixation seem implausible because of the very strong reducing potential needed and the requirement for a non-aqueous environment; furthermore, the Ti-based systems do not parallel N_2ase in other reactions; *e.g.*, ethylene interacts strongly with Ti-systems, such as Ziegler catalysts, while no similar interaction of C_2H_4 and N_2ase has been observed.

(2) Diazene and/or hydrazine

Four mechanistic proposals of this type will be discussed. The first three avoid the thermodynamically unfavorable free diazene intermediate while the fourth involves a metal and hydrogen-bond stabilized diazene. Formation of free diazene is endothermic to the extent of 48.7 kcal/mole and probably represents the most unfavorable stage in the energy profile for N_2 reduction (Fig. 4). Model systems exist for all proposals and support them to varying degrees. Three proposals are based almost exclusively on results with synthetic systems while the fourth is constructed on the basis of both biochemical and inorganic results from several laboratories.

The first mechanistic proposal of this type evolved from model studies of hydrogenase using cobalt(II) cyanide[11,197]. It is significant that it preceded the extraction of N_2ase from cells, formation of transition metal N_2 complexes and reduction of N_2 to NH_3 by synthetic homogeneous

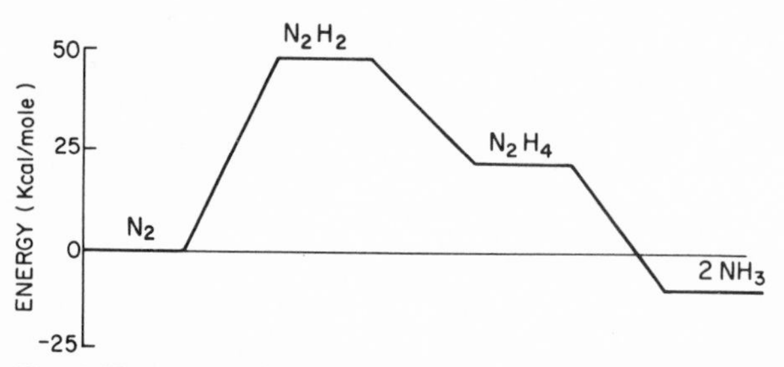

Fig. 4. Energy profile (free energies of formation) in the reduction of N_2.

catalysts. This mechanism proposed the following requirements for N_2ase: (i) a dinuclear site in an aqueous medium; (ii) two metal atoms of a type which can form either a double covalent bond, or a single covalent bond to dinitrogen and behave chemically like Co^{II} or Fe^{I} rather than Fe^{II}; (iii) a means of transferring H atoms to dinitrogen; (iv) provision of H atoms at an energy level at least equal to that of dihydrogen (a qualification met by the subsequently identified electron donors), and (v) protection of the catalyst against poisoning by H_2, O_2, cyanide, carbon monoxide, *etc.*, compounds with which N_2ase is now known to react. The specific mechanism invoked two theoretical mononuclear hydrogenase prosthetic groups juxtaposed on N_2ase 3–4 Å apart, *e.g.*,

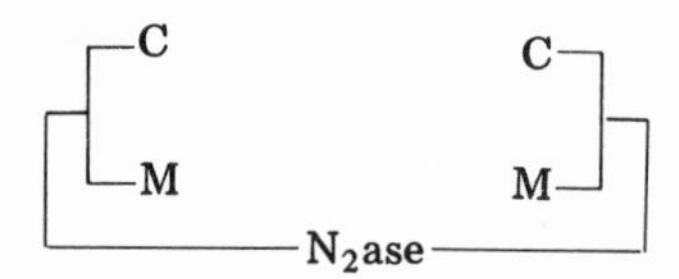

The C and M (metal) atoms are approximately planar parallel prosthetic groups, and the metal is chemically like Co^{II} or Fe^{I}. Based on the thermodynamics of N_2 chemisorption on iron, the following favorable reactions were suggested:

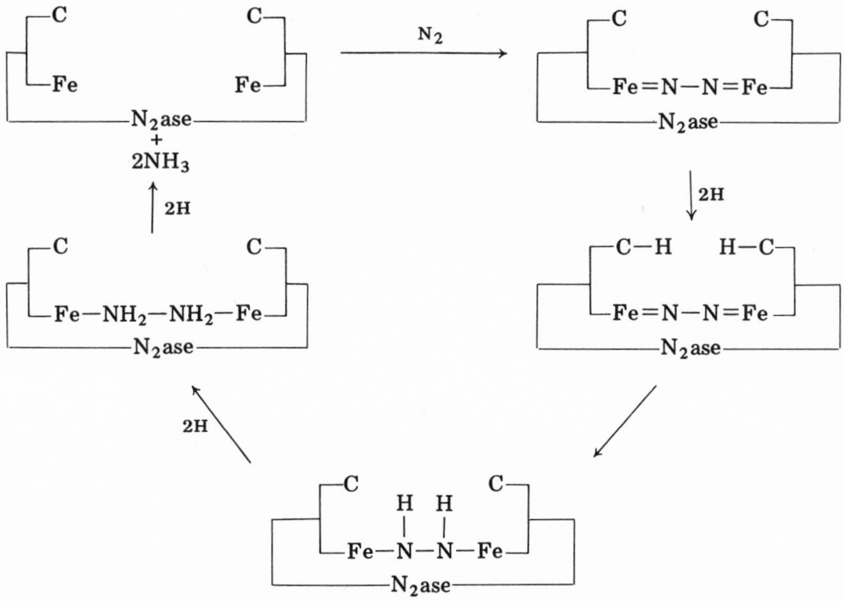

C C
Fe Fe
N_2ase
+
$2NH_3$

$\xrightarrow{N_2}$

C C
Fe=N—N=Fe
N_2ase

↓ 2H

C—H H—C
Fe=N—N=Fe
N_2ase

↙

C C
H H
| |
Fe—N—N—Fe
N_2ase

↖ 2H

C C
Fe—NH_2—NH_2—Fe
N_2ase

↑ 2H

The proposal contains many attractive concepts, including a dinuclear site, stepwise reduction, explanation of H_2 inhibition and avoidance of the energetically unfavorable diazene; its greatest limitations include inability to accommodate many of the known substrates of N_2ase or the ATP function of N_2ase and the absence of a conventional hydrogenase activity in purified N_2ase.

A more recent proposal[16,198] avoids the unfavorable thermodynamics of a diazene intermediate through a four electron addition from a dinuclear hydride to a dinuclear dinitrogen complex, *e.g.*:

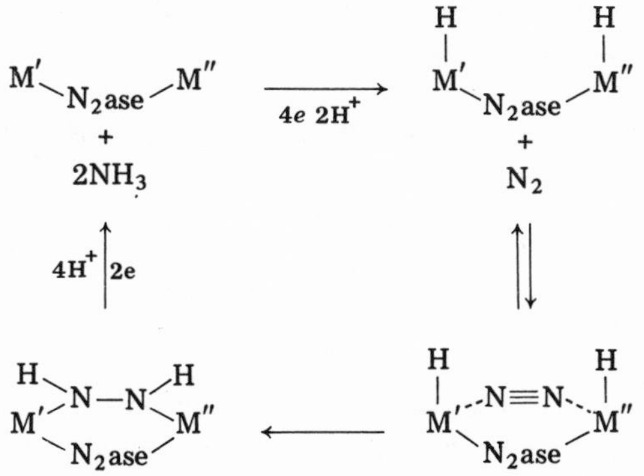

M′ M″
N_2ase
+
$2NH_3$

$\xrightarrow{4e\ 2H^+}$

H H
| |
M′ M″
N_2ase
+
N_2

⇅

H H
| |
M′ ··N≡N·· M″
N_2ase

⟵

H H
N—N
M′ M″
N_2ase

↑ $4H^+$ 2e

Subsequent addition of $4H^+$ and $2e$ convert the dinuclear hydrazine to $2NH_3$. Experimental basis[199] for this proposal is the formation of dinuclear dinitrogen complexes, hydrazine and ammonia from N_2 with Ti, Fe and Mo as catalysts and RMgX as reductants. However, N_2H_4 has never been detected as a product of biological N_2 fixation.

An earlier proposal[200] also avoided diazene through dinitrogen insertions into a bridged dihydride system:

$$M_{red}(\mu\text{-}H)_2M_{red} + N_2 \longrightarrow M(H)(H)(N{\equiv}N)M_{red} \longrightarrow M_{red}(HN{-}NH)M_{red}$$

$$\downarrow$$

$$M_{ox}(\mu\text{-}NH)_2M_{ox} \xleftarrow{2H^+} \left[M_{ox}(\mu\text{-}NH_2)_2M_{ox}\right]^{2+} \xleftarrow{2H^+} \ldots$$

$$2[M_{ox}]^{2+} + 2NH_3 \xleftarrow{2H^+} \left[M_{ox}(\mu\text{-}NH_2)_2M_{ox}\right]^{2+}$$

$$2[M_{ox}]^{2+} \xrightarrow{2e} 2[M_{red}]^{1+} \xrightarrow{4e,\ 2H^+} M_{red}(\mu\text{-}H)_2M_{red}$$

This mechanism was based on model studies with a Cp_2TiCl_2 + RMgX + N_2 system; however, e.s.r. studies of the model system have not substantiated the bridged dihydride interpretation[201].

The final theory for N_2 fixation invokes separate sites on N_2ase for electron activation and substrate complexation[41] and proposes a stepwise reduction with N_2ase-bound diazene and hydrazine as intermediates of N_2 fixation[9,169]. The basic hypotheses and support for this proposal are the following:

(1) The major role of ATP is electron activation to form a strong nucleophile, possibly a metal hydride, which evolves H_2 and/or reduced N_2ase substrates. The interdependent electron transfer and ATP utilization reactions of N_2ase provide biochemical support. The irreversible evolution of H_2 suggests energy-coupled electron transfer to yield a reduced species with a potential more negative than –0.43 V. A possible mechanism invokes

formation of an intermediate possessing a preferred leaving group which facilities reduction[10, 204] *e.g.*:

$$\mathrm{Mo{-}OH + ATP} \xrightarrow{-\,PO_4^{2-}} \mathrm{Mo{-}ADP} \xrightarrow[-\,ADP]{2e} \mathrm{Mo\!:\ or\ MoH}$$

$$\mathrm{Mo{-}OH + ATP} \xrightarrow{-\,ADP} \mathrm{Mo{-}PO_4^{2-}} \xrightarrow[-\,PO_4^{2-}]{2e} \mathrm{Mo\!:\ or\ MoH}$$

The stimulation by XTP of H_2 evolution from the molybdothiol-boro-hydride model of N_2ase supports this proposal[159].

(2) Electron activation occurs at one site while substrate complexation (*i.e.*, reducible substrates other than H_3O^+) occurs at a separate site.

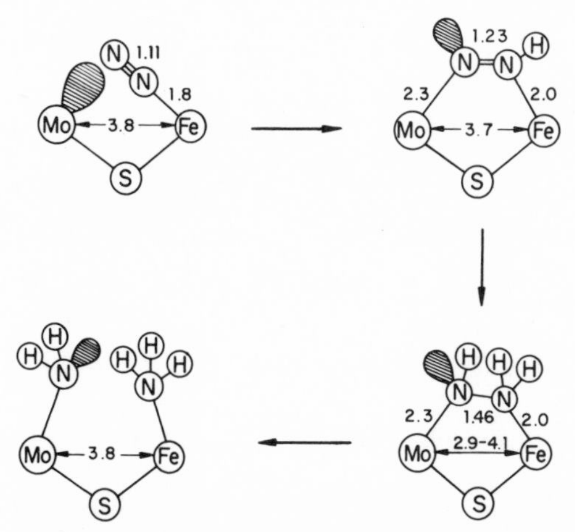

Fig. 5. Complexation and reduction sequence in proposed diazene/hydrazine mechanism for N_2ase[9]. Mo may be replaced by V. Interatomic distances are shown in Å. The shaded orbitals may be protonated.

The two sites are differentiated by CO inhibition of reactions only at the substrate complexation site.

(3) The substrate complexation site is dinuclear and is suggested to be composed of Mo and Fe bridged by a ligand such as S with the site situated in a pocket of N_2ase (Fig. 5). Dinitrogen complexes initially with Fe and other substrates with Mo. Inhibitor studies, metal replacement experiments and model substrate studies are compatible with this structure. Dihydrogen competitively inhibits only N_2 reduction, while CO inhibits the reduction of all substrates other than H_3O^+, indicating a CO-sensitive site for these substrates and in addition a H_2-sensitive site specifically for N_2 (Fig. 2). Comparison of reaction kinetics of V–N_2ase with Mo–N_2ase

References pp. 787–793

supports Mo as the common CO-sensitive site. Inability to reduce isobutyronitrile and dimethylacetylene, relatively low activity reductions of *trans*-crotononitrile and methacrylonitrile *vs.* *cis*-crotononitrile and differences between the product mixtures from ethyl isocyanide and methyl isocyanide support a dinuclear site in a sterically restrictive pocket on N_2ase. Reduction of a variety of N_2ase substrates, especially acetylene by the molybdothiol-borohydride system but not by an analogous iron system, provides model support for the proposed role of Mo.

(4) Dinitrogen reaction with Fe in a valence state $\leqslant 2$ to form a linear N≡N–Fe complex. All known $M \cdot N_2$ complexes are of the structure M–N≡N, not $M \leftarrow \overset{N}{\underset{N}{|||}}$. The ligand affinities of N_2ase and of various transition metals that form complexes with N_2 are similar, *e.g.*,

$$N_2ase - CO \sim NO \gg N_2 \sim C_2H_2 \sim H_2 > N_3^-, N_2O, RNC >$$

$$RCN \gg NH_3, H_2O, C_2H_4$$

$$(Ph_3P)_3Co(L)H - CO \gg N_2 \sim H_2 > C_2H_4, NH_3$$

indicating these $M \cdot N_2$ complexes are valid models of the initial reaction between N_2 and N_2ase. Theoretical calculations on dinitrogen reduction indicate that a complex containing N_2 with a bond order of 2 to 3 should be optimal for facile reduction[202].

(5) The $Fe \cdot N_2$ complex undergoes nucleophilic or hydride attack to form a dinuclear $Mo{-}N{=}\overset{H}{N}{-}Fe$ complex at the diazene level of reduction. Subsequent electron additions in one or two electron increments produce a dinuclear hydrazine complex followed by cleavage of the N–N bond to bound ammonia with low affinity for the site. Exchange occurs between H_2 and protons of the media via protons of the above intermediates. The excessively high activation energies of the formation of diazene and hydrazine are markedly reduced by coordination of these intermediates to metals (Fig. 4). It is proposed that the activation energy involved in adding the first two electrons is lowered from 48.7 to less than 15 kcal/mole by coordination and hydrogen bonding of diazene to N_2ase. Hydrogen bonding has been demonstrated in hydroxymethyldiazenes and is believed to contribute substantially to their stability[203]. Production of 2-, 4-, 6-, 8-, 10-, 12- and 14-electron addition products in N_2ase catalysis suggests electron transfer in 1- or 2-electron steps and supports diazene and hydrazine as intermediates.

Three model systems support this reductive sequence by N_2ase. The interaction of a benzenediazonium salt as an example of $N_2ase \cdot N_2$ with a platinum hydride as a model for the Mo: or MoH portion yields the proposed diazene and hydrazine intermediates[204]:

$$(PEt_3)_2ClPt{-}H + [N{\equiv}N{-}\phi]^+ \longrightarrow [(PEt_3)_2ClPt{-}\overset{H}{\overset{|}{N}}{=}N{-}\phi]^+$$

$$\downarrow H_2/Pt$$

$$(PEt_3)_2ClPtH + {}^+NH_3NH{-}\phi \xleftarrow{H_2/Pt} [(PEt_3)_2ClPt{-}NH_2{-}NH{-}\phi]^+$$

The reduction of the phenyldiazene complex to the hydrazine derivative may also be accomplished by $Na_2S_2O_4$ reduction or by polarographic reduction at −0.4 V. The similarity between the HD exchange reactions of these model diazene and hydrazine intermediates and the HD exchange in N_2ase-catalyzed N_2 fixation is the sole experimental evidence for enzyme-bound diazene and hydrazine as intermediates.

The first reductive step of the proposed N_2ase reaction (Fig. 5) is supported by the interaction of a $Re \cdot N_2$ complex, $[(PMe_2Ph)_4Re(N_2)Cl]$ with a variety of transition metal ions including Ti^{III}, Cr^{III}, Mo^{III} and Mo^{IV} [205]. It is interesting that Mo^{IV} is most effective in this reaction. The reaction is formulated as follows:

$$Mo^{IV} + (N_2)Re \rightarrow Mo{\cdots}N{\equiv}N{\cdots}Re.$$

The N≡N stretching frequency is reduced from 2331 cm^{-1} in free N_2 to 1992 cm^{-1} in $Re \cdot N_2$ to 1680 cm^{-1} in the dinuclear product, which is within 100–200 cm^{-1} of that of diazene. Unfortunately, the product has shown no tendency toward further reduction.

The molybdothiol–borohydride model[158,159] system appears to be an excellent representation of the Mo nucleophile proposed in this mechanism since it is stimulated by ATP, is not inhibited by H_2, and reduces all substrates of N_2ase except perhaps N_2. Dinitrogen itself is reported to be reduced in a reaction stimulated by Fe^{2+}. Convergence of model and biochemical approaches in the future will provide the basis for further definition of the scientifically intriguing mechanism of biological N_2 fixation.

D. Ancillary roles of metals in biological N_2 fixation

Besides the Mo–Fe protein and Fe protein of N_2ase, other metalloproteins and metals play roles in N_2 fixation. The role of Fd has already been described [Section B(1)], and the consistent association of hydrogenase and N_2ase suggests a possible ancillary role of hydrogenase in N_2 fixation. Leghemoglobin is the hemoprotein responsible for the intense red color of root nodules, the site of N_2 fixation in legumes. The natural form of LHb is high spin LHb^{II}, and like myoglobin it is able to

References pp. 787–793

undergo reversible oxygenation[206]. Leghemoglobin has the highest affinity of any known hemoprotein for O_2 and recent results show that it can facilitate O_2 diffusion[207]. An earlier supposed N_2 complex[208] of LHb which was suggested to be involved in N_2 transport or activation has not been substantiated[209]. The molecular weights, amino acid composition and absorption spectra of two crystalline LHb's from soybeans have been determined[210].

Leghemoglobin concentration is qualitatively correlated with N_2-fixing activity of legume nodules[8,211]. It is not found in free-living organisms but may occur in non-legume symbionts that fix N_2. Both LHb and N_2ase are products of the symbiosis between leguminous plants and *Rhizobia.* The heme portion is synthesized by the bacteria[212] and the globin by the plant[213]. The physiological role of LHb is unknown. It is not part of N_2ase and is localized outside of the bacteroids, which contain N_2ase. Leghemoglobin does not stimulate N_2 fixation by either isolated bacteroids or N_2ase[116,117]. Suggested LHb functions include facilitation of O_2 diffusion in actively respiring nodules or O_2-scavenging to produce a localized anaerobiosis for the O_2-sensitive N_2ase system.

Differences in the cytochromes and hemoproteins between *Rhizobia* bacteria and bacteroids have also been found, including the absence of cytochromes a_3 and o and a rhizobial leghemoglobin (antigenically unrelated to LHb of nodules) and occurrence of cytochromes 552, P-420 and P-450 in bacteroids[1]. These changes are of unknown physiological significance.

The nutritional requirement of Co for N_2-fixing organisms has been extensively explored[6]. All N_2-fixing organisms that have been thoroughly investigated require Co for growth and for synthesis of vitamin B_{12} compounds. The role of Co in N_2 fixation is obviously an indirect one since purified N_2ases do not contain significant quantities. Other metals, including Ca and Cu, are suggested to play ancillary roles in development of N_2 fixation in symbionts and/or free-living organisms[6].

IV. ABIOLOGICAL DINITROGEN FIXATION

The history of catalytic dinitrogen fixation parallels that of catalysis generally. Even before the first world war a heterogeneous catalyst to effect the reaction of N_2 and H_2 had been devised. The first commercial ammonia synthesis unit employing the Haber–Bosch process opened in 1913. This process (Section A) has undergone engineering development to the point that it is one of the most efficient commercial processes and makes NH_3 available at about $0.03 per kg on a 1000-ton/day scale.

In contrast, the homogeneous catalytic reactions of N_2, like those of olefins, have been studied only in recent years. The apparent chemical inertness of N_2 inhibited study of its interaction with metal complexes until the early 1960s. However two major developments in the solution chemistry of N_2, from other than specific studies of N_2 chemistry, have made N_2 fixation one of the most active current areas of inorganic chemistry.

The first of these developments was the report[214] in 1964 that several transition metal halides catalyze the reduction of N_2 under mild conditions. The second, reported in the following year, was the isolation of the first discrete complex of N_2 with a transition metal★. These two discoveries, described in more detail in Sections B and C, have had a major impact on inorganic chemistry and have stimulated much thought about the role of Fe and Mo in N_2ase.

A. Haber–Bosch process

Commercial dinitrogen fixation is effected by passing a 3:1 mixture of highly purified H_2 and N_2 (from methane reforming with limited air injection) over an iron catalyst at about 450°C and 200–300 atm.[215,216]. Roughly 20% of the reaction mixture is converted to ammonia on a single pass through the catalyst. The ammonia is condensed from the product stream and the gases are recycled to the reactor inlet. The chemical yield is essentially quantitative and the catalyst lifetime is measured in years. The catalyst is a porous iron which contains small amounts of Al_2O_3 to stabilize the crystallites and traces of K_2O which seem to facilitate the chemisorption of nitrogen.

The mechanism of reduction of N_2 on an iron catalyst is far from established, but a commonly proposed sequence involves the following steps: (1) chemisorption of N_2 as a surface species Fe–N≡N analogous to the well characterized molecular complexes, $[(NH_3)_5Ru(N_2)]^{2+}$ and $(Ph_3P)_3(N_2)CoH$; (2) dissociation of the diatomic species to a monoatomic species, perhaps a metal nitride like those postulated as intermediates in the homogeneous reduction of N_2 (this dissociation is generally accepted as the rate-determining step in the heterogeneous reduction); (3) hydrogenation of the monoatomic complex via species such as Fe–NH and $Fe–NH_2$ to give coordinated NH_3 as the final product; (4) dissociation of NH_3 from the catalyst. Other sequences also fit the reported kinetics of the reaction[217], but do not explain the isotope exchange between $^{14}N_2$ and $^{15}N_2$ observed on some catalysts.

B. Homogeneous catalysis

Of many transition metal salts tested as catalysts for the reduction of N_2 by organometallic compounds, maximum activity was found with Ti

★ An entertaining and instructive account of this discovery has been provided by Allen and Bottomley[15].

derivatives, although V, Cr, Fe, Mo and W compounds also gave significant reduction[214]. The combination of Cp_2TiCl_2 catalyst and an organometallic reducing agent has been studied extensively and several significant facts have emerged. The Ti complex appears to be reduced at least to the divalent state and probably even lower[218,219]. Some evidence for formally zerovalent Ti species has appeared, and e.s.r. studies of the reaction solutions indicate the presence of several paramagnetic species including[20,201]:

$$[Cp_2TiH_2]^- \text{ and } [Cp_2Ti\langle^{H}_{H}\rangle TiCp_2]^-.$$

However, the species that captures N_2 may be diamagnetic. Solutions of titanocene, usually written as $[Cp_2Ti]_2$,* react with N_2 slowly and reversibly to give a species with infra-red absorption probably assignable to a Ti–N≡N function although definitive experiments with $^{15}N_2$ have not been reported[220,221]. This transient N_2 complex is reduced by both Ti and additional reducing agent to a dark substance which liberates NH_3 on hydrolysis. This dark substance, probably a nitride, appears to contain a maximum of one N per Ti atom. Thus, the overall sequence may be:

$$2Ti^{IV} \xrightarrow{4e} 2Ti^{II} \underset{}{\overset{N_2}{\rightleftharpoons}} (Ti^{II})_2\text{-}N{\equiv}N \xrightarrow{2e} 2Ti^{IV}{\equiv}N \xrightarrow{H_2O} 2NH_3$$

In a typical experiment, Cp_2TiCl_2 was treated with a large excess of ethylmagnesium bromide in ether at 25°C under N_2. Hydrolysis of the mixture after nine hours gave 0.67 mole hours/NH_3 per mole of Ti complex[222]. Under 150 atm. N_2, yields approached 1.0.

Simple metal halides, in combination with stronger reducing agents and high N_2 pressures, produce more than one mole of NH_3 per mole of metal complex. For example, $TiCl_4$ in combination with a Mg/MgI_2 couple gives 1.25 NH_3/Ti[223]. The maximum yield of 2.0 NH_3/metal has been obtained by reduction with Li naphthalide in the presence of VCl_3[224]. Presumably, the reaction involves the conversion of coordinated N_2 to a mixture of nitrides:

$$V^0\text{-}N{\equiv}N + 3Li \rightarrow V^{III}{\equiv}N + Li_3N \xrightarrow{H_2O} 2NH_3$$

---

* Probably actually[154]

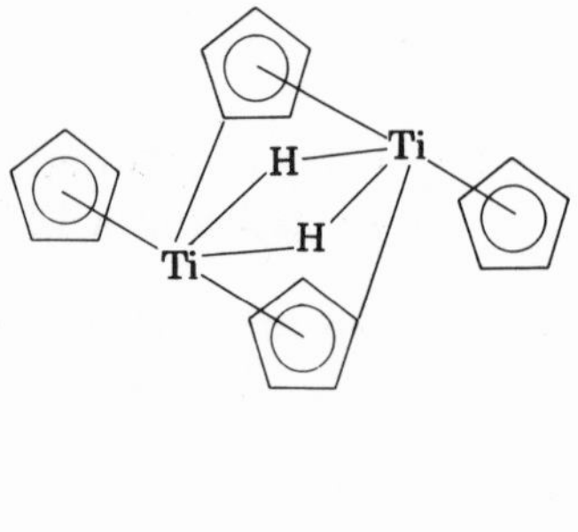

[Structure: two Cp₂Ti units bridged by two H atoms, each Ti bonded to one Cp ring and to a Cp ring of the other unit]

A catalytic reduction system that employs Al metal as the reducing agent has been developed[225]. A mixture of a catalytic amount of $TiCl_4$ and a gross amount of Al in molten $AlBr_3$ reacts with N_2 under pressure to give up to 284 moles of NH_3 per mole of $TiCl_4$ after hydrolysis. If the $TiCl_4$ is omitted, no reaction occurs. This system may well involve reduction of N_2 coordinated to Ti followed by reduction of the Ti≡N function, thus making the metal available for another cycle of operation.

$$TiCl_4 \xrightarrow{Al} Ti^0 \xrightarrow{N_2} TiN{\equiv}N \xrightarrow{Al} Ti{\equiv}N + AlN \xrightarrow{H_2O} NH_3$$

(Ti≡N $\xrightarrow{Al}$ AlN, regenerating Ti^0)

A further development in the use of Ti is the reaction cycle characterized by both electrolytic and chemical reduction[226]. An electrolysis cell fitted with an Al anode and a nichrome cathode contained titanium tetraisopropoxide, naphthalene, aluminum isopropoxide and tetrabutylammonium chloride in 1,2-dimethoxyethane. In the course of 11 days of electrolysis at 40 V under a stream of pure N_2, aluminum nitride accumulated to provide a 610% yield of NH_3 (based on Ti) when the mixture was subsequently treated with 8 *N* aqueous NaOH. In an earlier reaction system[196] near-stoichiometric reductions of N_2 to NH_3 (based on Ti) were obtained using Na metal as the ultimate electron source and Na naphthalide as the reducing species to maintain Ti as Ti^{II} and to reduce Ti-bound N_2. Hydrazine has since been identified as a product[227,228], and its formation is favored by a high ratio of fixing agent to reducing agent. The general characteristics of the reaction suggest that an intermediate at the hydrazine level of reduction (4 electrons per N_2) is formed, but that excess reductant converts it to a nitride which gives NH_3 on hydrolysis.

Despite speculation that Ti-based fixation may lead to a low pressure ammonia synthesis, it seems more likely to find practical application in the synthesis of organic amines. Dinitrogen has been successfully incorporated into organic carbonyl compounds according to the overall reaction:

$$R_2CO \xrightarrow{Cp_2TiCl_2 + Mg + THF + N_2} R_2CHNH_2 + (R_2CH)_2NH$$

With diethyl ketone, 25–50% of the N_2 taken up by the system was recovered in 3-pentylamine and di(3-pentyl)amine (2:1 ratio)[229]. In another system aromatic amines (along with NH_3) are formed[230]. The reduction of N_2 with excess phenyl lithium catalyzed by Cp_2TiCl_2 gave 0.03 mole aniline and 0.17 mole ammonia per g-atom Ti at 25°C and 1 atm. With *p*-tolyl lithium, the yield of aromatic amine was somewhat higher, 0.07 mole, but the *p*-toluidine was contaminated with *m*-toluidine.

References pp. 787–793

It is unclear whether this reaction represents N_2 insertion into a Ti–C bond or attack of Ti≡N by the aryl lithium reagent. The positional scrambling observed with the tolyl reagent is consistent with the presence of a benzyne intermediate. Indeed, an *o*-phenylene titanium species has been implicated in the decomposition of Cp_2TiPh_2 [231].

Of the other soluble systems for the reduction of N_2 to ammonia, greatest activity is found with Fe and Mo. Reduction of N_2 to NH_3 has been accomplished with an $FeCl_3$–lithium naphthalide system[232]. The reduction is stoichiometric, yielding a limit of 1 NH_3/Fe with 2,6-dimethylnaphthalene dianion as reductant, but not with naphthalides of weaker reducing activity. Four naphthalene moieties per Fe appear to be required for the critical reaction step; stoichiometric reductions were reported for reactions of 4 h duration at 1600 psi of N_2 and 60°C, while substantially longer times were indicated for conditions of one atm. and room temperature.

Iron(III) chloride is reduced by $(CH_3)_2CHMgCl$ in the presence of triphenylphosphine and N_2 to form an unstable dinitrogen complex[199]. With excess reducing agent at –50°C, the N_2 is reduced to the hydrazine level and, indeed, N_2H_4 is liberated on treatment with anhydrous HCl. At elevated temperatures (90°C) there occurs further reduction to a nitride species which yields NH_3 on hydrolysis. Similar behavior is reported for $MoCl_5$ and $(CH_3)_2CHMgCl$ with 4- and 6-electron reductions occurring at roughly equal rates[199].

Molybdenum-based systems are particularly interesting since they fix N_2 in water solutions. In addition to the molybdothiol–borohydride system[158] discussed earlier, an efficient Mo-catalyzed reduction of N_2 by Ti^{III} has been reported[189]. The latter reaction is carried out in water or aqueous ethanol at high pH (> 10.5) with stoichiometric quantities of $TiCl_3$ as reducing agent and a catalytic amount of $MoOCl_3$ or Na_2MoO_4. The chief product at 25°C is reported to be N_2H_4 together with some NH_3. The turnover number of 0.6 N_2 reduced/Mo · min is about 1% that of enzymic fixation. However, the system resembles N_2ase in that H_2 is evolved in competition with N_2 reduction, C_2H_2 is reduced to C_2H_4, CO is a competitive inhibitor and Mg^{2+} is a promoter.

The history of N_2 fixation by Mo-based catalysts is complex. Heterogeneous catalysts based on Mo are active in N_2 hydrogenation at lower temperatures (down to 200°C) than those based on Fe, but have not been commercialized because of operating deficiencies. A specially reduced molybdenum oxide in a slurry system is reliably reported to convert N_2 and H_2 to ammonia slowly at 25°C and 1 atm. pressure[233]. Two reports[234,235] of N_2 reduction by Zn catalyzed by MoO_4^{2-} in aqueous acid have been partly discredited[236]. A Mo-catalyzed photochemical reaction of N_2 with formaldehyde to give amino acids does not seem to have been verified[237].

C. *Transition metal complexes of* N_2

In the five years that have elapsed since the isolation of the first transition metal complex of N_2, well-confirmed complexes have been reported for ten of the transition metals. Historically, these complexes have been of two distinct types which are considered separately below. One group consists of cationic complexes of Ru and Os with ammonia or amines as stabilizing ligands. The other broader group is characterized by the presence of mono- or di-tertiary phosphine ligands and, usually, by the absence of charge. However, the distinctions have become blurred with the isolation of cationic, phosphine-containing complexes and by obvious parallelisms in the chemistry of the two classes.

(1) Amine-stabilized complexes

The first report[238] of an N_2 complex described the reduction of $RuCl_3$ by excess hydrazine in aqueous solution to give salts of $[(NH_3)_5Ru(N_2)]^{2+}$. Both the ammonia and the dinitrogen ligands were formed by the metal-catalyzed decomposition of N_2H_4. As in all the dinitrogen complexes reported to date, the presence of coordinated N_2 was evidenced by a strong infra-red absorption assignable to N≡N stretching. Initial skepticism about the identity of the complex was dispelled by the publication of a partial crystal structure determination[239]. Although fine points of the structure were lost due to disorder in the crystal, it was apparent that the complex contained five NH_3 ligands and an N_2 in an octahedral array about the metal atom. The N_2 was bound "end-on" like CO in a metal carbonyl rather than "side-on" like a coordinated acetylene.

Many different nitrogen compounds can serve as sources of the N_2 ligand in the preparation of $[(NH_3)_5Ru(N_2)]^{2+}$. This complex has been isolated from reactions in which N_2O, N_3^-, ammonia and, most importantly, N_2 itself provided the dinitrogen ligand.

The first indication that N_2 would take part in complex formation was the reported reduction of $RuCl_3$ with Zn under N_2 in aqueous THF to give an N_2 complex[240], subsequently identified as $RuCl_2(N_2)(H_2O)_2$-(THF)[241]. This discovery showed that N_2 could be absorbed from the gas phase even in competition with Lewis base ligands such as water and tetrahydrofuran.

A more elegant demonstration came from the following reaction sequence[242].

$$[(NH_3)_5RuCl]^{2+} \xrightarrow[H_2O]{Zn} [(NH_3)_5Ru(H_2O)]^{2+} \xrightarrow{N_2} [(NH_3)_5Ru(N_2)]^{2+}$$

In this reaction a preformed Ru^{III} complex is reduced by abstraction of chloride. The vacant coordination site on the Ru^{II} derivative is immediately filled by water and the aquo complex can be isolated in the absence of N_2.

References pp. 787–793

However, under an N_2 atmosphere, the water is rapidly displaced to give the N_2 complex.

This latter study also showed that the coordinated N_2 can complex with a second Ru atom[243]:

$$[(NH_3)_5Ru(N_2)]^{2+} + [(NH_3)_5Ru(H_2O)]^{2+} \rightleftharpoons [(NH_3)_5Ru\text{-}N{\equiv}N\text{-}Ru(NH_3)_5]^{4+}$$

The overall reaction of two equivalents of $[(NH_3)_5Ru(H_2O)]^{2+}$ with one mole of N_2 is exothermic to the extent of 22 kcal/mole[244]. The binuclear product has a linear Ru–N≡N–Ru structure as indicated by crystal structure determination[245]. The vibrational absorption is observable in the Raman spectrum rather than in the usual infra-red spectrum, an indication of a symmetrical structure. It also absorbs at a surprisingly high frequency, 2100 cm^{-1}, which suggests that the N–N bond order approaches a value of 3.

The N–N vibrational frequencies have been widely used (and sometimes misused) as criteria of bond character in dinitrogen complexes. The limiting value for a pure triple bond is probably well defined by that of molecular N_2, 2331 cm^{-1} (a Raman mode because of symmetry). The value of an NN double bond is probably in the range of 1400–1600 cm^{-1} as judged by the infra-red spectra of organic azo compounds. On this basis, all the cationic, amine-stabilized N_2 complexes of Ru and Os have bond orders near 3 since the infra-red absorptions are at 2000–2200 cm^{-1}. Thus, the character of the N_2 ligand in $[(NH_3)_5Ru(N_2)]^{2+}$ (ν_{NN} 2140 cm^{-1}) is not greatly perturbed from that of free N_2. This postulate is supported by the findings that the ligand is not activated toward reduction and that it reacts with a second mole of $[(NH_3)_5Ru(H_2O)]^{2+}$ just as readily as does N_2.

This class of N_2 complexes has been elaborated in many ways, such as the preparation of bis(dinitrogen) complexes and the use of organic amines as stabilizing ligands. However, the major interest from the viewpoint of bioinorganic chemistry is the interaction of these metals with substrates of N_2ase other than N_2. The aquo complex $[(NH_3)_5Ru(H_2O)]^{2+}$ reacts with CO to give stable carbonyl derivative[247]. Azide ion[168] and N_2O[172] react to give transient complexes which rapidly convert to the dinitrogen complex, $[(NH_3)_5Ru(N_2)]^{2+}$. Organic nitriles form stable complexes in which the ligand is attached by the nitrogen of the C≡N function[248]. This "end-on" attachment is consistent with that of N_2 itself. Isomerization studies of the $^{15}N^{14}N$ complex indicate little or no significant stability for attachment of N_2 "side-on"[249].

(2) Phosphine-stabilized complexes

Just as with the amine-stabilized complexes, the first N_2 complex in this class was prepared by an indirect synthesis. The reaction of

$[(C_6H_5)_3P]_2(CO)IrCl$ with an acyl azide gave a dinitrogen analog of the carbonyl complex[250].

$$Ir{-}C{\equiv}O + \phi{-}\overset{\overset{O}{\|}}{C}{-}N{=}N{=}N \longrightarrow Ir{-}N{\equiv}N + \phi{-}\overset{\overset{O}{\|}}{C}{-}N{=}C{=}O$$

Many structural modifications of the N_2 complex have been made by using other Ir carbonyl complexes[251] or by using Rh analogs[252]. However, none of the changes have significantly enhanced the stability of the products. The Ir-N_2 complex slowly and irreversibly loses N_2 on standing in solution at room temperature. The N_2 stretching vibrations are about 2100 cm^{-1} and do not vary greatly with the nature of the other ligands.

A major breakthrough in the chemistry of the phosphine-stabilized complexes was the simultaneous discovery by three research groups that $[(C_6H_5)_3P]_3(N_2)CoH$ could be prepared directly from N_2 [253-255]. The synthesis was the reduction of a Co^{III} salt with an alkylaluminum compound in the presence of the phosphine ligand, *e.g.*,

$$Co^{III} + 3(C_6H_5)_3P + (C_2H_5)_3Al + N_2 \rightarrow [(C_6H_5)_3P]_3(N_2)CoH + Al^{III} + 2C_2H_4 + C_2H_6$$

This reaction has since been developed into a general synthetic method for dinitrogen complexes of transition metals. The variety of compounds accessible by this method is indicated by Table IV. All the

TABLE IV

PHOSPHINE-STABILIZED DINITROGEN COMPLEXES
(Typical N_2 stretching frequencies in parentheses)

Group VI	Group VII	Group VIII		
No Cr	No Mn	$(R_3P)_3(N_2)FeH_2$ (2055–2060)	$(Ar_3P)_3(N_2)CoH$ (2033–2088)	$(R_3P)_2(N_2)Ni$ (2028) $[(R_3P)_2Ni]_2N_2$
$(\pi\text{-}C_6H_6)(Ar_3P)_2(N_2)Mo$ (2005) $(diphos)_2(N_2)_2Mo$ (1970)	No Tc	$(Ar_3P)_3(N_2)RuH_2$ (2147)	$(Ar_3P)_2(N_2)RhCl$ (2152)	No Pd
$(PhPMe_2)_4(N_2)_2W$ (1931, 1998)	$(R_3P)_4(N_2)ReCl$ (1920–1990)	$(R_3P)_3(N_2)OsH_2$ (2060–2090)	$(Ar_3P)_2(N_2)IrCl$ (2051–2113)	No Pt

compound types shown, except those of Rh and Ir, have been prepared by reduction of a metal salt in the presence of the phosphine ligand. (The Re and Os complexes were formed by the use of Zn instead of an alkylaluminum compound as the reducing agent.) Some interesting variations

References pp. 787–793

are the formation of a bis(N_2)Mo complex[256] and of a binuclear Ni complex[257], $[(C_6H_{11})_3P]_2Ni-N_2-Ni[(C_6H_{11})_3P]_2$. The latter is presumed to contain a linear M–N≡N–M arrangement like the amine-stabilized Ru complex.

The Co compound $[(C_6H_5)_3P]_3(N_2)CoH$ may be the best studied of the dinitrogen complexes. A careful crystal structure determination[258] has shown that the N–N bond length is 1.11 Å, a remarkably small increase over the 1.096 Å bond distance in N_2 itself. The N_2 ligand is bound "end-on" and the Co–N–N angle is approximately 180° (Fig. 6). Its reactions with

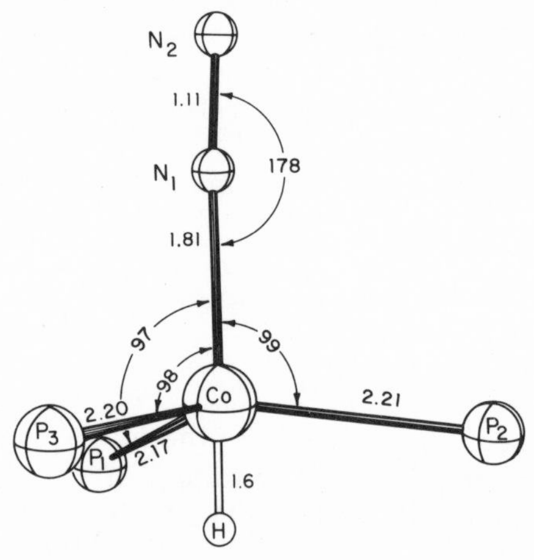

Fig. 6. Structure of $(Ph_3P)_3(N_2)CoH$[258]. The structural parameters are averaged from those reported for molecules in two sites in the crystal.

N_2ase substrates and inhibitors as discussed previously make it a useful model for the N_2-binding site.

The Fe[259,260] and Ru complexes[261,262] of the formula $(R_3P)_3(N_2)MH_2$ are strikingly similar to the Co complex in their synthesis and chemistry. The ligand affinity series for $[(C_6H_5)_3P]_3(N_2)RuH_2$, is like that for $[(C_6H_5)_3P]_3(N_2)CoH$. Cationic Fe and Ru complexes $[(DEPE)_2(N_2)MH]^+$ have been prepared[263] by abstraction of chloride from the neutral compounds $(DEPE)_2MHCl$ in the presence of N_2.

The isolation of dinitrogen complexes of Fe is significant not only because of the presence of Fe in N_2ase, but also because it permits the use of Mössbauer spectroscopy to study the metal–N_2 bond[263]. Preliminary results on $[(DEPE)_2(N_2)FeH]B(C_6H_5)_4$ and its carbonyl analog indicate that N_2 is a weaker ligand than CO, both as a σ-donor and a π-acceptor. This result is consistent with the bonding picture presented in Fig. 7.

Molybdenum, like Fe, has a special role in the bioinorganic chemistry of dinitrogen because of its presence in N_2ase. These two elements are likewise distinctive in showing activity in abiological N_2-fixing systems as well as giving isolable dinitrogen complexes. The first reported N_2 complex of Mo is similar in structure to one of the postulated intermediates in

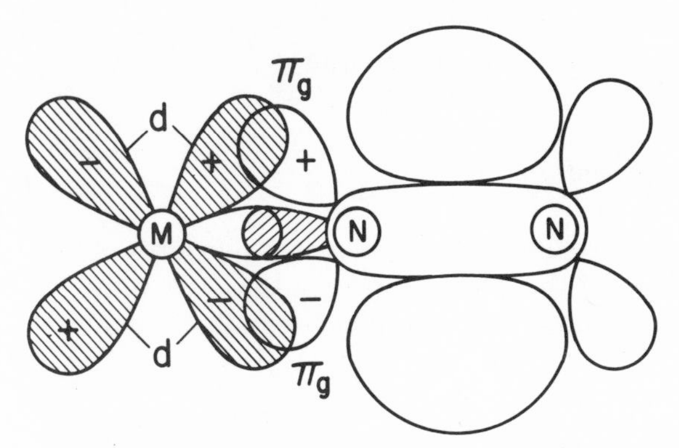

Fig. 7. Metal-dinitrogen bonding. The ligand-to-metal σ bond arises from overlap of a filled σ orbital ▨ (θ_4 of Fig. 1) of N_2 with a vacant σ orbital of the metal. The metal-to-ligand π bond involves a filled metal *d* orbital ▧ and a vacant π_g molecular orbital of dinitrogen.

abiological fixation. The reaction of Mo^{III} acetylacetonate with R_3Al and $(C_6H_5)_3P$ under N_2 gives a complex[264] best formulated as **I**. It seems very probable that a species such as **II** is present as an immediate precursor to

$(C_6H_5)_3P$, $(C_6H_5)_3P$ – Mo(η-C_6H_6) – N≡N $\qquad [(C_6H_5)_n$ – Mo(η-C_6H_6) – N≡N$]^{x-}$

(I) (II).

N_2 reduction in the $C_6H_5Li/MoCl_3$ system. Substantial e.s.r. evidence for analogous species in Li naphthalide reductions has been presented[20].

(3) The metal–N_2 bond

The best evidence from spectroscopic studies is that the stability of N_2 complexes arises from a combination of ligand-to-metal σ-bonding and metal-to-ligand π-bonding. This bond concept, illustrated in Fig. 7, is consistent both with molecular orbital calculations and with chemical evidence based on the effects of changes in stabilizing ligands.

In this view of the bond[27], a weakly antibonding but filled σ-orbital (θ_4) on a nitrogen atom overlaps a vacant metal orbital on close approach.

This weak σ-donation to the metal enhances its electron density and promotes the formation of a bond utilizing a filled *d* orbital (π-symmetry) of the metal and a vacant antibonding orbital of the N_2. The introduction of electron density into the antibonding orbital of the N_2 destabilizes the original orbital and makes it a better donor. Thus the two effects are synergistic, and it is difficult and probably unprofitable to separate the two.

D. Non-catalytic N_2 fixation reactions

Other N_2 fixation reactions are summarized below for purposes of completeness. With the exception of the Bucher process, these are non-catalytic systems. There is no apparent relationship between these and biological N_2 fixation.

1. Reaction of N_2 with metallic Li to give Li_3N which can be hydrolyzed to NH_3 [265].
2. Formation of calcium cyanamide by reaction of CaC_2 with N_2 at 1100°C.
3. The Bucher process[266] for manufacture of NaCN by reaction of Na_2CO_3 with powdered coal and N_2 in the presence of an iron catalyst.
4. Synthesis of NH_3 from mixtures of N_2 and H_2 by γ-irradiation, gas discharge, or sonolysis, but not by hydrated electrons[166].
5. Oxidation of N_2 by lightning[267] or in electric discharge[268] at ~1500°C, followed by quick quenching to give NO which can be oxidized to HNO_3 by conventional commerical processes.
6. Pyrolysis of methane and N_2 at 1200°C or higher to form HCN and H_2 [268].

E. Summary

It is apparent that abiological N_2-fixing systems and the isolable N_2 complexes have had separate historical developments, but the relationship between the two systems is becoming closer. The detection of an N_2 complex in the Ti-based systems for N_2 fixation and the isolation of a π-arene Mo dinitrogen complex point to this relation.

The efficacy of Fe and Mo in enzymic fixation is consistent with the inorganic chemistry of these two metals with N_2. Iron seems to be a slightly better choice as an N_2-binding site in the enzyme simply based on the special stability of dinitrogen complexes of iron subgroup (Fe, Ru, Os) metals. However, the direct involvement of Mo in the reduction of N_2 is supported by: (i) known examples of $Mo \cdot N_2$ complexes, (ii) homogeneous and heterogeneous catalysis of abiological fixation by Mo, and (iii) its occurrence in the best current inorganic models for N_2ase. The particular effectiveness of phosphines in stabilizing a wide variety of N_2 complexes points to involvement of a similar type of ligand in the dinitrogen-binding site of the enzyme. The best candidate among biologically available ligands is sulfur which, either as S^{2-} or as the thiolate group of cysteine, has electronic characteristics similar to that of a tertiary phosphine.

REFERENCES

1 F. J. Bergersen, *Proc. R. Soc., B172* (1969) 401.
2 F. J. Bergersen, *A. Rev. Pl. Physiol.*, 22 (1971) 121.
3 R. C. Burns and R. W. F. Hardy, *Nitrogen Fixation in Bacteria and Higher Plants*, Springer-Verlag, New York, in press.
4 R. H. Burris, *Proc. R. Soc.*, B172 (1969) 339.
5 R. H. Burris in J. Postgate, *The Chemistry and Biochemistry of Nitrogen Fixation*, Plenum Press, London, 1971, p. 106.
6 H. J. Evans and S. A. Russell, *ibid.*, p. 191.
7 R. W. F. Hardy and R. C. Burns, *A. Rev. Biochem.*, 37 (1968) 331.
8 R. W. F. Hardy, R. C. Burns, R. R. Hebert, R. D. Holsten and E. K. Jackson, *Plant and Soil*, Special Vol. (1971) 561.
9 R. W. F. Hardy, R. C. Burns and G. W. Parshall, *Adv. Chem.*, 100 (1971) 219.
10 R. W. F. Hardy and E. Knight, Jr., in L. Reinhold and Y. Liwschitz, *Progress in Phytochemistry*, Vol. I, Wiley, London, 1968, p. 407.
11 L. E. Mortenson, in I. C. Gunsalus and R. Y. Stanier, *The Bacteria*, Vol. 3, Academic Press, New York, 1962, p. 718.
12 L. E. Mortenson, *Surv. Progr. Chem.*, 4 (1968) 127.
13 J. R. Postgate, *Nature*, 226 (1970) 25.
14 A. D. Allen, *Adv. Chem.*, 100 (1971) 79.
15 A. D. Allen and F. Bottomley, *Accounts Chem. Res.*, 1 (1968) 360.
16 Yu. G. Borodko and A. E. Shilov, *Russ. Chem. Rev.*, 38 (1969) 355.
17 J. Chatt, *Proc. R. Soc.*, B172 (1969) 327.
18 J. Chatt and G. J. Leigh in E. J. Hewitt and C. V. Cutting, *Recent Aspects of Nitrogen Metabolism in Plants*, Academic Press, London, 1968, p. 3.
19 J. E. Ferguson and J. L. Love, *Rev. Pure Appl. Chem.*, 20 (1970) 33.
20 G. Henrici-Olive and S. Olive, *Angew. Chem. (Int. Edn)*, 8 (1969) 650.
21 K. Kuchynka, *Catalysis Rev.*, 3 (1969) 111.
22 R. Murray and D. C. Smith, *Coordination Chem. Rev.*, 3 (1968) 429.
23 E. E. van Tamelen, *Accounts. Chem. Res.*, 3 (1970) 361.
24 E. E. van Tamelen, *Adv. Chem.*, 100 (1971) 95.
25 W. L. Jolly, *The Inorganic Chemistry of Nitrogen*, W. A. Benjamin, New York, 1964, p. 115.
26 Natl. Bur. Standards Circ. 500, *Selected Values of Chemical Thermodynamic Properties*, 1952.
27 K. G. Caulton, R. L. DeKock and R. F. Fenske, *J. Am. Chem. Soc.*, 92 (1970) 515.
28 S. W. Benson, *J. Chem. Educ.*, 42 (1965) 502.
29 D. Burk, *Ergeb. Enzymforsch.*, 3 (1934) 23.
30 J. E. Carnahan, L. E. Mortenson, H. F. Mower and J. E. Castle, *Biochim. Biophys. Acta*, 44 (1960) 520.

31 L. E. Mortenson, H. F. Mower and J. E. Carnahan, *Bact. Rev.*, 26 (1962) 42.
32 L. E. Mortenson, R. C. Valentine and J. E. Carnahan, *J. Biol. Chem.*, 238 (1963) 794.
33 J. E. McNary and R. H. Burris, *J. Bact.*, 84 (1962) 598.
34 L. E. Mortenson, *Proc. Natn Acad. Sci. U.S.*, 52 (1964) 272.
35 R. W. F. Hardy and A. J. D'Eustachio, *Biochem. Biophys. Res. Commun.*, 15 (1964) 314.
36 A. J. D'Eustachio and R. W. F. Hardy, *ibid.*, 15 (1964) 319.
37 M. J. Dilworth, D. Subramanian, T. O. Munson and R. H. Burris, *Biochim. Biophys. Acta*, 99 (1965) 486.
38 L. E. Mortenson, in A. San Pietro, *Non-Heme Iron Proteins: Role in Energy Conversion*, Antioch Press, Yellow Springs, Ohio, 1965, p. 243.
39 R. C. Burns, *ibid.*, p. 289.
40 R. H. Burris, H. C. Winter, T. O. Munson and J. Garcia-Rivera, *ibid.*, p. 315.
41 R. W. F. Hardy, E. Knight, Jr. and A. J. D'Eustachio, *Biochem. Biophys. Res. Commun.*, 20 (1965) 539.
42 A. Lockshin and R. H. Burris, *Biochim. Biophys. Acta*, 111 (1965) 1.
43 R. W. F. Hardy, A. J. D'Eustachio and E. Knight, Jr., *Science*, 147 (1965) 310.
44 L. E. Mortenson, *Biochim. Biophys. Acta*, 127 (1966) 18.
45 E. Knight, Jr., A. J. D'Eustachio and R. W. F. Hardy, *ibid.*, 113 (1966) 626.
46 E. Knight, Jr. and R. W. F. Hardy, *J. Biol. Chem.*, 241 (1966) 2752.
47 E. Knight, Jr. and R. W. F. Hardy, *ibid.*, 242 (1967) 1370.
48 R. W. F. Hardy and E. Knight, Jr., *Biochem. Biophys. Res. Commun.*, 23 (1966) 409.
49 M. J. Dilworth, *Biochim. Biophys. Acta*, 127 (1966) 285.
50 R. Schöllhorn and R. H. Burris, *Proc. Natn. Acad. Sci. U.S.*, 57 (1967) 1317.
51 R. Schöllhorn and R. H. Burris, *ibid.*, 58 (1967) 213.
52 E. Moustafa and L. E. Mortenson, *Nature*, 216 (1967) 1241.
53 R. W. F. Hardy and E. Knight, Jr., *Biochim. Biophys. Acta*, 139 (1967) 69.
54 L. E. Mortenson, J. A. Morris and D. Y. Jeng, *ibid.*, 141 (1967) 516.
55 G. Daesch and L. E. Mortenson, *J. Bact.*, 96 (1967) 346.
56 R. W. Detroy, D. F. Witz, R. A. Parejko and P. W. Wilson, *Proc. Natn. Acad. Sci., U.S.*, 61 (1968) 537.
57 R. W. F. Hardy, R. D. Holsten, E. K. Jackson and R. C. Burns, *Pl. Physiol.*, 43 (1968) 1185.
58 H. Winter and R. H. Burris, *J. Biol. Chem.*, 243 (1968) 940.
59 P. T. Bui and L. E. Mortenson, *Proc. Natn. Acad. Sci. U.S.*, 61 (1968) 1021.
60 D. Y. Jeng, T. Devanathan and L. E. Mortenson, *Biochem. Biophys. Res. Commun.*, 35 (1969) 525.
61 E. Moustafa and L. E. Mortenson, *Biochim. Biophys. Acta*, 172 (1969) 106.
62 K. B. Taylor, *J. Biol. Chem.*, 244 (1969) 171.
63 P. T. Bui and L. E. Mortenson, *Biochemistry*, 8 (1969) 2462.
64 D. Y. Jeng, J. A. Morris and L. E. Mortenson, *J. Biol. Chem.*, 245 (1970) 2809.
65 J.-P. Vandecasteele and R. H. Burris, *J. Bact.*, 101 (1970) 794.
66 G. Nakos and L. E. Mortenson, *Biochemistry*, 10 (1971) 455.
67 G. Nakos and L. E. Mortenson, *Biochim. Biophys. Acta*, 229 (1971) 431.
68 L. E. Mortenson, in S. P. Colowick and N. O. Kaplan, *Methods in Enzymology*, Vol. 23, Academic Press, New York, 1972, in press.
69 W. A. Bulen, R. C. Burns and J. R. LeComte, *Biochem. Biophys. Res. Commun.*, 17 (1964) 265.
70 W. A. Buler, R. C. Burns and J. R. LeComte, *Proc. Natn. Acad. Sci. U.S.*, 53 (1965) 532.

71 R. C. Burns and W. A. Bulen, *Biochim. Biophys. Acta*, 105 (1965) 437.
72 W. A. Bulen, J. R. LeComte, R. C. Burns and J. Hinkson in A. San Pietro, *Non-Heme Iron Proteins: Role in Energy Conversion,* Antioch Press, Yellow Springs, Ohio, 1965, p. 261.
73 W. A. Bulen and J. R. LeComte, *Proc. Natn. Acad. Sci. U.S.*, 56 (1966) 979.
74 R. W. F. Hardy and E. Knight, Jr., *Biochim. Biophys. Acta*, 122 (1966) 520.
75 R. W. F. Hardy and E. K. Jackson, *Federation Proc.*, 26 (1967) 725.
76 M. Kelly, R. V. Klucas and R. H. Burris, *Biochem. J.*, 105 (1967) 3c.
77 M. Kelly, J. R. Postgate and R. L. Richards, *Biochem. J.*, 102 (1967) 1c.
78 G. W. Strandberg and P. W. Wilson, *Proc. Natn. Acad. Sci. U.S.*, 58 (1967) 1404.
79 J. C. Hwang and R. H. Burris, *Federation Proc.*, 27 (1968) 639.
80 E. K. Jackson, G. W. Parshall and R. W. F. Hardy, *J. Biol. Chem.*, 243 (1968) 4952.
81 M. Kelly, *Biochem. J.*, 107 (1968) 1.
82 R. J. Fisher and W. J. Brill, *Biochim. Biophys. Acta*, 184 (1969) 99.
83 J. R. Benemann, D. C. Yoch, R. C. Valentine and D. I. Arnon, *Proc. Natn. Acad. Sci. U.S.*, 64 (1969) 1079.
84 D. C. Yoch, J. R. Benemann, R. C. Valentine and D. I. Arnon, *ibid.*, 64 (1969) 1404.
85 R. I. Gvozdev, V. A. Yakovlev, V. R. Linde, L. V. Vorob'ev and E. Ya. Alfimova, *Isvestiya Biol. Ser.*, 2 (1969) 215.
86 S. Kajiyama, T. Matsuki and Y. Nosoh, *Biochem. Biophys. Res. Commun.*, 37 (1969) 711.
87 G. V. Novikov, L. A. Syrtsova, G. I. Likhtenshtein, V. A. Trukhtanov, V. F. Rachek and V. I. Gol'danskii, *Dokl. Akad. Nauk SSSR*, 181 (1969) 1170.
88 V. R. Linde, E. Ya. Alfimova, G. I. Slepko, A. M. Uzenskaia and G. I. Likhtenshtein, *Dokl. Phys. Chem. USSR*, 185 (1969) 208.
89 K. L. Hadfield and W. A. Bulen, *Biochemistry*, 8 (1969) 5103.
90 H. Dalton and J. R. Postgate, *J. Gen. Microbiol.*, 54 (1969) 463.
91 H. Dalton and J. R. Postgate, *ibid.*, 56 (1969) 307.
92 I. N. Ivleva, A. A. Medzhidov, G. I. Likhtenshtein, A. P. Sadkov and V. A. Yakovlev, *Biofizika*, 14 (1969) 639.
93 M. Kelly, *Biochim. Biophys. Acta*, 171 (1969) 9.
94 M. Kelly, *ibid.*, 191 (1969) 527.
95 R. W. F. Hardy and G. W. Parshall, *158th Natn. Mtg. Am. Chem. Soc.*, New York, Sept., 1969, Abstr. 226.
96 R. C. Burns, *Biochim. Biophys. Acta*, 171 (1969) 253.
97 J. Drozd and J. R. Postgate, *J. Gen. Microbiol.*, 60 (1970) 427.
98 M. G. Yates, *FEBS Lett.*, 8 (1970) 281.
99 M. G. Yates and R. M. Daniel, *Biochim. Biophys. Acta*, 197 (1970) 161.
100 M. G. Yates, *J. Gen. Microbiol.*, 60 (1970) 393.
101 J. Oppenheim, R. J. Fisher, P. W. Wilson and L. Marcus, *J. Bact.*, 101 (1970) 292.
102 J. Oppenheim and L. Marcus, *ibid.*, 101 (1970) 286.
103 W. H. Fuchsman and R. W. F. Hardy, *Bact. Proc.*, (1970) 148.
104 W. H. Fuchsman and R. W. F. Hardy, *Bioinorg. Chem.*, 1 (1972) 195.
105 R. C. Burns, R. D. Holsten and R. W. F. Hardy, *Biochem. Biophys. Res. Commun.*, 39 (1970) 90.
106 R. Silverstein and W. A. Bulen, *Biochemistry*, 9 (1970) 3809.
107 E. Moustafa, *Biochim. Biophys. Acta*, 206 (1970) 178.
108 G. J. Sorger and D. Trofimenkoff, *Proc. Natn. Acad. Sci. U.S.*, 65 (1970) 74.
109 W. A. Bulen and J. R. LeComte, in S. P. Colowick and N. O. Kaplan, *Methods in Enzymology*, Vol. 23, Academic Press, New York, 1972, in press.

110 R. C. Burns and R. W. F. Hardy, *ibid.*, in press.
111 R. C. Burns, K. T. Fry and R. W. F. Hardy, unpublished results.
112 R. C. Burns, W. H. Fuchsman and R. W. F. Hardy, *Biochem. Biophys. Res. Commun.*, 42 (1971) 353.
112a C. E. McKenna, J. R. Benemann and T. G. Traylor, *ibid.*, 41 (1970) 1501.
113 J. T. Stasny, R. C. Burns and R. W. F. Hardy, *Bact. Proc.*, (1971) 139.
114 D. R. Biggins and J. R. Postgate, *J. Gen. Microbiol.*, 56 (1969) 181.
115 F. J. Bergersen, *Biochim. Biophys. Acta*, 130 (1966) 304.
116 F. J. Bergersen and G. L. Turner, *ibid.*, 141 (1967) 507.
117 B. Koch, H. J. Evans and S. A. Russell, *Proc. Natn. Acad. Sci. U.S.*, 58 (1967) 1343.
118 B. Koch, H. J. Evans and S. A. Russell, *Pl. Physiol.*, 42 (1967) 466.
119 R. V. Klucas and H. J. Evans, *ibid.*, 43 (1968) 1458.
120 R. V. Klucas, B. Koch, S. A. Russell and H. J. Evans, *ibid.*, 43 (1968) 1906.
121 F. J. Bergersen and G. L. Turner, *J. Gen. Microbiol.*, 53 (1968) 205.
122 G. L. Turner and F. J. Bergersen, *Biochem. J.*, 115 (1969) 529.
123 F. J. Bergersen and G. L. Turner, *Biochim. Biophys. Acta*, 214 (1970) 28.
124 D. C. Yoch, J. R. Benemann, D. I. Arnon, R. C. Valentine and S. A. Russell, *Biochem. Biophys. Res. Commun.*, 38 (1970) 838.
125 B. Koch, P. Wong, S. A. Russell, R. Howard and H. J. Evans, *Biochem. J.*, 118 (1970) 773.
126 I. R. Kennedy, *Biochim. Biophys. Acta*, 222 (1970) 135.
127 H. J. Evans, B. Koch and R. V. Klucas, in S. P. Colowick and N. O. Kaplan, *Methods in Enzymology*, Vol. 23, Academic Press, New York, 1972, in press.
128 P. Wong, H. J. Evans, R. Klucas and S. A. Russell, *Plant and Soil*, Special Vol. (1971) 525.
129 M. C. Mahl and P. W. Wilson, *Can. J. Microbiol.*, 14 (1968) 33.
130 R. A. Parejko and P. W. Wilson, *ibid.*, 16 (1970) 681.
131 D. R. Biggins and M. Kelly, *Biochim. Biophys. Acta*, 205 (1970) 288.
132 M. Kelly and G. Lang, *ibid.*, 233 (1970) 86.
133 F. H. Grau and P. W. Wilson, *J. Bact.*, 85 (1963) 446.
134 D. F. Witz, R. W. Detroy and P. W. Wilson, *Arch. Mikrobiol.*, 55 (1967) 369.
135 R. J. Fisher and P. W. Wilson, *Biochem. J.*, 117 (1970) 1023.
136 H. C. Winter and D. I. Arnon, *Biochim. Biophys. Acta*, 197 (1970) 170.
137 D. C. Yoch and D. I. Arnon, *ibid.*, 197 (1970) 180.
138 R. C. Burns and W. A. Bulen, *Arch. Biochem. Biophys.*, 113 (1966) 461.
139 T. O. Munson and R. H. Burris, *J. Bact.*, 97 (1969) 1093.
140 R. V. Smith and M. C. W. Evans, *Nature*, 225 (1970) 1253.
141 W. D. P. Stewart and H. W. Pearson, *Proc. R. Soc.*, B175 (1970) 293.
142 W. D. P. Stewart, *Plant and Soil*, Special Vol. (1971) 377.
143 H. Dalton and L. E. Mortenson, *Bact. Proc.*, (1970) 148.
144 L. W. Gosser and C. A. Tolman, *Inorg. Chem.*, 9 (1970) 2350.
145 R. C. Burns and R. W. F. Hardy, *Federation Proc.*, 30 (1971) 1291.
146 J. A. Morris, H. Dalton and L. E. Mortenson, *Bact. Proc.*, (1969) 119.
147 R. D. Dua and R. H. Burris, *Proc. Natn. Acad. Sci. U.S.*, 50 (1963) 169.
148 D. Y. Jeng, T. Devanathan, E. Moustafa and L. E. Mortenson, *Bact. Proc.*, (1969) 119.
149 J. Weiher, R. C. Burns and R. W. F. Hardy, unpublished results.
150 D. C. Blomstrom, E. Knight, Jr., W. D. Phillips and J. Weiher, *Proc. Natn. Acad. Sci. U.S.*, 51 (1964) 1085.
151 K. Nauer, P. Hemmerich and J. D. W. Van Voorat, *Angew. Chem. (Int. Edn.)*, 6 (1967) 262.
152 B. A. Goldman and J. B. Raynor, *J. Chem. Soc. (A)*, (1970) 2038.

153 C. S. Yang and F. M. Huennekens, *Biochemistry*, 9 (1970) 2127.
154 F. N. Tebbe and G. W. Parshall, *J. Am. Chem. Soc.*, 93 (1971) 3793.
155 J. T. Spence, *Coordination Chem. Rev.*, 4 (1969) 475.
156 A. Kay and P. C. H. Mitchell, *J. Chem. Soc. (A)*, (1970) 2421.
157 L. R. Melby, *Inorg. Chem.*, 8 (1969) 349.
158 G. N. Schrauzer and G. Schlesinger, *J. Am. Chem. Soc.*, 92 (1970) 1808.
159 G. N. Schrauzer and P. A. Doemeny, *ibid.*, 93 (1971) 1608.
160 A. R. Dias, *Ph.D. Thesis*, Oxford University, 1970.
161 L. E. Mortenson, R. C. Valentine and J. E. Carnahan, *Biochem. Biophys. Res. Commun.*, 7 (1962) 448.
162 K. Tagawa and D. I. Arnon, *Nature*, 195 (1962) 537.
163 S. G. Mayhew, G. P. Foust and V. Massey, *J. Biol. Chem.*, 244 (1969) 803.
164 P. Wong, *Ph.D. Thesis*, Oregon State University, Corvallis, 1971.
165 I. R. Kennedy, J. A. Morris and L. E. Mortenson, *Biochim. Biophys. Acta*, 153 (1968) 777.
166 E. A. Shaede, B. P. Edwards and D. C. Walker, *J. Phys. Chem.*, 74 (1970) 3217.
167 P. K. Das, H. A. O. Hill, J. M. Pratt and R. J. P. Williams, *Biochim. Biophys. Acta*, 141 (1967) 644.
168 L. A. P. Kane-Maguire, P. S. Sheridan, F. Basolo and R. G. Pearson, *J. Am. Chem. Soc.*, 92 (1970) 5865.
169 G. E. Hoch, K. C. Schneider and R. H. Burris, *Biochim. Biophys. Acta*, 37 (1960) 273.
170 M. M. Mozen and R. H. Burris, *ibid.*, 14 (1954) 577.
171 R. G. S. Banks, R. J. Henderson and J. M. Pratt, *J. Chem. Soc. (A)*, (1968) 2886.
172 J. N. Armor and H. Taube, *J. Am. Chem. Soc.*, 91 (1969) 6874.
173 A. A. Diamantis and G. J. Sparrow, *Chem. Commun.*, (1970) 819.
174 A. A. Diamantis and G. J. Sparrow, *ibid.*, (1969) 469.
175 A. Yamamoto, S. Kitazume, T. S. Pu and S. Ikeda, *J. Am. Chem. Soc.*, 93 (1971) 371.
176 K. R. Laing and W. R. Roper, *J. Chem. Soc. (A)*, (1970) 2149.
177 K. R. Grundy, C. A. Reed and W. R. Roper, *Chem. Commun.*, (1970) 1501.
178 A. Misono, Y. Uchida, M. Hidai and T. Kuse, *ibid.*, (1969) 208.
179 P. C. Ford, *ibid.*, (1971) 7.
180 P. C. Ford, R. D. Foust, Jr. and R. E. Clarke, *Inorg. Chem.*, 9 (1970) 1933.
181 S. Otsuka, A. Nakamura and T. Yoshida, *J. Am. Chem. Soc.*, 91 (1969) 7196.
182 Y. Yamamoto and H. Yamazaki, *Bull. Chem. Soc. Japan*, 43 (1970) 2653.
183 Y. Yamamoto and H. Yamazaki, *J. Organometal. Chem.*, 24 (1970) 717.
184 B. Koch and H. J. Evans, *Pl. Physiol.*, 41 (1966) 1748.
185 C. Sloger and W. S. Silver, *Bact. Proc.*, (1967) 112.
186 W. D. P. Stewart, G. P. Fitzgerald and R. H. Burris, *Proc. Natn. Acad. Sci. U.S.*, 58 (1967) 2071.
187 J. P. Collman and J. W. Kang, *J. Am. Chem. Soc.*, 89 (1967) 844.
188 P. B. Tripathy and D. M. Roundhill, *ibid.*, 92 (1970) 3825.
189 N. T. Denisov, V. F. Shuvalov, N. I. Shuvalova, A. K. Shilova and A. E. Shilov, *Dokl. Akad. Nauk SSSR*, 195 (1970) 879,
190 C. Bradbeer and P. W. Wilson, in R. M. Hochster and J. H. Quastel, *Metabolic Inhibitors*, Vol. II, Academic Press, New York, 1963, p. 595.
191 P. W. Wilson, *Proc. R. Soc.*, B172 (1969) 319.
192 A. Yamamoto, L. S. Pu, S. Kitazume and S. Ikeda, *J. Am. Chem. Soc.*, 89 (1967) 3071.
193 M. Kelly, *Biochem. Biophys. Acta*, 109 (1968) 322.
194 D. C. Owsley and G. K. Helmkamp, *J. Am. Chem. Soc.*, 89 (1967) 4558.
195 J. Ellermann, F. Poersch, R. Kunstmann and R. Kramolowsky, *Angew. Chem. (Int. Edn.)*, 8 (1969) 203.

196 E. E. van Tamelen, G. Boche and R. Greeley, *J. Am. Chem. Soc.*, 90 (1968) 1677.
197 M. E. Winfield, *Rev. Pure Appl. Chem.*, 5 (1955) 217.
198 A. E. Shilov, *Kinet. Catalysis*, 11 (1970) 256.
199 M. O. Broitman, N. T. Denisov, N. I. Shuvalova and A. E. Shilov, *Kinet. Catalysis*, 12 (1971) 444.
200 H. Brintzinger, *Biochemistry*, 5 (1966) 3947.
201 H. Brintzinger, *J. Am. Chem. Soc.*, 89 (1967) 6871.
202 G. I. Likhtenshtein and A. E. Shilov, *Zh. Fiz. Khim.*, 44 (1970) 849.
203 S. Hünig and G. Büttner, *Angew. Chem. (Int. Edn.)*, 8 (1969) 451.
204 G. W. Parshall, *J. Am. Chem. Soc.*, 89 (1967) 1822.
205 J. Chatt, J. R. Dilworth, R. L. Richards and J. R. Sanders, *Nature*, 224 (1969) 1201.
206 C. A. Appleby, *Biochim. Biophys. Acta*, 189 (1969) 267.
207 J. D. Tjepkema and C. S. Yocum, *Pl. Physiol.*, 45 (1970) 44.
208 K. Abel, N. Bauer and J. T. Spence, *Arch. Biochem. Biophys.*, 100 (1963) 339.
209 C. A. Appleby, *Biochim. Biophys. Acta*, 180 (1969) 202.
210 N. Ellfolk and G. Sievers, *Acta Chem. Scand.*, 21 (1967) 1457.
211 F. J. Bergersen, *Biochim. Biophys. Acta*, 50 (1961) 576.
212 J. A. Cutting and H. M. Schulman, *ibid.*, 192 (1969) 486.
213 M. J. Dilworth, *ibid.*, 184 (1969) 432.
214 M. E. Volpin and V. B. Shur, *Dokl. Akad. Nauk SSSR*, 156 (1964) 1102.
215 A. Nielsen, *An Investigation of Promoted Iron Catalysts for the Synthesis of Ammonia*, Jul. Gjellerup Forlag, Copenhagen, 1968, p. 165.
216 G. W. Bridger, in *Catalyst Handbook*, Springer–Verlag, New York, 1970, p. 126.
217 S. Carra and R. Ugo, *J. Catalysis*, 15 (1969) 435.
218 R. Maskill and J. M. Pratt, *Chem. Commun.*, (1967) 950.
219 E. Bayer and V. Schurig, *Chem. Ber.*, 102 (1969) 3378.
220 E. E. van Tamelen, R. B. Fechter, S. W. Schneller, G. Boche, R. H. Greeley and B. Akermark, *J. Am. Chem. Soc.*, 91 (1969) 1551.
221 A. E. Shilov, A. K. Shilova and E. F. Kvashina, *Kinet. Catalysis*, 10 (1969) 1402.
222 M. E. Volpin, V. B. Shur and M. A. Ilatovskaya, *Izvest. Akad. Nauk, otdel. khim. nauk*, (1964) 1728.
223 M. E. Volpin, A. A. Belii and V. B. Shur, *Izvest. Akad. Nauk SSSR, Ser. Khim.*, (1965) 2196.
224 G. Henrici-Olive and S. Olive, *Angew. Chem. (Int. Edn.)*, 6 (1967) 873.
225 M. E. Volpin, M. A. Ilatovskaya, L. V. Kosyakova and V. B. Shur., *Chem. Commun.*, (1968) 1074.
226 E. E. van Tamelen and D. A. Seeley, *J. Am. Chem. Soc.*, 91 (1969) 5194.
227 E. E. van Tamelen, R. B. Fechter and S. W. Schneller, *ibid.*, 91 (1969) 7196.
228 A. E. Shilov and A. K. Shilova, *Zh. Fiz. Khim.*, 44 (1970) 288.
229 E. E. van Tamelen and H. Rudler, *J. Am. Chem. Soc.*, 92 (1970) 5253.
230 M. E. Volpin, V. B. Shur, R. V. Kudryavtsev and L. A. Prodayko, *Chem. Commun.*, (1968) 1038.
231 J. Dvorak, R. J. O'Brien and W. Santo, *ibid.*, (1970) 411.
232 L. G. Bell and H. Brintzinger, *J. Am. Chem. Soc.*, 92 (1970) 4464.
233 O. Glemser, Ger. Pat. 956,674 (1957).
234 G. P. Haight and R. Scott, *J. Am. Chem. Soc.*, 86 (1964) 743.
235 K. B. Yatsimirskii and V. K. Pavlova, *Dokl. Akad. Nauk SSSR*, 165 (1965) 130.
236 A. E. Shilov and A. K. Shilova, *Kinet. Catalysis*, 10 (1969) 1163.
237 K. Bahadur, S. Ranganayaki and L. Santamaria, *Nature*, 182 (1958) 1668.
238 A. D. Allen and C. Senoff, *Chem. Commun.*, (1965) 621.
239 F. Bottomley and S. C. Nyburg, *ibid.*, (1966) 897.
240 A. E. Shilov, A. K. Shilova and Yu. G. Borodko, *Kinet. Catalysis*, 7 (1966) 768.
241 A. K. Shilova and A. E. Shilov, *ibid.*, 10 (1969) 267.

242 D. E. Harrison and H. Taube, *J. Am. Chem. Soc.*, 89 (1967) 5706.
243 D. E. Harrison, E. Weissburger and H. Taube, *Science*, 159 (1968) 320.
244 E. L. Farquhar, L. Rusnock and S. J. Gill, *J. Am. Chem. Soc.*, 92 (1970) 416.
245 I. M. Treitel, M. T. Flood, R. E. Marsh and H. B. Gray, *ibid.*, 91 (1969) 6512.
246 J. Chatt, A. B. Nikolsky, R. L. Richards, J. R. Sanders, J. E. Ferguson and J. T. Love, *J. Chem. Soc. (A)*, (1970) 1479.
247 A. D. Allen, T. Eliades, R. O. Harris and P. Reinsalu, *Can. J. Chem.*, 47 (1969) 1605.
248 R. E. Clarke and P. C. Ford, *Inorg. Chem.*, 9 (1970) 227.
249 J. N. Armor and H. Taube, *J. Am. Chem.*, 92 (1970) 2560.
250 J. P. Collman and J. W. Kang, *ibid.*, 88 (1966) 3459.
251 J. Chatt, D. P. Melville and R. L. Richards, *J. Chem. Soc. (A)*, (1969) 2841.
252 L. Yu. Ukhin, Yu. A. Shvetsov and M. L. Khidekel, *Izvest. Akad. Nauk SSSR, Ser. Khim.*, (1967) 957.
253 A. Misono, Y. Uchida and T. Saito, *Bull. Chem. Soc. Japan*, 40 (1967) 700.
254 A. Yamamoto, S. Kitazume, L. S. Pu and S. Ikeda, *Chem. Commun.*, (1967) 79.
255 A. Sacco and M. Rossi, *ibid.*, (1967) 316.
256 M. Hidai, K. Tominari, Y. Uchida and A. Misono, *ibid.*, (1969) 1392.
257 P. W. Jolly and K. Jonas, *Angew. Chem. (Int. Edn.)*, 7 (1968) 731.
258 B. R. Davis, N. C. Payne and J. A. Ibers, *Inorg. Chem.*, 8 (1969) 2719.
259 A. Sacco and M. Aresta, *Chem. Commun.*, (1968) 1223.
260 C. H. Campbell, A. R. Dias, M. L. H. Green, T. Saito and M. G. Swanwick, *J. Organometal. Chem.*, 14 (1968) 349.
261 T. Ito, S. Kitazume, A. Yamamoto and S. Ikeda, *J. Am. Chem. Soc.*, 92 (1970) 3011.
262 W. H. Knoth, *ibid.*, 90 (1968) 7172.
263 G. M. Bancroft, M. J. Mays, B. E. Prater and F. P. Stefanini, *J. Chem. Soc. (A)*, (1970) 2146.
264 M. Hidai, K. Tominari, Y. Uchida and A. Misono, *Chem. Commun.*, (1969) 814.
265 C. C. Addison and B. M. Davies, *J. Chem. Soc. (A)*, (1969) 1822.
266 J. E. Bucher, *Ind. Eng. Chem.*, 9 (1917) 233.
267 E. E. Ferguson and W. F. Libby, *Nature*, 229 (1971) 37.
268 U. Landt, *Angew. Chem. (Int. End.)*, 9 (1970) 786.
269 R. D. Shannon and C. T. Prewitt, *Acta Cryst.*, 25 (1969) 925; B26 (1970) 1046.

PART VI

PORPHYRIN COMPOUNDS AND RELATED LIGANDS

Chapter 24

IRON PORPHYRINS—HEMES AND HEMINS

W. S. CAUGHEY

Department of Chemistry, Arizona State University, Tempe, Arizona, U.S.A.

INTRODUCTION

Iron porphyrins serve as prosthetic groups for an important class of proteins and enzymes known collectively as hemeproteins or hemoproteins[1-3]. These proteins can exhibit distinctly different functions which include reversible oxygen binding for transport (hemoglobins) or storage (myoglobins) of oxygen, oxygen reduction to the level of water (cytochrome *c* oxidase), mixed-function oxidation with oxygen (*e.g.*, cytochrome P_{450}), electron-transfer (cytochromes *b* and *c*), and hydrogen peroxide utilization (catalases and peroxidases). The prefixes *heme* and *hemo* are derived from the names for the iron porphyrin moieties—heme for iron(II) porphyrin and hemin for iron(III) porphyrin.

Hemeproteins differ in both polypeptide and heme components. Only a few differences in porphyrin structure are known whereas protein structures vary markedly as function and species change. Keen interest in structure–function relationships has accompanied the remarkably accelerating experimental advances. Here the hemes and hemins will be considered in terms of those structures, reactions, and physical properties that provide the bases for an understanding of hemeprotein function.

The unusually large number of observable variables of potential value for elucidation of structure and reaction mechanisms has made the hemeproteins particularly promising proteins for detailed study[4]. Thus variations found in the intensely colored, highly aromatic porphyrin ligand and in the iron atoms are characterized by changes in such properties as visible–u.v., Mössbauer, nuclear magnetic resonance (n.m.r.), electron paramagnetic resonance (e.p.r.), i.r., far i.r., and optical rotatory dispersion–circular dichroism (o.r.d.–c.d.) spectra and magnetic susceptibilities. However, successful interpretation of hemeprotein chemistry in terms of this array of independent variables is quite dependent upon reference to correlations of effects of porphyrin structure, axial ligand, oxidation state, and medium on similar properties of smaller, protein-free porphyrins where structure–property relationships can be established in greater detail.

References pp. 829–831

STRUCTURES OF PORPHYRINS OF NATURAL HEMES

Three particularly important hemes may be called heme *a*, heme *b*, and heme *c* since they are associated with the *a*, *b*, and *c* type cytochromes,

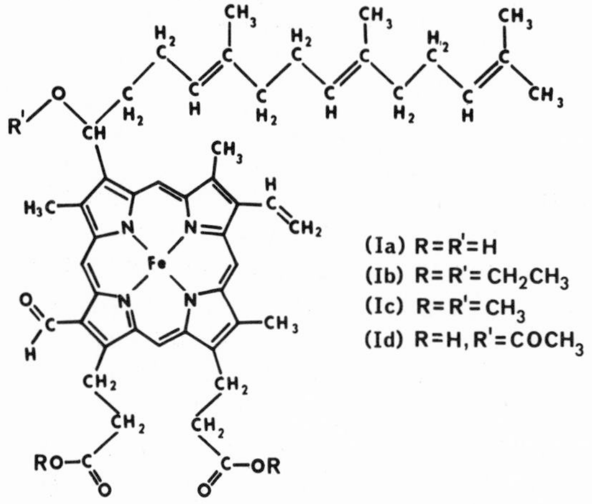

(Ia) R=R'=H
(Ib) R=R'= CH_2CH_3
(Ic) R=R'= CH_3
(Id) R=H, R'= $COCH_3$

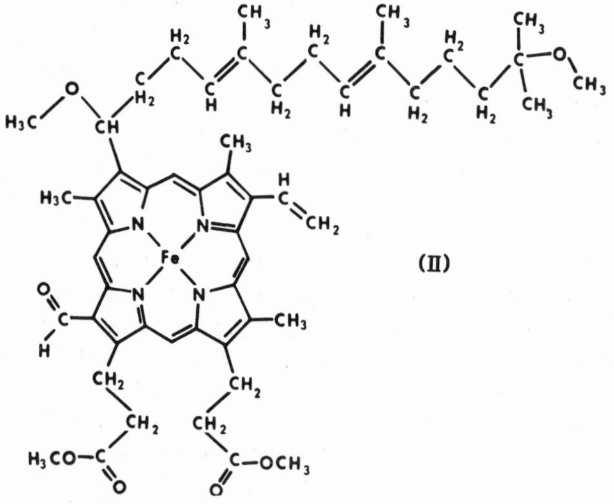

(II)

Fig. 1. Heme *a* (**Ia**) and derivatives. In **Ia**, R' is considered uncertain and could be designated X (see text).

respectively. In living systems essentially all the hemes found are combined with protein. That is, the concentrations of protein-free iron porphyrin are very low.

Heme *a* serves at the site of oxygen reduction in cytochrome *c* oxidase with a critical function in mitochondrial reduction of molecular oxygen for the production of cellular energy[1,2]. Although this heme may only be associated with one protein, cytochrome *c* oxidase (also known as cytochrome $a + a_3$), it is nevertheless extremely widely distributed over different species and tissues. The heme *a* structure (Fig. 1, Structure **Ia**) is based on our recent findings for heme *a* isolated from beef heart muscle[4–6]. This structure may also apply to hemes *a* in all species and tissues since visible and u.v. spectra for the *a*-type hemins in the mitochondria and in the cytochrome *c* oxidases are similar, but this possibility remains to be established. Such spectral evidence, while consistent with the presence of a formyl and probably a vinyl in each case, would not discriminate among other likely differences such as the length and degree of unsaturation of the long alkyl group. Present evidence also permits a difference in protein environment

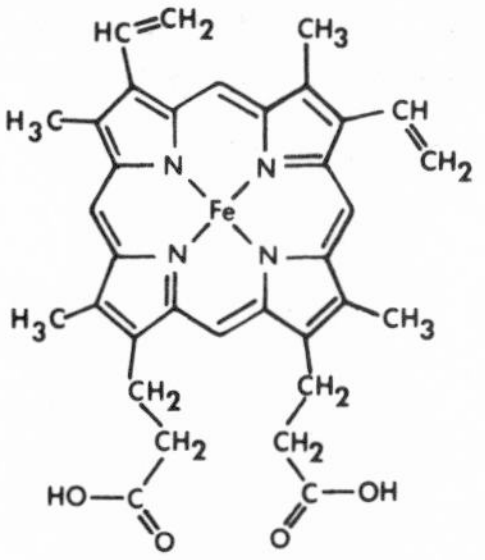

Fig. 2. Heme *b* or protoheme.

rather than a difference in porphyrin structure to serve as an explanation of the two types of heme *a* seen in the oxidase and classically ascribed to cytochrome *a* and cytochrome a_3[6]. Differences in bonding or environment at the 1′-hydroxy groups for the long alkyl group have been considered for the *a* and a_3 sites[6,7]. Indeed it is not clear what the 1′-substituent is in the intact oxidase and it is for this reason that we have frequently depicted the 1′-substituent as an OX group (see, *e.g.* p. 966, Chapter 27; also refs. 4 and 6). In any case the bonding between heme and aproprotein is sufficiently weak to allow extraction of the heme with reagents as mild as pyridine[5]. (The structure of heme *a* is discussed further on pp. 963–968, Chapter 27.)

Heme *b* is the most extensively studied natural heme. The structure (Fig. 2) has been known for more than 30 years[8]. It is readily isolated, commonly from the hemoglobin of beef blood, but is found in many proteins which include other hemoglobins, myoglobins, cytochrome P_{450}, catalases, and peroxidases as well as *b*-type cytochromes. The porphyrin ligand of heme *b* is protoporphyrin IX (Fig. 3); heme *b* is thus frequently called protoporphyrin IX iron(II), protoheme, or simply "heme". Similarly the oxidized or iron(III) form (hemin *b*) is called hemin or protohemin and, with a chloro axial ligand, hemin chloride, protohemin chloride or chloroprotoporphyrin IX iron(III) (Fig. 4). The "IX" refers to the specific

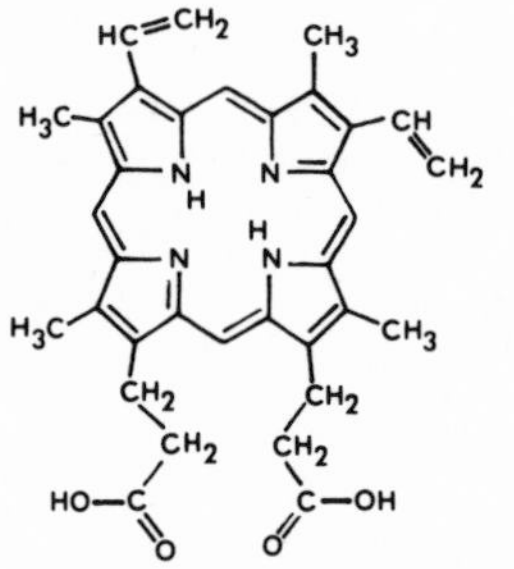

Fig. 3. Protoporphyrin IX (2,4-divinyldeuteroporphyrin IX).

arrangement of substituent groups on the porphyrin ring[8]. All other known natural hemins have many structural similarities to heme *b* presumably because protoporphyrin IX serves as precursor in the biosynthesis of the

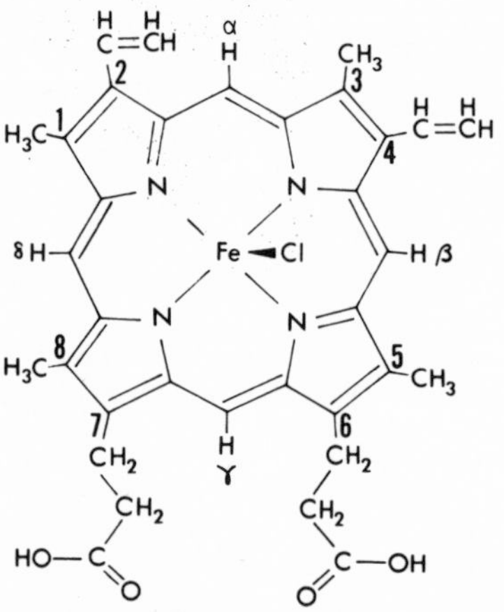

Fig. 4. Hemin *b* [chloroprotoporphyrin IX iron(III)]. The numbers 1,2,3, . . . 8 represent the β positions on the porphyrin ring, whereas α, β, γ, and δ are known as *meso* positions.

other porphyrin ligands. Heme *a* differs from heme *b* only by the 1-hydroxy-2-(*trans,trans*-farnesyl)ethyl and formyl groups at positions 2 and 8 in place of vinyl and methyl groups respectively.

Heme *c* (Fig. 5) has thioether linkages to cysteines at 1′-carbons of 2,4 ethyl groups with the remainder of the structure as in heme *b* (see also Fig. 2, Chapter 26). In cytochrome *c*, these two cysteine residues are incorporated in the polypeptide chains with two amino acid residues between the cysteines. The residues between and on either sides of the cysteines vary among cytochromes *c* from different species[9].

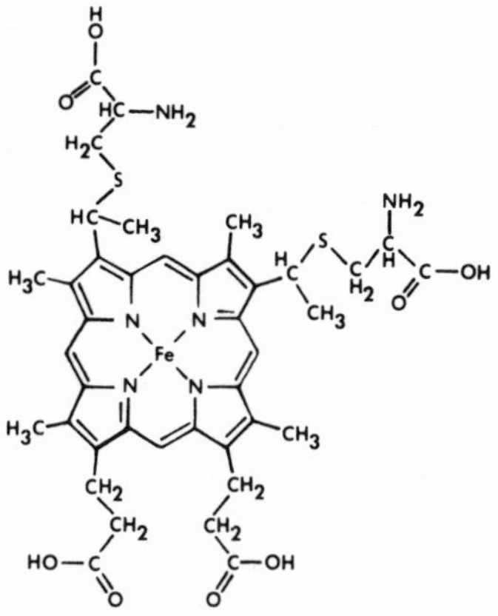

Fig. 5. Heme *c*.

Spirographis or chlorocruoro heme, which we shall call heme *s* is another well-characterized natural heme (Fig. 6). It is found in certain sea worms and differs from heme *b* only by a formyl in place of vinyl at position 2[8].

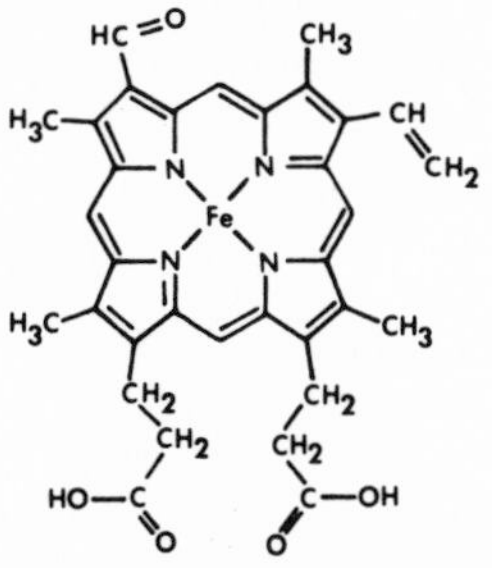

Fig. 6. Heme *s* (Spirographis heme).

Several other hemes are known, especially among micro-organisms, with still different structures or means of attachment to protein but structural details remain unclear.

DERIVATIVES OF NATURAL HEMES

Natural hemes have been modified at the carboxylic acid (*e.g.*, by esterification) and at other groups on the porphyrin ring as well as by removal of iron to give metal-free porphyrin or by replacement of iron by another metal[8]. Such modifications have been useful in structure elucidation, in altering solubility and aggregation characteristics, and in a general study of effects of structure on properties. The availability of these materials has permitted a more systematic evaluation of substituent effects than would be possible if studies were restricted to the limited number of natural porphyrins.

Metalloporphyrins have been prepared with many different metals[10,11]. However, present evidence suggests iron is the only metal other than magnesium with an important biological function as a porphyrin complex. Magnesium porphyrins do serve as intermediates on the biosynthetic pathway from protoporphyrin to the chlorophylls. Manganese porphyrins have been implicated, quite inconclusively, in photosynthesis. Copper uroporphyrin contributes to the color of feathers of Turaco and other birds but no biological function has been demonstrated for this copper complex. Nor do excreted zinc porphyrins appear to have an *in vivo* role. Crude petroleums contain vanadyl and nickel as well as iron porphyrins but there is no convincing evidence that either vanadium or nickel porphyrins have had or currently have a biological function.

Heme a derivatives

Derivatives that represent subtle modifications of native heme *a* were first obtained very recently. Compounds with the 6,7-propionic acids esterified and alkoxy groups at the 1′- or 1′- and 13′-positions on the long alkyl group proved easier to obtain in high purity and are more soluble in

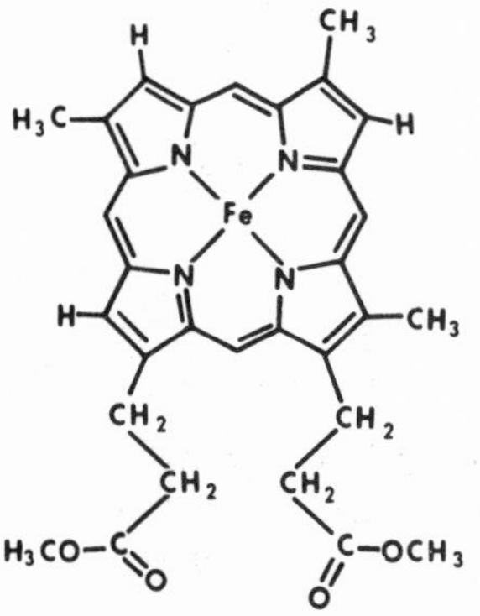

Fig. 7. Cytodeuteroheme dimethyl ester.

nonaqueous media than is heme *a* as isolated (Fig. 1)[4,6]. Such compounds are illustrated in structures **Ib–d** and **II** of Fig. 1. The 8-formyl group can be converted to an acetal during ester and ether formation[12]. The preparation of a number of other formyl derivatives and reduction products has been investigated but products of high purity have not been isolated[5,12-16]. An extensively modified derivative, cytodeuteroheme dimethyl ester (Fig. 7), has also been prepared from heme *a*[5,12-14,17].

Derivatives of heme b and heme s

Heme *b* and heme *s* have been converted to many other hemes and porphyrins which include deuteroheme (Fig. 8), mesoheme (Fig. 9), and

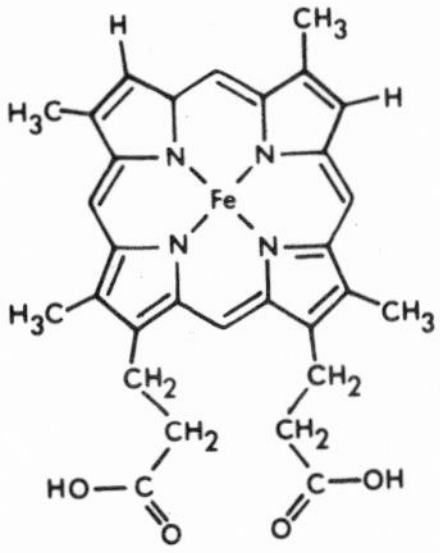

Fig. 8. Deuteroheme.

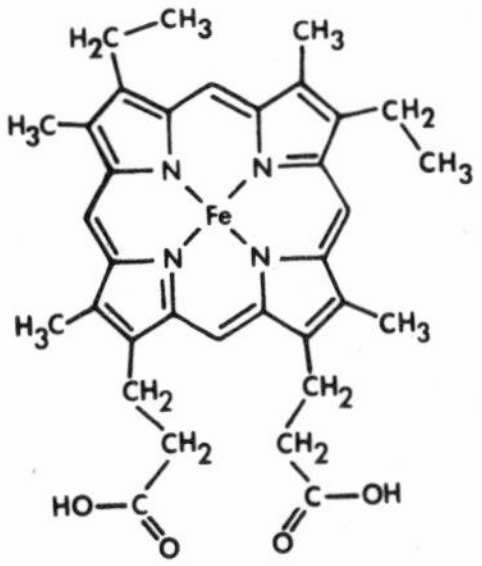

Fig. 9. Mesoheme.

hematoheme (Fig. 10) (see also Chapter 26, Fig. 2). Many derivatives have been prepared with different substituents at the 2,4,α and β positions on the porphyrin ring and at the propionic acid carboxyl group[8,18]. These compounds are conveniently termed derivatives of deuteroporphyrin IX. Thus, hemes *b* and *s* may be named as iron(II) complexes of 2,4-divinyldeuteroporphyrin IX and 2-formyl-4-vinyl-deuteroporphyrin IX respectively. The

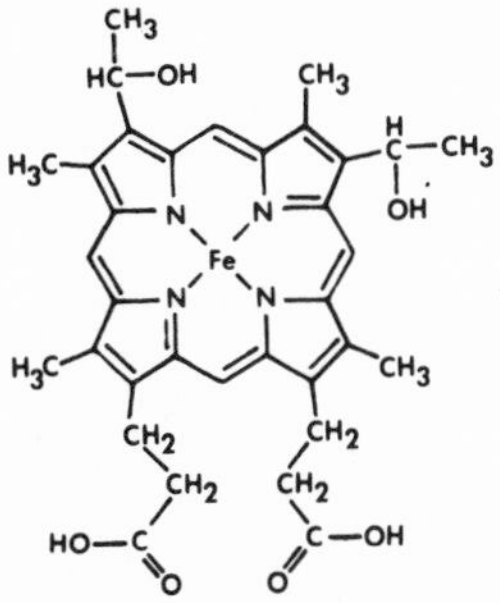

Fig. 10. Hematoheme.

prefix "deutero" was in use in porphyrin chemistry long before the discovery of ^{2}H and as used here does not indicate the presence of deuterium in the molecule.

Heme c derivatives

Heme *c* itself has not received much study. More attention has been given to heme peptides with many more amino acid residues which remain after degradation of the polypeptide of cytochrome *c*[19] (see Chapter 26, p. 904). Hematoheme (Fig. 10) in which all the amino acid residues are absent has been obtained directly from cytochrome *c* by cleavage of the thioether linkages[20].

Synthetic porphyrins

Porphyrins that bear close structural analogy to the natural porphyrins are generally more conveniently obtained via degradation of isolated natural

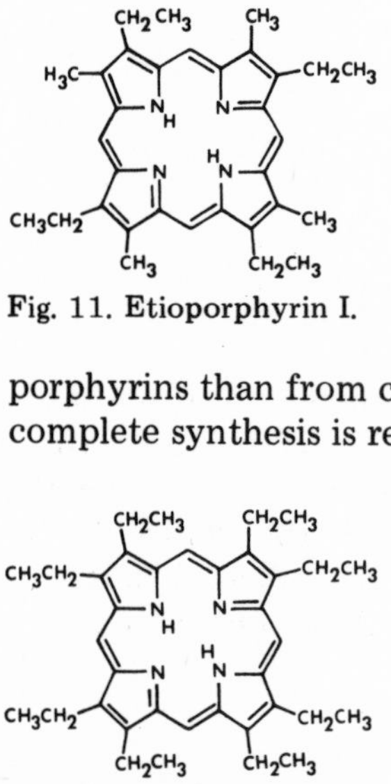

Fig. 11. Etioporphyrin I.

porphyrins than from complete synthesis. However, for several porphyrins complete synthesis is relatively simple and some of these compounds have

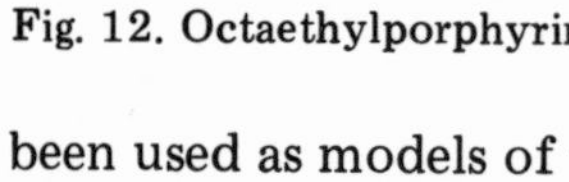

Fig. 12. Octaethylporphyrin.

been used as models of the natural systems. Examples are etioporphyrin I (Fig. 11), octaethylporphyrin (Fig. 12) and MS-tetraphenylporphin (TPP) (Fig. 13). A number of porphins with *meso* substituents other than the phenyls of TPP have been prepared[10,21]. Such *meso*-substituted porphins differ from the natural porphyrins in having substituents at the *meso* (*i.e.*, $\alpha, \beta, \gamma, \delta$) positions and no substituents at the β (*i.e.*, 1,2,3 . . . 8) positions

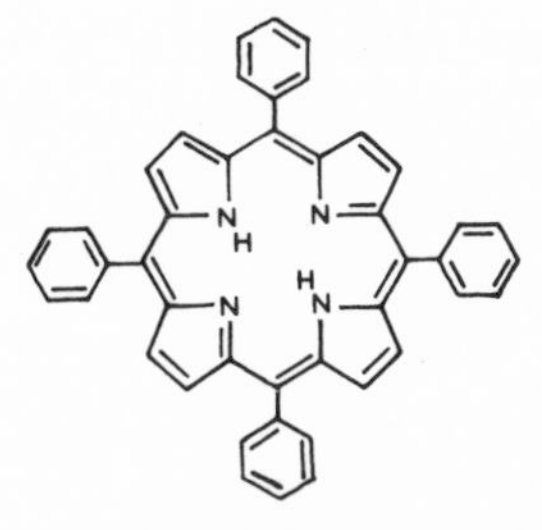

Fig. 13. MS-Tetraphenylporphin (α, β, γ, δ-tetraphenylporphine).

(see Fig. 4 for identification of positions). The natural porphyrins do not have *meso* substituents but do have β-substituents. Owing to both electronic and steric aspects of substituent effects it is usually desirable to use natural porphyrins or close structural analogs for studies where optimum biochemical relevance is desired.

EFFECTS OF PORPHYRIN STRUCTURE ON PROPERTIES

Differences in substituents on the periphery of the porphyrin ring of the type found among natural porphyrins have been shown to cause significant changes in the properties of iron and other metal porphyrins[3,4,22-25]. Electron (inductive) effects are now well established. Rather less evidence has been reported for steric effects although no doubt steric factors are frequently important especially in hemeproteins. The electron-withdrawing or electron-donating character of a substituent is effectively transmitted across the porphyrin to influence the basicity of the central nitrogens which in turn affects properties of the metal and exerts a *cis* effect upon axial ligands.

The basicity of the central nitrogens can be evaluated with metal-free porphyrins in terms of equilibria for protonation of the imine-type nitrogen of the neutral species (PH_2) to form the "acid salts" (PH_3 and PH_4).

N N N
H H

$$\text{NH}\quad\text{HN} + \text{H}^+ \underset{}{\overset{K_3}{\rightleftharpoons}} \text{NH} + \text{HN} + \text{H}^+ \underset{}{\overset{K_4}{\rightleftharpoons}} \text{NH} + + \text{HN}$$

H
N N N

PH_2 PH_3 PH_4

The determination of thermodynamically significant intrinsic equilibrium constants for the successive protonation steps (K_3 and K_4) for a series of differently substituted porphyrins has been complicated by aggregation, limited solubility, and field effects in aqueous media. However, consistent *relative* basicities have been observed in acetic acid, in chloroform, and in aqueous detergent solutions[23]. The aqueous detergent method gives pK_3 values presumably related to the equilibrium between neutral and mono-protonated species[24,25]. The more-effectively electron-withdrawing the substituents, the smaller is the pK_3 value (Table I). A similar order is found in other media[23].

TABLE I

BASICITY AND ABSORPTION MAXIMA DATA FOR SUBSTITUTED METAL-FREE DEUTEROPORPHYRIN IX DIMETHYL ESTERS[a]

Substituents	pK_3 ± 0.1[b]	λ_{max}(nm) in $CHCl_3$ at 30°C[c]	
2,4-ethyl	5.8	619	399.5
2,4-hydrogen	5.5	619	399
2,4-(2′-ethoxycarbonylcyclopropyl)[d]	4.8	623	403
2,4-vinyl	4.8	630	407.5
2,4-acetyloxime	4.5	625	406
2,4-oximino	4.3	641	418
2(and 4)-propionyl-4(and 2)hydrogen[d]	4.2	633	410.5
2(and 4)-formyl-4(and 2)vinyl[d]	3.75	642	420
2,4-acetyl	3.3	640	424.5
2,4-propionyl	3.2	637	422
α(and β)nitro-β(and α)hydrogen[d]	3.2	625	400
2,4-bromo	3.0	623	402
2,4-methoxycarbonyl	3.0	636	419.5
2,4-formyl	< 3.0	649	437
α,β-nitro	$\ll 3.0$	630	405.5

[a] From ref. 23.
[b] Determined in 2.5% aqueous sodium dodecylsulfate at 25°C by the method of Phillips[24]. See text for definition.
[c] Only two of several absorption maxima found in the visible and ultra-violet regions. Thus the 2,4-vinyl derivative (protoporphyrin IX dimethyl ester) also has maxima at 576, 541, 506 and 275 nm.
[d] Mixture of isomers, see ref. 18.

Cis effects on axial ligands follow the same substituent order found for nitrogen basicities. This first became clear in studies of equilibria between diamagnetic planar[26] nickel(II) porphyrins and pyridine or piperidine to give paramagnetic tetragonal dipyridine or dipiperidine species[27,28]. A series of 2,4-substituted nickel(II) deuteroporphyrin IX dimethyl esters in piperidine–chloroform solutions revealed striking differences in pK, ΔH, and ΔS owing to changes in porphyrin substituent (Table II)[28]. In general, the more basic is the porphyrin the smaller the proportion of liganded species found. However, the data (especially ΔS values) indicated inductive and steric effects on solvent interactions were important in addition to a

TABLE II

THERMODYNAMIC PARAMETERS FOR EQUILIBRIA FOR SUBSTITUTED DEUTEROPORPHYRIN IX DIMETHYL ESTER NICKEL(II), PIPERIDINE, AND DIPIPERIDINE NICKEL(II) COMPLEX[a]

2,4-Substituent	log K ± 0.02	ΔH ± 0.2 (kcal mole^{-1})	ΔS ± 0.7 (e.u.)
Ethyl	-2.04	-2.8	-18.8
Hydrogen	-1.69	-5.5	-26.2
2-Ethoxycarbonylcyclopropyl	-1.70	-4.3	-21.1
Vinyl	-1.39	-5.2	-23.8
Acetyl	-0.25	-7.8	-27.3
Formyl	+0.15	-7.8	-25.5

[a] In chloroform at 25°C. Data from ref. 28.

TABLE III

C-O STRETCH FREQUENCIES OF HEME CARBONYLS IN PYRIDINE-BROMOFORM AND IN RECONSTITUTED HEMOGLOBINS AND MYOGLOBINS

	ν_{CO} (cm^{-1})
In bromoform with 0.12 *M* pyridine:	
Substituted deuteroheme[30]	
2,4-ethyl	1973.0
2,4-hydrogen	1975.1
2,4-vinyl	1976.6
2,4-acetyl	1983.7
heme *a*[32]	1982.0
Hemoglobin A:	
Native[30]	1951.0
Reconstituted with substituted deuteroheme[a]	
2,4-vinyl	1951.0
2,4-hydrogen	1948.5
2,4-ethyl	1946.0
Myoglobin, sperm whale[33]:	
Native	1944 (~1933)
Reconstituted with substituted deuteroheme	
2,4-vinyl	1944 (~1933)
2,4-hydrogen	1941 (~1930)
2,4-ethyl	1939 (~1928)

[a] M. C. O'Toole, F. Wood, and W. S. Caughey, unpublished observations.

direct *cis* effect and, on occasion, substituent-axial ligand steric interactions are also important. Thus, although in these equilibria other factors (solvation, steric hindrance) are also significant, one important factor is the inverse relationship between porphyrin-to-metal and axial ligand-to-metal bonding; the stronger is one, the weaker is the other[22,27,28].

TABLE IV

QUADRUPOLE SPLITTING (ΔE) IN MÖSSBAUER SPECTRA FOR IRON 2,4-SUBSTITUTED DEUTEROPORPHYRIN IX DIMETHYL ESTERS

	Ligand	ΔE (mm sec^{-1})	
		4.6K[a]	298K[b]
Mesoporphyrin IX iron(III)	chloro	0.93	0.87
Deuteroporphyrin IX iron(III)	fluoro	0.78	0.59[c]
	chloro		0.91[c]
	bromo		1.12[c]
	iodo		1.30[c]
	μ-oxo		0.64[d]
	acetate	0.81	
	azido	0.83	
Protoporphyrin IX iron(III)	chloro	0.83	
Protoporphyrin IX iron(III)[e]	μ-oxo	0.62	
2,4-Diacetyldeuteroporphyrin iron(III)	chloro	0.89	0.79
	bromo	1.02	
Mesoporphyrin IX iron(II)	pyridine		1.27
Protoporphyrin IX iron(II)	pyridine	1.14	
2,4-Diacetyldeuteroporphyrin IX iron(II)	pyridine	1.17	1.22
2,4-Dibromodeuteroporphyrin IX iron(II)	pyridine		1.27

[a] Ref. 42. All values ±0.01 mm sec^{-1}. All samples enriched in ^{57}Fe.
[b] Determined in collaboration with Dr. J. Spijkerman on samples unenriched in ^{57}Fe. Values estimated ± 0.02.
[c] Ref. 22.
[d] Ref. 43.
[e] Chloroprotohemin in pyridine: 0.2 *N* aqueous NaOH (1:1, v/v) at 77K. (ref. 42). Subsequent observations by C. H. Barlow and W. S. Caughey, confirm the designation as a μ-oxo derivative.

Cis effects are also observed with iron porphyrins. For example, in heme carbonyls the CO stretching frequency (ν_{CO}) is sensitive to porphyrin structure[4,29-31]. As porphyrin basicity decreases, ν_{CO} increases (Table III). An increase in ν_{CO} indicates a higher C–O bond order, but a lower Fe–C bond order and therefore weaker CO bonding to heme; decreased porphyrin basicity renders iron(II) a poorer π-donor to CO. Similar effects on ν_{CO} are found in reconstituted myoglobins and hemoglobins A (Table III). Relative oxygen affinities of these reconstituted hemoglobins[34-36] and myoglobins[37] follow a similar order (vinyl < hydrogen < ethyl) which can be explained as an increase in strength of oxygen binding as the π-donor ability of the iron(II) is increased[3,29,31,38]. It was also found that in equilibria in which

TABLE V

OBSERVED ZERO-FIELD SPLITTINGS (Δ_1 AND Δ_2) AND DERIVED VALUES OF D AND λ FOR HEMINS, HEMOGLOBINS, AND MYOGLOBINS FROM FAR INFRA-RED MAGNETIC RESONANCE SPECTRA[a]

Compound	Ligand	Δ (cm^{-1})	D (cm^{-1})	λ
Deuteroporphyrin IX dimethyl ester iron(III)	Fluoro	Δ_1 = 11.1 ± 0.22	5.55 ± 0.11	~ 0
	Azido	Δ_1 = 14.8 ± 0.10	7.32± 0.05	0.036 ± 0.015
	Azido	Δ_2 = 29.2 ± 0.15		~ 0
	Chloro	Δ_1 = 17.9 ± 0.36	8.95 ± 0.18	~ 0
	Bromo	Δ_1 = 23.6 ± 0.46	11.8 ± 0.23	~ 0
	Iodo	Δ_1 = 32.8 ± 0.30	16.4 ± 0.15	~ 0
Protoporphyrin IX dimethyl ester iron(III)	Fluoro	Δ_1 = 10.0 ± 0.20	5.0 ± 0.10	~ 0
	Chloro	Δ_1 = 13.9 ± 0.28	6.95 ± 0.14	~ 0
	Azido	Δ_1 = 19.5 ± 0.30	9.10 ± 0.5	0.085 ± 0.025
		Δ_2 = 36.0 ± 0.75		
Methemoglobin (bovine)	Fluoro	Δ_1 = 12.60 ± 0.24	6.30 ± 0.12	
Metmyoglobin (sperm whale)	Fluoro	Δ_1 = 11.88 ± 0.16	5.94 ± 0.08	
2,4-Diacetyldeuteroporphyrin IX dimethyl ester iron(III)	Chloro	Δ_1 = 17.9 ± 0.36	8.9 ± 0.18	

[a]References 44 and 45. Temperatures 4.2 and 50K.

CO and pyridine compete for the iron(II) binding site, CO competes with pyridine more effectively as porphyrin basicity increases — a finding consistent with increased porphyrin basicity causing iron(II) to serve as a better π-donor to CO but a poorer σ-acceptor to pyridine[29,30]. Related

TABLE VI

ABSORPTION MAXIMA DATA FOR ELECTRONIC SPECTRA OF CHLORO IRON(III) 2,4-SUBSTITUTED DEUTEROPORPHYRIN IX DIMETHYL ESTERS IN CHLOROFORM AT 30°C

2,4-Substituents	λ_{max}, nm (ϵ_{mM})				
ethyl	918 (0.46)	635 (4.4)	534 (9.0)	507 (8.4)	379 (106)
vinyl	916 (0.55)	641 (5.0)	539 (9.9)	512 (10.0)	387 (100)
2-carboxyethyl-cyclopropyl	907 (0.48)	636 (4.7)	535 (9.7)	506 (9.9)	382 (95)
hydrogen	903 (0.47)	632 (4.1)	532 (8.9)	507 (8.6)	377 (89)
bromo	894 (0.50)	634 (4.7)	534 (9.5)	505 (9.6)	379 (97)
acetyl	890 (0.61)	643 (4.6)	545 (10.3)	516 (10.5)	419 (76)

explanations may be offered for slower rates with more electron-withdrawing substituents in autoxidation reactions of dipyridine hemes in solution and in thermal dissociation of pyridine from solid dipyridine hemes[29,39,40]. In

TABLE VII

ABSORPTION MAXIMA DATA FOR ELECTRONIC SPECTRA OF μ-OXO-BIS[2,4-SUBSTITUTED DEUTEROPORPHYRIN IX DIMETHYL ESTER IRON(III)] DERIVATIVES IN BENZENE AT 30°C[a]

2,4-Substituents	λ_{max}, nm (ϵ_{mM})[b]			
Hydrogen	s584 (10.1)	563 (12.9)	390 (120)	334 (69)
Ethyl	590 (11.0)	564 (13.4)	389 (121)	343 (80)
2′-Ethoxycarbonyl-cyclopropyl	592 (12.0)	568 (15.0)	393 (117)	344 (76)
Vinyl	599 (11.8)	573 (14.1)	397 (115)	s357 (75)
Propionyl	s626 (7.7)	580 (13.8)	415 (94.8)	340 (61)
Acetyl	s626 (8.2)	582 (13.9)	415 (95.5)	341 (63)

[a]Data of C. H. Barlow and W. S. Caughey, unpublished.
[b]s denotes "shoulder" where band maximum is only approximately determined.

the autoxidation reaction kinetic evidence indicates O_2, a π-acceptor as is CO, competes with pyridine for a site at the iron(II)[40] with this competition an important factor in overall rates. Substituent effects also appear to follow the same order for redox potentials[41]. Substituent effects in Mössbauer parameters[42,43] and zero-field splittings[44,45] have been noted but the range of data available is limited (Tables IV and V).

TABLE VIII

ABSORPTION MAXIMA DATA OF DIPYRIDINE IRON(II) 2,4-SUBSTITUTED DEUTEROPORPHYRIN IX DIMETHYL ESTERS IN PYRIDINE–BENZENE SOLUTIONS

2,4-Substituent	λ_{max}, nm (ϵ_{mM})			
	α-band	β-band	δ-band	Soret-band
Ethyl[39]	546 (31.0)	516 (16.3)	488 (11.7)	407 (131)
Hydrogen[39]	543 (21.3)	513.5 (13.6)	477 (10.4)	405 (120)
Vinyl[39]	555 (30.8)	523 (15.1)	473 (13.7)	419 (166)
Acetyl[39]	568 (18.5)	535 (15.1)		435 (150)
Bromo[23]	547 (1.0)[a]	518 (0.52)[a]	470 (0.38)[a]	409 (4.6)[a]
Hydrogen (α,β-nitro)[23]	549 (1.0)[a]	517 (0.93)[a]		407.5 (5.7)[a]

[a]Here number within parentheses represents the absorbance relative to the α band, *i.e.*, (A/A_α).

Electronic spectra serve as important criteria for porphyrin identification since these spectra vary with the nature and the relative ring positions of substituent groups (Tables I and VI–XI; Figs. 14 and 15)[8,11,23,46,47].

TABLE IX

ABSORPTION MAXIMA DATA FOR CARBONYLIRON(II) 2,4-SUBSTITUTED DEUTEROPORPHYRIN IX DIMETHYL ESTERS IN PYRIDINE–BENZENE SOLUTIONS[a]

2,4-Substituent	λ_{max}, nm (ϵ_{mM})		
	α-band	β-band	Soret band
Ethyl	557.5 (12.3)	528 (12.0)	408.5 (223)
Hydrogen	553.5 (8.8)	525.5 (10.0)	407 (202)
Vinyl	563.5 (14.3)	533.5 (13.8)	417 (172)
Acetyl		546 (12.5)	433.5 (141)

[a] Ref. 39.

However, shifts in electronic spectra do not necessarily correlate strictly with the extent of inductive interactions of substituents[23]. In many cases, but not always, electron-withdrawing substituents shift absorption maxima to longer wavelengths. Resonance or π-type substituent interactions affect

TABLE X

ABSORPTION MAXIMA DATA FOR PIPYRIDINE HEMES IN PYRIDINE–AQUEOUS SODIUM HYDROXIDE SOLUTIONS

Heme	λ_{max}, nm (ϵ_{mM})			
	α-band	β-band		Soret band
Heme *a*	587 (27.4)[a]			
	587 (28.5)[b]	535 (6.5)[b]		430 (131)[b]
Heme *b*	558 (30.6)[c]	526.5 (17.0)[c]	478 (12.3)[c]	419 (157)[c]
	558 (30.9)[d]	525 (16.3)[d]		418 (130)[d]
Heme *c*	551 (29.1)[e]	552 (18.6)[e]		
Heme *s*	580.5[f]	532–534[f]		434[f]
	583.1[g]	541.1[g]		

[a] Cytohemin chloride in pyridine–aq. NaOH–$Na_2S_2O_4$. Ref. 14.
[b] Hemin A chloride in pyridine–aq. NaOH–$Na_2S_2O_4$. Ref. 5.
[c] Ref. 79. [d] Ref. 80. [e] Ref. 81. [f] Ref. 82. [g] Ref. 83.

electronic transitions to a much greater extent than σ-type interactions[23]. For example, replacement of an ethyl by an acetyl causes a marked decrease in porphyrin basicity and in frequencies of absorption maxima whereas replacement of ethyl by bromo results in a comparable change in basicity but, in general, wavelengths of absorption maxima shift only slightly Tables I, VI, VIII and XI). This difference in behavior can be explained in

TABLE XI

WAVELENGTHS AND ABSORPTIVITIES OF SORET BAND MAXIMA OF NICKEL(II) 2,4-SUBSTITUTED DEUTEROPORPHYRIN IX DIMETHYL ESTERS IN CHLOROFORM AND IN PIPERIDINE[a]

2,4-Substituent	λ_{max}, nm (ϵ_{mM})	
	In chloroform[b]	In piperidine[c]
Ethyl	392 (207)	419 (259)
Hydrogen	391 (206)	418 (228)
Hydrogen (α,β-nitro)	397 ([d])	d
Bromo	395 ([d])	d
2-Ethoxycarbonylcyclopropyl	395.5 (212)	423 (224)
Vinyl	401 (192)	431 (254)
Acetyl	417.5 (132)	448 (175)
Formyl	428.5 (158)	459.5 (200)

[a] Refs. 23 and 28.
[b] Diamagnetic planar species.
[c] Paramagnetic tetragonal dipiperidine species.
[d] Not determined.

terms of the acetyl and bromo groups exerting comparable electron-withdrawing effects but bromo has much less π-interaction than does acetyl. Similarly, introduction of nitro groups at the α and β positions of deuteroporphyrins results in a much less basic porphyrin but only small wavelength shifts (Tables I, VIII and XI). Data for heme esters and natural hemes are in Tables VIII–X. Other data are available, *e.g.*, in Falk's treatise[46]. Throughout, the order of substituent effects upon wavelengths is generally similar for all derivatives — metal-free, iron(II), iron(III), or other metal. However, meaningful correlations or other use of reported spectral data require careful consideration of several factors which include solvent effects (Fig. 16), concentration (aggregation) dependent changes, the axial ligand(s) actually present, oxidation and spin states. All too frequently adequate proof of purity and identity of the species under consideration is absent in published accounts.

The theory of porphyrin electronic spectra has been discussed[47–49] but is beyond our scope here. The theory of iron(III) porphyrin spectra has been particularly difficult[49]. An aspect of bioinorganic interest is the effect of structure on the "near infra-red band" found between 1200 and 650 nm in high-spin iron(III) porphyrins (Tables VI and XII) and in heme proteins. This weak band has been ascribed, at least in part to electron transfer from porphyrin to iron(III). Indeed, the data of Table VI may be considered consistent with a process of electron-donation by the porphyrin since the energy of the transition increases with decreasing tendency of the 2,4-substituent to donate electrons. However, for this correlation to be

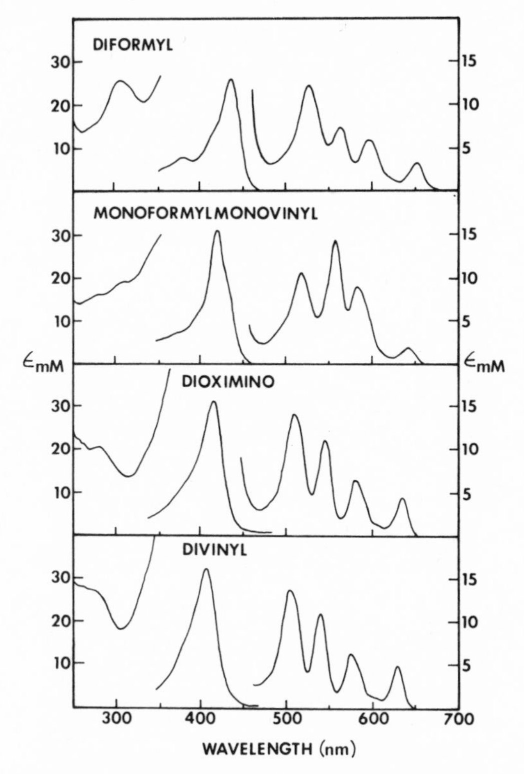

Fig. 14. Absorption spectra for dimethyl esters of 2,4-diformyldeuteroporphyrin IX, 2-(and 4-) formyl -4(and 2-) vinyldeuteroporphyrin IX, 2,4-dioximinodeuteroporphyrin IX, and protoporphyrin IX (top to bottom) in chloroform at 30°C. In the Soret region, the ϵ_{mM} range is 0–200. Ref. 23.

valid, it is necessary to assume that in an electron demand situation such as this the vinyl and cyclopropyl groups can serve as a better electron-donor than can hydrogen[3]. The wavelength shifts for band maxima other than the near infra-red band do not appear to follow this order.

Sufficient data have now been obtained on substituent effects with structural analogs to permit fairly accurate estimates of the chemical result from several of the structural differences found among the natural hemes *a*, *b*, *c* and *s*. Thus replacement of an 8-methyl or 2-vinyl by a formyl as is the case for heme *a* or heme *s*, respectively, compared with heme *b* has the effect of introducing a much more strongly electron-withdrawing substituent. Other reactions of the formyl (*e.g.* with protein) have been suggested but certainly not established. A particularly striking difference between hemes *a* and *b* is the unsaturated C_{17} side chain *vs.* a vinyl. Possible roles for heme *a* in the electron-transport and oxidative phosphorylation functions of cyto-

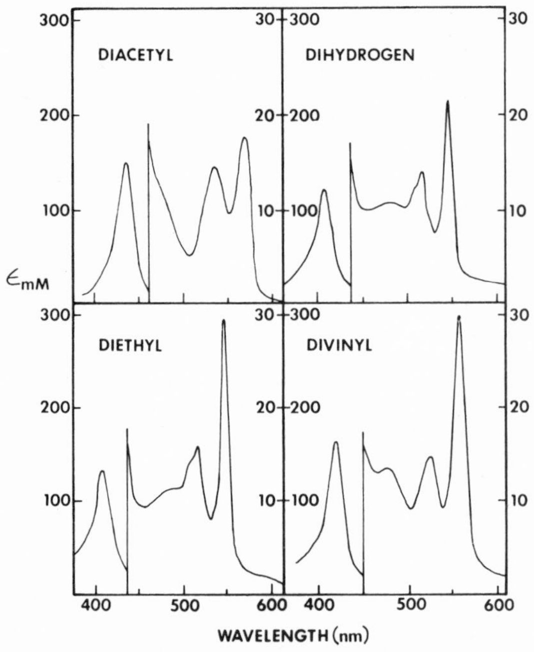

Fig. 15. Absorption spectra for dipyridine 2,4-substituted-deuteroporphyrin IX iron(II) dimethyl esters in benzene with 1*M* pyridine. (Ref. 39). Upper left: 2,4-acetyl. Upper right: 2,4-hydrogen (deuteroheme). Lower left: 2,4-ethyl (mesoheme). Lower right: 2,4-vinyl (protoheme).

chrome *c* oxidase merit consideration[4,6,7]. The *trans–trans*-farnesylethyl groups may form conformations in which the isolated double bonds can couple with each other and with porphyrin to serve in electron or energy transfer to or from the porphyrin ring or in which the terminal C_{12}–C_{13} double bond can participate in reactions at iron[4]. The inductive effect of the C_{17} group with either a free or bound OH at $C_{1'}$ is not clear. Similarly the inductive effect of 2,4-ethyl groups with thioether linkages as in heme *c* has not been well quantitated. It is clear from comparison of mesoheme and heme *c* that the presence of thioether results in little change in visible and Soret spectra compared with ethyl groups. Vinyl groups may serve more effectively as either donor or acceptor than an ethyl group.

Two common oxidation states for iron porphyrins are iron(II) and iron(III) with both high and low spin states known in each oxidation state. Also, iron(I) and iron(IV) species have been discussed as possible participants in certain redox reactions. High-spin iron(III) species have been studied most extensively.

The reactions and physical properties of iron porphyrins are greatly influenced by the nature of axial ligand(s) and of solvent medium. Studies

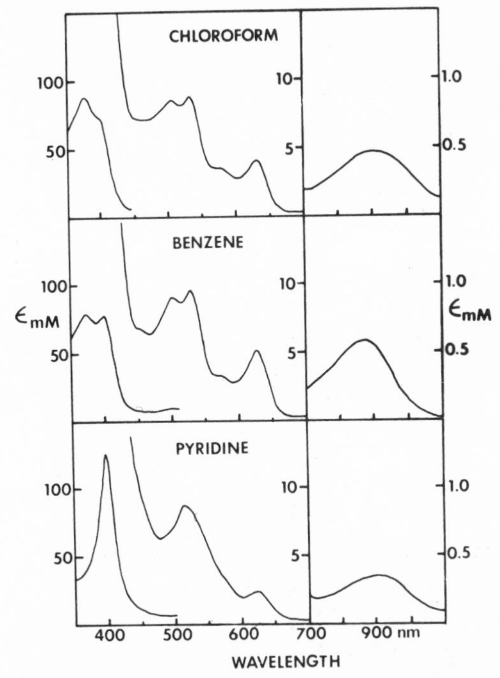

Fig. 16. Absorption spectra of chloro deuteroporphyrin IX dimethyl ester iron(III) in chloroform (top), benzene (middle) and pyridine (bottom).

of these properties in solution are often complicated by aggregation or association of monomers and by ligand dissociation or exchange. Since these phenomena can result in mixtures of species, it becomes necessary, though frequently difficult, to establish the structures of species actually present if meaningful structure–property correlations are to be established. Unfortunately, for many studies reported in the literature there remains a significant degree of uncertainty as to the species involved. Infra-red and n.m.r. spectroscopic approaches are most useful in the identification of ligands in porphyrins and in hemeproteins[4]. Identification of ligands is a critical problem with hemeproteins although association phenomena are rather different in the proteins compared with the protein-free iron porphyrins.

References pp. 829–831

TABLE XII

ABSORPTION MAXIMA DATA FOR ELECTRONIC SPECTRA OF DEUTEROPORPHYRIN IX DIMETHYL ESTER IRON(III) WITH DIFFERENT AXIAL LIGANDS IN BENZENE AT 30°C[a]

Ligand	λ_{max}, nm (ϵ_{mM})							
Phenoxo	769 (0.41)	588 (7.1)	564 (7.0)	s542 (6.5)[b]	477 (9.9)		392 (72)	337 (35)
Fluoro	776 (0.56)[c]	587 (9.0)	s557 (6.5)	s511 (7.8)	472 (10.8)		393 (98)	337 (36)
Azido	848 (0.55)	623 (4.9)	570 (4.6)	524 (9.65)	497 (11.3)	s451 (9.4)	394 (63)	362 (58)
Cyanato	867 (0.55)	625 (4.7)	570 (4.0)	527 (8.1)	499 (8.3)	s451 (8.3)	399 (82)	357 (53)
Thiophenoxo	872 (0.66)	629 (5.3)	s577 (3.5)	530 (10.0)	504 (9.7)	s454 (7.2)	400 (75)	373 (67)[e]
Chloro	877 (0.59)	628 (4.9)	s575 (3.3)	530 (9.2)	504 (8.7)	s453 (6.7)	401 (69)	373 (71)
Bromo	896 (0.53)	634 (4.2)	s583 (3.2)	533 (8.9)	503 (9.2)[d]	s457 (8.2)	383 (76)	s358 (56)
Iodo	920 (0.55)	641 (4.9)	s585 (4.3)	531 (10.8)	513 (10.7)	s461 (8.0)	s393 (67)	368 (76)
Thiocyanato	954 (0.55)	637 (4.0)	s583 (3.4)	s529 (13.9)	508 (15.5)	s460 (8.2)	387 (82)	s359 (56)
Selenocyanato	997 (0.49)	637 (6.5)	s583 (4.3)	540 (13.2)	s511 (10.6)	s458 (7.1)	393 (86)	s370 (65)

[a] Hemin conc. ~ 10^{-4} *M*. Data of H. Eberspaecher, C. H. Barlow, and W. S. Caughey, unpublished.
[b] s denotes "shoulder" where band maximum is only approximately determined.
[c] Also, s673 (0.48).
[d] Also, s486 (8.7).
[e] Also, s353 (61).

Association and ligand exchange phenomena which result in multiple and/or unknown species are particularly prevalent in aqueous media[46,50-53]. For example, several different species of protohemin are observed in aqueous media as pH and other solvents are varied[50]. The structures of many of these species remain poorly understood. Some forms are extensively aggregated; one species appears to be μ-oxo dimer. The formation of

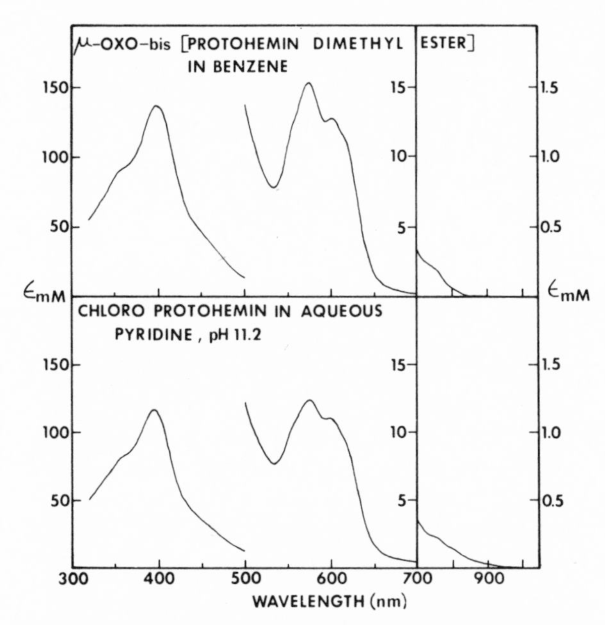

Fig. 17. Absorption spectra of μ-oxo-bis[protoporphyrin IX dimethyl ester iron(III)] (5×10^{-5} *M*) in benzene (upper curve) and of a solution of protohemin chloride (5×10^{-5} *M* as dimer) in aqueous sodium hydroxide with 10% pyridine at pH 11.2.

hemin dimers in aqueous alkali was recognized by Shack and Clark[50] who also reported a change in visible spectra upon addition of pyridine, a change they attributed to the formation of a new dimeric species for which an FeOFe linkage has been suggested[51]. The presence of such a linkage received some support from Mössbauer[42] and infra-red[43,52] spectra. Electronic spectra of authentic μ-oxo-bis[protohemin dimethyl ester] and of hemin dicarboxylic acid in aqueous alkaline pyridine compare favorably (Fig. 17) with any differences attributable to the fact that aqueous media tend to give somewhat broader and less well-resolved bands of lower intensity than do benzene solutions[54]. E.p.r. spectra at 77K are also similar. Aqueous alkaline solutions of chlorohemin at 77K exhibit typical high spin e.p.r. spectra which nearly disappear upon addition of pyridine[54]. Evidently the pyridine exerts a solvation or disaggregation effect and need not serve as a ligand to iron in that pyridines with substituents so bulky as to preclude nitrogen binding to metal are nevertheless effective in causing formation of the FeO–Fe species[53]. We have suggested double binuclear bridging with hydroxo ligands as a reasonable intermediate species (III) on the route from hemin hydroxide to μ-oxo dimer[43,72]:

$$2\,FeOH \rightleftharpoons Fe\underset{\underset{H}{O}}{\overset{\overset{H}{O}}{\diamond}}Fe \rightleftharpoons Fe{-}O{-}Fe + H_2O$$

(III)

Such species (III) are also likely to be present in alkaline hemin solutions. Hemin hydroxides (hematins) have not been demonstrated as pure monomeric solids although they may well be present in solutions to some extent. Associations involving bonding interactions of the donor–acceptor π–π type are well known in n.m.r. studies of diamagnetic porphyrins[4,55] and probably are also important with these iron porphyrins. The extent of donor–acceptor association is critically influenced by porphyrin structure[4,55] and appears important in heme–apoprotein interactions[3]. Since the nature of many species found in aqueous media remain unclear we shall restrict our discussion here mainly to systems where the only species, or at least the dominant species, present is well-characterized. Usually this means systems of iron porphyrin esters in non-aqueous media or of solids. However, non-aqueous systems are not as non-physiological as they may at first seem in that the environment of the heme moieties within hemeproteins appears to be largely that of neutral protein amino acid residues with only minor exposure to the outside aqueous medium.

Hemins with one axial ligand

With one axial ligand iron(III) porphyrins are square pyramidal structures. The iron atom in such structures may be expected to be out of the plane defined by the porphyrin nitrogens in the direction of the axial ligand. Crystal structures for chloro protohemin revealed the iron to be 0.47 Å out-of-plane toward the chlorine[56] and for methoxomesohemin dimethyl ester the distance was 0.45 Å toward the methoxide (Fig. 18)[57]★. In the solid and in solutions where the solvent does not serve as an additional (sixth) ligand, these compounds appear as essentially high-spin ($S = \frac{5}{2}$) compounds. Thus magnetic susceptibility measurements approach the spin-only value of five unpaired electrons[10,41], electron spin resonance spectra at 77K or below have a major absorption at $g \approx 6$ and a minor peak at $g \approx 2$[42,58], n.m.r. spectra exhibit broadening and paramagnetic

★See Chapter 25 for further discussion of the out-of-plane iron.

shifts[59,60], zero-field splittings directly measured in the far infra-red range from 0 to 50 cm^{-1} [44,45], and highly characteristic Mössbauer[42,61] and electronic spectra[8,46,55] are obtained. Each of these physical properties varies systematically with change in axial ligand.

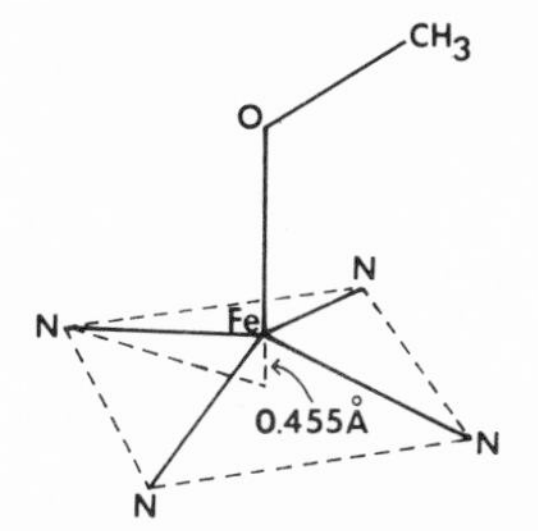

Fig. 18. Representation of the stereochemistry about iron in methoxo mesoporphyrin IX dimethyl ester iron(III). Ref. 57.

Proton n.m.r. spectra for high-spin hemins exhibit large paramagnetic shifts to high or low field with the extent of shift for most, but not all, protons significantly influenced by the nature of the axial ligand. Chemical

TABLE XIII

^{1}H CHEMICAL SHIFTS FOR HIGH-SPIN DEUTEROHEMIN ESTERS[a]

Ligand	2,4-H	Ring-CH_3[b]	6,7α-CH_2[c]	*Meso*-H[d]
Phenoxo	-75	-39	— — -31 (2)	38
Fluoro	-76	-43	— — -33 (2)	35
Azido	-72	-46	-41 (1) -39 (1) -35 (2)	46
Chloro	-75	-49	-44 (1) -42 (1) -39 (2)	57
Bromo	-72	-51	-45 (1) -42 (1) -38 (2)	57

[a] Ppm from $(CH_3)_4$ Si; negative values downfield, positive values upfield. In $CDCl_3$. Ref. 60.
[b] Chemical shift at maximum peak height.
[c] Numbers in parentheses represent proton counts.
[d] Chemical shifts for the broader *meso*-H absorptions were less accurately measured.

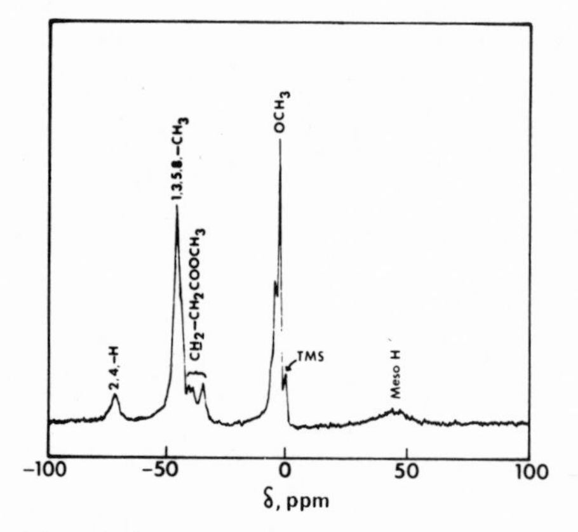

Fig. 19. ^{1}H n.m.r. spectra at 100 MHz for azido deuteroporphyrin IX dimethyl ester iron(III) in chloroform at 35°C.

TABLE XIV

^{1}H CHEMICAL SHIFTS FOR IRON(II) 2,4-SUBSTITUTED DEUTEROPORPHYRIN IX DIMETHYL ESTERS IN $CDCl_3$-N_2D_4[a]

2,4-Substituent	6,7αCH_2	Ring-CH_3	OCH_3	6,7-βCH_2	*Meso*-H
Ethyl	3.28	3.51	3.77	4.27	9.45 (γ) 9.48 (α,β,δ)
Vinyl	3.25	3.5 (5,8) 3.6 (1,3)	3.75	4.25	9.43 (γ) 9.56 (δ) 9.65 (β) 9.73 (α)
Hydrogen	3.25	3.5 (5,8) 3.55 (1,3)	3.75	4.25	9.4 (α,β,δ) 9.5 (δ)
Bromo	3.23	3.5 (5,8) 3.55 (1,3)	3.75	4.23	9.46 (γ) 9.53 (δ) 9.65 (β) 9.67 (α)
Propionyl	3.24	3.47 (5,8) 3.83 (1,3)	3.75	4.20	9.35 (γ) 9.56 (δ) 10.23 (β) 10.47 (α)
Acetyl	3.25	3.45 (5,8) 3.83 (1,3)	3.75	4.18	9.33 (γ) 9.56 (δ) 10.26 (β) 10.51 (α)

[a] M. C. McDaniel, D. H. O'Keeffe, and W. S. Caughey, unpublished.

shifts for a series of deuterohemin dimethyl esters with different axial ligands in $CDCl_3$ are given in Table XIII, a typical spectrum is in Fig. 19[60]. All porphyrin protons, except for the relatively isolated ester $-OCH_3$ and β-CH_2 of the 6,7-propionic acid ester groups, were shifted well outside the region for diamagnetic metal porphyrins (Table XIV and Fig. 20)[4,55].

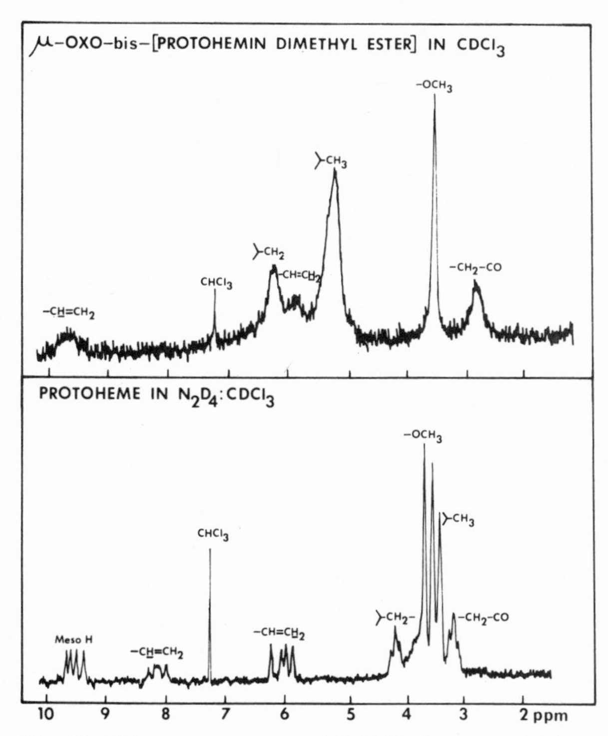

Fig. 20. ^1H n.m.r. spectra at 100 MHz for μ-oxo-bis[protoporphyrin IX dimethyl ester iron(III)] in $CDCl_3$ (upper curve) and for protoporphyrin IX dimethyl ester iron(II) in a mixture of $CDCl_3$ (0.4 ml) and $N_2D_4 \cdot D_2O$ (0.06 ml). Delta values downfield from $(CH_3)_4Si$.

Only *meso*-protons were shifted to high-field where they appeared as an especially broad absorption. With the phenoxo derivative protons of the axial ligand also experience large paramagnetic shifts with resonances at 97 and −83 ppm assigned to 2,6- and 3,5-phenyl protons respectively. The differences in chemical shift compared with diamagnetic porphyrins for ring methyl, α-CH_2, and *meso*-protons were particularly sensitive to a change in axial ligand. These data do not permit discrimination between contact and pseudo-contact shifts nor have possible association effects been evaluated but they are consistent with an important contribution from the transmission of isotropic hyperfine interactions from the iron to protons at the periphery of the porphyrin ring. The magnitude of resultant contact shifts is expected to depend upon the degree and type of bonding interactions between iron and the porphyrin nitrogens which in turn are influenced by differences in bonding between iron and axial ligand. That solvent and/or *trans* ligand effects may also be important is shown by data for chlorohemins in $(CD_3)_2SO$ where no high field shifts were found[59]; the

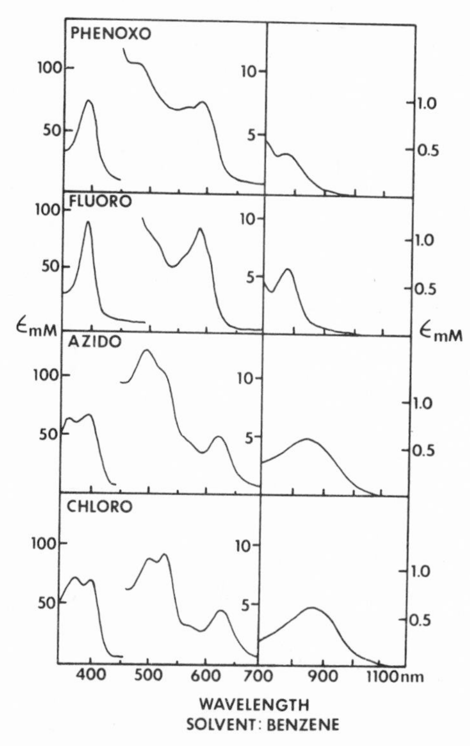

Fig. 21. Absorption spectra for deuteroporphyrin IX dimethyl ester iron(III) with different axial ligands in benzene.

difference between spectra in $CDCl_3$ and in $(CD_3)_2SO$ remains to be explained but it is likely the $(CD_3)_2SO$ serves as a ligand and/or displaces chloride whereas $CDCl_3$ does not.

Zero-field splittings (Δ_1) have been determined directly by far infra-red spectroscopy for several hemins (Table V)[44,45]. Mössbauer spectra have also given approximate values of Δ_1 as well as quadrupole splittings (ΔE) (Table IV) and isomer shifts[22,42]. Marked differences in zero-field and quadrupole splittings are found with a change in axial ligand while isomer shift is not much affected. Far infra-red data also show the iron environment in the fluorohemins and the fluoro hemeproteins (methemoglobin and metmyoglobin) to be nearly identical (Table V). These findings along with the similarly of other properties for hemin and hemeprotein fluorides suggest the protein environments of myoglobin and hemoglobin has little effect upon the fluoro protohemin and that the iron is out-of-plane (0.50 Å or greater) toward fluoride with only very weak, if any, bonding between iron

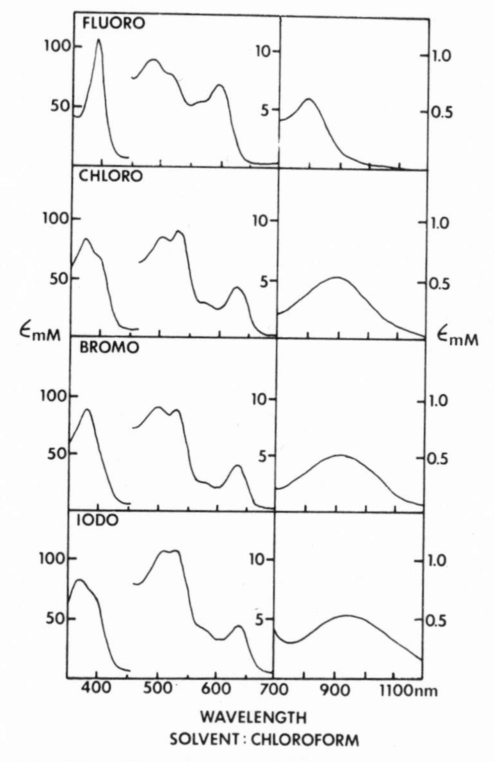

Fig. 22. Absorption spectra for deuteroporphyrin IX dimethyl ester iron(III) with different halogen axial ligands in chloroform.

and the proximal histidine. The Δ_1 and Δ_2 data for azido hemins (Table V) reveal deviations from axial symmetry (λ) in the case of non-linear Fe–N–N–N bonding.

Electronic absorption spectra are also markedly affected by differences in axial ligand as illustrated in Table XII and Fig. 21 for deuterohemin esters in benzene and in Fig. 22 in chloroform. The data presented in Table XII do not accurately reflect resolution of all the bands which contribute to visible and ultra-violet regions. Nevertheless the general trends are clear with the near infra-red and visible bands shifted in a direction opposite to the shift for Soret bands.

The ligand order followed for all of these physical properties is essentially the same, an order which corresponds to the spectrochemical series[4,10,22]. These data may be generalized most simply in terms of stronger interaction or bonding between iron and axial ligand resulting in a weaker interaction between porphyrin nitrogens and iron. Thus the asymmetry which gives rise to the quadrupole splitting of Mössbauer spectra and to zero field splittings is greater as the strength of bonding between the porphyrin nitrogens and iron increases. Related arguments serve in explanation of the magnitude of paramagnetic shifts in n.m.r. spectra and of electronic spectral shifts. Thus, the shifts in the λ_{max} of the near infra-red band can be explained in the following terms: the stronger is the bond between iron and axial ligand, the greater is the energy associated with donation of an electron from porphyrin to iron(III) because the iron becomes less electronegative toward the porphyrin. In contrast, the energy of the Soret band tends to decrease as the metal becomes less electronegative. This is also clearly shown in the spectra for nickel(II) porphyrins and their tetragonal complexes: upon addition of two nitrogenous bases as axial ligands (thereby reducing the electronegativity of Ni(II) to porphyrin), the Soret band shifts about 30 nm to the red[27,28].

The electronic spectra of hemins with one axial ligand are also dependent upon solvent. Figure 16 shows spectra for a chloro hemin ester in chloroform, benzene, and pyridine. In each case the species is high-spin with chloro ligand retained. The causes of such solvent effects are difficult to establish in detail. In addition to the aggregation and ligand dissociation that are frequently observed, differences in solvation of porphyrin, axial ligand, and metal can be expected. Some aspects of solvent interactions will be discussed below.

Hemins with two trans *axial ligands*

Hemins with two *trans* axial ligands are characteristically different in physical properties and stereochemistry from the high-spin penta-coordinated iron complexes we have just considered. It now appears that whenever an iron(III) porphyrin has two axial ligands bound at all strongly on each side of the porphyrin plane the iron atom will be essentially in the plane of the porphyrin nitrogens, the magnetic properties are of the "low-spin" type, and other properties are also characteristically different from the high-spin types. If one ligand to iron bond were long and very weak, a high-spin state may result though little evidence for this has been obtained with hemins. However, both acid methemoglobin and acid metmyoglobin are "high-spin"; crystal structure data indicate the iron atom is about 0.3 Å out-of-plane toward the proximal histidine which is *trans* to an oxygen containing ligand (hydroxo or hydrogen-bonded hydroxo)[3,62,63]. In this case the iron–oxygen bond is clearly much weaker than the iron–histidine bond. Mössbauer spectra support similar bonding for chloroprotohemin in aqueous pyridine[42].

Upon adjusting the medium to more alkaline pH levels met Hb and met Mb convert to low-spin forms ($S = \frac{1}{2}$) where the iron is undoubtedly in-plane or nearly in-plane with strong iron–oxygen bonding, with a true hydroxo or "deprotonated" hydroxo ligand present. Early proposals of in-plane iron stereochemistry for "low-spin" type iron(III) porphyrin systems[57] are now supported by a crystal structure for the low spin hemin-chloro-bis-imidazoletetraphenylporphin iron(III), where the iron was essentially in-plane with *trans* imidazole ligands oriented nearly perpendicular to each other to utilize different iron *d*-orbitals[64].

Low-spin species have long been recognized in solutions of hemins with cyanide present[41,46]. In fact no high spin cyano hemins are known, although pure solid compounds apparently have not been prepared. Both monocyano and dicyano species have been observed for hemins in aqueous pyridine–KCN solutions[50,65]. Early electronic spectral evidence for the two species now is confirmed by infra-red data for protohemin solutions where the dicyano, monocyano, and unbound cyanide exhibit ν_{CN} bands at 2112, 2125 and 2073 cm^{-1}, respectively[66,67]. Presumably pyridine is the *trans* ligand in the monocyano species; the *trans* effect of pyridine *vs.* cyano on ν_{CN} is considered reasonable. It is also of interest that cyano met Hb and cyano met Mb where monocyano and the *trans* imidazole of proximal histidine are the ligands also give ν_{CN} bands at 2125 cm^{-1} [66]. The n.m.r. spectra of cyanohemins are characterized by much narrower and smaller temperature dependent paramagnetic shifts than are found with high spin hemins[67-69]. The shifts are significantly less for dicyano than for monocyano species (*e.g.*, ~ 2.5 ppm less for ring methyl protons) as expected for increased iron–porphyrin interactions for monocyano compared with dicyano species[67].

Both high and low spin forms of azido hemins are well-established and these two forms exhibit characteristic spectral properties[66]. As pure solids without *trans* ligand, azidodeuterohemin esters are exclusively high-spin in terms of e.p.r. and Mössbauer spectra[42] and of Δ_1 and Δ_2 values from far infra-red spectra[44] (Table V) with a single infra-red band near 2060 cm^{-1} [66]. Also in chloroform solution, visible, near infra-red (Table XII) and e.p.r. spectra of the azide were similar to those for the chloro hemins, typical high-spin hemins. However, when dissolved in pyridine at room temperature both high and low spin forms are observed in n.m.r. and e.p.r. spectra as well as in the infra-red region where two bands at 2010 and 2045 cm^{-1} have been assigned to low and high-spin forms, respectively[66]. Presumably pyridine serves as a ligand *trans* to azide in the low-spin form and is not such a ligand in the high-spin form. However, the role of pyridine is not established. It is of interest that in MbN_3 infra-red bands at 2033 and 2045 cm^{-1} can be assigned to low- and high-spin forms, respectively. Also,

References pp. 829–831

with azidoprotohemin pyridine promotes the slow formation of a species with a 2100 cm^{-1} band consistent with a dinitrogen iron(II) ($Fe^{II}N_2$) complex[66].

In many cases, hemins remain essentially all high-spin in pure pyridine although electronic (Fig. 16) and Mössbauer spectra can differ somewhat from spectra obtained in solvents such as chloroform or benzene[42]. The extent of conversion of high-spin hemin to a low-spin form, with or without displacement of the original axial ligand by pyridine in pyridine solution depends greatly upon the nature of the axial ligand, porphyrin, solvent composition and temperature. Chloroprotohemin in pyridine was found as nearly all high-spin at room temperature by magnetic susceptibility and visible spectral criteria and at 77K exhibited a $g = 6$ band in an e.p.r. spectrum; a Mössbauer spectrum was interpreted as that of a high-spin compound with pyridine *trans* to the chloro ligand[42]. On the other hand, n.m.r. spectra for this hemin in CD_3OD with a large excess of substituted pyridines present at ~ 200K were interpreted in terms of only low-spin bis-pyridine iron(III) species being present[70]. The spectra consisted of fairly narrow resonances and the chemical shift of all protons, especially those of the peripheral methyl groups, were dependent upon the structure of the pyridines that served as axial ligands. For example, the paramagnetic shift of the methyl protons decreased as pyridine basicity increased (the same trend noted above for changes in axial ligand with high-spin hemin n.m.r. spectra). When the temperature was raised from 200K to room temperature, high-spin species were also present. Chlorohemins lost chloride to form bis-pyridine species to a greater extent in methanolic solutions than in pure pyridine and it has been shown that chlorohemins in the presence of methanol readily form high-spin methoxohemins which in turn appear to form low-spin bis-pyridine species more readily than do the chlorohemins[72].

μ-Oxo-bis-hemins

It was recently discovered that dimeric hemins with FeOFe bridging are readily prepared from monomeric hemins or from the autoxidation reactions of hemins[39,43,72]. Highly characteristic physical properties result from the presence of the FeOFe linkage[4]. Certain resonances in n.m.r. spectra exhibit broadening and small paramagnetic shifts (Fig. 20)[4,54,73]. However, the magnitude of broadening and shifts are much less than is found for the monomeric high-spin hemins as expected in view of the coupling through the oxygen bridge. The antiferromagnetic character of these compounds is illustrated in magnetic susceptibilities measured from 1.4 to 293K for μ-oxo-bis-[protoporphyrin IX dimethyl ester iron(III)] which were consistent with strong coupling ($2J = 380K$) between high-spin ($S = \frac{5}{2}$) iron(III) atoms[73]. Crystal structures for this compound[74] and for the related TPP compound[75,76] reveal a stereochemistry of the high-spin iron(III) type with each iron out-of-plane ~ 0.5 Å toward the bridging oxygen atom. Mössbauer[43,77] (Table IV) and electronic[39,43,73] (Fig. 17 and Table VII) spectra are also characteristic. The significance of such bridging in cytochrome *c* oxidase and possibly other hemeproteins is made more likely by the demonstration that a μ-oxo linkage can be formed from hemin *a*[4,78] and from 2,4-bis(ethoxycarbonylcyclopropyl) deuterohemin[54] despite the presence of such bulky groups on the porphyrin ring.

Monomeric hemins readily form dimers in the presence of water under neutral conditions presumably through formation of doubly bridged species as intermediates[43]. In the presence of $H_2{}^{18}O$, the ^{16}O of μ-oxo dimers are readily replaced by ^{18}O, again presumably via a doubly-bridged intermediate[43]. Since the FeOFe linkage is readily split in the presence of acids, these dimers represent conveniently obtained starting materials for the preparation of monomeric hemins.

Iron(II) porphyrins

Hemoglobin and myoglobin perform their major physiological role of reversible oxygen binding with the iron in the iron(II) oxidation state and many other hemeproteins function in this state at least part of the time[3]. Without a sixth ligand, *i.e.*, as "deoxy" forms, Hb and Mb are high-spin whereas with O_2 or CO ligands present they are low-spin, diamagnetic. Both low- and high-spin protein-free hemes are known in solution and to a lesser extent as pure solids. However, the iron(II) hemeproteins in several respects are more thoroughly studied than are the hemes which in many media tend to undergo autoxidation and aggregation reactions readily[46,50].

The most extensively studied or best understood hemes are the low spin compounds with two nitrogenous (*e.g.*, pyridine) axial ligands known as hemochromes[8,10,39,46]. The relatively sharp-banded hemochrome spectra obtained upon reducing hemins with sodium dithionite in aqueous alkaline pyridine[8,79] or with hydrazine[8] have long been used to characterize iron porphyrins (Tables VIII–X and Fig. 15). The aggregation phenomena which complicate interpretations of hemes—high- or low-spin—in aqueous media is less prevalent with hemochromes especially as heme esters in non-aqueous media[39,46,50,79]. For these reasons, and particularly because of relatively greater stability against autoxidation, the low-spin hemes have been studied more extensively than the high-spin hemes.

A number of dipyridine hemes have now been prepared as pure (in some cases crystalline) compounds[4,5,23,39,61]. As solids most dipyridine hemes are stable in the presence of oxygen until the temperature is raised to a level where pyridine dissociates to give a species without axial ligands which can form a 1:1, Fe:O_2 adduct upon exposure to molecular oxygen[39]. Oxyhemes are of particular interest as analogs of oxygenated hemeproteins. However, the oxyhemes studied thus far suggest that an $Fe(III)O_2{}^-$

description for the bonding situation is more important for the oxyheme than for oxyhemoglobin and oxymyoglobin.

Carbonyl hemes have only infrequently been prepared as solids but have long been studied in solution as have complexes with similar ligands (*e.g.*, NO, RNC, and R–NO)[3,4,10,31,46]. *Cis* effects due to changes in porphyrin structure were mentioned above. Also *trans* effects are shown by the increase in ν_{CO} with a decrease in basicity of *trans* pyridine[30]. Dicarbonyl hemes have not been reported although more than two species have been detected during formation of carbonyl hemes from dipyridine hemes[30].

Autooxidation reactions of hemes are relevant to the oxygen reduction of cytochrome *c* oxidase and mixed function oxidases. These reactions are also of interest in explanation of the lack of irreversible oxygen reduction by hemoglobin and myoglobin. The stoichiometry (4 hemes oxidized per O_2), kinetics, and products of reactions of dipyridine hemes with oxygen in benzene solutions are consistent with the following steps (I to V) where py = pyridine[39,40,43]:

I. py-Fe-py $\rightleftarrows$ py-Fe + py
II. py-Fe + O_2 $\rightleftarrows$ py-FeO_2
III. py-FeO_2 + Fe-py $\rightleftarrows$ py-FeO_2Fe-py
IV. py-FeO_2Fe-py $\rightarrow$ 2 py-FeO
V. py-FeO + py-Fe $\rightarrow$ FeOFe + 2 py

Other possibilities have been considered[39,40,84]. Discussion of solvent effects upon overall rates have been used to explain the stability of oxyhemoglobin and oxymyoglobin[84]. However, the kinetic data available gave no significant information on effects of medium on the iron oxidation steps *per se*[3,40]. For example, increasing the alcohol concentration in a pyridine–benzene solution will tend to promote step I (the dissociation of pyridine and formation of ligand-free site at iron II) and thus increase the rate of disappearance of dipyridine species. However, this is not evidence for increased rates for steps IV and V. Mechanisms with steps in which O_2^- (superoxide ion) dissociates from hemin have been proposed but pertinent experimental evidence has not been reported[84]. The data obtained cannot be considered significant for evaluation of the importance of the non-polar character of the heme environment of hemoglobin and myoglobin to reversible oxygen binding. However, present data do show the importance of steric restrictions which prevent access of potential electron donors to the O_2-heme site[3].

Insight into ligand exchange and redox reactions of iron and other porphyrins is limited with a great deal more work needed to clarify the mechanisms involved[10,43,85].

REFERENCES

1 B. Chance, R. W. Estabrook and T. Yonetani, *Hemes and Hemoproteins*, Academic Press, New York, 1966.
2 K. Okunuki, M. D. Kamen and I. Sekuzu, *Structure and Function of Cytochromes*, University of Tokyo Press, Tokyo, 1968.
3 W. S. Caughey, *A. Rev. Biochem.*, 36 (1967) 611.
4 W. S. Caughey, *Adv. Chem. Ser.*, 100 (1971) 248.
5 J. L. York, S. McCoy, D. N. Taylor and W. S. Caughey, *J. Biol. Chem.*, 242 (1967) 908.
6 G. A. Smythe and W. S. Caughey, *J. Chem. Soc. (D)*, (1970) 809.
7 R. A. Bayne, G. A. Smythe and W. S. Caughey in B. Chance, T. Yonetani and A. S. Mildvan, *Probes of Structure and Function of Macromolecules and Membranes*, Vol. II, Academic Press, New York, 1971, p. 613.
8 H. Fischer and H. Orth, *Die Chemie des Pyrrols*, Vol. 2, Part I, Akademische Verlagsgesellschafts, Leipzig, 1937.
9 M. O. Dayhoff, *Atlas of Protein Sequence and Structure*, Vol. 4, National Biomedical Research Foundation, Silver Spring, 1969, p. D-1.
10 P. Hambright, *Coordination Chem. Rev.*, 6 (1971) 247.
11 A. Treibs, *Liebigs Ann. Chem.*, 728 (1969) 115.
12 G. A. Smythe and W. S. Caughey, unpublished observation.
13 O. Warburg and H.-S. Gewitz, *Z. Physiol. Chem.*, 288 (1951) 1.
14 O. Warburg, H.-S. Gewitz and W. Volker, *Z. Naturforsch.*, 106 (1955) 541.
15 M. Grassl, G. Augsburg, U. Coy and F. Lynen, *Biochem. Z.*, 337 (1963) 35.
16 M. Grassl, U. Coy, R. Seyffert and F. Lynen, *Biochem. Z.*, 338 (1963) 771.
17 G. S. Marks, D. K. Dougall, E. Bullock and S. F. MacDonald, *J. Am. Chem. Soc.*, 81 (1959) 250; 82 (1960) 3183.
18 W. S. Caughey, J. O. Alben, W. Y. Fujimoto and J. L. York, *J. Org. Chem.*, 31 (1966) 2631.
19 H. A. Harbury and P. A. Loach, *J. Biol. Chem.*, 235 (1960) 3640.
20 K. G. Paul, *Acta Chem. Scand.*, 5 (1951) 389.
21 A. Treibs and N. Häberle, *Liebigs Ann. Chem.*, 718 (1968) 183.
22 W. S. Caughey, H. Eberspaecher, W. H. Fuchsman, S. McCoy and J. O. Alben, *Ann. N.Y. Acad. Sci.*, 153 (1969) 722.
23 W. S. Caughey, W. Y. Fujimoto and B. P. Johnson, *Biochemistry*, 5 (1966) 3830.
24 J. N. Phillips, *Rev. Pure Appl. Chem.*, 310 (1960) 35.
25 J. N. Phillips, in M. Florkin and E. H. Stotz, *Comprehensive Biochemistry*, Vol. 9, Elsevier, Amsterdam, 1963, p. 34.
26 T. A. Hamor, W. S. Caughey and J. L. Hoard, *J. Am. Chem. Soc.*, 87 (1965) 2305.
27 W. S. Caughey, R. M. Deal, B. D. McLees and J. O. Alben, *J. Am. Chem. Soc.*, 84 (1962) 1735.
28 B. D. McLees and W. S. Caughey, *Biochemistry*, 7 (1968) 642.
29 W. S. Caughey, J. O. Alben and C. A. Beaudreau, in T. E. King, H. S. Mason and M. Morrison, *Oxidases and Related Redox Systems*, John Wiley, New York, 1965, p. 97.
30 J. O. Alben and W. S. Caughey, *Biochemistry*, 7 (1968) 175.
31 W. S. Caughey, *Ann. N. Y. Acad. Sci.*, 174 (1970) 148.
32 W. S. Caughey, R. A. Bayne and S. McCoy, *J. Chem. Soc. (D)*, (1970) 950.
33 S. McCoy and W. S. Caughey, in B. Chance, T. Yonetani and A. S. Mildvan, *Probes of Structure and Function of Macromolecules and Membranes*, Vol. II, Academic Press, New York, 1971, p. 289.

34 A. Rossi-Fanelli and E. Antonini, *Arch. Biochem. Biophys.*, 80 (1959) 308.
35 A. Rossi-Fanelli, E. Antonini and A. Caputo, *Arch. Biochem. Biophys.*, 85 (1959) 37.
36 Y. Sugita and Y. Yoneyama, *J. Biol. Chem.*, 246 (1971) 389.
37 A. Rossi-Fanelli and E. Antonini, *Arch. Biochem. Biophys.*, 72 (1957) 243.
38 J. E. Falk, J. N. Phillips and E. A. Magnusson, *Nature*, 212 (1966) 1531.
39 J. O. Alben, W. H. Fuchsman, C. A. Beaudreau and W. S. Caughey, *Biochemistry*, 7 (1968) 624.
40 I. A. Cohen and W. S. Caughey, *Biochemistry*, 7 (1968) 636.
41 A. E. Martell and M. Calvin, *Chemistry of the Metal Chelate Compounds*, Prentice-Hall, New York, 1952, p. 373.
42 T. H. Moss, A. J. Bearden and W. S. Caughey, *J. Chem. Phys.*, 51 (1969) 2624.
43 N. Sadasivan, H. I. Eberspaecher, W. H. Fuchsman and W. S. Caughey, *Biochemistry*, 8 (1969) 534.
44 G. C. Brackett, P. L. Richards and W. S. Caughey, *J. Chem. Phys.*, 54 (1971) 4383.
45 P. L. Richards, W. S. Caughey, H. Eberspaecher, G. Feher and M. Malley, *J. Chem. Phys.*, 47 (1967) 1187.
46 J. E. Falk, *Porphyrins and Metalloporphyrins*, Elsevier, Amsterdam, 1964.
47 W. S. Caughey, R. M. Deal, C. Weiss and M. Gouterman, *J. Mol. Spectros.*, 16 (1965) 451.
48 M. Gouterman, *J. Mol. Spectros.*, 6 (1961) 138.
49 M. Zerner, M. Gouterman and H. Kobayashi, *Theor. Chim. Acta*, 6 (1967) 366.
50 J. Shack and W. M. Clark, *J. Biol. Chem.*, 171 (1947) 143.
51 W. Scheler, P. Mohr, K. Hecht and K. Gustav, *Z. Chem.*, 7 (1967) 303.
52 S. B. Brown, P. Jones and I. R. Lantzke, *Nature*, 223 (1969) 960.
53 P. Mohr and W. Scheler, *Eur. J. Biochem.*, 8 (1969) 444.
54 C. H. Barlow, D. O'Keeffe, G. A. Smythe and W. S. Caughey, unpublished.
55 W. S. Caughey, J. L. York and P. K. Iber in A. Ehrenberg, B. G. Malmström and T. Vänngård, *Magnetic Resonance in Biological Systems*, Pergamon Press, Oxford, 1967, p. 25.
56 D. F. Koenig, *Acta Cryst.*, 18 (1965) 663.
57 J. L. Hoard, M. J. Hamor, T. A. Hamor and W. S. Caughey, *J. Am. Chem. Soc.*, 87 (1965) 2312.
58 J. Peisach and W. E. Blumberg, in B. Chance, T. Yonetani and A. S. Mildvan, *Probes of Structure and Function of Macromolecules and Membranes*, Vol. II, Academic Press, New York, 1971, p. 231.
59 R. J. Kurland, D. G. Davis and C. Ho, *J. Am. Chem. Soc.*, 90 (1968) 2700.
60 W. S. Caughey and L. F. Johnson, *J. Chem. Soc. (D)*, (1969) 1362.
61 H. Kobayashi, Y. Maeda and Y. Yanagawa, *Bull. Chem. Soc. Japan*, 43 (1970) 2342J.
62 J. C. Kendrew, *Science*, 139 (1963) 1259.
63 M. F. Perutz, *Nature*, 228 (1970) 726, 734.
64 R. Countryman, D. M. Collins and J. L. Hoard, *J. Am. Chem. Soc.*, 91 (1969) 5166.
65 K. Kaziro, F. Uchimura, and G. Kikuchi, *J. Biochem. (Tokyo)*, 43 (1956) 539.
66 S. McCoy and W. S. Caughey, *Biochemistry*, 9 (1970) 2387.
67 D. H. O'Keeffe and W. S. Caughey, unpublished.
68 K. Wüthrich, R. G. Shulman, B. J. Wyluda and W. S. Caughey, *Proc. Natn. Acad. Sci. U.S.*, 62 (1969) 636.
69 K. Wüthrich, *Structure and Bonding*, 8 (1970) 53.
70 H. A. O. Hill and K. G. Morallee, *J. Chem. Soc. (D)*, (1970) 266.
71 W. S. Caughey, unpublished.
72 W. S. Caughey, J. L. Davies, W. H. Fuchsman and S. McCoy, in *Structure and Function of Cytochromes*, University of Tokyo Press, Tokyo, 1968, p. 20.

73 T. H. Moss, H. R. Lillienthal, C. Moleski, G. A. Smythe, M. C. McDaniel and W. S. Caughey, *J. Chem. Soc. (D)*, in press.
74 L. J. Radonovich, C. H. Barlow, W. S. Caughey and J. L. Hoard, unpublished.
75 E. B. Fleischer and T. S. Srivastava, *J. Am. Chem. Soc.*, 91 (1969) 2403.
76 A. B. Hoffman, D. M. Collins, V. W. Day, E. B. Fleischer, T. S. Srivastava and J. L. Hoard, personal communication.
77 I. A. Cohen, *J. Am. Chem. Soc.*, 91 (1969) 1980.
78 G. A. Smythe and W. S. Caughey, *Federation Proc.*, 29 (1970) 464.
79 W. A. Gallagher and W. B. Elliott, *Biochem. J.*, 97 (1965) 187.
80 D. L. Drabkin, *J. Biol. Chem.*, 146 (1942) 605.
81 R. K. Morton, *Rev. Pure and Appl. Chem.*, 8 (1958) 161.
82 P. S. Clezy and D. B. Morell, *Biochem. Biophys. Acta*, 71 (1963) 165.
83 R. Lemberg and J. E. Falk, *Biochem. J.*, 49 (1951) 674.
84 O. Kao and J. H. Wang, *Biochemistry*, 4 (1965) 342.
85 L. J. Boucher and H. K. Garber, *Inorg. Chem.*, 9 (1970) 2644.

Chapter 25

HEMOGLOBIN AND MYOGLOBIN

J. M. RIFKIND

Laboratory of Molecular Aging, Gerontology Research Center, National Institutes of Health, National Institute of Child Health and Human Development, Baltimore City Hospitals, Baltimore, Maryland 21224, U.S.A.

INTRODUCTION

Hemoglobin and myoglobin are hemoproteins designed to reversibly bind molecular oxygen to their hemes. Myoglobin is involved in the storage of oxygen in the muscle, while hemoglobin is the major component of the red blood cell and transports oxygen from the lungs to all parts of the organism.

The structures of hemoglobin[1,2] and myoglobin[3,4] have been determined by X-ray crystallography. The iron in both proteins is coordinated to the four pyrrole nitrogens of protoporphyrin IX (Fig. 3, Chapter 24) and an imidazole nitrogen of a histidine residue from the polypeptide or "globin" chain on the side of the porphyrin defined as the "proximal" side. The sixth coordination position on the other side of the porphyrin has a histidine too far away to coordinate with the iron and is, therefore, defined as the "distal" side. Oxygen and other small molecules can coordinate with the iron at this site.

Myoglobin[3-5] consists of one polypeptide or "globin" chain and one heme (Fig. 1; see also Fig. 4A, Chapter 7). Mammalian hemoglobin[1,2,5] consists of four polypeptide chains each containing a heme with two identical pairs that have been designated α and β chains. The four chains, or subunits, are in a tetrahedral arrangement forming a compact spheroidal molecule[6], in which each subunit can make contact with every other subunit (Fig. 2).

Sperm whale myoglobin contains 153 residues[7,8] (Fig. 3) and eight helical regions of varying length. The helical regions are denoted by letters from A nearest the amino end to H nearest the carboxyl end of the chain. The interhelical regions are denoted by the letters of the adjoining helical regions, and the non-helical regions at the amino and carboxyl ends are denoted by NA and HC, respectively. The residues within each helical and non-helical region are numbered from the residue closest to the amino end (Figs. 1 and 3). The amino acid sequence of the α and β chains of hemoglobin[7,9,10] are quite different from that of myoglobin. Nevertheless, the

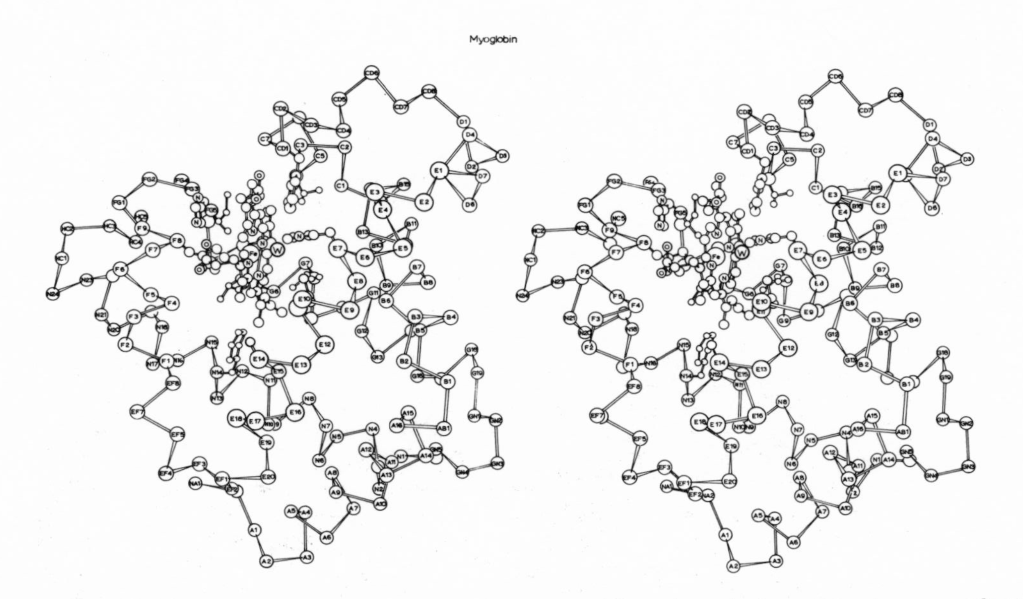

Fig. 1. A stereo drawing of sperm whale metmyoglobin based on the X-ray diffraction studies of Kendrew *et al.*[3]. The individual amino acid residues are designated by one or two letters and a number which indicate the position of the residue in the tertiary structure (see text for explanation). The water molecule and the imidazole side chain of histidine F8 coordinated to the iron are shown. Besides several amino acid residues in the region of the heme for which the side chains are included, only the polypeptide backbone is indicated. (From the stereo supplement to the *Structure and Action of Proteins* by permission of Harper and Row.)

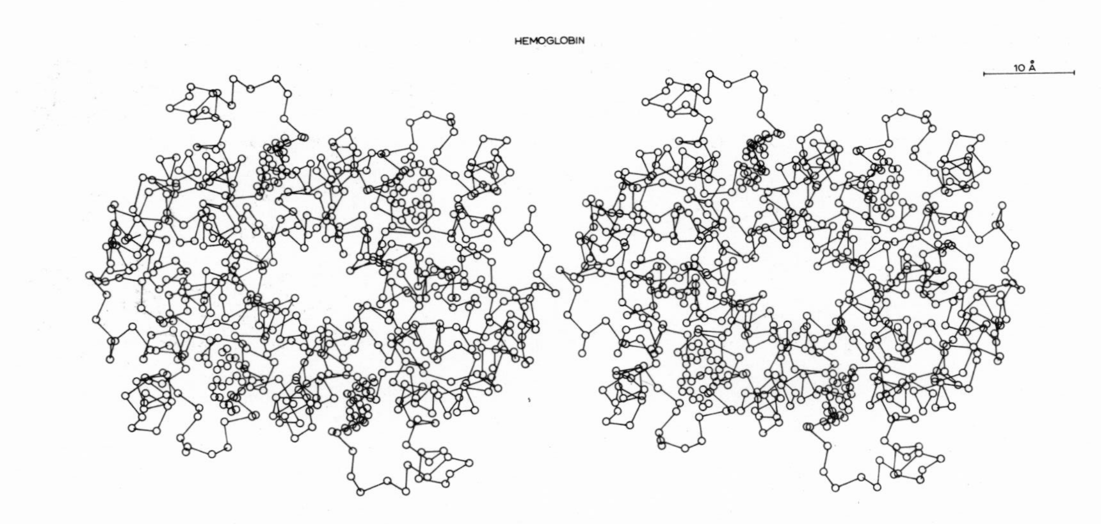

Fig. 2. A stereo drawing of liganded horse hemoglobin based on the X-ray diffraction studies of Perutz *et al.*[2]. Only every tenth residue from the amino terminus is numbered and the letters A and B indicate α and β chains respectively. Thus, "A3" indicates residue 30 of the α chain, and "B14" marks residue 140 of the β chain. The β-subunits are shown closer to the eyes of the viewer and the α-subunits are shown away from the eyes of the viewer. If the β-subunit to the right and toward the top of the page is designated β_1, then α_1 is the α-subunit away from the eyes of the viewer to the right and toward the bottom of the page. α_2 is away from the eyes of the viewer toward the top of the page and to the left. (From the stereo supplement to the *Structure and Action of Proteins* by permission of Harper and Row.)

NA1 A1 A5 A10 A16 AB1 B1 B5 B10 B16

Myoglobin (Sperm Whale): VAL-LEU — SER-GLU-GLY-GLU-TRP-GLN-LEU-VAL-LEU-HIS-VAL-TRP-ALA-LYS-VAL-GLU-ALA-ASP-VAL-ALA-GLY-HIS-GLY-GLN-ASP-ILE-LEU-ILE-ARG-LEU-PHE-LYS-SER (5, 10, 15, 20, 25, 30, 35)

Alpha Chains (Human): VAL-LEU — SER-PRO-ALA-ASP-LYS-THR-ASN-VAL-LYS-ALA-ALA-TRP-GLY-LYS-VAL-GLY-ALA-HIS-ALA-GLY-GLU-TYR-GLY-ALA-GLU-ALA-LEU-GLU-ARG-MET-PHE-LEU-SER (5, 10, 15, 20, 25, 30, 35)

Beta Chains (Human): VAL-HIS-LEU-THR-PRO-GLU-GLU-LYS-SER-ALA-VAL-THR-ALA-LEU-TRP-GLY-LYS-VAL — — ASN-VAL-ASP-GLU-VAL-GLY-GLY-GLU-ALA-LEU-GLY-ARG-LEU-LEU-VAL-VAL (5, 10, 15, 20, 25, 30)

C1 C5 C7 CD1 CD5 CD9 D1 D5 D7 E1 E5 E10 E15

HIS-PRO-GLU-THR-LEU-GLU-LYS-PHE-ASP-ARG-PHE-LYS-HIS-LEU-LYS — THR-GLU-ALA-GLU-MET-LYS-ALA-SER-GLU-ASP-LEU-LYS-LYS-HIS-GLY-VAL-THR-VAL-LEU-THR-ALA-LEU-GLY-ALA (40, 45, 50, 55, 60, 65, 70)

PHE-PRO-THR-THR-LYS-THR-TYR-PHE-PRO-HIS-PHE-ASP-LEU-SER-HIS-GLY — — — — — — — SER-ALA-GLN-VAL-LYS-GLY-HIS-GLY-LYS-LYS-VAL-ALA-ASP-ALA-LEU-THR-ASN (40, 45, 50, 55, 60, 65)

TYR-PRO-TRP-THR-GLN-ARG-PHE-PHE-GLU-SER-PHE-GLY-ASP-LEU-SER — THR-PRO-ASP-ALA-VAL-MET-GLY-ASN-PRO-LYS-VAL-LYS-ALA-HIS-GLY-LYS-LYS-VAL-LEU-GLY-ALA-PHE-SER-ASP (35, 40, 45, 50, 55, 60, 65, 70)

E20 EF1 EF5 EF8 F1 F5 F9 FG1 FG5 G1 G5 G10 G15

ILE-LEU-LYS-LYS-LYS-GLY-HIS-HIS-GLU-ALA-GLU-LEU-LYS-PRO-LEU-ALA-GLN-SER-HIS-ALA-THR-LYS-HIS-LYS-ILE-PRO-ILE-LYS-TYR-LEU-GLU-PHE-ILE-SER-GLU-ALA-ILE-ILE-HIS-VAL (75, 80, 85, 90, 95, 100, 105, 110)

ALA-VAL-ALA-HIS-VAL-ASP-ASP-MET-PRO-ASN-ALA-LEU-SER-ALA-LEU-SER-ASP-LEU-HIS-ALA-HIS-LYS-LEU-ARG-VAL-ASP-PRO-VAL-ASN-PHE-LYS-LEU-LEU-SER-HIS-CYS-LEU-LEU-VAL-THR (70, 75, 80, 85, 90, 95, 100, 105)

GLY-LEU-ALA-HIS-LEU-ASP-ASN-LEU-LYS-GLY-THR-PHE-ALA-THR-LEU-SER-GLU-LEU-HIS-CYS-ASP-LYS-LEU-HIS-VAL-ASP-PRO-GLU-ASN-PHE-ARG-LEU-LEU-GLY-ASN-VAL-LEU-VAL-CYS-VAL (75, 80, 85, 90, 95, 100, 105, 110)

G19 GH1 GH5 H1 H5 H10 H15 H20 H24 HC1 HC6

LEU-HIS-SER-ARG-HIS-PRO-GLY-ASN-PHE-GLY-ALA-ASP-ALA-GLN-GLY-ALA-MET-ASN-LYS-ALA-LEU-GLU-LEU-PHE-ARG-LYS-ASP-ILE-ALA-ALA-LYS-TYR-LYS-GLU-LEU-GLY-TYR-GLN-GLY (115, 120, 125, 130, 135, 140, 145, 150)

LEU-ALA-ALA-HIS-LEU-PRO-ALA-GLU-PHE-THR-PRO-ALA-VAL-HIS-ALA-SER-LEU-ASP-LYS-PHE-LEU-ALA-SER-VAL-SER-THR-VAL-LEU-THR-SER — — — LYS-TYR-ARG — — — (110, 115, 120, 125, 130, 135, 140)

LEU-ALA-HIS-HIS-PHE-GLY-LYS-GLU-PHE-THR-PRO-PRO-VAL-GLN-ALA-ALA-TYR-GLN-LYS-VAL-VAL-ALA-GLY-VAL-ALA-ASN-ALA-LEU-ALA-HIS — — — LYS-TYR-HIS — — — (115, 120, 125, 130, 135, 140, 145)

Fig. 3. The amino acid sequence of sperm whale myoglobin[8] and the α and β chains of human hemoglobin[9,10]. The amino acid residues are arranged to indicate the similarities in tertiary structure. Blanks are left where a particular position in the tertiary structure is missing in one of the polypeptide chains, *e.g.*, the D helix in the α chain. The helical segments are enclosed in boxes. The residue number starting from the amino end is indicated directly above the amino acid residues. The letters and numbers at the top indicate the position in the tertiary structure (see text for explanation).

References pp. 891–901

tertiary structure is quite similar[11], and as shown in Fig. 3 it is possible to align the amino acids in the helical and non-helical regions of the hemoglobin chains with those of myoglobin. Only a few additions and deletions are necessary.

The heme is located in a crevice between the E and F helices[12], the proximal histidine F8 being part of the F helix. The edge containing the polar propionic acid groups is near the surface and the rest of the heme is buried deep inside the globin (Fig. 1).

The functional forms of myoglobin and hemoglobin contain iron in the +2 state and in the absence of oxygen or other ligands are referred to as deoxymyoglobin and deoxyhemoglobin, respectively. When oxygen is bound they are called oxymyoglobin and oxyhemoglobin, respectively. The oxygen in the Fe(II) proteins can be replaced by other neutral ligands[13,14] such as CO[15-17], NO[18,19] and alkylisocyanides[20-22]. In the absence of such ligands the sixth coordination position of deoxyhemoglobin and deoxymyoglobin is not even occupied by H_2O; only five ligands are present[23-26]. Fe(II) in hemoglobin and myoglobin is readily oxidized to the Fe(III) state[27-30]. In this form the terminology ferrihemoglobin or methemoglobin and ferrimyoglobin or metmyoglobin is used. None of the above ligands, including oxygen, bind significantly to the oxidized proteins. Instead, other ligands[13,31-33] such as H_2O, OH^-[34], F^-[35], N_3^-[34,36], CN^-[35], SCN^-[35], and imidazole[37] can be bound.

The abbreviations Mb and Hb represent deoxymyoglobin and deoxyhemoglobin, respectively. In designating the various Fe(II) liganded proteins this notation is followed by the chemical formula of the sixth ligand, *i.e.*, MbO_2. For the Fe(III) proteins a superscript (+) is added, *i.e.*, Mb^+H_2O, Hb^+CN.

The "active site" where oxygen is bound to myoglobin and hemoglobin is on the heme and is not an intrinsic part of the protein. However, the "globin" was found to be necessary in order to reversibly bind oxygen without oxidizing the iron and to regulate the uptake and release of oxygen. The role of the "globin" in the oxygenation of hemoglobin is particularly crucial, since the effective transport of oxygen from the lungs to the cells requires highly cooperative binding[38]. Therefore, a structure is required in which the affinity of oxygen for one subunit changes when oxygen is bound to another subunit. This interaction between subunits has been referred to as "heme–heme interaction" and must involve large portions of the protein since the hemes are far removed from each other (Fig. 2).

The binding of oxygen and other ligands to hemoglobin and myoglobin has been thoroughly studied from various points of view. Various model compounds (Chapter 24) have been investigated. The state of the iron[39,40] has been elucidated by techniques such as the Mössbauer effect[41-43], electron paramagnetic resonance (e.p.r.)[44-48], and magnetic susceptibility[49-52]. In addition to X-ray diffraction studies, the protein conformation has been probed by chemical modification[53,54], spin labeling techniques[55–57] circular dichroism-optical rotatory dispersion (c.d.–o.r.d.)[58-61], nuclear magnetic resonance (n.m.r.)[62–64] and hydrogen exchange[65]. Detailed thermodynamics[17,66,67] and kinetics[68-70] of the binding of ligands have been elucidated.

In this chapter a picture will be developed of what is known about the functioning of these proteins by focusing first on the iron and the properties of the iron complex, then on the immediate ligands of the iron, then on the effect of the immediate protein environment and how it can influence the binding of oxygen, and finally on the total protein and the interactions between subunits. This approach should be particularly helpful for the inorganic chemist and perhaps place some ideas in a new perspective for the protein chemist. Such an approach emphasizes the role of the total protein in controlling the ligand properties of the metal complex. It will, of course, not be possible to be fully comprehensive in such a presentation; the choice of topics included is made on the basis of the organization described.

COORDINATION CHEMISTRY OF HEMOGLOBIN AND MYOGLOBIN

Geometry of the complex

When the sixth coordination position is occupied, the ligands for the iron in hemoglobin and myoglobin are to a first approximation in an octahedral geometry (Chapter 3). If the porphyrin is considered to be in the x–y plane of a cartesian coordinate system, the histidine and oxygen or other sixth ligand are on the z axis[39].

The major effect of the octahedral ligand field on the iron in hemoglobin and myoglobin is to separate the energy of the e_g orbitals directed toward the ligands, and that of the t_{2g} orbitals not directed toward the ligands (Fig. 4; see also Fig. 1, Chapter 3).

While the e_g orbitals directed toward the ligands always have higher energies than the t_{2g} orbitals, the magnitude of the splitting in hemoglobin and myoglobin is influenced by π-interactions[71,72]. The porphyrin, histidine, and many of the ligands found in the sixth coordination position possess electronic orbitals (p or π) perpendicular to the σ-bonds between the ligands and the iron. These ligand orbitals do not interact with the e_g orbitals involved in σ bonding with the ligands; however, π-interactions with the t_{2g} orbitals are quite possible. Such π-interactions can increase the octahedral splitting by lowering the energy of the t_{2g} orbitals, if double bonds involving the electrons in the t_{2g} orbitals and the ligand can be formed. If a double bond cannot be formed, the octahedral splitting can be decreased by the electrostatic repulsions between electrons in the t_{2g} orbitals and ligand electrons.

Deoxymyoglobin and deoxyhemoglobin with only five ligands are, of course, not octahedral complexes. However, even when a sixth ligand is present, the axial ligands are clearly different from the ligands in the x–y plane, and hemoglobin and myoglobin have considerable tetragonal distortion[71]. Thus, the d_{z^2} orbital directed toward the axial ligands in the z direction has lower energy than the $d_{x^2-y^2}$ orbital directed toward the porphyrin pyrroles in the x–y plane (Fig. 4).

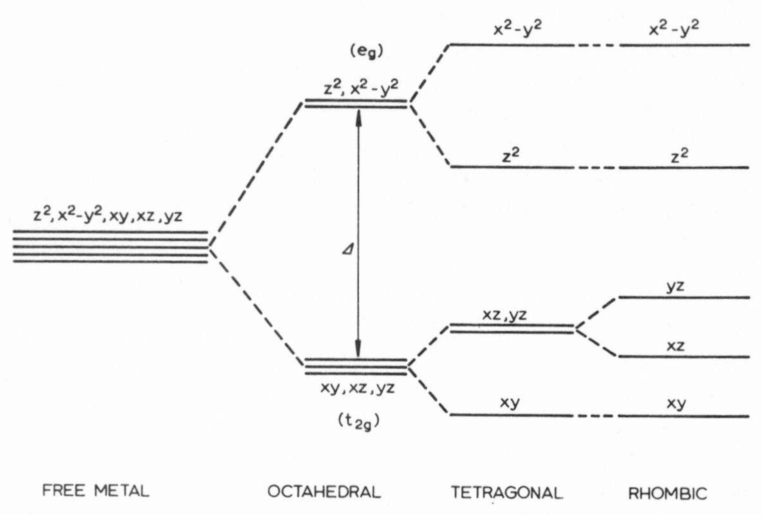

Fig. 4. The splitting of the d-orbitals in octahedral, tetragonal, and rhombic ligand fields.

The relative energies of the various t_{2g} orbitals in a tetragonal geometry are influenced by electrostatic interactions with the ligand electrons involved in σ-bonding. However, the dominant effect in hemoglobin and myoglobin involves the π-interactions between the t_{2g} orbitals and ligand orbitals. Since the d_{xy} orbital is in the plane of the porphyrin it is not properly positioned to interact with the π systems of the various ligands. Therefore π-interactions between the ligand and the iron will affect the energy of the d_{xz} and d_{yz} orbitals relative to the d_{xy} orbital in the same manner that the octahedral splitting is affected, *i.e.*, π-bonding lowers the energy of d_{xz} and d_{yz} and electrostatic repulsion raises the energy of these orbitals.

Simple porphyrin complexes maintain tetragonal symmetry[46] with degenerate d_{xz} and d_{yz} orbitals. However, many hemoglobin and myoglobin complexes have been shown to possess significant rhombic distortion[44,46,73], *i.e.*, the d_{xz} and d_{yz} orbitals are no longer degenerate (Fig. 4).

Rhombic distortion has been observed by e.p.r. spectroscopy[44,46]. Low-spin ferrimyoglobin and ferrihemoglobin complexes often possess three widely separated principal g values. For instance, with Hb^+N_3 crystals[44] it is found that $g_x = 1.72$, $g_y = 2.22$ and $g_z = 2.80$. High-spin complexes have $g_z = 2$ and $g_x \simeq g_y \simeq 6$ as expected for tetragonal symmetry. However, a careful analysis of the resonanace absorption derivative near $g = 6$ in the high-spin complexes indicates that in the protein it is either broadened or split into two resolvable g values, again indicative of rhombic distortion[46]. Rhombic distortion has also been observed by n.m.r.[73], c.d.[74], and polarization studies[75], and in certain cases the rhombic splitting seems to be comparable with the tetragonal splitting.

The rhombic distortion does not necessarily imply that the pyrrole nitrogen coordination sites in the x–y plane are arranged asymmetrically. Instead, the interaction of the globin with the heme in myoglobin and hemoglobin can asymmetrically orient the fifth[76] or sixth[77] ligand, thereby removing the degeneracy between the d_{xz} and d_{yz} orbitals. A similar effect can be produced by a distortion of the configuration of the porphyrin.

The spin state

The Fe(II) and Fe(III) complexes of myoglobin and hemoglobin can exist in high-spin or low-spin states (Fig. 2, Chapter 3). The state of a particular metal complex depends on the energy difference between the t_{2g} and e_g orbitals which results from the octahedral ligand field (Chapter 3). This energy must be large enough to counteract the electron pairing energy. Hemoglobin and myoglobin are unusual in that there is a close balance between these two energies so that a change at just one coordination position is able to push the complex from the high-spin to the low-spin state.

Thus, deoxyhemoglobin and deoxymyoglobin are both high-spin paramagnetic complexes[78,79] while complexing with any of the common ligands, i.e., O_2, CO, NO, and alkylisocyanides, produces low-spin diamagnetic complexes[78-80]. The ability of these ligands to push the complex into a low-spin state is thought to be related to π-bonding which further lowers the energy of the t_{2g} orbitals (see above). It has also been suggested that a sixth ligand will pull the iron closer to the plane of the porphyrin, thus increasing the ligand field of the porphyrin[81]. It is interesting that magnetic susceptibility measurements suggest that in the erythrocyte even deoxyhemoglobin may be partly low-spin[82].

An unusual situation exists for many of the Fe(III) myoglobin and hemoglobin complexes. The energy difference between the high-spin and low-spin states is so small that at room temperature they are in thermal equilibrium[49-52]. A mixture of high-spin and low-spin states was originally proposed for the OH^- complexes on the basis of magnetic susceptibility measurements at room temperature[76,78,83-86]. Hb^+OH has an effective Bohr

References pp. 891–901

magneton number of 4.5, which is clearly intermediate between 5.9 Bohr magnetons obtained for high-spin Hb^+F and 2.3–2.5 Bohr magnetons obtained for low-spin Hb^+CN[49,50,83,85,87].

A careful analysis of the temperature dependence of the susceptibility has shown that a thermal equilibrium between high-spin and low-spin states exists at 20°C for the H_2O, OH^-, OCN^-, and N_3^- complexes of myoglobin and hemoglobin and the imidazole complex of myoglobin[49–52,88]. The magnetic susceptibilities of these complexes vary inversely with temperature in the region of 77°K as predicted by the Curie law[51,52,88]. However, at higher temperatures significant deviations from the Curie law are observed. The hemoglobin complexes and the azide and imidazole complexes of myoglobin are in the low-spin state at 77°K and raising the temperature produces positive deviations from the Curie law, shifting the equilibrium toward the high-spin state. The Mb^+H_2O and Mb^+OCN complexes are in the high-spin state at 77°K, and raising the temperature produces negative deviations from the Curie law indicating that in these two complexes raising the temperature shifts the equilibrium towards the low-spin state. No deviations in the Curie law are observed for the F^- and CN^- complexes, which are in the high-spin and low-spin states, respectively.

These results have also been correlated with the temperature dependence of the visible spectrum of these complexes[40,49,50,52], which seem to be predominantly determined by the spin state of the complex. Thus, the spectra of low-spin CN^- and high-spin F^- complexes are different and the complexes with appreciable mixtures of both states have intermediate spectra. Furthermore, the spectra approach those of the F^- or CN^- complexes as the temperature increases, indicating, as found from the temperature dependence of the magnetic susceptibility, that temperature shifts the equilibrium toward the high-spin or low-spin state.

These results can be quantitatively described by assuming a Boltzmann distribution[40,51,52,88] among high-spin and low-spin states with the populations proportional to $W_H\, e^{-E_H/kt}$ and $W_L\, e^{-E_L/kt}$, where W_H and W_L are the probabilities for the high-spin and low-spin states, respectively and E_H and E_L are the energies for the high-spin and low-spin states, respectively. In terms of the Boltzmann distribution:

$$\Delta H^\circ = -N(E_H - E_L)$$

and $$\Delta S^\circ = -Nk \ln W_H/W_L$$

where N = Avogadro's number and

k = Boltzmann constant

Values of ΔH° and ΔS° for the transition from the low-spin to the high-spin states have been determined for all the hemoglobin and myoglobin complexes[49,52]. The ΔH° values indicate that for each protein the stability of the low-spin state relative to the high-spin state follows the order imidazole > N_3^- > OH^- > OCN^- > H_2O. This order agrees with the spectrochemical series.

The contribution of the globin to the spin equilibrium is indicated by the fact that the low-spin state is more stable for hemoglobin than myoglobin[52] for all the ligands studied. There is also a large ΔS° which most probably arises from the interaction with the globin. It has been suggested[89] that the ΔS° might be related to a cooperative breaking of van der Waals contacts between the porphyrin and globin.

Recently, it has been found by e.p.r. and visible spectroscopy of isolated α and β chains of ferrihemoglobin that the spin equilibrium in the β chain is influenced by association into tetrameric hemoglobin[48]. Thus, Hb^+H_2O and α^+H_2O are only 5% low-spin at 0° while β^+H_2O is 67% low-spin. This change in the spin equilibrium was attributed to a change in the interaction of the proximal histidine (F8) with the iron. A similar interaction may help to determine the spin state for different complexes.

The requirement for iron(II)

Many metallo-enzymes have been found to function quite adequately if the physiological metal ion is replaced by various other metal ions. In fact in the case of carboxypeptidase (Chapter 15) the replacement of Zn(II) by Co(II) increases the activity of the enzyme[90].

The inability of Fe(III) hemoglobin and myoglobin to bind oxygen has already been mentioned. This effect is thought to be related to the fact that Fe(III) cannot as readily donate electrons to oxygen from its t_{2g} orbitals as can Fe(II)[91].

Porphyrins (Chapter 24) form strong complexes with most metal ions[92]. However, the only functional protein in which Fe(II) of hemoglobin or myoglobin is replaced by another metal ion is coboglobin[93], which contains Co(II) in place of Fe(II). The affinity of coboglobin derived from hemoglobin for O_2 is about a factor of three lower than that of hemoglobin, and the heme-heme interaction, while appreciable, is less than that of hemoglobin (Fig. 5). Coboglobin derived from myoglobin also has a lower affinity for oxygen than myoglobin.

The reversible oxygenation of a variety of cobalt complexes (Chapter 20) has been known for some time[94-97]. It is, therefore, not completely surprising that cobalt should be able to replace iron in hemoglobin and myoglobin. In fact, although Fe(II) heme is oxidized to Fe(III) heme in the presence of oxygen, it is possible to reversibly bind O_2 to cobalt protoporphyrin[93] providing a base is bound to the fifth coordination position. In the absence of a base, cobalt, like iron, is oxidized. E.p.r. spectra indicate that the electronic structure of cobalt(II) protoporphyrin is very similar to that of coboglobin. However, the affinity for oxygen increases by a factor of 600 or more when the porphyrin is bound to the globin[93].

References pp. 891–901

Co(II) has 7 *d*-electrons compared to 6 *d*-electrons in Fe(II). Thus, octahedral Co(II) complexes are paramagnetic in both the low-spin and high-spin state. It is, therefore, possible to obtain e.p.r. spectra of low-spin oxygenated coboglobin, which is impossible for diamagnetic HbO_2 and

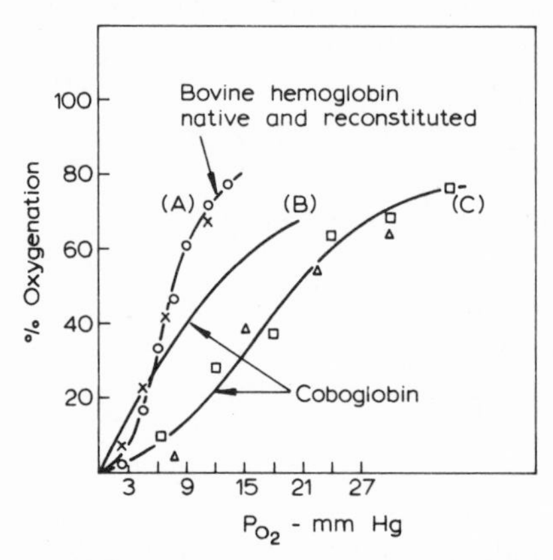

Fig. 5. The oxygenation of coboglobin in 0.2 *M* phosphate buffer, pH 7.0, at 14–15°C. (A), hemoglobin, native (○), and reconstituted (×); (B), coboglobin prepared from globin preparations which showed non-cooperative oxygenation curves when reconstituted with hematin; (C), coboglobin from two samples of globin which showed co-operative oxygenation curves when reconstituted with hematin; (□) is the same globin preparation as (×) in curve A. (From ref. 93 by permission of the authors.)

MbO_2. These spectra[93,96,98,99] indicate that the oxygen is bound in a bent orientation.

This configuration lifts the degeneracy of the $2p\pi^*$ orbitals of molecular oxygen and the two unpaired antibonding electrons are paired in one orbital. It is found that the remaining unpaired electron of the complex is localized mainly on the oxygen with the Co(II) formally oxidized to Co(III) and the oxygen reduced to the superoxide ion $O_2^{\cdot -}$.

In many respects the ligand reactions of coboglobin are similar to those of hemoglobin and myoglobin. Nevertheless, there are several important differences[93]. The oxygenation of coboglobin does not involve a change in spin state as found for hemoglobin and myoglobin, since both the deoxy and oxy forms of coboglobin are low-spin complexes. Furthermore, CO does not bind to coboglobin even though hemoglobin and myoglobin have a greater affinity for CO than O_2.

Modified hemoglobins or myoglobins where iron is replaced by Mn(II), Mn(III), Ni(II), Cu(II), or Zn(II) have also been prepared[100–104]. However, none of these derivatives reversibly binds oxygen. This behavior has been partly explained by the different number of d-electrons in these complexes[71,102,103]. Ni(II) and Cu(II) porphyrin complexes with eight and nine d-electrons, respectively, coordinate additional ligands with great difficulty, probably because of the repulsive effect of the filled d_{z^2} orbital. Planar Zn(II) porphyrin complexes with ten d-electrons are much weaker than other porphyrin complexes possibly because of the tendency for d^{10} complexes to be tetrahedral. Furthermore, these Zn(II) complexes can coordinate with only one additional ligand. It has been suggested that the zinc is then only coordinated to three porphyrin nitrogens. This hypothesis is consistent with o.r.d. and c.d. spectra of myoglobin containing Zn(II) instead of Fe(II). While the replacement by Cu(II) does not affect the conformation of the protein, replacement by Zn(II) has an appreciable influence on the protein conformation[103].

Role of the porphyrin

Interactions involving the porphyrin and the rest of the complex.
The highly conjugated porphyrin macrocycle interacts very strongly with the iron in myoglobin and hemoglobin. The four pyrrole nitrogens form σ-bonds with the iron while the π-system interacts with d-electrons in the t_{2g} orbitals.

Nuclear magnetic resonance spectroscopy of paramagnetic hemoglobin and myoglobin complexes indicates that the porphyrin proton resonances are shifted by Fermi contact interactions which arise from the delocalization of the unpaired electron spin[73,105-110]. In low-spin Fe(III) complexes with $S = \frac{1}{2}$ the unpaired electron is in the t_{2g} orbitals and spin delocalization occurs via the π-system of the porphyrin. For Mb^+CN and Hb^+CN it has been calculated that about 1% of the spin resides at each peripheral carbon of the porphyrin ring[105,107,108]. In the high-spin Fe(III) complexes[109,110] with $S = \frac{5}{2}$ unpaired electrons are found in all the d-orbitals, including the $d_{x^2-y^2}$ orbital involved in σ-bonding with the porphyrin. It has, therefore, been suggested[109] that some of the spin delocalization may take place through the σ-bonding system.

The interaction of the iron with the π-system of the porphyrin is also apparent from a study of the visible absorption and rotatory dispersion spectra of hemoglobin and myoglobin[39,58,59,61,111-123]. The electronic transitions in this region arise predominantly from porphyrin $\pi \rightarrow \pi^*$ bands[120-122] and these spectra are influenced by the oxidation state[112,115,120] and the spin state[49,50,114,118] of the iron (Fig. 6).

As discussed in Chapter 24, the structure of the porphyrin influences the axial ligands, and the axial ligands affect the properties of the porphyrin.

References pp. 891–901

Thus, the porphyrin spectra of hemoglobin complexes such as HbO_2 and HbCO which are in the same oxidation state and spin state depend on the axial ligand[111,112,114] (Fig. 6B). It has also been possible to study the effect of the porphyrin on the axial ligands of hemoglobin and myoglobin by using proteins reconstituted with different hemes. The carbon monoxide stretching frequency (Table III, Chapter 24) and the oxygen affinity[14,124-127] depend on the substituents at positions 2 and 4 of the porphyrin (Fig. 4,

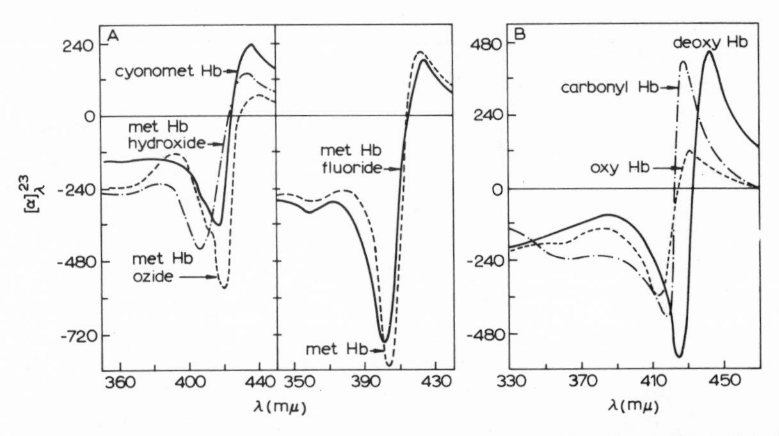

Fig. 6. The Soret optical rotatory spectra of hemoglobin derivatives. A-left: Fe(III) hemoglobin derivatives completely or partly low-spin. (—), Hb^+CN; (- - -), Hb^+N_3; (-·-·), Hb^+OH. A-right: Fe(III) hemoglobin derivatives predominantly high-spin. (—), Hb^+F; (- - -), Hb^+H_2O. B: Fe(II) hemoglobin derivatives. (—), Hb; (- - -), HbO_2; (-·-·), HbCO. (From ref. 114 by permission of the authors.)

Chapter 24). The oxygen affinity for hemoglobin with ethyl, hydrogen, hydroxyethyl, and vinyl substituents are in the ratio of 5:2:1.3:1[127]. This order is the reverse of that found for the negative induction effects of these groups, suggesting the importance of the π-electron density of the porphyrin in regulating the oxygen affinity of hemoglobin. It has, therefore, been proposed that the withdrawal of π-electron density from the porphyrin ring strengthens iron to porphyrin nitrogen π-bonds and thereby weakens the π-bonds of the iron to the axial ligands[127,128]. Such π-bonds are thought to play an important role in the binding of ligands[71,72,91,129] to the sixth coordination position particularly in Fe(II) hemoglobin and myoglobin complexes.

The studies of hemoglobin and myoglobin reconstituted with modified hemes have also helped to elucidate some of the interactions of the porphyrin with the globin[14,130,131]. Substitution of vinyl groups at positions 2 and 4 on the porphyrin for other groups decreases the cooperativity of oxygen binding significantly[124-127]. Modification of the heme at these positions also decreases the affinity for globin and enhances the denaturation of reconstituted hemoglobins[126,130]. Elimination of the negatively charged propionic acid side chains by esterification or replacement by an ethyl group was thought to affect the stability and oxygenation properties of reconstituted hemoglobins and myoglobins[14,132-134]. However, a recent study has found that the propionic acid side chain seems to have no effect on the oxygenation or spectral properties of hemoglobin[127].

Displacement of iron from the plane of the porphyrin

The metal in Ni(II), Cu(II), and Pd(II) metalloporphyrins lies in the plane of the four pyrrole nitrogens[81,135-137] to which it is coordinated. However, for almost all iron metalloporphyrins, "hemes", investigated[81,138-140] the iron is 0.4–0.5 Å out of the plane of the pyrrole nitrogens (Fig. 18, Chapter 24). The only exception seems to be the low-spin Bis(imidazole-$\alpha,\beta,\gamma,\delta$-tetraphenyl-porphinatoiron(III))[141] where the iron is essentially located in the plane of the nitrogens. Hoard[81] has explained this peculiar phenomenon in terms of the radius of the central hole of the porphyrin, Ct–N, which gives the distance of each pyrrole nitrogen from the center of the grouping. The radius of this cavity is limited by the cyclic structure of the porphyrin and has been found to be 2.015 ± 0.007 Å for all metalloporphyrins. The metal lies in the plane of the nitrogens only if the metal nitrogen bond distance can be equal to Ct–N. Consistent with the ionic radius for octahedral high-spin Fe(III)[142], the Fe–N bond distances for the high-spin hemes lie within the range of 2.061 ± 0.012 Å which is clearly larger than the radius of the central hole. Low-spin planar complexes are feasible because of the 1.91 Å Fe–N bond distance predicted for low-spin iron[142].

The protein X-ray studies can determine the plane of the porphyrin, but not the plane of the pyrrole nitrogens. The Fe is about 0.3 Å out of the plane of the porphyrin in all the Fe(II) and Fe(III) high-spin myoglobin derivatives[3,4,12,23,24,143,144] whose X-ray structures have been determined. The iron is thought to remain out of the plane in some low-spin complexes[145-148], but at least in Mb^+CN the iron seems to be closer to the plane of the porphyrin[149]. In hemoglobin the iron is displaced about 0.3 Å from the plane of the porphyrin toward the proximal histidine in the high-spin Fe(III) complexes[1,2,38]. However, for deoxyhemoglobin the displacement of the iron is about 0.75 Å[25,26,38,150,151], consistent with the even larger radius for high-spin Fe(II) than for high-spin Fe(III) and the lack of any ligand in the sixth position to pull the iron toward the plane of the porphyrin. At least in certain low-spin Fe(III) hemoglobins there is an indication that the iron may lie in the plane of the porphyrin[152-154]. These results imply that the iron may move as much as 0.75 Å when oxygen is bound to hemoglobin. It has been proposed that this movement may trigger heme–heme interaction in hemoglobin[81,140].

The out of plane geometry of iron in hemoglobin and myoglobin will lower the energy of the $d_{x^2-y^2}$ and d_{xy} orbitals, since the pyrrole nitrogens are below the x–y plane defined with the iron as the origin of the coordinate system. This effect will influence the relative energies of the $d_{x^2-y^2}$ and d_{z^2} orbitals; however, the effect on the relative energies of the more closely spaced t_{2g} orbitals is particularly significant. This geometry has been used to explain[39,155] why the d_{xy} orbital, which cannot form π-bonds, is the lowest orbital in $Hb^{+}N_3$[39,44,155,156] as determined by crystal e.p.r. spectroscopy.

Planarity of the porphyrin

The delocalization of π-electrons in the porphyrin is expected to favor a planar porphyrin structure. However, X-ray studies on various porphyrins indicate that the porphine skeleton is readily deformable into a ruffled[135,157] or domed[138,140] configuration. Ruffled structures have been observed in which some atoms are as much as 0.4 Å above the mean plane and others 0.4 Å below it. Deformations have been observed with and without the presence of a metal ion or axial ligands.

The explanation offered for these deformations[81,135] is that a planar configuration does not minimize the angular strains in the pattern of σ-bonding. The opposing forces of the σ strain and π planarity produce a quite flexible porphine skeleton which can bend to the configuration favored by the rather weak crystal packing forces.

The X-ray determination of hemoglobin and myoglobin structures are not refined to the point at which the detailed configuration of the porphyrin can be observed. However, these studies on models indicate that the numerous interactions between the globin chain and the porphyrin can control the configuration of the flexible heme and thereby the interactions between the porphyrin and iron. These interactions may contribute to the rhombic distortions observed when the heme is combined with globin.

Role of the fifth ligand

As illustrated by the interaction of the porphyrin with the sixth coordination site, all the ligands of a metal complex interact with each other. However, the mutual interaction between the two coordination sites on opposite sides of the metal, known as the "*trans* effect", is of particular importance. This effect is the result of both ligands interacting with the same *d*-orbitals.

In hemes the fifth and sixth ligands σ-bond with the d_{z^2} orbital and ligand π-electrons interact with metal electrons in the d_{xz} and d_{yz} orbitals. It is, therefore, not surprising that the binding of CO is very sensitive to the *trans* ligand. Thus, the mixed hemochrome in which one ligand is a base-like pyridine and the other ligand is CO is much more stable than the complexes in which the fifth and sixth ligands are identical[158–161]. In fact, CO cannot be bound to both axial positions. Furthermore, replacing H_2O by the more basic pyridine in the fifth position increases the affinity of heme for CO by an order of magnitude[159]. The effect of the basicity of the pyridine on the CO stretching frequency, ν_{CO}, of 4-substituted pyridine-2,4-diacetyldeuteroheme dimethylester carbonyls has also been investigated[162,163]. Changing the 4-substituent from cyano to amino and thus increasing the basicity decreases ν_{CO} from 1988.5 to 1971.1 cm^{-1}.

The decrease in ν_{CO} and the increase in affinity with the more basic *trans* ligand can be understood in terms of the partial double bond character of the bond between iron and the sixth ligand[91] (see below).

$$\mathrm{(N)_4{-}Fe{-}C{\equiv}O \rightleftharpoons (N)_4{-}Fe{=}C{=}O}$$

The formation of a double bond involves the donation of electrons from the d_{xz} and d_{yz} orbitals into π bonds made with p or π orbitals of the sixth ligand[158]. The stronger base polarizes these metal orbitals toward the sixth ligand strengthening the sixth ligand metal bond. However, stabilizing the Fe–C double bond for CO weakens the C–O bond thereby lowering ν_{CO}[128,129].

As mentioned above, a basic *trans* ligand is necessary to prevent cobalt protoporphyrin from being oxidized[93,95,96], thereby facilitating reversible oxygenation. This behavior can be understood in terms of a strengthening of the Fe–oxygen bond. However, too much strengthening of the Fe–ligand bond can produce irreversible binding of the sixth ligand. This phenomenon is illustrated by the iridium oxygen carrier, chloro-carbonyl-bis(triphenyl-phosphine) iridium, $IrCl(CO)(PPh_3)_2$ (Chapter 20)[164]. If the chlorine is replaced by the less electronegative, more basic, iodine the oxygen binds irreversibly[165].

The fifth ligand, which is *trans* to O_2 in HbO_2 and MbO_2, is a part of the polypeptide chain. The only proteins whose fifth ligand may not be an imidazole ring are two genetically abnormal hemoglobins[166,167] M Iwate[168,169] and M Hyde Park[170,171] (Table I). In both cases histidines of the α and β chains, respectively, are replaced by less basic tyrosine residues. The abnormal subunits do not reversibly bind oxygen and are oxidized instead. However, these effects cannot be completely attributed to the *trans* effect, since the replacement of histidine by the bulkier tyrosine will influence the structure of the protein. In fact, in hemoglobin M Iwate, but not hemoglobin M Hyde Park, the iron is coordinated to the distal histidine E7 (Fig. 1) as well as to the tyrosine, and the coordination of an external ligand requires the displacement of one of these internal ligands[185].

Since the fifth ligand is the only ligand covalently attached to the protein, changes in the complex as a result of binding ligands may be transmitted to the protein primarily through the fifth ligand. On the other hand, the protein conformation can be expected to influence the rest of the heme complex through the fifth ligand. The polypeptide chain can also be expected

TABLE I

ABNORMAL HEMOGLOBINS WHERE RESIDUES IN THE REGION OF THE HEME ARE SUBSTITUTED

Designation	Substitution	Effect on oxygenation and other properties	Ref.
M Iwate	His F8(87)α → Tyr	α-chains oxidized, β-chains low affinity	168,169
M Hyde Park	His F8(92)β → Tyr	α-chains normal affinity, β-chains oxidized	170,171
M Boston	His E7(58)α → Tyr	α-chains oxidized, β-chains low affinity	172
M Saskatoon	His E7(63)β → Tyr	α-chains normal affinity, β-chains oxidized	172
Zürich	His E7(63)β → Arg	Very high affinity, decreased cooperativity— Arginine does not fit in ligand pocket	173–175
M Milwaukee	Val E11(67)β → Glu	α-chains are oxidized— COO^- can form salt bridge with Fe(III)	172
Bristol	Val E11(67)β → Asp	Low affinity, COO^- can close pocket by forming salt bridge with His E7	176
Sydney	Val E11(67)β → Ala	Unstable, heme of β-chain readily lost	177
Hammersmith	Phe CD1(42)β → Ser	Low affinity — influences tilt of heme in liganded conformation	178,179
Bucuresti	Phe CD1(42)β → Leu	Low affinity — influences tilt of heme in liganded conformation	180
Santa Ana	Leu F4(88)β → Pro	No heme in β-chain	166
Borås	Leu F4(88)β → Arg	High affinity — gap on proximal side of heme	181,182
Köln	Val FG5(98)β → Met	High affinity	183,184

to orient the imidazole in a fixed configuration relative to the plane of the heme[186]. The degeneracy of the d_{xz} and d_{yz} orbitals will then be lifted producing rhombic distortion[39,156]. The e.p.r. spectra of abnormal hemoglobins in which the proximal histidine, F8, is replaced by tyrosine indicate an enhanced rhombic distortion[46,171,187-191] consistent with the contribution of the fifth ligand to this distortion.

THE IRON-OXYGEN BOND

Despite X-ray studies, and the use of e.p.r., n.m.r., and Mössbauer techniques, the controversy regarding the structure of the Fe–oxygen bond which began with Pauling[91,192] and Griffith[76] in the 1950s is still unresolved. In fact, additional hypotheses have been proposed in recent years[193,194].

Deoxyhemoglobin is in the high-spin paramagnetic state with four unpaired electrons[78], and there is one unpaired electron in each of two degenerate antibonding π^* orbitals of oxygen[195]. The oxygenation of hemoglobin produces a diamagnetic low-spin molecule with all the iron d-electrons in the t_{2g} orbitals[78]. It is generally accepted that the Fe–oxygen bond has considerable double bond character with the electrons in the d_{xz} and d_{yz} orbitals contributing to the bonding, as was mentioned earlier for HbCO and MbCO. The controversy revolves around the orientation of the oxygen relative to the heme plane and the extent to which electrons are transferred to the oxygen from the iron.

Orientation of oxygen relative to the heme

The first proposed structure of the O_2–iron bond involved a linear bond with the oxygen perpendicular to the plane of the porphyrin[192]. It was later recognized that such a structure would not lift the degeneracy between the two antibonding orbitals and could, therefore, not explain the diamagnetism of the complex. Later Pauling[91,196] proposed a similar structure with a 120° angle between the Fe and the oxygen.

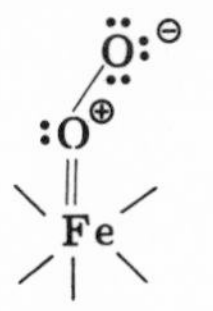

In this structure the two unpaired oxygen electrons are paired up into one of the antibonding orbitals and a double π-bond is formed involving these electrons and the d-electrons in the d_{xz} and d_{yz} orbitals.

Griffith[76] proposed an alternative structure with oxygen parallel to the plane of the heme.

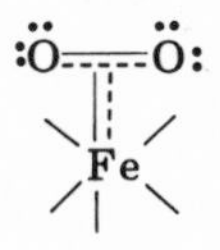

In this structure the two unpaired oxygen electrons are involved in σ-bonding to the Fe with additional π-bonding involving the d_{xz} and d_{yz} metal orbitals and other π orbitals on the oxygen molecule. Griffith based his argument on the fact that the unpaired antibonding electrons are more weakly bound to oxygen and more readily involved in binding to iron[76].

The Pauling structure is similar to that found for the Co–O_2 bond in coboglobin[93,95,96], and the Griffith structure is similar to the Ir–O_2 bond in the iridium oxygen carriers[164,165]. Model studies thus indicate that oxygen can bind in both ways.

Theoretical calculations for the Fe–oxygen bond are inconclusive. The molecular orbital calculations of Maggiura *et al.*[197] favor the Pauling structure, while the LCAO-MO calculations of Gouterman and Zerner[198,199] indicate that the Griffith structure should produce a stable low-spin complex.

Several attempts have been made to determine the orientation of oxygen in hemoglobin and myoglobin directly. A recent X-ray determination of MbO_2[200] suggests that the axis of the oxygen molecule may form an angle of 120° with the plane of the porphyrin, consistent with the Pauling structure. ^{17}O nuclear magnetic resonance[201-203] has been used in an attempt to differentiate between the identical environment of both oxygen atoms predicted by the Griffith structure, as opposed to the asymmetric environment predicted by the Pauling structure. It was at first thought that only one resonance line is observed[201]. However, it has recently been found that the only resonance detected is due to $H_2{}^{17}O$ and the bound oxygen cannot be detected[203].

Hoard[81] has calculated distances of the oxygen atoms from the pyrrole nitrogens for various geometries of oxygen binding and finds that the Griffith model produces an unfavorably short N–O distance of less than 2.60 Å, unless the Fe–imidazole or Fe–oxygen distance is increased to the extent that only very weak bonding to the fifth and sixth ligand is permitted.

Electron transfer from iron to oxygen

In addition to the question of the orientation of the oxygen iron bond, there has also been a controversy over the extent to which electron density is transferred from the iron to the oxygen[193,194,196,204-206]. In the Pauling[91,192,196] and Griffith[76] structures π electrons are shared without any significant electron transfer.

It has also been proposed that in oxyhemoglobin and oxymyoglobin one electron is transferred to oxygen from iron[193,206]; the binding can be described as a low-spin complex of Fe(III) with a superoxide anion radical ($\cdot OO^-$). In this model HbO_2 and MbO_2 have two unpaired electrons, one localized on the iron and one localized on the oxygen. The diamagnetism results from spin-coupling of the unpaired spins on adjacent atoms[206]. An analogous transfer of electrons from the metal to oxygen has been established by e.p.r. spectroscopy for the reversible binding of oxygen to various cobalt complexes[95,96] including coboglobin[93].

Molecular orbital (MO) calculations[205] and the Mössbauer effect[42,207] both indicate that electron density is transferred from iron to oxygen in HbO_2. The MO calculations suggest 0.5–1.0 electronic charge units[205] are transferred to oxygen, consistent with the formation of a superoxide.

Recently Gray[194] has questioned the superoxide model on the basis of the position of the superoxide ion in the spectrochemical series which suggests that a heme–superoxide complex should be high-spin, *i.e.*, HbO_2 should be high-spin. Furthermore, unsymmetrically bound O_2 or $O_2^{\cdot-}$ should be observable in the infra-red (i.r.) by an O–O stretching frequency. Since no such i.r. absorption is observed, Gray has suggested that oxygen binding to hemoglobin involves the transfer of two electrons from the iron to the oxygen resulting in a seven coordinate Fe(IV) complex with O_2^{2-}.

REVERSIBLE OXYGENATION

Fe(II) heme which is not attached to the globin cannot reversibly bind oxygen in aqueous solution[161], but instead is oxidized to the Fe(III) form which no longer binds oxygen. This is true even though in the absence of oxygen Fe(II) heme can combine reversibly with other ligands (Chapter 24) such as carbon monoxide, ammonia, pyridine, imidazole and cyanide. In myoglobin and hemoglobin autoxidation remains a problem[29,30], since these proteins are still slowly oxidized in the presence of oxygen. However, the rate of autoxidation is at least several orders of magnitude slower than in the absence of the globin.

Environment of the heme

The environment of the hemes in hemoglobin and myoglobin has been elucidated by X-ray diffraction studies[1-4,6,12,208,209]. As mentioned previously, the hemes are bound into crevices in each polypeptide chain and are far removed from each other in hemoglobin (Figs. 1 and 2). As in the case of other globular proteins, the interiors of these molecules contain closely packed non-polar side chains while the polar side chains lie on the surface[12]. The only polar side chains not on the surface are the proximal histidines bound to the iron, the distal histidines in the ligand pocket, and an internally hydrogen bonded threonine side chain[2].

In addition to the coordination bond between the iron and the proximal histidine F8 there are in hemoglobin and myoglobin about sixty interactions involving a contact of $\leqslant$ 4 Å between atoms of the globin chain and each heme[2,12,208]. All of these contacts, except one or two which involve the polar propionic acid groups at the surface of the molecule, are non-polar or hydrophobic. The environment of the heme in the β-chain which is similar to that of the α-chain and myoglobin is shown in Fig. 7.

The oxygen binding site on the distal side of the heme is also buried inside the molecule. Pointing into this pocket is histidine E7, valine E11 and slightly further away in contact with the porphyrin is phenylalanine CD1 (Figs. 1 and 7). The ligand pocket is therefore limited in size and hydrophobic.

Role of the protein in preventing autoxidation

Different aspects of the heme environment have been emphasized in attempts to explain the role of the protein in preventing autoxidation and

References pp. 891–901

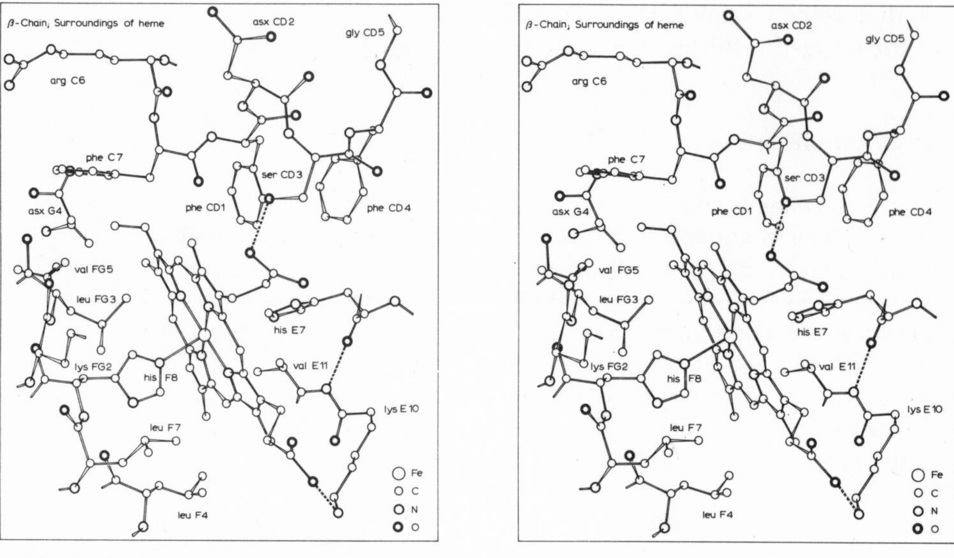

Fig. 7. A stereo drawing of the surroundings of the heme in the β chain of liganded horse hemoglobin. Broken lines indicate hydrogen bonds. (From ref. 166 by permission of the authors.)

thereby enabling hemoglobin and myoglobin to function as reversible oxygen carriers.

Even before the environment of the heme in myoglobin and hemoglobin had been determined by X-ray studies, Wang and co-workers[161] proposed that an environment of low dielectric constant could inhibit the charge separation which occurs upon oxidation and such an environment may be required for reversible oxygenation. This hypothesis was supported by the demonstration that it was possible to reversibly oxygenate heme embedded in a hydrophobic polystyrene matrix[210,211]. It was also shown that the rate of oxidation of dipyridine ferrohemochrome by molecular oxygen depends on the dielectric constant of the solvent[212].

Caughey and co-workers[213,214] (Chapter 24) found that the oxidation of hemes by oxygen often passes through an intermediate oxygen-bonded dimer. It was, therefore, suggested that the impossibility of forming such an intermediate with the heme buried inside the globin may be partly responsible for making reversible oxygenation possible.

N.m.r. and tritium incorporation experiments indicate that oxidation and reduction of hemes involve peripheral attack upon the porphyrin ring[215]. The fact that all parts of the periphery of the porphyrin except the propionic acid groups are inaccessible in hemoglobin and myoglobin may also help to explain the stability of hemoglobin and myoglobin to oxidation.

THE BINDING OF LIGANDS TO MYOGLOBIN

Myoglobin possesses one heme and one reactive ligand binding site (Fig. 1). Therefore, the ligand reactions are described by a simple hyperbolic binding curve[14,216-219] (Fig. 8) analogous to those obtained for hemes[159] or other model compounds[220]. It is, therefore, not immediately obvious whether or not the protein influences the ligand reactions. In fact, it has recently been reported[220] that even the thermodynamics for the binding of oxygen to bis(acetylacetone)-ethylenediiminecobalt(II) is quite similar to that of myoglobin. Nevertheless, a comparison of the ligand reactions of hemes with those of myoglobin indicates that the protein does influence the binding of ligands.

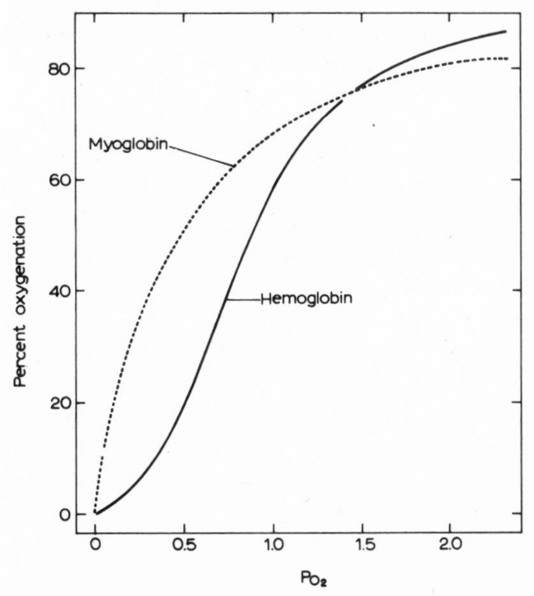

Fig. 8. The oxygenation of myoglobin and hemoglobin. (----), sperm whale myoglobin pH 8.6 in 0.005 *M* Tris tham chloride buffer at 25°C from Keyes[219]. (——), human hemoglobin pH 9.1 in 0.2 *M* borate buffer 19°C from Roughton and Lyster[67].

Role of the protein residues in the vicinity of the heme

The role of the low dielectric constant of the protein environment of the heme in regulating the affinity of myoglobin and hemoglobin for ligands is dramatized by a comparison of the binding of CN^- and CO[161]. These two ligands are isoelectric, almost identical in size and shape, and both bind to heme with a similar affinity[161]. However, neutral CO binds to myoglobin and hemoglobin with a greater affinity than to heme while negative CN^-

References pp. 891–901

essentially does not bind to the proteins in the Fe(II) forms. The affinity of CN^- for Fe(II) hemoglobin is at least five orders of magnitude smaller than for heme[161,221]. This difference has been attributed to the requirement for electro-neutrality[91] in the low dielectric heme pocket[161]. Thus, no ionic ligands bind Fe(II) myoglobin or hemoglobin. On the other hand, the affinity of CN^- for ferrimyoglobin and ferrihemoglobin, in which iron carries an extra positive charge, is very high and most of the strongly bound ligands are anions[91] which neutralize this extra charge.

The amino acid side chains in the region of the sixth ligand binding site not only provide a hydrophobic environment, but restrict the size and orientation of ligands bound to myoglobin and hemoglobin.

The X-ray structure of Mb^+N_3 [145] indicates that the azide is oriented away from the propionic acid side chains toward the α-*meso* proton of the heme (Fig. 4, Chapter 24) with an angle of 21° between the azide and the heme plane, *i.e.*, the Fe–azide bond angle is 111°. This orientation is dictated by steric considerations involving the histidine E7, phenylalanine CD1, and valine E11 side chains, which point into the ligand pocket and restrict the freedom of rotation of the ligand.

The ligand pocket also influences the Fe–CO bond. The CO stretching frequency, ν_{CO}, for protoheme mono-pyridine monocarbonyl is 1977 cm^{-1} and has a broad half-band width of 30 cm^{-1}, while for HbCO ν_{CO} is 1951 cm^{-1} with a narrow half-band width of 8 cm^{-1} [129,162,163,222], and for MbCO two bands are observed, a major narrow band at 1944 cm^{-1} and a minor band at 1931 cm^{-1} [222,223]. It has been suggested[129] that as a result of the close fitting pocket the Fe–CO bond is non-linear and the orientation of the CO in the pocket is fixed, as found for Mb^+N_3. The fixed orientation would decrease the band width while the bent structure is expected to enhance the double bond character of the Fe–CO bond[224], which should decrease the C–O stretching frequency[128,129]. The fact that ν_{CO} for myoglobin is even lower than for hemoglobin implies an even tighter ligand pocket in myoglobin.

The geometry of the sixth ligand, which can be determined by the protein environment of the heme, not only influences the Fe–ligand bond, but affects the structure of the whole complex. The bent azide bond is thought to explain the very large rhombic distortion of the Mb^+N_3 and Hb^+N_3 complexes[77]. The interaction of the ligand with the heme pocket also seems to be influenced by the spin state of the complex. Thus two i.r. bands are obtained for Mb^+N_3 [225,226] and the ratio of the areas agrees with the relative amounts of high-spin and low-spin complex found on the basis of magnetic susceptibility measurements[49,52].

Involvement of the total protein conformation

Myoglobin contains 153 amino acids on one polypeptide chain (Fig. 3)[8] which is wrapped around the heme (Fig. 1). Thus, residues 43, 64, and 68 (phenylalanine CD1, histidine E7, and valine E11) surround the distal side of the heme where the sixth ligand is bound (Fig. 1). Residue 93, the proximal histidine F8, is part of a rigid helical segment 9 residues long (Fig. 3) and directly coordinated to the iron (Fig. 1). In addition, close contact with the heme is made by residues 71 (alanine E14), 97 (histidine FG3), 99 (isoleucine FG5), 104 (leucine G5), 107 (isoleucine G8), and 138 (phenylalanine H15) (Fig. 1). It therefore seems that any interaction which influences the protein conformation almost anywhere in the molecule should be able to influence some of the residues in close proximity to the heme, and thereby influence the binding of ligands. In a reciprocal fashion, the binding of ligands which change the electronic structure of the iron complex and thereby the interaction of the iron and porphyrin with the portions of the protein adjacent to them, is expected to produce observable changes even in regions of the protein far removed from the heme. This type of "allosteric" linkage between the active site and other sites often far removed from the active site is fairly common in proteins[227-230]. The "rack" mechanism proposed by Lumry and Eyring[186,231] to explain the structure-function relationships of proteins encompassed such an inter-relationship between the conformation of the polypeptide and the electronic structure of the heme.

Such a reciprocal interaction between the protein conformation and the electronic structure of the heme is observed when apomyoglobin (myoglobin without the heme) recombines with heme[232,233]. A 20% increase in the rotatory strength in the far ultra-violet has been interpreted in terms of 15 additional protein residues in a helical configuration. At the same time the absorption and n.m.r. spectra of hemes[105-107] are perturbed by reaction with globin.

Ligand induced conformational changes

In view of these considerations, structural changes in myoglobin are expected, when the ligand at the sixth coordination site is changed. Nevertheless, the three dimensional X-ray diffraction studies of various myoglobin derivatives do not indicate any clear-cut structural changes in the protein conformation even at 1.4 Å resolution[3,4,12,23,24,143-148]. Some differences have been noticed when $Mb^{+}CN$ and $Mb^{+}OH$ are compared with the high-spin Fe(III)[143,144,147-149] and Fe(II)[23,24] derivatives; however, these changes are localized mainly in the immediate region of the heme, and some of the differences in $Mb^{+}OH$ have been attributed to partial denaturation of the protein[147,148].

Surprisingly, even the expected change in the position of the iron relative to the plane of the heme, when comparing high-spin and low-spin complexes, is not generally observed. In most of the derivatives investigated the position of the iron seems to be fixed at about 0.3 Å out of plane of the porphyrin on the side of the proximal histidine[4,23,24,144-146]. $Mb^{+}CN$ is the only derivative in which the iron seems to be displaced towards the plane of the heme as expected in the conversion from high-spin to low-spin complexes[149].

The failure to observe a change in the X-ray structure when ligands are bound does not necessarily imply that the protein is not involved in regulating the binding of ligands to myoglobin. In fact, the relative insensitivity of the position of the iron to the spin state of the complex in myoglobin, unlike model hemes, can probably be attributed to the protein. Thus, at least in the low-spin myoglobin complexes, the X-ray studies suggest that the protein affects the structure of the heme complex. Large rearrangements of the protein residues when different ligands are bound, are precluded by the X-ray studies. Nevertheless, it is indeed likely that there are small movements of protein residues not resolved by the X-ray technique, which may even be spread over large portions of the protein, when oxygen or other ligands are bound.

Recent n.m.r. spectroscopy studies indicate that such conformational changes do actually take place. Differences in the ring current shifts between Mb and MbO_2 clearly indicate that there are small changes of the atomic position of various aliphatic and aromatic protein residues at least in the region of the heme[62]. A specific resonance at –6.1 ppm in Mb is shifted at least 0.2 ppm upon oxygenation. These changes have been attributed to an aromatic residue, probably phenylalanine CD1, and imply the movement of this residue at least 0.2 Å away from the heme axis when ligands are bound. A slightly larger shift upon oxygenation of a resonance attributed to a methyl group of an aliphatic residue is interpreted in terms of a movement of at least 0.5 Å. The residue involved is probably valine E11 or leucine B10. Shifts have also been observed in the resonances of tryptophan residues[234], none of which are located near the heme (Figs. 1 and 3). Therefore, some conformational changes must be transmitted through relatively large regions of the molecule.

The changes in entropy for the binding of oxygen and carbon monoxide to myoglobin[218] are –60.0 and –65.7 e.u., respectively, with a standard state of 1 torr. This entropy change is quite large even when corrected to an organic solvent standard state more reasonable for the environment of the ligand in the protein. It therefore seems likely that the protein contributes significantly to the binding of ligands.

Allosteric linkage between the sixth coordination site and other binding sites

Oxygen and carbon monoxide binding studies have been performed in the presence of other small molecules, which bind to myoglobin at known sites other than the sixth coordination site of the heme. Since the affinity of the ligands is sensitive to quite small changes in free energy, these studies provide a very sensitive probe of ligand induced conformational changes in the region of the binding site for the other small molecule. X-ray crystallographic studies[147,235–237] of Mb, Mb^+H_2O, and Mb^+OH in the presence of xenon or cyclopropane indicate that xenon and cyclopropane bind at a hydrophobic binding site on the proximal side of the heme between histidine F8 and the pyrrole ring of the heme with the 2-vinyl substituent (Fig. 4, Chapter 24). N.m.r. spectroscopy indicates that the xenon and cyclopropane occupy this same site in Mb^+CN, even in solution[238]. In Mb^+OH Schoenborn[147] found a second site for xenon far removed from the heme between the AB and GH corners roughly equidistant from histidines GH1 and B5. This xenon atom makes van der Waals contact with 31 neighboring atoms. Schoenborn suggests that this site is available only at high pH, when the histidines lose a proton and cannot hydrogen bond to other groups.

Wishnia[239] has studied the binding of xenon as well as pentane and other alkanes to myoglobin and hemoglobin and has discovered that at least two molecules of these substances bind cooperatively. The site found near the heme by X-ray diffraction is undoubtedly involved. The second molecule may bind at a different site perhaps related to the high pH site found in Mb^+OH[147]. However, it is quite likely that both molecules bind at the same site near the heme.

A reciprocal interaction between the binding of these hydrophobic molecules and the binding of a ligand to the sixth coordination site has been observed. The binding constants for pentane change when deoxyhemoglobin is oxygenated[239], and there are at least two interacting xenon sites which, when occupied, increase the affinity of myoglobin for carbon monoxide (Fig. 9)[219,240,241]. It has also been observed that xenon doubles the rate for the recombination of myoglobin with carbon monoxide dissociated by flash photolysis[242]. N_2 and cyclopropane also seem to increase the affinity for CO (Fig. 9)[219,240]. However, they do not noticeably affect the rate of recombination with CO[242]. A comparison of the X-ray structures indicates that xenon is 0.8 Å closer to the proximal histidine F8 than cyclopropane[236,242]. This difference may explain the different rates for the recombination of myoglobin with CO in the presence of these two hydrophobes. The linkage between the affinity of myoglobin for carbon monoxide and the xenon binding to myoglobin is eliminated when sperm whale myoglobin is chromatographed on G-25 Sephadex (Fig. 9)[219]. An unidentified low molecular weight substance which binds strongly to myoglobin is removed by this procedure[218,219].

Certain divalent metal ions also influence the binding of O_2 and CO to myoglobin. Zn(II), Hg(II), Ni(II), Mn(II) and Cd(II) enhance the affinity, while Ca(II), Mg(II) and Co(II) have no observable effect[219,241]. Zn(II) produces the largest effect corresponding to an increase in the oxygen affinity of 67% at a metal concentration of 2×10^{-3} *M*. This concentration of Zn(II) is similar to that employed in the X-ray determination[243] of the highest affinity zinc binding site of ferrimyoglobin[244]. Therefore, the zinc binding

References pp. 891–901

site which affects the oxygenation is probably the site occupied in the X-ray studies, which is located between histidine GH1, lysine A14 and asparagine GH4 (Fig. 4B, Chapter 7). These residues are located near the surface of the molecule far removed from the heme. The effect of Cu(II), which is also known to bind in this region could not be investigated because it greatly enhances autoxidation. Interestingly the two tryptophan residues of myoglobin whose n.m.r. spectra are affected by oxygenation[234] are located at positions A5 and A12 in the same region of the molecule as the Zn(II) binding site (Fig. 4B, Chapter 7).

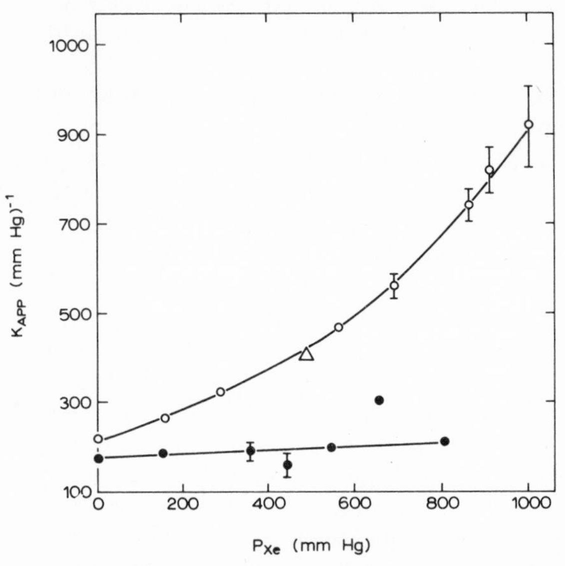

Fig. 9. The effect of xenon (or cyclopropane) on the equilibrium constant for the binding of carbon monoxide to sperm whale myoglobin. (○), xenon; (△), cyclopropane; (●), results with xenon and myoglobin rechromatographed on G-25 Sephadex. (From ref. 219 by permission of the author.)

THE BINDING OF LIGANDS TO HEMOGLOBIN

The physiological function of hemoglobin[245], unlike that of myoglobin, requires an efficient uptake and release of oxygen necessitating cooperative binding of oxygen and a mechanism for varying the oxygen affinity. The cooperative binding as shown in Fig. 8 is accomplished by a positive interaction between the four subunits (see below). Possible mechanisms for regulating the oxygen affinity involve pH[14,31,227,246] and organic phosphates[31,247-252] such as 2,3-diphosphoglyceric acid (2,3-DPG).

The variation of the oxygen affinity with pH known as the Bohr effect is shown in Fig. 10. The maximum between pH 6–6.5 corresponds to the pH range of lowest oxygen affinity. This pH sensitivity is thought to be produced by a small number of titratable amino acid residues whose environment changes when ligands are bound[253,254]. The affinity of the isolated subunits[255] is not affected by pH, as in the case of myoglobin[246] (Fig. 10).

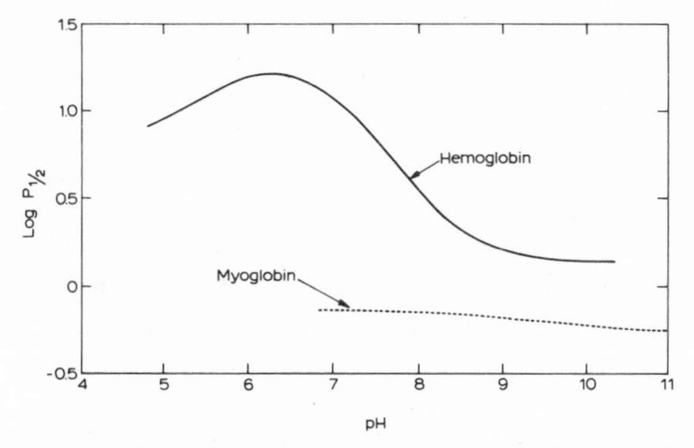

Fig. 10. The effect of pH on the oxygen affinity of hemoglobin and myoglobin. (——), human hemoglobin in 0.2 *M* phosphate, 0.4 *M* acetate, 0.05 *M* borate, or 0.4 *M* glycine-NaOH at 20°C; (- - - -), myoglobin in 0.2 *M* phosphate. (Modified from figure in ref. 246.)

It has been found that organic phosphates drastically lower the oxygen affinity of hemoglobin (Fig. 11)[250,252] and also do not affect the individual subunits[249] or myoglobin. This effect is correlated with the binding of these phosphates to hemoglobin[248,251,256]. Deoxyhemoglobin has a greater affinity for organic phosphates than oxyhemoglobin, consistent with a decreased oxygen affinity. The binding is dependent on ionic strength and under certain conditions, 2,3-DPG seems not to bind at all to oxyhemoglobin, while one mole is bound per tetramer of deoxyhemoglobin.

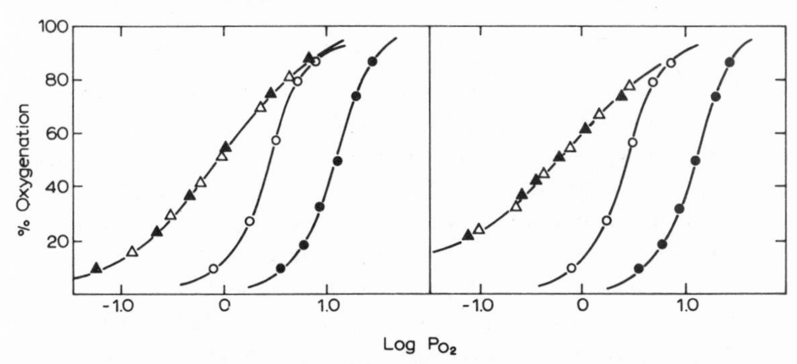

Fig. 11. The effect of 2,3-DPG on the oxygen equilibrium curves of human hemoglobin and its isolated subunits. (○, △), 0.3% protein in 0.01 *M* NaCl, pH 7.0 at 30°C; (●, ▲), 0.3% protein in 0.01 *M* NaCl and 1×10^{-4} *M*, 2,3-DPG, pH 7.0 at 30°C. Left: (△, ▲), α chains; (○, ●), human hemoglobin. Right: (△, ▲), β chains; (○, ●), human hemoglobin. (From ref. 249 by permission of the authors.)

The binding of ligands other than O_2 to Fe(II) hemoglobin is also cooperative and influenced by pH[14,31]. The major difference is in the affinity, which follows the order NO > CO > O_2 > alkyl isocyanides. On the other hand the binding of ligands to the non-physiological Fe(III) hemoglobin is essentially non-cooperative[14,31-37].

Conformational changes produced by the binding of ligands to hemoglobin

The changes in affinity for hemoglobin, though analogous to the effect of Zn(II) and xenon on myoglobin[241], are much larger. Furthermore, heme–heme interaction involves sites which are separated by 25–40 Å on different polypeptide chains[25]. It is, therefore, reasonable to expect greater structural changes when hemoglobin, rather than myoglobin, reacts with ligands.

The first direct evidence for the gross distortion of the hemoglobin structure by oxygen binding was reported by Haurowitz[257], who found that hemoglobin crystals fall apart when deoxygenated, *i.e.*, the entire crystal structure is changed. X-ray studies, even at 5.5 Å resolution[1,6,25,151,208,258], indicate that the structural change responsible for the change in crystal structure is the result of a rearrangement of the subunits of hemoglobin. A similar change in crystal structure occurs when other ligands are bound to deoxyhemoglobin[25]. In all these cases a 5-coordinated high-spin complex is transformed into a 6-coordinated low-spin complex. No similar change in structure is found when the 6-coordinated low-spin and high-spin Fe(III) crystals are compared. In fact, all the ferrihemoglobin crystals are isomorphous with oxyhemoglobin crystals, but not with those of deoxyhemoglobin. This behavior is consistent with the observation that essentially no heme–heme interaction is observed when ligands are bound to ferrihemoglobin.

X-ray

Three dimensional structures of various forms of mammalian hemoglobin have been determined by X-ray crystallography[1,2,6,25,26,258]. Unfortunately, an accurate determination of the structure of oxyhemoglobin has not yet been achieved[38]. Oxyhemoglobin crystals are oxidized in the course of collecting the X-ray data and the structure of Hb^+H_2O is obtained. Since HbO_2 crystals are isomorphous with those of Hb^+H_2O and not Hb, it has been assumed that the structural differences between deoxyhemoglobin and ferrihemoglobin are analogous to those between deoxyhemoglobin and oxyhemoglobin. Therefore, in the discussion of these differences, the "liganded form" of hemoglobin refers to the structure of any Fe(III) or Fe(II) hemoglobin with a ligand in the sixth position, while the "unliganded form" refers to deoxyhemoglobin.

X-ray studies on mammalian hemoglobins at 5.5 Å resolution[1,6,25,151,208,258] indicate that the tertiary structures of the α and β subunits are very similar to the tertiary structure of myoglobin (Figs. 1 and 2), even though there are numerous differences in the amino acid sequences[8-10] (Fig. 3). Furthermore, no changes in tertiary structure are resolved between the liganded and unliganded forms. No changes are resolved in myoglobin even at a 1.4 Å resolution (see above)[3,4,12,23,24,143-148]. However, ligand induced structural changes within the subunits of hemoglobin are delineated at 2.8 Å resolution[2,26,150]. The presence of such changes in hemoglobin, but not in myoglobin, can be explained by the fact that space filling models of the structures determined by X-ray crystallography indicate that the side chains in the interior of the myoglobin molecule are much more tightly packed than in the hemoglobin subunits[38].

In the α and β subunits of deoxyhemoglobin the iron is displaced 0.75 Å out of the plane[25,26,38,150,151] toward the proximal histidine F8 as compared to 0.3 Å in high-spin Hb^+H_2O[1,2,38]. The movement of the iron toward the plane of the porphyrin when binding a sixth ligand is expected to be at least as much for low-spin HbO_2, since the iron is thought to be in the plane of the porphyrin in low-spin hemoglobin complexes[152-154].

The ligand pocket[26,150] in the α-chain of the deoxy structure is large enough for small ligands such as H_2O, O_2, CO, and even ethyl isocyanide, if the ligand is oriented away from the protein side chains surrounding the pocket. Therefore, no rearrangement of the amino acid side chains in the ligand pocket is observed when the deoxy and liganded forms are compared[38] even though the pocket may orient the ligands, as in myoglobin[145]. However, some changes are observed in the region of the propionic side chains of the α-subunits indicating that their inclination is less steep in the liganded form. In the β-chain the ligand pocket is too small for even a small ligand like H_2O, and consequently valine E11(67) is displaced about 1 Å in the liganded form[38] to make room for the ligand; this displacement is accompanied by a decrease in the tilt of the heme in the liganded form.

Besides these changes in the region of the heme the major changes observed upon ligand binding occur on the proximal side of the porphyrins[26,38,150]. As a result of the movement of the iron, the F helix is moved inward toward the center of the molecule, narrowing the pocket between the F and H helices (Fig. 1). This pocket shrinks by 1.3 Å in the α- and by 2 Å in the β-subunits. The C-terminal residues, arginine HC3(141)α and histidine HC3(146)β, the terminal carboxyl groups, and the tyrosine residues adjacent to the C-termini (Fig. 3) are all clearly resolved in deoxyhemoglobin. Tyrosines HC2(140)α and HC2(145)β are located in the pockets between the F and H helices (Fig. 12) in deoxyhemoglobin. The phenolic hydroxyls are hydrogen bonded to the peptide carbonyls of Val FG5 whose side chain is in contact with the heme and is adjacent to residues forming part of the $\alpha_1\beta_2$ contact. In the β-subunit of deoxyhemoglobin there is also an internal salt bridge linking the imidazole of the terminal histidine HC3(146) with the α-carboxyl of aspartate FG1(94)β (Fig. 13).

References pp. 891–901

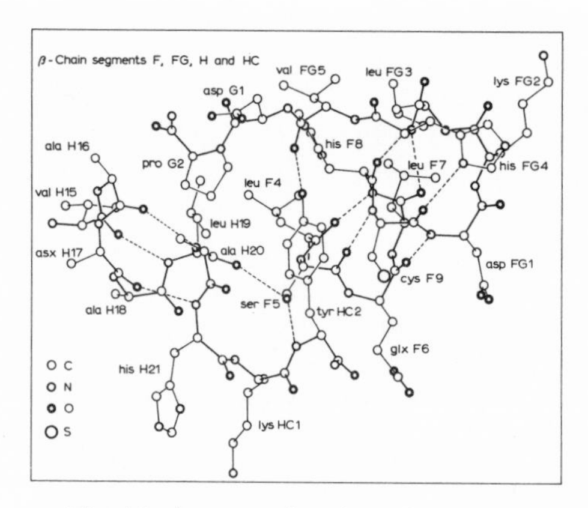

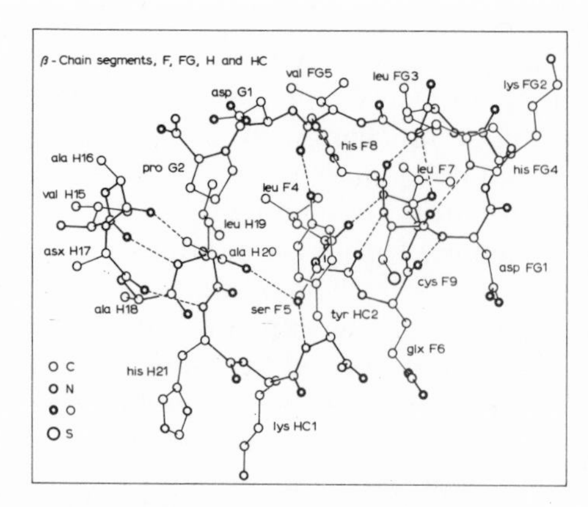

Fig. 12. A stereo drawing of segments F, FG, H, and HC of the β chain of horse hemoglobin showing the pocket between the F and H helices which contains tyrosine HC2. (From ref. 166 by permission of the authors.)

The narrowing of the pocket between the F and H helices in the liganded form of hemoglobin markedly decreases the electron density identified with the phenol ring of tyrosine[259] and the rest of the C-termini of the α- and β-subunits are not at all resolved[253]. The poor resolution in these regions of the molecule indicate a state of rapid conformational equilibrium in the liganded form at the C-termini[57].

The C-terminal residues of deoxyhemoglobin[26,150,253,254] are involved in eight salt bridges (Figs. 13 and 14). Two are within the β-subunits and six connect various subunits, 4 of them across the $\alpha_1\alpha_2$ interface and one each across the $\alpha_1\beta_2$ and $\alpha_2\beta_1$ interfaces. All of them are disrupted when ligands are bound and the C-termini are no longer in a fixed conformation[38,259]. A salt bridge connects the terminal COO^- of histidine HC3(146)β and lysine C5(40)α (Fig. 13). When ligands are bound and the tyrosine is no longer fixed in the pocket between the F and H helices (Fig. 12), this salt bridge as well as the internal salt bridge between histidine HC3(146)β and aspartate FG1(94)β are disrupted. The C-terminal arginine residues of the α-chains are involved in two salt bridges with their partner α-chains (Fig. 14). One bridge extends from the guanidinium group to aspartate H9(126) and the other from the terminal carboxyl group to the N-terminal amino group of the other chain, valine NA(1). These salt bridges are also disrupted when tyrosine HC2(140)α is no longer held in the pocket between the F and H helices in liganded hemoglobin.

The disruption of these salt bridges through ligand binding should change the pK values of the various titratable groups. It has, therefore, been proposed[253,254] that histidine HC3(146)β and valine NA(1)α which titrate in the necessary pH range, contribute to the alkaline Bohr effect (Fig. 10).

The fixed configuration of the C-terminal residues and the presence of the salt bridges connecting subunits in the deoxy conformation but not in

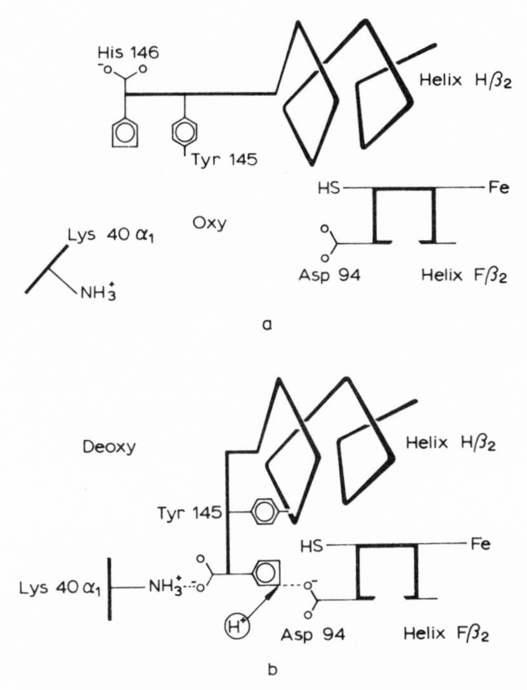

Fig. 13. Change in conformation of the C-terminal residues of the β chains on deoxygenation. a, in oxyhemoglobin or liganded hemoglobin histidine 146β, aspartate 94β, and lysine 40α are free and far from each other. The SH of cysteine 93β is accessible to sulfhydryl reagents; b, in deoxyhemoglobin or unliganded hemoglobin a salt bridge is formed between aspartate 94β and the imidazole of histidine 146β as well as the C-terminal carboxyl of the β_2 chain and lysine 40α_1. Access to the sulfhydryl of cysteine 93β is restricted. (From ref. 254 by permission of the authors.)

the liganded conformation suggests that the deoxy conformation is the constrained conformation[38]. This prediction is consistent with the much lower oxygen affinity of hemoglobin than of the isolated subunits[255] or of myoglobin[217,218] (Figs. 10 and 11).

The $\alpha_1\beta_1$ and $\alpha_2\beta_2$ contacts are very extensive (Table II)[2]. They consist of 34 residues of which 110 atoms approach within 4 Å of each other. The great majority of the interactions are non-polar. In the change from the deoxy to the liganded form[25,26,150,151] these subunits are rotated relative to each other 3.7° about a screw axis and moved 0.3 Å along the axis. The relative displacement of atoms at the contacts are only about 1 Å. This contact is almost identical in both forms of hemoglobin and the only reported difference involves a hydrogen bond in liganded hemoglobin between aspartate H9(126)α_1 and tyrosine C1(35)β_1 (Table II). In the deoxy form the tyrosine remains at the interface but the aspartate is bonded to the C-terminal arginine HC3(141) of the other α-chain (Fig. 14) instead of tyrosine C1(35)β_1.

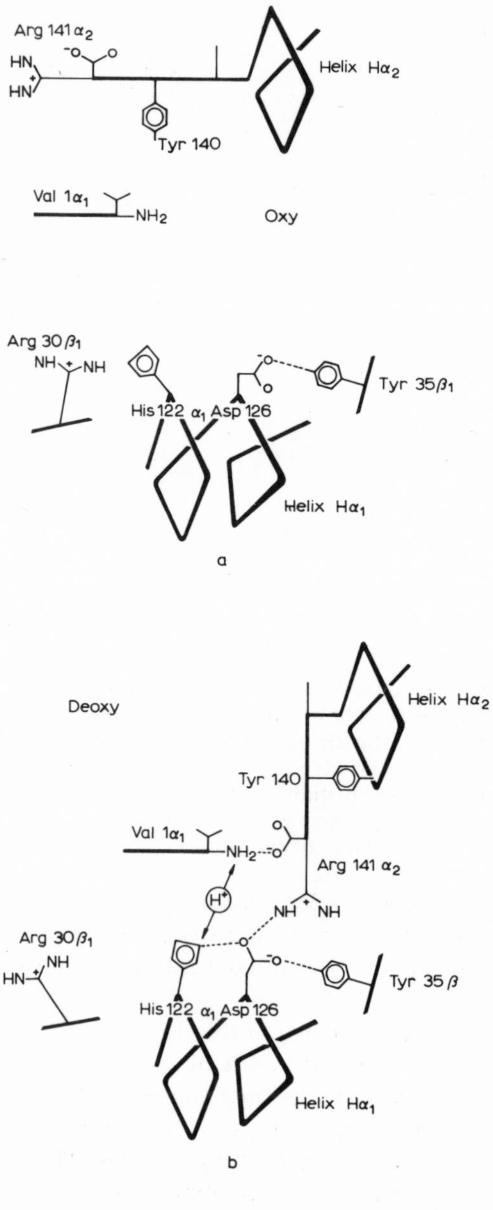

Fig. 14. Change in conformation of the terminal regions of the α chains on deoxygenation. a, in oxyhemoglobin or liganded hemoglobin valine 1α and arginine 141α are free, histidine $122\alpha_1$ is close to arginine $30\beta_1$; b, in deoxyhemoglobin or unliganded hemoglobin valine $1\alpha_1$ is in contact with the carboxyl group of arginine $141\alpha_2$, aspartate $126\alpha_1$ forms close contacts with histidine $122\alpha_1$, tyrosine $35\beta_1$, and arginine $141\alpha_2$. (From ref. 254 by permission of the author.)

TABLE II

THE $\alpha_1\beta_1$ CONTACT IN LIGANDED HEMOGLOBIN FROM PERUTZ *et al.*[2]

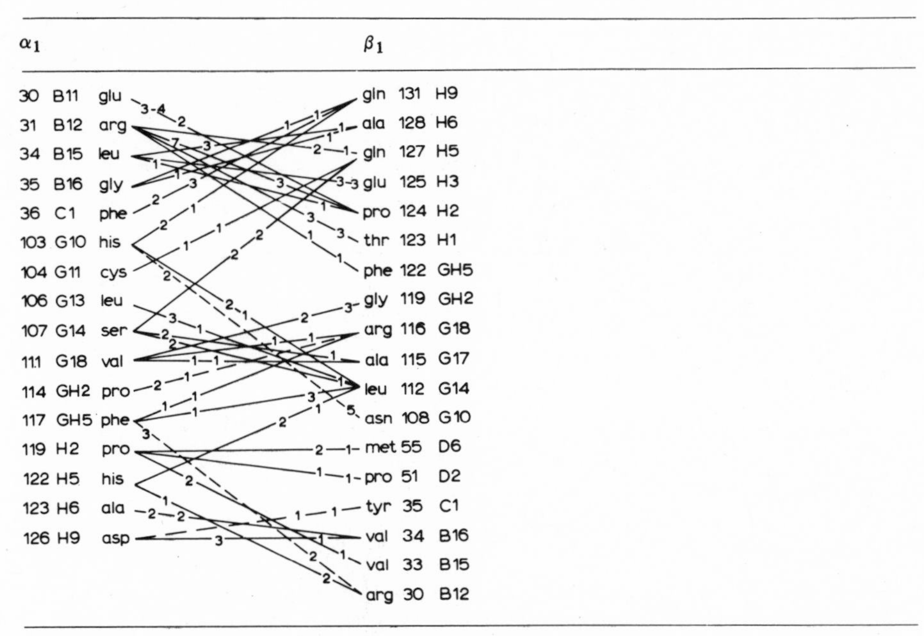

34 residues including about 110 atoms in contact at distances of 4 Å or less. Plain connecting lines indicate van der Waals contacts; broken ones indicate that the contact includes a hydrogen bond. The numbers on the lines give the number of atoms contributed to the contact by the residues on each side.

The $\alpha_1\beta_2$ and $\alpha_2\beta_1$ contacts consist of 19 residues of which 80 atoms approach within a distance of 4 Å in liganded hemoglobin (Table III)[2,38]. There is one and possibly a second hydrogen bond, while the rest of the interactions are non-polar. Residues in helices C and G and the FG non-helical segments of the α and β subunits are involved (Fig. 15). This contact is dovetailed so that the C region of one subunit fits into the FG and G region of the other subunit. The change from the liganded to unliganded conformation involves a rotation of these subunits relative to each other by 13.5° about a screw axis and a displacement along the screw axis of 1.9 Å. An appreciable displacement of atoms at the contact occurs whereby the ϵ-amino group of lysine C5(40)α moves about 7 Å (Fig. 13)[38] and other atoms move as much as 5.7 Å. This movement produces major changes in the contacts at the $\alpha_1\beta_2$ interface (Table III)[26,150]. In the liganded form threonine C3(38)α is in contact with valine FG5(98)β (Fig. 15). In the unliganded form the threonine C3(38)α is replaced by threonine C6(41)α, which is the residue protruding from the next turn of the C-helix. The hydrogen bond linking aspartate G1(94)α to asparagine G4(102)β in the

TABLE III

CONTACTS $\alpha_1\beta_2$ IN OXY- AND DEOXYHEMOGLOBIN FROM BOLTON AND PERUTZ[26]

Oxy α_1	Oxy β_2	Deoxy α_1	Deoxy β_2
38 C3 Thr	Asn 102 G4	37 C2 Pro	His 146 HC3
41 C6 Thr	Glu 101 G3	38 C3 Thr	Tyr 145 HC2
42 C7 Tyr	Asp 99 G1	40 C5 Lys	Glu 101 G3
91 FG3 Leu	Val 98 FG5	41 C6 Thr	Pro 100 G2
92 FG4 Arg	His 97 FG4	42 C7 Tyr	Asp 99 G1
93 FG5 Val	Arg 40 C6	44 CD2 Pro	Val 98 FG5
94 G1 Asp	Gln 39 C5	92 FG4 Arg	His 97 FG4
95 G2 Pro	Trp 37 C3	94 G1 Asp	Arg 40 C6
96 G3 Val	Pro 36 C2	95 G2 Pro	Trp 37 C3
140 HC2 Tyr		96 G3 Val	
		140 HC2 Tyr	

All contacts at a distance of 4 Å or less were counted. Plain connecting lines indicate van der Waals contacts; broken ones indicate that the contact includes a hydrogen bond. The numbers on the lines give the numbers of atoms contributed to the contact by the residues on each side.

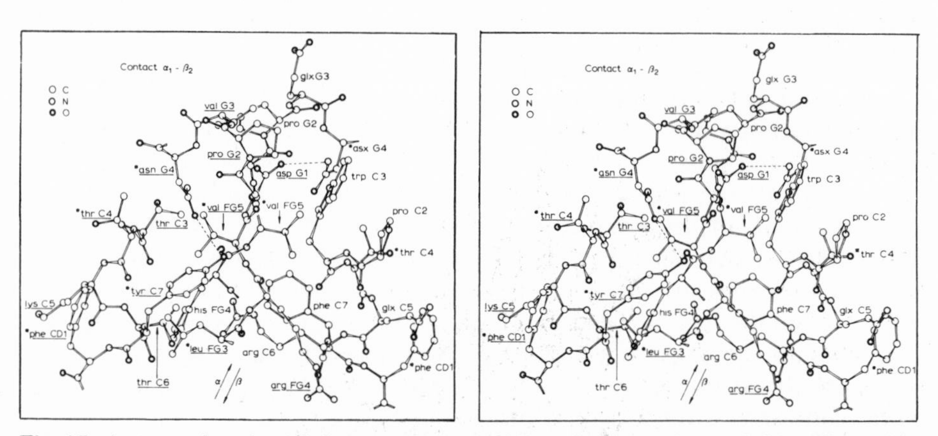

Fig. 15. A stereo drawing of the $\alpha_1\beta_2$ contact of liganded horse hemoglobin. The residues of the α chain are underlined and those of the β chain are not. All residues in this contact are included with the exception of tyrosine H23α which lies vertically above, and in van der Waals contact with tryptophan C3β. The heme groups are not shown, but would be seen approximately end-on, one of them on top of the lower left and the other underneath the lower right of the drawing. Residues which are in contact with the heme groups are starred. The arrows at the bottom indicate the relative movements of the subunits on deoxygenation. (From ref. 2 by permission of the authors.)

liganded form (Fig. 15) is replaced in the deoxy form by a hydrogen bond between tyrosine C7(42)α and aspartate G1(99)β. Lysine C5(40)α is free in solution in the liganded form, but is involved in a salt bridge with histidine HC3(146)β in deoxyhemoglobin (Fig. 13)[253,254]. There are also small adjustments of the G-helix in the α-chain and the C-helix in the β-chain.

The α-chains are not in contact in the liganded form[2]. However, in the deoxy form they are linked by four salt bridges (Fig. 14). The change in conformation from the liganded to the unliganded form results in a small

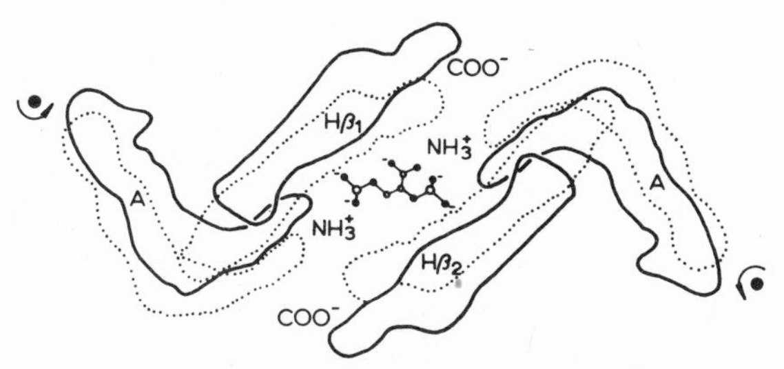

Fig. 16. Projection of the electron density maps of liganded (. . . .) and unliganded (——) hemoglobin at 5.5 Å resolution, showing the ββ interface consisting of the N and C terminal portions of both β chains. The likely position of 2,3-DPG in the central cavity is indicated. The figure shows that the phosphates of 2,3-DPG are close to the N-termini of the β chains in deoxyhemoglobin. On binding ligands the N-termini move apart and the H helices close up, thus expelling 2,3-DPG. The circular arrows indicate the positions of the axes around which the β-subunits rotate on deoxygenation. (From ref. 254 by permission of the author.)

rotation of the subunits relative to each other, and a decrease in the distance between subunits[25,26].

The β-chains are not in direct contact with each other in either form but the gap is much wider in the deoxy form and the rotation resulting from the change in quaternary structure moves the terminal amino groups closer together in the deoxy form[2,25,26] (Fig. 16). In deoxyhemoglobin they are 16 Å apart while in liganded hemoglobin they are 20 Å apart.

Organic phosphates such as 2,3-DPG are thought to bind at this interface. The five titratable acidic groups on 2,3-DPG can neutralize the positive terminal amino groups as well as positive lysine EF6(82)β and histidine H21(143)β, which also point into the cavity between the β-subunits[254]. The decreased size of the cavity and the greater distance between the terminal ε-amino groups in the liganded form relative to the deoxy form can account for the greater interaction of 2,3-DPG with deoxyhemoglobin[248,251,256]. It has recently been found[260,261] that EDTA, which has no phosphates but does have negatively charged carboxyls which can be arranged into a configuration very similar to that of the negative groups of 2,3-DPG, also markedly lowers the oxygen affinity.

References pp. 891–901

Spectroscopic methods

O.r.d. and c.d. spectroscopy provide sensitive conformational probes[262]. The use of these techniques to gain evidence for conformational changes within the globin moiety when ligands are bound is, however, complicated by the heme absorption bands, which are found in all regions of the spectrum[58,120-122,263,264]. These bands are, of course, affected by the binding of ligands (Fig. 6)[59,61,112-117]. It has nevertheless been possible to demonstrate that the polypeptide conformation is responsible for a change in the o.r.d. and c.d. spectra in the far u.v.[60,263,265]. The binding of O_2, CO, NO, or ethyl isocyanide to deoxyhemoglobin decreases the mean residue rotation at 233 nm by about 8%[60,265] and produces an analogous decrease in the c.d. band at 222 nm[263]. The decrease in the rotation is not observed for isolated chains or myoglobin and is the same for the different ligands. These results indicate that the protein conformation is influenced by the binding of ligands and that the change is not restricted to the heme or even the local environment of the ligand. The change can possibly be attributed to the perturbation of an aromatic residue[265]. However, the correspondence of the change with the known $n \rightarrow \pi^*$ transition of the peptide chromophore, which is extremely sensitive to conformational changes[266], suggests a perturbation of the polypeptide backbone. While a change in rotation is not observed for isolated hemoglobin subunits, modifications of the structure of hemoglobin, which eliminate the ligand induced changes in quaternary structure as well as heme–heme interaction without separating the subunits, do not eliminate this change in rotatory properties[265]. Thus, an analogous decrease in rotation at 233 nm is observed when hemoglobin is treated with carboxypeptidase A (CPA) or reacted with bis(*N*-maleimidomethyl)ether (BME), both of which completely eliminate heme–heme interaction (see below).

Spectral changes upon oxygenation attributed to the globin moiety have also been observed in the 280–290 nm region[59,267,268]. The absorbance increases and a negative c.d. band is eliminated when oxygen is bound. Again, no effect is observed with the isolated subunits. However, in this case, unlike the effect in the far u.v., the spectral changes are not observed when heme–heme interaction is eliminated by reaction with BME[268]. The changes in this region of the spectrum probably arise from the changed environment of an aromatic residue. It has been suggested[268] that the aromatic residues at the contacts between unlike chains, such as tyrosines C7(42)α and C1(35)β or tryptophan C3(37)β may be involved (Tables II and III). Therefore, this change can arise simply from the rearrangement of the subunits occurring when the quaternary structure is changed. However, the change in the 222 nm c.d. band occurs in the absence of a change in quaternary structure and is, therefore, evidence for a conformational change within the subunits.

N.m.r. spectroscopy has also been used to monitor changes in the protein conformation as a function of oxygenation for hemoglobin[234,269] as well as myoglobin[62,234]. A comparison of the spectra in H_2O and D_2O has been used to localize the changes involving exchangeable protons[234]. There are three tryptophans in human hemoglobin[9,10] (Fig. 3)—A12(14)α, A12(15)β and C3(37)β. The resonances of the indole NH exchangeable protons of these residues should be located in the region of -10 to -12 ppm from 3-(trimethylsilyl)-propane sulfonic acid sodium salt (DSS), which is a relatively clean region of the spectrum[62]. For the isolated α-subunits no

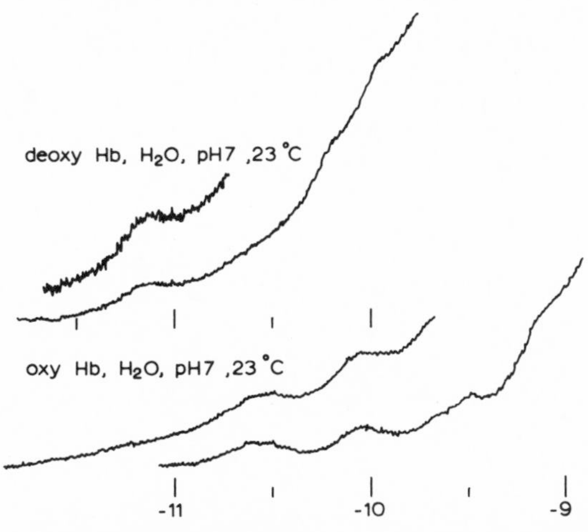

Fig. 17. 220 MHz proton n.m.r. spectra of the exchangeable tryptophan resonances of human HbO_2 and Hb between -9 and -12 ppm from DSS in H_2O, pH 7 and 23°C. From the areas under the peaks the HbO_2 resonance at -10.6 ppm and -10.1 ppm are attributed to 2–3 and 1 exchangeable protons/$\alpha\beta$ dimer, respectively. In the Hb spectrum there is approximately one exchangeable proton/$\alpha\beta$ dimer each at -11.2, -10.2, and -10.0 ppm. (From ref. 234 by permission of the authors.)

exchangeable protons with resonances from -10 to -15 ppm from DSS are observed[269]. For the isolated β-subunits the two resonances in this region are essentially not affected by the binding of ligands. On the other hand, for the cooperative tetramer all the resonances in the region from -10 to -15 ppm from DSS are influenced by binding oxygen. The resonances in the region from -10 to -12 ppm attributed to the tryptophans are shown in Fig. 17. It is interesting that while none of the tryptophans are near the heme the C3(37)β tryptophan is near the $\alpha_1\beta_2$ contact (Table III and Fig. 15) and the other tryptophans are on the other side of the molecule between the A and E helices. Ligand binding also influences the n.m.r. resonances of two exchangeable protons in the region from -12 to -15 ppm from DSS. These resonances are not observed in the monomer and arise from groups which are situated more than 10 Å from the heme[234]. The binding of ligands to hemoglobin also affects certain aliphatic side chains in close contact with the heme as well as certain other aromatic resonances[269].

Other methods

Hemoglobin has two reactive sulfhydryl groups located in the β-subunits at cysteine F9(93), the residue adjacent to the proximal histidine F8(92)β (Fig. 3). The reactions of these groups have been studied extensively[14,30,31,270-274] and it has been found that this site is much less reactive in deoxyhemoglobin than in any liganded hemoglobin (Fig. 18)[271,273,275].

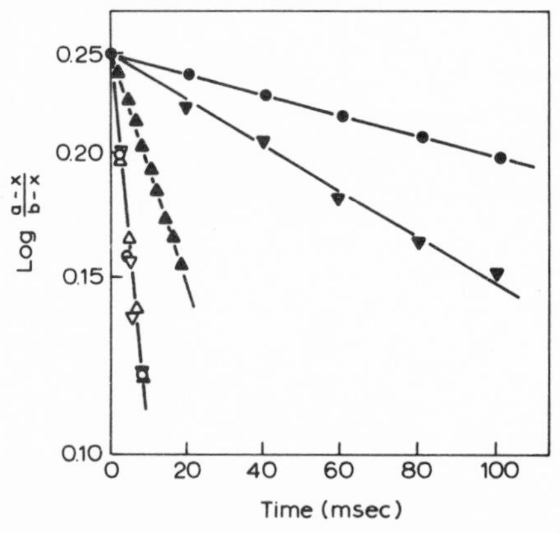

Fig. 18. Kinetic plots for the reaction of PMB with the F9(93)β sulfhydryl groups of human hemoglobin. a and b are the initial concentrations of the sulfhydryl groups and PMB, respectively, X is the concentration of the reacted sulfhydryl groups during time $t = 0$ to t. In all the experiments a is equal to 2.5×10^{-5} *M* and b is equal to 1×10^{-4} *M*. (●), deoxyhemoglobin; (○) oxyhemoglobin; (▼) deoxygenated hybrid hemoglobin, $\alpha^{+}\beta$; (▽) oxygenated hybrid hemoglobin, $\alpha^{+}\beta$; (▲), deoxygenated hybrid hemoglobin, $\alpha\beta^{+}$; (△), oxygenated hybrid hemoglobin, $\alpha\beta^{+}$. (From ref. 275 by permission of authors.)

It has been possible to investigate the conformational changes in the region of the protein near cysteine F9(93)β by using sulfhydryl reagents with a nitroso radical attached to it[55-57,276-281]. The e.p.r. spectrum of the nitroso radical is very sensitive to changes in its environment[57,279], and, therefore different e.p.r. spectra are observed with deoxyhemoglobin and oxyhemoglobin (Fig. 19). The changes in the e.p.r. spectrum as a function of oxygenation are reversible[56,278] for the iodoacetamide spin labels *N*-(1-oxyl-2,2,5,5-tetramethyl-3-pyrrolidinyl)iodoacetamide and *N*-(1-oxyl-2,2,6,6-tetramethyl-4-piperidinyl)iodoacetamide. The spectrum for deoxyhemoglobin corresponds to a weakly immobilized spin while for oxyhemoglobin the spectrum corresponds to a mixture of a weakly immobilized spin and a strongly immobilized spin in rapid conformational equilibrium[57].

The spin label e.p.r. spectrum of the isolated β-subunit is different from that of normal hemoglobin[56]. Nevertheless, changes are observed when oxygen is bound indicating that there are ligand induced conformational changes within the β-subunit.

In globular proteins such as hemoglobin, many of the exchangeable hydrogens are on the inside of the molecule, but they are nevertheless

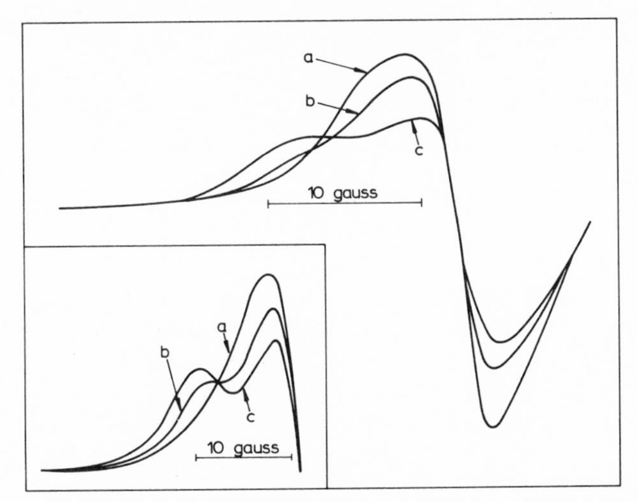

Fig. 19. Electron paramagnetic resonance spectra of 1% human hemoglobin labeled with the spin label *N*(1-oxyl-2,2,6,6-tetramethyl-4-piperidinyl) iodoacetamide. The hemoglobin was first treated with dithionite to reduce any methemoglobin. The spectra were run in 0.05 *M* phosphate buffer, pH 7.0 at 23°C. Insert: e.p.r. spectra of hemoglobin J Capetown reacted with the same spin label under the same conditions. (a), deoxyhemoglobin; (b), hemoglobin partially saturated with carbon monoxide; (c) HbCO. (Modified from ref. 281 by permission of the authors.)

exchanged with the solvent because of motility, or "breathing", of the protein[282]. Any small tightening or loosening of the protein is readily observed by following the exchange of hydrogens with deuterium or tritium. Frequently these changes do not affect the equilibrium conformation, and thus could not be resolved by X-ray studies. It is, therefore, not surprising that Englander[65] has observed that when oxygen is bound to hemoglobin 10 additional hydrogens exchange rapidly with the solvent, *i.e.*, oxygenation loosens up the hemoglobin structure.

Functionally significant conformational changes

The X-ray studies clearly delineate certain structural changes which occur when ligands are bound[26,38,150,253,254]. There are changes in the region of the heme on both the proximal side and, at least in the β-chain, on the distal side of the heme (Fig. 7). The changes on the proximal side of the heme seem to extend to the C-terminal residues (Figs. 13 and 14), the $\alpha_1\beta_2$ contact (Fig. 15, and Table III), and the $\alpha\alpha$ contact. The changes in this region of the β-subunit are also correlated with a change in the reactivity of cysteine F9(93)β[271,273,275] and the environment of nitroso spin labels attached to this residue[55-57,277-279].

However, n.m.r.[234,269] and c.d.[59,60,263,265-268] studies seem to indicate that conformational changes may actually be more widespread than suggested by the X-ray studies. As in myoglobin, X-ray studies do not necessarily resolve thermodynamically significant conformational rearrangements. In fact, the changes involving the largest spatial movement, which are observed by X-ray studies, need not be energetically the most significant changes.

Distal side of the heme

The functional significance of steric interactions with the polypeptide residues on the distal side of the heme was originally proposed by George and Pauling[21] on the basis of a comparison of the binding of ethyl, isopropyl and t-butyl alkyl isocyanides. They found that the binding constant for ethyl isocyanide is three times that of isopropyl isocyanide and 200 times that of t-butyl isocyanide. This was true even though the affinity for ferroheme itself is independent of the size of the alkyl group.

These results are consistent with the X-ray studies, since the t-butyl group cannot be accommodated into the ligand pocket of either subunit without significant perturbation of the protein conformation[26,38]. However, for smaller ligands like O_2 and CO the perturbations in normal hemoglobin seem limited to valine E11(67) in the β-chain.

As already discussed for myoglobin, the ligand pocket on the distal side of the heme determines the orientation of the ligand[129,145,162,163,222] and thereby its interaction with the iron and the rest of the heme[77,225,226], thus indirectly influencing the ligand affinity and heme–heme interaction. The importance of the distal side of the heme in determining the nature of the Fe–ligand bond has been investigated by studying the binding of CO to various hemoglobins having large differences in CO affinity[129,222]. The only variants studied that affected the carbonyl stretching frequency, ν_{CO}, were hemoglobin M Emory, which is the same as M Saskatoon[172], and hemoglobin Zürich[173-175] where histidine E7(63)β is replaced by tyrosine and arginine, respectively (Table I).

Various abnormal hemoglobins containing substituted amino acid residues in the ligand pocket (Fig. 7) have been isolated and studied (Table I). The oxygenation of these variants is usually markedly different from that of normal hemoglobin. However, it has been possible to explain most of these differences on the basis of the three-dimensional structures determined by X-ray crystallography[166,167].

The substitution of tyrosine for the distal histidine E7 in the α-chain (M Boston)[172] or the β-chain (M Saskatoon)[172] prevents the binding of oxygen to the abnormal chains because the iron is oxidized (Table I). It is thought that the Fe(III) heme is stabilized by the formation of an ionic bond between tyrosine and the iron[166]. The replacement of histidine E7(63)β by arginine, in hemoglobin Zürich (Table I), increases the oxygen affinity[173–175] because arginine is larger than tyrosine and cannot be accommodated into the ligand pocket[166,167]. The arginine must protrude at the surface, thus rendering the pocket more accessible.

When valine E11(67)β (Table I) is replaced by glutamic acid (M Milwaukee)[172] or aspartic acid (Bristol)[176], the carboxyl groups can form salt bridges with the iron and histidine E7, respectively[166,167]. The salt bridge with the iron then stabilizes the Fe(III) heme, preventing the binding of oxygen, while the salt bridge with the histidine E7 closes the heme pocket, drastically lowering the oxygen affinity.

Phenylalanine CD1(42)β acts as a spacer maintaining the heme in an inclined position[38,167] in liganded hemoglobin. Therefore, the substitution of phenylalanine by the smaller serine or leucine in hemoglobins Hammersmith[178,179] and Bucuresti[180], respectively, lowers the oxygen affinity by stabilizing the upright heme position of deoxyhemoglobin (Table I).

Proximal side of the heme

The importance of the proximal side of the heme in controlling the functioning of hemoglobin and myoglobin was originally proposed by Lumry and Eyring[231] as an example of a "Rack" mechanism[186,283]. It was thought that the electronic properties of the heme can influence the protein conformation via the Fe–proximal histidine bond in a reciprocal fashion. The proximal histidine is *trans* to the oxygen and utilizes the same metal orbitals for coordination. Therefore even a small distortion of the iron–imidazole bond produced by a change in protein conformation should be reflected in the binding of oxygen. On the other hand, binding oxygen should influence the Fe–proximal histidine bond, and the distortion should be transmitted directly to other parts of the protein. The X-ray studies indicate even larger changes in this region[26,38,150] than had been expected because of the movement of the iron and the attachment of the proximal histidine F8 to the rigid F helix (Figs. 1 and 3).

A strong case can be made for the functional role of the conformational changes observed by X-ray studies on the proximal side of the heme near the C-terminal.

The relationship between the C-terminal residues of hemoglobin and the oxygenation reaction has been known for some time. It was originally demonstrated by digestion of hemoglobin with carboxypeptidases A and B (CPA and CPB), which produce well defined homogeneous changes at the COO^- terminals of the various chains[14,284]. CPA removes histidine HC3(146) and tyrosine HC2(145) from the C-terminus of the β-chain (Fig. 3), thereby increasing the oxygen affinity, decreasing the Bohr effect and eliminating heme–heme interaction. CPB attacks the C-terminus of the α-chains, removing arginine HC3(141)α and in some cases tyrosine HC2(140)α and lysine HC1(139)α (Fig. 3), but does not attack the β-chain. In the case of CPB modification the binding remains cooperative; although the Bohr effect and the oxygen affinity are modified.

The requirement of tyrosine HC2(145)β in the pocket between the F and H helices of the β-subunit (Fig. 12) for the cooperative oxygen binding and the normal Bohr effect[38,254] has been recently demonstrated by the investigation of the functional properties and X-ray crystallographic studies on hemoglobins modified by various monofunctional and bifunctional reagents which react with cysteine F9(93)β[53,54,259,285-287]. It was found that even large monofunctional reagents produce derivatives which influence the oxygenation but do not completely eliminate heme–heme interaction and the Bohr effect[285]. However, bifunctional reagents such as bis(*N*-maleimido-methyl)ether (BME), which form covalent bridges between the SH group of cysteine F9(93)β and the imidazole of histidine FG4(97)β[259], completely eliminate heme–heme interaction and the Bohr effect[54]. X-ray studies[259] indicate that the formation of this internal crosslink displaces tyrosine HC2(145)β from its pocket (Fig. 12). The structural differences between this derivative and normal hemoglobin are largely confined to the section of the $\alpha_1\beta_2$ interface formed by contact between the C helix of the α-chain and the FG region of the β-chain (Table III and Fig. 15). The remainder of the $\alpha_1\beta_2$ interface (Table III) as well as the $\alpha_1\beta_1$ interface (Table II) and the environment of the heme groups (Fig. 7) are for the most part unaltered.

Equilibria and kinetics of the binding of ligands to hemoglobin

Equilibrium studies

The binding of ligands to hemoglobin containing four hemes can be described by the four equilibria[66,288]

$$\mathrm{Hb + X \overset{K_1}{\rightleftharpoons} HbX}$$

$$\mathrm{HbX + X \overset{K_2}{\rightleftharpoons} HbX_2}$$

$$\mathrm{HbX_2 + X \overset{K_3}{\rightleftharpoons} HbX_3}$$

$$\mathrm{HbX_3 + X \overset{K_4}{\rightleftharpoons} HbX_4}$$

where X represents the ligand, Hb, HbX, HbX_2, HbX_3, and HbX_4 are hemoglobin molecules coordinated with none, one, two, three and four ligand molecules respectively, and the Ks are equilibrium constants. The general equation proposed by Adair[288] for four stage binding is:

$$y = \frac{K_1X + 2K_1K_2X^2 + 3K_1K_2K_3X^3 + 4K_1K_2K_3K_4X^4}{4(1 + K_1X + K_1K_2X^2 + K_1K_2K_3X^3 + K_1K_2K_3K_4X^4)}$$

where y is the fractional saturation of hemoglobin.

Roughton and coworkers have obtained very precise gasometric binding data for the cooperative oxygenation of various mammalian hemoglobins[66,67,289,290]. From these data it has been possible to obtain values for the four stepwise constants, although there are quite large errors for K_2 and K_3 (Table IV). These results indicate that $K_4 \gg K_1$, as required for cooperative binding. Furthermore, K_4 is in the range of the binding constants for myoglobin and the isolated subunits[66,217,218].

TABLE IV

EQUILIBRIUM CONSTANTS[a] AND FREE ENERGIES OF INTERACTION[b] FOR THE BINDING OF OXYGEN TO HEMOGLOBIN

Species: pH:	Human[c] 7.0	Human[d] 9.1	Horse[c] 7.0	Horse[d] 9.1	Sheep[c] 7.1	Sheep[d] 9.1
K_1 (mmHg)$^{-1}$	0.0493	0.217	0.0715	0.239	0.063	0.109
K_2 (mmHg)$^{-1}$	0.0427	3.949	0.0041	2.732	0.025	0.218
K_3 (mmHg)$^{-1}$	0.221	0.551	1.473	0.773	0.015	0.185
K_4 (mmHg)$^{-1}$	0.320	2.116	0.270	2.136	0.583	2.000
ΔF_1 (cal/mole)	2700	2900	2400	2900	2900	3300
ΔF_2 (cal/mole)	2200	700	3500	900	2900	2300
ΔF_3 (cal/mole)	800	1400	-400	1200	2700	2000
ΔF_{Total} (cal/mole)	5700	5000	5500	5000	8500	7600

[a] The equilibrium constants are from Roughton, Otis and Lyster[66], and Roughton and Lyster[67].
[b] The free energies of interaction are unpublished results of J. Rifkind calculated directly from the equilibrium constants.
[c] 0.6 *M* phosphate buffer, 19°C.
[d] 0.2 *M* borate buffer, 19°C.

Most other binding data have been analyzed in terms of the Hill equation[291]

$$y = \frac{KX^n}{1 + KX^n}$$

where K is an equilibrium constant for the binding of n moles of ligand to one mole of hemoglobin in one step

$$\text{Hb} + n\text{X} \overset{K}{\rightleftharpoons} \text{HbX}_n$$

This equation was proposed before it was known that hemoglobin is a tetramer and no longer has any simple physical significance. However, it is useful for obtaining an empirical measure of the affinity and cooperativity[14]. In the Hill plot (Fig. 20) $\log y/(1-y)$ is plotted *vs.* log X. The value of X at 50% saturation, $X_{\frac{1}{2}}$, is equal to $1/K$, and the slope of the Hill plot in the middle region of the binding curve is equal to n, the effective number of

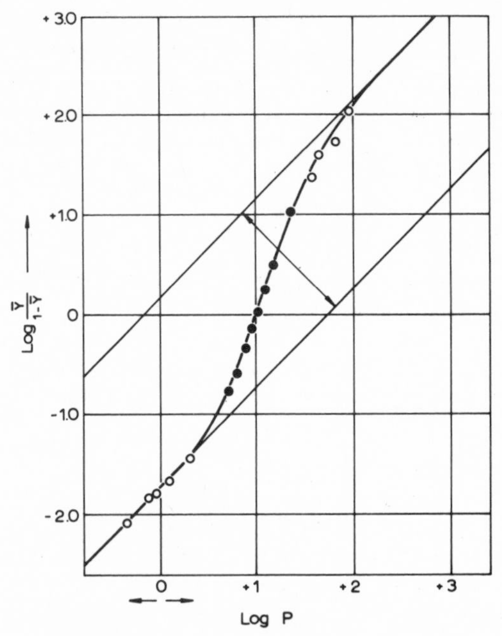

Fig. 20. Hill plot of oxygen equilibrium of horse hemoglobin in 0.6 M phosphate buffer pH 7.0 and 19°C[67]. Points are divided into three groups in accordance with different techniques employed for different portions of the binding curve[66]. The free energy of interaction per site is obtained by multiplying the perpendicular distance between asymptotes by $RT(2)^{\frac{1}{2}}$. From this data a free energy of interaction of 2600 cal is obtained with $n = 2.95 \pm 0.05$ at 50% saturation. (From ref. 227 by permission of the author.)

moles which bind cooperatively. The slope of the Hill plot n for the binding of oxygen to normal mammalian hemoglobin is about 3, indicating that the interaction between subunits is such as to produce the effect of 3 moles of oxygen binding simultaneously.

Wyman[227,228] has shown that the slope of the Hill plot extended to very low and very high degrees of saturation must approach unity (Fig. 20). From such a plot it is possible to obtain a value for the minimum free energy of interaction per site (ΔF_I), which is responsible for cooperative binding[227,228]. This method assumes that the interaction energy for binding each mole of oxygen is equivalent to that involved in binding the first molecule of oxygen to deoxyhemoglobin.

Alternatively, the stepwise Adair constants can be used to calculate apparent free energies of interaction[292]. The binding constant to an unconstrained hemoglobin molecule can be considered equal to K_4, the constant for binding the fourth ligand molecule[66]. It is then possible to calculate individual free energies of interaction for the reactions of the first three ligand molecules (Table IV), after taking into account statistical factors to correct for the different numbers of sites available at each step of the reaction[288]. These results indicate that the free energies of interaction of the various ligand molecules are not necessarily equivalent, and that the step at which the major change in interaction energy occurs is not unique and depends on the hemoglobin species and pH (Table IV).

Kinetic studies

The overall rate of reaction for the binding of ligands to hemoglobin[68] is much slower than the rate for myoglobin[293] and the subunits[255,294,295]. The slower rate seems to be related to the same constraints[38] which lower the equilibrium affinity for ligands[217,218,255]. Elimination of cooperative binding by such methods as the digestion with carboxypeptidase A markedly increases the reaction rate[31]. This hypothesis also explains the kinetics for the recombination of CO with hemoglobin after rapid dissociation of CO by flash photolysis at pH 9[31,296]. Under these conditions a fraction of the hemoglobin binds CO 20 to 30 times faster than deoxyhemoglobin. This behavior has been interpreted by assuming that when CO is rapidly removed hemoglobin retains the liganded conformation in which state it reacts rapidly with CO[296].

The simplest expression for describing the kinetics of binding ligands to hemoglobin is

$$\text{Hb} + \text{X} \underset{k}{\overset{k'}{\rightleftharpoons}} \text{HbX}$$

where $K = k'/k$ is equal to the equilibrium constant K. The kinetic data for the binding of ligands to myoglobin[14,293] and the isolated subunits[255,294,295] obey this simple expression, and in these cases values for the equilibrium constant obtained from kinetic data agree reasonably well with those obtained directly from equilibrium binding curves.

The kinetic data for hemoglobin are naturally more complicated, since the binding of each of the four ligand molecules has an "on" and "off" rate constant. Thus, the Adair mechanism[288] requires eight kinetic constants[68].

Thermodynamically, cooperative binding is attributed to an increased binding constant for the later stages of binding ligands[66]. Kinetically, this cooperativity can arise from either an increased "on" constant or a decreased

References pp. 891–901

"off" constant, or both, for the later stages of binding ligands[68-70]. The increased "on" constant has been demonstrated for the combination of CO with deoxyhemoglobin by the autocatalytic nature of the kinetic curve[68,297,298]. The "off" constants for oxygen are much larger than for CO so that an autocatalytic kinetic curve is usually not observed for O_2 binding[299]. However, by using an extremely fast stopped flow apparatus at

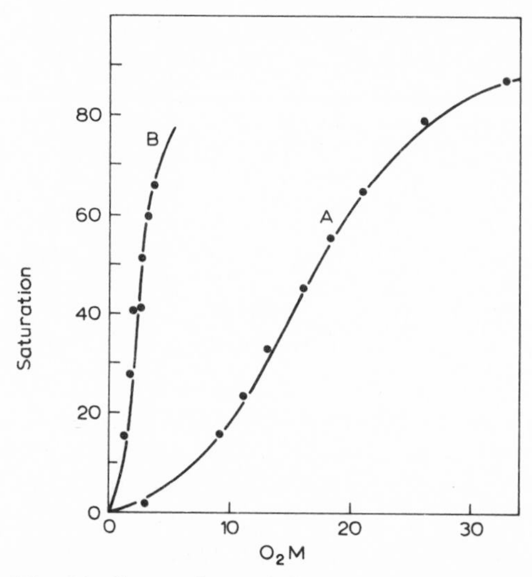

Fig. 21. Comparison of the experimentally observed oxygen equilibrium curves of human hemoglobin with those calculated from kinetic constants. A, in 0.1 *M* phosphate buffer, pH 7.0. (Points), from Roughton and Lyster[67]; (line) calculated from the kinetic constants k'_1 = 17.7, k'_2 = 33.2, k'_3 = 4.89, k'_4 = 33 per μM sec, k_1 = 1900, k_2 = 158, k_3 = 539, k_4 = 50 per sec. B, hemoglobin stripped of phosphates in 0.05 *M* 2-bis(hydroxy-methyl)-2,2′,2″-nitrilotriethanol–Tris buffer, pH 7.0. (Points), from spectrophotometric equilibrium binding curve; (line), calculated from the kinetic constants k'_1 = 14.7, k'_2 = 35.2, k'_3 = 15.8, k'_4 = 33 per μM sec and k_1 = 136, k_2 = 15.7, k_3 = 138, k_4 = 50 per sec. (From ref. 69 by permission of the author.)

high oxygen concentrations, the oxygen binding was shown to be autocatalytic, indicating that the "on" constants increase as oxygen is bound[300].

Recently, Gibson[69] has been able to fit the oxygenation kinetics for concentrated hemoglobin solutions with the eight kinetic constants of the Adair mechanism (Fig. 21). The agreement with the equilibrium data is reasonable, although significant discrepancies are still apparent. The constants obtained from these studies suggest that the cooperativity is attributed mainly to the "off" constants.

The non-equivalence of alpha and beta subunits

Thus far in discussing the binding of ligands it has been assumed that the intrinsic constants for all four sites are identical. However, the X-ray studies indicate that the environment of the heme is quite different in the two types of chains[26,38]. The hemes in isolated α- and β-subunits also have different n.m.r.[63], c.d. (Fig. 22)[59,113], e.p.r.[48], and absorption spectra[301,302] The difference in the n.m.r. spectra of the α and β hemes is also observed in tetrameric hemoglobin[303–305]. If the differences affect the affinity for ligands, it is necessary to include separate constants for each type of subunit in the analyses of equilibrium and kinetic studies. This heterogeneity will decrease the apparent heme–heme interaction[227,228].

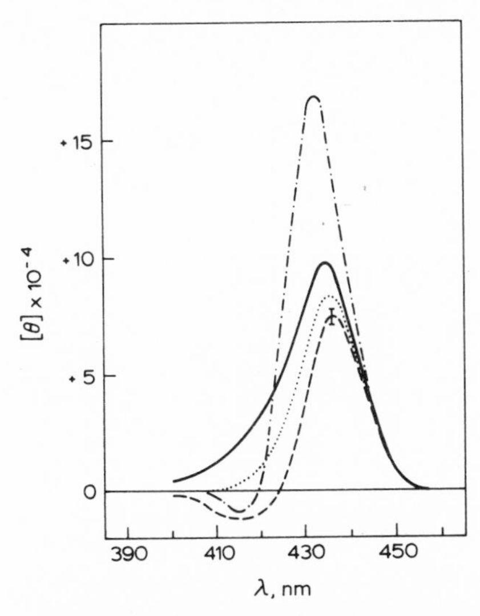

Fig. 22. Circular dichroism spectra in the Soret region of deoxy α chains (——), deoxy β chains (- - - -), recombined deoxy chains (–·–·–·) in 50 mM sodium phosphate (pH 7.0). (. . . .), the mean of the circular dichroism spectra of the isolated chains. (From ref. 113 by permission of the authors.)

The isolated β-subunit has a slightly higher oxygen affinity than the isolated α-subunit (Fig. 11)[255]. Recently it has been shown that even in tetrameric hemoglobin the binding of ligands to the α- and β-subunits can be quite different[306,307]. The non-cooperative binding of imidazole and CN^- to ferrihemoglobin cannot be explained by a single constant but can be fit by assuming two classes of independent and non-equivalent sites, *i.e.*, α- and β-chains.

In both kinetic[302] and thermodynamic[308] experiments it has been found that isosbestic points are not observed in the visible absorption spectra when n-butyl isocyanide reacts with Fe(II) hemoglobin. These results have been interpreted in terms of a preferential reaction with the β-subunits. Consistent with the difference in the absorption spectra of α and β-chains[302], a preferential reaction with one type of subunit can explain the lack of isosbestic points. The conclusion that the β-subunits have a higher affinity than the α-subunits has been corroborated by the faster decrease in the intensity of the n.m.r. spectra of the porphyrin methyl protons assigned to the β-chains than those assigned to the α-chains, as n-butyl isocyanide is added[309]. It has also been found that 2,3-DPG and inositol hexaphosphate (IHP), which drastically lower the oxygen affinity[247-252], also exaggerate the difference between the α- and β-subunits[308,309]. In fact, in the presence of IHP, the first 20% of n-butyl isocyanide is bound almost exclusively to the β-subunit[308].

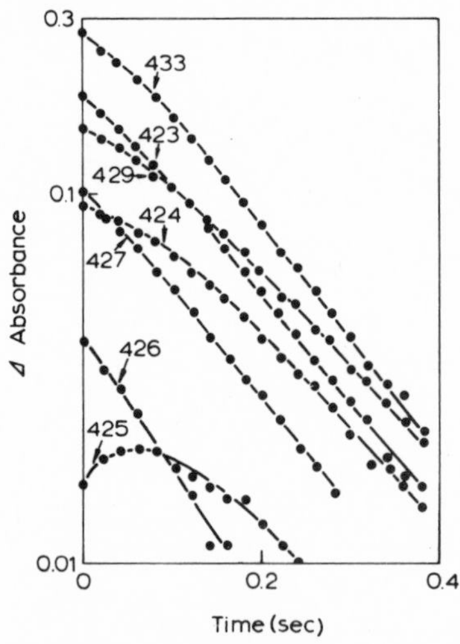

Fig. 23. Semi-logarithmic plot of the reaction of 3 μM human deoxyhemoglobin with 92 μM CO observed with light at various wavelengths between 423 nm and 433 nm. The buffer is 0.05 M bis-Tris, pH 7.0, containing 1 mM 2,3-diphosphoglycerate. (From ref. 310 by permission of the authors.)

Since n-butyl isocyanide is a very large ligand, differences between the subunits may be exaggerated. Such effects may, therefore, not be significant for ligands like oxygen. However, recent kinetic experiments[310] indicate that while in the absence of organic phosphates isosbestic points are obtained for the reaction of CO with hemoglobin, in the presence of organic phosphates isosbestic points are not observed and the time course of the reaction varies with the wavelength of the observing light (Fig. 23). These results also have been interpreted in terms of non-equivalent binding to α- and β-subunits.

Dissociation of hemoglobin

The dissociation of hemoglobin is another factor that is neglected in the simple four step mechanisms for the association of ligands with hemoglobin. Functional hemoglobin is a tetramer which potentially can dissociate into dimers and monomers[14,31]. Whenever hemoglobin is significantly dissociated, the binding expression must include the dissociation constants into dimers and monomers as well as the ligand binding constants for the dissociated species.

The functional behavior of monomers (Fig. 11) is well known from studies on isolated subunits[31,255]. They have a high affinity, no heme–heme interaction, and react rapidly with ligands (see above). However, under normal conditions, the concentration of monomers is relatively small if significant at all. Recent studies[311] indicate that in the presence of 1×10^{-3} *M* EDTA there is no dissociation of hemoglobin into monomers, even at a concentration of 6.2×10^{-8} *M* heme. The dissociation into monomers, when it does occur, is accompanied by pronounced spectral changes and is thought to be irreversible.

Understanding the dissociation into dimers and the functional characteristics of these dimers is, therefore, more significant. It has been observed that dilute hemoglobin solution, even in high concentrations of NaCl, possess functional characteristics similar to the tetramer[31,312-315], *i.e.*, the rate for the reaction of ligands is much slower than for monomers and the binding is cooperative. Since high ionic strength is thought to enhance the dissociation into dimers, it has been proposed that the equilibria[315] and kinetics[314] of binding ligands to dimers are very similar to tetramers and that the large change in functional behavior occurs when the dimers dissociate into monomers[31,312,313].

These conclusions require that liganded and unliganded hemoglobin must be significantly dissociated into dimers under the conditions of the binding experiments. While liganded hemoglobin is generally thought to be dissociated into dimers quite readily in high salt[316–318], it has been difficult to obtain reliable dissociation constants for deoxyhemoglobin[31,316,319,320]. Kellett[321] has recently found that in the presence of 1×10^{-3} *M* EDTA no dissociation of deoxyhemoglobin is observed in 2 *M* NaCl or 1 *M* NaI even at the very low concentration of 6.2×10^{-7} *M* heme.

These results[321] suggest that under the experimental conditions where dimers were thought to be present, deoxyhemoglobin may have been tetrameric, thus explaining the cooperative binding[312,315]. In fact, recent experiments[320,322-324] suggest that when deoxyhemoglobin is actually dissociated into dimers the kinetics for the reaction with ligands[320,323,324] is rapid and binding is non-cooperative[322,323] (Fig. 24), *i.e.*, dimers possess functional characteristics of monomers. These studies have been performed with normal hemoglobin in 2 *M* NaI[320,323] and with des-arginine HC3(141)α-hemoglobin in 0.9 *M* $MgCl_2$[322]. Under these conditions $n = 1$ at a concentration that produces dimers in both liganded and unliganded hemoglobin. In des-arginine HC3(141)α-hemoglobin, the C-terminal arginines of the α-chains are removed by CPB[284]. Since these amino acid residues are involved in ionic bonds linking the two α-chains in deoxyhemoglobin (Fig. 14), the

References pp. 891–901

dissociation of deoxyhemoglobin into $\alpha\beta$ dimers is enhanced. This modified hemoglobin exhibits cooperative binding in 0.2 *M* phosphate (Fig. 24) when the unliganded hemoglobin is not dissociated[322]. The removal of the

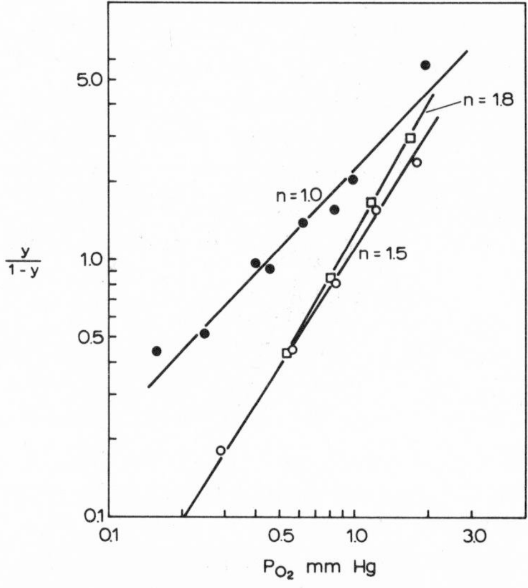

Fig. 24. Hill plots for the reaction of oxygen with human des-Arg hemoglobin (hemoglobin with the α terminal arginines removed). (•), 0.9 *M* $MgCl_2$–0.1 *M* BES, pH 7.2. Under these conditions the liganded and unliganded forms exist as dimers. (○), the sample used for determination of the equilibrium curve in $MgCl_2$ after removal of $MgCl_2$ by passing the sample through a Sephadex G-25 column equilibrated with 0.2 *M* phosphate buffer, pH 6.9. (□), the sample used for determination of the sedimentation velocity in $MgCl_2$ and then transferred to 0.2 *M* phosphate buffer as described above. The lower value of *n* for the sample previously used for the oxygen binding curve is attributed to the 17% methemoglobin present in this sample. (From ref. 322 by permission of the authors.)

C-terminal arginines are not expected to influence the binding properties of the dimers, because they are far removed from the $\alpha\beta$ contact and exposed to the solvent[322].

Other factors which enter into equilibria and kinetic studies

The binding of many other molecules can influence the binding of ligands. Organic phosphates have recently been studied most extensively (Fig. 11)[247-252]. CO_2 is also known to have an effect by reacting with the α-amino groups[31,325,326]. Many other substances such as metal ions, buffers, chloride, *etc.* can and probably do, also affect the oxygenation. However, in the analyses of the binding data such substances need be considered explicitly only if the extent to which they are reacted with hemoglobin changes significantly when oxygen or other ligands are bound.

In addition to all the above considerations, kinetic studies can be further complicated by conformational changes[31,296,327] which are slower than the rate for the reaction with ligands.

Interaction between subunits (heme–heme interaction)

If the binding of ligands to hemoglobin were an all-or-none process hemoglobin molecules would have either four ligands or no ligands. Then heme–heme interaction could be adequately understood on the basis of the structural differences between liganded and unliganded hemoglobin already discussed. However, significant concentrations of partly liganded hemoglobin[66–68,298] are present even when ligands bind cooperatively to hemoglobin. An understanding of the oxygenation of hemoglobin must, therefore, take into account these partly oxygenated intermediates. It is necessary to known at what stages in the oxygenation, between which subunits, and in what manner the structural changes within the subunits influence the oxygen affinity of other subunits.

The propagation of ligand-linked conformational changes is usually discussed in terms of two extreme points of view[31,328].

1. The allosteric model attributed to Monod, Wyman, and Changeux[229] considers an all-or-none equilibrium between two conformational forms of hemoglobin with different ligand affinities. The binding of ligands shifts this equilibrium from the low affinity, *i.e.* deoxy, conformation to the high affinity, *i.e.* liganded, conformation.
2. The induced-fit model of Koshland, Nemethy, and Filmer[329] considers changes in conformation within the subunits when ligands are bound. These conformational changes influence the energy necessary to undergo the ligand-induced conformational changes in adjacent subunits and thereby the affinity.

The difference between these models really lies in their emphasis, and they can both be considered special cases of a more general scheme[328] in which conformational changes are transmitted within and between subunits. Therefore, instead of discussing heme–heme interaction from the viewpoint of a particular model, the evidence for various types of interactions will be presented.

Effect on the heme

The cooperative binding of ligands requires that the binding of a ligand to a heme of one subunit increases the affinity for ligands at another subunit. Therefore, the interaction between subunits must exert some influence on the other heme.

References pp. 891–901

The absorption[308], c.d. (Fig. 22)[58,113], and n.m.r.[63] porphyrin spectra of high affinity isolated deoxygenated subunits are different from those of low affinity deoxyhemoglobin (Fig. 11). These results suggest that the spectra of the high affinity non-liganded subunits within partly oxygenated hemoglobin should be different from those for the low affinity completely deoxygenated tetramer[330]. It has usually been considered that this is not true, primarily because of the good isosbestic points obtained during the oxygenation of hemoglobin[14]. This problem has recently been approached by comparing spectroscopic and gasometric oxygen binding data on the same hemoglobin preparations[330]. The results suggest that intermediate spectra may exist for partly oxygenated hemoglobin. Furthermore, recent studies with CO[310] and n-butyl isocyanide[302,308,309] indicate that isosbestic points, at least under certain conditions, are not observed. While these results were interpreted in terms of non-equivalence of the α- and β-subunits, the contribution of intermediate spectral species cannot be ruled out.

In the case of abnormal hemoglobin M's (Table I)[168-172] the iron of the abnormal subunits is readily oxidized to produce valence hybrids with only the abnormal subunits in the nonfunctional Fe(III) state. Since e.p.r. spectra are observed only for these abnormal chains[46,171,187-191], it is possible to examine the effect of binding ligands to the normal chains on the e.p.r. spectra of the abnormal chain[46,189]. In most cases the binding of oxygen does not influence the e.p.r. spectrum[189]. However, effects have been observed (Table I) for hemoglobin M Hyde Park, in which the proximal histidine of the β-chain is replaced by tyrosine[189] and M Milwaukee in which valine in the ligand pocket of the β-chain is replaced by glutamic acid[46]. Deoxygenating the α-chains of M Hyde Park increases the rhombic distortion of the abnormal β heme, while deoxygenating the α-chains of hemoglobin M Milwaukee decreases the rhombic distortion of the abnormal β heme[46]. X-ray studies[185] on hemoglobin M Hyde Park indicate that binding oxygen to the normal α-chains changes the quaternary structure from that of deoxyhemoglobin to liganded hemoglobin. Hemoglobin M Iwate in which the proximal histidine of the α-chain, instead of that of the β-chain, is replaced by tyrosine does not undergo a change in quaternary structure when ligands are bound to the normal chains[188]. This behavior may explain the difference between hemoglobin M Hyde Park and other hemoglobin M's.

In recent years[31,331-333] techniques have been developed to prepare normal hemoglobin valence hybrids containing either the α-chains or β-chains in the Fe(III) state and, therefore, only one type of chain that can react with O_2 or CO. It has, thus, been possible to directly observe the effect of binding ligands to one type of subunit on the other types of subunits in normal hemoglobins. E.p.r. experiments[46] on the $\alpha^+\beta$ hybrid indicate that when the β-subunits are oxygenated the same e.p.r. spectrum is obtained as for fully oxidized hemoglobin. However, when the β-subunits are deoxygenated a change is produced indicating a greater rhombic distortion of the α heme.

Similar studies have also been performed with spin labels attached to the Fe(III) heme[334,335]. In these studies the $\alpha^{+}\beta$ hybrid was influenced to a greater extent than the $\alpha\beta^{+}$ hybrid. Smaller effects were observed when the Fe(III) chains had CN^- bound instead of F^- [335].

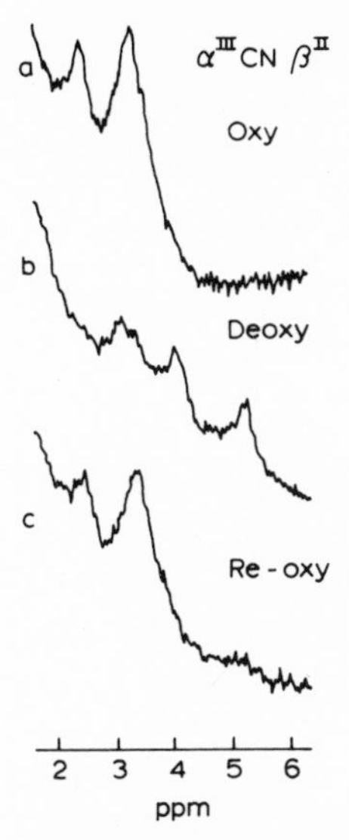

Fig. 25. N.m.r. spectra of the $\alpha^{+CN}\beta$ human hemoglobin hybrid in phosphate buffer at pH 7.1, 15°C. (a), the oxygenated hybrid; (b), the deoxygenated hybrid; (c) the deoxygenated hybrid reoxygenated. The high field region from 2 to 6 ppm is shown. The resonance at + 2.4 ppm comes from the oxygenated β subunit and disappears upon deoxygenation. The other resonances are attributed to the α^{+CN} heme. (From ref. 336 by permission of the authors.)

Shulman and coworkers[63,336,337] have investigated the n.m.r. spectra of $\alpha\beta^{+}$ and $\alpha^{+}\beta$ hemoglobin hybrids in which cyanide or H_2O are bound to the oxidized chains. In the original experiments[63] at pD $\leqslant$ 7, in the absence of the phosphates, it was reported that the n.m.r. spectra of the porphyrin of the oxidized subunits were insensitive to deoxygenation of the Fe(II) subunits. Since these techniques should be sensitive to small changes in the electronic state of the hemes, it was suggested that binding ligands to one heme does not significantly influence other hemes. In later experiments[336,337] conditions were found under which the n.m.r. spectra of the Fe(III) chains were influenced by the binding of ligands to the Fe(II) chains (Fig. 25). The effect was enhanced by phosphates and is most pronounced in the presence of inositol hexaphosphate (IHP). The enhanced n.m.r. sensitivity to ligand binding in the presence of IHP was correlated with changes in the kinetics for the reaction of CO with the Fe(II) subunits[337]. In the absence of organic phosphates, slowly reacting and rapidly reacting kinetic species are observed. The former reacts at a rate similar to that of deoxyhemoglobin, while the latter reacts at a rate similar to that of isolated subunits. The addition of organic phosphates increases the fraction of slowly reacting hemoglobin. These results have been interpreted in terms of a change in the equilibrium between the unliganded and liganded hemoglobin conformations[336,337]. The phosphates lower the oxygen affinity[247-252] and presumably stabilize the deoxy conformation. It has been suggested that the changed heme environment may not be mediated directly by binding a ligand to another subunit, but may depend on the change in quaternary structure[336,337].

Effect on the protein

We have already discussed how the binding of ligands produces a change in the conformation of at least the subunit where the ligand is bound. This change in tertiary structure, at a minimum, leads to stresses at the interfaces between various subunits which ultimately triggers a change in quaternary structure, *i.e.*, the arrangement of the subunits in the hemoglobin tetramer. There is also evidence that the binding of ligands to one subunit influences the protein conformation of other subunits. The reaction of cysteine F9(93)β with sulfhydrul reagents is much slower in the unliganded than in the liganded conformation (Fig. 18)[273,275]. By using hybrids with oxidized β-chains ($\alpha\beta^+$)[275,333], it is also possible to investigate the effect of binding oxygen to the α-chains on the environment of the F9(93)β sulfhydryl group. When CN^- is bound to the oxidized β-chains[333], the deoxygenation of the α-chains was found not to affect the reaction of the F9(93)β sulfhydryl with *p*-mercuribenzoate (PMB). However, when the oxidized chains were liganded with H_2O it was found that the rate for reaction of the F9(93)β sulfhydryl depends on the state of both subunits[275]. Thus, different rates are obtained for $\alpha\beta$(Hb), $\alpha^+\beta$, $\alpha\beta^+$, and $\alpha^+\beta^+$(Hb$^+$), when the Fe(II) subunits present are unliganded (Fig. 18). When all subunits are liganded the same rate is always obtained whether the subunits are in the Fe(II) or Fe(III) state. These results were interpreted in terms of four different conformations for the β-chains depending on whether ligands are bound to α and/or β-chains[275].

The environment of the spin label with a 6-membered ring, *N*-(1-oxyl-2,2,6,6-tetramethyl-4-piperidinyl)iodoacetamide[56] attached to cysteine F9(93)β of $\alpha\beta^{+CN}$ is influenced by binding oxygen or carbon monoxide to the α-chains[278,338]. This change in conformation of the β-subunits by binding ligands to the α subunits destroys the isosbestic points[278,280,281] for the e.p.r. spectra compared at different partial pressures of oxygen or carbon monoxide (Fig. 19). This behavior suggests the existence of an intermediate protein conformation for partly liganded hemoglobin[338].

The subunit contacts involved

The four subunits in hemoglobin are arranged in a tetrahedron[6] and each subunit is in contact with the other three subunits (Fig. 2). However, as discussed previously there are four different types of contacts ($\alpha_1\beta_2$ = $\alpha_2\beta_1$, $\alpha_1\beta_1$ = $\alpha_2\beta_2$, $\alpha\alpha$ and $\beta\beta$)[2,26], and binding a ligand to one subunit will not uniformly influence the other three subunits.

The X-ray structures indicate that the $\alpha\beta$ contacts are much more extensive than the $\alpha\alpha$ or $\beta\beta$ contacts (Tables II and III). Even though the $\alpha_1\beta_1$ contacts are more extensive than the $\alpha_1\beta_2$ contacts, a convincing argument can be made for the primary importance of the $\alpha_1\beta_2$ contacts. The change in quaternary structure influences the $\alpha_1\beta_2$ contact (Table III) to a much larger extent than the $\alpha_1\beta_1$ contact[26]. Any modification of amino acids in the former region (Table V) produce significant changes in the

TABLE V

ABNORMAL HEMOGLOBINS WHERE RESIDUES AT CONTACTS BETWEEN SUBUNITS ARE SUBSTITUTED

Designation	Substitution	Effect on oxygenation and other properties	Contact	Ref.
J Capetown	Arg FG4(92)$\alpha \rightarrow$ Gln	High affinity — diminished cooperativity	$\alpha_1\beta_2$	339,340
Chesapeake	Arg FG4(92)$\alpha \rightarrow$ Leu	High affinity — diminished cooperativity	$\alpha_1\beta_2$	341,342
Kansas	Asn G4(102)$\beta \rightarrow$ Thr	Very low affinity — diminished cooperativity dissociates into dimers	$\alpha_1\beta_2$	343,344
Yakima	Asp G1(99)$\beta \rightarrow$ His	Very high affinity, non-cooperative	$\alpha_1\beta_2$	345,346
Kempsey	Asp G1(99)$\beta \rightarrow$ Asn	Very high affinity	$\alpha_1\beta_2$	347
Malmö	His FG4(97)$\beta \rightarrow$ Gln	Very high affinity	$\alpha_1\beta_2$	167,348
Hirose	Trp C3(37)$\beta \rightarrow$ Ser	Very high affinity, diminished cooperativity	$\alpha_1\beta_2$	349
G Chinese	Glu B11(30)$\alpha \rightarrow$ Gln	None	$\alpha_1\beta_1$	350
Chiapas	Pro GH2(114)$\alpha \rightarrow$ Arg	None	$\alpha_1\beta_1$	351
Yoshizuka	Asn G10(108)$\beta \rightarrow$ Asp	Low affinity	$\alpha_1\beta_1$	352
Philly	Tyr C1(35)$\beta \rightarrow$ Phe	Weakens $\alpha_1\beta_1$ contact so that all cysteines react with PMB. Dissociates into monomers in presence of PMB	$\alpha_1\beta_1$	166
Tacoma	Arg B12(30)$\beta \rightarrow$ Ser	High affinity	$\alpha_1\beta_1$	353

oxygenation of hemoglobin, while certain changes at the $\alpha_1\beta_1$ contact produce no noticeable effect (Table V)[38,166,167]. The shortest distance between hemes is across this contact[208] and part of the β-chain involved in this contact is close to the region for which the largest change in tertiary structure is observed[26,38].

The modification of hemoglobin by bifunctional maleimides[53,54,259,285-287] like BME completely eliminates heme-heme interaction[54]. The major change in the X-ray structure produced by the modification[259] is in the portion of the $\alpha_1\beta_2$ contact where part of the BME molecule is located. The $\alpha_1\beta_1$ contact, the environment of the hemes, and the rest of the $\alpha_1\beta_2$ contact are essentially unaltered.

The role of the $\alpha_1\beta_2$ contact in transmitting conformational changes between subunits has been studied using abnormal hemoglobins in which residues at this contact are changed[64,280,281,305]. In Hb Chesapeake[341,342] and Hb J. Capetown[339,340], arginine FG4(92)α[354] is replaced by leucine and glutamine, respectively, while in Hb Yakima[345,346] and Hb Kempsey[347], aspartic acid G1(99)β is replaced by histidine and asparagine, respectively (Tables III and V).

The n.m.r. spectra of these abnormal hemoglobins in the deoxy form show alterations in the hyperfine shifted protons of both the α and β hemes, irrespective of whether the abnormality is on the α or β chain[64,269,305]. On the other hand, in the case of hemoglobins with modified β chains, such as Hb Zurich[173-175] in which histidine E7(63)β is replaced by arginine (Table I), fetal hemoglobin[355] in which the β chains are replaced by γ chains, hemoglobin treated with CPA[284], and hemoglobin reacted with a spin label on the β subunits[56], the n.m.r. spectrum shows alterations predominantly for the β heme.

It has also been shown that the failure to obtain isosbestic points for the e.p.r. spectra of the spin label *N*-(1-oxyl-2,2,6,6-tetramethyl-4-piperidinyl) iodoacetamide[278] when CO binds depends on the presence of an intact $\alpha_1\beta_2$ contact (Fig. 19)[280,281]. Thus, isosbestic points are obtained with Hb J. Capetown, Hb Chesapeake, Hb Yakima (Table V) and hemoglobin treated with CPA[284] but no isosbestic points are obtained with other hemoglobins which do not affect the $\alpha_1\beta_2$ contact, even if the oxygenation is quite different. These results indicate that the existence of an intermediate environment of the spin label on the β chain for partly liganded hemoglobin (see above) requires an unaltered normal $\alpha_1\beta_2$ contact.

It should be pointed out that although any modification at the $\alpha_1\beta_2$ contact influences the oxygen affinity and decreases the heme-heme interaction (Table V), in at least a few cases a significant amount of heme-heme interaction remains[354], as in the case of Hb J. Capetown (n = 2.2)[357] or Hb Chesapeake (n = 1.3)[342]. Furthermore, heme-heme interaction can be reduced without affecting the $\alpha_1\beta_2$ contact (Table I), as found for Hb Zurich (n = 1.8)[174].

The dissociation of hemoglobin into dimers at low concentrations in high salt is thought to result in $\alpha_1\beta_1$ and $\alpha_2\beta_2$ dimers, *i.e.*, the $\alpha_1\beta_2$ and $\alpha_2\beta_1$ contacts are broken[38,274,355]. Thus, the constant for dissociation into dimers of the ligand form of hemoglobin Kansas[343,344], in which asparagine G4(102)β is replaced by threonine (Tables III and V) is two orders of magnitude greater than that of normal human hemoglobin[343]. In the liganded form asparagine G4(102)β is hydrogen bonded with aspartic acid G1(94)α in the $\alpha_1\beta_2$ contact (Table III and Fig. 15)[357]. This hydrogen bond is almost the only polar interaction at the $\alpha_1\beta_2$ interface in liganded hemoglobin. Furthermore, the reaction of hemoglobin with one mole of PMB at the reactive F9(93)β sulfhydryl, which is near the $\alpha_1\beta_2$ contact, enhances dissociation into dimers[274].

In order to split human hemoglobin into monomers it is necessary to react cysteines G11(104)α and G14(112)β[274], which are in the region of the $\alpha_1\beta_1$ contact (Table II), with a sulfhydryl reagent such as PMB. The abnormal hemoglobin Philly (tyrosine C1(35)$\beta \rightarrow$ phenylalanine), in which a polar hydrogen bond at the $\alpha_1\beta_1$ contact in liganded hemoglobin cannot be formed (Table II), reacts readily with six moles of PMB and dissociates into monomers instead of dimers (Table V).

The studies on the functional behavior of the dimers discussed earlier, therefore, relate to the importance of the $\alpha_1\beta_1$ contacts in heme–heme interaction. On the basis of experiments suggesting that the ligand reactions of the dimer are similar to the tetramer[31,312-315], a "dimer hypothesis"[31,312,313] has been proposed which considers the $\alpha\beta$ dimer as the important functional unit even at the tetramer level. The observed n of about 3 in normal hemoglobin requires some interaction between dimers. This phenomenon can be explained by a strong coupling of 2800 cal between the α- and β-subunits together with a much weaker coupling, as small as 600 cal, between the dimers[228].

The validity of the dimer hypothesis has been questioned from various points of view. The X-ray structure seems to suggest that the $\alpha_1\beta_2$ contact is more important than the $\alpha_1\beta_1$ contact[38,26]. Furthermore, the ionic bonds across the $\alpha\alpha$ contacts and the $\alpha_1\beta_2$ contacts in deoxyhemoglobin (Figs. 13 and 14) as well as the proposed binding of 2,3-DPG between β-subunits (Fig. 16)[254] suggest that the functional tetramer is required for cooperative binding to occur. The kinetic[69,358] and equilibrium[66,67] studies, which indicate that a large change in the constants occurs after the third ligand is bound (Table IV and Fig. 21), are also inconsistent with a dimer model. Finally, as mentioned earlier, there is evidence[320,322-324] that deoxyhemoglobin is undissociated under conditions that are thought to produce the cooperative dimers, and that the ligand reactions of the deoxy dimer resemble those of the monomer (Fig. 24) indicating that the tetramer is required for cooperative binding.

The role of the $\alpha\alpha$ contact in heme–heme interaction has been elucidated by studies on hemoglobin which has the terminal arginine HC3(141)α removed by CPB[322]. These arginines are responsible for the four salt bridges linking the α-chains in deoxyhemoglobin[26,38,253,254]. The decrease in n from 3 to 2 by the removal of these residues (Figs. 20 and 24), when the hemoglobin is not dissociated, is therefore a measure of the contribution of the $\alpha\alpha$ contact to heme–heme interaction.

While the β-subunits are not in contact with each other in hemoglobin, an indication of the influence of this interface on heme–heme interaction can be acquired by investigating the effect of organic phosphates thought to link these subunits (Fig. 16)[254], on heme–heme interaction. It has recently been reported that 2,3-DPG increases the free energy of interaction by 1400 cal[359] and IHP increases the free energy of interaction by 600 cal[360].

Thus, the linkage between the β-subunits does contribute to heme–heme interaction.

Another indication of interaction between like subunits which must be transmitted across these contacts has been obtained from the oxygenation properties of the various hybrid hemoglobins incorporating non-functional subunits that either have no porphyrin ($\alpha^{\circ}\beta$, $\alpha\beta^{\circ}$)[361-363], contain Fe(III) porphyrins ($\alpha^{+}\beta$, $\alpha\beta^{+}$)[331-333,364] or contain Mn(III) porphyrins ($\alpha^{Mn}\beta$, $\alpha\beta^{Mn}$)[100]. The oxygen affinities of all of these hybrids are intermediate between that of normal hemoglobin and that of isolated chains, and the heme–heme interaction is reduced. The hybrids with functional β-subunits have a lower oxygen affinity than those with functional α-subunits, and in most cases the value of n is about 1.3[100,331,333,363,364]. Such a value of n corresponds to an interaction energy between the β-subunits of 700–800 cal, which is not insignificant relative to the 3000 cal for tetrameric hemoglobin (Table IV)[364]. For the valence hybrids with functional α-subunits the interaction between α-chains is dependent on whether CN^- is bound to the oxidized hemes. With CN^- bound, there does not seem to be any interaction between α chains[333,364]. Without CN^- bound to the Fe(III) β-chains, the shape of the oxygenation curve is dependent on pH[331], perhaps indicating that the interaction between subunits depends on the presence of H_2O or OH^- at the sixth coordination site of the Fe(III) hemes. At pH $\geqslant$ 8.2 the value of n is 1.1 which corresponds to an almost negligible interaction energy of about 70 cal. At lower pH values the binding curves are asymmetric, becoming steeper as the oxygenation proceeds. The interaction energy calculated for these curves corresponds to about 800 cal[331].

The interaction energy observed in these hybrids between like subunits is small relative to the total heme–heme interaction in normal hemoglobin. However, the oxidized subunits are presumably in the liganded state[38] and these hybrids are, therefore, models for the binding of the last two molecules of oxygen (or CO) to hemoglobin. Changing the Fe(III) ligand, while maintaining the liganded state, influences the interaction between like chains[331-333,364]. Therefore, the interaction between the like chains at early stages of oxygen binding can be quite different from that observed in these hybrids.

Source of the interaction energy

Can any conclusions be reached regarding the source of energy responsible for heme-heme interaction? Equilibrium studies furnish an estimate of the free energy of interaction (Table IV)[227,228,292] and both equilibrium[66,67,292] and kinetic studies[69,358] suggest what stages in the binding of ligands are involved in the strongest interactions. Changes are observed at the hemes[38,46,189,330,335], within the subunits[38,275,338], and at the contacts between the hemes[26,38]. However, these changes do not necessarily reflect the thermodynamic source for the energy which the tetrameric protein uses to regulate the binding of ligands at sites far removed from each other.

Shulman and coworkers[63,336,337], on the basis of their n.m.r. studies suggest that the contribution of the hemes is relatively small[63]. They suggest that the interaction energy arises from the change in quaternary structure[336,337]

Perutz has suggested[38] that the six salt bridges which are disrupted on the conversion from the unliganded to the liganded conformation constitute the primary source of the interaction energy. The extensive non-polar contacts between the α- and β-subunits are not felt to be of primary importance since there are actually fewer van der Waals contacts between the subunits in the more constrained unliganded conformation.

Recent studies on proteins[365,366] indicate that solvent interactions may play a dominant role. The interactions of water with the different types of ionic, polar, and non-polar protein side chains are quite different. The tertiary structure of proteins is determined in large part by the requirement to place non-polar side chains on the inside, away from the polar solvent, and polar side chains in contact with the polar solvent. The interactions with water usually seem to involve large changes in ΔH and ΔS which vary linearly and compensate each other, thus producing much smaller changes in free energy[365].

Such compensation phenomena have been observed in studies on the spin equilibrium[52,89] and binding of ligands to ferrihemoglobins[33-37], suggesting the importance of solvent interactions in hemoglobin. The rearrangement of the subunits observed in the X-ray studies[25,26] probably influences the interaction of the protein with the solvent. This interaction can therefore account for a significant fraction of the interaction energy.

ACKNOWLEDGEMENT

I wish to thank Dr. James Butzow and Dr. Yong Ae Shin for valuable discussions and Jacqueline Blake for assistance in the preparation of this chapter.

REFERENCES

1 M. F. Perutz, M. G. Rossman, A. F. Cullis, H. Muirhead, G. Will and A. C. T. North, *Nature*, 185 (1960) 416.
2 M. F. Perutz, H. Muirhead, J. M. Cox and L. C. G. Goaman, *Nature*, 219 (1968) 131.
3 J. C. Kendrew, R. E. Dickerson, B. E. Strandberg, *et al.*, *Nature*, 185 (1960) 422.
4 H. C. Watson, *Progress in Stereochemistry*, Vol. 4, Butterworths, London, 1968.

5 R. E. Dickerson and I. Geis, *The Stereo Supplement to the Structure and Action of Proteins*, Harper and Row, New York, 1969.
6 M. F. Perutz, *Science*, 140 (1963) 863.
7 M. O. Dayhoff, *Atlas of Protein Sequence and Structure*, Vol. 4, The National Biomedical Research Foundation, Silver Spring, 1969.
8 A. B. Edmundson, *Nature*, 205 (1965) 883.
9 G. Braunitzer, R. Gehring-Muller, M. Hilschmann, *et al.*, *Z. Physiol. Chem.*, 325 (1961) 283.
10 R. J. Hill and W. Konigsberg, *J. Biol. Chem.*, 237 (1962) 3151.
11 M. F. Perutz, J. C. Kendrew and H. C. Watson, *J. Mol. Biol.*, 13 (1965) 669.
12 J. C. Kendrew, *Science*, 139 (1963) 1259.
13 R. Lemberg and J. W. Legge, *Haematin Enzymes and Bile Pigments*, Interscience, New York, 1949.
14 A. Rossi-Fanelli, E. Antonini and A. Caputo, *Adv. Protein Chem.*, 19 (1964) 73.
15 H. Theorell, *Biochem. Z.*, 268 (1934) 46.
16 C. G. Douglas, J. S. Haldane and J. B. S. Haldane, *J. Physiol. (London)*, 44 (1912) 275.
17 F. J. W. Roughton, *Ann. N.Y. Acad. Sci.*, 174 (1970) 177.
18 Q. H. Gibson and F. J. W. Roughton, *J. Physiol. (London)*, 136 (1957) 507.
19 Q. H. Gibson and F. J. W. Roughton, *Proc. R. Soc. (B)*, 147 (1957) 44.
20 A. Lein and L. Pauling, *Proc. Natn. Acad. Sci. U.S.*, 42 (1956) 51.
21 R. C. C. St. George and L. Pauling, *Science*, 114 (1951) 629.
22 B. Talbot, M. Brunori, E. Antonini and J. Wyman, *J. Mol. Biol.*, 58 (1971) 261.
23 C. L. Nobbs, in B. Chance, R. W. Estabrook and T. Yonetani, *Hemes and Hemoproteins*, Academic Press, New York, 1966, p. 143.
24 C. L. Nobbs, H. C. Watson and J. C. Kendrew, *Nature*, 209 (1966) 339.
25 H. Muirhead, J. M. Cox, L. Mazzarella and M. F. Perutz, *J. Mol. Biol.*, 28 (1967) 117.
26 W. Bolton and M. F. Perutz, *Nature*, 228 (1970) 551.
27 J. Brooks, *Proc. R. Soc. (B)*, 109 (1931) 35.
28 W. N. M. Ramsay, in F. J. W. Roughton and J. C. Kendrew, *Haemoglobin*, Butterworths, London, 1949, p. 231.
29 W. D. Brown and L. B. Mebine, *J. Biol. Chem.*, 244 (1969) 6696.
30 J. Rifkind, *Biochim. Biophys. Acta*, 273 (1972) 30.
31 E. Antonini and M. Brunori, *A. Rev. Biochem.*, 39 (1970) 977.
32 P. George, in D. E. Green, *Currents in Biochemical Research*, Interscience, New York, 1956, p. 338.
33 J. G. Beetlestone and D. H. Irvine, *J. Chem. Soc. (A)*, (1969) 735.
34 J. E. Bailey, J. G. Beetlestone and D. H. Irvine, *J. Chem. Soc. (A)*, (1969) 241.
35 A. C. Anusiem, J. G. Beetlestone and D. H. Irvine, *J. Chem. Soc. (A)*, (1968) 960.
36 A. C. Anusiem, J. G. Beetlestone and D. H. Irvine, *J. Chem. Soc. (A)*, (1968) 1337.
37 J. G. Beetlestone, A. A. Epega and D. H. Irvine, *J. Chem. Soc. (A)*, (1968) 1346.
38 M. F. Perutz, *Nature*, 228 (1970) 726.
39 M. Weissbluth, in C. K. Jørgensen, *et al.*, *Structure and Bonding*, Vol. 2, Springer-Verlag, New York, 1967, p. 1.
40 M. Kotani, *Ann. N.Y. Acad. Sci.*, 158 (1969) 20.
41 U. Gonser and R. W. Grant, *Biophys. J.*, 5 (1965) 823.
42 G. Lang and W. Marshall, *J. Mol. Biol.*, 18 (1966) 385.
43 A. Trautwein, H. Eichler, A. Mayer, *et al.*, *J. Chem. Phys.*, 53 (1970) 963.
44 J. F. Gibson and D. J. E. Ingram, *Nature*, 180 (1957) 29.
45 J. F. Gibson, D. J. E. Ingram and D. Schonland, *Discuss. Faraday Soc.*, 26 (1958) 72.

46 J. Peisach, W. E. Blumberg, S. Ogawa, E. A. Rachmilewitz and R. Oltzik, *J. Biol. Chem.*, 246 (1971) 3342.
47 J. S. Griffith, in B. Pullman and M. Weissbluth, *Molecular Biophysics*, Academic Press, New York, 1965, p. 191.
48 R. Banerjee, Y. Alpert, F. Leterrier and R. J. P. Williams, *Biochemistry*, 8 (1969) 2862.
49 J. Beetlestone and P. George, *Biochemistry*, 3 (1964) 707.
50 P. George, J. Beetlestone and J. S. Griffith, in J. E. Falk, R. Lemberg and R. K. Morton, *Haematin Enzymes*, Pergamon Press, Oxford, 1961, p. 105.
51 T. Iizuka and M. Kotani, *Biochim. Biophys. Acta*, 181 (1969) 275.
52 T. Iizuka and M. Kotani, *Biochim. Biophys. Acta*, 194 (1969) 351.
53 S. R. Simon and W. H. Konigsberg, *Proc. Natn. Acad. Sci. U.S.*, 56 (1966) 749.
54 J. K. Moffat, S. R. Simon and W. H. Konigsberg, *J. Mol. Biol.*, 58 (1971) 89.
55 J. C. A. Boeyens and H. M. McConnell, *Proc. Natn. Acad. Sci. U.S.*, 56 (1966) 22.
56 S. Ogawa and H. M. McConnell, *Proc. Natn. Acad. Sci. U.S.*, 58 (1967) 19.
57 J. K. Moffat, *J. Mol. Biol.*, 55 (1971) 135.
58 G. L. Eichhorn and A. J. Osbahr, *Adv. Chem. Coordination Compounds*, (1961) 216.
59 S. Beychok, I. Tyuma, R. E. Benesch and R. Benesch, *J. Biol. Chem.*, 242 (1967) 2460.
60 M. Brunori, J. Engel and T. M. Schuster, *J. Biol. Chem.*, 242 (1967) 773.
61 M. Nagai, Y. Sugita and Y. Yoneyama, *J. Biol. Chem.*, 244 (1969) 1651.
62 R. G. Shulman, K. Wüthrich, T. Yamane, D. J. Patel and W. E. Blumberg, *J. Mol. Biol.*, 53 (1970) 143.
63 R. G. Shulman, S. Ogawa, K. Wüthrich, T. Yamane, J. Peisach and W. E. Blumberg, *Science*, 165 (1969) 251.
64 D. G. Davis, T. R. Lindstrom, N. H. Mock, *et al.*, *J. Mol. Biol.*, 60 (1971) 101.
65 S. W. Englander, in B. Chance, T. Yonetani and A. S. Mildvan, *Probes of Structure and Function of Macromolecules and Membranes*, Vol. 2, Academic Press, New York, 1971, p. 389.
66 F. J. W. Roughton, A. B. Otis and R. L. J. Lyster, *Proc. R. Soc. (B)*, 144 (1955) 29.
67 F. J. W. Roughton and R. L. J. Lyster, *Hvalrådets Skrifter Nr*, 48 (1961) 185.
68 Q. H. Gibson, *Progr. Biophys. Biophys. Chem.*, 9 (1959) 1.
69 Q. H. Gibson, *J. Biol. Chem.*, 245 (1970) 3285.
70 R. A. B. Holland, *Ann. N.Y. Acad. Sci.*, 174 (1970) 154.
71 J. E. Falk, *Porphyrins and Metalloporphyrins*, Elsevier, Amsterdam, 1964, p. 53.
72 L. E. Orgel, in J. E. Falk, R. Lemberg and R. K. Morton, *Haematin Enzymes*, Vol. 1, Pergamon Press, Oxford, 1961, p. 1.
73 R. G. Shulman, S. H. Glarum and M. Karplus, *J. Mol. Biol.*, 57 (1971) 93.
74 M. C. Hsu and R. W. Woody, *J. Am. Chem. Soc.*, 91 (1969) 3679.
75 W. A. Eaton and R. M. Hochstrasser, *J. Chem. Phys.*, 49 (1968) 985.
76 J. S. Griffith, *Proc. R. Soc. (A)*, 235 (1956) 23.
77 M. Kotani, *Biopolymers Symp.*, 1 (1964) 67.
78 C. D. Coryell, F. Stitt and L. Pauling, *J. Am. Chem. Soc.*, 59 (1937) 633.
79 N. Nakano, J. Otsuka and A. Tasaki, *Biochim. Biophys. Acta*, 236 (1971) 222.
80 C. D. Coryell, L. Pauling and R. W. Dodson, *J. Phys. Chem.*, 43 (1939) 825.
81 J. L. Hoard, in A. Rich and N. Davidson, *Structural Chemistry and Molecular Biology*, W. H. Freeman, San Francisco, 1968, p. 573.
82 A. Tasaki, *J. Appl. Phys.*, 41 (1970) 1000.
83 H. Theorell and A. Ehrenberg, *Acta Chem. Scand.*, 5 (1951) 823.
84 H. Taube, *Chem. Rev.*, 50 (1952) 69.

85 W. Scheler, G. Schoffa and F. Jung, *Biochem. Z.*, 329 (1957) 232.
86 R. Havemann and W. Haberditzl, *Z. Phys. Chem.*, 209 (1958) 135.
87 D. Keilin and E. F. Hartree, *Biochem. J.*, 49 (1951) 88.
88 T. Iizuka and M. Kotani, *Biochim. Biophys. Acta*, 154 (1968) 417.
89 J. Otsuka, *Biochim. Biophys. Acta*, 214 (1970) 233.
90 J. E. Coleman and B. L. Vallee, *J. Biol. Chem.*, 236 (1961) 2244.
91 L. Pauling, in F. J. W. Roughton and J. C. Kendrew, *Haemoglobin*, Butterworths, London (1949) p. 57.
92 J. E. Falk, *Porphyrins and Metalloporphyrins*, Elsevier, Amsterdam, 1964, p. 30.
93 B. M. Hoffman and D. H. Petering, *Proc. Natn. Acad. Sci. U.S.*, 67 (1970) 637.
94 R. G. Wilkins, in R. F. Gould, *Bioinorganic Chemistry*, Advances in Chemistry Series, Washington, D.C., American Chemical Society, 1971, p. 111.
95 A. L. Crumbliss and F. Basolo, *J. Am. Chem. Soc.*, 92 (1970) 55.
96 B. M. Hoffman, D. Diemente and F. Basolo, *J. Am. Chem. Soc.*, 92 (1970) 61.
97 B. C. Wang and W. D. Schaefer, *Science*, 166 (1969) 1404.
98 J. H. Bayston, N. K. King, F. D. Looney and M. E. Winfield, *J. Am. Chem. Soc.*, 91 (1969) 2775.
99 A. Misono and S. Koda, *Bull. Chem. Soc. Japan*, 42 (1969) 3048.
100 M. R. Waterman and T. Yonetani, *J. Biol. Chem.*, 245 (1970) 5847.
101 T. Yonetani, H. R. Drott, J. S. Leigh, Jr., G. H. Reed, M. R. Waterman and T. Asakura, *J. Biol. Chem.*, 245 (1970) 2998.
102 M. Z. Atassi, *Biochem. J.*, 103 (1967) 29.
103 S. F. Andres and M. Z. Atassi, *Biochemistry*, 9 (1970) 2268.
104 T. L. Fabry, C. Simo and K. Jayaherian, *Biochim. Biophys. Acta*, 160 (1968) 118.
105 K. Wüthrich, R. G. Shulman and J. Peisach, *Proc. Natn. Acad. Sci. U.S.*, 60 (1968) 373.
106 K. Wüthrich, R. G. Shulman, B. J. Wyluda and W. S. Caughey, *Proc. Natn. Acad. Sci. U.S.*, 62 (1969) 636.
107 K. Wüthrich, R. G. Shulman and T. Yamane, *Proc. Natn. Acad. Sci. U.S.*, 61 (1968) 1199.
108 R. G. Shulman, K. Wüthrich, T. Yamane, E. Antonini and M. Brunori, *Proc. Natn. Acad. Sci. U.S.*, 63 (1969) 623.
109 R. J. Kurland, R. G. Little, D. G. Davis and C. Ho, *Biochemistry*, 10 (1971) 2237.
110 R. J. Kurland, D. G. Davis and C. Ho, *J. Am. Chem. Soc.*, 90 (1968) 2700.
111 A. J. Osbahr and G. L. Eichhorn, *J. Biol. Chem.*, 237 (1962) 1820.
112 T. Samejima and K. Masako, *J. Biochem. (Tokyo)*, 65 (1969) 759.
113 G. Geraci and T.-K. Li, *Biochemistry*, 8 (1969) 1848.
114 T.-K. Li and B. P. Johnson, *Biochemistry*, 8 (1969) 3638.
115 Y. Sugita, M. Nagai and Y. Yoneyama, *J. Biol. Chem.*, 246 (1971) 383.
116 Y. Ueda, T. Shiga and I. Tyuma, *Biochim. Biophys. Acta*, 207 (1970) 18.
117 M. Brunori, E. Antonini, J. Wyman and S. R. Anderson, *J. Mol. Biol.*, 34 (1968) 357.
118 N. M. Rumen and B. Chance, *Biochim. Biophys. Acta*, 175 (1969) 242.
119 D. L. Drabkin, in J. E. Falk, R. Lemberg and R. K. Morton, *Haematin Enzymes*, Vol. 1, Pergamon Press, Oxford, 1961, p. 142.
120 R. J. P. Williams, in J. E. Falk, R. Lemberg and R. K. Morton, *Haematin Enzymes*, Vol. 1, Pergamon Press, Oxford, 1961, p. 41.
121 R. J. P. Williams, in B. Chance, R. W. Estabrook and T. Yonetani, *Hemes and Hemoproteins*, Academic Press, New York, 1966, p. 557.
122 R. J. P. Williams, *Chem. Rev.*, 56 (1956) 299.
123 J. E. Falk and D. D. Perrin, in J. E. Falk, R. Lemberg and R. K. Morton, *Haematin Enzymes*, Vol. 1, Pergamon Press, Oxford, 1961, p. 56.

124 A. Rossi-Fanelli and E. Antonini, *Arch. Biochem. Biophys.*, 80 (1959) 308.
125 A. Rossi-Fanelli, E. Antonini and A. Caputo, *Arch. Biochem. Biophys.*, 85 (1959) 37.
126 E. Antonini, M. Brunori, A. Caputo, E. Chiancone, A. Rossi-Fanelli and J. Wyman, *Biochim. Biophys. Acta*, 79 (1964) 284.
127 Y. Sugita and Y. Yoneyama, *J. Biol. Chem.*, 246 (1971) 389.
128 W. S. Caughey, H. Eberspaecher, W. H. Fuchsman, S. McCoy and J. O. Alben, *Ann. N.Y. Acad. Sci.*, 153 (1969) 722.
129 W. S. Caughey, *Ann. N.Y. Acad. Sci.*, 174 (1970) 148.
130 A. Rossi-Fanelli and E. Antonini, *J. Biol. Chem.*, 235 (1960) PC 4.
131 Q. H. Gibson and E. Antonini, *J. Biol. Chem.*, 238 (1963) 1384.
132 J. E. O'Hagan, *Biochem. J.*, 74 (1960) 417.
133 J. E. O'Hagen and P. George, *Biochem. J.*, 74 (1960) 424.
134 T. Asakura, H. Yoshikawa and K. Imahori, *J. Biochem. (Tokyo)*, 64 (1968) 515.
135 E. B. Fleischer, C. K. Miller and L. E. Webb, *J. Am. Chem. Soc.*, 86 (1964) 2342.
136 E. B. Fleischer, *J. Am. Chem. Soc.*, 85 (1963) 146.
137 T. A. Hamor, W. S. Caughey and J. L. Hoard, *J. Am. Chem. Soc.*, 87 (1965) 2305.
138 D. F. Koenig, *Acta Cryst.*, 18 (1965) 665.
139 J. L. Hoard, G. H. Cohen and M. D. Glick, *J. Am. Chem. Soc.*, 89 (1967) 1992.
140 J. L. Hoard, M. J. Hamor, T. A. Hamor and W. S. Caughey, *J. Am. Chem. Soc.*, 87 (1965) 2312.
141 R. Countryman, D. M. Collins and J. L. Hoard, *J. Am. Chem. Soc.*, 91 (1969) 5166.
142 L. Pauling, *The Nature of the Chemical Bond*, 3rd Edn., Cornell University Press, Ithaca, 1960.
143 B. Chance, in B. Chance, T. Yonetani and A. S. Mildvan, *Probes of Structure and Function of Macromolecules and Membranes*, Vol. 2, Academic Press, New York, 1971, p. 321.
144 H. C. Watson and B. Chance, in B. Chance, R. W. Estabrook and T. Yonetani, *Hemes and Hemoproteins*, Academic Press, New York, 1966, p. 149.
145 L. Stryer, J. C. Kendrew and H. C. Watson, *J. Mol. Biol.*, 8 (1964) 96.
146 C. L. Nobbs, *J. Mol. Biol.*, 13 (1965) 325.
147 B. P. Shoenborn, *J. Mol. Biol.*, 45 (1969) 297.
148 B. P. Schoenborn, in B. Chance, T. Yonetani and A. S. Mildvan, *Probes of Structure and Function of Macromolecules and Membranes*, Vol. 2, Academic Press, New York, 1971, p. 181.
149 P. A. Bretscher, *Nature*, 219 (1968) 606.
150 H. Muirhead and J. Greer, *Nature*, 228 (1970) 516.
151 W. Bolton, J. M. Cox and M. F. Perutz, *J. Mol. Biol.*, 33 (1968) 283.
152 E. A. Padlan and W. E. Love, in B. Chance, T. Yonetani and A. S. Mildvan, *Probes of Structure and Function of Macromolecules and Membranes*, Vol. 2, Academic Press, New York, 1971, p. 187.
153 E. A. Padlan and W. E. Love, *Nature*, 220 (1968) 376.
154 W. E. Love, *Abstr. Eighth Int. Congr. Biochem.*, (1970) 2.
155 J. S. Griffith, *Biopolymers Symp.*, 1 (1964) 35.
156 J. S. Griffith, *Nature*, 180 (1957) 30.
157 M. T. Hamor, T. A. Hamor and J. L. Hoard, *J. Am. Chem. Soc.*, 86 (1964) 1938.
158 J. H. Wang, in J. E. Falk, R. Lemberg and R. K. Morton, *Haematin Enzymes*, Vol. 1, Pergamon Press, Oxford, 1961, p. 76.
159 A. Nakahara and J. H. Wang, *J. Am. Chem. Soc.*, 80 (1958) 6526.
160 J. N. Phillips, in J. E. Falk, R. Lemberg and R. K. Morton, *Haematin Enzymes*, Vol. 1, Pergamon Press, Oxford, 1961, p. 78.

161 J. H. Wang, A. Nakahara and E. B. Fleischer, *J. Am. Chem. Soc.*, 80 (1958) 1109.
162 J. O. Alben and W. S. Caughey, in B. Chance, R. W. Estabrook and T. Yonetani, *Hemes and Hemoproteins*, Academic Press, New York, 1966, p. 139.
163 J. O. Alben and W. S. Caughey, *Biochemistry*, 7 (1968) 175.
164 L. Vaska, *Science*, 140 (1963) 809.
165 J. A. McGinnety, R. J. Doedens and J. A. Ibers, *Science*, 155 (1967) 709.
166 J. F. Perutz and H. Lehmann, *Nature*, 219 (1968) 902.
167 H. Morimoto, H. Lehmann and M. F. Perutz, *Nature*, 232 (1971) 408.
168 T. Miyaji, I. Iuchi, S. Shibata, I. Takeda and A. Tamura, *Acta Haematol. Japan*, 26 (1963) 538.
169 N. Hayashi, Y. Mutokawa and G. Kikuchi, *J. Biol. Chem.*, 241 (1966) 79.
170 P. Heller, R. D. Coleman and V. Yakulis, *J. Proc. 11th Congr. Int. Soc., Haematol.*, Govt. Printer, Sydney, 1966, p. 427.
171 A. Hayashi, T. Suzuki, A. Shimizu, K. Imai, *et al.*, *Arch. Biochem. Biophys.*, 125 (1968) 895.
172 D. S. Gerald and M. L. Efron, *Proc. Natn. Acad. Sci. U.S.*, 47 (1958) 1758.
173 C. J. Muller and S. Kingma, *Biochim. Biophys. Acta*, 50 (1961) 595.
174 K. H. Winterhalter, N. M. Anderson, G. Amiconi, E. Antonini and M. Brunori, *Eur. J. Biochem.*, 11 (1969) 435.
175 W. H. Hitzig, P. C. Frick, K. Betke and T. H. J. Huisman, *Helv. Paediat. Acta*, 15 (1960) 499.
176 J. H. Steadman, A. Yates and E. R. Huehns, *Br. J. Haematol.*, 18 (1970) 435.
177 R. W. Carrell, H. Lehmann, P. A. Lorkin, E. Raik and E. Hunter, *Nature*, 215 (1967) 626.
178 J. V. Dacie, N. K. Shinton, P. J. Gaffney, R. W. Carrell and H. Lehmann, *Nature*, 216 (1967) 663.
179 A. J. Grimes, A. Meisler and J. V. Dacie, *Br. J. Haematol.*, 10 (1964) 281.
180 V. Bratu, P. A. Lorkin, H. Lehmann and C. Predescu, *Biochim. Biophys. Acta*, 251 (1971) 1.
181 A. Hollender, P. A. Lorkin, H. Lehmann and B. Svensson, *Nature*, 222 (1969) 953.
182 B. Svensson and L. Strand, *Scand. J. Haematol.*, 4 (1967) 241.
183 R. W. Carrell, H. Lehmann and H. E. Hutchison, *Nature*, 210 (1966) 915.
184 R. Vaughan-Jones, A. J. Grimes, R. W. Carrell and H. Lehmann, *Br. J. Haematol.*, 13 (1967) 394.
185 J. Greer, *J. Mol. Biol.*, 59 (1971) 107.
186 R. Lumry, *Biophysics (Japan)*, 1 (1961) 138.
187 A. Hayashi, A. Shimizu, Y. Yamamura and H. Watari, *Biochim. Biophys. Acta*, 102 (1965) 626.
188 A. Hayashi, A. Shimizu, Y. Yamamura and H. Watari, *Science*, 152 (1966) 207.
189 A. Hayashi, T. Suzuki, A. Shimizu, H. Morimoto and H. Watari, *Biochim. Biophys. Acta*, 147 (1967) 407.
190 G. Bemski and R. L. Nagel, *Biochim. Biophys. Acta*, 154 (1968) 592.
191 H. Watari, A. Hayashi, H. Morimoto and M. Kotani, in S. Fujiwara and L. H. Piette, *Recent Developments of Magnetic Resonance in Biological Systems*, Hirokawa Publishing Co., Tokyo, 1968, p. 128.
192 L. Pauling and C. P. Coryell, *Proc. Natn. Acad. Sci. U.S.*, 22 (1936) 210.
193 J. J. Weiss, *Nature*, 202 (1964) 83.
194 H. B. Gray, in R. F. Gould, *Bioinorganic Chemistry*, Advances in Chemistry Series, Washington, D.C., American Chemical Society, 1971, p. 365.
195 C. A. Coulson, *Valence*, Clarendon Press, Oxford (1952).
196 L. Pauling, *Nature*, 203 (1964) 182.

197 G. M. Maggiora, R. O. Viale and L. L. Ingraham, in T. E. King, H. S. Mason and M. Morrison, *Oxidases and Related Redox Systems,* Vol. 1, John Wiley, New York, 1965, p. 88.
198 M. Gouterman and M. Zerner, in B. Chance, R. W. Estabrook and T. Yonetani, *Hemes and Hemoproteins,* Academic Press, New York, 1966, p. 589.
199 M. Zerner and M. Gouterman, *Theoret. Chim. Acta,* 4 (1966) 44.
200 H. C. Watson and C. L. Nobbs, *Coll. Ges. Biol. Chem.*, 19th Mosbach, Springer-Verlag, Berlin, 1968.
201 S. Maričić, J. S. Leigh, and D. E. Sunko, *Nature,* 214 (1967) 462.
202 G. Pifat, S. Maričić, M. Petrinovic, V. Kramer, J. Marcel and K. Bonhard, *Croat. Chem. Acta,* 41 (1969) 195.
203 C. S. Irving and A. Lapidot, *Nature (New Biology),* 230 (1971) 224.
204 J. J. Weiss, *Nature,* 203 (1964) 183.
205 R. O. Viale, G. M. Maggiora and L. L. Ingraham, *Nature,* 203 (1964) 183.
206 J. B. Wittenberg, B. A. Wittenberg, J. Peisach and W. E. Blumberg, *Proc. Natn. Acad. Sci. U.S.*, 67 (1970) 1846.
207 G. Lang and W. Marshall, in B. Chance, R. W. Estabrook and T. Yonetani, *Hemes and Hemoproteins,* Academic Press, New York, 1966, p. 115.
208 M. F. Perutz, *J. Mol. Biol.*, 13 (1965) 646.
209 H. C. Watson, in B. Chance, R. W. Estabrook and T. Yonetani, *Hemes and Hemoproteins,* Academic Press, New York, 1966, p. 63.
210 J. H. Wang, *J. Am. Chem. Soc.*, 80 (1958) 3168.
211 J. H. Wang, in J. E. Falk, R. Lemberg and R. K. Morton, *Haematin Enzymes,* Vol. 1, Pergamon Press, Oxford, 1961, p. 98.
212 O. H. Kao and J. H. Wang, *Biochemistry,* 4 (1965) 342.
213 J. O. Alben, W. H. Fuchsman, C. A. Beaudreau and W. S. Caughey, *Biochemistry,* 7 (1968) 624.
214 I. A. Cohen and W. S. Caughey, *Biochemistry,* 7 (1968) 636.
215 C. E. Castro and H. F. Davis, *J. Am. Chem. Soc.*, 91 (1969) 5405.
216 H. Theorell, *Biochem. Z.*, 268 (1934) 73.
217 A. Rossi-Fanelli and E. Antonini, *Arch. Biochem. Biophys.*, 77 (1958) 478.
218 M. H. Keyes, M. Falley and R. Lumry, *J. Am. Chem. Soc.*, 93 (1971) 2035.
219 M. H. Keyes, *Ph.D. Dissertation,* University of Minnesota, Minneapolis, 1968.
220 G. Amiconi, M. Brunori, E. Antonini, G. Tauzher and G. Costa, *Nature,* 228 (1970) 549.
221 F. Stitt and C. D. Coryell, *J. Am. Chem. Soc.*, 61 (1939) 1263.
222 W. S. Caughey, J. O. Alben, S. McCoy, S. H. Boyer, S. Charache and P. Hathaway, *Biochemistry,* 8 (1969) 59.
223 S. McCoy and W. S. Caughey, in B. Chance, T. Yonetani and A. S. Mildvan, *Probes of Structure and Function of Macromolecules and Membranes,* Vol. 2, Academic Press, New York, 1971, p. 289.
224 W. B. Rippon, *Ph.D. Dissertation,* University of Newcastle, New South Wales, 1969.
225 S. McCoy and W. S. Caughey, *Biochemistry,* 9 (1970) 2387.
226 S. McCoy and W. S. Caughey, in B. Chance, T. Yonetani and A. S. Mildvan, *Probes of Structure and Function of Macromolecules and Membranes,* Vol. 2, Academic Press, New York, 1971, p. 295.
227 J. Wyman, *Adv. Protein Chem.*, 19 (1964) 223.
228 J. Wyman, *J. Am. Chem. Soc.*, 89 (1967) 2202.
229 J. Monod, J. Wyman and J. P. Changeux, *J. Mol. Biol.*, 12 (1965) 88.
230 J. Wyman, *J. Mol. Biol.*, 11 (1965) 631.
231 R. Lumry and H. Eyring, *J. Phys. Chem.*, 58 (1954) 110.

232 E. Breslow, S. Beychok, K. D. Hardman and F. R. N. Gurd, *J. Biol. Chem.*, 240 (1965) 304.
233 S. C. Harrison and E. R. Blout, *J. Biol. Chem.*, 240 (1965) 299.
234 D. J. Patel, L. Kampa, R. G. Shulman, T. Yamane and M. Fujiwara, *Biochem. Biophys. Res. Commun.*, 40 (1970) 1224.
235 B. P. Shoenborn, H. C. Watson and J. C. Kendrew, *Nature*, 207 (1965) 28.
236 B. P. Shoenborn and C. L. Nobbs, *J. Mol. Pharmacol.*, 2 (1966) 491.
237 B. P. Shoenborn, *Nature*, 214 (1967) 1120.
238 R. G. Shulman, J. Peisach and B. J. Wyluda, *J. Mol. Biol.*, 48 (1970) 517.
239 A. Wishnia, *Biochemistry*, 8 (1969) 5064.
240 M. Keyes and R. Lumry, *Federation Proc.*, 27 (1968) 895.
241 M. Keyes and R. Lumry, in B. Chance, T. Yonetani and A. S. Mildvan, *Probes of Structure and Function of Macromolecules and Membranes*, Vol. 2, Academic Press, New York, 1971, p. 343.
242 T. Reed, *Ann. N.Y. Acad. Sci.*, 174 (1970) 172.
243 L. Banaszak, H. C. Watson and J. C. Kendrew, *J. Mol. Biol.*, 12 (1965) 130.
244 E. Breslow and F. R. N. Gurd, *J. Biol. Chem.*, 238 (1963) 1332.
245 C. Bishop and D. M. Surgenor, *The Red Blood Cell*, Academic Press, New York, 1964.
246 E. Antonini, J. Wyman, A. Rossi-Fanelli and A. Caputo, *J. Biol. Chem.*, 237 (1962) 2773.
247 R. Benesch and R. E. Benesch, *Nature*, 221 (1969) 618.
248 R. Benesch, R. E. Benesch and C. I. Yu, *Proc. Natn. Acad. Sci. U.S.*, 59 (1968) 526.
249 R. Benesch, R. E. Benesch and Y. Enoki, *Proc. Natn. Acad. Sci. U.S.*, 61 (1968) 1102.
250 R. E. Benesch, R. Benesch and C. I. Yu, *Biochemistry*, 8 (1969) 2567.
251 A. Chanutin and R. R. Curnish, *Arch. Biochem. Biophys.*, 121 (1967) 96.
252 A. Chanutin and E. Hermann, *Arch. Biochem. Biophys.*, 131 (1969) 180.
253 M. F. Perutz, H. Muirhead, L. Mazzarella, R. A. Crowther, J. Greer and J. V. Kilmartin, *Nature*, 222 (1969) 1240.
254 M. F. Perutz, *Nature*, 228 (1970) 734.
255 E. Antonini, E. Bucci, C. Fronticelli, J. Wyman and A. Rossi-Fanelli, *J. Mol. Biol.*, 12 (1965) 375.
256 L. Garby, G. Gerber and C.-H. deVerdier, *Eur. J. Biochem.*, 10 (1969) 110.
257 F. Haurowitz, *Z. Physiol. Chem.*, 254 (1938) 268.
258 M. F. Perutz and F. S. Mathews, *J. Mol. Biol.*, 21 (1966) 199.
259 J. K. Moffat, *J. Mol. Biol.*, 58 (1971) 79.
260 J. Rifkind, Paper presented at 160th National Meeting of the American Chemical Society, Chicago, 1970.
261 J. Rifkind, to be submitted.
262 C. Djerrasi, *Optical Rotatory Dispersion*, McGraw-Hill, New York, 1960.
263 D. W. Urry, *J. Biol. Chem.*, 242 (1967) 4441.
264 A. S. Brill and H. E. Sandberg, *Proc. Natn. Acad. Sci. U.S.*, 57 (1967) 136.
265 G. Hänisch, J. Engel, M. Brunori and H. Fasold, *Eur. J. Biochem.*, 9 (1969) 335.
266 J. A. Schellman and P. Oriel, *J. Chem. Phys.*, 37 (1962) 2114.
267 Y. Enoki and I. Tyuma, *Jap. J. Physiol.*, 14 (1964) 280.
268 S. R. Simon and C. R. Cantor, *Proc. Natn. Acad. Sci. U.S.*, 63 (1969) 205.
269 C. Ho, D. G. Davis, N. H. Mock, T. R. Lindstrom and S. Charache, *Biochem. Biophys. Res. Commun.*, 38 (1970) 779.
270 J. F. Taylor, E. Antonini, M. Brunori and J. Wyman, *J. Biol. Chem.*, 241 (1966) 241.

271 A. Riggs, *J. Biol. Chem.*, 236 (1961) 1948.
272 M. Brunori, J. F. Taylor, E. Antonini, J. Wyman and A. Rossi-Fanelli, *J. Biol. Chem.*, 242 (1967) 2295.
273 G. Guidotti, *J. Biol. Chem.*, 242 (1967) 3673.
274 M. A. Rosemeyer and E. R. Huehns, *J. Mol. Biol.*, 25 (1967) 253.
275 T. Maeda and S. Ohnishi, *Biochemistry*, 10 (1971) 1177.
276 S. Ohnishi, J. C. A. Boeyens and H. M. McConnell, *Proc. Natn. Acad. Sci. U.S.*, 56 (1966) 809.
277 H. M. McConnell and C. L. Hamilton, *Proc. Natn. Acad. Sci. U.S.*, 60 (1968) 776.
278 S. Ogawa, H. M. McConnell and A. Horwitz, *Proc. Natn. Acad. Sci. U.S.*, 61 (1968) 401.
279 H. M. McConnell, W. Deal and R. T. Ogata, *Biochemistry*, 8 (1969) 2580.
280 C. Ho, J. J. Baldassare and S. Charache, *Proc. Natn. Acad. Sci. U.S.*, 66 (1970) 722.
281 J. J. Baldassare, S. Charache, R. T. Jones and C. Ho, *Biochemistry*, 9 (1970) 4707.
282 A. Hvidt and S. O. Nielsen, *Adv. Protein Chem.*, 21 (1966) 287.
283 R. Lumry, in T. King and M. Klingenberg, *Electron and Coupled Energy Transfer in Biological Systems*, Marcel Dekker, New York, 1971, p. 1.
284 R. Zito, E. Antonini and J. Wyman, *J. Biol. Chem.*, 239 (1964) 1804.
285 S. R. Simon, D. J. Arndt and W. H. Konigsberg, *J. Mol. Biol.*, 58 (1971) 69.
286 D. J. Arndt and W. H. Konigsberg, *J. Biol. Chem.*, 246 (1971) 2594.
287 D. J. Arndt, S. R. Simon, T. Maita and W. H. Konigsberg, *J. Biol. Chem.*, (1971) 2602.
288 G. S. Adair, *J. Biol. Chem.*, 63 (1925) 529.
289 W. Paul and F. J. W. Roughton, *J. Physiol. (London)*, 113 (1951) 23.
290 F. J. W. Roughton, *J. Physiol. (London)*, 126 (1954) 359.
291 A. V. J. Hill, *J. Physiol. (London)*, 40 (1910) iv-vii.
292 J. Rifkind, unpublished results.
293 G. A. Mullikan, *Proc. R. Soc. (B)*, 120 (1936) 366.
294 M. Brunori, R. W. Noble, E. Antonini and J. Wyman, *J. Biol. Chem.*, 241 (1966) 5238.
295 M. Brunori and T. M. Schuster, *J. Biol. Chem.*, 244 (1969) 4046.
296 Q. H. Gibson, *Biochem. J.*, 71 (1959) 293.
297 F. J. W. Roughton, in F. J. W. Roughton and J. C. Kendrew, *Haemoglobin*, Butterworths, London, 1949, p. 83.
298 Q. H. Gibson and F. J. W. Roughton, *Proc. R. Soc. (B)*, 146 (1957) 206.
299 Q. H. Gibson, in B. Chance, T. Yonetani and A. S. Mildvan, *Probes of Structure and Function of Macromolecules and Membranes*, Vol. 2, Academic Press, New York, 1971, p. 407.
300 R. L. Berger, E. Antonini, M. Brunori, J. Wyman and A. Rossi-Fanelli, *J. Biol. Chem.*, 242 (1967) 4841.
301 Y. Henry and R. Banerjee, *J. Mol. Biol.*, 50 (1970) 99.
302 J. S. Olson and Q. H. Gibson, *J. Biol. Chem.*, 246 (1971) 5241.
303 D. G. Davis, S. Charache and C. Ho, *Proc. Natn. Acad. Sci. U.S.*, 63 (1969) 1403.
304 D. G. Davis, N. H. Mock, V. R. Laman and C. Ho, *J. Mol. Biol.*, 40 (1969) 311.
305 D. G. Davis, N. H. Mock, T. R. Lindstrom, S. Charache and C. Ho, *Biochem. Biophys. Res. Commun.*, 40 (1970) 343.
306 H. Uchida, J. Heystek, and M. H. Klapper, *J. Biol. Chem.*, 246 (1971) 2031.
307 M. H. Klapper and H. Uchida, *J. Biol. Chem.*, 246 (1971) 6849.
308 J. S. Olson and Q. H. Gibson, *Biochem. Biophys. Res. Commun.*, 41 (1970) 421.
309 T. R. Lindstrom, J. S. Olson, N. H. Mock, Q. H. Gibson and C. Ho, *Biochem. Biophys. Res. Commun.*, 45 (1971) 22.

310 R. D. Gray and Q. H. Gibson, *J. Biol. Chem.*, 246 (1971) 5176.
311 G. L. Kellett and H. K. Shachman, *J. Mol. Biol.*, 59 (1971) 387.
312 E. Antonini, *Science*, 158 (1967) 1417.
313 G. Guidotti, *J. Biol. Chem.*, 242 (1967) 3685.
314 E. Antonini and M. Brunori, *J. Biol. Chem.*, 245 (1970) 5412.
315 N. M. Anderson, E. Antonini, M. Brunori and J. Wyman, *J. Mol. Biol.*, 47 (1970) 205.
316 G. Guidotti, *J. Biol. Chem.*, 242 (1967) 3673.
317 A. G. Kirschner and C. Tanford, *Biochemistry*, 3 (1964) 291.
318 I. B. E. Norén, C. Ho and E. F. Casassa, *Biochemistry*, 10 (1971) 3222.
319 S. J. Edelstein, M. J. Rehmar, J. S. Olson and Q. H. Gibson, *J. Biol. Chem.*, 245 (1970) 4372.
320 G. L. Kellett and H. Gutfreund, *Nature*, 227 (1970) 921.
321 G. L. Kellett, *J. Mol. Biol.*, 59 (1971) 401.
322 J. A. Hewitt, J. V. Kilmartin, L. F. Ten Eyck and M. F. Perutz, *Proc. Natn. Acad. Sci. U.S.*, 69 (1972) 203.
323 G. L. Kellett, *Nature (New Biology)*, 234 (1971) 189.
324 M. E. Andersen, J. K. Moffat and Q. H. Gibson, *J. Biol. Chem.*, 246 (1971) 2796.
325 L. Rossi-Bernardi and F. J. W. Roughton, *J. Physiol. (London)*, 189 (1967) 1.
326 J. V. Kilmartin and L. Rossi-Bernardi, *Nature*, 222 (1969) 1243.
327 T. M. Shuster and G. Ilgenfritz, in B. Chance, T. Yonetani and A. S. Mildvan, *Probes of Structure and Function of Macromolecules and Membranes*, Vol. 2, Academic Press, New York, 1971, p. 367.
328 M. Eigen, in B. Chance, T. Yonetani and A. S. Mildvan, *Probes of Structure and Function of Macromolecules and Membranes*, Vol. 2, Academic Press, New York, 1971, p. 439.
329 D. E. Koshland, G. Nemethy and D. Filmer, *Biochemistry*, 5 (1966) 365.
330 J. Rifkind and R. Lumry, *Federation Proc.*, 26 (1967) 673.
331 R. Banerjee and R. Cassoly, *J. Mol. Biol.*, 42 (1969) 351.
332 Y. Enoki and S. Tomita, *J. Mol. Biol.*, 32 (1968) 121.
333 M. Brunori, G. Amiconi, E. Antonini and J. Wyman, *J. Mol. Biol.*, 49 (1970) 461.
334 T. Asakura, J. S. Leigh, H. R. Drott, T. Yonetani and B. Chance, *Proc. Natn. Acad. Sci. U.S.*, 68 (1971) 861.
335 T. Asakura and H. R. Drott, *Biochem. Biophys. Res. Commun.*, 44 (1971) 1199.
336 S. Ogawa and R. G. Shulman, *Biochem. Biophys. Res. Commun.*, 42 (1971) 9.
337 R. Cassoly, Q. H. Gibson, S. Ogawa and R. G. Shulman, *Biochem. Biophys. Res. Commun.*, 44 (1971) 1015.
338 H. M. McConnell, S. Ogawa and A. Horwitz, *Nature*, 220 (1968) 787.
339 M. C. Botha, D. Beale, W. A. Isaacs and H. Lehmann, *Nature*, 212 (1966) 792.
340 J. G. Lines and R. McIntosh, *Nature*, 215 (1967) 297.
341 J. B. Clegg, M. A. Naughton and D. J. Weatherall, *J. Mol. Biol.*, 19 (1966) 91.
342 R. L. Nagel, Q. H. Gibson and S. Charache, *Biochemistry*, 6 (1967) 2395.
343 J. Bonaventura and A. Riggs, *J. Biol. Chem.*, 243 (1968) 980.
344 K. R. Reissman, W. E. Ruth and T. Nomura, *J. Clin. Invest.*, 40 (1961) 1826.
345 R. T. Jones, E. E. Osgood, B. Brimhall and R. D. Koler, *J. Clin. Invest.*, 46 (1967) 1840.
346 M. J. Novy, M. J. Edwards and J. Metcalfe, *J. Clin. Invest.*, 46 (1967) 1848.
347 C. S. Reed, R. Hampson, S. Gordon, R. T. Jones, *et al.*, *Blood*, 31 (1968) 623.
348 P. A. Lorkin, H. Lehmann, V. Fairbanks, G. Berglund and T. Leonhardt, *Biochem. J.*, 119 (1970) 68p.
349 T. Yanase, M. Hanada, M. Seita, Y. Ohta, *et al.*, *Jap. J. Human Genet.*, 13 (1968) 40.

350 R. T. Swenson, R. L. Hill, H. Lehmann and R. T. S. Jim, *J. Biol. Chem.*, 237 (1962) 1517.
351 R. T. Jones, B. Brimhall and R. Lisker, *Biochim. Biophys. Acta*, 154 (1968) 488.
352 T. Imamura, S. Fujita, Y. Ohta, M. Hanada and T. Yanase, *J. Clin. Invest.*, 48 (1969) 2341.
353 E. W. Baur and A. G. Motulsky, *Humangenetik*, 1 (1965) 621.
354 J. Greer, *J. Mol. Biol.*, 62 (1971) 241.
355 E. Antonini, J. Wyman, M. Brunori, *et al.*, *Arch. Biochem. Biophys.*, 108 (1964) 569.
356 R. L. Nagel, Q. H. Gibson and T. Jenkins, *J. Mol. Biol.*, 58 (1971) 643.
357 J. Greer, *J. Mol. Biol.*, 59 (1971) 99.
358 Q. H. Gibson and L. J. Parkhurst, *J. Biol. Chem.*, 243 (1968) 5521.
359 I. Tyuma, K. Shimizu and K. Imai, *Biochem. Biophys. Res. Commun.*, 43 (1971) 423.
360 I. Tyuma, K. Imai and K. Shimizu, *Biochem. Biophys. Res. Commun.*, 44 (1971) 682.
361 K. H. Winterhalter, G. Amiconi and E. Antonini, *Biochemistry*, 7 (1968) 2228.
362 M. R. Waterman, R. Gondko and T. Yonetani, *Arch. Biochem. Biophys.*, 145 (1971) 448.
363 R. Cassoly and R. Banerjee, *Eur. J. Biochem.*, 19 (1971) 514.
364 J. E. Haber and D. E. Koshland, *Biochim. Biophys. Acta*, 194 (1969) 339.
365 R. Lumry and S. Rajender, *Biopolymers*, 9 (1970) 1125.
366 R. Lumry, in B. Chance, T. Yonetani and A. S. Mildvan, *Probes of Structure and Function of Macromolecules and Membranes*, Vol. 2, Academic Press, New York, 1971, p. 353.

Chapter 26

CYTOCHROMES *b* AND *c*

HENRY A. HARBURY AND RICHARD H. L. MARKS

Section of Biochemistry and Molecular Biology, Department of Biological Sciences, University of California, Santa Barbara, California 93106, U.S.A.

The cytochromes, a widely distributed series of intracellular oxidation–reduction catalysts, were first studied in the 1880s, when MacMunn[1–3] provided spectroscopic evidence for the existence of pigments he referred to as histo- and myohematins, and proposed that such compounds fulfill a respiratory function. However, it was not until 1925, when Keilin[4,5] discovered these substances anew, that more detailed investigations were undertaken. Working with a variety of tissues and micro-organisms, Keilin demonstrated that each contained at least three heme proteins, which he named cytochromes *a*, *b*, and *c*. These compounds, corresponding to MacMunn's histo- and myohematins, could be reduced by the addition of substances such as succinate, and were found to be reoxidized when oxygen was admitted to the preparations. Subsequent studies led to the recognition of additional components[6,7], and, in time, the nature and significance of electron-transfer chains of the type shown in Fig. 1 became apparent. In

Substrates → Non-cytochrome components → Cytochrome *b* →

Cytochrome c_1 → Cytochrome *c* → Cytochrome $(a + a_3)$ → Oxygen

Fig. 1. Simplified version of the mitochondrial electron-transfer chain.

accordance with the fact that the constituents of such chains undergo cyclic oxidation–reduction, the definition of the cytochromes adopted by the Commission on Enzymes of the International Union of Biochemistry states that "a cytochrome is a heme protein whose characteristic mode of action is electron and/or hydrogen transport by virtue of a reversible valency change of its heme iron"[8].

In eukaryotic cells, the components indicated in Fig. 1 occur in the mitochondria, where they are deployed along the inner membrane and form part of the machinery whereby oxidative energy is conserved through respiratory-chain phosphorylation[9–11]. Other representatives, serving more specialized functions, have been found in the endoplasmic reticulum and chlorophyll-containing organelles, and a very broad array of cytochromes has been encountered among the prokaryotic organisms. These prokaryotic cytochromes, which are arranged along the cell membrane, can often be obtained in aqueous solution more easily than their counterparts from eukaryotic sources and, because of their great diversity, offer potentially excellent opportunities for the study of structure–function relationships[12].

The present review is limited to the cytochromes *b* and *c*, and will focus on observations concerning these proteins in the isolated state, with emphasis on the nature and role of the central coordination complex. The cytochromes *c* of the "mammalian type", about which the most is known in this context, will be discussed in some detail, while the other *c*-type cytochromes and the *b*-type molecules, which can usefully be discussed only much more briefly at this juncture, will be considered in terms of selected examples. As will become apparent, the available information is still very incomplete. Fortunately, however, the pace of recent progress suggests this will not much longer remain the case.

CYTOCHROMES *c*

The cytochromes *c* are defined as those cytochromes in which the heme group is bound covalently to the protein via side chains of the porphyrin[8]. Studies of preparations from a broad range of source materials have shown that the members of this family obtained from mammalian tissues and many other eukaryotic cells have similar structures and properties, and are functionally equivalent to one another. On the other hand, some eukaryotic proteins of the *c*-type exhibit appreciably different features, and resemble more nearly some of the molecules found among prokaryotic organisms.

The eukaryotic proteins with similar properties share the following characteristics: (i) molecular weights near 12,000; (ii) a single heme group per molecule; (iii) $E_{m,7}$ values* near 0.25 V; (iv) isoelectric points near pH 10; and (v) high rates of reaction with mammalian cytochrome oxidase preparations[14,15]. Since the first such molecules to be examined in detail were obtained from mammalian tissues, the various members of this group, regardless of the source material, have commonly been referred to as being of the "mammalian type". In view of the wide distribution of such proteins among non-mammalian organisms, this is an awkward designation, and during the past few years the term "eukaryotic cytochromes *c*" has come to be used in this context. However, the latter description presents the difficulty that a number of eukaryotic organisms contain *c*-type proteins which, in terms of the criteria just indicated, are not of the "mammalian

*$E_{m,7}$: oxidation–reduction potential of the half-reduced system (midpoint potential) at pH 7[13]. The values quoted in this chapter were obtained at 20–30°C.

References pp. 947–954

type". For example, certain of the fungi and algae contain *c*-type molecules with isoelectric points in the acidic or neutral pH range, and such proteins are oxidized only slowly in the presence of the mammalian cytochrome oxidases. To avoid confusion, the term "eukaryotic cytochromes *c*" will be employed here solely to designate the *c*-type cytochromes of the eukaryotes collectively, and, in referring to the more restricted "mammalian-type" category, the original usage will be retained. Although a new term could be introduced in the latter connection, it seems preferable to defer such changes until sufficient information is at hand to permit a structural basis to be adopted.

"Mammalian-type" cytochromes c

Primary structure

The "mammalian-type" cytochromes *c* consist in each case of a single heme group and peptide chain. The heme group, a derivative of iron protoporphyrin IX, is joined through thioether bonds to two cysteine residues of the molecule. Hydrolysis with hydrochloric or sulfuric acid results in the release of porphyrin *c*[16–19] (Fig. 2), while treatment with silver salts leads to cleavage of the thioether linkages and the formation of hematoporphyrin[20] (see also Chapter 24, p. 803 (6)). Upon enzymatic hydrolysis of the protein, heme peptides are obtained in which the thioether bridges remain intact, and studies of such products have contributed significantly to knowledge concerning the parent protein systems.

The primary structure varies with the source material. Contributions from a number of laboratories, most notably those of Margoliash and E. Smith, have led to knowledge of the complete amino acid sequences for preparations from over 30 different species[15,21,22]. The first sequence to be determined, that of cytochrome *c* from horse heart[23], is shown in Fig. 3, and above this sequence are indicated the substitutions which occur at each position among the other proteins thus far examined. The chains range in length from 103 residues for tuna cytochrome *c* to 112 residues for the protein from wheat germ and, as may be seen, 33 of the residues are invariant. This number may decrease somewhat as additional sequences become known, but is unlikely to drop much below 32[53,54].

Nearly half of the constant residues occur in positions 67 through 84, while no more than two successive side chains remain constant elsewhere along the chain. The two cysteine residues which are joined to the heme group are invariant features in positions 14 and 17 of the sequences, and position 18 is taken by a constant histidine residue. Of the aromatic amino acids of the horse heart molecule, the single tryptophan residue, two of the four tyrosine residues, and two of the four phenylalanine residues occur in each of the proteins thus far examined. Other invariant groups include the arginine residues in positions 38 and 91, several lysine residues, and some two-thirds of the glycine residues of the molecule. No residues with free carboxylate groups are constant.

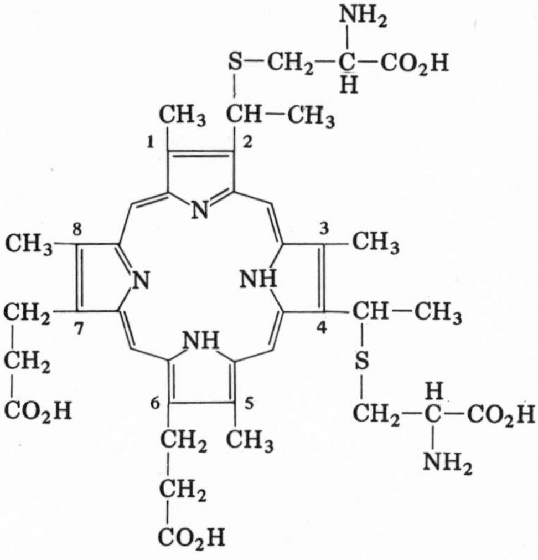

Porphyrin *c*

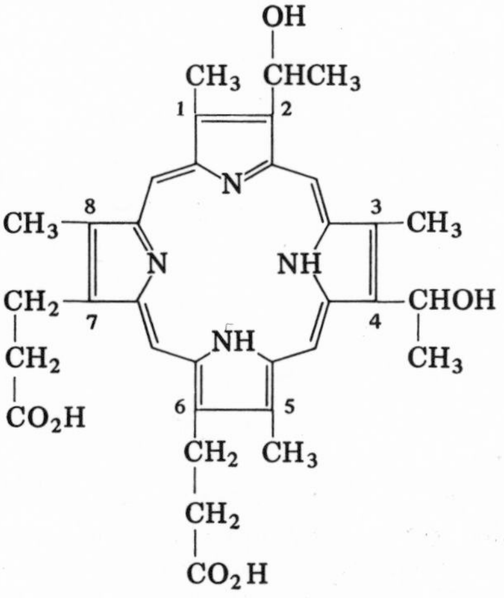

Hematoporphyrin

Fig. 2. Structures of porphyrin *c* and hematoporphyrin.

A number of the substitutions indicated in Fig. 3 are conservative ones. For example, seven of the eight positions in which there are aromatic amino acid residues in the horse heart protein are also taken by aromatic groups in the other sequences. In other instances, even though more radical

Acetyl	Ala	Lys Ser	Gly Pro Phe	Gly Asn Thr Ser Ala	Ser Pro Glu Gly	Phe Val Ala	Lys Glu Pro Ser	Gln Pro Ala		Asn Ser	Ile Pro Ala Ser	Thr Ala Asp Lys	Thr Ser Asn Ala
···	···	···	···	···	···	···	···	ACETYL	GLY ·	ASP ·	VAL ·	GLU ·	LYS ·
				−5					1				5

	Glu Ala	Thr Asn	Val Leu Thr		Thr Lys Ile	Met Thr	Arg		Glu Ser	Glu Leu			Gly	Glu Cys Ile
GLY ·	LYS ·	LYS ·	ILE ·	PHE ·	VAL ·	GLN ·	LYS ·	CYS ·	ALA ·	GLN ·	CYS ·	HIS ·	THR ·	VAL ·
				10				└	15 HEME		┘			20

Ala Gly Asp	Gly Ala Asn	Ala Asn	Val Leu Ala	Ser Thr Pro Gly	Gln		Val Ile Gln			Ala		Gln Asn Ser Tyr Trp		Phe Ile
GLU ·	LYS ·	GLY ·	GLY ·	LYS ·	HIS ·	LYS ·	THR ·	GLY ·	PRO ·	ASN ·	LEU ·	HIS ·	GLY ·	LEU ·
				25					30					35

Ile Tyr	Ser		His Gln	Ser		Thr Ser	Thr Val	Asp Ala Gln Glu Val		Tyr	Ala Ser		Ser	Glu Ala Asn
PHE ·	GLY ·	ARG ·	LYS ·	THR ·	GLY ·	GLN ·	ALA ·	PRO ·	GLY ·	PHE ·	THR ·	TYR ·	THR ·	ASP ·
				40					45					50

		Ile	Ala Ser Gln Arg Lys	Ala	Asn Ala	Val	Gln Glu Leu Ile Val		Asn Gln Glu Ala Asp Gly	Gln Asn Asp	Lys Pro Asn Asp	Asn	Met	Arg Ser Tyr Phe
ALA ·	ASN ·	LYS ·	ASN ·	LYS ·	GLY ·	ILE ·	THR ·	TRP ·	LYS ·	GLU ·	GLU ·	THR ·	LEU ·	MET ·
				55					60					65

Val Ile Asp			Thr Leu			TLy		Phe						
GLU ·	TYR ·	LEU ·	GLU ·	ASN ·	PRO ·	LYS ·	LYS ·	TYR ·	ILE ·	PRO ·	GLY ·	THR ·	LYS ·	MET ·
				70					75					80

Ala Val		Val Thr Pro Gly		Leu	TLy Ser		Asn Thr Glu Pro Asp Ala	Ser Ala Asp Gly Gln Glu Asn Lys	Asp		Lys Asp Gln Thr Ala Gly Asn Val	Asn	Ile	Val
ILE ·	PHE ·	ALA ·	GLY ·	ILE ·	LYS ·	LYS ·	LYS ·	THR ·	GLU ·	ARG ·	GLU ·	ASP ·	LEU ·	ILE ·
				85					90					95

Thr	Phe	Met	Leu	Gln Thr Asp Glu Ser	Thr Lys Ser	(D) Ala Cys	(D) Lys Ser Ala Cys	(D) Lys Ala Ser
ALA ·	TYR ·	LEU ·	LYS ·	LYS ·	ALA ·	THR ·	ASN ·	GLU
				100				104

Fig. 3. Variable and constant amino acid residues of the "mammalian-type" cytochromes *c*. The continuous sequence is that of cytochrome *c* from horse heart[23]. Trimethyllysine is abbreviated as TLy, and the letter D indicates an amino acid deletion in a given position. The residues shown above the continuous sequence are those encountered among the cytochromes *c* of man[24], chimpanzee[21,25], Rhesus monkey[26], donkey[21,27], cow[28], pig[29], sheep[21,30], dog[31], rabbit[32], whale[33], kangaroo[34], chicken[35], turkey[36], pigeon[21,36], duck[21,36], rattlesnake[37], snapping turtle[38], bullfrog[21,39,40], tuna[41,42], dogfish[43], Pacific lamprey[21,44], *Samia cynthia*[45], tobacco hornworm moth[21,46], *Drosophila melanogaster*[21,47], screwworm fly[21,48], *Neurospora crassa*[49], bakers' yeast (iso-1)[50], *Candida krusei*[51], *Humicola lanuginosa* (J. P. Riehm, personal communication), wheat germ[52] and sesame, sunflower, mung bean, and castor bean (D. Boulter, personal communication).

substitutions may be involved, the region in which the substitution occurs is kept relatively constant in terms of its overall properties. Significant conservatism is exhibited, for instance, in connection with the distribution of hydrophobic residues along the chains[15,55-58]. Such residues are arranged in clusters, and, although substitutions occur at individual positions, the hydrophobic character of the segment as a unit is carefully preserved. Similarly, there are sets of positively charged* groups and regions with negatively charged groups which remain constant[15,56,57], and a high degree of conservatism is displayed with regard to the charge of the molecule as a whole. Although the numbers of positively and negatively charged residues vary considerably, all the "mammalian-type" cytochromes *c* are basic proteins with isoelectric points near pH 10[59,60]. The similarities in composition and sequence are thus substantially greater than might be inferred from the fact that fewer than a third of the residues have been found to be invariant. Indeed, since the "mammalian-type" proteins appear to be functionally interchangeable with one another (L. Smith, personal communication), the primary structures are generally assumed to be sufficiently similar to allow essentially equivalent three-dimensional folding patterns to be adopted.

Ligand groups

The prosthetic group and the protein are bound to one another not only via the thioether bridges which have been mentioned, but also, of course, through coordination to the heme iron atom. The latter type of interaction is of major structural and functional importance for the molecule, and has been a focal point of interest in cytochrome *c* research since the early and influential reports on this topic by Theorell and Åkesson in 1941[59,61-63].

As noted initially by Keilin[64] and Dixon *et al.*[65], the "mammalian-type" cytochromes *c* do not react with oxygen or carbon monoxide at neutral pH, and exhibit absorption spectra that are typical of ferri- and ferroporphyrin complexes in which the fifth and sixth coordination positions about the iron atom are occupied by strong-field ligands (Fig. 4). Theorell and Åkesson, working with the molecule from beef heart, suggested that these fifth and sixth ligands are the imidazole groups of two histidine side chains. This hypothesis was based largely on data obtained through acid–base titrations[62,63]. Although the protein from beef heart contains three histidine residues, it was found that over the pH range 5.5–8.5 a total of only two equivalents could be titrated. Furthermore, determinations of the apparent heat of dissociation yielded values which increased continuously over this interval. This was taken to indicate that the groups undergoing titration were of more than one kind, and could include no more than one histidine

* References to charge type pertain to pH 7.

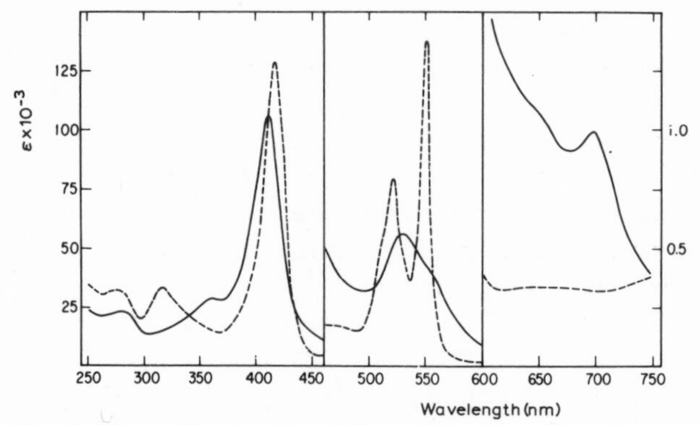

Fig. 4. Absorption spectra of horse heart cytochrome *c* at pH 7. —, oxidized form; -----, reduced form. Center panel, *read* ordinate on right x 20.

side chain. The two remaining histidine residues of the molecule were assumed to be coordinated to the iron atom, and therefore subject to protonation only under more acidic conditions. Indeed, studies of the absorption spectra[61] and magnetic susceptibility[63] of the molecule showed that the low-spin (hemochrome)* properties observed in neutral solution were maintained to very low values of pH. In the case of the reduced form of the compound, no high-spin character could be noted even at pH 1, and the oxidized form was found to undergo conversion to a high-spin species (Fig. 5) only with a midpoint of change near pH 2.5. Two hydrogen ions

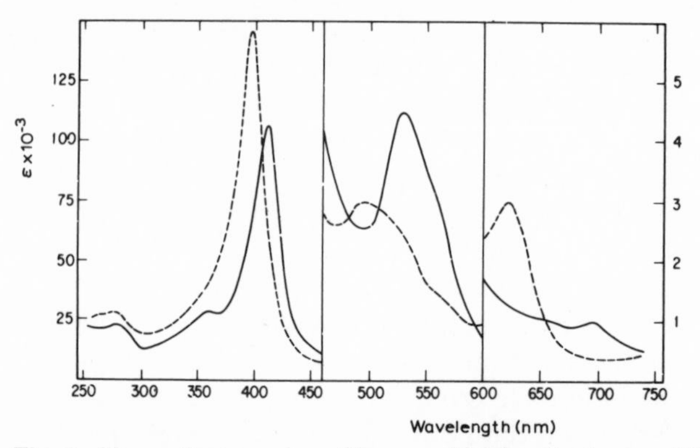

Fig. 5. Absorption spectra of low- and high-spin forms of horse heart ferricytochrome *c*. —, pH 7 (low-spin form); -----, pH 1 (high-spin form). Center panel, *read* ordinate on right x 2.5.

---

* The term hemochrome[66] has commonly been used to designate low-spin iron porphyrin complexes with nitrogenous ligands in coordination positions 5 and 6. It is employed here in the broader context of low-spin iron porphyrin complexes in which the strong-field ligands in positions 5 and 6 may be either nitrogenous or non-nitrogenous. The prefixes ferri and ferro are used to indicate the oxidation state of the complex.

per molecule were required for this transition[67], which was inferred to reflect protonation of the coordinated histidine residues to the non-coordinated imidazolium form.

It is firmly established today that the side chain of a histidine residue does in fact occupy one of the positions about the heme iron atom. Spectrophotometric, potentiometric and magnetometric studies of small heme peptides containing histidine residue 18 have shown that this group serves as a ligand in both the oxidized and reduced forms of such derivatives[68–72], and X-ray diffraction data for ferricytochrome *c* from horse heart have confirmed the occurrence of this linkage in the intact protein[73]. On the other hand, it has been found that the second coordinated group is provided not by a histidine residue, but by the thioether group of a methionine side chain; *i.e.*, the "mammalian-type" cytochromes *c* are mixed hemochromes, with a nitrogen and a sulfur atom bound to the heme.

That there can be only one coordinated imidazole group became apparent once amino acid sequence data began to accumulate. Ambler[74,75], in work with cytochrome *c*-551* from *Pseudomonas aeruginosa***, a molecule with an oxidation–reduction potential at pH 7 near that of the "mammalian-type" systems[77–79], found this protein to contain just a single histidine side chain. When it was shown subsequently that this cytochrome remains a hemochrome to the same low values of pH as does the molecule from horse heart, it began to seem likely that there is no difference in the nature of the groups coordinated to the iron in the two cases, and that the "mammalian-type" proteins, like the *Pseudomonas* molecule, have but one heme-bound imidazole group[80,81]. Further and more direct evidence to this effect became available in the form of some of the sequence data given in Fig. 3. Whereas most of the "mammalian-type" cytochromes *c* examined to date contain histidine residues in positions 18, 26, and 33, some, as observed initially by Kreil[41], have such groups only in positions 18 and 26. Similarly, as shown by Heller and Smith[49] and by Stewart *et al.*[82], there are some which have no hisitidine residue in position 26. Thus, only histidine residue 18, adjacent to the second of the cysteine bridges, is invariant among these proteins. If it is assumed that the groups functioning as ligands in the different molecules do not vary either in kind or position, it follows that they can include but this single histidine side chain.

Evidence that a methionine residue is coordinated to the iron atom

* Many of the *b*- and *c*-type cytochromes are identified by a number indicating the position of the longer-wavelength hemochrome band (α-band) in the visible absorption spectrum of the reduced species.

** The organism was identified originally as a strain of *Pseudomonas fluorescens*, but is more probably a *Pseudomonas aeruginosa*[76].

arose in the course of chemical modification studies and experiments with model heme peptides. When horse cytochrome *c* is treated with bromoacetate at pH 7, no changes in properties are observed. However, as demonstrated by Schejter and George[83], if the bromoacetate is applied in the presence of cyanide, and the cyanide then is removed, very major alterations are seen. The product obtained in this manner fails to remain a hemochrome in acid solution, and, in the reduced form, is partly in the high-spin state even at pH 7. Since the reaction of proteins with bromoacetate is known to lead to the carboxymethylation of available histidine residues at neutral pH, and since studies of the pH-dependence of the interaction of cytochrome *c* with ligands such as cyanide and azide had been inferred to indicate that these groups displace a coordinated histidine side chain[84,85], a simple interpretation was entertained at first: *i.e.*, that the cyanide replaces a coordinated imidazole group, which then becomes subject to alkylation, and, once modified, cannot resume its normal role when the cyanide is withdrawn[83]. However, studies by Harbury *et al.*[80,86-88] and by Stellwagen[89] have shown that the only histidine residue alkylated under these conditions is histidine 33, and that the modification of this group occurs in the absence as well as in the presence of cyanide. When the experiment is performed with tuna cytochrome *c*, which contains histidine residues only in positions 18 and 26, the same spectral changes are observed as with the horse heart molecule, but without attendant histidine modification. For both proteins, the analytical data indicate, rather, a specific effect of cyanide on the susceptibility to carboxymethylation of methionine residue 80[80,86-89]. In the presence of cyanide, treatment with bromoacetate at pH 7 leads to the alkylation of this residue, while in the absence of cyanide, no alkylation is observed at this position.

Parallel investigations by Ando *et al.*[90-93] and Tsai and Williams[94,95] have established that residue 80 can also be converted to the carboxymethyl derivative by treatment with iodoacetic acid at pH 3, and that such modification results in the elimination of electron-transfer activity in the cytochrome oxidase and succinate oxidase systems, increased rates of autoxidation, and other changes in properties indicative of major changes in the structure of the molecule. Tsai and Williams, finding that residues 65 and 80 react at about the same rate as free methionine, concluded that both groups are on the surface of the protein. On the other hand, Ando *et al.*, noting that at pH 5–6 the modification of residue 80 proceeds more slowly than that of residue 65, and that the rate becomes slower still when the reaction is attempted with the reduced protein, inferred that residue 80 is less exposed than methionine 65 and becomes more inaccessible upon reduction of the molecule. Both groups of investigators interpreted the changes in properties introduced by the modification of residue 80 to be the result of the conversion of a hydrophobic group to a hydrophilic one[92-94]. The substitution of a charged carboxymethylsulfonium group for a hydrophobic thioether side chain in the largely hydrophobic sequence of residues encompassing methionine residue 80 was taken to alter the three-dimensional structure sufficiently to preclude maintenance of the normal properties of the molecule.

Harbury *et al.*[80,86,87,96], in a different interpretation, suggested that the side chain of methionine 80, and the corresponding side chain in *Pseudomonas* cytochrome *c*-551, might in fact provide the sought-for sixth ligand in the proteins under study. This suggestion was based in part on observations concerning the interaction of *N*-acetylmethionine methyl ester and simpler thioethers with a heme octapeptide in which the fifth coordination position was known to be occupied by the imidazole group of the single histidine residue of the molecule[72,97]:

┌—Heme——┐
Cys · Ala · Gln · Cys · His · Thr · Val · Glu ·

It was found that thioethers, like nitrogenous ligands, readily form hemochromes with this peptide[80]. Furthermore, in contrast to imidazole or primary amines[72], the thioethers coordinate far more firmly to the reduced than to the oxidized form of the heme peptide, and remain bound to the iron to very low values of pH. These are, of course, attractive features in connection with such key properties of the parent proteins as: (i) the retention of low-spin character in acid solution; (ii) the high oxidation–reduction potentials relative to those of free heme[98] and the heme octapeptide in the absence of added ligands[72]; and (iii) the generally greater structural stability of the reduced as compared to the oxidized proteins (pp. 920–921).

Further evidence for a heme–methionine linkage was provided by carboxymethylation studies of the cytochrome from *Pseudomonas*[80,81,86,87,96]. It was found that treatment with bromoacetate at pH 7, in the absence of cyanide, leads to the alkylation of only one of the two methionine residues of the molecule, residue 22, and is without effect on the visible absorption spectra. On the other hand, in the presence of cyanide, both residue 22 and methionine 61 are modified, and the curves are altered greatly. Spectra comparable to those of the reduced form of the parent protein are obtained only at alkaline pH, and when the carboxymethylation is carried out with a preparation in which the amino groups are first trifluoroacetylated, the reduced species fails to develop hemochrome character even at pH 10. As in the case of the "mammalian-type" cytochromes *c*, the loss of hemochrome character hinges on the modification of a single specific residue. Furthermore, this residue can be alkylated only upon the disruption of the central coordination complex by procedures such as the addition of an extrinsic ligand or adjustment of the pH to the range where transition to a high-spin structure is known to occur. Since the amino acid sequence of the *Pseudomonas* molecule[75] differs markedly from the sequences of the "mammalian-type" proteins, and especially so in the vicinity of the relevant methionine residue (Fig. 6), these observations support the view that the iron is bound, in the case of *Pseudomonas* cytochrome *c*-551, to histidine residue 16 and methionine residue 61, and, in the "mammalian-type" molecules, to histidine 18 and methionine 80.

```
                                                     ┌──── HEME ────┐
Glu·Asp·Pro·Glu·Val·Leu·Phe·Lys·Asn·Lys·Gly·Cys·Val·Ala·Cys·HIS·Ala·Ile·
 1               5                   10                  15

Asp·Thr·Lys·Met·Val·Gly·Pro·Ala·Tyr·Lys·Asp·Val·Ala·Ala·Lys·Phe·Ala·Gly·
    20                  25                  30                  35

Gln·Ala·Gly·Ala·Glu·Ala·Glu·Leu·Ala·Gln·Arg·Ile·Lys·Asn·Gly·Ser·Gln·Gly·
            40                  45                  50

Val·Trp·Gly·Pro·Ile·Pro·MET·Pro·Pro·Asn·Ala·Val·Ser·Asp·Asp·Glu·Ala·Gln·
 55                 60                  65                  70

Thr·Leu·Ala·Lys·Trp·Val·Leu·Ser·Gln·Lys
        75                  80
```

Fig. 6. Amino acid sequence of cytochrome *c*-551 from *Pseudomonas aeruginosa*[75]. The residues which provide the groups coordinated to the iron atom are indicated by capital letters.

Final confirmation of the nature of the ligand groups has come from the X-ray diffraction studies being conducted by Dickerson *et al.*[73]. A preliminary examination of horse ferricytochrome *c* at 4 Å resolution yielded data consistent with a folding pattern that would bring residue 80 into the position requisite for coordination, but did not establish the nature of the sixth ligand with certainty[54,99,100]. The results subsequently obtained at 2.8 Å resolution show clearly, however, that the iron is linked to the thioether sulfur atom of methionine 80 on one side of the heme plane, and to the imidazole group of histidine 18 on the other side[73]. In addition, it has been demonstrated by difference mapping that the same arrangement holds for ferricytochrome *c* from bonito[73].

Although crystallographic data are not yet available for the reduced form of the protein, it seems likely from the heme peptide data and the results of the chemical modification studies that the same two residues will be found to be coordinated to the prosthetic group under those conditions. Indeed, the heme–methionine linkage should be intrinsically a firmer one than in the case of the oxidized form[80]. Recently, in a proton magnetic resonance study of ferrocytochrome *c* preparations from nine different species, McDonald *et al.*[101] have concluded that, in each instance, the methyl- and γ-methylene-group protons of a methionine residue are arranged close to the face of the heme group, as would be expected were the methionine coordinated via the thioether sulfur atom.

Propionate linkages

The heme group is joined to the protein not only through the iron atom and the two cysteine bridges, but also via at least one of the propionate groups of the porphyrin. Studies of preparations of horse cytochrome *c* in which the carboxylate groups were blocked by esterification or coupling with glycine ethyl ester have shown that, in all likelihood, one or more of these groups play an important role in maintenance of the normal structure of the molecule[102,103]. Whereas conversion of the positively charged ϵ-ammonium groups to the uncharged trifluoroacetyl derivative[104] results in only minor alterations in structure[105,106], modification of the negatively charged carboxylate groups leads to changes in the protein–heme interaction so marked as to suggest that linkages directly involving the prosthetic group are being affected. This interpretation is borne out by the X-ray diffraction data, which show that the propionate groups of horse ferricytochrome *c* do not extend outward, as in hemoglobin and myoglobin, but toward the bottom of the crevice into which the heme group is built, and are joined there through hydrogen bonds to residues of the protein[73]. The side chain which is buried the more deeply of the two is hydrogen-bonded to tyrosine residue 48, tryptophan 59, and threonine 40, while the more exposed chain is thought to be bound to threonine 49 and/or asparagine 52.

Heme crevice

Since the heme group is coordinated to two side chains of the protein, it is evident that it must be partly or wholly enclosed by the latter in a crevice-like arrangement[63,107,108]. Solvent perturbation studies by Stellwagen[109] have been taken to indicate that only a portion of the heme is accessible to molecules of the medium, and the X-ray diffraction data for horse ferricytochrome *c*[73,99,100] show that the prosthetic group is embedded in such a manner that only an edge is exposed to the surroundings, with no part of the heme disc projecting beyond the surface of the protein. The crevice is lined with the side chains of about 16 hydrophobic amino acid residues. Approximately half of these are invariant among the proteins thus far examined, and the remainder are subject to conservative substitution only. As might be expected, the hydrophobic groups are concentrated especially along the inner reaches of the pocket.

The prosthetic group is positioned in the crevice with the edge from porphyrin positions 4 to 5 at the opening. The bond to cysteine residue 17, and the propionate side chain at position 6 of the porphyrin thus occur near the surface of the molecule. The propionate group at position 7, on the other hand, is located in the highly hydrophobic interior of the crevice. As has been mentioned, it is linked to the protein via three hydrogen bonds.

Overall structure (see note added in proof, p. 946)

The X-ray diffraction studies of Dickerson *et al.*[54,73,99,100] have established the complete three-dimensional structure for ferricytochrome *c* from horse heart. In general, the features agree well with what would be predicted on the basis of a simple "oil drop" model. The peptide chain is arranged in such a manner that the charged groups are on the outside of the molecule, while hydrophobic groups are for the most part in the interior, clustered about the heme group. The folding pattern is obviously determined in major measure by the prosthetic group, with the peptide chain winding around it to provide a covering essentially a single layer thick. Although there are a number of short segments with Ramachandran angles similar to those of an α-helix, true α-helicity is seen only between residues 92 and 102 near the carboxyl terminus. Most of the chain is folded back and forth, in extended form, to bring the hydrophobic groups into position around the heme.

The overall shape of the molecule, including the side chains, is that of a prolate spheroid, 30 x 34 x 34 Å. When the heme group is placed in a near-vertical plane, with the cysteine bridges at the top and the open end of the crevice facing the viewer, residues 1–47 and 92–104 compose the right half of the molecule, and residues 48–91 the left half (Fig. 7). The chain begins at the top, crosses from right to left below the crevice, and then, following residue 91, crosses back to the right along the top of the crevice, with residues 92–104 continuing down the right-hand surface to complete the molecule. The sulfur atoms of the two cysteine bridges are accessible from the exterior, especially so in the case of the linkage between residue 17 and position 4 of the porphyrin ring at the upper front of the molecule. The imidazole group of histidine 18, extending to the heme from the right half of the molecule, and the sulfur atom of methionine 80, coordinating from the left side, are, on the other hand, located in somewhat more shielded positions.

In each half of the molecule, a region can be charted from the heme group to the surface in which only hydrophobic side chains are encountered. Clusters of positively charged lysine side chains adjoin these hydrophobic regions where they meet the surface, and, between these clusters, towards the rear of the molecule, a large number of negatively charged side chains are congregated. It has been suggested that each of these charged areas fulfills a binding function. The positively charged groups may play a role in the binding of cytochrome c to the negatively charged cytochrome c_1 and cytochrome oxidase molecules with which it reacts in the mitochondrial electron-transfer chain, while the area with negatively charged groups may be involved in the interaction of the protein with other mitochondrial

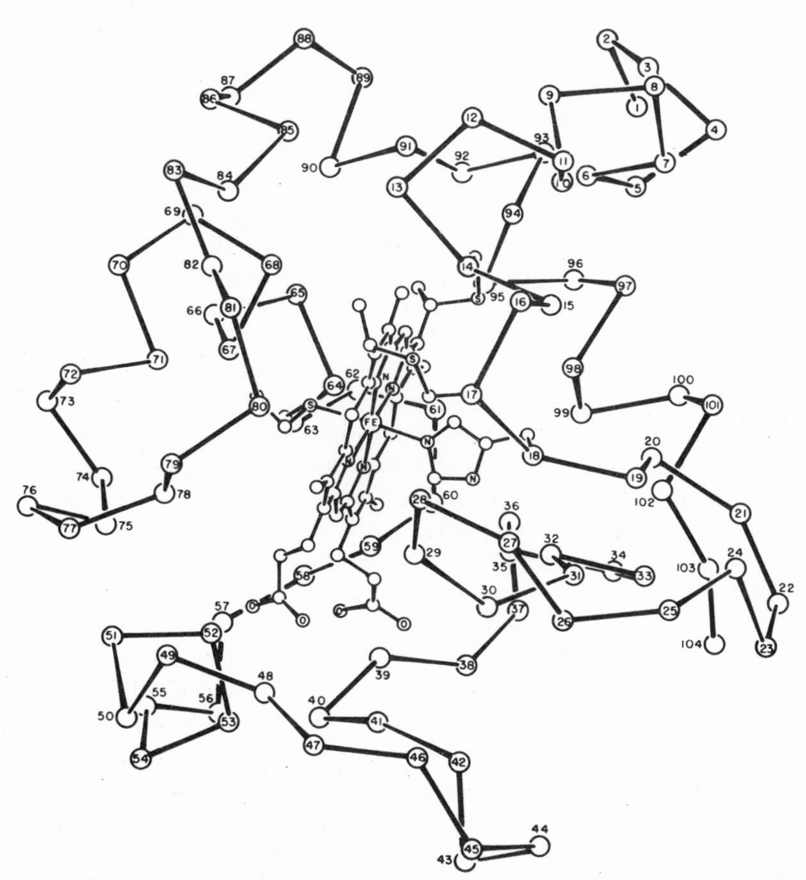

Fig. 7. α-Carbon diagram of horse heart ferricytochrome c. (Provided by R. E. Dickerson[73]).

components. In view of the fact that the hydrophobic regions meet the surface near the presumed binding sites for the electron donor and acceptor molecules, the possibility has been raised that they might serve a function in electron and/or energy transfer to and from the heme group. It has been noted, in this connection, that the aromatic groups of tyrosine residue 74 and tryptophan residue 59, which form part of the left-hand region, are arranged in planes that are approximately parallel to one another, and that the plane of the ring of tyrosine residue 67, in the same region, is roughly

References pp. 947–954

parallel to that of the heme group. Similarly, on the right side, the imidazole group of histidine 18, and the aromatic groups of phenylalanine residue 10 and tyrosine residue 97, are approximately parallel to one another.

There are, of course, numerous hydrogen bonds in the molecule. Through one such linkage, the non-coordinated nitrogen atom of the imidazole group of histidine residue 18 is joined to the carbonyl group of proline residue 30, in an arrangement analogous to that found for the heme-bound histidine side chain of myoglobin. The invariant arginine side chains are hydrogen bonded to glutamine residues; arginine 91 to glutamine 12 in a linkage which helps to join the front and back sections of the protein at the top of the crevice, and arginine 38 to glutamine 42 along the bottom of the molecule. The indole group of the constant tryptophan residue is linked to the posterior propionate side chain of the heme group, as is the invariant tyrosine side chain in position 48, and the second constant tyrosine residue, in position 67, is bound to threonine residue 78.

The three-dimensional structure readily accounts for most of the features noted upon amino acid sequence comparisons. The many invariant glycine residues occur principally in positions where there is insufficient room for a side chain, while groups which vary widely occur on the surface in regions where they can extend without consequence into the medium. The largely constant sequence between positions 67 and 82 forms the important part of the molecule adjoining the left face of the heme group. The reasons for the constancy or conservative variance of other residues of the molecule are in many instances easily appreciated, as is the basis for the preservation of positively charged, negatively charged and hydrophobic clusters along the chain. Although the differences in the amino acid sequences of the "mammalian-type" proteins will cause the individual structures to differ in points of detail, there is every reason to anticipate that the structure established for the horse heart molecule will be found to hold in its essentials for the family as a whole. Indeed, this has already been shown to be the case for ferricytochrome *c* from bonito[73]. The crystals of this protein are isomorphous with those from horse, and it has been demonstrated by difference mapping that the two vary in structure only at the expected side chain positions, with at most minor adjustments of the main chain.

It will be of great interest to have comparative data for the molecules in the reduced state. Studies directed to this end are currently in progress (R. E. Dickerson, personal communication), and preliminary results should soon be available.

Protein–prosthetic group interaction

The heme group of the cytochromes *c* profoundly influences the structure and properties of the protein component, and the latter, in turn, greatly affects the behavior of the prosthetic group. This interplay lies at the heart of cytochrome chemistry, but remains to be analyzed in detail.

As has been mentioned, the "mammalian-type" molecules retain their hemochrome properties to unusually low values of pH. Horse ferrocytochrome *c* remains in the low-spin state even at pH 1, and the oxidized form undergoes transition to the high-spin state only with a midpoint of change near pH 2.5[61,67]. Since the imidazole group of histidine residue 18 does not remain coordinated to the heme iron at such low values of pH in any of the model heme peptide complexes which have been studied[68-72,80], it is apparent that cooperative effects beyond those attributable to the presence of the two cysteine bridges must come into play. A simple interpretation would be that in the intact protein molecule, the protonation of the histidine residue to the imidazolium form is rendered unfavorable by the hydrophobic environment about the central coordination complex. This view would be consistent with the shifts observed in the midpoint of the spin-state transition when denaturing agents are added to the solution. In the presence of 6 *M* guanidine hydrochloride or 10 *M* urea, horse ferricytochrome *c* loses its low-spin character with a midpoint of change near pH 5.6, while in the presence of 10^{-3} *M* sodium dodecyl sulfate a value of 7.2 is found[110].

The coordination of methionine residue 80 provides a further illustration of the summed effects encountered. Thioethers are bound only weakly to the oxidized form of the small heme peptides that have been examined[80], and, in the case of the cytochromes *c*, such coordination is dependent upon conditions imposed by the folding pattern of the molecule. Chemical modification studies have shown that acetylation of all the tyrosine residues of horse ferricytochrome *c* does not result in a change in spin state[111,112]. Similarly, trifluoroacetylation[104] or acetylation[113] of all the lysine residues is without effect on the hemochrome properties of the molecule. However, when both sets of side chains are acetylated, the oxidized form exhibits high-spin character even at pH 7[112,113]. Presumably, hydrogen bonds formed by tyrosine residues, and the state of charge of some of the amino groups, are collectively of importance for the retention of a folding pattern compatible with maintenance of the heme–methionine linkage. When the fully acetylated preparation is reduced, and the methionine residue, as judged from model studies, becomes bound more firmly, the system reverts to the low-spin state at neutral pH.

Upon modification of the methionine residue itself, the product remains in the low-spin form so long as another strong-field group of the protein can enter the sixth coordination position. For instance, carboxymethylation of the methionine residues of horse cytochrome *c* and *Pseudomonas* cytochrome *c*-551 has been found to yield derivatives which are in the high-spin state at pH 4, but fully low-spin at values of pH sufficiently alkaline for a histidine–heme–lysine complex to be formed. On

References pp. 947–954

the other hand, when the *Pseudomonas* molecule, with its single histidine residue, is carboxymethylated after the amino groups are first blocked by trifluoroacetylation, the reduced form fails to exhibit low-spin character even at pH 10[87,96]. In general, the results obtained in experiments with modified preparations and heme peptide derivatives of the cytochromes *c* indicate that there is a great tendency for such molecules to adopt a structure in which the heme is maintained in the low-spin state. Indeed, a single strong-field group potentially capable of binding in the sixth position is usually sufficient for this condition to be fulfilled. In other words, the normal folding pattern of the parent proteins at pH 7 provides not only for the coordination of a group with a relatively low binding constant[80], but does so to the exclusion of other potential ligands which readily become bound to the heme iron when the methionine residue is modified or deleted.

This important condition is found not to be maintained when solutions of the oxidized form of the protein are examined at alkaline pH. The structure which predominates at pH 7 yields an absorption band, attributable to a transition polarized perpendicularly to the plane of the porphyrin[114], at 695 nm (Fig. 4). This band disappears when the molecule is heated, treated with denaturing agents, or subjected to complex formation with extrinsic ligands[115-118], and has been observed in experiments with model heme peptide systems only upon the coordination of a thioether group[119]. It is thus thought to reflect the heme-methionine 80 linkage present in the "mammalian-type" proteins at neutral pH. As first noted by Theorell and Åkesson[61], increase in the alkalinity of a solution of horse or beef ferricytochrome *c* leads to loss of the 695 nm band with a pK' of about 9.3, and marked alterations indicative of a change in the protein–heme interaction occur in the optical rotatory dispersion (o.r.d.) and circular dichroism (c.d.) spectra over this part of the pH range[105,120-124]. The simplest interpretation is that the coordinated methionine side chain is replaced as a ligand by a nitrogenous group of the protein, and, in keeping with the observations concerning the carboxymethylated-trifluoroacetylated preparations, this could well be the amino group of a lysine residue. Consistent with this possibility is the fact that guanidination of the lysine side chains raises the midpoint for loss of the 695 nm band of the horse heart protein to 9.7[106].

The pK' value for the transition which leads to loss of the 695 nm band, like the midpoint of the transition which results in the loss of low-spin character in acid solution, reflects summed effects which in large part remain to be delineated. Upon nitration of one to two of the tyrosine residues of horse ferricytochrome *c* by treatment of the molecule with tetranitromethane, the loss of the 695 nm band occurs with a pK' of about 6 instead of 9.3[125,126]. Since the nitration procedure reduces the pK' value for the ionization of the modified tyrosine side chains to approximately this extent, it is tempting to attribute this result to the disruption of structurally important hydrogen bonds, and, indeed, it has been found that the residues which are nitrated first are those known to be involved in such bonding, *i.e.*, residues 48 and 67[125,126]. However, the band near 695 nm has been observed in other studies to be retained at pH 7 upon *O*-acetylation of these two residues[106,112]. The results of nitration, like the change in spin state mentioned earlier, can thus not be accounted for solely in terms of elimination of a hydrogen-donor role for the phenolic hydroxyl groups. A more plausible interpretation would be that the generation of a phenolate anion in the hydrophobic interior of the molecule, or the introduction of the nitro group *per se*, contributes to the loss of the 695 nm band at the lower values of pH. Such an effect is, however, unlikely to be a factor in the transition with a pK' of 9.3 displayed by the unmodified protein, since the interior tyrosine residues would be expected to ionize with midpoints appreciably higher than this[127–129]. In fact, recent observations by Aviram and Schejter[130] suggest that other, as yet unspecified, factors remain to be taken into account. It has been found by these authors that loss of the 695 nm band of bakers' yeast iso-1-cytochrome *c* occurs with a midpoint more than a pH unit lower than that which holds for the protein from horse heart. Since comparisons of the stability of the two proteins[131], their susceptibility to the action of proteolytic enzymes[132,133], and the enthalpy changes for complex formation with cyanide[85,130,134] all indicate that the structure of the yeast molecule is more easily altered than that of the horse protein, this difference in midpoints is readily interpretable in overall terms. However, the summed effects which are involved are likely to be complex, and detailed deductions must await more complete information than is currently available. In this connection, it may well be useful to consider again the cytochrome *c*-551 from *Pseudomonas.* Although this molecule, like yeast iso-1-cytochrome *c*, is less stable than the protein from horse heart, the band at 695 nm is lost only at substantially more alkaline values of pH; the midpoint in this instance occurs at pH 10.9, near the upper end of the pH range within which the observed changes remain reversible[106]. In contrast to the horse and yeast proteins, the *Pseudomonas* molecule contains just a single tyrosine residue, and does not have a nitrogenous group capable of coordination to the heme iron in the sequence position preceding the coordinated methionine side chain. The latter factor alone could account for the retention of the 695 nm band to higher pH values, since it is probable that a more extensive rearrangement would be required to bring a ligand into play from elsewhere along the chain. Clarification of this point may aid significantly in the analysis of the effects operative in the more complex "mammalian-type" systems.

The influence of the protein on the central coordination complex has a counterpart, of course, in effects of the prosthetic group on the peptide chain of the molecule. The X-ray diffraction data for horse ferricytochrome *c* indicate that residues 92–102 are arranged in the form of an α-helix, while residues 9–13, 14–18, 49–54, 62–70, and 71–75 form segments which are gently helical, and in part have Ramachandran angles close to those of an α-helix[73]. The presence of such ordered structure is reflected in the o.r.d. and c.d. spectra recorded over the range of pH in which the molecule remains a hemochrome[102,105,120–124,135–139]. However, upon disruption of the central coordination complex in strongly acid solution, a concomitant transition to a structure with the features of a random coil is observed to take place[102,121,124]. Since a similar effect is obtained with fully esterified horse ferricytochrome *c*[102], where there is no change in net charge as the pH is decreased, it is apparent that maintenance of the ordered state at low pH requires that the coordinated histidine and methionine residues be retained in positions 5 and 6 about the heme iron. Upon dissociation of the metal–ligand bonds under these conditions, the conformation is altered sufficiently that not only the residues near the heme group, but also those in the more distant segment of the chain near the carboxyl terminus (residues 92–102), are affected.

A still more striking example of the relationship between ligand bonding and the conformation of the peptide chain has been encountered in experiments with the cytochrome from *Pseudomonas*[121]. At neutral pH, in the presence of 8 *M* urea, the oxidized form of this protein is a high-spin species, and the o.r.d. and c.d. spectra indicate that the chain is in the form of a random coil. However, when the molecule is subjected to reduction, the absorption spectrum reverts to that of a hemochrome, and the chain regains the ordered structure observed in the absence of urea. Presumably, the firmer bonding of methionine residue 61 to the reduced than to the oxidized prosthetic group results in the restoration of this residue to the sixth coordination position, with attendant reversal of the order–disorder transition, even in 8 *M* urea.

Structural effects introduced by alteration of the oxidation state of the cytochromes are, of course, of potentially great interest in connection with the mode of action of these compounds. In the absence of denaturing agents, the reduction of cytochrome *c*-551[121,140] and the "mammalian-type" proteins[137-139] appears not to lead to a change in helix content at neutral pH, but there are indications that there may be significant differences between the reduced and oxidized forms with regard to other structural features. It has long been known, for example, that the oxidation state affects the chromatographic behavior[141-143] and surface properties[144] of the "mammalian-type" proteins, and that the reduced form is less susceptible to enzymatic hydrolysis than the corresponding oxidized form[132,133,145,146]. There are also differences in terms of antigenic characteristics[147], rates of hydrogen–deuterium exchange[148,149], and the chemical reactivity of various side-chain groups. As might be expected in view of the intrinsically firmer coordination of thioethers to the reduced than to the oxidized prosthetic group[80], the carboxymethylation of methionine residue 80 proceeds much more slowly with ferro- than with ferricytochrome *c*[91,93]. In addition, however, differential reactivity has also been observed with respect to the tyrosine residues[112,150] and carboxylate groups[103] of the molecule. Spectrophotometric titration data indicate an effect of the oxidation state on the environment of the phenolic hydroxyl groups at alkaline values of pH[127], and the o.r.d. and c.d. spectra for horse ferri- and ferrocytochrome *c* show that oxidation-state dependent differences affecting at least one of the aromatic side chains of the molecule also prevail at neutral pH[102,105,120,124,135-139]. Still other indications of a change in structure are provided by the markedly different proton magnetic resonance spectra of the two forms[101,151-153], as well as the fact that a decrease in entropy accompanies the reaction[154,155]:

$$\text{ferricytochrome } c + \tfrac{1}{2}\,H_2 \rightarrow \text{ ferrocytochrome } c + H^+ \qquad (1)$$

In general, the results suggest a tighter structure for the reduced species, and this is consistent with the greater resistance of the latter to structural change upon increase in the temperature[156], or exposure of the molecule to guanidine hydrochloride or urea[120,121,124].

It may be that the structural differences between the ferri- and ferrocytochromes *c* will be found to involve primarily the half of the molecule containing the coordinated methionine residue (Fig. 7, left side). If it is assumed that this residue, as in the case of the heme peptide systems, interacts more strongly with the reduced than with the oxidized prosthetic group, changes in structure in this part of the protein would have an obvious basis. Moreover, the X-ray diffraction data for horse ferricytochrome *c* show that, in the vicinity of this residue, the molecule is less rigidly constructed than is the part to the right side of the crevice[73]. From an inspection of models of the molecule, it appears that changes could be introduced in the side-chain positions and/or folding pattern of the left part of the protein that would be wholly consistent with the known differences in properties of the two states. Such changes need not, of course, be very extensive. Indeed, the lack of a change in helix content[137-139], and the observation that reaction (1) is attended by only a small change in apparent heat capacity[154], suggest that the structural effects may be more limited than has sometimes been assumed. The crystallographic studies of the reduced molecule, currently in progress, should before long provide more complete information on this point.

The heme–methionine linkage is, of course, not the only factor of importance in the binding of the prosthetic group to the protein that is altered upon change in the oxidation state. This is emphasized by observations concerning the hydrophobic interaction of the heme group with the peptide chain. Whereas compounds such as guanidine hydrochloride and

References pp. 947–954

urea more readily alter the structure of ferri- than of ferrocytochrome *c*[120,121,124], the reverse holds true for the action of sodium dodecyl sulfate on the two forms[110]. The latter agent converts horse ferrocytochrome *c* to a high-spin species in neutral solution, and, upon increase in the pH at a 10^{-3} *M* concentration of the denaturant, low-spin properties emerge only with a midpoint of change near pH 10. Conversion of the oxidized form of the molecule to the high-spin state requires greater concentrations of sodium dodecyl sulfate than are needed for the reduced form, and, at a sodium dodecyl sulfate concentration of 10^{-3} *M*, the pH-dependent change in spin state occurs in this instance with a midpoint of 7.2. Apparently, hydrophobic interaction of the prosthetic group with this detergent is favored when the heme is in the reduced rather than the oxidized form, and a corresponding difference is likely to apply to the interaction of the prosthetic group with hydrophobic groups of the protein. The effects of alteration of the oxidation state would then reflect, in addition to the change in ligand binding, a difference in hydrophobic interaction.

Electron transfer (see also Chapter 19 and Chapter 27, pp. 977–979)

Little is known at this time concerning the mechanisms operative upon oxidation and reduction of the cytochromes *c*. It has been suggested that cytochrome c_1 and cytochrome oxidase, the electron donor and electron acceptor for cytochrome *c* in the mitochondrial electron-transfer chain, are bound to the latter at the regions of positive charge of the molecule[73,100]. However, the effects which then come into play, in the course of electron transfer from the donor to the acceptor heme group, remain to be elucidated.

DeVault and Chance[157–159], in studies of the kinetics of electron transfer between cytochrome *c*-552 and light-oxidized chlorophyll in the photosynthetic bacterium *Chromatium* D, have shown that, below 100°K, the rate of this reaction is temperature-independent. The process thus requires no thermal motion under these conditions. Moreover, it is sufficiently rapid (half-time of 2 msec at 34°K) that it is assumed that no bond-breaking steps or changes in conformation are involved[158]. Movement of the electron from the heme group to the surface of the molecule is thought to proceed very rapidly under these circumstances, with the rate-limiting step of the overall transfer reaction involving intermolecular "tunneling" of the electron from the donor to the acceptor system.

That electron transfer from the heme group to an acceptor molecule can also occur very rapidly under solution conditions is indicated by the rate at which cytochrome *c* is oxidized by hydrogen peroxide in the presence of yeast cytochrome *c* peroxidase. Yonetani and Ray[160] have reported a rate constant of $5 \times 10^8\ M^{-1}\ \mathrm{sec}^{-1}$ for this reaction, a value which approaches the limit set by the protein–protein collision frequency.

The reduction and reoxidation of cytochrome *c* in the mitochondrial electron-transfer chain proceeds considerably more slowly than this, especially under conditions of oxidative phosphorylation, and the possibility has been widely discussed that cyclic changes in the structure of the protein may assume major significance in connection with respiratory-rate control and energy-coupling mechanisms[158,161]. The information currently at hand is, however, insufficient to allow adequate analysis of this possibility. In fact, much remains to be established even concerning the events which attend the interaction of the cytochromes *c* with relatively simple oxidizing and reducing agents such as ferricyanide or ascorbate.

At neutral pH, the "mammalian-type" cytochromes *c* react with ferricyanide at rates comparable to those obtained upon the oxidation of cytochrome *c* by cytochrome oxidase. At temperatures from 5 to 40°C, pH 7.0, and $\mu = 0.2$, a rate constant of $7 \times 10^6\ M^{-1}\ sec^{-1}$ is obtained[76,162,163], which compares with a value of approximately $10^7\ M^{-1}\ sec^{-1}$ for the reaction with cytochrome oxidase under generally similar conditions[164]. In accordance with the fact that ferricyanide and the "mammalian-type" cytochromes *c* are oppositely charged, the rate diminishes with increasing ionic strength[76]. Conversely, in the case of the fully trifluoroacetylated derivative, which is net negatively charged at neutral pH, the rate of the reaction with ferricyanide increases with increasing ionic strength, and the same is true for the rate of oxidation of the negatively charged cytochrome *c*-551 molecule[106]. At pH 7.0 and $\mu = 0.2$, the latter reaction yields a rate constant of $10^5\ M^{-1}\ sec^{-1}$.

The reverse process, in which the protein is reduced by ferrocyanide, proceeds more slowly. At 20°C, pH 7.0, and $\mu = 0.2$, the rate constants are about $10^4\ M^{-1}\ sec^{-1}$ for the cytochrome from horse heart, and $10^2\ M^{-1}\ sec^{-1}$ for the *Pseudomonas* preparation[106,162,163]. However, if the pH is increased to 9.5, the kinetics for the reduction of the "mammalian-type" proteins become more complex. As noted initially by Greenwood and Palmer[165] in studies of the reaction of cytochrome *c* with ascorbic acid and tetrachlorohydroquinone, the reduction at this pH proceeds in a biphasic manner. This has been taken to indicate the existence of a reactive and an unreactive form of ferricytochrome *c* under these conditions, the two forms corresponding, respectively, to the species with and without an absorption band at 695 nm. In keeping with this interpretation, the oxidation of ferrocytochrome *c* by ferricyanide results, at pH 9.5, in the rapid formation of a product with a 695 nm band, followed by partial conversion of this species to a structure lacking this feature[106]. Furthermore, *Pseudomonas* cytochrome *c*-551, which loses its 695 nm band only at more alkaline values of pH ($pK' = 10.9$ *vs.* $pK' = 9.3$ for the horse heart molecule), gives no evidence of an unreactive form at pH 9.5; *i.e.*, the reduction of the *Pseudomonas* protein by ascorbate or ferrocyanide, and its reoxidation by ferricyanide, are monophasic processes at this pH[106]. The simplest inference is that, in the "mammalian-type" proteins, the previously discussed displacement of the coordinated methionine residue by a lysine side chain yields, in alkaline solution, a structure which can no longer be reduced by the agents employed. Since there is, in contrast, no loss of reducibility when the coordinated methionine residue is displaced by extrinsic imidazole[115,118,166], it would appear that it is the nature of the group substituted in the sixth coordination position, more than the alteration in protein structure accompanying such substitution, that governs the effects obtained. Indeed, the available o.r.d. and c.d. data suggest that the overall changes in structure which occur upon complex formation with imidazole differ little from those introduced by increase in the pH[105,124]. These observations are of interest, of course, in connection with the possibility that electron transfer to the heme group may proceed under normal conditions via the coordinated thioether sulfur atom.

There have been a variety of suggestions over the years concerning the manner in which electrons might be passed to and from the prosthetic group of the cytochromes *c* within the mitochondrial electron transfer chain (see also Chapter 19, p. 648). The proposed mechanisms, which in part are not mutually exclusive, include: (a) transfer via π-orbital overlap of the heme group with that of the electron donor or acceptor molecule; (b) transfer via the protein groups occupying coordination positions 5 and/or 6 about the heme iron; (c) transfer via the thioether groups linking the protein to porphyrin side chains 2 and 4; (d) transfer via the methine linkages of the porphyrin ring; (e) transfer via one or more aromatic side chain of the protein in π-orbital overlap with one another and/or the heme group; (f) transfer via segments of the peptide chain; (g) transfer via bound solvent or small solute molecules; (h) transfer via quantum-mechanical tunneling; and (i) appropriate combinations of the above. Some of these proposals seem unlikely in the light of present information. For instance, the dimensions of the crevice and the arrangement of the heme group within it[73,99,100] render it improbable that the first mechanism prevails. Similarly, the structure of ferricytochrome *c* in the vicinity of residues 80 and 18 makes it unlikely that the reduction of cytochrome *c* proceeds via a direct interaction of the electron donor with the coordinated thioether or imidazole group. Such interaction could, however, occur in an indirect fashion, and the hypothesis that electron transfer occurs via the ligand groups, first advanced by Theorell[63], remains an attractive one. In fact, were reduction to proceed by way of one of the ligands, and oxidation via the other, this would offer some interesting possibilities with respect to respiratory-control and energy-conservation mechanisms. The existence of what may be separate binding sites for cytochrome c_1 and cytochrome oxidase, and the hydrophobic regions with roughly parallel aromatic groups which have been described, would, of course, be consistent with an arrangement such as this.

In view of the growing body of structural information which has become available for the cytochromes *c*, the effects operative upon electron and energy transfer will undoubtedly be receiving intensive attention during the next few years. A useful step would be to examine reactions such as those of the cytochromes *c* with cytochrome c_1, cytochrome oxidase and other heme proteins under conditions which permit the structure of the donor and acceptor molecules to be varied in systematic fashion. In this connection, a study has been undertaken recently of the factors governing the rate of electron transfer between horse cytochrome *c* and *Pseudomonas* cytochrome *c*-551[76]. The *Pseudomonas* molecule yields absorption spectra sufficiently different from those of the "mammalian-type" proteins to allow the electron-transfer process to be followed spectrophotometrically, and offers the advantage that the properties of a number of useful derivatives have been delineated. In the experiments completed to date, it has been found that, although the reactants differ greatly in primary structure and are oppositely charged at neutral pH (Figs. 3 and 6), the rate of electron transfer at pH 7 and 20°C corresponds closely to that estimated from nuclear magnetic resonance data for the electron exchange reaction between horse ferri- and ferro-cytochrome *c* at this pH and temperature ($5 \times 10^4\ M^{-1}\ sec^{-1}$)[76,152].

Other cytochromes c of eukaryotes

As mentioned earlier, a number of the *c*-type cytochromes obtained from eukaryotic cells have physical and chemical properties differing from those of the "mammalian-type" proteins. Detailed investigations of these preparations remain to be undertaken, and only a few characteristics will be indicated here.

The firmly bound cytochrome c_1 component of the mitochondrial electron-transfer chain (Fig. 1) exhibits absorption bands at 411 and 523 nm in the oxidized form, and 418, 523, and 553 nm in the reduced state[167-169]. Purified preparations from beef heart have a weight of about 37,000 per heme group[169,170], but are subject to extensive aggregation[169] and yield average particle weights of about 1.2×10^6. The protein is acidic, with an isoelectric point near pH 3.6. The $E_{m,7}$ value is 0.22 V[171], and the molecule is non-autoxidizable[170].

Two *c*-type cytochromes have been found in studies of *Euglena gracilis*[172-176]; both have non-alkaline isoelectric points and $E_{m,7}$ values near 0.30 V[175]. One of these proteins, with an α-band maximum at 552 nm, is not found in dark-grown cells, and appears to be located in the chloroplasts[176]. The other, reported to have an α-band maximum at 558 nm, with a shoulder at 556 nm, is thought to fulfill a respiratory function[176].

The cytochromes *c* of *Ustilago sphaerogena*[177] and *Aspergillus oryzae*[178], two fungi, have a number of properties in common with the "mammalian-type" molecules. For instance, the protein from *Ustilago* yields o.r.d. spectra which closely resemble those of the "mammalian-type" preparations[121], and displays electron-transfer activity in the rat liver succinate oxidase system[177]. On the other hand, the cytochrome from *Aspergillus* is inactive in the mammalian cytochrome oxidase system[179], and both proteins have been found to have non-alkaline isoelectric points[177,178].

The cytochromes commonly given the designation *f* contain covalently bound heme, and are thus, by definition[8], of the *c*-type. These proteins, which occur in the chloroplasts of higher plants[180], have acidic isoelectric points, and have been reported to have molecular weights as high as 245,000[181]. At neutral pH, the main visible absorption band of the reduced species (α-band) is located at 552 to 555 nm[182], as compared to 550 nm for the "mammalian-type" molecules, and there is a major difference in terms of the oxidation-reduction potentials at pH 7: $E_{m,7} \cong 0.35$ V[182], similar to the potentials of the cytochromes c_2 of photosynthetic bacteria (Table I), *vs.* values near 0.25 V for the "mammalian-type" systems.

Acidic, high-potential cytochromes of the *c*-type have also been found among the algae[183-190]. Like the cytochromes *f* of higher plants, these are associated with the plastids[185,186,189,190], and appear to be of importance in connection with photosynthesis.

Cytochromes c of prokaryotes

The cytochromes *c* of the prokaryotes are far more varied in structure and properties than are the "mammalian-type" proteins, and, as has been stressed particularly by Kamen[12], provide unusually good opportunities for investigations of structure–function relationships. Information concerning this broad array of compounds, although still limited, has increased rapidly in recent years[191,192].

A few of the properties exhibited by this group of proteins are summarized in Table I. The $E_{m,7}$ values range from 0.38 V to -0.22 V, and there are marked differences in the positions of the characteristic α-band maxima. Some of the preparations have neutral or acidic isoelectric points, while others are basic. The molecular weights range from less than 10,000 to greater than 70,000. Most of the molecules have a single heme group, but some contain two or three hemes, and in several cases there is also a flavin group present. In general, little or no electron-transfer activity is observed in assays conducted with mitochondrial cytochrome oxidase or succinate oxidase systems[179].

One of the more extensively studied preparations, the cytochrome *c*-551 of *Pseudomonas aeruginosa*, has already been commented upon in the course of discussion of the "mammalian-type" proteins. Although it differs greatly from the latter in amino acid composition and sequence[74,75],

TABLE I

PROPERTIES OF SOME CYTOCHROMES *c* FROM PROKARYOTES

Cytochrome	Source	Molecular weight	Isoelectric point	α-Band maximum (nm)	Number of heme groups	$E_{m,7}$ (V)	Reference
c	*Nitrobacter*			550		0.28	193
	Rhodospirillum molischianum	10,000	9.4	550	1	0.38	194,195
c_2	*Rhodospirillum rubrum*	13,000	6.1–6.4	550	1	0.32	196,197,198
	Rhodopseudomonas spheroides	13,000	7.9	550	1	0.35	199
	Rhodopseudomonas palustris	< 20,000	> 9.0	552	1	0.35	200
	Rhodospirillum molischianum	13,400	> 9.4	550	1	0.29	194,195
	Rhodospirillum vannielii		~ 7	550		0.30	201
c_3	*Desulfovibrio desulfuricans*	~ 13,000	7.2	552-553	3	-0.20	202-204
	Desulfovibrio gigas		5.2	553		-0.22	205
	Desulfovibrio vulgaris	~ 13,000	10.0	552-553	3		203,204,206
	Desulfovibrio salexigens	14,000	10.8	552	3	-0.21	204,207
c_4	*Azotobacter vinelandii*	24,000	4.5	551	2	0.30	208–210
c-550	*Bacillus subtilis*	12,500	8.65	550	1	0.21	211,212
	Spirillum itersonii	11,000	9.86	550	1	0.30	213
c-551	*Pseudomonas aeruginosa*	9,000	4.7	551	1	0.29	74,77,78

References pp. 947–954

	Chlorobium thiosulfatophilum	45,000–60,000	6.0	551	2	0.14	214
c-552	*Pseudomonas stutzeri* (I)	8,400	4.0–6.6	552	1	0.28	215
	Pseudomonas stutzeri (II)	20,000	4.0–6.6	552	2	0.28	215
	Escherichia coli	12,000[b]	4.1–4.7	552	1	~ -0.2	216,217
	Chromatium	72,000	5.1	552	2[a]	0.01	218,219
c-553	*Chlorobium thiosulfatophilum*	50,000	6.7	553	1[a]	0.10	214
	Pseudomonas denitrificans	45,000		553	2	~ -0.1	220
	Rhodopseudomonas spheroides	25,000		553	1	0.12	221
c-554	*Bacillus subtilis*	14,000	4.4	554	1	-0.08	211,212
	Azotobacter vinelandii	24,000	4.3	554	2	0.32	196,208,210
c-555	*Chlorobium thiosulfatophilum*	10,000	~ 10.5	555	1	0.14	214,222,223
	Crithidia fasciculata	12,000	9.9	555	1	0.28	224
c-550,553	*Chromatium*	13,000[b]	4.38	550,553		0.33	226
c-552,558	*Pseudomonas stutzeri*	70,000–74,000		552,558	2	~ 0	225
c′	*Rhodopseudomonas palustris*	< 20,000	> 9.0	550–560	1	0.10	200
cc′	*Rhodospirillum rubrum*	27,000–30,000	~ 5.5	550–560	2	~ 0	196,198,227
	Chromatium	27,000–30,000	5.5	550–560	2	~ 0	196,227
	Pseudomonas denitrificans	28,000		550–560	2	0.12	228

[a] Also contains one flavin prosthetic group.
[b] Obtained in aggregated form; the molecular weight indicated is per heme group.

this molecule has the same types of groups coordinated to the iron atom[80,81,86,87,96]. Indeed, it has been suggested recently that, in spite of the major differences in amino acid sequence, the *Pseudomonas* and "mammalian-type" proteins have similar spatial configurations (R. E. Dickerson, personal communication).

Structural similarities may also hold for many of the other prokaryotic proteins. However, it is evident from Table I that, in a number of instances, fundamental distinctions will be encountered. In the following sections, some properties of three representative subgroups will briefly be indicated.

Cytochromes c_2

The members of this family, studied primarily by Kamen and his colleagues, bear a closer resemblance to the "mammalian-type" proteins than do some of the other prokaryotic preparations. The molecular weights appear to be of the order of 13,000, and the only amino acid sequence thus far determined, that of the molecule from *Rhodospirillum rubrum*[229,230], exhibits a significant degree of correspondence to the sequence of the protein from horse heart. On the other hand, the isoelectric points vary considerably, and the $E_{m,7}$ values are appreciably higher than those of the "mammalian-type" systems.

The circular dichroism spectra of cytochrome c_2 from *Rhodospirillum molischianum*[195] closely resemble those of the horse heart protein, suggesting that the environment about the heme group is essentially equivalent in the two cases. *R. rubrum* cytochrome c_2, on the other hand, yields circular dichroism patterns[138] in the Soret and near ultra-violet regions which differ markedly from the curves for the "mammalian-type" systems (Fig. 8). In the Soret pattern for the oxidized form, a single positive band (perhaps consisting of more than one component) replaces the overlapping negative and positive bands displayed by the horse heart protein. In addition, the band of the latter at 263 nm, postulated to reflect a heme–histidine transition[138], is lacking in the *R. rubrum* spectrum. In these respects, the c.d. spectrum of *R. rubrum* ferricytochrome c_2 resembles more nearly the pattern for *Pseudomonas* ferricytochrome *c*-551 than it does that for cytochrome c_2 from *R. molischianum*; both the negative band in the Soret region and the positive band at 263 nm are missing from the c.d. spectrum of the oxidized *Pseudomonas* protein (Fig. 8)[140,231]. Although the *Pseudomonas* and horse heart systems possess the same ligand groups, the environment about the heme group clearly differs in the two cases[121], as would be expected in view of the very different amino acid sequences. Presumably, such a difference also prevails between *R. rubrum* cytochrome c_2 and the "mammalian-type" proteins. It will be of much interest to obtain crystallographic data allowing the structures of the *R. rubrum*, *Pseudomonas*, and "mammalian-type" proteins to be compared in detail in this connection.

References pp. 947–954

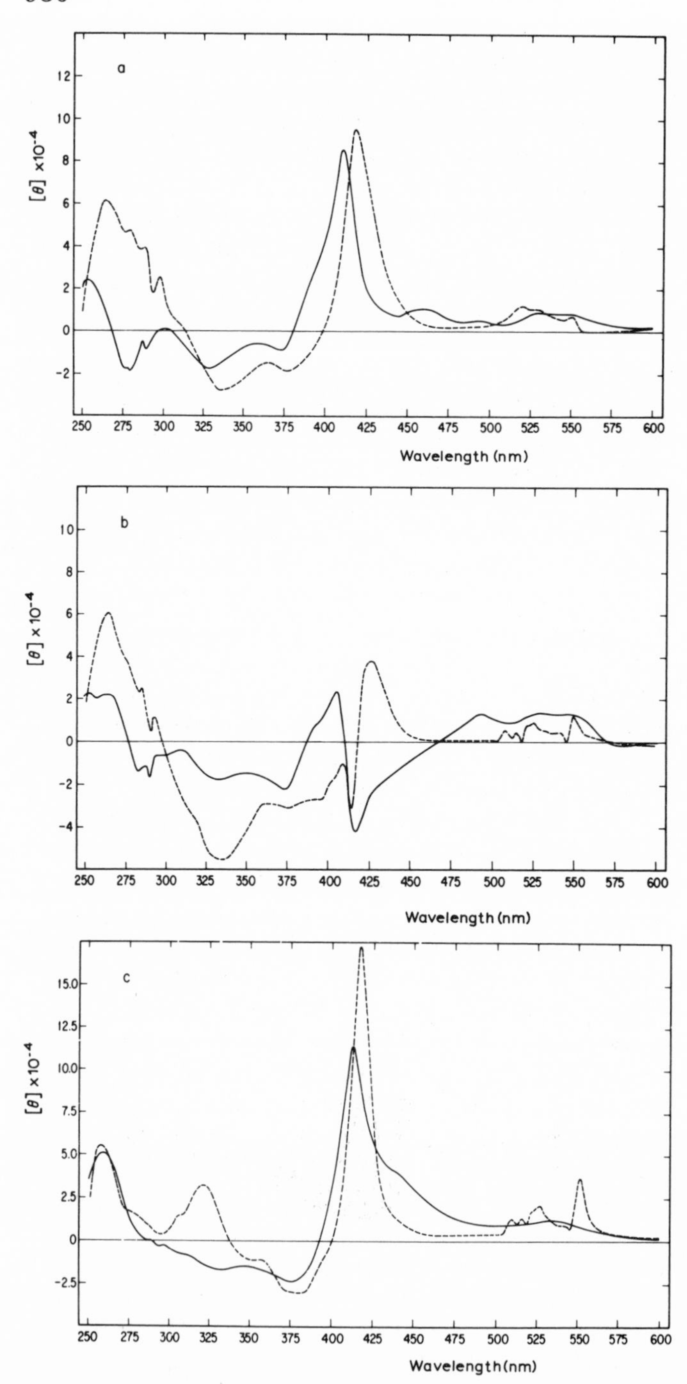

Fig. 8. Circular dichroism spectra of *c*-type cytochromes. (a) *Rhodospirillum rubrum* cytochrome c_2[138]; (b) horse heart cytochrome *c*[137,139,140]; (c) *Pseudomonas aeruginosa* cytochrome *c*-551[140,231]. —, oxidized form; -----, reduced form.

In the first X-ray diffraction study of a prokaryotic cytochrome, it has been observed by Kraut *et al.*[232] that the oxidized and reduced forms of the *R. rubrum* protein are isomorphous. There is thus no major change in conformation upon alteration of the oxidation state of the molecule. This is consistent with what has been inferred concerning the horse heart molecule. As noted previously, although the crystals obtained for the oxidized and reduced forms of the latter are non-isomorphous, the overall differences in molecular structure need not be extensive.

Cytochromes c_3

As observed initially by Postgate[202] in studies of a preparation from *Desulfovibrio desulfuricans*, the cytochromes in this category have low oxidation-reduction potentials and are subject to rapid autoxidation; the $E_{m,7}$ values are on the order of −0.2 V (Table I). The molecular weights correspond closely to those of the "mammalian-type" proteins, and the α-band maxima are at 552–553 nm at neutral pH. Three of the four preparations thus far examined (those from *Desulfovibrio desulfuricans*, *D. vulgaris*, and *D. salexigens*) are basic proteins, while the fourth (from *D. gigas*) is acidic. Each of these cytochromes contains three heme groups[204,207].

Studies by Ambler[206] have shown that the amino acid sequence of the protein from *D. vulgaris* contains only two regions corresponding to those which commonly provide the cysteine bridges and coordinated histidine residue of the cytochromes *c*: -Cys–A–B–Cys–His- (Fig. 3). However, there are also two regions of the type -Cys–A–B–C–D–Cys–His- in the chain. Circular dichroism studies of this protein, as well as of the molecules from *D. desulfuricans* and *D. salexigens*, suggest that the heme groups are in each case located in substantially equivalent environments[233].

Upon decrease of the pH, the oxidized form of the preparation from *D. vulgaris* undergoes a low-spin to high-spin conversion with a midpoint of change near pH 2.7[233], essentially the same value as that obtained with *Pseudomonas* cytochrome *c*-551 and the "mammalian-type" proteins. Two protons per heme group are involved in this transition.

It is assumed that, at neutral pH, each of the hemes has a histidine side chain bound to the iron atom, but little is known concerning the nature of the group in the *trans* position. In view of the low oxidation-reduction potentials, this ligand would be unlikely to be a thioether group, and, in fact, it has been found that in at least one of the c_3-cytochromes there are no methionine residues present[234]. Possibly the complex corresponds to the histidine–heme–histidine arrangement considered initially in connection with the "mammalian-type" proteins. Such a structure would be consistent with the $E_{m,7}$ values and other properties exhibited by model heme peptide complexes in which the iron is bound to two imidazole groups[70–72], but until further data become available for the c_3-systems, no firm conclusions can be drawn.

References pp. 947–954

Cytochromes c′ and cc′

Some of the more atypical *c*-type proteins thus far encountered are those given the designations *c′* and *cc′*. These molecules have covalently bound heme groups, but lack the familiar hemochrome properties of the cytochromes *c*.

Much of what is known about these systems has come from the work of Kamen and Horio and their colleagues. The cytochromes *c′* are single-chain proteins, with one heme group per molecule. The preparation from *Rhodopseudomonas palustris*, which has been examined the most extensively, has a molecular weight of less than 20,000, an alkaline isoelectric point, and an $E_{m,7}$ value of 0.1 V[200,235]. The cytochromes *cc′* are also single-chain proteins, but contain two heme groups each[196,227,236]. They have molecular weights in the range of 27,000–30,000, isoelectric points near pH 5.5, and $E_{m,7}$ values of about 0–0.1 V[198,227]. Sequence studies of a diheme peptide of 27 residues, obtained from the *cc′*-cytochrome of *Chromatium*, suggest that the heme groups are bound to the peptide chain rather near one another[237]. One binding site is thought to involve the sequence -Cys–A–B–Cys–His- in positions 5–9 of the peptide. The other may involve the sequence -Lys–Cys–His- in positions 19–21. No other cysteine residues are available, and it has been suggested that the second heme group may be joined to the peptide through only a single thioether linkage, or, perhaps, a thioether bridge and an ether linkage involving a threonine side chain[237]. Of course, the location of the heme groups in terms of sequence positions does not necessarily imply spatial proximity. In fact, c.d. and o.r.d. studies have provided no evidence of heme–heme interaction[238,239].

Although they differ in many of their properties, the cytochromes *c′* and *cc′* exhibit similar absorption, o.r.d., and c.d. spectra[200,235,238–241]. The absorption curves of the reduced forms are devoid of hemochrome character, and have a split band in the Soret region. The oxidized species, on the other hand, yield spectra indicative of a mixed high-spin and low-spin character reminiscent of that exhibited by the alkaline forms of ferrihemoglobin and ferrimyoglobin. However, whereas the latter exist as thermal mixtures of high- and low-spin species, for which the degree of high-spin character increases with the temperature, the high-spin character of the ferricytochromes *cc′* diminishes with increasing temperature[241]. The results of Mössbauer spectroscopy suggest that the two heme groups of the cytochromes *cc′* of *Chromatium* and *R. rubrum* are each subject to similar ligand effects, and that, as in ferrihemoglobin, the high-spin oxidized species has a highly distorted electronic configuration[242]. Interestingly, the electric-field gradients for the *cc′*-proteins and ferrihemoglobin are of opposite sign.

At high values of pH, both oxidation states of the *c′* and *cc′* preparations exhibit fully low-spin properties[196,235,240]. Upon the addition of certain denaturing agents, such conversion to the low-spin state can also be effected in neutral solution. High ionic strengths, on the other hand, prevent the emergence of low-spin character even at alkaline pH[238,239]. Circular dichroism studies have shown that the transition to the low-spin state is in both instances paralleled by the formation of a random coil. This loss of ordered structure upon conversion to the low-spin form contrasts markedly with the relationships between spin state and peptide-chain conformation observed in the case of *Pseudomonas* cytochrome *c*-551 and the "mammalian-type" cytochromes *c*[102,121,124].

CYTOCHROMES *b*

The cytochromes *b* are characterized by (a) a protoheme prosthetic group (Fig. 9) which is not bound covalently to the protein via side chains of the porphyrin; and (b) an α-band maximum at 556–558 nm when the

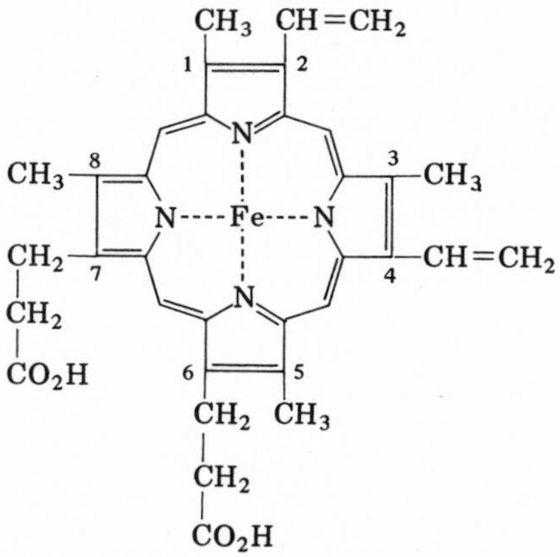

Fig. 9. Structure of protoheme.

spectrum of the reduced form is determined in the presence of pyridine at alkaline pH[8]. The members of this group are of widespread occurrence, and, particularly in the case of the bacterial representatives[191,192], vary considerably in structure and properties. Many occur in tightly bound form, and are difficult to dissolve in aqueous media. Others, such as the cytochrome b_5 of microsomes and a number of the prokaryotic *b*-proteins, are more readily soluble and have been studied in greater detail thus far.

Cytochromes b

The *b*-type cytochromes were first observed in work with heart muscle preparations[4], and have been the object of much interest in connection with the role which they play as components of the mitochondrial electron-transfer chain. Forceful measures are required to obtain these mitochondrial proteins in purified form, and monomeric dispersions have been obtained only with the use of cationic detergents. The molecular weight has been found to be approximately 28,000 under these conditions[243].

The purified preparations exhibit low-spin absorption spectra (Table II), and the protein from beef heart has been reported to have an $E_{m,7}$ value of -0.34 V[244]. In contrast, the corresponding particulate preparation has been found to have a potential of approximately 0.08 V[245]. The difference in the two values could be the result of (a) the difference in dielectric constant of the two environments and (b) differential binding of the reduced and oxidized forms to the membrane and/or other mitochondrial components.

TABLE II

ABSORPTION MAXIMA OF SOME CYTOCHROMES *b*

Cytochrome	Wavelength of maximum absorbance (nm)				
	Reduced form			Oxidized form	
	Visible		Soret	Visible	Soret
b	562	532	429	538	418
b-562	562	531	427	531	418
b_5	556	526	423	530	413
b-555	555	528	424	530	414
b_2	556	528	424	530–560	413

The prosthetic group can be removed from the protein by treatment with acid–acetone, and has been shown to be iron protoporphyrin IX[246]. It has not been established what the ligand groups are, and there is little other structural information available at this time.

Cytochromes b_5

Probably the best characterized of the *b*-type cytochromes are those in the b_5 category. These are found in the endoplasmic reticulum of liver and other tissues, and have been studied extensively by Strittmatter and his co-workers.

The protein is firmly bound in the cells, and has been obtained in soluble form only through the use of enzymatic digestion procedures. The entity which results may of course differ in composition and/or structure from that present originally[247]. However, the purified preparations have been found to exhibit properties which, on the whole, are similar to those observed in studies of microsomal suspensions. Indeed, no changes in spectroscopic and electron-transfer characteristics have been noted even upon treatment of the product from calf liver with trypsin to yield a derivative with 81 rather than 93 amino acid residues[248].

At pH 7 and 20°C, the purified preparations are in the low-spin state[248,249] (Table II). A small amount of high-spin character has been noted in electron-spin resonance studies of the oxidized form at 88 to 173°K, and, at room temperature, the absorption spectrum of the latter becomes predominantly that of a high-spin species when the pH is increased to above 11.

There is a single heme group per molecule[248,250], and the purified preparation from calf liver has an $E_{m,7}$ value of 0.02 V[251]. The corresponding potential for cytochrome b_5 as a component of microsomal suspensions has been estimated to be 0.1 V lower than this[252,253]. Whether the difference reflects primarily effects of the purification procedure on the molecule *per se*, or a differential interaction of the oxidized and reduced forms with components of the particulate material, remains uncertain at this juncture.

Upon removal of the prosthetic group by treatment with acid–acetone, metalloporphyrins other than protoheme can be bound to the protein[254,255]. With the exception of the derivative formed with hematoheme, the complexes in which the metal ion is hexacoordinate are sufficiently stable that the metalloporphyrin is not displaced from the protein when protoheme is readded to the solution. Metal-free porphyrins and metalloporphyrins in which the metal is not hexacoordinate are, on the other hand, bound only weakly. All of the iron–porphyrin complexes which have been examined can be reduced by NADH–cytochrome b_5 reductase, while the complexes formed with cobalt and manganese porphyrins have been found to serve as competitive inhibitors of the electron-transfer reactions.

In keeping with the view that the protein-prosthetic group interaction plays an important role in the determination of the overall structure of the molecule, the metalloporphyrin complexes are less sensitive to the effects of urea, low pH, and the action of trypsin than is the apoprotein.

The cytochromes b_5, like the *b*-type cytochromes generally, react with neither carbon monoxide nor cyanide. However, these ligands do bind to the iron of the hematoheme–protein complex. Although such binding is presumed to involve the displacement of a coordinated protein group, it has been reported that the cyanide-containing system can still react with NADH-cytochrome b_5 reductase[255].

On the basis of the pH-dependence of the binding of protoheme to the protein, it has been proposed that the side chain of at least one histidine residue is linked to the iron atom[256]. This would be consistent with the fact that the binding of the heme group can be prevented by treatment of the protein with diazotized sulfanilic acid under conditions leading to the

References pp. 947–954

modification of a single imidazole group. In addition, it has been established that: (i) a thioether or thiol group cannot be involved in the interaction with the heme group, since there are no methionine or cysteine residues in the calf liver protein; (ii) the ϵ-amino groups of lysine residues cannot be involved, since the binding reaction is unaffected by acetylation of the protein; and (iii) the tyrosine residues can all be iodinated without affecting the binding of the prosthetic group[248,255-257].

Succinylation of the eight lysine residues and the α-amino group of the protein, like the acetylation of these groups, is without effect on the spectral and enzymatic properties of the molecule: despite the change from positively to negatively charged groups at neutral pH, the fully modified apocytochrome b_5 recombines with protoheme to form a product with properties equivalent to those of untreated preparations[257]. On the other hand, upon trinitrophenylation of all of the amino groups of the apoprotein a derivative is obtained which binds protoheme only weakly to yield a complex which no longer can be reduced with NADH–cytochrome b_5 reductase. Interestingly, although the lysine residues of cytochrome b_5 are not difficult to modify chemically, the molecule is relatively resistant to the action of trypsin[248].

The amino acid sequences of preparations from calf and rabbit liver have been determined (Fig. 10)[258-261]. The two are fairly similar overall, with the main differences occurring near the ends of the chains. It has

Rabbit: (Gln,Ala,Ala,Ser,Asp,Lys)Asp Lys Ala Val Lys Tyr Tyr Thr Leu
Calf: SER·LYS·ALA·VAL·LYS·TYR·TYR·THR·LEU·
5

Glu Glu Ile Lys Lys His Asn His Ser Lys Ser Thr Trp Leu Ile Leu His His
GLU·GLN·GLU·ILE·LYS·HIS·ASN·ASN·SER·LYS·SER·THR·TRP·LEU·ILE·LEU·HIS·TYR·
10 15 20 25

Lys Val Tyr Asp Leu Thr Lys Phe Leu Glu Glu His Pro Gly Gly Glu Glu Val
LYS·VAL·TYR·ASP·LEU·THR·LYS·PHE·LEU·GLU·GLU·HIS·PRO·GLY·GLY·GLU·GLU·VAL·
30 35 40 45

Leu Arg Glu Gln Ala Gly Gly Asp Ala Thr Glu Asp Phe Glu Asp Val Gly His
LEU·ARG·GLU·GLN·ALA·GLY·GLY·ASP·ALA·THR·GLU·ASP·PHE·GLU·ASP·VAL·GLY·HIS·
50 55 60

Ser Thr Asp Ala Arg Glu Leu Ser Lys Thr Phe Ile Ile Gly Glu Leu His Pro
SER·THR·ASP·ALA·ARG·GLU·LEU·SER·LYS·THR·PHE·ILE·ILE·GLY·GLU·LEU·HIS·PRO·
65 70 75 80

Asp Asp Arg Ser Lys
ASP·ASP·ARG·SER·LYS·ILE·THR·LYS·PRO·SER·GLU·SER
85 90 93

Fig. 10. Amino acid sequences of cytochrome b_5 from calf liver[261] and rabbit liver[260].

been suggested by Strittmatter that segments 63–76 and 53–57 resemble those in the vicinity of the proximal and distal heme-linked[262] histidine residues of hemoglobin and myoglobin, and that the tertiary structures near the heme groups of the three proteins may be rather similar to one another[258,259]. In keeping with earlier indications[255-257], it could be that, at neutral pH, the central coordination complex of cytochrome b_5 is one in which the fifth and sixth positions are both occupied by imidazole groups. X-ray diffraction studies are under way to explore this and other points more fully[263,264].

Cytochrome b-555

A soluble cytochrome with properties generally comparable to those of cytochrome b_5 has been obtained by Okada and Okunuki[265] from the larvae of the housefly, *Musca domestica* L. (Table II). The preparation can be reduced with rat liver NADH–cytochrome b_5 reductase, and has an $E_{m,7}$ value of 0.01 V. There is no evidence to suggest that the material is a fragment of a larger molecule present *in vivo.*

Determination of the amino acid composition has shown that there are only 65 residues present, as compared to 93 residues in the solubilized preparation from calf liver[258-261]. Comparison of the c.d. curves[266] with those of yeast cytochrome b_2 [267] and a second cytochrome of *Musca domestica* L., cytochrome *b*-563[268], has led Okada and Okunuki to postulate some empirical rules for the c.d. spectra of the *b*-type proteins generally[266,268]. It may be worthy of note, in connection with attempts at more detailed interpretations, that the patterns for the *b*-555 and *b*-563 molecules bear a roughly mirror-image relationship to those for *Pseudomonas* cytochrome *c*-551 (Fig. 11).

Cytochrome b-562

A water-soluble preparation designated as cytochrome *b*-562 has been obtained from *Escherichia coli* by Itagaki and Hager[269]. The oxidation–reduction potential at pH 7 is 0.11 V, the highest value thus far recorded for a *b*-type cytochrome. The positions of the principal absorption bands in the Soret and visible regions are indicated in Table II.

The prosthetic group can be removed by means of the conventional acid–acetone procedure, and, when protoheme is readded to the apoprotein in a ratio of 1:1, a product with properties similar to those of the original *b*-562 molecule is obtained[270]. The kinetics of the recombination process are complex, and suggest that several intermediates are involved. The rate-limiting step becomes more rapid with increase in the pH over the range of pH 4–9. Treatment of the apoprotein with acetic anhydride or diazotized sulfanilic acid yields, in each instance, a derivative which can still

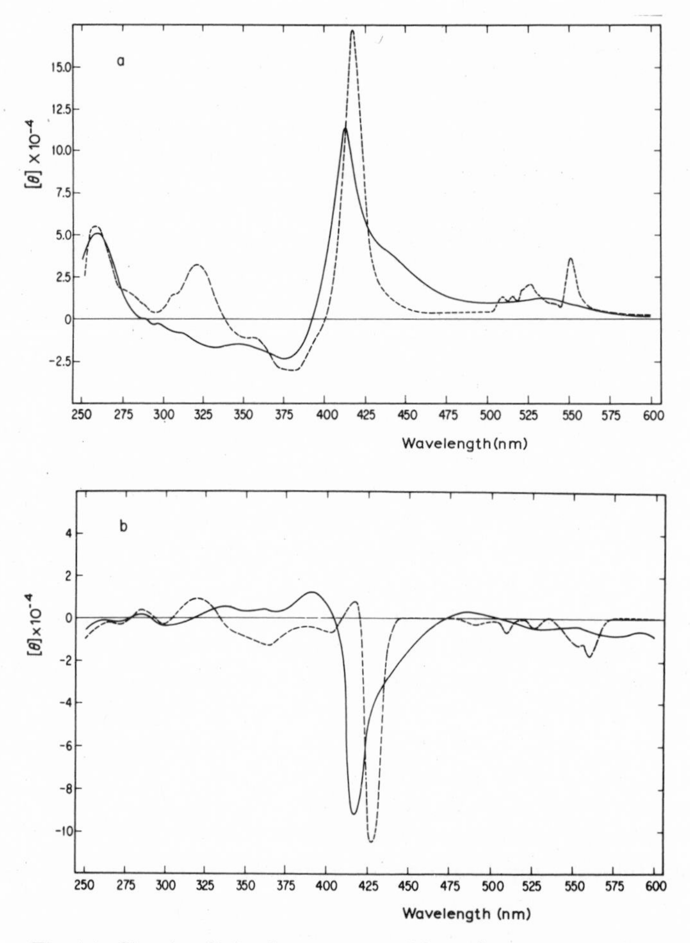

Fig. 11. Circular dichroism spectra of *b*- and *c*-type cytochromes. (a) cytochrome *c*-551 from *Pseudomonas aeruginosa*[140,231]; (b) cytochrome *b*-555 from the larvae of the housefly *Musca domestica* L.[266]. ——, oxidized form; -----, reduced form.

combine with heme, but the oxidized form of the product so obtained has spectral properties which differ markedly from those of the oxidized form of the unmodified molecule. When such complexes are reduced, however, the spectra of the modified and unmodified systems become similar to one another. Treatment of the apoprotein with iodine has a more far-reaching effect, and completely inhibits the reconstitution reaction.

The amino acid sequence[271] is given in Fig. 12. There are similarities to the sequences of hemoglobin and myoglobin, especially in the vicinity of the two histidine residues of the cytochrome, and it has been suggested that these may be the groups coordinated to the heme iron atom.

```
Ala·Asp·Leu·Glu·Asp·Asp·Met·Gln·Thr·Leu·Asn·Asp·Asn·Leu·Lys·Val·Ile·Glu·Lys·
                5                   10                  15

Ala·(Asx,Asx,Glx)·Lys·Ala·Asn·Asp·Ala·Ala·Gln·Val·Lys·Leu·Lys·Met·Arg·Ala·
 20                    25                  30                  35

Ala·Ala·Leu·Asn·Ala·Gln·Lys·Lys·Ala·Thr·Pro·Pro·Lys·Leu·Glu·Asp·Lys·Ser·Pro·
    40                  45                  50                  55

Asn·Ser·Gln·Pro·Met·Lys·Asp·Phe·Arg·His·Gly·Phe·Asp·Ile·Leu·Val·Gly·Glu·Ile·
            60                  65                  70                  75

Asp·Asp·Ala·Leu·Lys·Leu·Ala·Asn·Glu·Gly·Lys·Val·Lys·Glu·Ala·Gln·Ala·Ala·Glu·
                80                  85                  90

Ala·Gln·Leu·Lys·Thr·Thr·Arg·Asn·Ala·Tyr·Lys·His·Gln·Lys·Tyr·Arg
 95                 100                 105                 110
```

Fig. 12. Amino acid sequence of cytochrome *b*-562 from *Escherichia coli*[271].

Cytochrome b_2

A *b*-type molecule which has received considerable attention is that known as cytochrome b_2, obtained from yeast. This protein differs markedly from the other cytochromes of the *b*-type in that it contains both protoheme and flavin mononucleotide (FMN) as a prosthetic group. The system serves as an electron-transfer agent and also as the lactate dehydrogenase of yeast, thus combining the functions of its mammalian counterparts, L-lactate dehydrogenase and cytochrome *b*, in a single catalytic entity. Ferricytochrome *c* acts as the electron acceptor. Some isolation procedures result in a product with about 15 deoxyribonucleotide units per heme group[272]. However, this DNA can be removed by electrophoresis[273] and appears to be without effect on the catalytic and spectral properties of the preparation.

The molecular weight of the enzyme is about 235,000, and there are four subunits, each containing an FMN group as well as a heme group[274-276]. Upon the addition of urea or guanidine hydrochloride to the system, smaller units with weights of 17,000 and 34,000 are obtained[277].

Of the 10 cysteine residues per heme group, 5 can be titrated with *p*-chloromercuribenzene sulfonate in the case of the oxidized form of the enzyme, and 6 if the molecule is in the reduced form[278]. Such titration yields a derivative which slowly loses FMN and activity, presumably as a result of a conformation change. The catalytic activity can be partly restored by the addition of an excess of FMN to the product in the presence of lactate, but the reconstituted enzyme so obtained is less stable than the original molecule.

The FMN can also be removed by treatment of the protein with ammonium sulfate in acid solution[278-280]. The flavin-free preparation obtained in this manner has spectral properties similar to those of cytochrome b_2, except for the absence of the bands contributed by the flavin moiety. Upon readdition of FMN to this derivative, greater enzymatic activity is observed when cytochrome *c* serves as the electron acceptor than when ferricyanide is substituted in this capacity. This has been taken to suggest that the binding of cytochrome *c* may help to stabilize the reconstituted preparation[279,280], and also has been cited as support for the view that the reduction of cytochrome *c* and ferricyanide involves different mechanisms[281,282].

The flavin-free derivative obtained through the action of mercurials bears a number of similarities to the "mammalian-type" cytochromes *c*[280,283]: (i) it is active in the cytochrome *c* reductase and oxidase systems, and is no longer autoxidizable; (ii) the absorption spectra are similar; (iii) the oxidation–reduction potential at pH 7 is 0.28 V, as compared to 0.25 V for the cytochromes *c* and 0.12 V for cytochrome b_2; and (iv) the heme group cannot be removed by treatment with acid–acetone. In view of these properties, it has been proposed that a cytochrome *c*-like molecule is created by the reaction of one or more sulfhydryl groups with the vinyl side chain(s) of the porphyrin ring to form one or two thioether linkages between the protein and the prosthetic group.

To convert cytochrome b_2 to a heme-free flavoprotein, mild conditions have to be used in order to protect the flavin moiety. In a procedure developed by Morton, the heme is removed with cold acetone in the presence of cyanide, and the product then stabilized by the addition of lactate[280]. The flavin group of the preparation obtained in this fashion, unlike that of the intact molecule, is fluorescent. The product is catalytically inactive, and remains so upon readdition of heme to the solution.

A heme peptide with a molecular weight of about 11,000 per heme group, and which displays absorption spectra in the visible region similar to those of the parent molecule, can be obtained by treatment of cytochrome b_2 with trypsin[284,285]. The heme group of this derivative can be removed with acid–acetone to yield a product which displays tyrosine and tryptophan fluorescence. Upon the addition of an equivalent of protoheme, this fluorescence is quenched as the heme peptide is reconstituted[285]. The $E_{m,7}$ value is only 0.03 V more negative than that of the intact protein[284]. It may thus be that the environment about the heme group in cytochrome b_2 is largely conserved in the heme peptide derivative, and further studies of this fragment could be most useful.

The absorption spectra of cytochrome b_2 are those of a low-spin heme protein, with additional contributions attributable to the flavin moiety. The principal bands are at 556.5, 528, 424, 330, and 265 nm for the reduced form, and at 530–560 (broad band), 413, 359, and 265 nm in the case of the oxidized form[286]. Optical rotatory dispersion and circular dichroism data indicate an α-helix content of 15–40%; the curves in the wavelength range from 200–250 nm are the same for the oxidized and reduced forms of the molecule, and no change is seen upon removal of the prosthetic group components[267,287-289]. The patterns in the region from 250–350 nm are affected by FMN binding. A negative dichroic band at 270 nm, which is not obtained in the case of the flavin-free derivative[267], is attributed to interaction of the FMN with aromatic groups of the protein. Upon reduction of the molecule with lactate, the dispersion spectrum in the visible range includes a Cotton effect in the α-band region, but such a Cotton effect has not been found upon reduction of the protein with hydrogen or thiols[288]; the significance of this is uncertain at present. The most striking observations are those pertaining to the Soret region. The Cotton effect in this range is essentially inverted upon removal of the flavin moiety[287,288], and can be restored to its original form by the readdition of FMN. Whether this is indicative of a direct interaction between the flavin and the heme group, as suggested by Sturtevant and Tsong[288], or an indirect effect involving the induction of a conformational change upon FMN binding, as postulated by Iwatsubo and DiFranco[287], remains to be determined.

It is thought that, in the course of functional activity, electron transfer proceeds from the lactate to the flavin to the heme group[281,282,290]. Not surprisingly, the interplay of the FMN with the heme group results in a significant degree of flavin semiquinone formation[281,282,290,291]. Little is known, however, concerning the details of the various transfer steps, especially with respect to those involved in the reoxidation of the molecule. Indeed, it is not known definitely at this time whether such reactions proceed in each case via the heme group. Although the electron transfer to ferricytochrome *c* undoubtedly occurs in this manner, it has been postulated that when ferricyanide is the acceptor molecule, the transfer sequence becomes one in which the heme group is bypassed. This suggestion is based on the observation that ferrocyanide can serve to lower the rate of heme reduction without exerting a corresponding effect on the rate of flavin reduction or the overall turnover number of the enzyme[290]. In view of findings such as this, the possibility has been considered that the electron transfer between the flavin and the heme group proceeds relatively slowly for an intramolecular electron-transfer process[282,290]. However, other evidence has been taken to indicate that this reaction is a rapid one[281], and the matter remains open to a number of uncertainties at this time.

References pp. 947–954

Microsomal suspensions reduced in the presence of carbon monoxide yield a difference spectrum with a major absorption band near 450 nm[292]. Representatives of the class of heme proteins responsible for this band, the cytochromes P-450[293], have also been found in adrenal mitochondria and in bacteria. These cytochromes serve as catalysts of mixed-function oxidase reactions of the type[294]:

$$RH + DH + H^+ + O_2^* \rightarrow RO^*H + D^+ + H_2O \qquad (2)$$

where DH is an appropriate electron donor such as NADPH. Most of the investigations to date have focused on the microsomal members of this

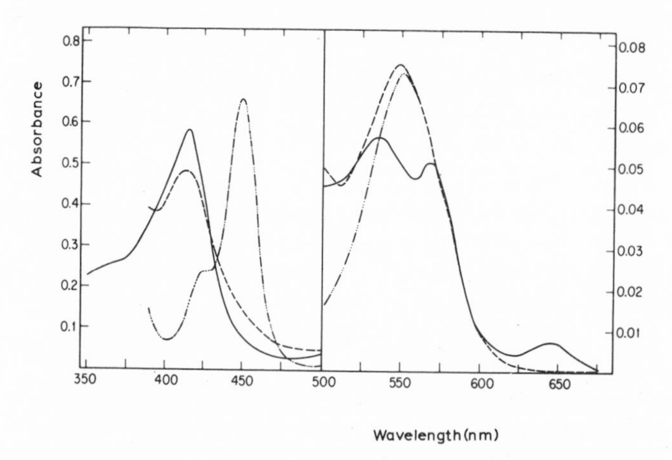

Fig. 13. Absorption spectra of P-450 preparation from rabbit liver microsomes. ——, oxidized form; -----, reduced form; — · · ·, reduced form in the presence of carbon monoxide. (From the data of Nishibayashi and Sato[298]).

family, which have generally been studied in the form of suspensions. Only recently has a satisfactory solubilized product been obtained[295–297].

Figure 13 indicates the Soret and visible absorption spectra for a particulate preparation from rabbit liver microsomes. At neutral pH, the oxidized and reduced forms yield Soret bands at 415 and 412 nm, respectively, while the carbon monoxide complex of the reduced species has a band at 449 nm, with a shoulder at 424 nm. As is evident from the curves in the visible region (Fig. 13), the reduced protein in the absence of carbon monoxide is in the high-spin state at room temperature. On the other hand, the oxidized preparation exhibits both low- and high-spin features, with the low-spin character predominating under the conditions indicated.

Addition of substrates to the oxidized form affects the absorption spectrum in one of two ways, depending on the nature of the substrate[299,300]. Hexobarbital and 3-methylcholanthrene, among others, induce a "Type I" change (Fig. 14), in which there is an increase in the absorbance at 390 nm and a decrease at 420 nm. The magnitude of the change is half-maximal at the concentrations of substrate required for half-maximal change in the rate of microsomal metabolism of the latter. A nearly opposite change, with a decrease in absorbance at 390 nm and an increase at 430 nm ("Type II" effect), is obtained when substances such as aniline are added to the P-450. In this case, greater than stoichiometric quantities of substrate are required for the full transition to be observed. The shift to 390 nm obtained with Type I substances is accompanied by changes in the visible region which indicate an increase in the high-spin character of the system, while the shift to longer wavelengths is attended by changes indicative of an increase in low-spin character.

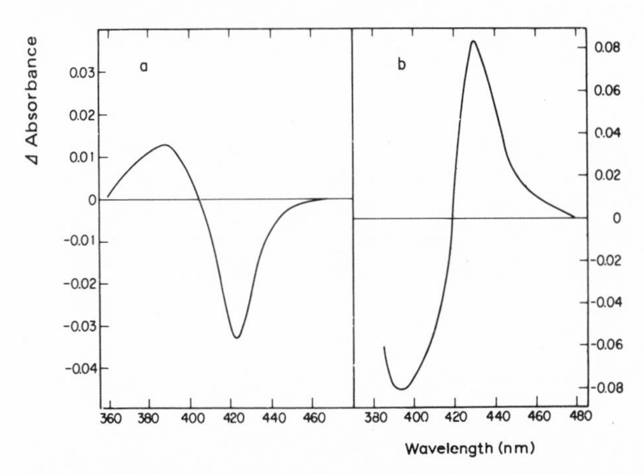

Fig. 14. Changes in absorption spectra upon addition of substrates to rat liver microsomes. (a) "Type I" change, induced by hexobarbital; (b) "Type II" change, effected by aniline. (From the data of Remmer *et al.*[299]).

A variety of agents, including urea, guanidine hydrochloride, detergents, and phospholipases, serve to convert the P-450 to an inactive form, P-420, which yields a carbon monoxide complex with a Soret maximum at 420 nm[293,301–303]. High concentrations of neutral salts and changes in the pH to either side of neutrality can also bring about this conversion[301]. The P-420 displays low-spin spectra characteristic of the *b*-type cytochromes.

The changes in properties involved in the P-450 to P-420 conversion, and the nature of the substances which bring about this transition[293,301], have led to the reasonable inference that the heme group of P-450 is bound in a region which is strongly hydrophobic. Less is known concerning the nature of the groups coordinated to the iron atom. The oxidized form of P-450 yields an electron paramagnetic resonance spectrum with features characteristic of low-spin ferriheme systems[302,304], and the g values which have been obtained (g = 1.91, 2.25, and 2.40) are similar to ones observed in model studies of complexes of thiols with ferriheme, ferrihemoglobin, and ferrimyoglobin[305,306]. In view of the fact that the low-spin character and enzymatic activity are lost when sulfhydryl reagents such as *p*-chloromercuribenzoate are allowed to act on the molecule, it is generally assumed that, as first suggested by Mason *et al.*[302], a mercaptide group of a cysteine residue is a ligand in P-450. Although the evidence is not conclusive (*e.g.*, comparable *g* values have been obtained with heme systems not involving mercaptide coordination, and the effects of the sulfhydryl reagents could be the result of a conformation change which affects the protein–heme interaction secondarily), it seems likely that this interpretation will turn out to be a correct one. Little information is available, however, concerning the group in the *trans* coordination position. It has been proposed, primarily on the basis of the similarity of the electron paramagnetic resonance data for oxidized P-450 and the thiol–ferrimyoglobin complexes, that the iron of the former is bound to the imidazole group of a histidine residue to yield a structure in which the available coordination positions are taken by a sulfur and a nitrogen atom, respectively[305]. However, the electron paramagnetic resonance spectrum of the oxidized form of P-450 closely resembles that of oxidized P-420, even though the group in the *trans* position is thought to differ in the two cases[307]. It is apparent, therefore, that firm conclusions cannot be drawn until additional data are at hand.

In the reduced state, P-450 is a high-spin complex, and it has been proposed that, in this form of the molecule, the fifth and sixth coordination positions are vacant or taken by groups which are bound to the iron only loosely[306]. Such a structure is in sharp contrast to that thought to hold for P-420 in the reduced state. As in the case of oxidized P-420, the latter is a low-spin species with what are assumed to be two strong-field groups coordinated to the iron.

In studies of the absorption spectra of complexes of ferroheme with ethyl isocyanide[308,309] and with carbon monoxide and pyridine[310], data have been obtained which indicate that, under conditions favoring aggregation, such model systems yield Soret bands near that of the carbon monoxide complex of P-450. These results have prompted suggestions that some of the characteristic spectral properties of P-450 are a consequence of heme–heme interaction in the protein molecule, but this seems unlikely on a number of grounds, including the observation that the preparation from *Pseudomonas putida* contains only one heme group per molecular weight of approximately 40,000[311]. A more plausible view[305-307] is that the spectral behavior of P-450 is attributable to changes in the heme–ligand group distances and/or the composition of the coordination complex. In an analysis advanced by Hill *et al.*[306], it is postulated that the increase in high-spin character observed upon the addition of Type I substances to oxidized P-450 reflects a conformationally induced change in which the distance between the iron and the ligand in the position opposite the mercaptide group is lengthened, with the metal assuming a position out-of-plane with respect to the porphyrin, towards the coordinated sulfur atom. Type II substances, on the other hand, are thought to become bound directly to the iron, through displacement of the coordinated protein group in the position opposite the cysteine residue, to yield a complex with enhanced low-spin character.

Jefcoate and Gaylor[305] and Tsai *et al.*[307], in another interpretation of the effects of Type I substances on the oxidized form of P-450, also invoke conformationally governed changes in the nature of the coordination complex, but conclude that the metal–sulfur distance increases rather than diminishes in the presence of such agents. On the basis of the currently available data, this would appear to be the more likely possibility of the two.

In the case of the reduced form of the molecule, it has been proposed that complex formation with carbon monoxide occurs at the position taken by the mercaptide group in the oxidized species[306]. The resulting structure is pictured as one in which the metal is out-of-plane towards the carbon monoxide, with either no group or a weakly linked ligand in the *trans* position. Presumably, in the course of the normal enzymatic activity of the system, oxygen would be bound to the heme in this manner.

As work on the microsomal and mitochondrial systems progresses, important benefits are beginning to accrue from studies of P-450 preparations from prokaryotic organisms. Recently, Gunsalus and his colleagues[311], working with *Pseudomonas putida* grown on camphor, have obtained a readily soluble protein, referred to as P-450_{cam}, which yields spectra similar to those of the P-450 systems of eukaryotic cells. Electron paramagnetic resonance studies of this preparation have been conducted in the presence and absence of the substrate, (+)-camphor[307]. The g values of 1.91, 2.26, and 2.45 found for the oxidized form in the absence of substrate agree well with those exhibited by the microsomal preparations. Only below 20°K does a high-spin contribution begin to emerge ($< 7\%$; $g = 1.8, 4, 8$). Upon the addition of camphor, a marked increase in the degree of high-spin character, and a corresponding diminution of the low-spin signal, are observed. These results provide further support for the interpretations arrived at on the basis of the absorption spectra of the various preparations.

In summary, the information currently available for this group of cytochromes has led to the following inferences: (i) the heme group of P-450 is located in a hydrophobic environment; (ii) at neutral pH and in the absence of substrates, oxidized P-450 exists predominantly in the low-spin state; (iii) in this low-spin state, a mercaptide group of a cysteine residue is bound to the iron; (iv) the remaining coordination position of the oxidized form is occupied by some other strong-field group of the protein, perhaps the imidazole group of a histidine residue; (v) in the

References pp. 947–954

presence of Type I substrates, a conformationally induced transition results in conversion of the oxidized form to a high-spin species; (vi) the reduced form of P-450 is also a high-spin species, whereas the oxidized and reduced forms of P-420 are both low-spin systems.

NOTE ADDED IN PROOF

X-ray diffraction data have become available for cytochrome *c* in the reduced form (tuna)[312]. As anticipated, the structure under these conditions differs significantly from that of the oxidized species. Some of the more striking changes occur in the vicinity of methionine residue 80. Upon reduction of the molecule, the side chain of phenylalanine 82, which in ferricytochrome *c* projects into the medium, moves inward to a position near the heme group, while isoleucine 81 undergoes a change in the opposite direction, from a position in which its side chain is arranged along the surface to one where it is extended into the surrounding solution. These shifts are accompanied by a rearrangement of some of the residues following histidine 18, with glutamate 21 moving towards phenylalanine 10 and tyrosine 97. Still other changes are seen in the orientation of tryptophan 59 and tyrosine residues 67 and 74.

Examination of the data for ferricytochrome *c* in the light of the results for the reduced form has prompted a revision[312] (R. E. Dickerson, personal communication) of the structure advanced earlier for the oxidized species[73]. It is thought that the oxidized as well as the reduced form of the molecule contains two regions of α-helix (residues 1–12 and 89–102), instead of the single such segment (residues 92–102) postulated previously. Other modifications include an altered deployment of arginine 91 and changes in the arrangement of some of the aromatic groups, so that, for example, the planes of the side chains of tyrosine 74 and tryptophan 59 are less nearly parallel to one another than they were before.

X-ray diffraction data have also become available for a *b*-type cytochrome[313].

ACKNOWLEDGEMENT

We are indebted to Drs. D. Boulter, R. E. Dickerson, J. P. Riehm and L. Smith for permission to cite unpublished material. Unpublished data from the authors' laboratory were obtained in the course of investigations supported by grants from the National Institutes of Health (GM-15379) and the National Science Foundation (GB-6995).

REFERENCES

1 C. A. MacMunn, *Phil. Trans. R. Soc.*, 177 (1886) 267.
2 C. A. MacMunn, *J. Physiol. London*, 8 (1887) 57.
3 C. A. MacMunn, *Spectrum Analysis Applied to Biology and Medicine*, Longmans, Green and Co., London, 1914.
4 D. Keilin, *Proc. R. Soc. (B)*, 98 (1925) 312.
5 D. Keilin, *The History of Cell Respiration and Cytochrome*, Cambridge University Press, New York, 1966.
6 D. Keilin and E. F. Hartree, *Proc. R. Soc. (B)*, 127 (1939) 167.
7 E. Yakushiji and K. Okunuki, *Proc. Imp. Acad. Tokyo*, 16 (1940) 299.
8 M. Florkin and E. H. Stotz, (Eds.), *Comprehensive Biochemistry*, Vol. 13, Elsevier, Amsterdam, 2nd edn., 1965, p. 18.
9 A. L. Lehninger, *The Mitochondrion*, W. A. Benjamin, New York, 1965.
10 D. E. Green and H. Baum, *Energy and the Mitochondrion*, Academic Press, New York, 1969.
11 E. Racker, *Membranes of Mitochondria and Chloroplasts*, Van Nostrand, Reinhold, New York, 1970.
12 M. D. Kamen, *Proc. Int. Symp. Enzyme Chem. Japan*, Maruzen, Tokyo, 1958, p. 245.
13 W. M. Clark, *Oxidation–Reduction Potentials of Organic Systems*, Williams and Wilkins, Baltimore, 1960.
14 E. Margoliash, *Brookhaven Symp. Biol.*, 15 (1962) 266.
15 E. Margoliash and A. Schejter, *Adv. Protein Chem.*, 21 (1966) 113.
16 R. Hill and D. Keilin, *Proc. R. Soc. (B)*, 107 (1930) 286.
17 H. Theorell, *Enzymologia*, 4 (1937) 192.
18 H. Theorell, *Biochem. Z.*, 298 (1938) 242.
19 K. Zeile and H. Meyer, *Naturwissenschaften*, 27 (1939) 598.
20 K. G. Paul, *Acta Chem. Scand.*, 5 (1951) 389.
21 M. O. Dayhoff, (Ed.), *Atlas of Protein Sequence and Structure*, National Biomedical Research Foundation, Silver Spring, 1969, p. D-7.
22 H. A. Sober, (Ed.), *Handbook of Biochemistry*, Chemical Rubber Company, Cleveland, 1968, p. C-156.
23 E. Margoliash, E. L. Smith, G. Kreil and H. Tuppy, *Nature*, 192 (1961) 1125.
24 H. Matsubara and E. L. Smith, *J. Biol. Chem.*, 238 (1963) 2732.
25 S. B. Needleman, unpublished experiments.
26 J. A. Rothfus and E. L. Smith, *J. Biol. Chem.*, 240 (1965) 4277.
27 O. F. Walasek and E. Margoliash, unpublished experiments.
28 T. Nakashima, H. Higa, H. Matsubara, A. M. Benson and K. T. Yasunobu, *J. Biol. Chem.*, 241 (1966) 1166.
29 J. W. Stewart and E. Margoliash, *Can. J. Biochem.*, 43 (1965) 1187.
30 S. K. Chan, S. B. Needleman, J. W. Stewart and E. Margoliash, unpublished experiments.
31 M. A. McDowell and E. L. Smith, *J. Biol. Chem.*, 240 (1965) 4635.
32 S. B. Needleman and E. Margoliash, *J. Biol. Chem.*, 241 (1966) 853.
33 A. Goldstone and E. L. Smith, *J. Biol. Chem.*, 241 (1966) 4480.
34 C. Nolan and E. Margoliash, *J. Biol. Chem.*, 241 (1966) 1049.
35 S. K. Chan and E. Margoliash, *J. Biol. Chem.*, 241 (1966) 507.
36 S. K. Chan, I. Tulloss and E. Margoliash, unpublished experiments.
37 O. P. Bahl and E. L. Smith, *J. Biol. Chem.*, 240 (1965) 3585.
38 S. K. Chan, I. Tulloss and E. Margoliash, *Biochemistry*, 5 (1966) 2586.

39 S. K. Chan, O. F. Walasek, G. H. Barlow and E. Margoliash, *Federation Proc.*, 26 (1967) 723.
40 S. K. Chan, O. F. Walasek, G. H. Barlow and E. Margoliash, unpublished experiments.
41 G. Kreil, *Z. Physiol. Chem.*, 334 (1963) 154.
42 G. Kreil, *Z. Physiol. Chem.*, 340 (1965) 86.
43 A. Goldstone and E. L. Smith, *J. Biol. Chem.*, 242 (1967) 4702.
44 C. Nolan, T. Uzzell, L. J. Weiss, W. M. Fitch and E. Margoliash, unpublished experiments.
45 S. K. Chan and E. Margoliash, *J. Biol. Chem.*, 241 (1966) 335.
46 S. K. Chan, unpublished experiments.
47 C. Nolan, L. J. Weiss, J. J. Adams and E. Margoliash, unpublished experiments.
48 S. K. Chan, I. Tulloss and E. Margoliash, unpublished experiments.
49 J. Heller and E. L. Smith, *Proc. Natn. Acad. Sci. U.S.*, 54 (1965) 1621.
50 Y. Yaoi, K. Titani and K. Narita, *J. Biochem. Tokyo*, 59 (1966) 247.
51 K. Narita and K. Titani, *Proc. Japan Acad.*, 41 (1965) 831.
52 F. Stevens, A. N. Glazer and E. L. Smith, *J. Biol. Chem.*, 242 (1967) 2764.
53 W. M. Fitch and E. Margoliash, *Biochem. Genet.*, 1 (1967) 65.
54 E. Margoliash, W. M. Fitch and R. E. Dickerson, *Brookhaven Symp. Biol.*, 21 (1968) 259.
55 E. L. Smith and E. Margoliash, *Federation Proc.*, 23 (1964) 1243.
56 E. Margoliash and E. L. Smith, in V. Bryson and H. J. Vogel, *Evolving Genes and Proteins*, Academic Press, New York, 1965, p. 221.
57 E. Margoliash, in B. Chance, R. W. Estabrook and T. Yonetani, *Hemes and Hemoproteins*, Academic Press, New York, 1966, p. 371.
58 E. L. Smith, in K. Okunuki, M. D. Kamen and I. Sekuzu, *Structure and Function of Cytochromes*, University of Tokyo Press, Tokyo, 1968, p. 282.
59 H. Theorell and Å. Åkesson, *J. Am. Chem. Soc.*, 63 (1941) 1804.
60 G. H. Barlow and E. Margoliash, *J. Biol. Chem.*, 241 (1966) 1473.
61 H. Theorell and Å. Åkesson, *J. Am. Chem. Soc.*, 63 (1941) 1812.
62 H. Theorell and Å. Åkesson, *J. Am. Chem. Soc.*, 63 (1941) 1818.
63 H. Theorell, *J. Am. Chem. Soc.*, 63 (1941) 1820.
64 D. Keilin, *Proc. R. Soc. (B)*, 106 (1930) 418.
65 M. Dixon, R. Hill and D. Keilin, *Proc. R. Soc. (B)*, 109 (1931) 29.
66 R. Lemberg and J. W. Legge, *Hematin Compounds and Bile Pigments*, Interscience, New York, 1949, p. 164.
67 E. Boeri, A. Ehrenberg, K. G. Paul and H. Theorell, *Biochim. Biophys. Acta*, 12 (1953) 273.
68 S. Paléus, A. Ehrenberg and H. Tuppy, *Acta Chem. Scand.*, 9 (1955) 365.
69 E. Margoliash, N. Frohwirt and E. Wiener, *Biochem. J.*, 71 (1959) 559.
70 H. A. Harbury and P. A. Loach, *Proc. Natn. Acad. Sci. U.S.*, 45 (1959) 1344.
71 H. A. Harbury and P. A. Loach, *J. Biol. Chem.*, 235 (1960) 3640.
72 H. A. Harbury and P. A. Loach, *J. Biol. Chem.*, 235 (1960) 3646.
73 R. E. Dickerson, T. Takano, D. Eisenberg, O. B. Kallai, L. Samson, A. Cooper and E. Margoliash, *J. Biol. Chem.*, in press.
74 R. P. Ambler, *Biochem. J.*, 89 (1963) 341.
75 R. P. Ambler, *Biochem. J.*, 89 (1963) 349.
76 R. A. Morton, J. Overnell and H. A. Harbury, *J. Biol. Chem.*, 245 (1970) 4653.
77 T. Horio, *J. Biochem. Tokyo*, 45 (1958) 267.
78 T. Horio, T. Higashi, M. Sasagawa, *et al.*, *Biochem. J.*, 77 (1960) 194.
79 F. L. Rodkey and E. G. Ball, *J. Biol. Chem.*, 182 (1950) 17.
80 H. A. Harbury, J. R. Cronin, M. W. Fanger, *et al.*, *Proc. Natn. Acad. Sci. U.S.*, 54 (1965) 1658.

81 S. N. Vinogradov and H. A. Harbury, *Biochemistry*, 6 (1967) 709.
82 J. W. Stewart, E. Margoliash and F. Sherman, *Federation Proc.*, 25 (1966) 647.
83 A. Schejter and P. George, *Nature*, 206 (1965) 1150.
84 P. George, S. C. Glauser and A. Schejter, *Proc. 5th Int. Conf. Biochemistry, Moscow*, 4 (1961) 192.
85 P. George, S. C. Glauser and A. Schejter, *J. Biol. Chem.*, 242 (1967) 1690.
86 T. P. Hettinger, M. W. Fanger, S. N. Vinogradov and H. A. Harbury, *Federation Proc.*, 25 (1966) 648.
87 H. A. Harbury, in B. Chance, R. W. Estabrook and T. Yonetani, *Hemes and Hemoproteins*, Academic Press, New York, 1966, p. 391.
88 T. P. Hettinger, *Ph.D. Thesis*, Yale University, New Haven, 1966.
89 E. Stellwagen, *Biochemistry*, 7 (1968) 2496.
90 K. Ando, H. Matsubara and K. Okunuki, *Proc. Japan Acad.*, 41 (1965) 79.
91 H. Matsubara, K. Ando and K. Okunuki, *Proc. Japan Acad.*, 41 (1965) 408.
92 K. Ando, H. Matsubara and K. Okunuki, *Biochim. Biophys. Acta*, 118 (1966) 240.
93 K. Ando, H. Matsubara and K. Okunuki, *Biochim. Biophys. Acta*, 118 (1966) 256.
94 H. J. Tsai and G. R. Williams, *Can. J. Biochem.*, 43 (1965) 1409.
95 H. J. Tsai, H. Tsai and G. R. Williams, *Can. J. Biochem.*, 43 (1965) 1995.
96 M. W. Fanger, T. P. Hettinger and H. A. Harbury, *Biochemistry*, 6 (1967) 713.
97 H. A. Harbury, Y. P. Myer and P. A. Loach, *Abstr. 141st National Meeting Am. Chem. Soc.*, Washington, 1962, p. 32C.
98 J. Shack and W. M. Clark, *J. Biol. Chem.*, 171 (1947) 143.
99 R. E. Dickerson, M. L. Kopka, J. E. Weinzierl, J. Varnum, D. Eisenberg and E. Margoliash, *J. Biol. Chem.*, 242 (1967) 3015.
100 R. E. Dickerson, M. L. Kopka, J. E. Weinzierl, *et al.*, in K. Okunuki, M. D. Kamen and I. Sekuzu, *Structure and Function of Cytochromes*, University of Tokyo Press, Tokyo, 1968, p. 225.
101 C. C. McDonald, W. D. Phillips and S. N. Vinogradov, *Biochem. Biophys. Res. Commun.*, 36 (1969) 442.
102 Y. P. Myer, A. J. Murphy and H. A. Harbury, *J. Biol. Chem.*, 241 (1966) 5370.
103 A. J. Murphy and H. A. Harbury, *Proc. 7th Int. Congr. Biochemistry*, Tokyo, 1967, A111.
104 M. W. Fanger and H. A. Harbury, *Biochemistry*, 4 (1965) 2541.
105 Y. P. Myer and H. A. Harbury, *Proc. Natn. Acad. Sci. U.S.*, 54 (1965) 1391.
106 A. F. Esser, J. Overnell and H. A. Harbury, unpublished experiments.
107 H. Theorell, *Adv. Enzymol.*, 7 (1947) 265.
108 P. George and R. L. J. Lyster, *Proc. Natn. Acad. Sci. U.S.*, 44 (1958) 1013.
109 E. Stellwagen, *J. Biol. Chem.*, 242 (1967) 602.
110 Y. Orii, C. Montagner and H. A. Harbury, unpublished experiments.
111 J. R. Cronin and H. A. Harbury, *Biochem. Biophys. Res. Commun.*, 20 (1965) 503.
112 J. R. Cronin, K. M. Ivanetich, J. R. Maynard and H. A. Harbury, unpublished experiments.
113 K. Wada and K. Okunuki, *J. Biochem. Tokyo*, 64 (1968) 667.
114 W. A. Eaton and R. M. Hochstrasser, *J. Chem. Phys.*, 46 (1967) 2533.
115 A. Schejter and P. George, *Biochemistry*, 3 (1964) 1045.
116 B. L. Horecker and A. Kornberg, *J. Biol. Chem.*, 165 (1946) 11.
117 B. L. Horecker and J. N. Stannard, *J. Biol. Chem.*, 172 (1948) 589.
118 A. Schejter and I. Aviram, *Biochemistry*, 8 (1969) 149.
119 E. Shechter and P. Saludjian, *Biopolymers*, 5 (1967) 788.
120 D. W. Urry, *Proc. Natn. Acad. Sci. U.S.*, 54 (1965) 640.

121 H. A. Harbury, Y. P. Myer, A. J. Murphy and S. N. Vinogradov, in B. Chance, R. W. Estabrook and T. Yonetani, *Hemes and Hemoproteins*, Academic Press, New York, 1966, p. 415.
122 R. Mirsky and P. George, *Proc. Natn. Acad. Sci. U.S.*, 56 (1966) 222.
123 P. Saludjian and E. Shechter, *Biopolymers*, 5 (1967) 561.
124 Y. P. Myer, *Biochemistry*, 7 (1968) 765.
125 K. Skov, T. Hofmann and G. R. Williams, *Can. J. Biochem.*, 47 (1969) 750.
126 A. Schejter and M. Sokolovsky, *FEBS Lett.*, 4 (1969) 269.
127 E. Stellwagen, *Biochemistry*, 3 (1964) 919.
128 J. A. Rupley, *Biochemistry*, 3 (1964) 1648.
129 T. Flatmark, *Acta Chem. Scand.*, 18 (1964) 1796.
130 I. Aviram and A. Schejter, *J. Biol. Chem.*, 244 (1969) 3773.
131 J. M. Armstrong, J. H. Coates and R. K. Morton, *Aust. J. Sci.*, 21 (1958) 119.
132 M. Nozaki, H. Mizushima, T. Horio and K. Okunuki, *J. Biochem. Tokyo*, 45 (1958) 815.
133 T. Yamanaka, H. Mizushima, M. Nozaki, T. Horio and K. Okunuki, *J. Biochem. Tokyo*, 46 (1959) 121.
134 P. George and C. L. Tsou, *Biochem. J.*, 50 (1952) 440.
135 D. W. Urry and P. Doty, *J. Am. Chem. Soc.*, 87 (1965) 2756.
136 D. D. Ulmer, *Biochemistry*, 4 (1965) 902.
137 Y. P. Myer, *J. Biol. Chem.*, 243 (1968) 2115.
138 T. Flatmark and A. B. Robinson, in K. Okunuki, M. D. Kamen and I. Sekuzu, *Structure and Function of Cytochromes*, University of Tokyo Press, Tokyo, 1968, p. 318.
139 R. Zand and S. Vinogradov, *Arch. Biochem. Biophys.*, 125 (1968) 94.
140 R. H. L. Marks and H. A. Harbury, unpublished experiments.
141 K. Zeile and F. Reuter, *Z. Physiol. Chem.*, 221 (1933) 101.
142 S. Paléus and J. B. Neilands, *Acta Chem. Scand.*, 4 (1950) 1024.
143 E. Margoliash, *Biochem. J.*, 56 (1954) 535.
144 J. H. P. Jonxis, *Biochem. J.*, 33 (1939) 1743.
145 M. Nozaki, T. Yamanaka, T. Horio and K. Okunuki, *J. Biochem. Tokyo*, 44 (1957) 453.
146 H. Mizushima, M. Nozaki, T. Horio and K. Okunuki, *J. Biochem. Tokyo*, 45 (1958) 845.
147 E. Margoliash, M. Reichlin and A. Nisonoff, *Science*, 158 (1967) 531.
148 D. D. Ulmer and J. H. R. Kägi, *Biochemistry*, 7 (1968) 2710.
149 J. H. R. Kägi and D. D. Ulmer, *Biochemistry*, 7 (1968) 2718.
150 D. D. Ulmer, *Biochemistry*, 5 (1966) 1886.
151 A. Kowalsky, *J. Biol. Chem.*, 237 (1962) 1807.
152 A. Kowalsky, *Biochemistry*, 4 (1965) 2382.
153 K. Wüthrich, *Proc. Natn. Acad. Sci. U.S.*, 63 (1969) 1071.
154 G. D. Watt and J. M. Sturtevant, *Biochemistry*, 8 (1969) 4567.
155 R. Margalit and A. Schejter, *FEBS Lett.*, 6 (1970) 278.
156 W. D. Butt and D. Keilin, *Proc. R. Soc. (B)*, 156 (1962) 429.
157 D. de Vault, J. H. Parkes and B. Chance, *Nature*, 215 (1967) 642.
158 B. Chance, *Biochem. J.*, 103 (1967) 1.
159 D. de Vault, in K. Okunuki, M. D. Kamen and I. Sekuzu, *Structure and Function of Cytochromes*, University of Tokyo Press, Tokyo, 1968, p. 488.
160 T. Yonetani and G. S. Ray, *J. Biol. Chem.*, 241 (1966) 700.
161 B. Chance, C. P. Lee, L. Mela and D. de Vault, in K. Okunuki, M. D. Kamen and I. Sekuzu, *Structure and Function of Cytochromes*, University of Tokyo Press, Tokyo, 1968, p. 475.
162 K. G. Brandt, P. C. Parks, G. H. Czerlinski and G. P. Hess, *J. Biol. Chem.*, 241 (1966) 4180.

163 B. H. Havsteen, *Acta Chem. Scand.*, 19 (1965) 1227.
164 Q. H. Gibson, C. Greenwood, D. C. Wharton and G. Palmer, *J. Biol. Chem.*, 240 (1965) 888.
165 C. Greenwood and G. Palmer, *J. Biol. Chem.*, 240 (1965) 3660.
166 K. Skov and G. R. Williams, in K. Okunuki, M. D. Kamen and I. Sekuzu, *Structure and Function of Cytochromes*, University of Tokyo Press, Tokyo, 1968, p. 349.
167 D. E. Green, J. Järnefelt and H. D. Tisdale, *Biochim. Biophys. Acta*, 31 (1959) 34.
168 I. Sekuzu, Y. Orii and K. Okunuki, *J. Biochem. Tokyo*, 48 (1960) 214.
169 Y. Orii and K. Okunuki, *A. Rep. Biol. Works, Faculty of Science, Osaka Univ.*, 17 (1969) 1.
170 R. Bomstein, R. Goldberger and H. Tisdale, *Biochim. Biophys. Acta*, 50 (1961) 527.
171 D. E. Green, J. Järnefelt and H. Tisdale, *Biochim. Biophys. Acta*, 38 (1960) 160.
172 M. Nishimura, *J. Biochem. Tokyo*, 46 (1959) 219.
173 J. A. Gross and J. J. Wolken, *Science*, 132 (1960) 357.
174 J. J. Wolken and J. A. Gross, *J. Protozool.*, 10 (1963) 189.
175 F. Perini, M. D. Kamen and J. A. Schiff, *Biochim. Biophys. Acta*, 88 (1964) 74.
176 F. Perini, J. A. Schiff and M. D. Kamen, *Biochim. Biophys. Acta*, 88 (1964) 91.
177 J. B. Neilands, *J. Biol. Chem.*, 197 (1952) 701.
178 T. Yamanaka, T. Nishimura and K. Okunuki, *J. Biochem. Tokyo*, 54 (1963) 161.
179 T. Yamanaka and K. Okunuki, *J. Biol. Chem.*, 239 (1964) 1813.
180 R. Hill and R. Scarisbrick, *New Phytologist*, 50 (1951) 98.
181 G. Forti, M. L. Bertole and G. Zanetti, *Biochim. Biophys. Acta*, 109 (1965) 33.
182 D. S. Bendall and R. Hill, *A. Rev. Plant Physiol.*, 19 (1968) 167.
183 E. Yakushiji, *Acta Phytochim.*, 8 (1935) 325.
184 S. Katoh, *J. Biochem. Tokyo*, 46 (1956) 629.
185 S. Katoh, *Plant Cell Physiol.*, 1 (1959) 29.
186 S. Katoh, *Plant Cell Physiol.*, 1 (1959) 91.
187 E. Yakushiji, Y. Sugimura, I. Sekuzu, I. Morikawa and K. Okunuki, *Nature*, 185 (1960) 105.
188 S. Katoh, *Nature*, 186 (1960) 138.
189 Y. Sugimura and E. Yakushiji, *J. Biochem. Tokyo*, 63 (1963) 261.
190 M. Nishimura and A. Takamiya, *Biochim. Biophys. Acta*, 120 (1966) 45.
191 M. D. Kamen and T. Horio, *A. Rev. Biochem.*, 39 (1970) 673.
192 R. G. Bartsch, *A. Rev. Microbiol.*, 22 (1968) 181.
193 P. A. Ketchum, H. K. Sanders, J. W. Gryder and A. Nason, *Biochim. Biophys. Acta*, 189 (1969) 360.
194 K. Dus, T. Flatmark, H. de Klerk and M. D. Kamen, *Biochemistry*, 9 (1970) 1984.
195 T. Flatmark, K. Dus, H. de Klerk and M. D. Kamen, *Biochemistry*, 9 (1970) 1991.
196 T. Horio and M. D. Kamen. *Biochim. Biophys. Acta*, 48 (1961) 266.
197 K. Sletten, K. Dus, H. de Klerk and M. D. Kamen, *J. Biol. Chem.*, 243 (1968) 5492.
198 K. Sletten and M. D. Kamen, in K. Okunuki, M. D. Kamen and I. Sekuzu, *Structure and Function of Cytochromes*, University of Tokyo Press, Tokyo, 1968, p. 422.
199 S. Morita, *Bot. Mag. Tokyo*, 79 (1966) 630.
200 H. de Klerk, R. G. Bartsch and M. D. Kamen, *Biochim. Biophys. Acta*, 97 (1965) 275.
201 S. Morita and S. F. Conti, *Arch. Biochem. Biophys.*, 100 (1963) 302.
202 J. R. Postgate, *J. Gen. Microbiol.*, 14 (1956) 545.
203 H. Drucker and L. L. Campbell, *J. Bact.*, 100 (1969) 358.

204 H. Drucker, E. B. Trousil, L. L. Campbell, G. H. Barlow and E. Margoliash, *Biochemistry*, 9 (1970) 1515.
205 J. le Gall, G. Mazza and N. Dragoni, *Biochim. Biophys. Acta*, 99 (1965) 385.
206 R. P. Ambler, *Biochem. J.*, 109 (1968) 47P.
207 H. Drucker, E. B. Trousil and L. L. Campbell, *Biochemistry*, 9 (1970) 3395.
208 A. Tissieres and R. H. Burris, *Biochim. Biophys. Acta*, 20 (1956) 436.
209 B. F. van Gelder, D. W. Urry and H. Beinert, in K. Okunuki, M. D. Kamen and I. Sekuzu, *Structure and Function of Cytochromes*, University of Tokyo Press, Tokyo, 1968, p. 335.
210 R. T. Swank and R. H. Burris, *Biochim. Biophys. Acta*, 180 (1969) 473.
211 K. Miki and K. Okunuki, *J. Biochem. Tokyo*, 66 (1969) 831.
212 K. Miki and K. Okunuki, *J. Biochem. Tokyo*, 66 (1969) 845.
213 G. D. Clark-Walker and J. Lascelles, *Arch. Biochem. Biophys.*, 136 (1970) 153.
214 T. E. Meyer, R. G. Bartsch, M. A. Cusanovich and J. H. Mathewson, *Biochim. Biophys. Acta*, 153 (1968) 854.
215 T. Kodama and S. Shidara, *J. Biochem. Tokyo*, 65 (1969) 351.
216 T. Fujita, *J. Biochem. Tokyo*, 60 (1966) 204.
217 J. Barrett and P. Sinclair, *Biochim. Biophys. Acta*, 143 (1967) 279.
218 R. G. Bartsch, T. E. Meyer and A. B. Robinson, in K. Okunuki, M. D. Kamen and I. Sekuzu, *Structure and Function of Cytochromes*, University of Tokyo Press, Tokyo, 1968, p. 443.
219 R. G. Bartsch and M. D. Kamen, *J. Biol. Chem.*, 235 (1960) 825.
220 H. Iwasaki and S. Shidara, *J. Biochem. Tokyo*, 66 (1969) 775.
221 J. A. Orlando, *Biochim. Biophys. Acta*, 57 (1962) 373.
222 T. Yamanaka and K. Okunuki, *J. Biochem. Tokyo*, 63 (1968) 341.
223 J. Gibson, *Biochem. J.*, 79 (1961) 151.
224 J. P. Kusel, J. R. Suriano and M. M. Weber, *Arch. Biochem. Biophys.*, 133 (1969) 293.
225 T. Kodama and T. Mori, *J. Biochem. Tokyo*, 65 (1969) 621.
226 M. A. Cusanovich and R. G. Bartsch, *Biochim. Biophys. Acta*, 189 (1969) 245.
227 R. G. Bartsch and M. D. Kamen, *J. Biol. Chem.*, 230 (1958) 41.
228 H. Iwasaki and S. Shidara, *Plant Cell Physiol.*, 10 (1969) 291.
229 K. Dus and K. Sletten, in K. Okunuki, M. D. Kamen and I. Sekuzu, *Structure and Function of Cytochromes*, University of Tokyo Press, Tokyo, 1968, p. 293.
230 K. Dus, K. Sletten and M. D. Kamen, *J. Biol. Chem.*, 243 (1968) 5507.
231 R. Zand and S. Vinogradov, *Biochem. Biophys. Res. Commun.*, 26 (1967) 121.
232 J. Kraut, S. Singh and R. A. Alden, in K. Okunuki, M. D. Kamen and I. Sekuzu, *Structure and Function of Cytochromes*, University of Tokyo Press, Tokyo, 1968, p. 252.
233 H. Drucker, L. L. Campbell and R. W. Woody, *Biochemistry*, 9 (1970) 1519.
234 M. Bruschi-Heriaud and J. le Gall, *Bull. Soc. Chim. Biol.*, 49 (1967) 753.
235 K. Dus, H. de Klerk, R. G. Bartsch, T. Horio and M. D. Kamen, *Proc. Natn. Acad. Sci. U.S.*, 57 (1967) 367.
236 L. P. Vernon and M. D. Kamen, *J. Biol. Chem.*, 211 (1954) 643.
237 K. Dus, R. G. Bartsch and M. D. Kamen, *J. Biol. Chem.*, 237 (1962) 3083.
238 F. C. Yong and T. E. King, *J. Biol. Chem.*, 245 (1970) 2457.
239 Y. Imai, K. Imai, K. Ikeda, K. Hamaguchi and T. Horio, *J. Biochem. Tokyo*, 65 (1969) 629.
240 Y. Imai, K. Imai, R. Sato and T. Horio, *J. Biochem. Tokyo*, 65 (1969) 225.
241 A. Ehrenberg and M. D. Kamen, *Biochim. Biophys. Acta*, 102 (1965) 333.
242 T. H. Moss, A. J. Bearden, R. G. Bartsch and M. A. Cusanovich, *Biochemistry*, 7 (1968) 1583.
243 R. Goldberger, A. L. Smith, H. Tisdale and R. Bomstein, *J. Biol. Chem.*, 236 (1961) 2788.

244 R. Goldberger, A. Pumphrey and A. Smith, *Biochim. Biophys. Acta,* 58 (1962) 307.
245 F. A. Holton and J. Colpa-Boonstra, *Biochem. J.,* 76 (1960) 179.
246 G. Hübscher, M. Kiese and R. Nicolas, *Biochem. Z.,* 325 (1954) 223.
247 T. Kajihara and B. Hagihara, in K. Okunuki, M. D. Kamen and I. Sekuzu, *Structure and Function of Cytochromes,* University of Tokyo Press, Tokyo, 1968, p. 581.
248 P. Strittmatter and J. Ozols, *J. Biol. Chem.,* 241 (1966) 4787.
249 R. Bois-Poltoratsky and A. Ehrenberg, *Eur. J. Biochem.,* 2 (1967) 361.
250 P. Strittmatter and S. F. Velick, *J. Biol. Chem.,* 221 (1956) 253.
251 S. F. Velick and P. Strittmatter, *J. Biol. Chem.,* 221 (1956) 265.
252 H. Yoshikawa, *J. Biochem. Tokyo,* 38 (1951) 1.
253 C. F. Strittmatter and E. G. Ball, *Proc. Natn. Acad. Sci. U.S.,* 38 (1952) 19.
254 J. Ozols and P. Strittmatter, *J. Biol. Chem.,* 239 (1964) 1018.
255 P. Strittmatter and J. Ozols, in B. Chance, R. W. Estabrook and T. Yonetani, *Hemes and Hemoproteins,* Academic Press, New York, 1966, p. 447.
256 P. Strittmatter, *J. Biol. Chem.,* 235 (1960) 2492.
257 J. Ozols and P. Strittmatter, *J. Biol. Chem.,* 241 (1966) 4793.
258 J. Ozols and P. Strittmatter, *Proc. Natn. Acad. Sci. U.S.,* 58 (1967) 264.
259 J. Ozols and P. Strittmatter, in K. Okunuki, M. D. Kamen and I. Sekuzu, *Structure and Function of Cytochromes,* University of Tokyo Press, Tokyo, 1968, p. 576.
260 A. Tsugita, M. Kobayashi, T. Kajihara and B. Hagihara, *J. Biochem. Tokyo,* 64 (1968) 727.
261 J. Ozols and P. Strittmatter, *J. Biol. Chem.,* 244 (1969) 6617.
262 J. Wyman, Jr., *Adv. Protein Chem.,* 4 (1948) 410.
263 F. S. Mathews and P. Strittmatter, *J. Mol. Biol.,* 41 (1969) 295.
264 R. H. Kretsinger, B. Hagihara and A. Tsugita, *Biochim. Biophys. Acta,* 200 (1970) 421.
265 Y. Okada and K. Okunuki, *J. Biochem. Tokyo,* 65 (1969) 581.
266 Y. Okada and K. Okunuki, *J. Biochem. Tokyo,* 67 (1970) 487.
267 J. M. Sturtevant and T. Y. Tsong, *J. Biol. Chem.,* 244 (1969) 4942.
268 Y. Okada and K. Okunuki, *J. Biochem. Tokyo,* 67 (1970) 603.
269 E. Itagaki and L. P. Hager, *J. Biol. Chem.,* 241 (1966) 3687.
270 E. Itagaki, G. Palmer and L. P. Hager, *J. Biol. Chem.,* 242 (1967) 2272.
271 E. Itagaki and L. P. Hager, *Biochem. Biophys. Res. Commun.,* 32 (1968) 1013.
272 C. A. Appleby and R. K. Morton, *Biochem. J.,* 75 (1960) 258.
273 R. K. Morton and K. Shepley, *Biochem. J.,* 89 (1963) 257.
274 C. Jacq and F. Lederer, *Eur. J. Biochem.,* 12 (1970) 154.
275 P. Pajot and O. Groudinsky, *Eur. J. Biochem.,* 12 (1970) 158.
276 C. Monteilhet and J. L. Risler, *Eur. J. Biochem.,* 12 (1970) 165.
277 A. Baudras and F. Labeyrie, in K. Okunuki, M. D. Kamen and I. Sekuzu, *Structure and Function of Cytochromes,* University of Tokyo Press, Tokyo, 1968, p. 601.
278 P. Pajot, K. S. Pell and J. M. Sturtevant, *J. Biol. Chem.,* 242 (1967) 3555.
279 A. Baudras, *Biochem. Biophys. Res. Commun.,* 7 (1962) 310.
280 R. K. Morton, *Nature,* 192 (1961) 727.
281 H. Suzuki and Y. Ogura, *J. Biochem. Tokyo,* 67 (1970) 277.
282 M. Iwatsubo, A. Baudras, A. di Franco, C. Capeillere and F. Labeyrie, in K. Yagi, *Flavins and Flavoproteins, Proc. 2nd Conf.,* University of Tokyo Press, Tokyo, 1967, p. 41.
283 R. K. Morton and K. Shepley, *Biochim. Biophys. Acta,* 96 (1965) 349.
284 F. Labeyrie, O. Groudinsky, Y. Jacquot-Armand and L. Naslin, *Biochim. Biophys. Acta,* 128 (1966) 492.
285 F. Labeyrie, A. di Franco, M. Iwatsubo and A. Baudras, *Biochemistry,* 6 (1967) 1791.
286 C. A. Appleby and R. K. Morton, *Biochem. J.,* 73 (1959) 539.

287 M. Iwatsubo and A. di Franco, in K. Okunuki, M. D. Kamen and I. Sekuzu, *Structure and Function of Cytochromes*, University of Tokyo Press, Tokyo, 1968, p. 613.
288 J. M. Sturtevant and T. Y. Tsong, *J. Biol. Chem.*, 243 (1968) 2359.
289 M. Iwatsubo and J. L. Risler, *Eur. J. Biochem.*, 9 (1969) 280.
290 R. K. Morton and J. M. Sturtevant, *J. Biol. Chem.*, 239 (1964) 1614.
291 K. Hiromi and J. M. Sturtevant, *J. Biol. Chem.*, 240 (1965) 4662.
292 M. Klingenberg, *Arch. Biochem. Biophys.*, 75 (1958) 376.
293 T. Omura and R. Sato, *J. Biol. Chem.*, 239 (1964) 2370.
294 T. Omura, R. Sato, D. Y. Cooper, O. Rosenthal and R. W. Estabrook, *Federation Proc.*, 24 (1965) 1181.
295 F. Mitani and S. Horie, *J. Biochem. Tokyo*, 65 (1969) 269.
296 F. Mitani and S. Horie, *J. Biochem. Tokyo*, 66 (1969) 139.
297 Y. Miyake, K. Mori and T. Yamano, *Arch. Biochem. Biophys.*, 133 (1969) 318.
298 H. Nishibayashi and R. Sato, *J. Biochem. Tokyo*, 63 (1968) 766.
299 H. Remmer, J. Schenkman, R. W. Estabrook, *et al.*, *Mol. Pharmacol.*, 2 (1966) 187.
300 J. B. Schenkman, H. Remmer and R. W. Estabrook, *Mol. Pharmacol.*, 3 (1967) 113.
301 Y. Imai and R. Sato, *Eur. J. Biochem.*, 1 (1967) 419.
302 H. S. Mason, J. C. North and M. Vanneste, *Federation Proc.*, 24 (1965) 1172.
303 T. Omura and R. Sato, *J. Biol. Chem.*, 239 (1964) 2379.
304 K. Murakami and H. S. Mason, *J. Biol. Chem.*, 242 (1967) 1102.
305 C. R. E. Jefcoate and J. L. Gaylor, *Biochemistry*, 8 (1969) 3464.
306 H. A. O. Hill, A. Röder and R. J. P. Williams, *Naturwissenschaften*, 57 (1970) 69.
307 R. Tsai, C. A. Yu, I. C. Gunsalus, *et al.*, *Proc. Natn. Acad. Sci. U.S.*, 66 (1970) 1157.
308 Y. Imai and R. Sato, in K. Okunuki, M. D. Kamen and I. Sekuzu, *Structure and Function of Cytochromes*, University of Tokyo Press, Tokyo, 1968, p. 626.
309 Y. Imai and R. Sato, *J. Biochem. Tokyo*, 64 (1968) 147.
310 C. R. E. Jefcoate and J. L. Gaylor, *J. Am. Chem. Soc.*, 91 (1969) 4610.
311 M. Katagiri, B. N. Ganguli and I. C. Gunsalus, *J. Biol. Chem.*, 243 (1968) 3543.
312 T. Takano, R. Swanson, O. B. Kallai and R. E. Dickerson, *Cold Spring Harbor Symp. Quant. Biol.*, 36 (1972) 397.
313 F. S. Mathews, P. Argos and M. Levine, *Cold Spring Harbor Symp. Quant. Biol.*, 36 (1972) 387.

Chapter 27

CYTOCHROME OXIDASE

D. C. WHARTON

Section of Biochemistry and Molecular Biology, Division of Biological Sciences, Cornell University, Ithaca, New York, U.S.A.

INTRODUCTION

What is now known as cytochrome oxidase was first observed in 1884 by C. A. MacMunn, a British physician, as an absorption band in the spectrum of slices of various tissues[1-3]. This cytochrome and the others whose presence was indicated by four absorption bands were referred to collectively as myohematin or histohematin, depending upon whether they occurred in muscle or other tissues, respectively. Using reducing and oxidizing agents, MacMunn demonstrated the reversible oxidoreduction of these pigments which led him to postulate that they serve a respiratory function. Unfortunately, MacMunn failed to convince his contemporaries and the significance of his work was never appreciated during his lifetime.

Following the work of MacMunn, little significant information of terminal cellular respiration appeared until the investigations of Warburg and his associates who showed that respiration could be imitated by iron-containing charcoals which catalyze the oxidation of certain amino acids[4-6]. They demonstrated that this oxidation is inhibited by small amounts of cyanide not unlike the inhibition of cellular respiration. Since Warburg was aware that all cells contain iron in significant concentrations, he believed that it was involved in the reaction with oxygen and that it underwent valence changes in the course of this reaction. Warburg referred to the cellular catalyst as the "Atmungsferment", or respiratory enzyme, and speculated that it was allied to hematin. Later, in some classic experiments with Negelein[7,8], Warburg demonstrated the heme character of the "Atmungsferment". These experiments were based on the fact that the inhibition of respiration by carbon monoxide is reversed by light. By showing that light of certain wavelengths is more efficient in reversing the inhibition of respiration in yeast, Warburg and Negelein were able to construct a photochemical action spectrum of the phenomenon. Warburg recognized the relationship of this action spectrum, characterized by absorption bands at 590 nm and 433 nm, with that of carboxyhemoglobin.

The work of Warburg was overlapped in part and was followed by the brilliant researches of Keilin who, over a period of almost forty years,

References pp. 983–987

studied the compounds he named cytochromes★. It was Keilin[9-12] who demonstrated conclusively that the cytochromes participate in cellular respiration and that their prosthetic group is heme. Keilin recognized the relationship between the respiratory pigments he was studying and the myohaematin and histohaematin reported by MacMunn. However, since the four-banded absorption spectrum observed in the tissues actually belongs to several structurally distinct heme compounds and because of confusion with myoglobin and hemoglobin, Keilin chose to name the substances cytochromes. Each band in the four-banded spectrum was given a letter designation (*a*, *b*, *c*, *d*) beginning with the band at the longest wavelength. Since these pioneering studies it has been shown that what Keilin called cytochrome *d* is in reality the β-band of the other cytochromes and not that of a separate entity. Furthermore, it has been demonstrated that band *a* is composed of two functionally distinct cytochromes, *a* and a_3, while band *c* is composed of two physically distinct cytochromes, *c* and c_1. The Soret bands of the various cytochromes could not be observed in Keilin's microspectroscope at that time.

In subsequent investigations[13-15], Keilin and his longtime colleague Hartree, demonstrated that cytochrome *a* is more closely associated with the reaction with oxygen than cytochromes *b* or *c* and that in reality, cytochrome *a* consists of two heme components, one of which reacts with cyanide and carbon monoxide, and probably oxygen, while the other does not react with these substances. Keilin called the latter compound cytochrome *a* but, because other cytochromes found in bacteria had already been named cytochromes a_1 and a_2, Keilin named the former compound cytochrome a_3. Since the carbon monoxide complex of cytochrome a_3 was observed to have absorption bands at 590 nm and 432 nm its relationship to the "Atmungsferment" of Warburg was established. Furthermore, Keilin found that cytochrome $a + a_3$ directly oxidized cytochrome *c* and, therefore, could be named cytochrome oxidase.

The concept of cytochrome oxidase as a complex of two different cytochromes has often been questioned[16-18] primarily because they have never been separated physically. However, at present, only Okunuki[19] in Japan and Wainio[20] in the United States remain as the chief advocates of the "unitarian theory" from which it is concluded that only one cytochrome, cytochrome *a*, is contained in cytochrome oxidase. They argue that the anomalous reactions of the enzyme can be explained without invoking the participation of a second cytochrome. These arguments will be discussed in more detail later.

★A detailed account of the evolution of our knowledge of cytochromes and respiration may be found in Keilin's book, *The History of Cell Respiration and Cytochrome*, published by Cambridge University Press in 1966.

Keilin and Hartree[14] found significant amounts of copper in their heart muscle respiratory particles and speculated for a time that copper is oxidized directly by oxygen. However, when they demonstrated that cytochrome a_3 satisfied the criteria of a terminal oxidase[15], they rejected their hypothesis about copper.

It should now be apparent that cytochrome oxidase is the terminal enzyme of cellular respiration, coupling the entire process to oxidation by molecular oxygen. The chief purpose of this form of respiration is to gather the energy resulting from the oxidation of foodstuffs and conserve it in the high energy phosphate bond of adenosine triphosphate (ATP) whereupon it can be utilized for various cellular requirements.

In plants, animals, and fungi cytochrome oxidase as well as its parent electron transport system is a constituent of subcellular organelles called mitochondria. The mitochondrion is an ellipsoid to elongated bag, 3–10 μm in length and about 1 μm in diameter, composed of an outer membraneous envelope and an invaginated inner membrane. Cytochrome oxidase is a constituent of the inner membrane. Details of mitochondrial structure and function have been discussed in a number of excellent reviews and the interested reader is advised to consult these sources[21-23].

PURIFIED PREPARATIONS

Methods

Because of its membraneous lipoprotein character, cytochrome oxidase is not readily purified in a water-soluble form. Most purified preparations have been obtained by extracting the membranes of beef heart mitochondria with bile salts such as deoxycholate or cholate[24-32] or with such synthetic detergents as Triton X-114 and Triton X-100[33,34]. Heart has been the organ of choice as a source because of its high concentration of cytochrome oxidase and because of the stability of its mitochondria. Although these preparations differ from one another in their degree of purity and in certain other respects, they are probably all quite similar in their mechanism of action. In order to maintain the solubility of the oxidase in aqueous solutions, detergents must always be present. This requirement is not without complications since many of these inhibit enzymic activity.

Molecular weight

The size of the cytochrome oxidase molecule has always been open to some doubt because of its property of aggregation and disaggregation, because of the questionable purity of available preparations, and the presence of detergents. Takemori *et al.*[35] found that in the presence of cholate, their oxidase preparation is polydispersed while in the presence of the non-ionic detergent Emasol 1130, it is monodispersed. They reported that the latter has a sedimentation coefficient of 21.9*S* from which they calculated a molecular weight of 530,000.

Tzagoloff *et al.*[36] calculated a sedimentation coefficient of 12*S* for their preparation of cytochrome oxidase which was stabilized as a monodispersed protein with 0.1% taurodeoxycholate. Using light-scattering measurements they calculated a molecular weight of about 300,000.

When the results of Takemori *et al.*[35] and of Tzagoloff *et al.*[36] are corrected for their lipid content the molecular weights are about 400,000 and 200,000, respectively. On the basis of a minimum molecular weight of 100,000 to 130,000, calculated from the concentration of heme *a* in the protein, the preparation of Takemori *et al.* contains four subunits while that of Tzagoloff and his associates contains two subunits.

Orii and Okunuki[37] have depolymerized the preparation studied by Takemori *et al.* using guanidine hydrochloride and sodium dodecyl sulfate (SDS). At a concentration of 0.30% SDS they found particles of increased oxidase activity with a sedimentation coefficient of 16.6*S*. Thus, they calculated that the original particles of 530,000 molecular weight depolymerize into particles of 290,000–330,000 molecular weight which are similar in size to those reported by Tzagoloff *et al.*[36]. When Orii and Okunuki[37] treated their oxidase preparation with higher concentrations of SDS they found smaller but inactive units of 5.7*S* corresponding to a molecular weight of about 67,000. The latter value is in good agreement with the earlier figure of Criddle and Bock[38] who calculated a molecular weight of 72,000 from data collected by an approach to equilibrium of their inactive "monomer" which had been obtained with the aid of SDS.

Wharton (unpublished results) has found that under certain conditions[39] cytochrome oxidase depolymerizes on dilution and this is accompanied by an increase in activity. He has obtained a sedimentation velocity constant of less than 9*S* from which a molecular weight of at least 140,000 can be calculated. Uncertainties in the extrapolation of data to zero concentration have precluded a determination of the minimal size in these preparations.

Chan and Stotz (unpublished results) have depolymerized a preparation of cytochrome oxidase by treatment with dilute alkali at pH 10–11. They report a molecular weight of about 200,000 before treatment and 100,000 after treatment. They claim that the latter material retains significant oxidase activity.

Although these data are of interest and amply demonstrate the polymeric character of oxidase preparations, their value in determining the more or less precise molecular weight of the minimum active unit of cytochrome oxidase remains somewhat limited because of the uncertainties stemming from the use of detergents and of preparations of unknown purity.

Composition

Matsubara *et al.*[40] have reported on the amino acid composition of a purified cytochrome oxidase preparation. They found some 820 amino acid residues with an aggregate weight of 93,000. They reported the following composition: Lys_{39}, His_{29-30}, Arg_{30-31}, Tyr_{30}, Asp_{58-61}, Thr_{42-43}, Ser_{53-55}, Glu_{59-60}, Gly_{58-59}, Pro_{46}, Ala_{60-64}, Cys_{7}, Val_{49-52}, $Ileu_{43}$, Leu_{86-88}, Met_{35}, Phe_{46-47}, Tyr_{32-33}, $(NH_3)_{56}$. Although these results are of interest as a first step, particularly the low concentration of cysteine, they must be viewed with some caution because of the unknown purity of the preparations.

As noted earlier, cytochrome oxidase is a lipid-containing hemoprotein. Some preparations have been reported which contain less than 5% lipid but in these cases the lipid is replaced with detergents. The composition of one type of purified preparation is shown in Table I. The iron in the preparation

TABLE I

COMPOSITION OF CYTOCHROME OXIDASE PURIFIED FROM BEEF HEART (GRIFFITHS AND WHARTON[28])

Component	1/mg protein
Heme *a*	8.1–9.2 nmoles
Iron	8.2–9.4 nmoles
Copper	9.2–10.6 nmoles
Lipid	0.20–0.28 mg
Deoxycholate	0.8–1.4 mg

is almost totally restricted to the heme while copper is present in an amount equimolar with the iron. Of the lipid contained in most purified oxidase preparations over 90% is phospholipid and this is composed of cardiolipin, lecithin, phosphatidylethanolamine, and phosphatidyl inositol[41]. The deoxycholate present in the preparation shown in Table I was introduced as a means of solubilizing the enzyme. For the student of inorganic biochemistry it is the iron and copper which make cytochrome oxidase of such vital interest. Accordingly, discussion of these metals will constitute the bulk of this chapter.

SPECTRAL PROPERTIES

Absorption spectra

As indicated earlier, the spectral characteristics of cytochrome oxidase have played a key role in its discovery and characterization. The absorption spectra of the oxidized and reduced forms of the enzyme are shown in Fig. 1. The spectrum of the oxidized form features an α-band at 599 nm and a γ- or Soret band which may vary between 417 and 424 nm. The reason for this variation is not known but may be related to the age of the preparation, to contact with certain detergents, to differences in protein conformation which affect the heme, or more likely, to some other, as yet unknown factor. On reduction the α-band shifts to 605 nm, the Soret band to 444 nm, and a weak β-band appears at 517 nm. The position of the Soret band in the reduced oxidase does not vary in wavelength as it does in the oxidized enzyme.

The addition of carbon monoxide to the reduced form of cytochrome oxidase results in the appearance of new absorption bands illustrated in Fig. 2A and seen as a shoulder at 590 nm and as a peak at 431 nm. This

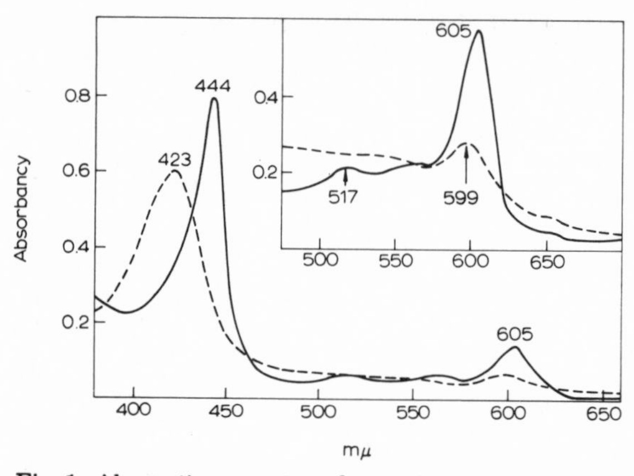

Fig. 1. Absorption spectra of cytochrome oxidase. The oxidized preparation, ----; the reduced preparation, ——. 0.85 mg of protein per ml.

spectrum also retains the absorption bands of the unreacted component at 605 nm and 444 nm. The absorption bands of the carbon monoxide complex are illustrated better by a difference spectrum between the carbon monoxide complex and the reduced oxidase as shown in Fig. 2B. The wavelengths of the maxima at 590 nm and 431 nm are identical with those described by Warburg and Negelein[7,8] for their photochemical action spectrum. Tzagoloff and Wharton[42] have shown that a carbon monoxide complex can be formed by exposing oxidized cytochrome oxidase to carbon monoxide anaerobically. Apparently under these conditions, the component which reacts with carbon monoxide is reduced and combines with the ligand while the unreactive component remains oxidized. Conversely, Horie and Morrison[43] and Tzagoloff and Wharton[42] have found that if reduced cytochrome oxidase complexed with carbon monoxide is oxidized by ferricyanide anaerobically, the non-liganded component becomes oxidized while the complex remains intact. Under both circumstances the difference spectra

(CO-complex minus oxidized) are identical with sharp peaks at 590 nm and 431 nm.

These results constitute one of the most compelling arguments for the presence of two heme components differing in their reactivity and represented by the cytochromes *a* and a_3. If at this point one assumes

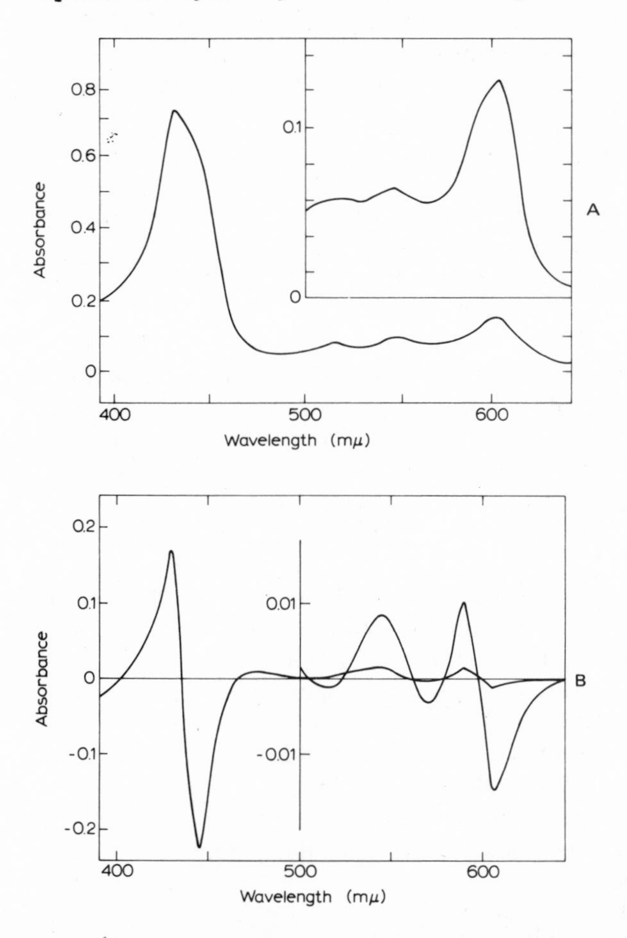

Fig. 2. Absorption spectra of the carbon monoxide derivative of cytochrome oxidase. A, direct spectrum; B, difference spectrum of carbon monoxide derivative minus reduced preparation.

equivalent amounts of cytochromes *a* and a_3, then one can visualize a minimum unit of cytochrome oxidase as having two cytochromes and two atoms of copper.

In addition to carbon monoxide, other reagents which bind to cytochrome oxidase also produce changes in its absorption spectrum. Of these cyanide has been studied most intensively[32,44–48]. When added to the oxidized enzyme, cyanide causes a slight shift of the Soret band to a longer wavelength but the α-band does not change. Its addition to the reduced oxidase causes a more significant effect with a shoulder at 590 nm on the α-band as well as a decrease in the absorbance of the Soret band. Difference spectra reveal that the reduced cytochrome component combined with cyanide has an α-band at 590 nm and a Soret band at 444 nm. If cyanide is added to the oxidized enzyme it inhibits the reduction of a portion of the oxidase. On the basis of difference spectra the inhibited component has been identified with cytochrome a_3.

These difference spectra with either carbon monoxide or cyanide have been utilized to determine the relative contributions of cytochrome *a* and cytochrome a_3 to the absorption bands and to calculate their extinction coefficients at each wavelength. Thus, Yonetani[45] concluded that cytochrome *a* constitutes 72% of the α-band and 49% of the Soret band while cytochrome a_3 contributes the remainder. VanGelder[49] calculated that each cytochrome contributes 50% to the difference spectrum at 445 nm while cytochrome *a* contributes 81% to the α-band. He has also calculated extinction coefficients of 19.4 and 4.6 mM^{-1} cm^{-1} for cytochrome *a* and cytochrome a_3, respectively in the difference spectrum at 605 nm while at 445 nm he has derived an extinction coefficient of 82 mM^{-1} cm^{-1} for each cytochrome. Vanneste and Vanneste[50] have determined similar extinction coefficients for the α-region of the oxidase but obtained coefficients in the Soret region of 57.2 and 112 mM^{-1} cm^{-1} for cytochromes *a* and a_3, respectively. Van Gelder[49] attributed the differences in their results to the lack of a correction for the presence of non-reducible cytochromes *a* and a_3 in the Vanneste preparation. However, it is of interest that Greenwood and Gibson[51] determined from rapid-reaction kinetic experiments that the ratio of cytochrome a_3 to cytochrome *a* at 445 nm is 2.3:1, a value much closer to that of Vanneste and not influenced by non-reducible components.

In addition to spectral changes induced by carbon monoxide and cyanide, other inhibitors of cytochrome oxidase such as nitric oxide[44,52], azide[53,54], and fluoride[54] have been observed to alter its absorption spectrum to varying degrees.

Circular dichroism and optical rotatory dispersion

A beginning has already been made in studying the properties of cytochrome oxidase, particularly the interaction of components, by circular dichroism (c.d.) and optical rotatory dispersion (o.r.d.). Urry and his coworkers[55,56] have found by c.d. that the oxidized enzyme has a relatively simple and positive Soret band while the reduced enzyme is characterized by a larger complex Soret band of greater molar ellipticity than is found in other hemoproteins. They conclude that the anisotropies, the ratio of rotational strength to dipole strength, for cytochrome oxidase are over twice the magnitude of those observed in deoxyhemoglobin and deoxy-myoglobin but are similar to those found in the case of certain heme peptides. It is their opinion as a result of comparing these data with those of known systems that the hemes in reduced cytochrome oxidase are juxtaposed although they rightly urge caution in this interpretation since this is the first heme *a* protein to be studied. However, this interpretation is strengthened by the results of their c.d. experiments with the carbon monoxide complex, in which the rotational strength of the carbon monoxide–cytochrome a_3 complex is negative and that of the unreacted cytochrome *a* is positive. These reciprocal relations would be expected from the dissymmetric interaction of juxtaposed groups.

Yong and King[57] have recently proposed on the basis of c.d. as well as absorption spectroscopy that the oxidation state of cytochrome *a* affects cytochrome a_3 although they retain their earlier conclusion[58] that there is no direct heme–heme interaction. They suggest that the interaction occurs by means of the copper moiety which interacts with cytochrome a_3.

Several studies[59-64] have applied the technique of o.r.d. to cytochrome oxidase with the aim of resolving the stereochemistry of both the heme and protein components. King and his collaborators[63] have observed three extrinsic Cotton effects in the oxidized enzyme with centers at 432, 423.7, and 376 nm while in the reduced oxidase they have found two such effects with centers at 446.5 and 444 nm. The presence of double Cotton effects in the Soret region may arise from the x–y polarization of the porphyrin π-system which leads to splitting of the absorption bands. King *et al.* have calculated from the o.r.d. spectra that the effective helical content of the reduced oxidase is greater than that of the oxidized form. It can be concluded from these early studies that the electronic relationship between the prosthetic group and the apoprotein is dependent on the oxidation state of the enzyme although the source of this dependence has yet to be deduced.

PROSTHETIC GROUPS

Heme a

The heme prosthetic group of cytochrome oxidase can be removed readily from the protein by the use of solvents such as acetone–HCl but its purification has been made difficult by its association with lipids[65]. More recently, however, York *et al.*[66] have reported a simple column chromatographic purification of heme *a* after removal of the lipids. Except for a recent report[67] which requires corroboration, studies of the purified heme indicate that it is a single component[68]. Its absorption spectrum as the pyridine hemochrome is shown in Fig. 3. As seen in Fig. 4, heme *a* is a

References pp. 983–987

tetrapyrrole in which the pyrrole groups are linked together by methene bridges, while the single iron atom is coordinated through the nitrogen atoms of each pyrrole (see also Fig. 1, Chapter 24). Each pyrrole group differs from the others either in the nature or position of its substituent

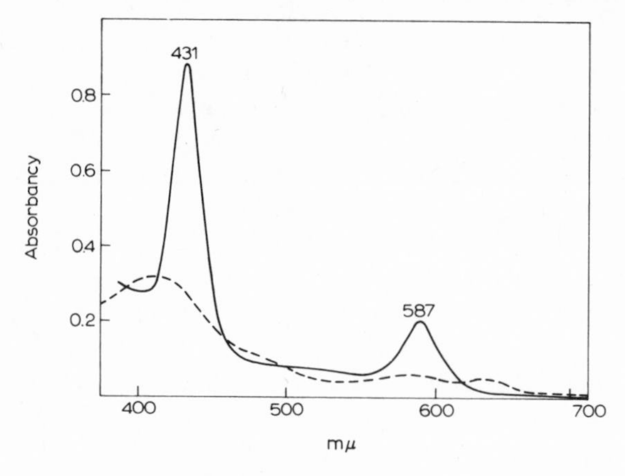

Fig. 3. Absorption spectra of pyridine hemochrome of heme *a*. The oxidized preparation, - - - -; the reduced preparation, —.

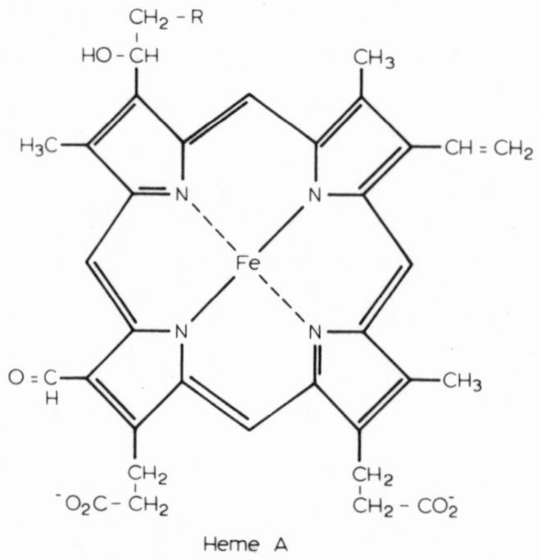

Fig. 4. Structural diagram of heme *a*. R represents the long alkyl sidechain described in the text.

groups. It is the nature of these peripheral groups which has provided the greatest challenge to an elucidation of the structure of heme *a*. The structure of all but one of these has now been demonstrated conclusively. Only the structure of the long alkyl sidechain on position 2 remains to be clarified and this only in certain details.

The presence of a formyl group on heme *a* was first suggested by Warburg[69] on the basis of the similarity of the photochemical action spectrum of the "Atmungsferment" with the absorption spectrum of chlorocruorin which was known to contain a formyl group. This hypothesis has since been confirmed by a number of studies including reactions specific for carbonyl groups[70]. One formyl group has been indicated by infra-red and nuclear magnetic resonance (n.m.r.) spectroscopy[71] while the assignment of this group to position 8 has been made by Nicolaus and Mangoni[72] and by Piattelli[73] from a study of pyrrole-α, α-dicarboxylic acids obtained by permanganate oxidation of porphyrin *a* derivatives. This conclusion is supported by spectral data[74] which show that the formyl group is attached to the pyrrole diagonally opposite that containing the vinyl side-chain. The presence of the latter was suggested by the work of Warburg and Gewitz[75] who subjected hemin *a* to the resorcinol melt. Lemberg *et al.*[74] provided evidence that the group on position 4 is vinyl while Lynen and his associates[76] concluded that an unsubstituted vinyl group is present by the identification of methylethylmaleimide following chromic acid oxidation of hydrogenated hemin *a* chloride. This conclusion has been confirmed by Caughey *et al.*[77] using a similar reaction on their pyridine hemochrome *a* as well as by infra-red and n.m.r. spectra. The synthesis of cytodeuteroporphyrin by Marks *et al.*[78], who showed that it is identical with a derivative of hemin *a*, established the presence of two propionic acid groups on positions 6 and 7. Their presence has also been indicated by the preparation of disodium salts of hemin *a* and of dimethyl esters of metal-free porphyrin derivatives of heme *a*[77]. The synthesis of cytodeuteroporphyrin[78] (see Fig. 7, Chapter 24, for structure of the methylated heme derivative) also documented the presence of the methyl groups on positions 1, 3, and 5 that had been suggested earlier by the work of Warburg[75]. N.m.r. spectra by Caughey and co-workers[79] confirmed the presence of three methyl groups and also demonstrated the presence of four meso protons, one on each of the methene bridges. A long alkyl side chain on position 2 was originally suggested by elemental analysis and by the solubility properties of heme *a*[74,75]. The possibility that this side-chain is an α-hydroxyalkyl group was prompted by Barrett's finding[80] that the chromatographic behavior of porphyrin *a* is similar to that of hematoporphyrin which is known to contain an α-hydroxyalkyl group. The exact structure of this side-chain is still uncertain.

It has been the contention of Lynen and his associates[76,82] with support from Lemberg[81] that the alkyl side-chain is a 1-hydroxy-5,9,13-trimethyltetradecyl group ($C_{17}H_{31}O$):

$$\underset{\text{Ring}}{\overset{\text{H}}{}}\!>\!C\!<\!\underset{\text{OH}}{\overset{CH_2-[CH_2-CH_2-\overset{\displaystyle CH_3}{\overset{|}{C}H}-CH_2]_3H.}{}}$$

References pp. 983–987

They cite several lines of evidence to support this structure. They propose that heme *a* is synthesized by the alkylation of a vinyl group on position 2 of protohemin by farnesyl pyrophosphate or some other alkyl pyrophosphate:

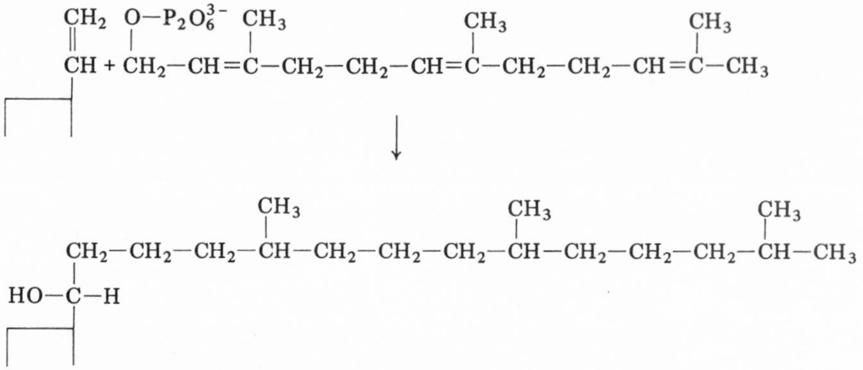

$$
\begin{array}{l}
\overset{CH_2}{\overset{\|}{CH}} + \overset{O-P_2O_6^{3-}}{\overset{|}{CH_2}}-CH=\overset{CH_3}{\overset{|}{C}}-CH_2-CH_2-CH=\overset{CH_3}{\overset{|}{C}}-CH_2-CH_2-CH=\overset{CH_3}{\overset{|}{C}}-CH_3 \\
\downarrow \\
CH_2-CH_2-CH_2-\overset{CH_3}{\overset{|}{CH}}-CH_2-CH_2-CH_2-\overset{CH_3}{\overset{|}{CH}}-CH_2-CH_2-CH_2-\overset{CH_3}{\overset{|}{CH}}-CH_3 \\
\quad | \\
HO-C-H
\end{array}
$$

The alkylation of the vinyl group on position 2 is favored over that on position 4 because the former is more electronegative. Degradative studies[82,83] in Lynen's laboratory, where hemin *a* was converted to its hydrogenated porphyrin derivative followed by oxidation of the reduced porphyrin with chromic acid, yielded a lipophilic methyl-alkyl-maleimide. The latter compound has been reported to compare favorably with a synthetic methyl-5,9,13-trimethyltetradecylmaleimide on the basis of infrared spectroscopy, mass spectrocopy, gas–liquid chromatography, and thin-layer chromatography. Their mass spectra and gas–liquid chromatograms also indicate that the sidechain is terminated with an isopropyl group. The chromatographic behavior of the ester of porphyrin *a* as observed by Clezy and Barrett[84] further supports the presence of a hydroxyl group in the side-chain.

Caughey and his associates[79] concluded from their n.m.r. spectra that heme *a* differs in several respects from the structure containing a 1-hydroxy-5,9,13-trimethyltetradecyl group as proposed by Lynen and his co-workers. Their most recent data (unpublished results) suggest the following structure:

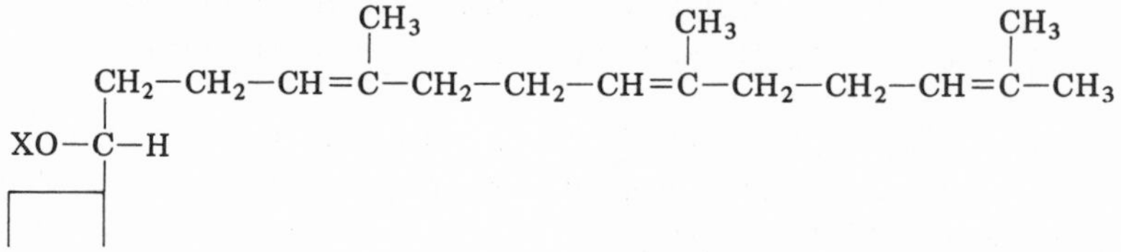

$$
\begin{array}{l}
CH_2-CH_2-CH=\overset{CH_3}{\overset{|}{C}}-CH_2-CH_2-CH=\overset{CH_3}{\overset{|}{C}}-CH_2-CH_2-CH=\overset{CH_3}{\overset{|}{C}}-CH_3 \\
\quad | \\
XO-C-H
\end{array}
$$

Lynen and Lemberg[81] have not argued against the inclusion of unsaturated bonds in the side-chain since the procedure employed in Lynen's laboratory[82] removes all such bonds by hydrogenation.

Whether the oxygen present in the native heme exists as a hydroxyl or an ether has been debated. Caughey[66], Lemberg[81], and Seyffert[83] have discussed the possibility that the alkyl side-chain contains either a closed pyrane or furane ring:

```
                CH2                                H2C——CH2
            CH2     CH2                            /      \
     H\      |       |  /R                    H\  /        \
        C            C                          C           CH—CH—R
Pyrrole ring/   \O/     \CH3          Pyrrole ring/  \O/          |
                                                                 CH3
```

Upon treatment with acid or conversion of the heme to porphyrin the ring opens and the oxygen could be converted to a free α-hydroxyl group:

```
             H    CH2—CH2
              \  /      \
               C         CH
              / \        //
Pyrrole ring    OH      C—R
                        |
                        CH3
```

However, Caughey (personal communication) presently is of the opinion that a ring structure involving this particular oxygen is not present in the intact enzyme.

The major controversial aspect of the structure of heme *a* at this time is the argument of Caughey and coworkers[67,79] that a nitrogen-containing moiety ($C_6H_{11}NO_4$), indicated by the letter X in their structural representations, is attached to the 1-carbon of the long alkyl sidechain. They cite elemental analyses, solubility characteristics, and infra-red spectra as evidence to support the presence of this structure which they speculate may be an amino sugar in a pyranose form. This moiety could form a bridge between the alkyl sidechain and the iron of the heme. Caughey *et al.* have indicated that conversion of their heme *a* to porphyrin *a* or acid-treatment of the heme in the presence of chloride ion results in a loss of the group. On the other hand, the presence of such a moiety in heme *a* has been disputed by Lemberg[68,81,85] and by Morrison[86]. Lemberg[68] reported that, in his hands, Caughey's hemochrome does not contain, nor does it lose on acid treatment, any hexosamine or carbohydrate. He also has argued against the thesis that a bridge extends from the sidechain to the iron for stereochemical reasons.

Caughey and his associates[67] have suggested the presence of two chemically distinct hemes in cytochrome oxidase as a means of explaining the differences between their experimental results and those of Lynen and Lemberg. They reported the chromatographic separation of heme *a* into two fractions of equal amounts to support this thesis. Confirmation of two different hemes would of course strengthen further the concept of two distinct cytochromes in cytochromes oxidase. This evidence, however, has

References pp. 983–987

been questioned by Morrison[86] on the grounds that the second component was created from the first during the manipulations.

The final resolution of the structure of the alkyl side-chain and of the question of a hexosamine group must await the clarification of the current conflicting evidence.

As noted earlier the iron of heme *a* is coordinated through the four nitrogen atoms of the tetrapyrrole. A significant contribution of covalent bonding is indicated by the high degree of stability of the iron in the heme.

Although the mode of binding of heme *a* to the protein has not been ascertained there is some evidence that a histidine imidazole on the protein is involved. Lemberg *et al.*[87], Horie[88] and Vanderkooi and Stotz[89] have studied the spectral properties of the reaction products of heme *a* with histidine, imidazole, and 4(5)-methylimidazole. The absorption spectrum of the reduced imidazole complex has an α-band near 593 nm and a Soret band near 438 nm which is not unlike that of alkali- or urea-denatured cytochrome oxidase.

Lemberg[68] has postulated a difference in the type of binding as a means of explaining the difference between cytochrome *a* and cytochrome a_3. According to this hypothesis, heme *a* in cytochrome *a* is bound coordinately in both the 5 and 6 positions of the iron to protein-N, perhaps by two histidine imidazole groups. An added possibility is that one of the positions is coordinated to a methionine *S* as is the case in cytochrome *c*. Lemberg cites as evidence for this hypothesis the inability of cytochrome *a* to react with carbon monoxide, the low ratio of its Soret to α maxima, the relatively small difference between the wavelength of the Soret maxima of the oxidized and reduced forms, and the absence of a charge transfer band in ferric cytochrome *a*. He postulates that heme *a* in cytochrome a_3 is probably coordinated only through the 5 position perhaps to the N of a histidine imidazole. To support this argument, Lemberg notes the reaction of cytochrome a_3 with carbon monoxide and cyanide, the high ratio of its Soret to α-band maxima, and the larger difference between the wavelength of the Soret bands of the oxidized and reduced forms.

Properties of iron

Ehrenberg and Yonetani[90] conducted extensive magnetic studies of various forms of cytochrome oxidase as well as of heme *a* and hematin *a*. Although their values of molar susceptibility may be in error due to significant copper contamination, their results indicate that iron in cytochrome *a* is low-spin in the Fe(III) state and is diamagnetic in the Fe(II) state. On the other hand, their evidence indicates that iron in cytochrome a_3 is high-spin in the Fe(II) state and is a thermal mixture of high-spin and low-spin in the Fe(III) state. The carbon monoxide and cyanide complexes of ferrocytochrome a_3 are both diamagnetic.

Beinert and his associates[91,92] have employed electron paramagnetic resonance (e.p.r.) spectroscopy to study the spin states of the heme iron at various stages of oxidoreduction. The e.p.r. spectrum of oxidized cytochrome oxidase is shown in Fig. 5A and is characterized by one dominant heme signal of a low-spin Fe(III) type with g values at 3.0, 2, and 1.5. The signal at $g = 2.26$ is all but obscured by the intense copper signal in that region, but can be observed near the temperature of liquid helium. The signal at $g = 4.3$ is weak and is thought to result from some iron contaminant. The presence of these signals has been confirmed by Tsudzuki and Okunuki[93]. When the enzyme is partly reduced by reducing agents, the copper signal and the Fe(III) signal at $g = 3.0$ decrease and there is a significant increase

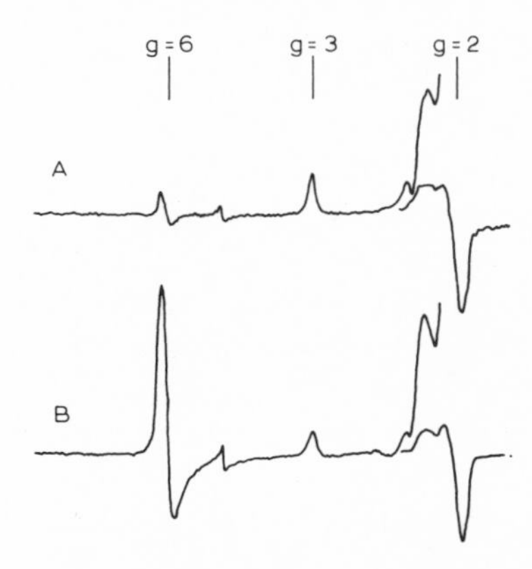

Fig. 5. E.p.r. spectra of cytochrome oxidase in the low field region. A, the oxidized preparation; B, half-reduced preparation. The conditions of e.p.r. spectroscopy were: microwave power, 27 mW; modulation amplitude, 6 gauss; scanning rate, 1000 gauss/min; time constant, 0.5 sec.; temperature, 39K. The copper signal at $g = 2$ was recorded at one-fifth the amplification used for the remainder of the spectrum.

in the formerly very weak signal at $g = 6.0$. The latter is typical of high-spin Fe(III) heme compounds. The concentration of the component responsible for this signal is judged to be about the same as that responsible for the low-spin signals in the oxidized enzyme. As the signal at $g = 6.0$ increases there is also an increase in signals of a low-spin Fe(III) heme type with g values at 2.6, 2.2, and 1.87. These are illustrated in Fig. 5B. It is argued[92], though a relationship is not precluded, that the latter high-spin and low-spin signals do not belong to an equilibrium mixture of the same heme component since the high-spin species reaches a maximum when the system is about one-half reduced whereas the low-spin signals do not reach a maximum until the system is about three-fourths reduced. The observation that the intensity of the latter signals is not proportional to that of the high-spin signal would also seem to preclude the possibility that a thermally

References pp. 983–987

balanced mixture of spin states in a single component is responsible. Both types of these signals, after reaching a maximum amplitude, decrease on further reduction of the system. Quite recently Beinert (unpublished results) has observed that some active preparations of the enzyme do not contain any of the low-spin signals that appear on partial reduction. He concludes that these signals result from damage to some of the heme iron.

Although Beinert and his associates have observed the appearance of the high-spin Fe(III) form on rapid (~ 10 msec) reduction of the enzyme by ferrocytochrome *c* or by dithionite they have been unable to observe the signal during rapid oxidation of the reduced oxidase. Also, they have found that strong ligands such as cyanide or azide block the appearance of the high-spin Fe(III) signal.

On the basis of their results, Beinert and his co-workers conclude that two different heme components of approximately equal concentration are present in their enzyme. They assign the low-spin Fe(III) signals seen in the oxidized form to cytochrome *a* and conclude that the high-spin Fe(III) signal seen only at intermediate stages of oxidation can be attributed to cytochrome a_3. They interpret these results to mean that in the oxidized enzyme antiferromagnetic coupling occurs between cytochrome a_3 and another component of the system, possibly another molecule of cytochrome a_3 or a molecule of copper, resulting in a marked decrease in the paramagnetism of the system. One attractive interpretation of the reductive titration results is based on an interaction between cytochrome a_3 and copper. As the first one or two electrons enter the system the copper, which incidentally would also be without significant net paramagnetism, would become reduced with a consequent cessation of the heme–copper interaction and the appearance of the a_3 high-spin Fe(III) signal. They conclude from the remainder of the titration that the introduction of additional electrons reduces first the cytochrome *a* and then the e.p.r.-detectable copper and cytochrome a_3. However, it seems difficult to reconcile on thermodynamic grounds the reduction of cytochrome *a* before cytochrome a_3 since under equilibrium conditions one would expect cytochrome a_3 to have a more positive oxidation–reduction potential. Perhaps, as these investigators suggest, the conditions of low temperature used in the e.p.r. experiments result in a distribution of electrons not encountered in an enzyme unit at higher temperatures. Their observations are supported, however, by the results of Ehrenberg and Vanneste[94,95] who have found that during reductive titrations at room temperature the magnetic susceptibility of cytochrome oxidase increases to a maximum between 50 and 75% reduction. The latter result also supports the hypothesis that an antiferromagnetic interaction is responsible for the reduction of net paramagnetism in the oxidized enzyme. The two other likely explanations, line broadening due to an increase in the electron spin relaxation rate or a magnetic dipole interaction between neighboring electron spins, would still result in a measurable bulk magnetic susceptibility.

Another hypothesis to explain the appearance of the high-spin ferric signals is that the introduction of electrons into the system causes a conformational change which alters the antiferromagnetic interaction of the iron in the heme groups. This is supported by the observation that the low-spin Fe(III) signals present in the oxidized enzyme decrease while the high-spin Fe(III) signals increase although, admittedly, it is not certain whether or not these changes are synchronized. This hypothesis does not require that the high-spin signal result from conversion of the low-spin signal, a reaction that has been rendered unlikely particularly by the observation[92] that denaturation of the enzyme results in the appearance of the high-spin signal without any decline of the low-spin signal.

Properties of copper

The function of copper in cytochrome oxidase has been one of its most controversial aspects. Keilin[14,15] was among the first to consider it as a functional component of terminal respiration but discarded this possibility when he demonstrated that cytochrome a_3 satisfactorily fulfills that role. Wainio and his associates[24] detected significant concentrations of copper in their purified oxidase preparations and briefly considered the possibility that the heme in cytochrome oxidase is really a cuproporphyrin. However, they discounted that hypothesis when they clearly demonstrated that iron is coordinated to the porphyrin. The case for a functional role of copper in cytochrome oxidase was strengthened by the results of Green *et al.*[97], Mackler and Penn[98] and of Wainio and co-workers[99] who showed a parallel increase of copper, cytochrome oxidase activity, and heme *a* on partial purification. Further support came from nutritional studies[100-104] which showed that the concentration of cytochrome oxidase is decreased in animals fed copper-deficient diets. Suggestive as these results were they did not supply sufficient evidence to justify naming copper as a functional constituent of cytochrome oxidase. It is only within the past ten years that this evidence has been gathered.

In order to ascertain that a metal is an active prosthetic group in an enzyme that undergoes oxidoreduction, a number of requirements must be met. In the case of copper in cytochrome oxidase, these may be enumerated as follows: (a) there must be a stoichiometric relationship between copper and heme *a*; (b) the concentration of copper must increase concomitantly with that of the heme during purification; (c) the copper should not be easily displaced from the protein; (d) reagents which complex copper so as to block changes in its oxido-reduction state should cause an inhibition of enzymic activity; (e) removal of the copper must result in a parallel decrease in enzymic activity; (f) replacement of copper to a copper-depleted oxidase must result in a parallel restoration of activity; (g) the copper in the completely oxidized or completely reduced enzyme must have the same oxidation state as that of the heme iron and the rate of its oxidation and reduction must approximate that of the heme components.

As already noted, Mackler and Penn[98] and Wainio *et al.*[99] showed that some increase in the concentration of copper occurred as cytochrome oxidase is purified. Later Griffiths and Wharton[28,105] clearly demonstrated a parallel increase of copper and heme *a* during purification of their highly purified preparation. Furthermore, they established that a 1:1 molar ratio of copper and heme *a* is present at all stages of purification. These results have since been confirmed by a number of investigators[106–110].

Most reagents which chelate copper and which inhibit other copper enzymes do not inhibit cytochrome oxidase[105]. This has led to some confusion as to the role of copper in cytochrome oxidase. However, it has been amply demonstrated by both chemical and e.p.r. methods that reagents such as bathocuproine disulfonate (BCS) do not combine with the copper of cytochrome oxidase[105,111]. Apparently, the copper is sufficiently shielded by the hemoprotein so as to prevent its reaction with externally added complexing reagents. There is evidence to indicate that cyanide, which does inhibit cytochrome oxidase, combines with a part of the copper in the enzyme as well as reacting with the heme iron of cytochrome a_3[112].

The copper of cytochrome oxidase is not readily displaced from the protein. Griffiths and Wharton[113] could detect no exchange between added ^{64}Cu and enzymic copper during purification of their enzyme or during turnover experiments. Wharton and Tzagoloff[114] succeeded in removing copper from the oxidase by dialyzing the enzyme against high concentrations of cyanide or against lower concentrations of cyanide in moderately alkaline buffers of relatively high ionic strength. Copper was also removed by treatment of the oxidase with BCS at pH 4–5. In each case, removal of the copper is accompanied by a parallel decrease of oxidase activity. Attempts by these workers to reconstitute the copper-depleted enzyme were unsuccessful. These results have been interpreted to indicate that in order for copper to react with and be removed by chelators the protein must unfold in the vicinity of the copper. Thus, it is not surprising that the reverse process may be difficult since, on addition of copper, the protein would have to be unfolded with the copper-binding site exposed and then refolded to its original state following binding of the copper.

The earlier reports by Vander Wende and Wainio[115] of the successful reconstitution of a copper-depleted oxidase have not been confirmed. The results of Nair and Mason[116] which indicated the successful removal and reconstitution of copper with a parallel loss and restoration of activity, have been shown recently to be erroneous due to artifacts in their copper assay procedure[117]. Thus, the reconstitution of the enzyme with copper remains the most serious unsolved problem related to its functional role in cytochrome oxidase.

Although some earlier reports[114] on the valence state of copper were in error due to its reduction by protein-bound SH-groups, chemical studies using valence-specific chelating agents by Takemori *et al.*[117] and by Griffiths and Wharton[105] indicate that copper in the oxidized enzyme is in the cupric state. Reduction of the cytochromes is accompanied by a parallel reduction of the copper. Similar results using e.p.r. spectroscopy have been obtained by Beinert and his associates[111,120].

A broad absorption band centered at 830 nm (cf. Fig. 6) was discovered in ferricytochrome oxidase by Griffiths and Wharton[28]. This band with a millimolar extinction coefficient of about 1.4 disappears when the oxidase is reduced and reappears on reoxidation. Cyanide does not interfere with the reduction of the band but blocks its reoxidation. These results have since been confirmed in several laboratories[32,50,120]. Because of its similarity to the absorption band of many copper proteins, Griffiths and Wharton suggested that it might be an absorption band of copper in the oxidase.

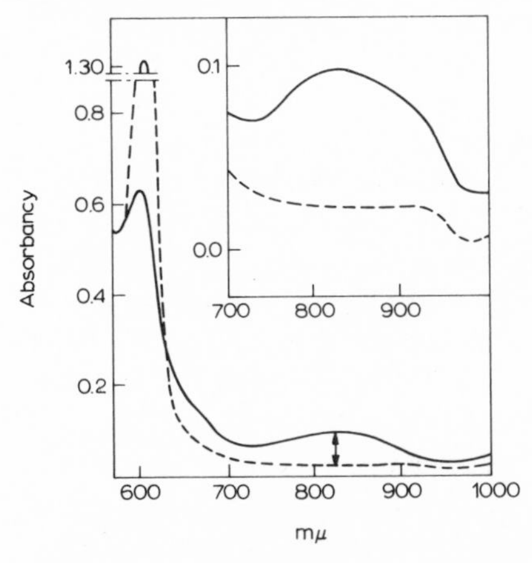

Fig. 6. Absorption spectrum of cytochrome oxidase from 580 to 900 nm. The oxidized preparation, ——; the reduced preparation, - - - -. 8.0 mg of protein per ml.

This suggestion was reinforced by the copper removal studies of Wharton and Tzagoloff[114] in which a concomitant loss of copper and the 830 nm band was observed.

Whether the 830 nm band represents all of the enzymic copper in cytochrome oxidase has not been determined with certainty. Work of vanGelder, Muijsers *et al.*[49,54,111] suggests that all of the enzymic copper contributes to the band. This conclusion is based on the observation that in the presence of cyanide only half the copper is reduced by substrate and the absorbance of 830 nm is decreased by about 50%.

The hypothesis that the 830 nm band is due to a copper chromophore has been criticized by Caughey[120] who contends that it may be a band of heme *a* since several ferric hematins are known to have absorption bands in the near infra-red region. This argument is partly blunted by the demonstration that the 830 nm chromophore can be reoxidized completely by ferricyanide while cytochrome a_3 remains blocked in the reduced form by carbon monoxide[122]. Gilmour[123] has also presented evidence that the 830 nm band is not due to cytochrome *a* by showing that the band reappears before cytochrome *a* is oxidized following autoxidation of an oxidase preparation in which the 830 nm chromophore and cytochrome *a*, but not cytochrome a_3, had been reduced. The independence of the 830 nm chromophore from cytochromes *a* and a_3, has also been supported by the titration experiments of vanGelder and associates[54,55,112] and by the experiments of Gibson *et al.*[124,125] which indicate that the 830 nm band is kinetically independent of the cytochrome bands in the visible region.

Although the available evidence supports the conclusion that the 830 nm band is an absorption band of copper the question could finally be settled by the successful reconstitution of a copper-depleted preparation.

A copper protein has been isolated from cytochrome oxidase by Tzagoloff and MacLennan[126] using Sephadex fractionation following succinylation of acetone-treated oxidase. This protein contains 2 copper atoms per 25,000 molecular weight. However, there is not sufficient evidence to conclude that such a protein component exists in the original enzyme preparation.

Little is known about the binding of copper to the protein in cytochrome oxidase although Hemmerich[127,128], on theoretical grounds as well as on the basis of some model studies with Beinert, has suggested the involvement of a sulfur ligand. Tsudzuki *et al.*[129] have studied the effect of titration with *p*-chloromercuribenzoate (PCMB) on the reaction of bathocuproine disulfonate (BCS) with enzymic copper in an oxidase preparation. They found that the PCMB could induce a reaction with BCS only when the enzyme is denatured with sodium dodecyl sulfate and suggested that the copper is bound to the protein through a sulfhydryl group or groups. This postulate requires a rigorous test, however, since it is well known that the titration with PCMB could result in a reaction between copper and BCS by other than a direct effect.

E.p.r. studies on copper

The e.p.r. spectrum of copper in cytochrome oxidase was first recorded by Sands and Beinert[130] who also observed that the cupric signal of the oxidized enzyme disappears upon the addition of substrate. In more detailed experiments Beinert and his associates[111] found that the e.p.r. spectrum, shown in Fig. 7, accounts for only about 40% of the total copper present in the enzyme and is characterized by a significant lack of hyperfine structure. Beinert *et al.* suggested that these spectral characteristics could be explained by an exchange interaction between a Cu^{2+}–Cu^{2+} or a Cu^{+}–Cu^{2+} pair as well as between a Cu^{2+}–Fe^{3+} pair. Such interactions can lead to a diminution of the signal of the magnitude observed with cytochrome oxidase even if all the copper is present as Cu^{2+}. Although other possible explanations are not rigorously excluded, an interaction with antiferromagnetic coupling is also supported by the observation that on denaturation of the oxidase by heat or urea in the presence of PCMS, all the copper in the protein can be accounted for quantitatively. Furthermore, as seen in Fig. 8 a considerable amount of hyperfine structure appears. Thus, two types of copper in cytochrome oxidase must be considered: one which has an e.p.r. signal and the other which has no e.p.r. signal.

Beinert and co-workers[111] have also demonstrated that the oxidation state of the copper closely parallels that of the heme although their low time resolution merely suggests a functional relationship. At about the same time that these experiments were being conducted, Ehrenberg and

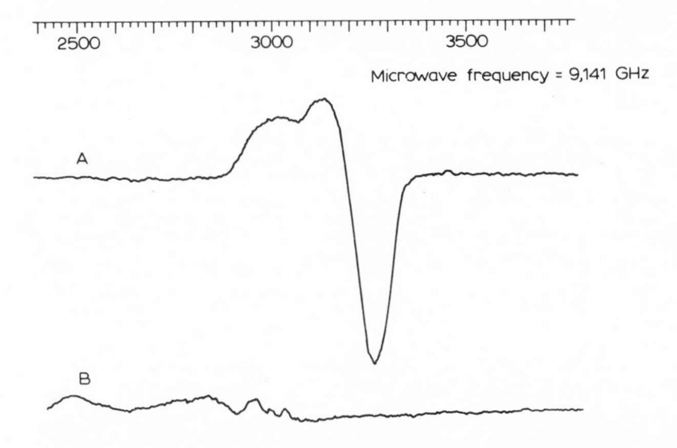

Fig. 7. E.p.r. spectra of copper in cytochrome oxidase. A, the oxidized preparation; B, the preparation reduced with cytochrome *c* and ascorbate. The conditions of e.p.r. spectroscopy were: microwave power, 25 mW; modulation amplitude, 20 gauss; scanning rate, 250 gauss/min; time constant, 1.0 sec; temperature, 97K.

Yonetani[90] reported that the copper present in cytochrome oxidase is not reduced significantly by ferrocytochrome *c* despite the fact that the cytochrome components are easily reduced. Thus, they concluded that the copper present in the oxidase is not a functional constituent of the enzyme. This controversy was clarified by Beinert and his associates[111] who pointed out that the oxidase preparation used by Ehrenberg and Yonetani had a copper to heme ratio of about two, whereas the ratio in preparations used in Beinert's laboratory approached unity. The considerable hyperfine structure present in the former preparation, which Ehrenberg and Yonetani had considered to be a property of enzymic copper, was shown to be a feature of damaged oxidase or of preparations containing adventitiously bound copper and was no doubt attributable to the excess copper present in their preparation. Beinert and Palmer[119] studied the effects of microwave power on the saturation of the e.p.r. signals and showed that the signal responsible for the hyperfine structure is readily saturated whereas that signal without significant hyperfine structure is not power-saturated at -170°C. The former signal is not diminished by substrates while the latter is readily reduced. Furthermore, Ehrenberg and Yonetani used a low microwave power in their experiments which accentuates the copper species with the hyperfine structure relative to the other as compared to spectra taken at high power.

Using a rapid freezing technique[131] Beinert and Palmer demonstrated that the oxidation state of the e.p.r.-detectable copper changes at a rate which is parallel to that of the cytochromes. Though this evidence further suggests a participatory role for copper the time resolution of the rapid-

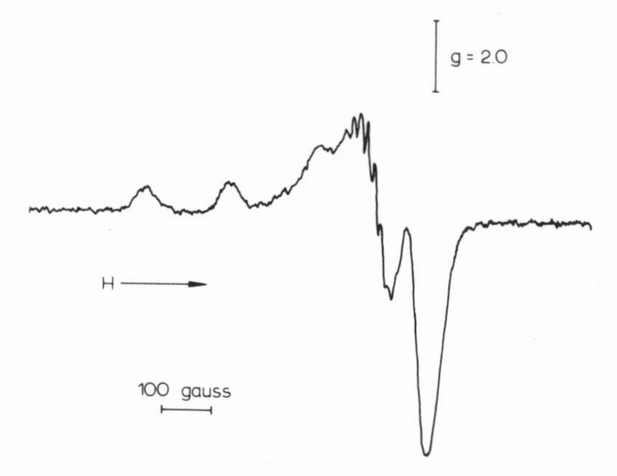

Fig. 8. E.p.r. spectrum of copper in cytochrome oxidase after treatment of the enzyme with 6.5 *M* urea and 15 m*M* *p*-chloromercuriphenyl sulfonate. The conditions of e.p.r. spectroscopy were: microwave power, 25 mW; modulation amplitude, 5 gauss; scanning rate, 250 gauss/min; time constant, 0.5 sec.; temperature, 103K.

freezing technique of 10 msec is not sufficiently fast during the oxidation of the enzyme to conclude positively that this is so.

Oxidation–reduction potentials

Values for the mid-point potential of cytochrome oxidase have appeared from time to time[132-134]. These reports with values near 290 mV have been based on measurements of the α-band of the oxidase and, thus, include both cytochromes a and a_3 in the calculation. Tzagoloff and Wharton[42] were the first to measure the midpoint potential of cytochrome a distinct from cytochrome a_3 by blocking the latter in the reduced form with carbon monoxide. By oxidizing the reduced cytochrome a anaerobically with ferricyanide they obtained a midpoint potential of 250 mV. However, it is unknown whether the presence of carbon monoxide modified the potential of the cytochrome a.

Horio and Ohkawa[135] have obtained values for three different heme *a* groups with mid-point potentials of 338, 300 and 208 mV, respectively. However, these values were calculated using non-integral values of N in the Nernst equation. Cusanovich and Wharton (unpublished results) have determined values of about 360 mV and 210 mV for the mid-point potentials of cytochrome a_3 and cytochrome *a*, respectively, in a purified preparation using integral values of N in a computer program to give a line of best fit with experimental results. A reasonable fit was obtained on the basis of two heme components.

Wilson and Dutton[136] have calculated mid-point potentials of 190 mV for cytochrome *a* and 395 mV for cytochrome a_3 in rat liver mitochondria under phosphorylating conditions. The addition of ATP to the system does not affect the value for cytochrome *a* but does change that of cytochrome a_3 to about 300 mV. These results contrast with that of Caswell[137] who, in the same laboratory, obtained a value of 330 mV for cytochrome *a* in cyanide-inhibited rat liver mitochondria. The reason for this difference has not been determined.

The oxidation–reduction potential of copper in purified cytochrome oxidase has been determined utilizing the 830 nm band. Tzagoloff and MacLennan[126] have calculated a mid-point potential of 284 mV during reductive titrations with ascorbate in the presence of cytochrome *c*. Using a similar reducing system, Wharton and Cusanovich[138] found the same mid-point potential while with dithionite as a reductant, they determined a mid-point potential of 278 mV. These values were repeated on oxidation by ferricyanide. Unfortunately, these measurements do not discriminate between one or two copper groups contributing to the absorption band at 830 nm. However, by measuring changes in the e.p.r.-copper signal on reduction by ascorbate and cytochrome *c*, Wharton and Cusanovich (unpublished results) have calculated a mid-point potential similar to that obtained using the near infra-red band.

These results of thermodynamic equilibrium seem to suggest a sequence of:

cytochrome $a \rightarrow$ copper $\rightarrow$ cytochrome a_3,

for the potential electron carriers in cytochrome oxidase.

REACTION OF CYTOCHROME OXIDASE

In the course of its work cytochrome oxidase takes an electron away from ferrocytochrome *c* and passes it along to a molecule of oxygen whereupon the latter, after its acceptance of four such electrons and four protons, becomes converted into two molecules of water. Although some information, particularly the rates of these reactions and their inhibition by certain reagents, is known, the mechanisms by which these reactions occur remain unknown. Of course, not even considered here is the fact that the oxidation of cytochrome *c* through cytochrome oxidase is accompanied by the synthesis of ATP, the mechanism of which is an even greater mystery.

Reaction with cytochrome c

There is some evidence[139-141] to suggest that the initial reaction between cytochrome oxidase and cytochrome *c* occurs by an electrostatic interaction between the acidic oxidase and the basic substrate. Experiments by Okunuki and his collaborators[140,141] have shown that acetylation and succinylation of cytochrome *c* results in an inhibition of cytochrome oxidase activity. Succinylation, which introduces an acidic group as well as neutralizing a basic group on the protein, results in a more potent inhibition than acetylation, which only does the latter.

A stable enzyme–substrate complex is suggested by the work of King *et al.*[142], of Nicholls[143], and of Orii *et al.*[144]. King's group has obtained a complex of cytochrome *c* and cytochrome oxidase by chromatographing a mixture of these proteins through a column of Sephadex G-200. The resulting complex has a heme *c*:heme *a* ratio of 1:1 and is stable to centrifugation although it can be separated in solutions of high ionic strength. Acetylation or succinylation of the cytochrome *c* inhibits the formation of this complex. Nicholls has also observed a stable cytochrome *c*–cytochrome oxidase complex where the ratio of heme *a* to heme *c* is 2:1 and which can be cleaved in solutions of high ionic strength. Orii *et al.* have obtained a complex by paper chromatography with a ratio of heme *a* to heme *c* of 1:1. The reason for the difference in the ratios of hemes is not clear at this time.

In the course of their studies on the reactions of cytochrome oxidase Wharton and Griffiths[145] obtained a K_m of about 10^{-5} M for cytochrome *c*. This is in agreement with the results of McGuiness and Wainio[146] but is slightly different from the results of Yonetani and Ray[147] who determined values near 8×10^{-6} M for the purified enzyme.

Gibson and his associates[148] have studied the reaction between cytochrome *c* and cytochrome oxidase by means of rapid reaction techniques[149]. They have found that the addition of reduced cytochrome *c* to oxidized cytochrome oxidase produces a very rapid reduction of cytochrome *a* with a rate constant of about 4×10^7 M^{-1} sec^{-1}. Their evidence indicates that this fast reaction involves only cytochrome *a* and not cytochrome a_3; the reduction of the latter follows at a much slower rate. This observation is somewhat surprising since earlier Gibson and Greenwood[150] had observed in oxidation experiments that the rate of the reaction between cytochrome *a* and cytochrome a_3 is at least 750 sec^{-1}. One explanation advanced to explain these results is that cytochrome *c* forms an inhibitory complex with cytochrome oxidase whereby turnover of cytochrome *a* depends upon the slow dissociation of ferricytochrome *c* from the complex. This interpretation is supported by the work of Smith *et al.*[139,151] and of McGuiness and Wainio[146] who have found that increasing concentrations of cytochrome *c* inhibit the activity of cytochrome oxidase. Yonetani and Ray[147] have determined that the oxidized cytochrome *c* is the inhibitory species. The objections of Lemberg[68] to the existence of an inhibitory cytochrome *c*–cytochrome oxidase complex seem unwarranted in view of the positive evidence in its favor.

Experiments in Gibson's laboratory[148] indicate an apparent heat of activation of 15 kcal per mole for the oxidase–cytochrome *c* reaction. The rate constant is somewhat larger than would be expected to correspond to this value thus indicating the possible role of an intermediate complex in the reaction.

Reaction with oxygen

The terminal step in the cytochrome oxidase reaction is the transfer of electrons to oxygen. Oxygen has been shown to have a very high affinity for cytochrome oxidase[152,153]. The turnover decreases only when the concentration of oxygen is below 10^{-6} *M*. Although H_2O has been shown to be the end product of these reactions[154] no partially reduced intermediates have ever been detected. This is hardly surprising in the light of the high speed of this reaction as measured by Gibson and his associates.

Gibson and his colleagues[52,150] have studied the reaction of reduced cytochrome oxidase with oxygen under a variety of conditions. The principal method used, called flow-flash, involves mixing the carbon monoxide complex of the reduced enzyme with buffer containing oxygen and initiating the reaction by the photochemical removal of the inihibitory carbon monoxide. Under the proper conditions first order rate constants up to 10^5 sec^{-1} can be measured by this technique. Thus Gibson *et al.*[52] have found that oxygen enters the oxidase unit at a diffusion-controlled rate (1×10^8 M^{-1} sec^{-1}) followed by the first order oxidation of cytochrome a_3 with a rate of 3×10^4 sec^{-1} and of cytochrome *a* at 7×10^2 sec^{-1}. As seen elsewhere in this discussion the copper component, as represented by the absorption band at 830 nm, is oxidized at a rate of 7×10^3 sec^{-1}. From these data Greenwood and Gibson[52] concluded that the intermediates of the reaction must have half-lives of 10 μsec or less.

The assignment of the reaction rates to cytochrome *a* or cytochrome a_3 in these studies is based on the biphasic nature of the absorbance changes in the spectral bands at 605 nm and 444 nm, cytochrome a_3 contributing proportionately less in the α-region than in the Soret region. This technique also permits the spectrokinetic separation of the oxidized minus reduced difference spectra of cytochromes *a* and a_3. These results are essentially in agreement with those obtained from static experiments except, as mentioned

References pp. 983–987

previously, that the ratio of the contribution of cytochrome a_3 to that of cytochrome a at 445 nm is 2.3:1 instead of 1:1 as obtained in the static method.

The oxidation of reduced cytochrome oxidase observed in a spectrophotometer on a slow time scale is accompanied by the appearance of a different spectral form with absorption bands at 601 nm and 428 nm[155-157]. It is most commonly observed after oxidation of a dithionite-reduced enzyme although it, or a very related species, can be formed following reaction of the oxidized or reduced enzyme with hydrogen peroxide[49,158,159] or after prolonged treatment of the oxidized enzyme with formamidine-sulfinic acid[160,161]. The product obtained after oxidation of the dithionite-reduced oxidase reverts gradually to the spectral form of the oxidized enzyme although this conversion can be speeded significantly by the addition of cytochrome c[162]. Although this observation has been confirmed[159] it is difficult to see how it supports the argument of Okunuki[155] and of Wainio and their colleagues[20] that the presence of cytochrome c is required for the oxidation of cytochrome oxidase, particularly in view of the myriad of observations[68,16'] that the dithionite-reduced oxidase is rapidly oxidized in the absence of cytochrome c. It has been demonstrated that this new species is not a mixture of oxidized and reduced enzyme[49]. During reduction by ferrocytochrome c or dithionite this species follows a less biphasic reaction than does the oxidized enzyme[160]. This spectral form has created a lively controversy as to its origin and function and is most commonly referred to as the oxygenated form or complex although its association with oxygen has yet to be demonstrated. Okunuki and his associates[155,156,162], Wainio and his co-workers[20,110,157], and more recently, Lemberg[68,163] have advocated for this species a role as an intermediate in the oxidation of cytochrome oxidase. On the other hand, Wharton and Gibson[161] have demonstrated, by using rapid-reaction methods, that the spectral form that appears within a few msec after reaction of the reduced enzyme with oxygen corresponds to the oxidized enzyme which in turn is converted over a period of seconds to what corresponds spectrally with the so-called oxygenated form. Wharton and Gibson argue that the latter can not be treated as an intermediate in the oxidation of cytochrome oxidase since it appears many seconds after oxidation whereas expected intermediates should have half-lives in the order of μsec.

Lemberg, who earlier was critical[68] of the conclusions of Wharton and Gibson, has more recently approached their point of view. Thus, Lemberg and Cutler[163] have proposed that a form of ferricytochrome a_3 occurs on oxidation which is conformationally distinct from the ferricytochrome a_3 in the static oxidized enzyme, although differentiation is difficult since both forms have a Soret band at 418 nm. They interpret the species with a Soret band at 428 nm as being due to a ferryl form of cytochrome a_3, on the basis that this form consumes more electrons on titration with ferrocytochrome c than does the static oxidized species. However, equally likely is the formation of sulfhydryl groups during the reduction as might be expected following a conformational change, the latter being suggested by the c.d. studies of the so-called oxygenated form by Yong, Meyer, and King (unpublished results). In any event, it is difficult at this time to conclude from available evidence that the 428 nm species occupies a functional role as an intermediate in the oxidation reaction.

Various schemes have appeared to describe the reaction of oxygen with the oxidase but due to a lack of experimental evidence they will not be discussed here. Those who wish to pursue them may find some included in several reviews[68,154,164–166].

Reaction of copper

Gibson and his co-workers[124,125] have studied the rate of oxidation of copper in reduced cytochrome oxidase by following the absorbance changes of the near infra-red band using the flow-flash technique. They have concluded that the increase in absorbance observed at 830 nm occurs at a rate of about 7×10^3 sec^{-1} which is different from those found in the case of the oxidation of cytochrome a and cytochrome a_3; 7×10^2 sec^{-1} and 3×10^4 sec^{-1}, respectively. Thus, if the assignment of the 830 nm band is correct, a strong case can be made in favor of copper as an integral electron carrier in cytochrome oxidase and the following reaction sequence assigned: cytochrome $a \rightarrow$ copper $\rightarrow$ cytochrome $a_3 \rightarrow O_2$.

Since it is uncertain whether all or only part of the total enzymic copper contributes to the absorption at 830 nm it is impossible to assign a place to a hypothetical invisible copper atom with relation to the cytochromes. However, it is an intriguing possibility that this invisible copper, if present, reacts directly with oxygen. This receives some support from the rapid kinetic measurements of Greenwood and Gibson[52] who found a considerable interval between the time when oxygen diffuses into the system and when cytochrome a_3 becomes oxidized. An invisible copper might be expected to react directly with oxygen and then in turn with cytochrome a_3 at the observed first order rate of 3×10^4 sec^{-1}. However, this remains highly conjectural.

Reactions with inhibitors

The spectral observation that carbon monoxide reacts with only part of the total heme a in cytochrome oxidase has been a prime reason for postulating the existence of cytochromes a and a_3. The carbon monoxide binding capacity of the oxidase has been measured in purified preparations and in respiratory particles derived from beef heart mitochondria[150,167-171]. Using ^{14}C-labeled carbon monoxide Gibson *et al.*[167] found that in the particles close to 50% of the total heme *a* becomes liganded and concluded that a 1 to 1 ratio of cytochromes *a* and a_3 exists in cytochrome oxidase. With purified preparations these workers and others[169-171] have usually obtained data which indicate a ratio of $a{:}a_3$ of between 1 and 2 to 1. The reason for the discrepancy may be due to an alteration of cytochrome a_3 in the purified preparations resulting in a loss of reactivity. The presence of low-spin Fe(III) forms as intermediates during the reduction of purified cytochrome oxidase[92] may be an expression of altered cytochrome a_3.

Wald and Allen[172] measured an affinity constant of 3.5×10^{-7} *M* for the reaction between carbon monoxide and cytochrome oxidase in heart muscle.

Cyanide has also been shown to bind to cytochrome a_3 on the basis of spectral studies. However, unlike carbon monoxide cyanide has been observed to bind two sites on the enzyme[173]. The reduction of the oxidase in the presence of cyanide[112] also suggests two sites of binding. Although there is no overpowering evidence to support the conclusion, copper has been suggested as the second binding site.

CONCLUSIONS

Cytochrome oxidase, a mitochondrial enzyme which oxidizes ferrocytochrome *c* and in turn is itself oxidized by molecular oxygen, is a lipoprotein containing prosthetic groups of heme *a* and copper in an equimolar ratio. The enzyme most probably reacts with cytochrome *c* through an electrostatic interaction while oxygen appears to approach the reaction site in a diffusion-controlled reaction. Evidence supports the conclusion that there are two functionally distinct cytochromes present, named *a* and a_3, from which it can be postulated that a minimum functional unit is composed of one cytochrome *a*, one cytochrome a_3, and two atoms of copper. One atom of copper may be associated with each of the cytochromes. The distinction between cytochrome *a* and cytochrome a_3 may be related to the binding of heme *a* to the protein but equally possible is a difference arising from the relationship of copper to each heme group or from a conformational change which affects the spin-state of the heme iron. The available evidence suggests a sequence of components from cytochrome *c* to cytochrome *a*–copper–cytochrome a_3–O_2 in which all the prosthetic groups undergo rapid changes in their oxidation state. Unfortunately, the mechanisms of these reactions remain unknown.

ACKNOWLEDGEMENTS

The author wishes to express his appreciation to Dr. Quentin Gibson for his helpful comments on the manuscript and to Dr. Helmut Beinert for providing the figure of e.p.r. spectra of heme iron. The author is the recipient of a USPHS Career Development Award HE 42378.

REFERENCES

1 C. A. MacMunn, *J. Physiol. London*, 5 (1884) xxiv.
2 C. A. MacMunn, *Phil. Trans. R. Soc.*, 177 (1886) 267.
3 C. A. MacMunn, *J. Physiol. London*, 8 (1887) 51.
4 O. Warburg and E. Negelein, *Biochem. Z.*, 113 (1921) 257.
5 O. Warburg, *Biochem. Z.*, 119 (1921) 134.
6 O. Warburg, *Biochem. Z.*, (1924) 479.
7 O. Warburg and E. Negelein, *Biochem. Z.*, 193 (1928) 339.
8 O. Warburg and E. Negelein, *Biochem. Z.*, 200 (1928) 414.
9 D. Keilin, *Proc. R. Soc. (B)*, 98 (1925) 312.
10 D. Keilin, *Proc. R. Soc. (B)*, 100 (1926) 129.
11 D. Keilin, *Proc. R. Soc. (B)*, 104 (1929) 206.
12 D. Keilin, *Proc. R. Soc. (B)*, 106 (1930) 418.
13 D. Keilin and E. F. Hartree, *Proc. R. Soc. (B)*, 125 (1938) 171.
14 D. Keilin and E. F. Hartree, *Nature*, 141 (1938) 870.
15 D. Keilin and E. F. Hartree, *Proc. R. Soc. (B)*, 127 (1939) 167.
16 D. E. Green and D. C. Wharton, *Biochem. Z.*, 338 (1963) 335.
17 W. W. Wainio, in J. E. Falk, R. Lemberg and R. K. Morton, *Haematin Enzymes*, Pergamon Press, New York, 1961, p. 281.
18 T. Horio, I. Sekuzu, T. Higashi and K. Okunuki, in J. E. Falk, R. Lemberg and R. K. Morton, *Haematin Enzymes*, Pergamon Press, New York, 1961, p. 302.
19 K. Okunuki, I. Sekuzu, Y. Orii, T. Tsudzuki and Y. Matsumara, in K. Okunuki, M. D. Kamen and I. Sekuzu, *Structure and Function of Cytochromes*, University Park Press, Baltimore, 1968, p. 35.
20 W. W. Wainio, D. Grebner and H. O'Farrell, in K. Okunuki, M. D. Kamen and I. Sekuzu, *Structure and Function of Cytochromes*, University Park Press, Baltimore, 1968, p. 66.
21 A. L. Lehninger, *The Mitochondrion*, W. A. Benjamin, New York, 1965.
22 D. E. Green and H. Baum, *Energy and the Mitochondrion*, Academic Press, New York, 1969.
23 E. Racker, *Membranes of Mitochondria and Chloroplasts*, Van Nostrand Reinhold, New York, 1970.
24 B. Eichel, W. W. Wainio, P. Person and S. J. Cooperstein, *J. Biol. Chem.*, 183 (1950) 89.
25 L. Smith and E. Stotz, *J. Biol. Chem.*, 209 (1954) 819.
26 K. Okunuki, I. Sekuzu, T. Yonetani and S. Takemori, *J. Biochem. Tokyo*, 45 (1958) 847.
27 T. Yonetani, *J. Biol. Chem.*, 236 (1961) 1680.
28 D. E. Griffiths and D. C. Wharton, *J. Biol. Chem.*, 236 (1961) 1850.
29 L. R. Fowler, S. H. Richardson and Y. Hatefi, *Biochim. Biophys. Acta*, 64 (1962) 170.
30 S. Horie and M. Morrison, *J. Biol. Chem.*, 238 (1963) 1855.

31 W. W. Wainio, *J. Biol. Chem.*, 239 (1964) 1402.
32 R. Lemberg, T. B. G. Pilger, N. Newton and L. Clark, *Proc. R. Soc. (B)*, 159 (1964) 405.
33 E. E. Jacobs, E. C. Andrews, W. Cunningham and F. L. Crane, *Biochem. Biophys. Res. Commun.*, 25 (1966) 87.
34 F. F. Sun and E. E. Jacobs, *Biochim. Biophys. Acta*, 143 (1967) 639.
35 S. Takemori, I. Sekuzu and K. Okunuki, *Biochim. Biophys. Acta*, 51 (1961) 464.
36 A. Tzagoloff, P. S. Yang, D. C. Wharton and J. S. Rieske, *Biochim. Biophys. Acta*, 96 (1965) 1.
37 Y. Orii and K. Okunuki, *J. Biochem. Tokyo*, 61 (1967) 388.
38 R. S. Criddle and R. M. Bock, *Biochem. Biophys. Res. Commun.*, 1 (1959) 138.
39 H. S. Mason and K. Ganapathy, *J. Biol. Chem.*, 245 (1970) 230.
40 H. Matsubara, Y. Orii and K. Okunuki, *Biochim. Biophys. Acta*, 97 (1965) 61.
41 S. Fleischer, H. Klouwen and G. Brierley, *J. Biol. Chem.*, 236 (1961) 2936.
42 A. Tzagoloff and D. C. Wharton, *J. Biol. Chem.*, 240 (1965) 2628.
43 S. Horie and M. Morrison, *J. Biol. Chem.*, 238 (1963) 2859.
44 W. W. Wainio, *J. Biol. Chem.*, 212 (1955) 723.
45 T. Yonetani, *J. Biol. Chem.*, 235 (1960) 845.
46 Y. Orii and K. Okunuki, *J. Biochem. Tokyo*, 55 (1964) 37.
47 S. Horie and M. Morrison, *J. Biol. Chem.*, 238 (1963) 2859.
48 R. Lemberg and G. E. Mansley, *Biochim. Biophys. Acta*, 118 (1966) 19.
49 B. F. vanGelder, *Biochim. Biophys. Acta*, 118 (1966) 36.
50 W. H. Vanneste and M. T. Vanneste, *Biochem. Biophys. Res. Commun.*, 19 (1965) 182.
51 C. Greenwood and Q. H. Gibson, *J. Biol. Chem.*, 242 (1967) 1782.
52 I. Sekuzu, S. Takemori, T. Yonetani and K. Okunuki, *J. Biochem. Tokyo*, 46 (1959) 43.
53 A. O. Muijsers, E. C. Slater and K. J. H. van Buuren, in K. Okunuki, M. D. Kamen and I. Sekuzu, *Structure and Function of Cytochromes*, University Park Press, Baltimore, 1968, p. 129.
54 A. O. Muijsers, B. F. vanGelder and E. C. Slater, in B. Chance, R. W. Estabrook and T. Yonetani, *Hemes and Hemoproteins*, Academic Press, New York, 1966, p. 467.
55 D. W. Urry, W. W. Wainio and D. Grebner, *Biochem. Biophys. Res. Commun.*, 27 (1967) 625.
56 D. W. Urry and B. F. van Gelder, in K. Okunuki, M. D. Kamen and I. Sekuzu, *Structure and Function of Cytochromes*, University Park Press, Baltimore, 1968, p. 210.
57 F. C. Yong and T. E. King, *Biochem. Biophys. Res. Commun.*, 38 (1970) 940.
58 Y. P. Myer and T. E. King, *Biochem. Biophys. Res. Commun.*, 34 (1969) 170.
59 V. Shashoua, in B. Chance, R. W. Estabrook and T. Yonetani, *Hemes and Hemoproteins*, Academic Press, New York, 1966, p. 503.
60 M. Morrison, in B. Chance, R. W. Estabrook and T. Yonetani, *Hemes and Hemoproteins*, Academic Press, New York, 1966, p. 503.
61 T. E. King, in B. Chance, R. W. Estabrook and T. Yonetani, *Hemes and Hemoproteins*, Academic Press, New York, 1966, p. 505.
62 J. A. Schellman and T. E. King, in B. Chance, R. W. Estabrook and T. Yonetani, *Hemes and Hemoproteins*, Academic Press, New York, 1966, p. 507.
63 T. E. King, F. C. Yong and P. M. Bayley, in K. Okunuki, M. D. Kamen and I. Sekuzu, *Structure and Function of Cytochromes*, University Park Press, Baltimore, 1968, p. 204.
64 T. E. King and J. A. Schellman, *Federation Proc.*, 25 (1966) 411.
65 R. Lemberg, *Adv. Enzymol.*, 23 (1961) 265.

66 J. L. York, S. McCoy, D. N. Taylor and W. S. Caughey, *J. Biol. Chem.*, 242 (1967) 908.
67 W. S. Caughey, J. L. Davies, W. H. Fuchsman and S. McCoy, in K. Okunuki, M. D. Kamen and I. Sekuzu, *Structure and Function of Cytochromes*, University Park Press, Baltimore, 1968, p. 20.
68 R. Lemberg, *Physiol. Rev.*, 49 (1969) 48.
69 O. Warburg, *Heavy Metal Prosthetic Groups*, Oxford University Press, Oxford, 1949, p. 165.
70 J. E. Falk, *Porphyrins and Metalloporphyrins*, Elsevier, Amsterdam, 1964, p. 98.
71 W. S. Caughey and J. L. York, *J. Biol. Chem.*, 237 (1962) PC 2414.
72 R. A. Nicolaus and L. Mangoni, *Ann. Chim. Rome*, 48 (1958) 400.
73 M. Piattelli, *Tetrahedron*, 8 (1960) 266.
74 R. Lemberg, P. Clezy and J. Barrett, in J. E. Falk, R. Lemberg and R. K. Morton, *Haematin Enzymes*, Vol. 1, Pergamon Press, New York, 1961, p. 344.
75 O. Warburg and H.-S. Gewitz, *Z. Physiol. Chem.*, 292 (1953) 174.
76 M. Grassl, U. Coy, R. Seyffert and F. Lynen, *Biochem. Z.*, 337 (1963) 35.
77 W. S. Caughey, S. McCoy and J. L. York, *Federation Proc.*, 25 (1966) 647.
78 G. S. Marks, D. K. Dougall, E. Bullock and S. F. MacDonald, *J. Am. Chem. Soc.*, 82 (1960) 3183.
79 W. S. Caughey, J. L. York, S. McCoy and D. P. Hollis, in B. Chance, R. W. Estabrook and T. Yonetani, *Hemes and Hemoproteins*, Academic Press, New York, 1966, p. 25.
80 J. Barrett, *Nature*, 183 (1959) 1185.
81 R. Lemberg, in B. Chance, R. W. Estabrook and T. Yonetani, *Hemes and Hemoproteins*, Academic Press, New York, 1966, p. 37.
82 M. Grassl, U. Coy, R. Seyffert and F. Lynen, *Biochem. Z.*, 338 (1963) 771.
83 R. Seyffert, M. Grassl and F. Lynen, in B. Chance, R. W. Estabrook and T. Yonetani, *Hemes and Hemoproteins*, Academic Press, New York, 1966, p. 45.
84 P. S. Clezy and J. Barrett, *Biochim. Biophys. Acta*, 33 (1959) 584.
85 R. Lemberg, in discussion of K. Okunuki, M. D. Kamen and I. Sekuzu, *Structure and Function of Cytochromes*, University Park Press, Baltimore, 1968, p. 32.
86 M. Morrison, in discussion of K. Okunuki, M. D. Kamen and I. Sekuzu, *Structure and Function of Cytochromes*, University Park Press, Baltimore, 1968, p. 32.
87 R. Lemberg, D. B. Morell, N. Newton and J. E. O'Hagan, *Proc. R. Soc. (B)*, 155 (1961) 339.
88 S. Horie, *J. Biochem. Tokyo*, 57 (1965) 650.
89 G. Vanderkooi and E. Stotz, *J. Biol. Chem.*, 241 (1966) 2260.
90 A. Ehrenberg and T. Yonetani, *Acta Chem. Scand.*, 15 (1961) 1071.
91 B. F. vanGelder, W. H. Orme-Johnson, R. E. Hansen and H. Beinert, *Proc. Natn. Acad. Sci., U.S.*, 58 (1967) 1073.
92 B. F. vanGelder and H. Beinert, *Biochim. Biophys. Acta*, 183 (1969) 1.
93 T. Tsudzuki and K. Okunuki, *J. Biochem. Tokyo*, 66 (1969) 281.
94 A. Ehrenberg and W. H. Vanneste, in *19th Colloq. Ges. Biol. Chem., Mosbach*, 1968, Springer, New York, 1968, p. 127.
95 A. Ehrenberg, in *3rd Int. Conf. Magnetic Resonance Biol. Systems, Warrenton, Va.*, 1968.
96 P. Person, W. W. Wainio and B. Eichel, *J. Biol. Chem.*, 202 (1953) 369.
97 D. E. Green, R. E. Basford and B. Mackler, in W. D. McElroy and H. B. Glass, *Symposium on Nitrogen Metabolism*, The Johns Hopkins Press, Baltimore, 1956, p. 125.
98 B. Mackler and N. Penn, *Biochim. Biophys. Acta*, 24 (1957) 294.
99 W. W. Wainio, C. VanderWende and N. F. Shimp, *J. Biol. Chem.*, 234 (1959) 2433.
100 E. Cohen and C. A. Elvehjem, *J. Biol. Chem.*, 107 (1934) 97.
101 M. O. Schultze, *J. Biol. Chem.*, 138 (1941) 219.

102 C. H. Gallagher, J. D. Judah and K. R. Rees, *Proc. R. Soc. (B)*, 145 (1956) 134.
103 C. J. Gubler, G. E. Cartwright and M. M. Wintrobe, *J. Biol. Chem.*, 224 (1957) 553.
104 J. M. Howell and A. N. Davison, *Biochem. J.*, 72 (1959) 365.
105 D. E. Griffiths and D. C. Wharton, *J. Biol. Chem.*, 236 (1961) 1857.
106 H. Beinert, in J. Peisach, P. Aisen and W. E. Blumberg, *The Biochemistry of Copper*, Academic Press, New York, 1966, p. 213.
107 M. Morrison, S. Horie and H. S. Mason, *J. Biol. Chem.*, 238 (1963) 2220.
108 K. Okunuki, in O. Hayaishi, *The Oxygenases*, Academic Press, New York, 1962, p. 409.
109 E. C. Slater, B. F. van Gelder and K. Minnaert, in T. E. King, H. S. Mason and M. Morrison, *Oxidases and Related Redox Systems*, John Wiley, New York, 1965, p. 667.
110 W. W. Wainio, in T. E. King, H. S. Mason and M. Morrison, *Oxidases and Related Redox Systems*, John Wiley, New York, 1965, p. 622.
111 H. Beinert, D. E. Griffiths, D. C. Wharton and R. H. Sands, *J. Biol. Chem.*, 237 (1962) 2337.
112 B. F. van Gelder and A. O. Muijsers, *Biochim. Biophys. Acta*, 118 (1966) 47.
113 D. E. Griffiths and D. C. Wharton, *Biochem. Biophys. Res. Commun.*, 4 (1961) 199.
114 D. C. Wharton and A. Tzagoloff, *J. Biol. Chem.*, 239 (1964) 2036.
115 C. VanderWende and W. W. Wainio, *J. Biol. Chem.*, 235 (1960) PC 11.
116 P. M. Nair and H. S. Mason, *J. Biol. Chem.*, 242 (1967) 1406.
117 H. Beinert, C. R. Hartzell, B. F. van Gelder, *et al.*, *J. Biol. Chem.*, 245 (1970) 225.
118 S. Takemori, I. Sekuzu and K. Okunuki, *Biochim. Biophys. Acta*, 38 (1960) 158.
119 H. Beinert and G. Palmer, *J. Biol. Chem.*, 239 (1964) 1221.
120 H. S. Mason, P. M. Nair and K. Ganapathy, in K. Okunuki, M. D. Kamen and I. Sekuzu, *Structure and Function of Cytochromes*, University Park Press, Baltimore, 1968, p. 138.
121 W. S. Caughey and S. McCoy, in J. Peisach, P. Aisen and W. E. Blumberg, *The Biochemistry of Copper*, Academic Press, New York, 1966, p. 271.
122 D. C. Wharton, *Biochim. Biophys. Acta*, 92 (1964) 607.
123 M. V. Gilmour, *Federation Proc.*, 26 (1967) 455.
124 Q. H. Gibson and C. Greenwood, *J. Biol. Chem.*, 240 (1965) 2694.
125 D. C. Wharton and Q. H. Gibson, in J. Peisach, P. Aisen and W. E. Blumberg, *The Biochemistry of Copper*, Academic Press, New York, 1966, p. 235.
126 A. Tzagoloff and D. H. MacLennan, in J. Peisach, P. Aisen and W. E. Blumberg, *The Biochemistry of Copper*, Academic Press, New York, 1966, p. 253.
127 P. Hemmerich, in J. Peisach, P. Aisen and W. E. Blumberg, *The Biochemistry of Copper*, Academic Press, New York, 1966, p. 15.
128 P. Hemmerich, in J. Peisach, P. Aisen and W. E. Blumberg, *The Biochemistry of Copper*, Academic Press, New York, 1966, p. 269.
129 T. Tsudzuki, Y. Orii and K. Okunuki, *J. Biochem. Tokyo*, 62 (1967) 37.
130 R. H. Sands and H. Beinert, *Biochem. Biophys. Res. Commun.*, 1 (1959) 175.
131 R. C. Bray, *Biochem. J.*, 81 (1961) 189.
132 E. G. Ball, *Biochem. Z.*, 295 (1938) 262.
133 W. W. Wainio, *J. Biol. Chem.*, 216 (1955) 593.
134 K. Minnaert, *Biochim. Biophys. Acta*, 110 (1965) 42.
135 T. Horio and J. Ohkawa, *J. Biochem. Tokyo*, 64 (1968) 393.
136 D. F. Wilson and P. L. Dutton, *Arch. Biochem. Biophys.*, 136 (1970) 583.
137 A. H. Caswell, *J. Biol. Chem.*, 243 (1968) 5827.
138 D. C. Wharton and M. A. Cusanovich, *Biochem. Biophys. Res. Commun.*, 37 (1969) 111.
139 L. Smith and H. E. Conrad, *Arch. Biochem. Biophys.*, 63 (1956) 403.

140 S. Takemori, K. Wada, M. Ando, *et al.*, *J. Biochem. Tokyo*, 52 (1962) 28.
141 K. Wada, *A. Rep. Sci. Works, Faculty Science, Osaka Univ.*, 12 (1964) 19.
142 T. E. King, M. Kuboyama and S. Takemori, in T. E. King, H. S. Mason and M. Morrison, *Oxidases and Related Redox Systems*, John Wiley, New York, 1966, p. 707.
143 P. Nicholls, *Arch. Biochem. Biophys.*, 106 (1964) 25.
144 Y. Orii, I. Sekuzu and K. Okunuki, *J. Biochem. Tokyo*, 51 (1962) 204.
145 D. C. Wharton and D. E. Griffiths, *Arch. Biochem. Biophys.*, 96 (1962) 103.
146 E. T. McGuiness and W. W. Wainio, *J. Biol. Chem.*, 237 (1962) 3273.
147 T. Yonetani and G. S. Ray, *J. Biol. Chem.*, 240 (1965) 3392.
148 Q. H. Gibson, C. Greenwood, D. C. Wharton and G. Palmer, *J. Biol. Chem.*, 240 (1965) 888.
149 Q. H. Gibson and L. Milnes, *Biochem. J.*, 91 (1964) 161.
150 Q. H. Gibson and C. Greenwood, *Biochem. J.*, 86 (1963) 541.
151 H. C. Davies, L. Smith and A. R. Wasserman, *Biochim. Biophys. Acta*, 85 (1964) 238.
152 I. S. Longmuir, *Biochem. J.*, 57 (1954) 81.
153 B. Chance and F. Schindler, in T. E. King, H. S. Mason and M. Morrison, *Oxidases and Related Redox Systems*, John Wiley, New York, 1966, p. 921.
154 T. Yonetani, in P. D. Boyer, H. Lardy and K. Myrbäck, *The Enzymes*, Vol. 8, Academic Press, New York, 1963, p. 41.
155 K. Okunuki, in M. Florkin and E. Stotz, *Comprehensive Biochemistry*, Vol.14, Elsevier, Amsterdam, 1966, p. 232.
156 I. Sekuzu, S. Takemori, T. Yonetani and K. Okunuki, *J. Biochem. Tokyo*, 46 (1959) 43.
157 A. N. Davison and W. W. Wainio, *Federation Proc.*, 23 (1964) 323.
158 Y. Orii and K. Okunuki, *J. Biochem. Tokyo*, 54 (1963) 207.
159 R. Lemberg, M. V. Gilmour and J. T. Stanbury, *Federation Proc.*, 25 (1966) 647.
160 R. Lemberg and M. V. Gilmour, *Biochim. Biophys. Acta*, 143 (1957) 500.
161 D. C. Wharton and Q. H. Gibson, *J. Biol. Chem.*, 243 (1968) 742.
162 Y. Orii and K. Okunuki, *J. Biochem. Tokyo*, 53 (1963) 498.
163 R. Lemberg and M. E. Cutler, *Biochim. Biophys. Acta*, 197 (1970) 1.
164 P. George and J. S. Griffiths, in P. D. Boyer, H. Lardy and R. Myrbäck, *The Enzymes*, Vol. 1, Academic Press, New York, 1959, p. 347.
165 H. S. Mason, *Adv. Enzymol.*, 19 (1957) 79.
166 M. E. Winfield, in T. E. King, H. S. Mason and M. Morrison, *Oxidases and Related Redox Systems*, John Wiley, New York, 1966, p. 115.
167 Q. H. Gibson, G. Palmer and D. C. Wharton, *J. Biol. Chem.*, 240 (1965) 915.
168 M. Morrison and S. Horie, *J. Biol. Chem.*, 240 (1965) 1359.
169 W. H. Vanneste, *Biochem. Biophys. Res. Commun.*, 18 (1965) 563.
170 W. H. Vanneste, *Biochemistry*, 5 (1966) 838.
171 G. E. Mansley, J. T. Stanbury and R. Lemberg, *Biochim. Biophys. Acta*, 113 (1966) 33.
172 G. Wald and D. W. Allen, *J. Gen. Physiol.*, 40 (1957) 593.
173 W. W. Wainio and J. T. Greenlees, *Arch. Biochem. Biophys.*, 90 (1960) 18.

Chapter 28

PEROXIDASES AND CATALASES

B. C. SAUNDERS

University Chemical Laboratory, Lensfield Road, Cambridge, CB2 1EW (Gt. Britain)

INTRODUCTION

It is appropriate that a chapter on peroxidases and catalases should appear in a book on inorganic biochemistry. Both enzymes contain iron and are concerned primarily with the utilization of hydrogen peroxide. Moreover many attempts have been made to simulate the action of peroxidase and catalase by purely inorganic substances. For example Fenton's reagent (Fe^{2+} plus H_2O_2 discussed later pp. 995, 996, 1012, sometimes appears to act as an artificial peroxidase and it is well known that manganese dioxide or colloidal platinum will initiate decomposition of hydrogen peroxide in a catalase-like manner.

Few enzymes have attracted more attention than has peroxidase, which is widely distributed in nature and has proved to be of interest to inorganic, physical, and organic chemists, biochemists and physiologists.

Many workers have been attracted to study peroxidase-catalyzed reactions because highly coloured compounds are often produced in such changes. Others have paid more attention to the kinetics and mechanisms of the reactions. In view of the diverse approaches and of the enormous number of relevant communications that have appeared, comprehensive coverage is not possible. An attempt, however, will be made to demonstrate the specificity of peroxidase action and, paradoxically, its versatility and some account is given of catalatic activity.

PEROXIDASE

Historical

In 1855 Schönbein[1] recorded that the oxidation of certain organic compounds by dilute solutions of hydrogen peroxide could be catalyzed by "substances" occurring in plants and animals. He thought, however, that activation and decomposition of hydrogen peroxide ($2H_2O_2 = 2H_2O + O_2$) were due to the same "substance". Later workers showed that the decomposition of hydrogen peroxide was promoted by catalase, whereas the peroxide-activating substance was a distinct enzyme called peroxidase. Nevertheless, over many years confusion existed. Except where stated otherwise, the peroxidase system will be used to indicate "a system consisting of dilute aqueous hydrogen peroxide solution together with the enzyme peroxidase, the latter being free from oxidase activity".★ The name peroxidase was first given by Linossier[2], who isolated from pus an oxidase-free preparation of the enzyme.

Hydrogen peroxide oxidations can often be catalyzed by substances other than true peroxidase, but frequently with different results. Such reactions are usually less specific and less clearly defined. It is, therefore, of the utmost importance that highly purified materials and enzymes should be used.

Work on peroxidase may be conveniently divided into three periods.

(a) Up to 1918. This period covers the researches of Bach and Chodat[3], who noted the wide occurrence of peroxidase in plants and showed that peroxidases could be obtained from horseradish (free from oxidases, catalase, amylase, invertase, emulsin and proteolytic enzymes). In 1903 they described qualitative experiments which indicated that the peroxidase system could oxidize pyrogallol, gallic acid and certain amines, but no attempt was made to isolate the coloured products, nor to investigate the reactions involved. Bach[4] showed in 1904 that peroxidase action was inhibited by an excess of hydrogen peroxide. In 1907, Ernest and Berger[5] made a preparation of peroxidase from beetroot. Battelli and Stern[6], in 1908, showed the presence of peroxidase in many animal tissues.

(b) 1918–1931. These years cover the work of Willstätter[7] on the purification of peroxidase and determination of its activity by oxidizing pyrogallol to purpurogallin. He introduced and defined the *purpurogallin number* (PN) as the number of mg of purpurogallin formed by 1 mg of peroxidase preparation, when allowed to act upon 5 g of pyrogallol in 2 l of water for 5 min at 20°C in the presence of 10 ml of 0.5% H_2O_2.

(c) 1931 onwards. This period includes the investigations of Chance, Keilin, Theorell, Mason, Paul and George on the nature of peroxidase and the kinetics of peroxidase action and the work of Saunders and colleagues on the detailed examination of the products obtained in such reactions.

OCCURRENCE

Peroxidases are widely distributed among higher plants; the richest known sources are the sap of the fig tree and root of the horseradish. Peroxidases are present in many animal tissues. They occur in human saliva[8], the adrenal medulla[9], liver, kidney and in most organs. A per-

---

★ A true oxidase catalyzes oxidations involving molecular oxygen.

oxidase occurs in leucocytes. The presence in blood plasma of a peroxidase that utilizes alkyl hydrogen peroxide, but not hydrogen peroxide, has been claimed[10]. Peroxidase reactions of micro-organisms have been recorded. Yeast contains cytochrome-*c* peroxidase.

Among the several peroxidases that have been obtained crystalline are: (i) verdoperoxidase (myeloperoxidase) from leucocytes[11]; (ii) lacto-peroxidase from milk[11]; (iii) cytochrome-*c* peroxidase from yeast[12]. This enzyme utilizes hydrogen peroxide to oxidize reduced cytochrome-*c* to oxidized cytochrome-*c*; (iv) Horseradish peroxidase[13]. Identical crystalline products are obtained from horseradish and turnips. This is the enzyme employed in the investigations described in this chapter. It reacts with ROOH as well as with HOOH, but not with ROOR. These peroxidases have many properties in common and often catalyze similar reactions. Occasionally, however, they appear to have somewhat different reactions with a given substrate.

Isozymes

The term isozyme is now frequently appearing in the literature, but Dixon and Webb[14] prefer the term isoenzyme. By the process of zone electro-phoresis Hosoya[15] has separated turnip peroxidase electrophoretically into four forms which differ only slightly in pK of dissociation of active centre groups.

Shannon and co-workers[16] have recently isolated from horseradish seven peroxidase isozymes by chromatography, ultracentrifugation and disc electrophoresis. No interconversions among the enzymes were detected and the two most abundant isoenzymes crystallized. Both of these oxidized mesidine in 95% yield (p. 996)[17]. Each isozyme contained protohaemin IX as the prosthetic group. All peroxidase isoenzymes appear to have almost identical catalytic properties, but differ in physico-chemical and kinetic properties.

Commission on Enzymes of the International Union of Biochemistry

The Report of the Commission on Enzymes of the International Union of Biochemistry[18] classifies peroxidase as follows: "1.11.1.7. Donor: H_2O_2 oxidoreductase.

Reaction: Donor + $2H_2O_2$ = oxidized donor + $2H_2O$

Specificity: A group of haemoprotein enzymes. A great variety of substances can act as donors".

PROPERTIES OF HORSERADISH PEROXIDASE (HRP)

HRP, which is brown, has a molecular weight[19] of about 40,000 and contains, C, 47.0; H, 7.35; N, 13.2; S, 0.43; Fe, 0.127%. The protein contains some 16 amino acids, which include cysteine and methionine, thus accounting for the presence of sulphur. Peroxidase contains carbohydrate probably as uronic acid.

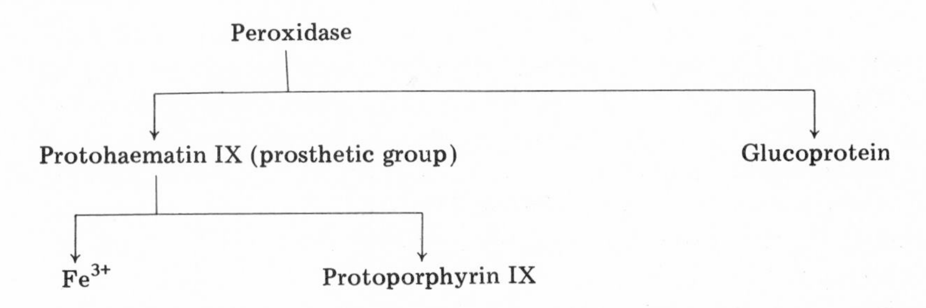

The iron has six co-ordination positions, of which four are occupied by porphyrin nitrogen atoms and a fifth by a protein group; the sixth can be variously occupied by H_2O, CN^-, etc. Peroxidase appears to operate by exchange of groups in this 6-position.

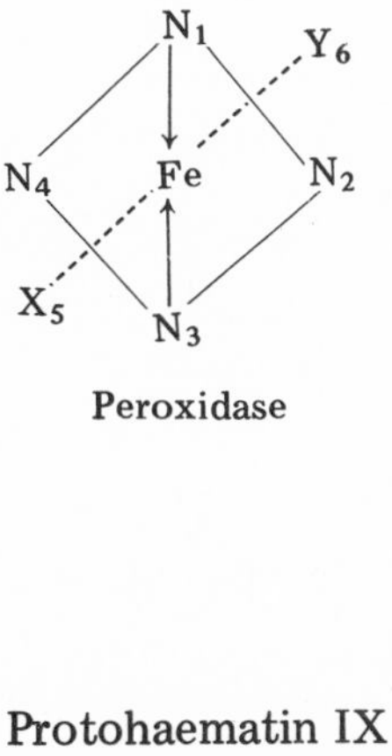

Peroxidase

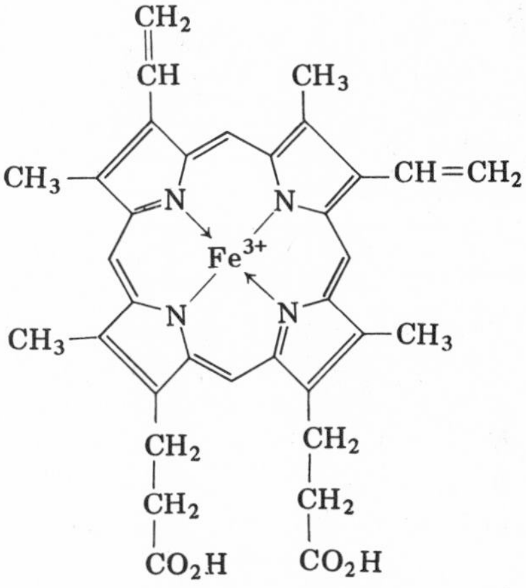

Protohaematin IX

Protohaematin IX can be easily and reversibly detached from the colourless protein by acetone and hydrochloric acid below 0°C. Ferriperoxidase reacts reversibly with cyanide, sulphide, fluoride, azide, hydroxylamine and nitric oxide forming compounds that can be detected spectroscopically[19,20]. The enzyme is inhibited by cyanide and sulphide at concentrations of 10^{-5}-10^{-6} *M*, and by fluoride at rather higher concentrations. Pure peroxidase has a PN of about 1200.

HRP bears some resemblance to methaemoglobin (containing the same protohaematin, but with the protein globin). Peroxidase and methaemoglobin show similarity in colour and absorption spectra. Both change from brown to red at higher pH and are reduced by $Na_2S_2O_4$ to the iron(II) state. Both form compounds with peroxidase inhibitors (*e.g.* CN^- and CO) and with hydrogen peroxide.

COMPOUNDS OF HORSERADISH PEROXIDASE (HRP) AND HYDROGEN PEROXIDE

Compound I (green) is formed from 1 mole of HRP and 1 mole of H_2O_2. It was first observed by Theorell[21]. It has absorption maxima at 658 and 407 nm, is unstable and readily passes into:

Compound II (red)[20], which is also formed from 1 mole of HRP and 1 mole of H_2O_2 (absorption maxima: 561, 530 and 417 nm).

Compound III (red)[20] is formed with a large excess of H_2O_2 (absorption maxima at 583, 545 and 417 nm).

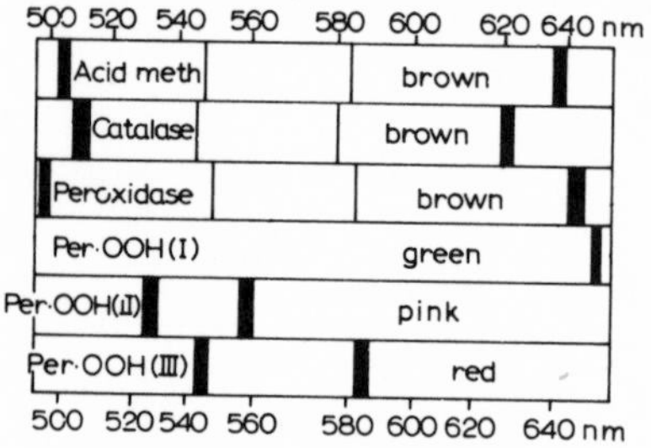

Chance studied the kinetics of the formation of these compounds[22]. The results were summarized thus:

(Per · OH = peroxidase)

$$\text{Per} \cdot \text{OH} + \text{HOOR} \underset{K_2}{\overset{K_1}{\rightleftharpoons}} \text{Per} \cdot \text{OOR(I)} + H_2O$$

$$\text{Per} \cdot \text{OOR(I)} \xrightarrow{K_5} \text{Per} \cdot \text{OOR(II)}$$

Decomposition of (II) in presence of hydrogen donor AH_2 (*e.g.* ascorbic acid);

$$\text{Per} \cdot \text{OOR(II)} \xrightarrow{K_4 = K_3/(AH_2)} \text{Per} \cdot \text{OH} + \text{ROH} + \text{A}$$

At pH = 4.7 the following results were obtained for ascorbic acid and ROOH = H_2O_2: $K_1 = 0.9 \times 10^7\ M^{-1}\ \text{sec}^{-1}$; $K_5 = 4\ \text{sec}^{-1}$; $K_4 = 1.2 \times 10^{-4}\ M^{-1}\ \text{sec}^{-1}$. The rate-determining step is the reaction of compound (II) with AH_2. Thus compound (II) was considered as the classical enzyme-substrate complex of Michaelis and Menten, although (II) does not decompose in a unimolecular manner.

George[23,24] believed that compound II underwent a one-equivalent reduction to ferriperoxidase with agents such as ferrocyanide, ferrocytochrome-*c* and Fe(II) ions. (Ferriperoxidase is the initial oxidation state of the enzyme.) He proposed a scheme along the following lines.

$$\underset{\text{(HRP)}}{Fe_p{}^{3+}H_2O} + H_2O \xrightarrow[\text{oxidation}]{\text{two equiv.}} Fe_p{}^{V} \cdot O^{3+} + 2H_2O$$

$$\underset{\text{(I)}}{Fe_p{}^{V} \cdot O^{3+}} + AH_2 \xrightarrow[\text{transfer}]{\text{one electron}} Fe_p{}^{IV} \cdot O^{2+} + AH^{\bullet} + H^+$$

$$\underset{\text{(II)}}{Fe_p{}^{IV} \cdot O^{2+}} + AH_2 + H^+ \xrightarrow[\text{or } H^{\bullet} + H^+]{\text{electron } +2H^+} Fe_p \cdot {}^{3+}H_2O + AH^{\bullet}$$

In other words, the protoporphyrin iron (Fe_p) was considered to have effective oxidation states of +5 and +4 (probably as ferryl ions). George based his theory partly on spectroscopic evidence that other oxidizing agents, *e.g.* HOCl, can replace hydrogen peroxide in the peroxidase system (but see below). Recent Mössbauer spectroscopy[25] indicates that the structures of compounds I and II are compatible with Fe^{IV}.

The theory of George postulates that the enzyme acts in virtue of higher oxidation states of the protohaematin. In this connection, however, there seem to be conditions[26] in which hydrogen peroxide is not interchangeable with hypochlorous acid in the peroxidase system. For example, peroxidase and H_2O_2 convert mesidine to the purple anil (p. 996). However, when HOCl replaces H_2O_2 at comparable concentrations no purple compound is produced[26]. Under these conditions the enzyme is not destroyed, for when H_2O_2 is now added to the mixture the oxidation proceeds with the usual rapidity. Hypochlorous acid will, however, at high concentrations oxidize mesidine, but under these conditions the reaction is not a catalytic one and it proceeds just as well in the absence of peroxidase. Other similar experiments indicate that hydrogen peroxide is not necessarily interchangeable with other oxidizing agents.

PEROXIDASE OXIDATION OF AMINES

Aniline

The production of colours by the action of peroxidase on certain amino and phenolic compounds has been recorded from time to time, but such reactions have usually been regarded as tests for peroxidase, the nature of the products being rarely determined. The wide distribution of the enzyme suggested that it could be of biological importance. However, the role that it plays in metabolic processes is not clear. From the work of Elliott[27], it appears that the enzyme is not concerned with general oxidative catabolism since it does not attack the majority of amino acids, fatty acids and carbohydrates.

For these reasons Saunders proposed to investigate in detail the action of peroxidase on a range of compounds, and from a consideration of the change brought about to suggest possible mechanisms of reaction and to undertake the synthesis of more complex compounds.

The substrate★ chosen for the initial investigation was aniline[28]. Identical oxidation products were obtained with HRP and turnip peroxidase. In control experiments no action on aniline was shown by the active enzyme in the absence of hydrogen peroxide or by heat-inactive enzyme in the presence of hydrogen peroxide. It is therefore justifiable to conclude that the product was the result of peroxidase action.

Reactions were carried out at room temperature with a solution of aniline in dilute acetic acid (pH = 4.5). A few mg of the peroxidase preparation (PN = 100) were added and dilute hydrogen peroxide was added at intervals.

This intermittent addition of hydrogen peroxide is necessary since excess of it inhibits peroxidase action[4]. The first effect was a blue-violet colour which rapidly changed to brown. A brown solid separated and was shown to be a mixture of 2,5-dianilino-*p*-benzoquinone imideanil **(I)**, pseudo-mauveine **(II)**, induline **(III)** and aniline black **(IV)**. It is important to note that benzoquinone, nitrosobenzene and nitrobenzene were not produced. This enzymic oxidation of aniline probably proceeded thus:

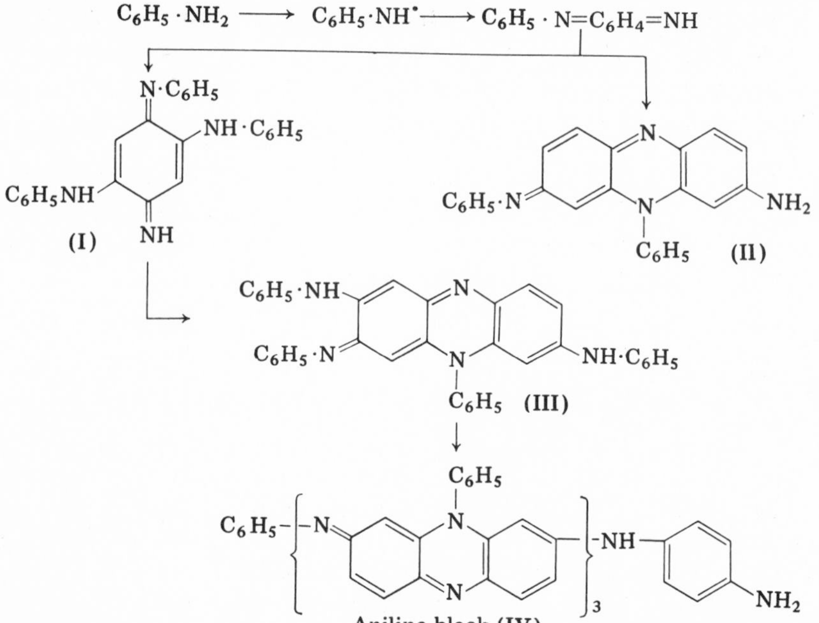

$$C_6H_5\cdot NH_2 \longrightarrow C_6H_5\cdot NH^{\bullet} \longrightarrow C_6H_5\cdot N{=}C_6H_4{=}NH$$

(I) (II)

(III)

Aniline black (IV)

★ Some workers prefer to describe H_2O_2 as the substrate, but there are advantages in assigning this term to the hydrogen donor.

Inorganic catalysts

It is important to know to what extent the action of the enzyme can be imitated by inorganic catalysts. Accordingly the oxidation of aniline in acetic acid was carried out using dilute hydrogen peroxide and iron(II) sulphate (Fenton's reagent)[28]. No blue-violet colour was produced and a brown precipitate separated, consisting of azobenzene, aminoanilinoquinone monoanil (**V**), and 2,5-dianilino-benzoquinone monoanil (**VI**). The absence of polycyclic compounds such as ψ-mauveine, induline and aniline black is of interest. It is evident therefore that Fenton's reagent behaves differently from the enzyme system.

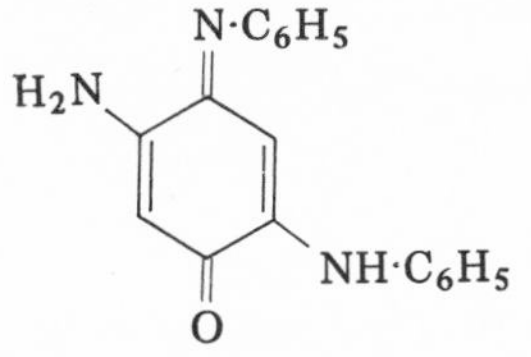

$N{\cdot}C_6H_5$ H_2N $NH{\cdot}C_6H_5$ O

(V)

$N{\cdot}C_6H_5$ $C_6H_5{\cdot}NH$ $NH{\cdot}C_6H_5$ O

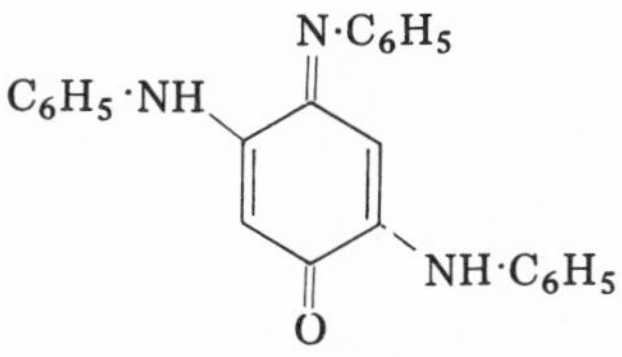

(VI)

p-Toluidine

Oxidations[29] were carried out as for aniline. A reddish-brown solid gradually separated. This was a mixture of 4-amino-2,5-toluquinonebis-*p*-tolylimine (**VII**), 4-*p*-toluidino-2,5-toluquinonebis-*p*-tolylimine (**VIII**), 4,4′-dimethyldiphenylamine, a small quantity of 4,4′-dimethylazobenzene and traces of 4-amino-2,5-toluquinone-2-*p*-tolylimine (**IX**) and 4-*p*-toluidino-2,5-toluquinone-2-*p*-tolylimine (**X**).

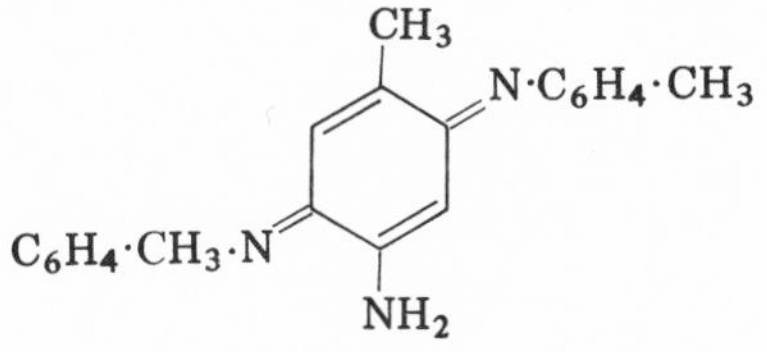

CH_3 $N{\cdot}C_6H_4{\cdot}CH_3$ $C_6H_4{\cdot}CH_3{\cdot}N$ NH_2

(VII)

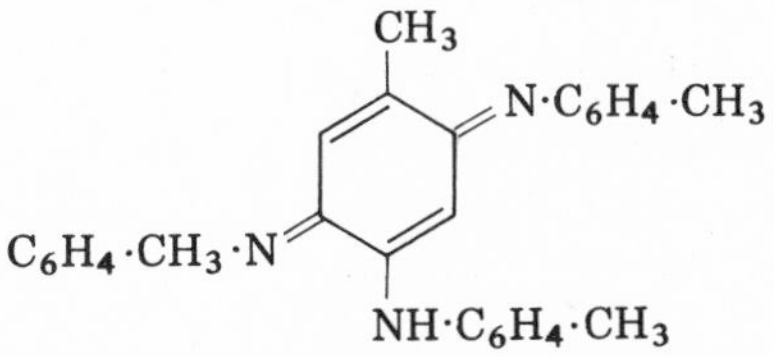

CH_3 $N{\cdot}C_6H_4{\cdot}CH_3$ $C_6H_4{\cdot}CH_3{\cdot}N$ $NH{\cdot}C_6H_4{\cdot}CH_3$

(VIII)

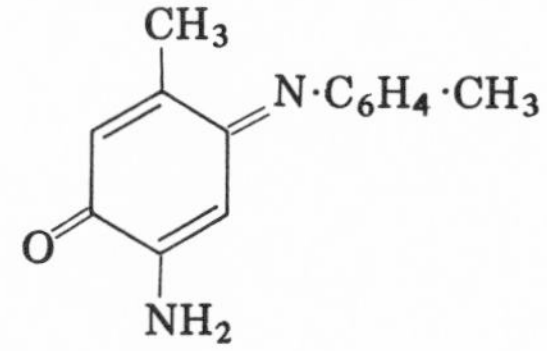

CH_3 $N{\cdot}C_6H_4{\cdot}CH_3$ O NH_2

(IX)

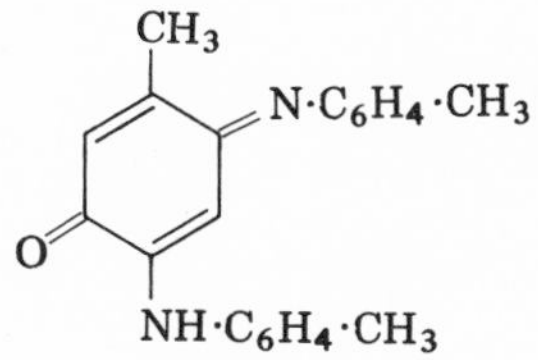

CH_3 $N{\cdot}C_6H_4{\cdot}CH_3$ O $NH{\cdot}C_6H_4{\cdot}CH_3$

(X)

References pp. 1019–1021

Inorganic catalysts and p-toluidine

In contrast to the above, it was shown that if iron(II) sulphate were substituted for peroxidase, the reaction led mainly to amorphous material. From this, however, a compound was isolated, not obtained in the peroxidase oxidation.

Syntheses by two independent routes proved it to be 2,7-dimethyl-3-*p*-toluidinophenazine[30] (XI).

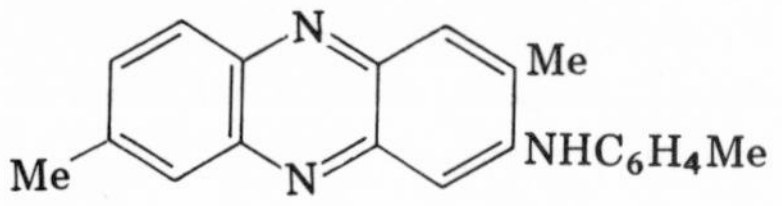

Me NHC$_6$H$_4$Me N N Me

(XI)

Mesidine

The oxidations of aniline and of *p*-toluidine were complex because of the several "free-points" in the molecule at which oxidation and oxidative coupling could take place. The work was therefore extended to mesidine as substrate[31]. Since this compound has only two free nuclear hydrogen atoms, it seemed likely that only a limited number of compounds could be produced and that a knowledge of their constitution might throw more light on the mechanisms of peroxidase oxidation. Mesidine in dilute acetic acid was oxidized at pH = 4.5, and a crystalline purple solid separated. The filtrate contained formaldehyde and ammonia.

With highly purified enzyme the yield of the purple crystalline compound was over 95%. Neither nitroso- nor nitro-compounds were detected, an observation consistent with the enzymic oxidations of aniline and of *p*-toluidine.

This enzymic oxidation can also be carried out in a dilution of one part of amine in 50,000 parts of solution with the same results.

The purple crystalline compound had the molecular formula $C_{17}H_{19}ON$, indicating its formation from two molecules of mesidine with elimination of a carbon atom (as a methyl group). This was confirmed by hydrolysis; treating it with dilute sulphuric acid mesidine and 2,6-dimethylbenzoquinone were produced.

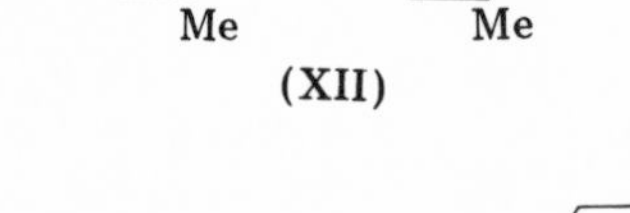

Me Me Me N =O Me Me

(XII)

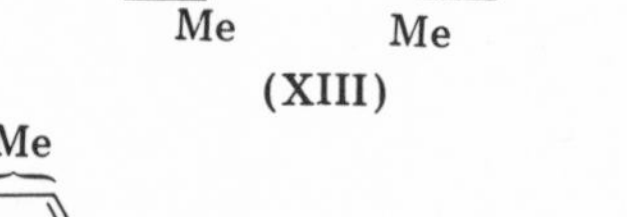

Me Me Me N =O Me Me

(XIII)

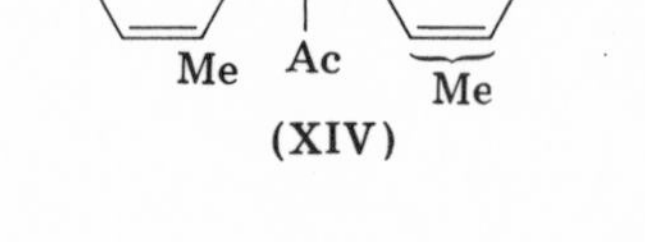

Me Me Me N OAc Ac Me Me

(XIV)

Thus the purple compound was either (**XII**), or (**XIII**). Reductive acetylation gave the colourless neutral diacetyl derivative of 4-hydroxy-3,5,2′,4′,6′ (or 2,6,2′,4′,6′)-pentamethyldiphenylamine (**XIV**).

The purple compound was synthesized by an independent and unambiguous method and shown to have the constitution (**XII**).

Other catalysts

Manometric experiments showed that mesidine was not oxidized by indophenol oxidase, catechol oxidase or laccase.

4-Methoxy-2,6-dimethylaniline

In the previous reaction, the elimination of a methyl group at room temperature was surprising. The next step was to examine the peroxidase oxidation of a related amine in which the *p*-substituent was again "chemically" stable.

For this purpose 4-methoxy-2,6-dimethylaniline[32] was selected. A crystalline purple compound was obtained. Reactions and independent syntheses proved it to have the formula (XV):

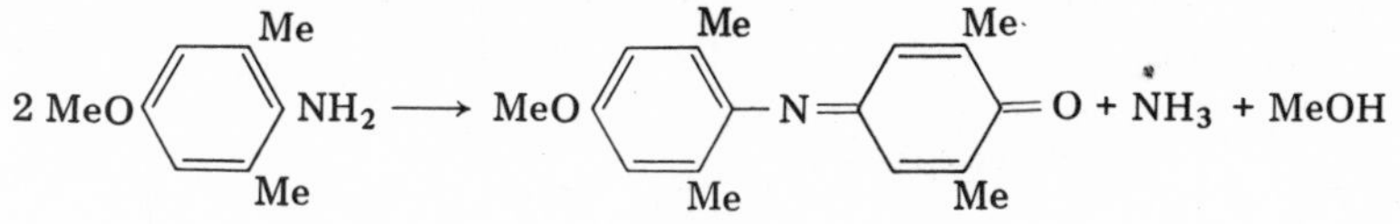

$$2\ \text{MeO}-C_6H_2(\text{Me})_2-\text{NH}_2 \longrightarrow \text{MeO}-C_6H_2(\text{Me})_2-\text{N}{=}C_6H_2(\text{Me})_2{=}\text{O} + \text{NH}_3 + \text{MeOH}$$

(XV)

Ammonia in quantitative yield was produced, and the "stable" ether grouping was readily eliminated as methanol.

Dimethylaniline

Dimethylaniline[33] was used as substrate to discover the effect of replacing the N-hydrogen atoms by methyl groups. The enzyme and H_2O_2 were added to the amine solution (in acetic acid, pH = 4.5). The mixture became yellow (stage 1), and quickly changed to intense green (stage 2). A light blue precipitate separated and the green colour gradually changed to blue-purple.

When a portion of the green solution (stage 2) was treated with excess of hydrogen peroxide and peroxidase it became orange (stage 3), but usually rapidly changed back to green.

At the end of the oxidation, the solid was filtered off, the main constituent was *N,N,N′,N′*-tetramethylbenzidine obtained pure by chromatography. This procedure showed traces of many coloured materials. The filtrate did not contain dimethylaniline oxide—an observation of importance.

The formation of *N,N,N′,N′*-tetramethylbenzidine (**XVII**) from dimethylaniline involves dehydrogenation in the *p*-position. It seems that

References pp. 1019–1021

free radicals (**XVI**) are produced which combine to give the benzidine (**XVII**), some of which separates, the remainder being further oxidized. (Alternatively the first stage may be the removal of an electron from dimethylaniline giving the ion-radical (**XVIa**). Such ion-radicals would combine in pairs to give (**XVII**) with elimination of two protons.) It is probable that the transient yellow colour (stage 1) is concerned with this initial process.

The action of peroxidase on *N,N,N',N'*-tetramethylbenzidine in dilute acetic acid was also examined. A greenish-yellow colour gave way to intense green. Further oxidation gave an orange-coloured solution. These green and orange colours correspond with those observed in the oxidation of dimethylaniline itself—stages 2 and 3, respectively.

The green compound (in solution) was a free radical (**XVIII**), formed by the removal of one electron from tetramethylbenzidine.

Further oxidation gives the quinonoid structure (**XIX**) by loss of a second electron. This is only feebly coloured because of the reduced possibility of resonance, and it is considered to be the compound responsible for the orange colour. In the oxidation of dimethylaniline traces of other coloured products are formed via this quinonoid compound (**XIX**). Some of these compounds are of the electron-transfer type.

The main changes are represented in outline thus:

Also XIX + XVII = 2 XVIII

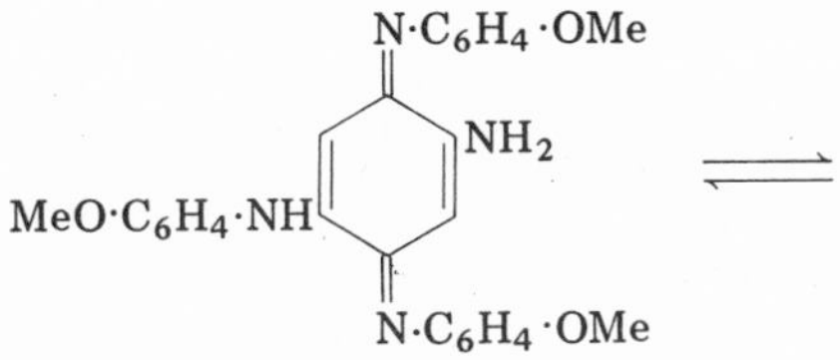

$N{\cdot}C_6H_4{\cdot}OMe$ / NH_2 / $MeO{\cdot}C_6H_4{\cdot}NH$ / $N{\cdot}C_6H_4{\cdot}OMe$

(XX)

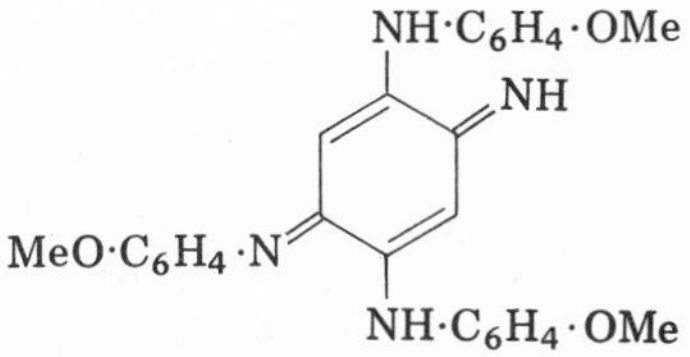

$\rightleftharpoons$

$NH{\cdot}C_6H_4{\cdot}OMe$ / NH / $MeO{\cdot}C_6H_4{\cdot}N$ / $NH{\cdot}C_6H_4{\cdot}OMe$

(XXI)

$N{\cdot}C_6H_4{\cdot}OMe$ / $NH{\cdot}C_6H_4{\cdot}OMe$ / $MeO{\cdot}C_6H_4{\cdot}NH$ / $N{\cdot}C_6H_4{\cdot}OMe$

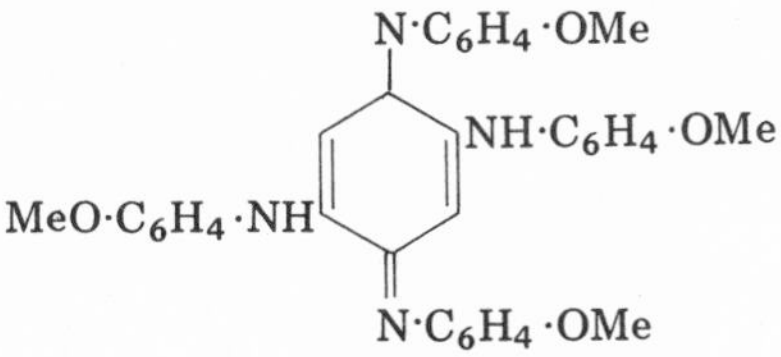

(XXII)

p-Anisidine

The oxidation of *p*-anisidine[34] might be expected to give products analogous to those obtained by the peroxidase oxidation of *p*-toluidine. This, however, was not so, nor was the reaction parallel with the oxidation of 4-methoxy-2,6-dimethylaniline.

With the enzyme system a deep violet colour was produced immediately, and a red-brown solid separated. Three compounds were isolated: (a) 4,4′-dimethoxyazobenzene in very small quantity; (b) dark red needles (**XXII**) representing 5% of the product; (c) lustrous cerise-red leaflets (**XX**), being 80% of the product.

Compound (**XX**) was produced by condensation of four molecules of *p*-anisidine with elimination of one methoxyl group. Hence compound (**XX**) was 2-amino-5-*p*-anisidinobenzoquinone di-*p*-methoxyphenylimine (for which the alternative formula (**XXI**) is possible).

Compound (**XXII**) was $C_{34}H_{32}O_4N_4$ derived from five *p*-anisidine residues with elimination of one methoxyl and one amino group. Moreover, it contained four methoxyl groups per mol. If the structure assigned to compound (**XX**) is correct, then the structure of (**XXII**) is 2,5-di-*p*-anisidinobenzoquinone di-*p*-methoxyphenylimine (or tetra-*p*-methoxyazophenine).

p-Chloroaniline[35]

The stability of the chlorine atom in *p*-chloroaniline prompted the investigation of the action of the enzyme system on this substrate particularly as *p*-toluidine is readily oxidized by the enzyme without elimination of the *p* methyl group. Throughout the oxidation of *p*-toluidine in dilute

References pp. 1019–1021

acetic acid, the pH remained constant at pH = 4.5 even in absence of buffer solution.

In exploratory experiments, hydrogen peroxide was added to a solution of *p*-chloroaniline in aqueous acetic acid (pH = 4.5), and the oxidation started by addition of enzyme, whereupon a deep red colour was produced. Only a trace of red-brown solid oxidation product was obtained and the pH fell to 3.5. This fall was due to accumulation of hydrochloric acid, and the low yield was attributed in part to the inhibition of the reaction by the lowering of pH and in part to retardation of the enzymic oxidation by the accumulation of hydrogen peroxide relative to the enzyme.

Therefore in later experiments the oxidation was carried out in acetate buffer at pH = 4.5, and high yields of a red-brown solid were obtained. This consisted of: (a) a dark-red tetra-azophenine (**XXIV**); (b) traces of 4,4′-dichloroazobenzene; and (c) 2-amino-5-*p*-chloroanilinobenzoquinone di-*p*-chloroanil (**XXIII**).

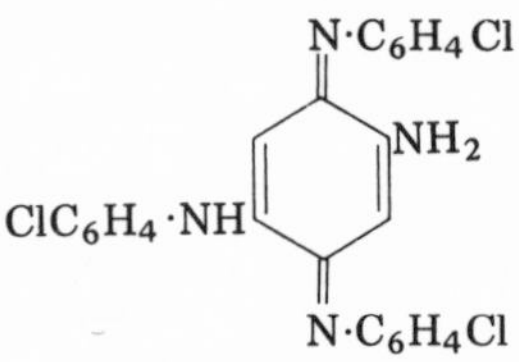

$N{\cdot}C_6H_4Cl$; NH_2 ; $ClC_6H_4{\cdot}NH$; $N{\cdot}C_6H_4Cl$

(XXIII)

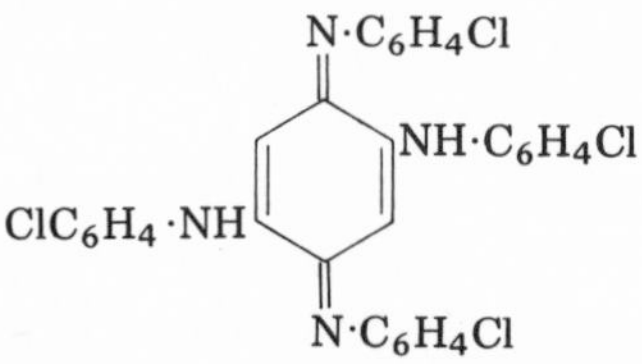

$N{\cdot}C_6H_4Cl$; $NH{\cdot}C_6H_4Cl$; $ClC_6H_4{\cdot}NH$; $N{\cdot}C_6H_4Cl$

(XXIV)

The aminochloroanil (**XXIII**) could be produced from four molecules of *p*-chloroaniline provided that dechlorination of one molecule of the latter takes place. The filtrate contained the amount of chloride ion theoretically required for the formation of (**XXIII**).

Enzymic rupture of a C–F Bond

In extending oxidations of this type, the enzymic cleavage of the stable C–F bond in *p*-fluoroaniline was effected[36]. In acetate buffer (pH = 4.5), at room temperature, the amine was oxidized to give the red crystalline 2-amino-5-*p*-fluoroanilinobenzoquinone di-*p*-fluoroanil (**XXV**; X = F).

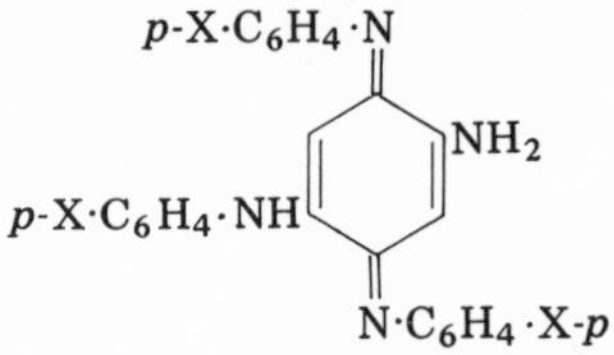

p-$X{\cdot}C_6H_4{\cdot}N$; NH_2 ; p-$X{\cdot}C_6H_4{\cdot}NH$; $N{\cdot}C_6H_4{\cdot}X$-p

(XXV)

The formation of (**XXV**; X = F) requires elimination of one fluorine atom per four molecules of amine. The fluorine was expelled as F^-, and since enzyme activity is retarded by F^-, the process is self-poisoning. As expected the reaction was not complete: it stopped at 30% completion and F^- was present. This oxidation of *p*-fluoroaniline provided the first recorded case of an enzymic cleavage of a C–F bond.

It was also shown that *p*-bromoaniline and *p*-iodoaniline gave mainly (**XXV**), X being Br and I respectively.

The eliminated bromine was produced quantitatively as Br^-, whereas the corresponding I^- was oxidized completely to I_2 (an oxidation accelerated by peroxidase), which iodinated unchanged iodoaniline giving 2,4-diiodoaniline. Thus we have an example of deiodination followed by *reiodination.* In other words, peroxidase is responsible for the *transport of iodine.*

TRANSIODINATION AND RELATED PROCESSES[37]

This principle of de-iodination followed by reiodination was extended so that a compound other than the original substrate became the "iodine acceptor". In this way the peroxidase system is able to effect a process that we may call *transiodination.*

Accordingly investigations of reactions involving this intermolecular transfer of iodine under the influence of the peroxidase system were carried out. Particular attention was paid to reactions with compounds similar to those present in the thyroid gland. Results are shown in Table I. We see for example how thyroxin can be synthesized by peroxidatic transiodination.

TABLE I

RESULTS OF TRANSIODINATION REACTIONS

Iodine donor	Iodine acceptor	Transiodinated product
p-Iodoaniline	Pyrrole	Tetraiodopyrrole
4-Iodo-2,6-dimethylphenol	Indole	3-Iodoindole
4-Iodo-2,6-dimethylaniline	Phenol	Tri-iodophenol
4-Iodo-2,6-dimethylaniline	*p*-Hydroxyphenylacetic acid	3,5-Di-iodo-4-hydroxyphenylacetic acid
4-Iodo-2,6-dimethylphenol	Tyrosine	Mono- and di-iodotyrosines
Di-iodotyrosine	Thyronine	3′-Iodothyronine
Di-iodotyrosine	3,5-Di-iodothyronine	Thyroxine
3-Iodotyrosine	3-Iodotyrosine	3,5-Di-iodotyrosine

References pp. 1019–1021

Reactions were carried out at room temperature by addition of peroxidase and hydrogen peroxide to mixtures of iodine donor and acceptor in weakly acid buffer solutions.

Other reactions in which peroxidase systems in the presence of iodine might be utilized were then investigated. It is known that quantitative yields of phosphoroiodidates are obtained by the action of iodine on trialkyl phosphites[38]:

$$(RO)_3P + I_2 = (RO)_2POI + RI$$

However, when dialkyl hydrogen phosphites are used in place of trialkyl phosphites, negligible reaction occurs. The hydrogen iodide first formed interferes with the formation of phosphoroiodidate, perhaps by dealkylation

$$(RO)_2POH + I_2 \nrightarrow (RO)_2POI + HI$$

of the hydrogen phosphite by the hydrogen iodide formed[39]. If the reaction between iodine (or an iodine donor) and a dialkyl phosphite could be carried out in the presence of peroxidase to remove the hydrogen iodide, then a high yield of phosphoroiodidate should result. This was found to be the case.

Since phosphoroiodidates react with amines to give dialkyl phosphoramidates and hydrogen iodide[38], it was thought worthwhile to investigate the possible formation of phosphorus–nitrogen bonds, in the presence of the enzyme. Such a system was indeed found to give good yields of phosphoramidate.

It is known that the controlled hydrolysis of phosphorochloridates yield tetra-alkyl pyrophosphates and halide[40]. Here again it seemed that the enzyme system might be used in the formation of P–O–P links. Preliminary experiments showed that careful hydrolysis of diethyl phosphoroiodidate gave tetra-ethyl pyrophosphate (TEPP). Separate experiments showed that addition of iodine to aqueous solutions of tricyclohexyl phosphite gave dicyclohexyl hydrogen phosphate, formed by hydrolysis of the phosphoroiodidate. On the other hand, TEPP could be obtained by careful addition of peroxidase and peroxide to solutions containing triethyl phosphite and iodide in the presence of a limited amount of water. In this latter reaction, iodide was continually oxidized by the enzyme to iodine which could re-enter the cycle:

$$(RO)_3P + I_2 = (RO)_2POI + RI$$

$$(RO)_2POI + H_2O = (RO)_2P(O)OH + HI$$

$$(RO)_2POI + (RO)_2P(O)OH = (RO)_2P(O)OP(O)(OR)_2 + HI$$

$$2HI + H_2O_2 \xrightarrow{\text{peroxidase}} I_2$$

Phosphorylation reactions involving dialkyl phosphites were also considered. As explained above, hydrogen iodide could be removed by oxidation with the peroxidase system. In the scheme given below, a cyclic formation and oxidation of iodide continually occur; thus the peroxidase system is able to effect the formation of phosphates, pyrophosphates and phosphoramidates in the presence of only catalytic amounts of iodine:

$$(RO)_2POH + I_2 = (RO)_2POI + HI$$

$$(RO)_2POI + H_2O = (RO)_2P(O)OH + HI$$

$$(RO)_2POI + (RO)_2P(O)OH = (RO)_2P(O)OP(O)(OR)_2 + HI$$

$$(RO)_2POI + R'NH_2 = (RO)_2P(O)NHR' + HI$$

$$4HI + 2H_2O_2 \xrightarrow{\text{peroxidase}} 2I_2$$

The formation of disulphide linkages from sulphydryl groups by the action of peroxidase and iodine was also studied. Here again, hydrogen iodide is formed, oxidized to iodine and returned as iodine to an earlier part of the cycle:

$$2RSH + I_2 = R{-}S{-}S{-}R + 2HI$$

$$2HI + H_2O_2 \xrightarrow{\text{peroxidase}} I_2 + 2H_2O$$

Thus cysteine was readily converted to cystine by the action of peroxidase and hydrogen peroxide in the presence of diiodotyrosine as iodine donor. This oxidation is of interest in connection with the description by Mills[41] of a "glutathione peroxidase" and his suggestion that this enzyme may protect haemoglobin from oxidative breakdown.

It is now convenient to summarize the action of iodine in phosphorylation and oxidation processes. It is known that some 8 μg of iodine are present in 100 ml of systemic blood, and this work may serve to indicate possible functions of this circulating iodine. Furthermore, we emphasize the triple function of peroxidase in this connection: (a) release of iodine from an iodine donor; (b) actual transfer of iodine — "transiodination"; (c) oxidation of iodide to iodine, hence the circulation of iodine is ensured.

All these processes are represented by the annexed scheme:

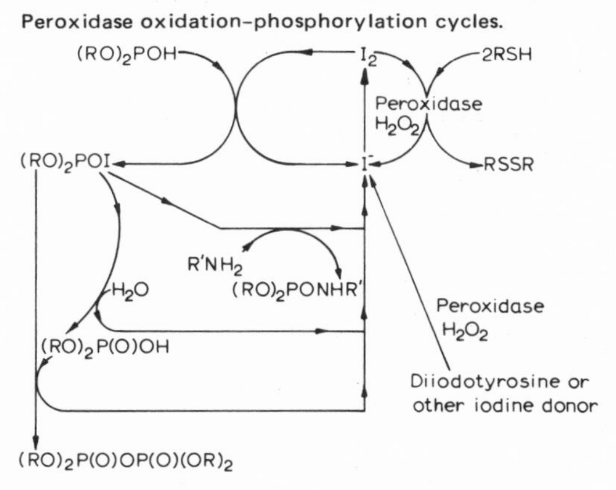

In general, in the transiodinations described above the yields were low, but these *in vitro* reactions were carried out in aqueous media, whereas the multi-phase structure of the animal cell may well permit a more efficient iodine transport process.

TRANSALKYLATION REACTIONS

It has recently been shown that the enzymic oxidation of 2,4-dimethylaniline (**XXVI**) does not give rise to a product analogous to that obtained from mesidine in which a methyl group is readily eliminated.

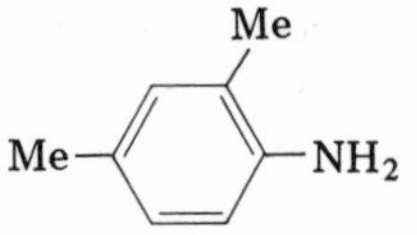

(**XXVI**)

Surprisingly compound (**XXVII**) is obtained, indicating that migration[42] of a methyl group must have taken place during oxidation.

The mechanism of formation of 2,5-dimethyl-*p*-benzoquinonebis-(2′,4′-dimethyl)anil (**XXVII**) is believed to proceed via 2,4,2′,4′-tetramethyldiphenylamine (**XXVIII**), produced from two molecules of 2,4-dimethylaniline with elimination of ammonia, and then according to the following scheme (**XXIX** to **XXXI**).

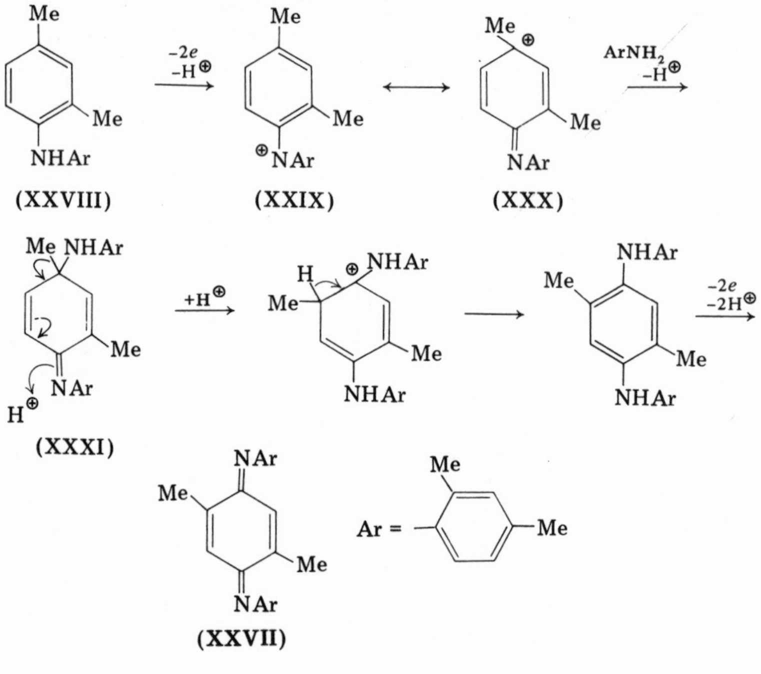

(**XXVIII**) (**XXIX**) (**XXX**)

(**XXXI**)

(**XXVII**)

In this connection it is of particular interest to note that 4-t-butyl-2,6-dimethylaniline (**XXXII**) gives the expected anil (**XXXIII**) and, in addition, by migration of the t-group[43] it gives 3-t-butyl-2,6-dimethyl-*p*-benzoquinone-4-(4′-t-butyl-2′,6′-dimethyl)anil (**XXXIV**).

The migrations of a methyl group in the oxidation of 2,3,4-trimethylaniline and of a cyclohexyl group in the oxidation of 4-cyclohexyl-2,6-dimethylaniline have been noted[44].

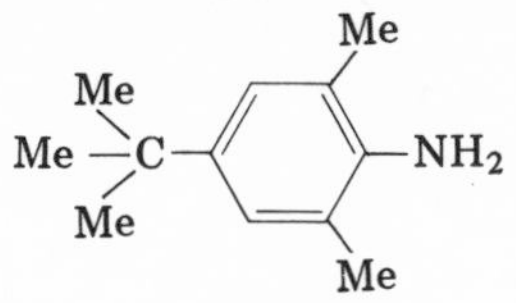

(XXXII)

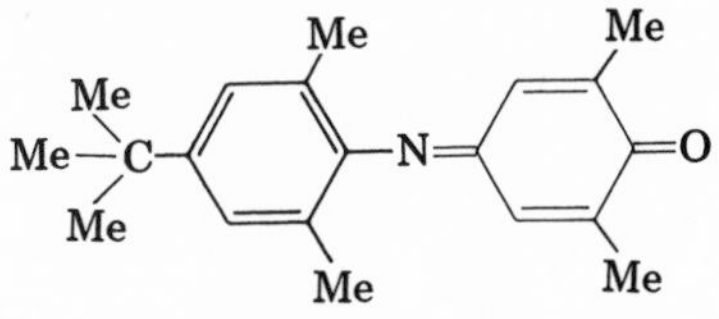

(XXXIII)

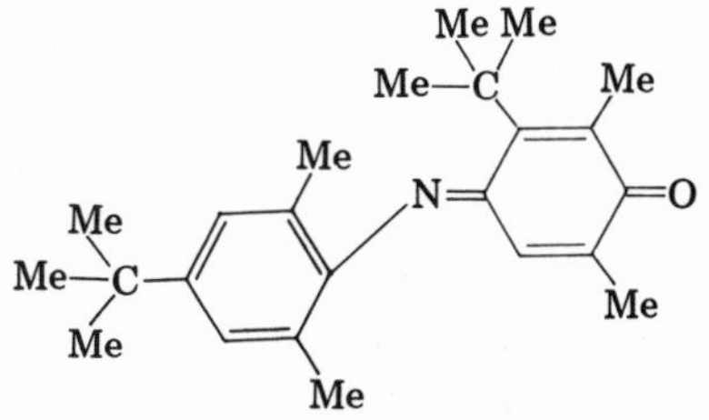

(XXXIV)

INVESTIGATIONS INTO THE BENZIDINE BLOOD TEST

As explained above *N,N,N′,N′*-tetramethylbenzidine (from *N,N*-dimethylaniline) gives an intense deep green colour with the peroxidase system. Benzidine (**XXXV**) itself gives a blue coloration with the peroxidase system and the main coloured product can be formulated as (**XXXVI**) or (**XXXVII**). Further oxidation gives a less highly coloured product benzidine brown (**XXXVIII** or **XXXIX**). Blood has a peroxidase-like action and in the presence of hydrogen peroxide gives all the colour reactions with the substrates (aniline, mesidine *etc.*) so far mentioned in this chapter. In particular, benzidine has been known for many years as an extremely sensitive and specific test for blood.

It has recently been shown that the intense blue colour given by the interaction of peroxidase, hydrogen peroxide and benzidine has an absorption spectrum almost identical with that obtained by the interaction of blood, hydrogen peroxide and benzidine[45].

The main disadvantage of benzidine as a test for blood in forensic work is that benzidine is extremely carcinogenic and therefore constitutes a severe potential hazard to workers employing the test.

References pp. 1019–1021

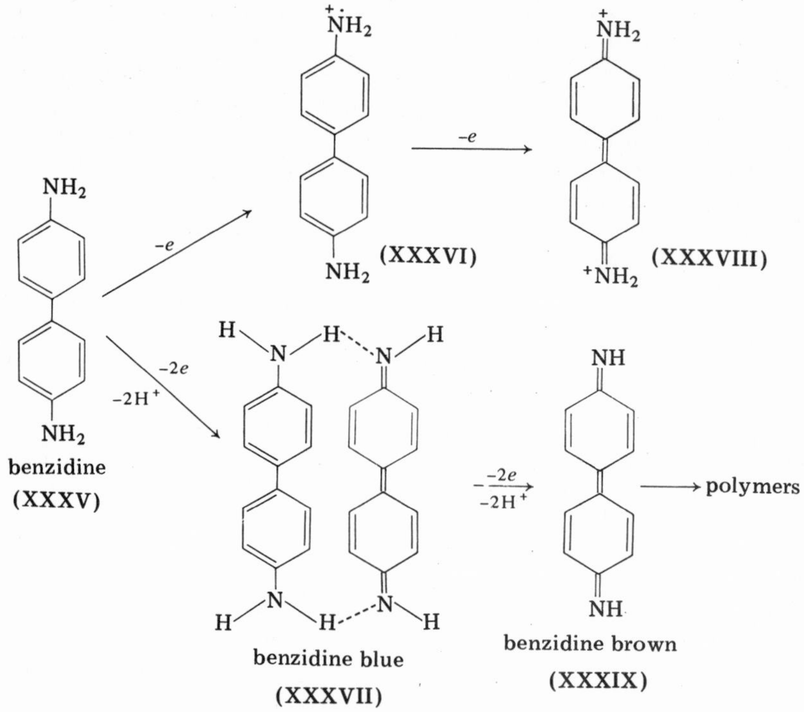

NH_2

benzidine
(XXXV)

(XXXVI)

(XXXVIII)

benzidine blue
(XXXVII)

benzidine brown
(XXXIX)

→ polymers

The use of *o*-tolidine (**XL**) has been suggested[46] as a sensitive test for blood but this compound, although not so carcinogenic as benzidine, is still dangerous. It has been suggested[47] that an aromatic amine is likely to be carcinogenic if the positions ortho to the amino-group can undergo metabolic hydroxylation. In other words the effective carcinogen is an *ortho*-aminophenol.

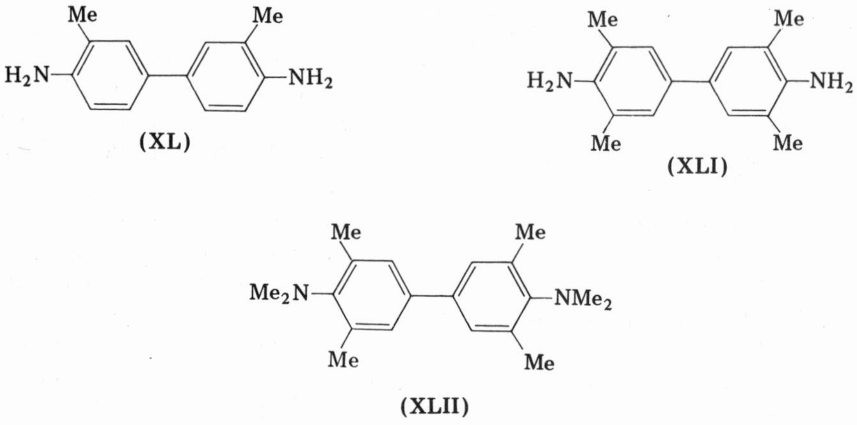

(XL)

(XLI)

(XLII)

One was therefore led to believe that 3,5,3′,5′-tetramethylbenzidine (**XLI**) would be non-carcinogenic and would still give an intensely coloured "peroxidase" test with blood. Work in progress[48] confirms these anticipations.

Substitution at both N and *ortho*-carbon results in steric strain in this benzidine system (**LXII**), sufficient to twist the dialkylamino group out of the plane of the ring and prevent any π orbital overlap. This precludes the possibility of any highly conjugated coloured species being

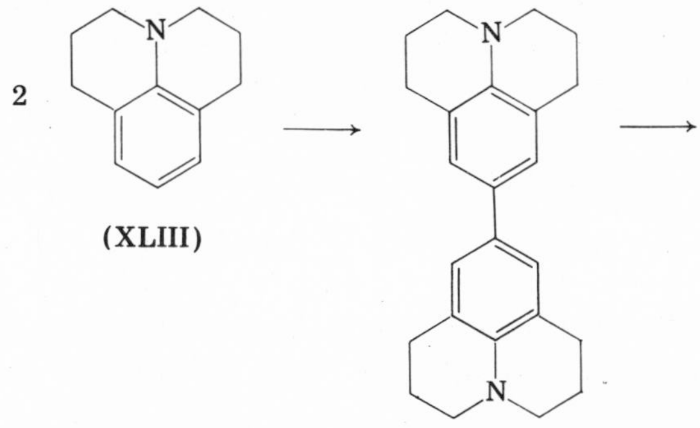

2 (XLIII) ⟶ (XLIV) ⟶

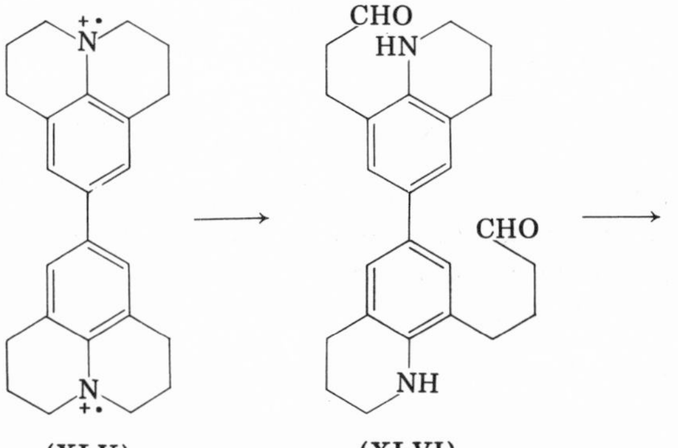

(XLV) ⟶ (XLVI) ⟶

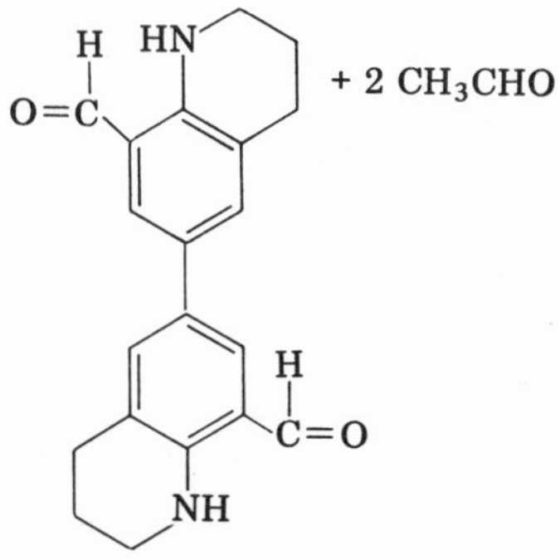

(XLVII) + 2 CH_3CHO

References pp. 1019–1021

formed. The twisting process may be prevented by the introduction of extra rings, as exemplified by the julolidine (**XLIII**) system[49]. This compound on oxidation gives the colourless dimer (**XLIV**) which goes to the highly coloured radical-ion (**XLV**) on further oxidation. The production of this colour constitutes a test for peroxidase and, of course, for blood. Further peroxidatic oxidation disrupts the molecule giving (**XLVI**) and then (**XLVII**). Compound (**XLIV**) appears to be non-carcinogenic. Quite apart from any possible forensic applications of the peroxidatic oxidation of substituted benzidines and related compounds to give coloured products, the production of compounds such as (**XLVI**) and (**XLVII**) provides the chemist with a valuable synthetic tool.

PEROXIDATIC OXIDATION OF MIXTURES

The peroxidase oxidation of two amines together has been investigated and in the case of aniline plus *p*-toluidine[50] none of the products given by each amine separately is obtained (apart from traces of 4,4′-dimethylazobenzene), but instead compounds of "intermediate" structure are formed, *e.g.*:

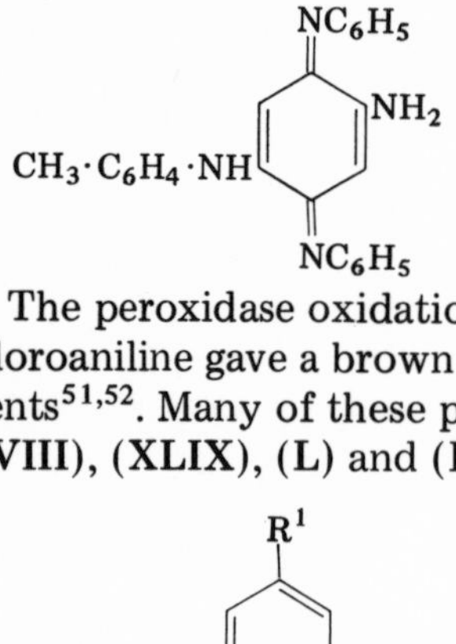

NC_6H_5
NH_2
$CH_3 \cdot C_6H_4 \cdot NH$
NC_6H_5

The peroxidase oxidation of an equimolar mixture of *p*-toluidine and 4-chloroaniline gave a brown solid consisting of at least fourteen components[51,52]. Many of these possessed "intermediate" structures *e.g.* (**XLVIII**), (**XLIX**), (**L**) and (**LI**).

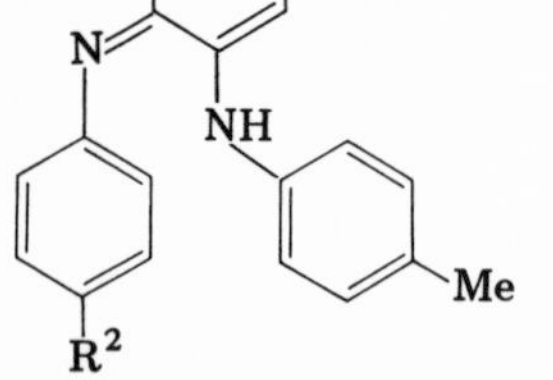

(**XLVII**; R^1=Cl. R^2=Me)
(**XLIX**; R^1=Me. R^2=Cl)

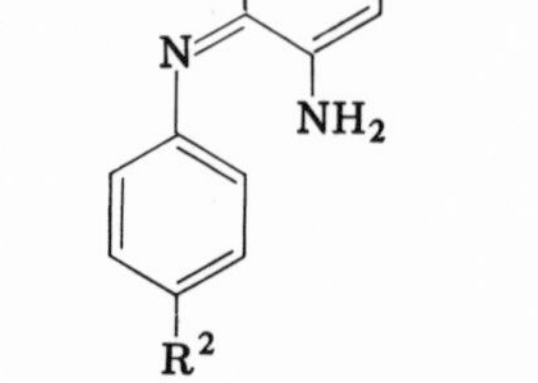

(**L**; R^1=Cl. R^2=Me)
(**LI**; R^1=Me. R^2=Cl)

Enzymic oxidation of $-CH_3$ *to -CHO*

It has been known since 1900 that phenols are oxidized by the peroxidase system, but most investigators have merely recorded colour changes. In a few cases however, the products have been characterized. Vanillin yields "dehydrodivanillin" and pyrogallol gives purpurogallin.

As the oxidation of mesidine resulted in the loss of a methyl group with the formation of 2,6-dimethylbenzoquinone-4-(2′,4′,6′-trimethyl)anil it seemed that similar considerations might apply to the oxidation of substituted phenols and therefore the oxidations of mesitol[53] and durenol[54] were investigated.

An aqueous solution of mesitol (**LII**), with hydrogen peroxide and peroxidase, became yellow and a pale yellow solid separated. The yellow filtrate contained 4-hydroxy-3,5-dimethylbenzaldehyde (**LIV**) and 2,6-dimethylbenzoquinone (**LV**). The yellow amorphous solid was fractionally sublimed in a high vacuum and gave 4-hydroxy-3,5-dimethylbenzyl alcohol (**LIII**).

An aqueous solution of the alcohol was rapidly oxidized to the corresponding aldehyde by the enzyme system. It was also noted that 4-hydroxy-3,5-dimethylbenzaldehyde was attacked by the system, giving 2,6-dimethyl-*p*-benzoquinone (**LV**).

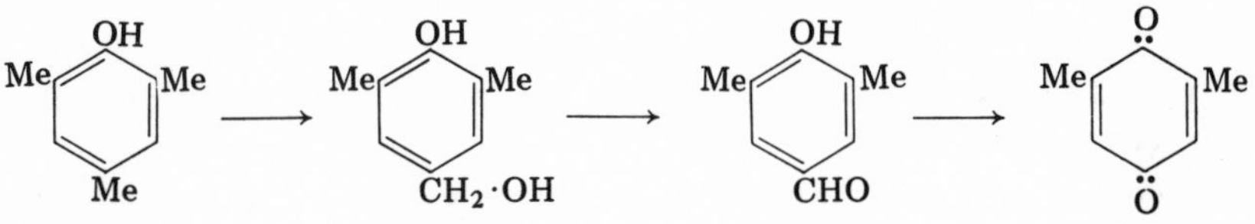

OH, Me, Me, Me (**LII**) → OH, Me, Me, $CH_2 \cdot OH$ (**LIII**) → OH, Me, Me, CHO (**LIV**) → O, Me, Me, O (**LV**)

Thus we have evidence for the enzymic oxidation of mesitol according to the above scheme (it should be noted that the carboxylic acid is not formed). Qualitative experiments showed that the quinone was unaffected by the enzyme system and was thus an end-product of the oxidation.

This was the first recorded peroxidase oxidation of a $\cdot CH_3$ group to a $\cdot$CHO group[53].

Durenol

Durenol being slightly soluble in water, was oxidized in fine suspension. Hydrogen peroxide and peroxidase were added intermittently. No colour changes were observed. The product consisted of colourless 4,4′-dihydroxy-octamethylbiphenyl (**LVI**).

References pp. 1019–1021

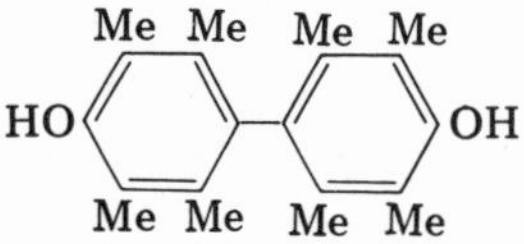

Me Me Me Me

HO OH

Me Me Me Me

(LVI)

Guaiacol

Many workers, *e.g.* Bach and Chodat[55] have noted that guaiacol is oxidized by the peroxidase system to a red substance. Ucko and Bansi[56] gave the optimum pH for the reaction as 5–5.2; Chance[57], however, found that the velocity of reaction of "compound II" with guaiacol was little affected by pH in the range 3.4–5.3. Many workers have used this reaction with guaiacol to investigate the kinetics of peroxidase action.

It was therefore important to make a careful investigation of this much-used reaction[54]. When a solution of guaiacol in acetate buffer (pH = 5) was treated with hydrogen peroxide and peroxidase, the product was a red-brown solid with a metallic lustre.

When the crude product was sublimed in high vacuum a white solid was isolated, which crystallized in colourless plates.

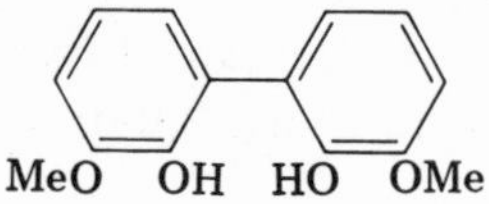

MeO OH HO OMe

(LVII)

An independent synthesis proved it to be 2,2′-dihydroxy-3,3′-dimethoxybiphenyl (**LVII**).

When the crude oxidation product from guaiacol was heated with acetic anhydride, zinc dust and sodium acetate, a white amorphous solid was obtained. Fractional sublimation gave colourless crystals of 4,4′-diacetoxy-3,3′-dimethoxybiphenyl. The isolation of this biphenyl by reductive acetylation of the crude oxidation product shows that the latter contains 3,3′-dimethoxydipheno-4,4′-quinone (**LVIII**).

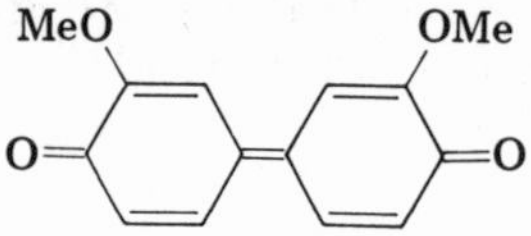

MeO OMe

O O

(LVIII)

The characteristic red-brown of the crude guaiacol oxidation product is undoubtedly due largely to this quinone. The crude product may also contain small quantities of the isomeric 3,3′-dimethoxydipheno-2,2′-quinone by further oxidation of the dihydroxylbiphenyl (**LVII**).

The fact that oxidation of durenol gives a colourless product is due to the steric effects of the four *o*-methyl groups. Further oxidation of the biphenol (**LVI**) to a coloured diphenoquinone would require the production of a double bond between the two rings, and this necessitates a shortening of the link. A model indicates that such shortening, with retention of a planar structure, is not likely.

HYDROXYL RADICALS AND THE PEROXIDASE–HYDROGEN PEROXIDE SYSTEM

We have seen that the peroxidase system and dilute iron(II) sulphate–hydrogen peroxide frequently give different products with a given substrate. It is known that the iron(II) sulphate–hydrogen peroxide system contains hydroxyl radicals:

$$Fe^{2+} + HOOH = Fe^{3+} + HO^{\bullet} + HO^{-}$$

It is important to know whether the peroxidase–hydrogen peroxide system also contains hydroxyl radicals or is capable of oxidation through double bond hydroxylation. Dainton seems to have indicated the absence of free hydroxyl radicals in the peroxidase system if acrylic esters are used as a monomer source. Apart from this there is little convincing evidence either way, but attention may be drawn to the following experiment[58]. Styrene was used in place of acrylic esters and search made for possible changes at the double bond. Bromine titration for unsaturation was chosen for following the reaction because it detects any one of the following changes in the monomer peroxidase–hydrogen peroxide system: (1) formation of low molecular weight polymers not separable by filtration; (2) catalyzed hydroxylation, by hydrogen peroxide, of the double bond; (3) epoxide formation; (4) formation of high-molecular-weight polymers.

Care was taken to obtain highly purified reagents. In outline the technique was as follows.

Charges A, B and C were made up in special flasks and the pH of each solution was adjusted to 4.8: (A) 50 ml of distilled water; 1 ml of 0.8 *M* hydrogen peroxide; 1 ml of enzyme of PN 50; 2 ml of styrene; (B) 50 ml of distilled water; 1 ml of 0.8 *M* hydrogen peroxide; 50 mg of $FeSO_4 \cdot 7H_2O$; 2 ml of styrene; (C) 50 ml of distilled water; 1 ml of 0.8 *M* hydrogen peroxide; 2 ml of styrene. The contents of each flask were frozen in dry ice–acetone and air was pumped out to a pressure of 1×10^{-4} mm. Then each flask was sealed under vacuum, and shaken in the dark for 24 hours. There was no evidence for polymerization in C, the styrene remaining in discrete droplets. In B there were no longer droplets and polymer was clearly visible as "fish-like" structures. Vigorous shaking caused the contents of A to emulsify, but there was no visual evidence of polymerization. The flasks were opened and each product was extracted with chloroform.

References pp. 1019–1021

The chloroform layer from a particular reaction was titrated with a 3% solution of bromine in glacial acetic acid. The titres (in ml) of a typical series for each of the reactions A, B and C were: (A) 31.2; (B) 24.4; (C) 31.1. Direct titration on 2 ml of styrene = 32 ml. It was also demonstrated that the enzyme system of A had not been inactivated during the treatment.

Conclusions

1. There is no evidence for the presence of hydroxyl radicals in the enzyme system.
2. There is no detectable hydroxylation★ nor other oxidative attack of the double bond of styrene.
3. The experimental results are in agreement with the observations of Saunders *et al.* that Fenton's reagent and the peroxidase system are not necessarily interchangeable.
4. These results do not preclude the possibility of hydroxyl radicals being passed directly to the substrate simultaneously with hydrogen atom removal (see p. 1013). The present experimental method, however, would not detect such a mechanism, since styrene is not likely to function as a hydrogen donor.

MECHANISM OF PEROXIDASE ACTION

At present there seems to be no single mechanism which will explain adequately all recorded peroxidatic oxidations. Suggested mechanisms in particular cases are given throughout this chapter (pp. 994, 998, 1004, 1007), (see also Chapter 20) and certain generalizations emerge.

Saunders, Holmes-Siedle and Stark have suggested several possible mechanisms[60]. Among such theories special note ought to be made of the unsymmetrical dimerization of radicals. Oxidation of an amine

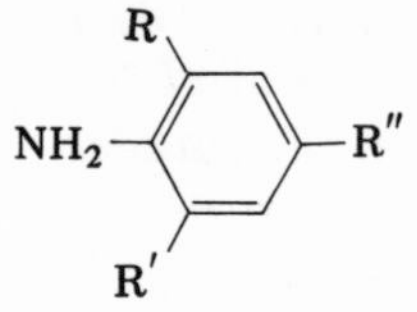

R

NH_2 — — R″

R′

will give the free radical (**LIXa**). Symmetrical pairing of two such radicals would give hydrazo-compounds which would be oxidized to the corresponding azo-compounds (actually produced in small yield in certain reactions,

---

★ Mason[59] reports hydroxylations in presence of dihydroxyfumaric acid.

see p. 1000). Unsymmetrical dimerization of the radicals (**LIXa**) and (**LIXb**) would give the quinone di-imine (**LX**) by elimination of R″. Indeed compounds of the type (**LX**) are often produced and frequently the C=NH is subsequently hydrolyzed to C=O. When R″ = F, Cl, Br or I and R′=R=H (compound **LX**) would give, by a series of established additive and oxidative reactions, products of the type (**XXV**) p. 1000.

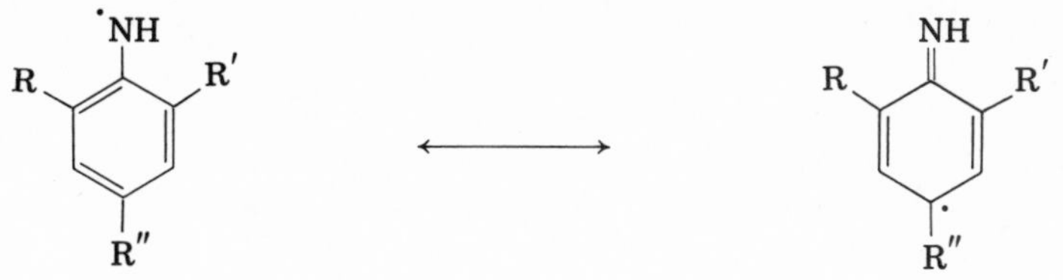

(**LIXa**) (**LIXb**)

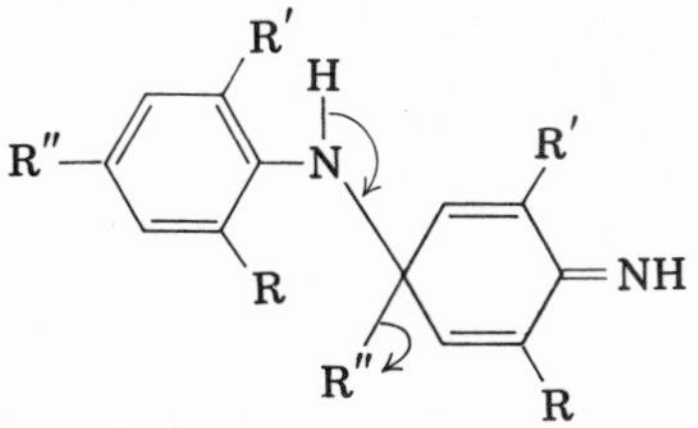

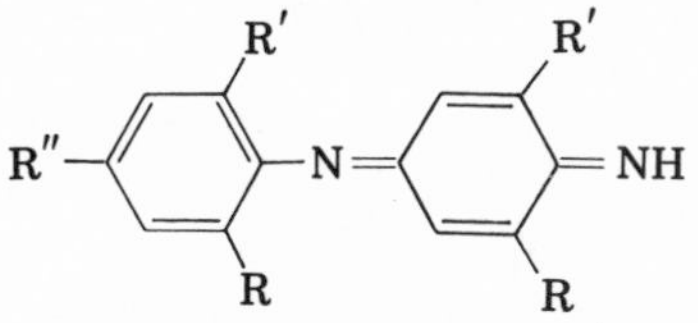

(**LX**)

It has been suggested (1950 onwards)[60] that the reaction of the amine with compound II might involve an activated complex such as (**LXI**). This idea does not necessitate the existence of free hydroxyl radicals. The complex would then give, by elimination and condensation reactions, a typical anil.

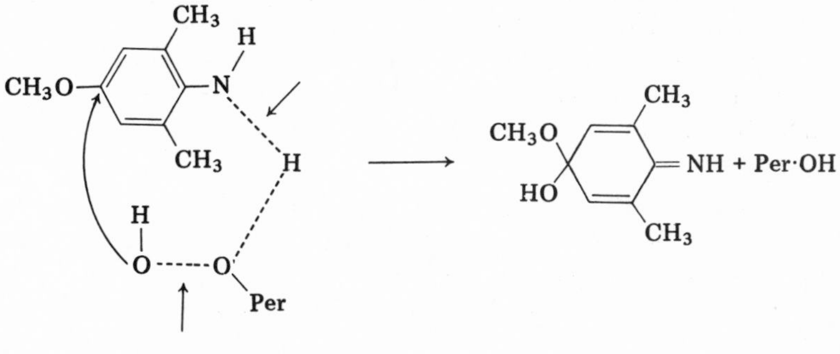

(**LXI**)

Other alternative mechanisms are clearly possible, but present work seems to confirm an initial ternary complex between the substrate, hydrogen peroxide and peroxidase[61,62].

We emphasize that the above suggestions merely serve as convenient working hypotheses for this type of oxidation. In this connection it is interesting to note that methyl 4-amino-3,5-dimethylbenzoate fails to give a coloration with hydrogen peroxide and peroxidase. Such a compound, containing a strong electron-attracting substituent, is less likely either (a) to react directly with hydrogen peroxide and peroxidase as above, or (b) to permit removal of an electron as in the suggested alternative mechanism. In fact the oxidizability of a compound by the peroxidase system depends primarily upon the tendency to such electronic changes; *p*-nitroaniline and *p*-aminoacetophenone for example, are not oxidized by peroxidase.

PEROXIDASE OXIDATIONS INVOLVING OTHER INORGANIC MATERIALS

Recorded oxidations are too numerous to quote, but the following should be noted.

Manganese

Kenten and Mann[63–67] observed that various peroxidase–hydrogen peroxide systems can, in the presence of certain phenols, oxidize manganous ion to higher valency states, and they suggested[63] that a cyclic mechanism operates as follows:

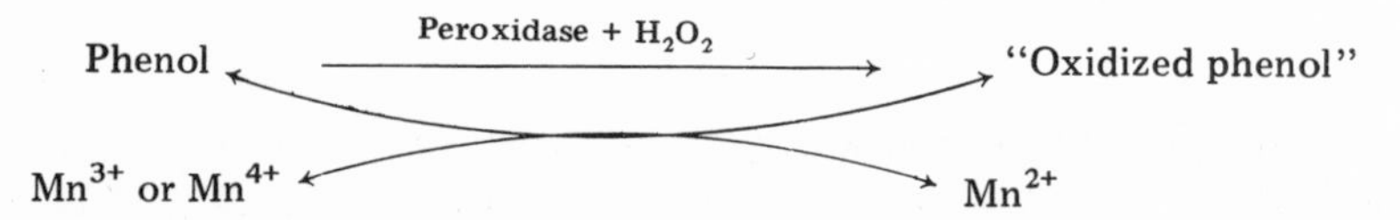

These workers also found[65] that such a peroxidase–phenol system could oxidize ferrocyanide, and also, possibly molybdate, tungstate and vanadate ions, the rate of oxidation of Mn^{2+} by the enzymic system being increased by addition of these last three ions. Even in the absence of phenols, peroxidases will catalyze the oxidation of ferrocyanide by hydrogen peroxide, but peroxidases do not appear to catalyze the oxidations of molybdate, tungstate or vanadate[65].

The elevation of Mn^{2+} to higher oxidation levels is probably responsible for the oxidation of various compounds (which are not themselves hydrogen donors in normal peroxidase reactions) by systems containing both peroxidase and manganese. For example a horseradish peroxidase–Mn^{2+} system oxidizes phenylacetaldehyde to benzaldehyde and formic acid[68]. Such a system can also oxidize certain dicarboxylic acids, including oxalic acid, oxaloacetic acid, ketomalonic acid, dihydroxytartaric acid, and dihydroxyfumaric acid, especially if a phenol is present[69].

Chelated iron

It is claimed that iron salts of ethylene-diaminetetraacetic acid have "peroxidase" activity towards ethanol[70], ascorbic acid[71], and *p*-amino-diethylaniline[72]. Anan[73] has reported that certain complex metal salts (Fe^{3+}, Cu^{2+}, Mg^{2+} and Mn^{2+}) behave as "model peroxidases" towards pyrogallol.

It should be emphasized that many of these oxidations may indeed not be true peroxidase reactions as defined on p. 989. In our view probably the most diagnostic test for peroxidase is its ability to oxidize mesidine to the purple anil (p. 996) in the presence of hydrogen peroxide in a yield of 95% or better.

Further recent applications of peroxidase

Lehrer[74] has reported that peroxidase systems may contribute to antifungal defence mechanisms in human beings.

Sutton and Becker[75] have reached the conclusion that HRP is a promising protein for specifically carrying cytotoxic therapeutic agents into certain types of tumour cells.

CATALASES

From a structural and functional point of view catalases are related to peroxidases. They are haem enzymes with four Fe atoms per molecule attached to protein and chelated to protoporphyrin IX. Each of the four haem groups is part of a separate polypeptide chain[75a]. The most characteristic reaction of catalase is the rapid catalysis of the reaction of hydrogen peroxide with itself, *i.e.* an oxidation–reduction process to oxygen and water ($K \simeq 10^7\,M^{-1}\,\text{sec}^{-1}$). Another, and perhaps less important, reaction of catalases is also known. This is slower ($K \simeq 10^3\,M^{-1}\,\text{sec}^{-1}$) and catalyzes oxidation by peroxides. This is sometimes called the "peroxidatic" action of catalase.

Historical

Thénard, the discoverer of hydrogen peroxide, was the first to note catalatic activity in animal tissues although it was not until 1907 that Loew showed that the action was due to a single enzyme which he called "catalase". In 1937 Sumner and Dounce[76] crystallized catalase obtained from beef liver.

It seems that catalase is always present in systems where cytochromes operate and where oxygen is reduced to hydrogen peroxide which is toxic to the cell and therefore must be removed. Catalases are found in all plants and in most aerobic (but usually not anaerobic) bacteria.

It was originally thought that the only important function of catalase was to destroy hydrogen peroxide, but it probably functions in additional and more complicated ways. For example, Keilin and Hartree[77] found that hydrogen peroxide produced by oxidases *in vitro* could be used by catalase to oxidize alcohols and certain other donors.

It is interesting to note that there exists in animals a hereditary condition known as actalasaemia. It has also been found that human beings who are deficient in erythrocyte catalase have a predisposition to oral gangrene.

Bonnichsen[78] describes a typical preparation of catalase from beef liver. An extract of ground liver is treated with chloroform and alcohol to precipitate the crude enzyme which, on dialysis and fractional precipitation with ammonium sulphate, gives the crystalline enzyme. Similar crystalline products are obtained from human, horse and guinea-pig liver and from human and beef erythrocytes.

The molecular weight of catalases from different sources are of the same order, *viz.* 225,000 to 251,000[79] with four atoms of iron per molecule. It seems that catalases from different animals differ in their protein content as indicated by the production of anticatalase antibodies in rabbits after injection of catalase obtained from other animals[80].

The Commission on Enzymes[18] classifies catalase as follows: "1.11.16. H_2O_2 oxidoreductase.

Reaction: $H_2O_2 + H_2O_2 = O_2 + 2H_2O$

Specificity: A haemoprotein enzyme. Several organic substances, especially ethanol can act as hydrogen donor."

Mechanism of catalase action (see also Chapter 20, p. 684)

It is probably true to say that the detailed mechanism of catalase action is not yet definitely established, but a large amount of physicochemical evidence on the course of the reaction has been published.

Bonnichsen, Chance and Theorell[81] showed that the catalatic decomposition of hydrogen peroxide is not a chain process. There is no induction period and chain-breaking reagents are ineffective. The catalase molecule can decompose 6×10^6 molecules of peroxide per second; the reaction between peroxide and bacterial catalase has been described by Chance as the "fastest biochemical event that has yet been measured directly"[82].

Hydrogen peroxide forms three spectroscopically distinguishable compounds with catalase, only one of which is active. Decomposition of peroxide takes place[83] via green "Compound I" (λ_{max} = 405, 655 nm), similar to peroxidase compound I. Compound II (λ_{max} = 429, 536, 568 nm), which is red, is formed with excess of peroxide. Its activity is about 1/10,000 that of compound I.

Keilin and Hartree[84] detected inactive compound III (λ_{max} = 416, 545, 585 nm) in the presence of high concentrations of peroxide.

Chance's interpretations of a large amount of kinetic spectrophotometric data[83] are as follows:

Decomposition of hydroperoxide to oxygen occurs on collision of catalase compound I with a second molecule of hydrogen peroxide or alkyl hydroperoxide. Compound I exhibits two oxidizing equivalents and probably reacts with the colliding peroxide by virtue of a shift of four electrons; this results in a two-electron transfer by the free peroxide to the bound peroxide, giving water and oxygen:

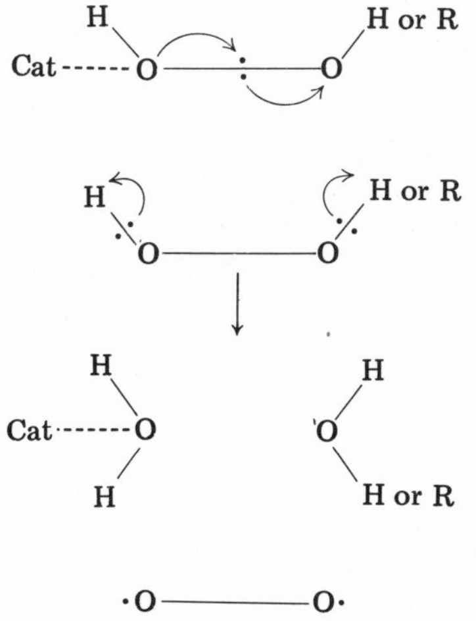

```
 H                      H or R
  \                    /
Cat------O ----- : ---- O

 H                      H or R
  \:                  /.
     O --------------O
            |
            v
 H                       H
  \                     /
Cat------O             O
        /               \
       H                 H or R

     ·O ------------- O·
```

The oxygen molecule is stable in the form shown. A hydrogen donor other than hydrogen peroxide (*e.g.* ethanol) can react with compound I. A similar mechanism probably holds, in which H—C—O—H (with C bearing two further bonds) takes the place of H—O—O—H; the diradical produced passes to the stable carbonyl group. The rate of oxidation of alcohols by compound I falls as the alkyl group increases in size. This may have some relevance to the steric arrangement of the active site in catalase. Only one of the four haem groups is required for catalatic action.

Keilin and Hartree[85] suggested that the oxidation–reduction of hydrogen peroxide by catalase proceeds through a ferrous form of catalase.

The chemical nature of catalase compound II is not as clear as that of the corresponding peroxidase compound. Unlike peroxidase, there is no requirement for a secondary compound one oxidation level lower than the primary, since the evidence suggests that the transfer of the two electrons from the colliding donor peroxide molecule to the bound acceptor peroxide molecule is virtually simultaneous[86]. Spectra and magnetometric data for compound II indicate covalent bonding of the haem ligands. It may be that the other three haem groups (or some of them) are linked to peroxide and that the resultant changes in bonding render the compound "inactive".

At neutral pH, free catalase, like peroxidase, has a hydroxyl group attached to the free haem position. Most anions can compete with this hydroxyl group for combination with the haem. CN^-, N_3^-, F^-, NH_2OH, N_2H_4, NO and H_2S, inhibit catalase action, but CO does not.

Some inorganic catalases

The iron(III) complex of triethylenetetramine (Chapter 20, p. 685) has catalatic activity[87,88]. It seems that in this chelate, two secondary amine nitrogen atoms are coplanar with the Fe^{3+} atom, the two primary amine atoms lying, one above, and one below, the iron atom. Wang[87,88] suggested that when peroxide is added, both oxygen atoms in the peroxide become coordinated with the iron atom, so that the O–O bond is stretched, thus lowering the activation energy for fission of the bond. Consequently, the O and the OH of the peroxide, in an intramolecular reaction, become two separate ligands with the iron, and react with a second molecule of peroxide to yield O_2, regenerating the original complex. It is claimed that a cobalt/magnesium carbonate system behaves both as a model peroxidase and a model catalase[89].

A recently recorded joint action of catalase and peroxidase

It is interesting to note that the explosive ejection of the defensive secretion from the carrion beetle[90] is considered to be due to an enzymic oxidation of constituent hydroquinones by hydrogen peroxide to quinones and free oxygen, the latter acting as a propellant for the former. Electrophoretic studies[90] indicate that 66% of the enzymes are of the catalase and 33% of the peroxidase type.

CONCLUSION

In conclusion it can be said that the enzyme peroxidase is remarkably specific in its action, but nevertheless (paradoxically) exhibits great versatility in the variety of reactions that it can promote, once the initial stages of oxidation have been effected. It is of considerable interest in plant and animal biology, and it is now proving a very valuable synthetic tool in the hands of the organic chemist, who can, by its use, perform selective reactions often under extremely mild conditions (*e.g.* the oxidation of a methyl group, the elimination of a methyl group and the migration of alkyl groups). The effect of peroxidase and of catalase on cell growth may be of far-reaching importance. Even the production of colours, aesthetically pleasing, has given peroxidase a most important place among biological catalysts.

ACKNOWLEDGEMENT

The author wishes to thank his colleagues Dr. V. R. Holland, Dr. P. B. Baker and Mrs. J. M. Ridgewell who have kindly read the manuscript and made valuable suggestions.

REFERENCES

1 C. F. Schönbein, *Verh. naturf. Ges. Basel, 1* (1855) 339.
2 M. G. C. R. Linossier, *Soc. Biol., Paris,* 50 (1898) 373; cf. O. Loew, *Rep. U.S. Dept. Agric.*, (68) (1901) 47.
3 A. N. Bach and R. Chodat, *Ber. Dtsch. chem. Ges.*, 36B (1903) 606.
4 A. N. Bach, *Ber. Dtsch. chem. Ges.*, 37 (1904) 3787; A. N. Bach, *Collected Works in Chemistry and Biochemistry*, Academy of Science, Moscow, 1950.
5 A. Ernest and H. Berger, *Ber. Dtsch. chem. Ges.*, 40 (1907) 4671.
6 F. Battelli and L. Stern, *Biochem. Z.*, 13 (1908) 44.
7 R. Willstätter and A. Pollinger, *Liebigs Annln.*, 430 (1923) 269; R. Willstätter and A. Pollinger, *Z. Physiol. Chem.*, 130 (1932) 281.
8 W. Mosimann and J. B. Sumner, *Arch. Biochem. Biophys.*, 33 (1951) 487.
9 I. Huszak, *Biochem. Z.*, 312 (1942) 330.
10 M. Polonovski and M. F. Jayle, *Bull. Soc. Chim. Biol.*, 21 (1939) 66.
11 K. Agner, *Acta Chem. Scand.*, 12 (1958) 89.
12 A. M. Altschul, R. Abrams and T. R. Hogness, *J. Biol. Chem.*, 142 (1942) 303.
13 H. Theorell, *Enzymologia,* 10 (1942) 250.
14 M. Dixon and E. C. Webb, *Enzymes,* 2nd Edn., Longmans, London, 1964, p. 696.
15 T. Hosoya, *J. Biochem. Tokyo,* 47 (1960) 369.
16 L. M. Shannon *et al. J. Biol. Chem.*, 241 (1966) 2166; 242 (1967) 2470; 243 (1968) 3560. See also Y. Morita, *Biochem. J.*, 66 (1969) 191 and K. G. Paul and T. Stigbrand, *Acta Chem. Scand.*, 24 (1970) 3607.
17 P. B. Baker, Dissertation, Cambridge University, 1970.
18 *Commission on Enzymes,* Pergamon Press, New York and Oxford, 1961.
19 D. Keilin and E. F. Hartree, *Biochem. J.*, 49 (1951) 88.
20 D. Keilin and T. Mann, *Proc. R. Soc. (B)*, 122 (1937) 119.
21 H. Theorell, *Ark. Kemi Min. Geol.*, 15B (1942) No. 24.
22 B. Chance, *Arch. Biochem. Biophys.*, 22 (1949) 224.
23 P. George, *Biochem. J.*, 54 (1953) 267.
24 P. George and D. H. Irvine, *Br. J. Radiol.*, 27 (1954) 131.
25 T. H. Moss, H. Ehrenberg and A. J. Bearden, *Biochemistry,* 8 (1969) 4519.
26 A. G. Holmes-Siedle and B. C. Saunders, unpublished work.
27 K. A. C. Elliott, *Biochem. J.*, 26 (1932) 10, 128.
28 P. J. G. Mann and B. C. Saunders, *Proc. R. Soc. (B)*, 119 (1935) 47.
29 B. C. Saunders and P. J. G. Mann, *J. Chem. Soc.*, (1940) 769.
30 D. G. H. Daniels, F. T. Naylor and B. C. Saunders, *J. Chem. Soc.*, (1951) 3433.

31 N. B. Chapman and B. C. Saunders, *J. Chem. Soc.*, (1941) 496.
32 B. C. Saunders and G. H. R. Watson, *Biochem. J.*, 46 (1950) 629.
33 F. T. Naylor and B. C. Saunders, *J. Chem. Soc.*, (1950) 3519.
34 D. G. H. Daniels and B. C. Saunders, *J. Chem. Soc.*, (1951) 2112.
35 D. G. H. Daniels and B. C. Saunders, *J. Chem. Soc.*, (1953) 822.
36 G. M. K. Hughes and B. C. Saunders, *Chem. Ind.*, (1954) 1265; *J. Chem. Soc.*, (1954) 4630.
37 B. C. Saunders and B. P. Stark, *Tetrahedron*, 4 (1958) 169.
38 H. McCombie, B. C. Saunders and G. I. Stacey, *J. Chem. Soc.*, (1945) 921.
39 W. Gerrard and G. J. Jeacocke, *J. Chem. Soc.*, (1954) 3647.
40 A. D. F. Toy, *J. Am. Chem. Soc.*, 70 (1948) 3882.
41 G. C. Mills, *J. Biol. Chem.*, 229 (1957) 189.
42 V. R. Holland, B. M. Roberts and B. C. Saunders, *Tetrahedron*, 25 (1969) 2291.
43 P. B. Baker, V. R. Holland and B. C. Saunders, (1972) in press.
44 P. B. Baker and B. C. Saunders (1972) in press.
45 V. R. Holland and B. C. Saunders, *Report to Home Office*, London (1969).
46 D. Quaglino and R. Flemens, *Lancet*, 2 (1958) 1020.
47 A. L. Walpole, M. U. C. Williams and D. C. Roberts, *Br. J. Ind. Med.*, 9 (1952) 255.
48 V. R. Holland and B. C. Saunders, (1972) in press.
49 V. R. Holland and B. C. Saunders, *Tetrahedron*, 27 (1971) 2851.
50 G. M. K. Hughes and B. C. Saunders, *J. Chem. Soc.*, (1956) 3814.
51 V. R. Holland and B. C. Saunders, *Tetrahedron*, 25 (1969) 4153.
52 D. G. H. Daniels and B. C. Saunders, *J. Chem. Soc.*, (1953) 822.
53 H. Booth and B. C. Saunders, *Nature*, 165 (1950) 567.
54 H. Booth and B. C. Saunders, *J. Chem. Soc.*, (1956) 940.
55 A. N. Bach and R. Chodat, *Arch. Sci. Phys. Natn.*, 42 (1916) 56.
56 H. Ucko and H. W. Bansi, *Z. Physiol. Chem.*, 159 (1926) 235.
57 B. Chance, *Arch. Biochem. Biophys.*, 24 (1949) 410.
58 A. A. Patchett and B. C. Saunders, unpublished work.
59 H. S. Mason, *Adv. Enzymol.*, 19 (1957) 147.
60 B. C. Saunders, A. G. Holmes-Siedle and B. P. Stark, *Peroxidase*, Butterworths, London, 1964, p. 28.
61 W. Heimann and K. Wirrer, *Nahrung*, 12(1) (1968) 45.
62 P. B. Baker, Dissertation, Cambridge University, 1970.
63 R. H. Kenten and P. J. G. Mann, *Biochem. J.*, 46 (1950) 67.
64 R. H. Kenten and P. J. G. Mann, *Biochem. J.*, 45 (1949) 255.
65 R. H. Kenten and P. J. G. Mann, *Biochem. J.*, 50 (1951) 29.
66 R. H. Kenten and P. J. G. Mann, *Biochem. J.*, 61 (1955) 279.
67 R. H. Kenten and P. J. G. Mann, *Biochem. J.*, 65 (1957) 179.
68 R. H. Kenten, *Biochem. J.*, 55 (1953) 350.
69 R. H. Kenten and P. J. G. Mann, *Biochem. J.*, 53 (1953) 498.
70 B. Matyska and F. P. Doucek, *Chem. Listy*, 51 (1957) 1791.
71 R. R. Grinstead, *J. Am. Chem. Soc.*, 82 (1960) 3464.
72 P. M. Mader, *J. Am. Chem. Soc.*, 82 (1960) 2956.
73 K. Anan, *J. Japan Biochem. Soc.*, 23 (1951) 105.
74 R. K. Lehrer, *Clin. Res.*, 17(2) (1969) 370.
75 C. H. Sutton and N. H. Becker, *Ann. N.Y. Acad. Sci.*, 159(2) (1969) 497.
75a K. Agner, *Arkiv. Kemi, Mineral. Geol.*, 16A (6) (1942).
76 J. B. Sumner and A. L. Dounce, *J. Biol. Chem.*, 121 (1937) 417.
77 D. Keilin and E. F. Hartree, *Biochem. J.*, 39 (1945) 148.
78 R. K. Bonnichsen, *Acta Chem. Scand.*, 2 (1948) 561.
79 A. C. Maehly, in D. Glick, *Methods of Biochemical Analysis*, Vol. 1, Interscience, New York, 1954, p. 357.

80 E. Tria, *J. Biol. Chem.*, 129 (1939) 385.
81 R. K. Bonnichsen, B. Chance and H. Theorell, *Acta Chem. Scand.*, 1 (1947) 685.
82 B. Chance, in J. B. Sumner and K. Myrbäck, *The Enzymes*, Vol. 2, Pt. 1, Academic Press, New York, 1951, p. 446.
83 B. Chance, *J. Biol. Chem.*, 179 (1949) 1331.
84 D. Keilin and E. F. Hartree, *Biochem. J.*, 49 (1951) 88.
85 D. Keilin and E. F. Hartree, *Biochem. J.*, 39 (1945) 148.
86 B. Chance, in W. D. McElroy and B. Glass, *The Mechanism of Enzyme Action*, Johns Hopkins Press, Baltimore, 1954, p. 396.
87 J. H. Wang, *J. Am. Chem. Soc.*, 77 (1955) 822.
88 J. H. Wang, *J. Am. Chem. Soc.*, 77 (1955) 4715.
89 W. Waneste and R. Vercauteren, *Naturwiss.*, 12 (1963) 413.
90 H. Schildknecht *et al. Z. Naturforsch.*, 23B (9) (1968).

Chapter 29

CHLOROPHYLL

J. J. KATZ

Chemistry Division, Argonne National Laboratory, Argonne, Illinois 60439, U.S.A.

1.0 INTRODUCTION

1.1 General

The chlorophylls are a group of closely related pigments found in organisms capable of photosynthesis. These substances absorb visible light strongly, and, because of their universal distribution in plants, have long been considered to act as photoreceptors in the primary light conversion act in photosynthesis. Useful introductions to the light conversion problem in photosynthesis have been provided by Kamen[1] and by Clayton[2]. The number of chlorophylls in nature is small. Chlorophyll *a* (Chl *a*) is the most abundant and the most widely distributed. It is present in all plants that produce oxygen by photosynthesis, and it is not suprising that it is by far the most intensively studied of the chlorophylls. In this chapter, the abbreviation, Chl, without further qualification refers to chlorophyll *a*. The other chlorophylls include chlorophyll *b* (Chl *b*), a subsidiary chlorophyll in green algae and higher plants; chlorophylls c_1 and c_2 (Chl c_1 and c_2)[3–5], minor chlorophylls in diatoms and some marine brown algae; chlorophyll *d* (Chl *d*), a minor chlorophyll in some marine red algae[6]; protochlorophyll (Pchl), widely distributed in plants but found only in low concentrations as an intermediate in the biosynthesis of chlorophylls *a* and *b*; bacterio-chlorophyll, (BChl), the major chlorophyll of purple photosynthetic bacteria; and the chlorobium chlorophylls, number and structure[7–9] still uncertain, the principal photosynthetic pigments of green photosynthetic bacteria. The extraction, separation, isolation and purification of the chlorophylls from natural sources have been described by Strain and Svec[10]. It must be emphasized that a large number of chlorophyll alteration products can be and have been isolated from plant material[11-13]. These substances arise from enzymatic or non-enzymatic chemical reactions on chlorophyll, which result in de- or trans-esterification[14], oxidation[15], allomerization[16], demetallation, and ring opening. As a result, great caution must be used before the presence of new pigments significant for photosynthesis can be considered established. At present there appears to be no cogent reason to suppose that any of the many chlorophyll alteration products that have been detected in plants extracts are anything but *post mortem* artifacts of no particular physiological significance. Numerous better defined derivatives of chlorophyll can be prepared in the laboratory, one of which, pyrochlorophyll (Pyrochl *a* and *b*)[14] is finding application in physical chemical studies on chlorophyll. Structural formulae for the chlorophylls and chlorophyll derivatives important for the subsequent discussion are shown in Fig. 1.

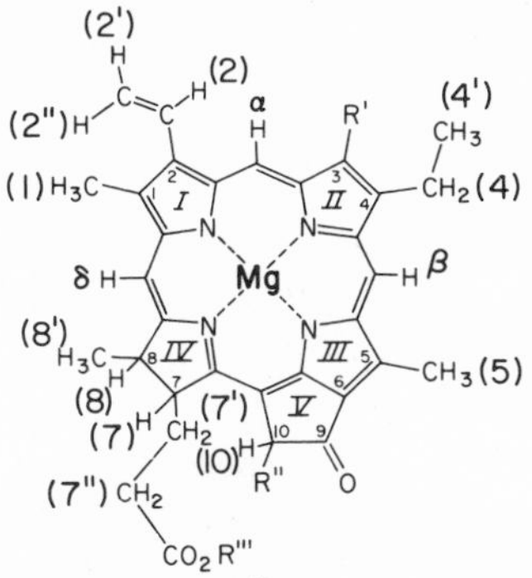

I

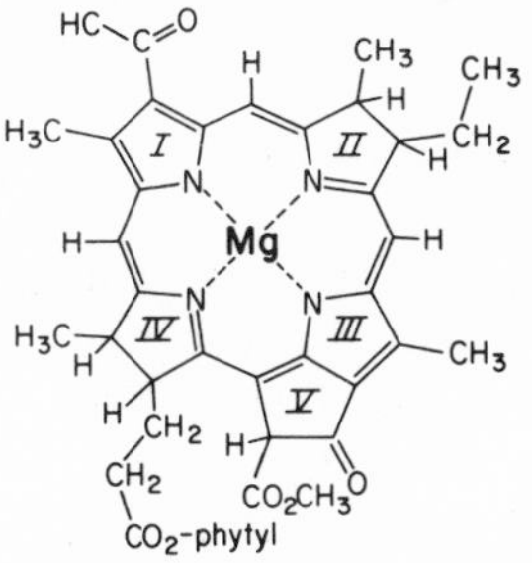

II

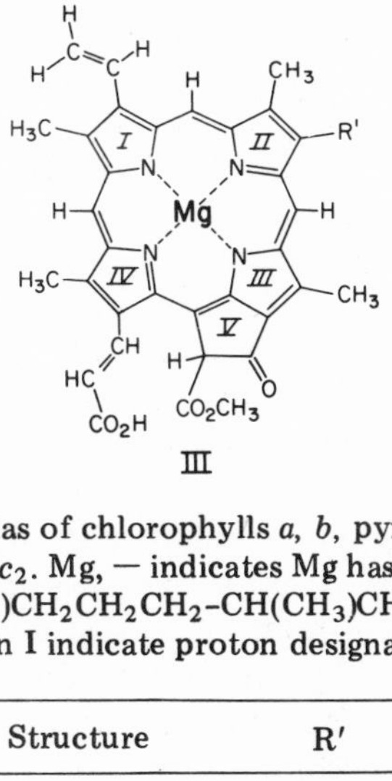

III

Fig. 1. Structural formulas of chlorophylls *a*, *b*, pyrochlorophyll *d*, bacteriochlorophyll, and chlorophylls c_1 and c_2. Mg, — indicates Mg has been removed and replaced by 2H. Phytyl, $-CH_2CH{=}C(CN_3)CH_2CH_2CH_2{-}CH(CH_3)CH_2CH_2CH_2CH(CH_3)CH_2CH_2CH_2CH(CH_3)_2$. The bold-face numbers in I indicate proton designations used in n.m.r.

	Structure	R′	R″	R‴	Mg
Chlorophyll *a*	I	CH_3-	$-COCH_3$	phytyl	+
Chlorophyll *b*	I	-CHO	$-COCH_3$	phytyl	+
Pyrochlorophyll *a*	I	CH_3-	-H	phytyl	+
Pheophytin *a*	I	CH_3-	$-COCH_3$	phytyl	-
Methylpheophorbide *a*	I	CH_3-	$-COCH_3$	$-CH_3$	-
Bacteriochlorophyll	II				+
Chlorophyll c_1	III	CH_3CH_2-			+
Chlorophyll c_2	III	$CH_2{=}CH$-			+
Protochlorophyllide	III	CH_3CH_2-			+

References pp. 1063–1066

1.2 Literature

The chlorophyll literature is vast. Still indispensable are the classic works of Willstätter and Stoll[17] and Fischer and Stern[18]. Useful as keys to the more modern literature are the reviews by Aronoff[19] and by French[20], and the comprehensive book on chlorophyll by Vernon and Seely[21]. The treatment of the subject matter of this chapter is necessarily selective, and is not intended to provide a complete key to the literature.

1.3 Properties of chlorophyll

The chlorophylls (Fig. 1) are cyclic tetrapyrroles. They thus belong to the porphyrin family of which many other biologically important substances are also members. The porphyrins and the chlorophylls have both important similarities and differences. The side-chains present in the chlorophylls, *i.e.*, methyl, ethyl, vinyl and propionic acid, are those commonly found in the porphyrins. The side-chain positions in the chlorophylls and porphyrins are for the most part identical, probably a result of similar biosynthetic pathways[22]. Unique to the chlorophylls, however, is the alicyclic Ring V, which contains a keto carbonyl function at position C-9 and a carbomethoxy group and an H atom at position C-10. This unique structural feature qualifies a compound as a chlorophyll, and this is the practice adopted here. The proton at C-10 is part of a β-keto ester system, and as such can enolize; under ordinary conditions the keto $\rightleftarrows$ enol equilibrium is displaced very much to the keto side. Nevertheless, enolization provides a mechanism for the stereochemical inversion and hydrogen exchange observed at the C-10 position. The absolute configurations of chlorophylls *a*, *b*, and bacteriochlorophyll have been established by Fleming[23] and by Brockmann[24]. The C-10 epimers of Chl *a*, *b*, and BChl have been detected and characterized by nuclear magnetic resonance (n.m.r.) spectroscopy[25]. Hydrogen exchange occurs at C-10 in both the chlorophylls and the Mg-free pheophytins, but exchange at the δ-methine position is observed to occur at an appreciable rate only in the Mg-containing chlorophylls. The rate of exchange of the δ-methine proton is about 100-fold slower than that of the C-10 proton. In BChl, all of the methine protons exchange readily[26].

The macrocycle in the chlorophylls constitutes a highly aromatic system which produces strong ring current effects in nuclear magnetic resonance. The π system of the chlorophylls differs, however, from that of the porphyrins. Chls *a* and *b* possess two extra hydrogen atoms in Ring IV, and BChl an additional two in Ring II; thus, Chls *a* and *b* are dihydroporphyrins, and BChl is a tetrahydroporphyrin. Unlike many of the porphyrins proper, the chlorophylls have no symmetry elements at all. It should be noted that Chl c_1 and c_2 and PChl are in the same oxidation state as the true porphyrins, but are saved for this chapter by possession of the alicyclic Ring V. Chlorophylls c_1 and c_2 have an acrylic acid side-chain at position 7, where a propionic acid side-chain is usually found.

Chlorophylls *a*, *b*, and BChl are di-basic acids esterified by two different aliphatic alcohols. The carboxylic acid function at C-10 is always esterified with methanol, and the propionic acid side-chain at position 7 by the long-chain, unsaturated, aliphatic alcohol phytol, $C_{20}H_{39}OH$ (3, D-7, D-11, 15-tetramethyl hexadec-*trans*-2-en-1-ol). This large aliphatic residue results in a molecule that has a polar region (the macrocycle) and a non-polar region (the phytyl moiety). Although the phytyl-bearing chlorophylls are highly water insoluble, the polar head-non-polar tail organization suggests that these compounds can function as surfactants in nonaqueous media. Both the phytyl and methyl ester functions, as will be seen below, are important in the interaction of chlorophyll with water in non-aqueous systems.

1.4 Coordination properties of magnesium in chlorophyll

All chlorophylls active in photosynthesis are chelate compounds of magnesium. Whereas the porphyrins that constitute the prosthetic groups of such biologically important substances as the hemoglobins, cytochromes, catalases and peroxidases are complexes of transition metal ions, the chlorophylls have had to be content with a regular metal ion, Mg(II), whose chemistry, at least in aqueous solutions, is prosaic by contrast with that of the transition elements. The central Mg atom of chlorophyll can be readily replaced by 2H by even weak acids[27], or by Cu(II), Ni(II), Co(II), Fe(II) and Zn(II). Zinc(II) mimics the behavior of Mg(II) to a certain extent[28], but none of the other metallo-chlorophylls behave like the Mg(II) compound insofar as coordination reactions are concerned. The coordination behavior of the central Mg atom in chlorophyll, however, is a crucial factor in chlorophyll behavior, and no other metal ion that can be inserted into chlorophyll duplicates the coordination behavior of Mg(II).

Although no structural information is available on chlorophyll or any of its Mg-containing derivatives, the structures of a number of related compounds have been determined recently by X-ray crystallography. A low resolution study of a phyllochlorin ester (a dihydroporphyrin lacking Ring V) has been reported[29], and the structure of vanadyl deoxophylloerythroetioporphyrin (a porphyrin that has a Ring V) has been determined[30]. Fischer *et al.*[31] have determined the structure of methyl pheophorbide *a*, a compound derived from Chl *a* by removal of the Mg atom and replacement of the phytyl ester by a methyl ester function at the propionic acid side-chain. The dihydroporphyrin ring is nearly flat, but the departures from planarity are significant. Ring IV is not planar; relative to the plane of the 4 pyrrole nitrogen atoms, Rings I and II are tipped up, while Rings III, IV, and V are tipped down. Timkovich and Tulinsky[32] have solved the crystal and molecular structure of mono-aquomagnesiumtetraphenylporphyrin by X-ray crystallographic techniques. The Mg is hydrated and possesses approximate square-pyramidal coordination. The Mg atom is no less than 0.273 Å out of the plane of the nitrogen atoms. A similar finding by Fischer *et al.*[31] places the Mg atom in monohydrated di-pyridinated Mg phthalocyanin 0.496 Å out of the plane of the central nitrogen atoms. If the Mg atom in chlorophyll itself is displaced from the dihydroporphyrin plane to the same extent, the tendency displayed by Mg in chlorophyll to assume a coordination number of 5 finds a reasonable explanation.

The structural formula of chlorophyll as usually written (Fig. 1) assigns the Mg atom a coordination number of 4. In many coordination compounds, to be sure, magnesium has a coordination number of 4, and assumes a tetrahedral configuration. Parry[33] and Eichhorn[34] have considered some possible consequences of the distortion of the usual magnesium configuration that might result from the more or less planar configuration of the chlorophyll molecule. However, it now appears that the Mg atom is not in the plane of the macrocycle, a circumstance that facilitates assumption of the coordination number 5. Further, magnesium halides, MgX_2, are known to form compounds with alcohols, ethers, ketones, esters and amides in which the magnesium is six-coordinated[35]. Magnesium(II) in chelates resembling chlorophyll in structure has long been known to assume a coordination number larger than 4. Mg(II) phthalocyanin shows a strong tendency to solvation by water[36], and the same has been shown to be true for Mg tetraphenylporphyrin[32]. Wei *et al.*[37] found magnesium-containing porphyrins form stable dipyridine complexes similar to hemichromes, in which the magnesium has octahedral coordination. With the exception of Evstigneev *et al.*[38], however, who concluded that magnesium was somehow involved in the activation of chlorophyll fluorescence by Lewis bases, the possibility that the central Mg atom of chlorophyll could assume coordination numbers larger than 4 was not seriously entertained until recently[39-42] to explain the unusual spectral properties of the chlorophylls.

1.5 Spectroscopy of chlorophyll

Other than its structural formula, most of what is known about chlorophyll in photosynthesis is derived from absorption spectroscopy in the visible. It has been known for some time that chlorophyll in the plant has anomalous spectral characteristics. Chlorophyll *a* in an acetone, diethyl ether, or pyridine solution is blue to the eye, with an absorption maximum in the red near 662 nm, and the solutions are strongly fluorescent. In the plant, the absorption maximum of chlorophyll *a* is shifted to the red and the chlorophyll is essentially non-fluorescent. From the characteristics of the red absorption peak, it has been decided that chlorophyll *a* in the plant exists in anywhere from two[43] to six different forms[44,45]. Nevertheless, when chlorophyll *a* is extracted from plants, only one chlorophyll *a* has ever been found. The nature of the long wavelength forms of chlorophyll has thus been the focal point of investigations on the role of chlorophyll in the primary light conversion act in photosynthesis. This has been particularly true since 1960, when Kok[46] found evidence for the participation in the light conversion act in photosynthesis of a form of chlorophyll *a* that absorbs light near 700 nm. The long wavelength forms of chlorophyll *a* have been presumed to arise by complex formation of chlorophyll with protein[47], or to be simply a result of chlorophyll "aggregation"[48]. According to the theory of McRae and Kasha[49], pigment aggregation leads to the splitting of absorption bands with the formation of lower-lying electronic transitions, which manifest themselves by a red shift in the absorption spectrum. Thus, chlorophyll "aggregation" became the watch-word. Newer investigations, however, make it possible and in fact require a more precise definition of this term than has been customary in the past, and closer examination of the red envelope (Section 4.0) raises serious questions about the multiplicity of chlorophyll forms in the plant.

This chapter will for the most part be concerned with laboratory studies on chlorophyll, and will describe infra-red (i.r.), n.m.r. and electron spin resonance (e.s.r.), as well as electron transition (e.t.) spectroscopic studies on well-defined laboratory chlorophyll systems. The spectral anomalies evident in the plant are to a considerable extent encountered also in the laboratory. In the laboratory, however, a variety of spectroscopic techniques can be brought to bear on the problem, greatly facilitating deductions about the nature of the chlorophyll species giving rise to the observed spectra. Some of the chlorophyll species characterized in the laboratory give every evidence of being important ones in the plant. It will be the principal purpose of this chapter to describe the chlorophyll species that have been characterized by spectroscopic techniques *in vitro*, and to relate these species to chlorophyll as it occurs in the plant.

1.6 Chlorophyll as electron donor–acceptor

In the interpretation of chlorophyll spectra, the coordination properties of the magnesium turn out to be of central importance. The interpretation advanced in this chapter to account for the observed solvent-dependence of the spectra of chlorophyll is based on the premise that magnesium in chlorophyll with a coordination number of 4 is coordinatively unsaturated, and that at least one of the axial Mg positions must always be filled with an electron donor group[42]. In polar solvents, the solvent acts as electron donor, and chlorophyll exists as the monomer, with solvent molecules in one or both axial positions, $Chl \cdot L$ or $Chl \cdot L_2$. In non-polar solvents, in the absence of extraneous nucleophiles or electron donor molecules, the coor-

References pp. 1063–1066

dination unsaturation of the magnesium can be relieved only by another chlorophyll molecule acting as electron donor. The keto C=O function in Ring V appears to serve as the principal chlorophyll electron donor. The interaction of the keto C=O function of one chlorophyll molecule with the Mg atom of another generates dimers, or, depending on the nature of the non-polar solvent, oligomers, in which the binding force is also a keto C=O---Mg interaction. Extraneous nucleophiles can compete for the coordination site in such systems. Chlorophyll, then, is distinguished by its ability to form donor–acceptor complexes either by self-interaction, or by interaction with extraneous donors or ligands. The chlorophyll molecule has an acceptor center at magnesium and a donor center at the keto C=O function in Ring V. Although compounds have been described that act both as donor and acceptor in charge transfer complexes, the chlorophyll-molecule appears to have unique donor–acceptor properties, and no exactly analogous compounds seem to have been described[50,51]. It is convenient to refer to the self-interactions of chlorophyll as *endogamous* and to chlorophyll-nucleophile interactions as *exogamous.* These interactions determine the state of aggregation of chlorophyll and, as the donor–acceptor behavior of chlorophyll is strongly solvent dependent, also largely account for the solvent dependency of the spectra of chlorophyll systems.

2.0 CHLOROPHYLL–CHLOROPHYLL (ENDOGAMOUS) INTERACTIONS

That chlorophyll can form a donor–acceptor molecular complex in which chlorophyll acts as both donor and acceptor can be deduced from the solvent dependence of i.r. and n.m.r. spectra. The applications of these spectroscopic techniques to chlorophyll have been reviewed by Katz *et al.*[52].

2.1 Infra-red spectra in the 1800–1600 cm^{-1} (carbonyl) region

I.r. spectra arise from the stretching, bending, and twisting motions of atoms in molecules, and these may be expected to be very complicated and essentially uninterpretable for a molecule composed of as many atoms as is chlorophyll. That region of the spectrum that is related to carbonyl and C=C stretch frequencies, however, is open to relatively straightforward interpretation and yields valuable information about the state of aggregation of the chlorophyll.

In polar solvents such as tetrahydrofuran, Chl *a* shows three absorption maxima in the 1800–1600 cm^{-1} region, whereas in CCl_4 benzene or other non-polar solvents, four peaks are present (Fig. 2). Earlier workers[53,54] attributed this phenomenon to keto–enol tautomerism involving the β-keto ester system in Ring V, but subsequent work[39,41] has shown that interpretation in terms of electron donor–acceptor interactions is more suitable. The three-banded spectrum in tetrahydrofuran or in non-polar solvents when tetrahydrofuran is present in large molar excess can readily be assigned in terms of the structural formula: the peak at 1735 cm^{-1} to the two ester C=O functions; the peak near 1695 cm^{-1} to the free keto C=O of Ring V;

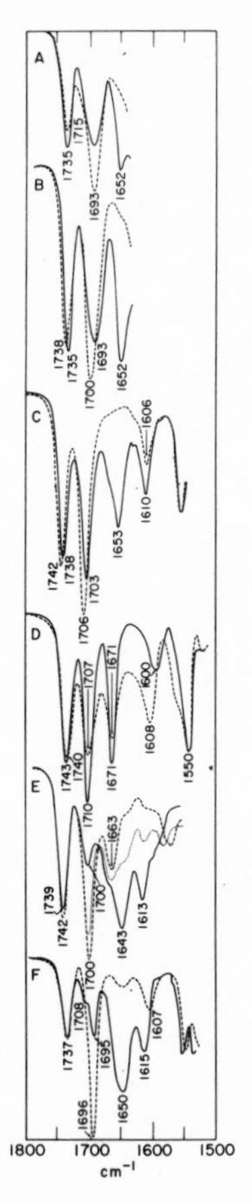

Fig. 2. Infra-red spectrum of chlorophylls in polar and non-polar media in the 1800–1500 cm^{-1} region. A, Chl *a*, (—), CCl_4, (- - -), CCl_4 + CH_3OH; B, Chl *a*, (—), CCl_4, (- - -), CCl_4 + tetrahydrofuran (THF); C, Chl *a*, (—), n-butylcyclohexane (NBC), (- - -) NBC + THF; D, Chl *b*, (—), NBC, (- - -), NBC + THF; E, BChl, (—), NBC, (- - -), NBC + THF; F, pyrochlorophyll *a* (—), NBC, (- - -), NBC + THF. Chlorophyll concentration $\sim 10^{-3}$ *M*. In the disaggregated spectra, the base is present in a ten-fold molar excess. For further details, see ref. 55.

and the absorption maximum near 1600 cm^{-1} to the skeletal C=C, C=N vibrations. In the four-banded spectrum, the ester C=O absorptions (~ 1735 cm^{-1}) and the skeletal vibrations (1610 cm^{-1}) are relatively unchanged, but the free keto C=O absorption is diminished in intensity, and a prominent new peak at 1652 cm^{-1} is observed. This peak can be seen

References pp. 1063–1066

in chlorophyll and its derivatives only when Mg is present and extraneous nucleophiles are absent. Pheophytin *a*, for instance, has i.r. spectra independent of the electron donor properties of the solvent. A reasonable assignment for the origin of the 1652 cm^{-1} absorption peak then, is to the keto C=O function of one chlorophyll molecule coordinated to the central magnesium atom of another: keto C=O---Mg. This interaction results in the formation of dimers in which, on the average, half of the keto C=O groups are free and half are coordinated to Mg. The 1652 cm^{-1} maximum can now be referred to as a chlorophyll *aggregation peak.*

Chl *b* and Bchl exhibit a similar solvent dependency[55] (Fig. 2). Because Chl *b* and BChl both possess additional carbonyl functions (Fig. 1), *b* in the form of an aldehyde, -CHO, function at position 3, and Bchl as an acetyl group, -$COCH_3$, at position 2, additional modes of carbonyl–magnesium interaction are available, and the interpretation of the i.r. spectra of these chlorophylls is therefore more ambiguous. In both cases, absorption maxima are observed that appear to be composite: in Chl *b*, the peak at 1608 cm^{-1} contains the skeletal vibrations but also has a contribution from ald C=O---Mg, and the keto C=O---Mg absorption peak at 1671 cm^{-1} also contains the free ald C=O absorption. Similarly in Bchl, the skeletal vibration at 1613 cm^{-1} contains a contribution from acetyl C=O---Mg, and free acetyl C=O contributes to the 1665 and 1643 cm^{-1} absorption peaks. Assignments are thus difficult to make, but there is reason to hope that the application of deconvolution techniques to these spectra may improve the situation in the future. For convenience, assignments of the i.r. absorption maxima in the 1800–1600 cm^{-1} region are collected in ref. 52. Far infra-red spectra of chlorophyll in the 340–160 cm^{-1} region show an analogous solvent dependence that can also be related to the state of aggregation of the chlorophyll. Bands at 311–303 cm^{-1} in chlorophyll *a* solutions in benzene or cyclohexane can be attributed to intermolecular coordination involving Mg and oxygen, and strong absorption bands at 292–296 and 195 cm^{-1} in monomeric chlorophyll solutions can be assigned to Mg–N vibrational modes. The peak at 312–306 cm^{-1} shows the same solvent dependence as does the keto C=O---Mg aggregation peak in the carbonyl stretch region of the i.r.[56].

For carefully dried solutions of chlorophyll in non-polar solvents, very little absorption in the O–H stretch region (3800–3000 cm^{-1}) can be detected. In concentrated dry solutions of chlorophyll *a* in benzene or CCl_4, in agreement with the observations of Sidorov and Terenin[57] and of Karyakin and Chibisov[54], a weak absorption maximum at 3538 cm^{-1} can be observed. This peak disappears rapidly on addition of D_2O, establishing its identity as an O–H group. This absorption peak can also be seen in concentrated solutions of pheophytin in CCl_4. The O–H absorption is thus assigned to the enol form of chlorophyll *a*[55]. The enol concentration in chlorophyll *a* must be very small, probably considerably less than 1% as judged by the fact that very concentrated solutions are required for the 3538 cm^{-1} peak to be seen at all. From n.m.r. (see below), it is evident that the chlorophylls all occur in solution predominantly in the keto form, an observation also in agreement with the results of i.r. in the O–H stretch region.

2.2 Nuclear magnetic resonance spectra

The n.m.r. spectra of chlorophyll exhibit an entirely analogous solvent dependency. High resolution n.m.r., moreover, provides information about the structure of the chlorophyll–chlorophyll dimer in non-polar solvents. Chlorophyll lends itself very well to n.m.r. studies, as the tetrapyrrole macrocycle is an aromatic system with a delocalized π system that generates a strong ring current in a magnetic field. Ring current effects on proton chemical shifts can be expected to be prominent whenever a hydrogen atom is positioned above or below the plane of the macrocycle, or in the plane of the macrocycle, to an extent determined by the closeness of the proton to the ring. Thus, valuable structural information can be extracted from ring current effects on the proton chemical shifts of chlorophyll itself or on ligands bound to chlorophyll.

Closs *et al.*[40] assigned essentially all of the protons on the dihydroporphyrin ring of chlorophylls *a* and *b* and additional n.m.r. data for these and other chlorophylls are to be found in ref. 52. In recent years, Inhoffen and co-workers[58] have provided a large amount of useful n.m.r. data on porphyrins and dihydroporphyrins related to the chlorophylls[58]. Figure 3 shows n.m.r. spectra of various chlorophylls in polar solvents. The methine protons in the plane of the ring are at unusually low field because of the deshielding action of the ring current. The vinyl protons form a readily recognized AMX pattern, and the C-10 proton is also clearly visible. This proton is slowly exchangeable with CH_3OD, and is very close in area to 1 proton by integration. This provides the basis for the conclusion that chlorophyll *a* exists in solution very largely in the keto form. The chemical shifts of the CH_3– groups directly attached to the conjugated system, *i.e.*, the CH_3– groups at positions 1, 3 and 5, and the CH_3 group of the carbomethoxy group at position 10 constitute a well-resolved set of 4 peaks. Some of these low field methyl resonances are very sensitive to the state of aggregation of the chlorophyll, and together with the C-10 proton, furnish considerable information about the intermolecular interactions of chlorophyll. The more aliphatic hydrogen atoms in the molecule, *i.e.*, the CH_3 groups at positions 8′ and 4′, the propionic acid -CH_2– groups, and the phytyl hydrogens show resonances at higher field, as expected. Some of these resonances are overlapped by the phytyl resonances and can be seen clearly only in phytyl-free derivatives of chlorophyll. The n.m.r. spectrum of chlorophyll *a* in tetrahydrofuran or pyridine, then, is entirely compatible with the structural formula shown in Fig. 1. In non-polar solvents, however, the situation is quite different.

The n.m.r. spectrum of Chl *a* in various non-polar solvents are shown in Fig. 4A. It can be seen at once that the spectra in non-polar solvents are considerably different from those in polar solvents and differ from one

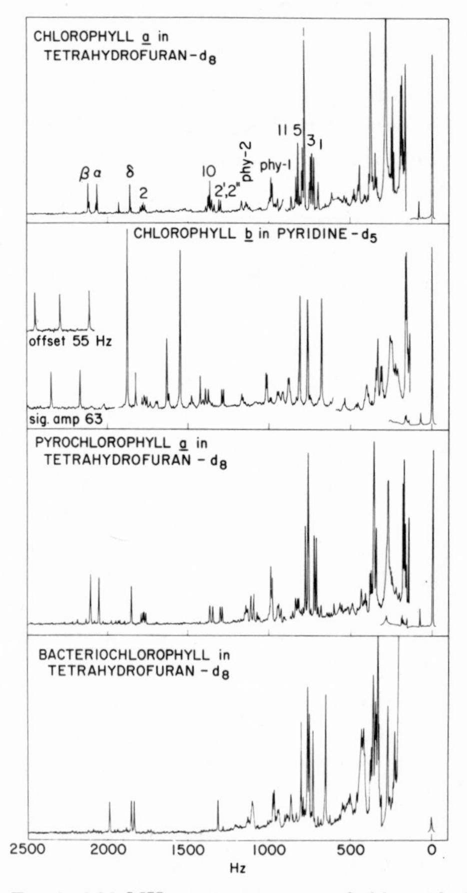

Fig. 3. 220 MHz n.m.r. spectra of chlorophylls in polar solvents. See I in Fig. 1 for proton numbering. Chemical shifts are in Hz, from internal hexamethyl disiloxane (HMS).

non-polar solvent to the other. The spectra in non-polar solvents vary from quite well-defined spectra in $CDCl_3$ and benzene to those observed in aliphatic hydrocarbon solvents, where distortion and shifting of the resonances leads to very poorly resolved spectra. Even in $CDCl_3$ or benzene, where the spectra are reasonably well resolved, significant differences are evident when the spectra are compared to those recorded in polar solvents. The C-10 proton is nowhere to be seen; the low field methyl resonances appear to be shifted about in such a way as to destroy the correlation between the number of peaks in the low-field methyl region and the structural formula. The phytyl resonances, however, are relatively unaffected. The gross appearance of the spectra, particularly in aliphatic hydrocarbon solvents, is such as to suggest that much larger entities are responsible for the spectra than is the case for chlorophyll dissolved in polar solvents. The

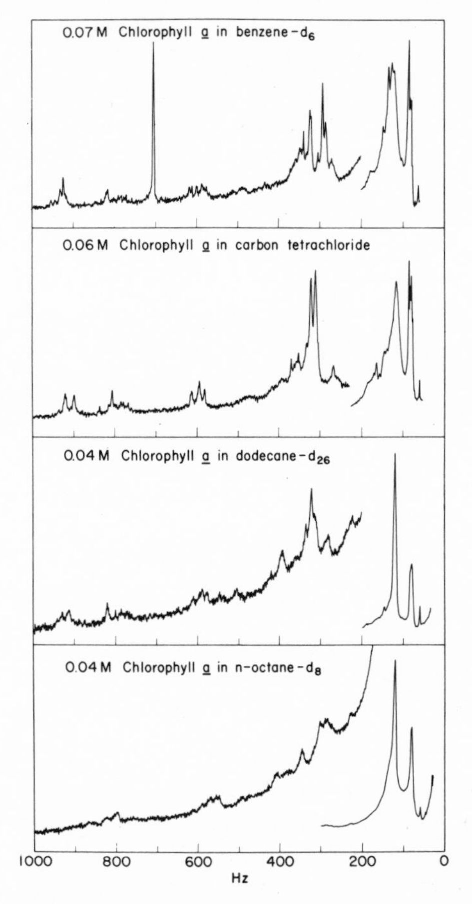

Fig. 4A. 100 MHz n.m.r. spectra of Chl *a* in non-polar solvents. The high-field portions of the spectra are shown at lower gain.

correlation time appears to be larger (and the tumbling rate smaller), with the result that the natural line widths are considerably increased, as is the case for polymers or other large, anisotropic, slowly moving species. Raising the temperature sharpens the spectra (Fig. 4B), presumably by dissociating the aggregates with a resulting improvement in rotational performance. Thus, the n.m.r. spectra in non-polar solvents suggest qualitatively that chlorophyll-chlorophyll interactions are present, just as in the case of the i.r. spectra.

References pp. 1063–1066

The n.m.r. spectra, however, can be used to provide a more precise picture of the nature of the chlorophyll–chlorophyll interaction. If chlorophyll aggregates form in these media, the macrocyclic rings will be brought into closer proximity than is the case for even a concentrated solution of monomer, and ring current effects can be used to make deductions about the structure of the aggregates. Protons of one chlorophyll molecule covered by the ring of another will necessarily experience an upfield shift. Closs *et al.*[40] showed that a close examination of these ring current effects can give a quite detailed picture of the way in which the aggregates are put together. Addition of base to a chlorophyll solution in a non-polar solvent causes changes in the n.m.r. spectra, and with a sufficiently large amount of

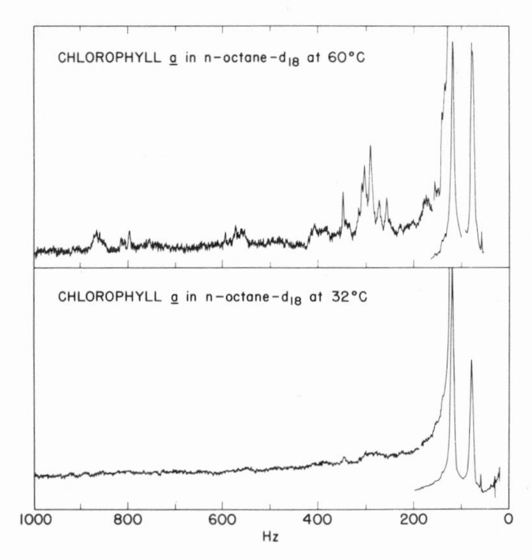

Fig. 4B. Effect of temperature on n.m.r. spectrum of chlorophyll *a* in a non-polar solvent.

base, the spectra become identical with those observed in polar solvents. Thus, incremental addition of base converts the chlorophyll aggregates into $Chl \cdot L_1$, and from the chemical shift changes as the disaggregation proceeds, the structure of the aggregate can be deduced. It is important to remember that on the n.m.r. time scale, the dynamic equilibrium between chlorophyll aggregates and the monomer · monosolvate is fast, and that only one set of lines with chemical shifts having the weighted average value can be observed at ambient temperatures. The necessary data can thus be obtained in a titration experiment. To chlorophyll dissolved in a non-polar solvent, *e.g.* CCl_4 or benzene, successive increments of a base, *e.g.* CH_3OH or pyridine, are added. The chemical shifts of the chlorophyll protons are then measured. At any point in the titration, the chemical shift of the various chlorophyll protons will be determined by the relative amounts of chlorophyll aggregate and Chl · L_1. In the limit, for a sufficiently large excess of base, the monomer spectrum of the chlorophyll is observed, and as the resonances in this spectrum are fully assigned, the location of many of the peaks in the aggregated spectrum can be inferred (Fig. 5).

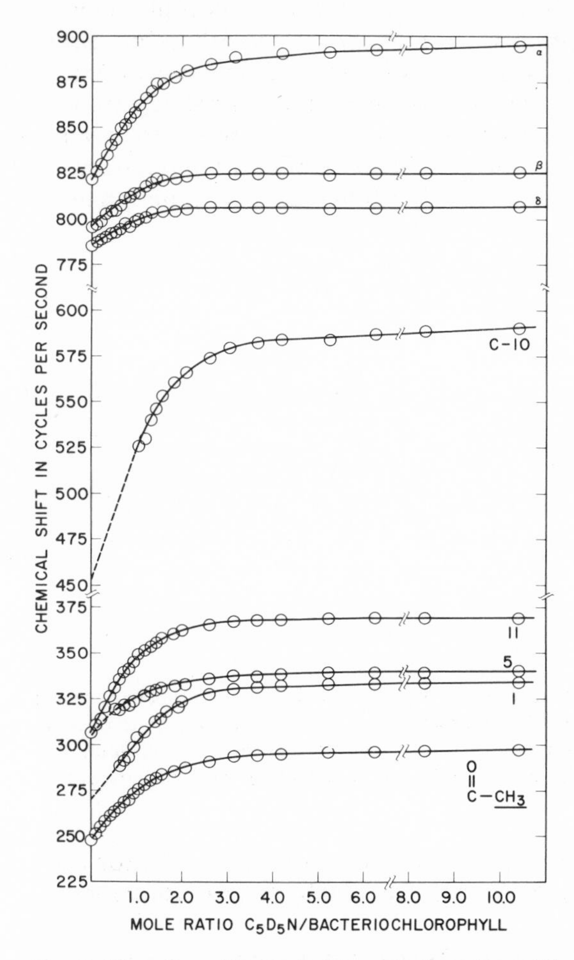

Fig. 5. Titration of bacteriochlorophyll with pyridine-d_5. Chemical shifts (in Hz, from internal HMS) are plotted as a function of mole ratio (see ref. 40). See Fig. 1 for proton numbering.

In such a titration experiment, the chlorophyll protons divide themselves up into two groups. The first contains the vinyl protons, the methine protons, and the methyl groups at positions 1 and 3. These protons have chemical shifts that are relatively insensitive to the basicity of the medium, *i.e.*, only small changes in chemical shift result from the addition of base. The other group of protons, which includes the C-10 proton, and the methyl groups at positions 5 and 11, shows a large paramagnetic (downfield) shift as base is added. From ring current considerations it can be concluded that the C-10 proton and the protons in the 5 and 11 methyl groups are in that portion of the chlorophyll molecule subject to strong shielding by another chlorophyll ring. That is to say, the chlorophyll molecules in the dimer are not positioned directly above each other, but rather are displaced in such a way that only a part of each molecule is eclipsed. The chemical shift data indicate that the regions of mutual overlap are in the vicinity of Ring V. The titration data of Fig. 5 can be presented graphically in the form of an "aggregation map" (Fig. 6). For chlorophyll *b*, an additional region of overlap is found near the aldehyde C=O function at

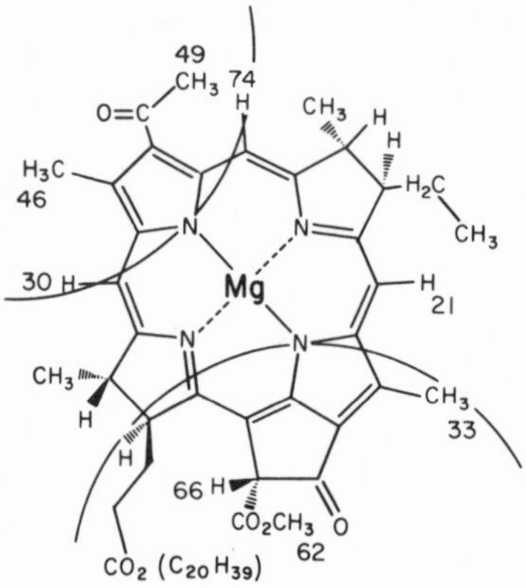

Fig. 6. Aggregation map of bacteriochlorophyll from chemical shift differences between aggregated and monomeric chlorophyll[40]. The numbers on the figure give the maximum chemical shift differences for the indicated protons as deduced from titration data. The semi-circles indicate region of overlap and show that both the acetyl and keto carbonyl functions are coordinated to Mg.

position 3, involving the α methine proton and the ethyl group protons at position 4. Thus, in Chl *b*, both the ketone and aldehyde C=O function as electron donors.

Although many porphyrin spectra show a strong concentration dependence of chemical shifts that has been interpreted to indicate a monomer $\rightleftarrows$ dimer equilibrium, with the dimer held together by weak π–π interactions[59], chlorophyll n.m.r. spectra in non-polar solvents show very little concentration dependence. Mg-free derivatives of chlorophyll, however, show highly selective dilution shifts of some resonances, but are unaffected by the addition of base. Thus, a distinction must be made between π–π interactions, which are weak and readily reversed by dilution, and coordination interactions between keto or aldehyde C=O and Mg which are strong, and which form dimers that are not dissociated by dilution. In the case of pheophytin *a*, the largest chemical shift changes on dilution are found in the vicinity of Ring II, and this then is the region of maximum overlap of the monomer units in the pheophytin dimer. Presence or absence of Mg, then, causes profound differences in the way self-interaction occurs. Zinc chlorophyll *a* shows behavior intermediate between Chl *a* and pheophytin *a*, *i.e.*, both (weak) coordination and π–π interactions. The extent of coordination interaction is small, and the results of titration and dilution experiments indicate that two kinds of aggregates occur in zinc chlorophyll dissolved in non-polar solvents[28].

In Fig. 6, the region of overlap in the bacteriochlorophyll is defined by the semi-circles, which represent other monomer units. When the keto C=O function of one chlorophyll molecule is introduced into the coordination sphere of the Mg atom of another, then the keto C=O function of the acceptor chlorophyll is positioned under the Mg atom of the donor chlorophyll molecule. Only a small motion is thus required to interchange the donor and acceptor chlorophylls, and the dimer can be viewed as a dynamic system in which a particular chlorophyll molecule on the average is donor one-half the time, and acceptor for the other half. The chemical shift data in Chl *a* dimer can best be explained in terms of a small angle between the ring planes. Circular dichroism (c.d.) measurements by Houssier and Sauer[60] suggest an angle of 45° between the ring planes. The c.d. measurements have been interpreted to indicate that the structure of Chl *a* and Pyrochl *a* dimers is very similar. N.m.r. data for Pyrochl *a*, however, are quite emphatic in suggesting a smaller angle between the ring-planes of the Pyrochl *a* dimer[14], than is the case for (Chl $a)_2$. The n.m.r. data are probably more sensitive to orientation than are the c.d. measurements or at least can be more readily interpreted. The absence of the carbomethoxy group in Pyrochl *a* should in any event reduce steric interferences in dimer formation. It has also been suggested that the propionic ester C=O is involved in dimer formation[60], but this conclusion is at variance with both i.r. and n.m.r. data. If the propionic acid ester group were involved in dimer formation a large ring current effect on the propionic $-CH_2-$ proton chemical shifts would be expected, and this is not observed. Likewise, in methyl chlorophyllide *a*, the chemical shift of the CH_3- group in the propionic ester shows that these protons are not subject to a ring current to any appreciable extent[40].

2.3 Chlorophyll–chlorophyll oligomers

In the chlorophyll dimer described here, the central Mg atom has the average coordination number 4.5. There is reason to suppose that the Mg atom in chlorophyll is displaced from the plane of the macrocycle[31,32]. A coordination number of 6 might therefore be difficult to achieve, but there is no obvious reason why a coordination number of 5 should not be attained. Additional keto C=O---Mg interactions first between dimers to form tetramers, then between tetramers to form oligomers, and so forth, may easily be visualized to occur. The oligomer found by this process will contain Mg with a coordination number 5. Close examination indicates that the equilibrium $n(Chl_2) \rightleftarrows (Chl_2)_n$ does in fact hold, but the fact that it is strongly solvent dependent is responsible for the reason that this process was not detected until recently. In non-polar solvents with good solvation characteristics for the macrocycle portion of the chlorophyll molecule, such as CCl_4 or benzene, n is small, and even over a large range of concentrations, chlorophyll *a* occurs mainly as the dimer. In non-polar solvents with poor solvation characteristics, principally aliphatic or cyclo-aliphatic hydrocarbons, n may be larger than 10, and chlorophyll *a* occurs in these solvations as oligomers with molecular weights in excess of 20,000. Thus, all non-polar solvents cannot be placed into one category, and it must be recognized that major differences in the state of aggregation of chlorophyll may occur in going from one non-polar solvent to another.

If the infra-red spectrum of chlorophyll *a* in the 1800–1600 cm^{-1} region is studied in CCl_4 as a function of concentration, it rapidly becomes apparent that only small differences are observed even over a considerable concentration range[61], implying no major changes in aggregation in this solvent as the concentration increases. In hexane, the situation is quite different (Fig. 7). As the concentration increases the keto C=O---Mg aggregation peak found at 1652 cm^{-1} in dilute solution shifts to 1660 cm^{-1} in a 0.01 *M* solution. The shift to higher frequency of the aggregation peak in the oligomer may be considered to indicate that the first chlorophyll–chlorophyll interaction to form dimer is energetically more favorable than subsequent interactions between dimers, tetramers, *etc.*, and that the higher frequency is indicative of a lower average stability for this interaction in the oligomer. It may be conjectured that the tendency to form oligomer in aliphatic hydrocarbon solvents is as much a matter of solvent pressure to reject an unfavorable solute as it is a manifestation of the drive for the Mg to achieve a higher coordination number. Competition between solvation by the solvent and the stabilization afforded by keto C=O---Mg interactions seems adequate to account for the solvent dependence of the aggregation process in non-polar solvents.

N.m.r. data also indicate that the state of aggregation of chlorophyll varies considerably in non-polar media. These observations, however, are not easy to interpret, involving as they do anisotropic particles of high molecular weight. Molecular weight measurements yield more direct and less ambiguous information. Vapor phase osmometry was first introduced for chlorophyll studies by Aronoff[62], and this technique has proved itself very useful for studying the aggregation of chlorophyll. Ballschmiter *et al.*[61] have made the most detailed studies by this technique. In dry CCl_4, benzene, or 1,2-dichloroethane, the aggregation number of chlorophyll (defined as the ratio, R, of the calculated to the observed molarity) approaches but never exceeds 4, even in solutions as concentrated as 0.1 M. Over much of the concentration range in these solvents, the predominant form of chlorophyll *a* is the dimer (Fig. 8), accompanied by variable amounts of oligomer. Aggregation behavior in CCl_4 and benzene is very similar. Extrapolation of the molecular weight data in these solvents to infinite dilution suggests that the basic chlorophyll *a* unit is the dimer.

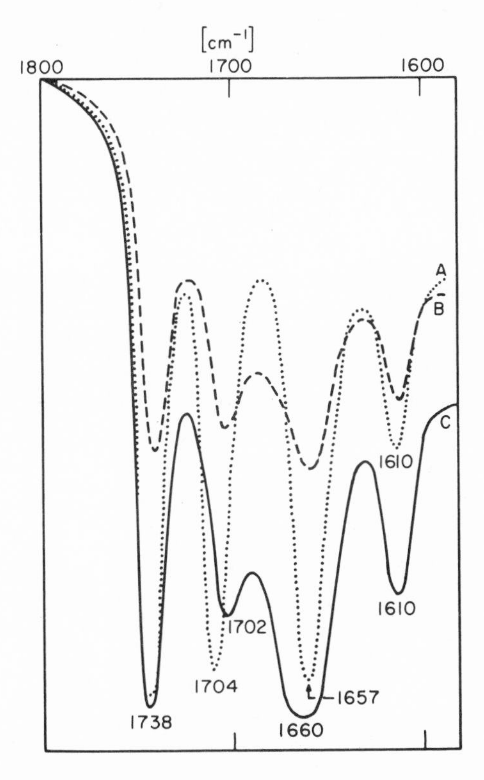

Fig. 7. Infra-red spectra of chlorophyll *a* in hexane as a function of aggregation number. A, 0.0017 M, R (see text) about 4; B, 0.04 M, R about 10; C, 0.08 M, R about 16[61].

In cycloaliphatic or aliphatic hydrocarbon solvents, Chl *a* is much more highly aggregated[72]. In cyclohexane solution at 25°C, 0.1 M chlorophyll *a* has an aggregation number of 10, and a comparable solution in hexane an aggregation number in excess of 20. Extrapolation of the molecular weight data to infinite dilution suggests that at 25°C dry Chl *a* in hexane or octane is a tetramer. As is to be expected from the mobile equilibria involved, raising the temperature shifts the equilibrium in the direction of disaggregation, and conversely, lowering the temperature increases the degree of aggregation. The changes in aggregation with temperature must surely be

References pp. 1063–1066

important in the interpretation of low-temperature spectral studies. When spectroscopists lower the temperature, they do a good deal more than just sharpen the lines of the chlorophyll electronic transition spectra.

Because both Chl *b* and Bchl each have two electron donor groups, the aggregation behavior of these chlorophylls can be expected to be more

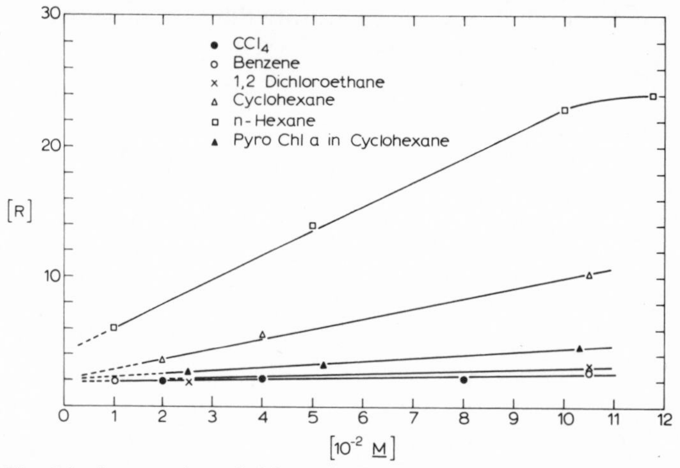

Fig. 8A. Aggregation of chlorophyll *a* in non-polar solvents from molecular weight measurements by vapor phase osmometry.

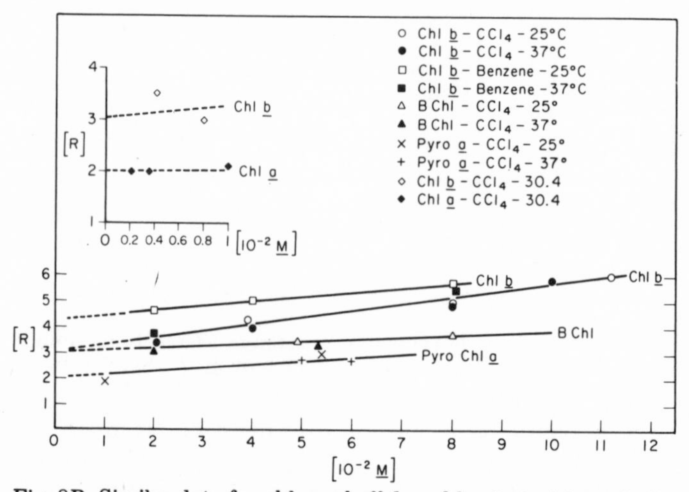

Fig. 8B. Similar data for chlorophyll *b* and bacteriochlorophyll.

complex than is the case for chlorophylls with only one donor center. The infra-red spectra in the carbonyl stretch region are also not as useful because of overlap of the absorption maxima. From molecular weight determinations, it appears that the basic unit for both Chl *b* and Bchl is the trimer (Fig. 8), and the assumption made in the past in the interpretation of electronic

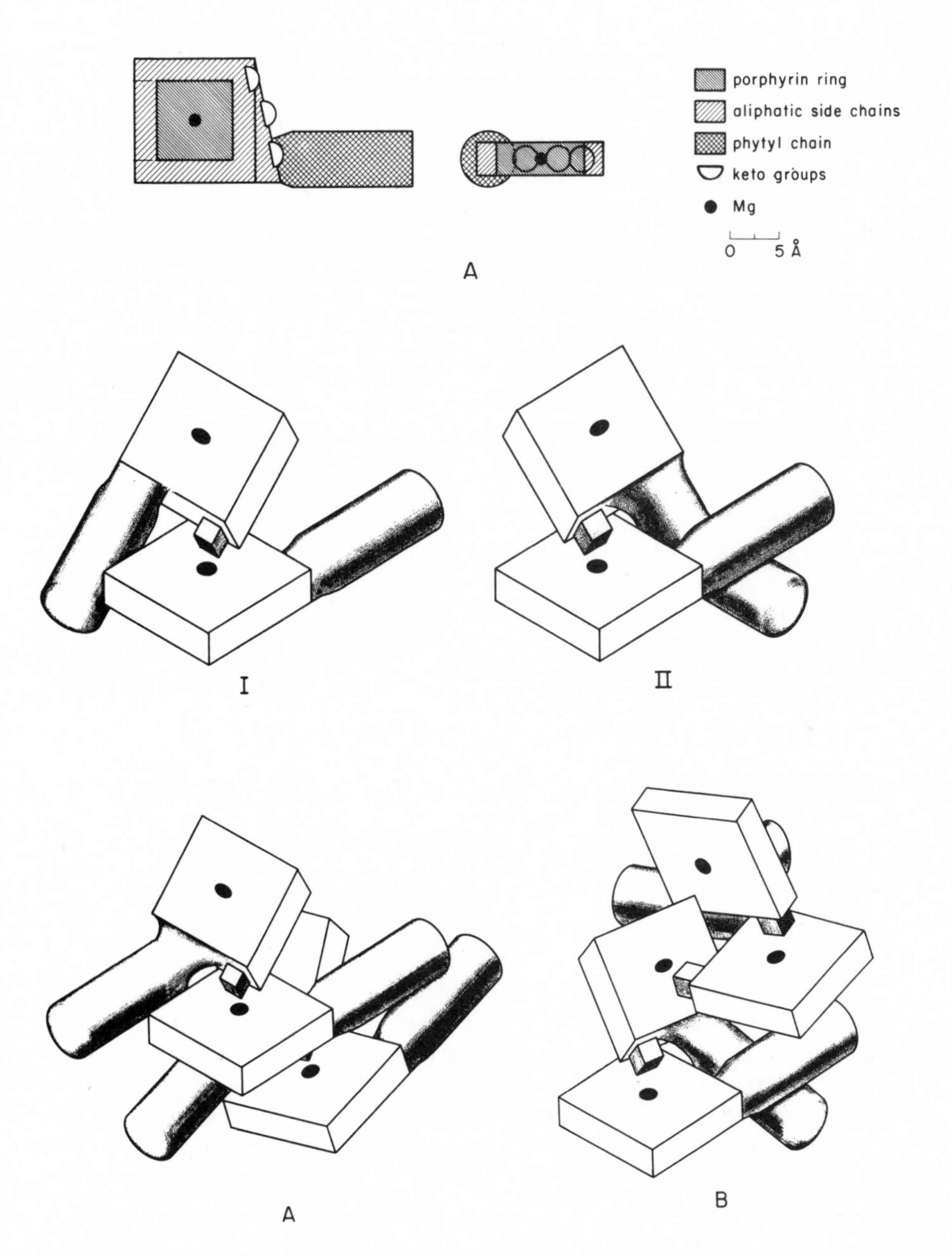

Fig. 9. Dimer (A) and oligomer (B) structures deduced from i.r. and n.m.r. data. Types I and II dimers represent limiting forms of dimer that can be interconverted by rotation. The oligomer structures are derived from the different dimer types. Whereas the dimers are readily interconverted, this may not be the case for the oligomers. The dimensions are taken from Courtauld space-filling models.

References pp. 1063–1066

transition spectra that all chlorophylls occur in dilute solution at room temperature as dimer is no longer valid. Molecular weight measurements also indicate that Chl *b* and BChl, particularly the former, are already aggregated significantly beyond the trimer stage even in CCl_4 or benzene solution. Whereas carefully dried BChl can be dissolved in aliphatic hydrocarbon solvents to form solutions as concentrated as 0.1 *M*, the solubility of Chl *b* in aliphatic hydrocarbons appears to be much more limited. Whether this reflects a greater tendency to aggregation on the part of Chl *b* in these media, or is a consequence of the difficulty of removing the last traces of water from the system (see below) is still uncertain.

Molecular weight determinations confirm conclusions arrived at from i.r. and n.m.r. that Mg is indispensable for chlorophyll aggregation in non-polar solvents. Both pheophytin *a* and pyropheophytin *a* occur as monomers in CCl_4 solutions more dilute than 0.005 *M*. With increasing concentration, the aggregation number increases, but the aggregation follows a monomer $\rightleftarrows$ dimer equilibrium, suggestive of π–π interactions. Thus, the behavior of the Mg-free chlorophyll derivatives is qualitatively different from that of the chlorophylls.

Representations of chlorophyll *a* dimers and oligomers are shown in Fig. 9.

3.0 CHLOROPHYLL-LIGAND (EXOGAMOUS) INTERACTIONS

The chlorophyll–chlorophyll interactions described above are the result of electron donor–acceptor interactions. Any extraneous donor molecule can compete for the coordination site at Mg, and, depending on the basicity of the nucleophile, can displace the chlorophyll acting as donor. Thus, addition of bases (ketones, ethers, alcohols, tertiary bases, *etc.*) to chlorophyll dissolved in a non-polar solvent forms monomeric chlorophyll species of the general type $Chl \cdot L_1$ and $Chl \cdot L_2$. The early observations of Livingston[63] on the activation of chlorophyll fluorescence by nucleophiles such as water, alcohols, amines, and ethers finds an explanation along these lines; it is the $Chl \cdot L_1$ and $Chl \cdot L_2$ species that are strongly fluorescent. It is obvious that competition between extraneous nucleophiles and chlorophyll as a donor is generally important only in non-polar media, because in polar media mass action considerations appear to prevent effective competition by either chlorophyll or small amounts of an additional nucleophile.

Chlorophyll–nucleophile interactions can be followed by a variety of spectroscopic techniques. Both n.m.r. and i.r., particularly the former, lend themselves very well to this purpose. Visible absorption spectroscopy has also been used, but of the spectroscopic methods available, is probably of the most limited utility. The intense light absorption by chlorophyll requires the use of highly dilute solutions, so that adventitious nucleophiles pose very serious experimental problems.

3.1 Chlorophyll–ligand interactions from infra-red spectroscopy

When nucleophiles are added to chlorophyll solutions in non-polar solvents, disruption of keto C=O---Mg interactions can readily be followed by i.r. observations in the carbonyl stretch region. As the chlorophyll dimers, trimers, or oligomers are disaggregated by addition of base, the free keto C=O absorption increases and the keto C=O---Mg aggregation peak decreases in intensity and area. An examination of the spectra indicates that the gain in area of the free keto C=O absorption with disaggregation is much smaller than the area associated with the aggregation peak

TABLE I

EQUILIBRIUM CONSTANTS FOR CHLOROPHYLL-LIGAND INTERACTIONS FROM i.r.[55] AND n.m.r.[64] DATA

$$Chl_2 + 2L \underset{}{\overset{K_6}{\rightleftharpoons}} 2Chl \cdot L$$

Chlorophyll	Solvent[a]	Ligand[b]	K_6 (mol l^{-1})	Method
Chl *a*	CCl_4	CH_2OH	45	i.r.
^{2}H-Chl *a*	CCl_4	CH_2OH	50	n.m.r.
^{2}H-Chl *a*	CH_2Cl_3	CH_2OH	40	n.m.r.
^{2}H-Chl *a*	CCl_4	CH_2CH_2OH	40	n.m.r.
Chl *a*	CCl_4	THF	22	i.r.
Chl *a*	NBC	THF	0.3	i.r.
Pchl *a*	NBC	THF	0.03	i.r.

[a] NBC, n-butyl cyclohexane
[b] THF, tetrahydrofuran

which is lost. The extinction coefficients for the free keto C=O must be different in (Chl_2) and Chl · L_1, and caution is therefore indicated in the use of peak areas in deducing equilibrium constants. Nevertheless, it is possible to make some semi-quantitative deductions from such titration data by using only the decrease in area of the keto C=O---Mg absorption to determine the concentration of dimer. Equilibrium constants deduced in this way are included in Table I.

3.2 Chlorophyll–ligand interactions from n.m.r. spectroscopy

The large ring current effects arising from the π system of the chlorophyll macrocycle make n.m.r. spectroscopy of great value in the study of

References pp. 1063–1066

chlorophyll–ligand interactions. A proton positioned above the plane of the macrocycle will experience a large diamagnetic shift from the ring current, the magnitude of which is determined by the geometry of the Chl · L_1 interaction product. The protons near the coordination center of a nucleophile coordinated to the Mg atom of chlorophyll will thus experience an up-field chemical shift. The displacement of the proton resonances in the nucleophile can be more readily followed if fully deuterated chlorophylls[64], readily available from fully deuterated algae, are used. As the deuterium

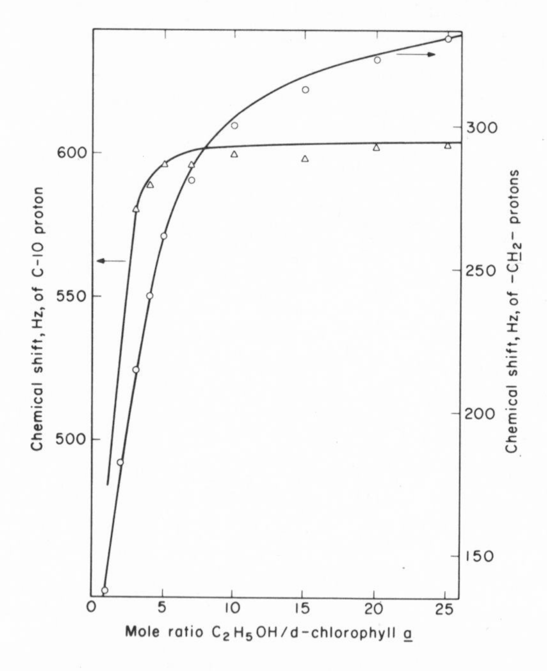

Fig. 10. ^{2}H-chlorophyll *a*–ethanol interaction in CCl_4 solution (0.068 *M*), showing chemical shifts of C-10 (△) and methylene $-CH_2-$ protons (○) as a function of ethanol/ Chl ratio. The solid lines are calculated values[64]. The equilibrium constant deduced for this interaction from these data is given in Table I.

atoms are invisible in an n.m.r. spectrometer adjusted to detect protons, the chemical shifts of the ligands can be followed with no complications introduced by the complex spectrum of resonances produced by the protons of ordinary chlorophyll. Fully deuterated ligands, particularly complicated ones of biological importance can also be usefully employed, in which case the n.m.r. spectrum of ordinary ^{1}H-chlorophyll is observed[42]. Closs[40] first observed the large up-field shift in methanol coordinated to the Mg of methyl chlorophyllide *a*. Studies using the same principle for studying ligand interactions with Mg-containing porphyrins and chlorophylls have also been reported[65].

The experiment consists in adding incremental amounts of base to a solution of ^{2}H-chlorophyll in a non-polar solvent, and recording the chemical shifts of the ligand as a function of ligand/chlorophyll ratio. As the C-10 and δ protons of chlorophyll are exchangeable[26], the deuterio-chlorophyll with ^{1}H in the C-10 and δ positions is employed, and these chemical shifts are also recorded. The chemical shifts of the ligand protons will be the weighted average of bound and free ligand, and the chemical shift of the C-10 proton is the weighted average in dimer and $Chl \cdot L_1$ (or $Chl \cdot L_2$). Results of a typical n.m.r. titration experiment are shown in Fig. 10. The ligand and the C-10 proton chemical shifts suffice to determine the concentrations of the species related by the equilibrium $Chl_2 + 2L \rightleftarrows 2Chl \cdot L$ and a computer analysis makes it possible to deduce self-consistent values of the equilibrium constant[64]. The equilibrium constants deduced by n.m.r. are further discussed in Section 3.6 and are listed in Table I.

The geometry of the chlorophyll–ligand interaction can be inferred from ring current calculations, using the procedure introduced by Abraham *et al.*[66]. The ring current effect of the chlorophyll macrocycle can be related to that of benzene by their anisotropic magnetic susceptibilities, and the observed chemical shifts of the ligand protons can then be used to deduce the distance and position of these protons from the center of the macrocycle. A distance (for alcohols) of 3.1 Å from the center of the macrocycle is deduced. If it is assumed that the Mg atom is displaced by 0.5 Å from the center of the macrocycle[31], then the Mg---O distance is estimated to be 2.6 Å, not much different from a conventional Mg–O bond.

3.3 Chlorophyll interactions with bifunctional ligands

As has been pointed out in Section 3.1, interaction of chlorophyll dimers or oligomers with monofunctional bases results in the formation of monomeric chlorophyll species. With bifunctional ligands the results may be very different. It has long been known that dioxane interacts with chlorophyll to form chlorophyll–dioxane adducts that have visible absorption spectra that are red-shifted relative to the dimer (Chl_2)[67–69]. Many other bifunctional ligands interact in essentially the same way: pyrazine, 4,4′-bipyridine, 2,2′-bipyrimidine, and 1,4-diazabicyclo (2,2,2) octane all form adducts with chlorophyll *a* that show long wavelength absorption[70]. Sherman and Fujimori[69], on the basis of infra-red spectra, appear to believe that all the keto C=O---Mg interactions are disrupted in the chlorophyll-dioxane adduct, which therefore has the structure -Chl dioxane Chl–dioxane Chl---. Such a structure would probably be blue shifted rather than red shifted. Closer examination of the infra-red spectra indicates that all the keto C=O---Mg interactions may not be disrupted, and that the structure of the adducts may be better represented by -Chl Chl–L–Chl Chl–L–Chl

Chl–. That is, it is the dimer units that are crosslinked by the bifunctional adduct coordinated to two Mg atoms. This structure would be expected to be red shifted. The c.d. studies by Sherman[71] on the chlorophyll–dioxane adduct appear on the whole to support this interpretation. The optical spectra of adducts of chlorophyll with bifunctional ligands are related to the size of the bifunctional ligand; with 4,4′-bipyridine, for example, where the chlorophyll dimers are relatively far apart, the red-shift is small. Pyrazine, 1,2-diazabicyclo(2,2,2)octane, 4,4′-bipyridine and 2,2′-bipyrimidine all are able to precipitate chlorophyll from decane solution, so strong is the interaction[70]. The insolubility of the adducts in aliphatic hydrocarbons may be taken to indicate a large aggregate. Where geometry or basicity is inappropriate, only disaggregation of chlorophyll dimers may occur. Such is the case for pyrimidine, which forms a (chlorophyll $\cdot x$ pyrimidine)$_n$ adduct at –20° C, but which forms monomeric (Chl · pyrimidine) in decane at room temperature. Pyrazine and pteridine, probably because of low basicity, and 2,2′-bipyridine, probably because of steric complications, interact with (Chl_2) weakly if at all[70]. There are many compounds in nature that could conceivably act as bifunctional ligands for chlorophyll, particularly nitrogen bases such as pyrimidine or pteridine. Although there is no reason to believe coordination compounds between chlorophyll and such ligands are of biological significance, this possibility can not be excluded at this time.

3.4 Chlorophyll–water interactions[55,72]

This is one of the more confusing episodes in the annals of chlorophyll chemistry. The older literature (reviewed in ref. 72) gave many reasons to suppose that the chlorophyll–water interaction was of unusual significance. Thus, Livingston and Weil[73] noted that water activated chlorophyll fluorescence in hydrocarbon solvents. The pioneer work of Jacobs *et al.*[74] on "crystalline" chlorophyll is probably most responsible for the wide-spread impression that chlorophyll and water interact in some special fashion. These workers found that water was required to produce chlorophyll that appeared crystalline by X-ray diffraction. Because their product also showed a strong red shift, Jacobs *et al.*[74] concluded that in contrast to ordinary chlorophyll, their chlorophyll was crystalline. Many workers subsequently observed changes in i.r. spectra[52,75-79] and e.s.r. spectra of chlorophyll[80-83] in the presence of water. Because the basic facts about the hydration, dehydration, and hydrates of chlorophyll were not established, and because no analytical procedures for water in chlorophyll were available, earlier work was carried out on ill-defined systems of unknown water content, the observations were uninterpretable, and the relationships between various effects were not perceived. A new, simple, sensitive vapor phase chromatographic analytical procedure[84] for water in chlorophyll has greatly clarified the situation. Perhaps more important, the donor–acceptor properties of chlorophyll were not recognized and the tenacity with which water is coordinated to solid chlorophyll was not appreciated. Above all, the extraordinary solvent dependence of the chlorophyll–water interaction was not taken into account, nor was it realized that the presence of water would have very different consequences in very dilute chlorophyll solutions in CCl_4 or benzene (such as are used in visible absorption spectrophotometry) and on a concentrated solution in aliphatic hydrocarbons or in films.

Water appears to be a unique nucleophile for chlorophyll. By virtue of its small size, its ability to act as an electron donor (via oxygen), and its capacity to form two hydrogen bonds simultaneously, the chlorophyll–water interaction is without a close parallel. Figure 11 shows the nature of

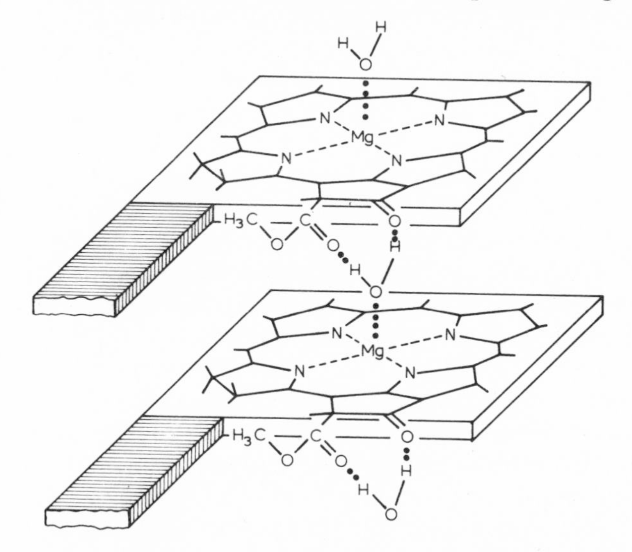

Fig. 11. Structure illustrating the nature of the chlorophyll–water interaction. The dimensions of the phytyl chain are not to scale[55].

the situation. Exposure of a $(Chl_2)_n$ oligomer solution in an aliphatic hydrocarbon solvent, or a chlorophyll film[55,75] to water vapor causes a striking change in the i.r. spectrum. The keto C=O---Mg aggregation peak at 1652–1660 cm^{-1} is replaced by an intense peak at 1638 cm^{-1}, the residual free keto C=O absorption near 1700 cm^{-1} vanishes, and the ester C=O absorption maxima at 1735 cm^{-1} are split and appear in the hydrated species at 1745 and 1727 cm^{-1}. Examination of the i.r. spectra in the O–H stretch region likewise reveals striking changes on hydration of chlorophyll. Anhydrous chlorophyll shows very little absorption in this region. A hydrated Chl *a* film or oligomer, however, possesses maxima at 3590, 3460 and 3240 cm^{-1}. Monomeric water in non-polar solvents shows the antisymmetric OH stretch fundamental ν_1 at 3615 cm^{-1}. Consequently, the OH absorptions seen in hydrated chlorophyll indicate the water to be bound, and the absorption maxima in both the carbonyl and OH stretch regions can be given a consistent interpretation. When chlorophyll oligomers (in solution or as a film) are hydrated, the keto C=O---Mg interactions are disrupted completely, and a water molecule is coordinated to the central Mg atom of the chlorophyll. From models, it can be seen that such a water molecule can hydrogen bond simultaneously to the two carbonyl functions in Ring V. The peak at 1638 cm^{-1} is due to a hydrogen bond interaction: keto C=O---HO(H)Mg. The ester C=O splitting is consistent with another hydrogen bond to the ester C=O in the carbomethoxy group at C-10. The three OH absorption maxima then correspond to water coordination to Mg, Mg---OH_2 (3600 cm^{-1}); Mg coordinated water that is hydrogen bonded to keto C=O: Mg---(H)OH---O=C keto (3240 cm^{-1}); and Mg coordinated water hydrogen bonded to ester carbonyl, Mg---(H)OH---O=C ester (3460 cm^{-1}). The overall interaction can then be formulated:

```
      ┌─────────────────────┐
ester C=O---H—O—H---O=C keto
                :
              —Mg—
```

The absorption maximum at 3240 cm^{-1} correlates with the absorption peak at 1638 cm^{-1}, and the presence of one of these peaks implies the presence of the other. Analysis shows the Chl:H_2O ratio to be 1:1 as required by this formulation[72].

The chlorophyll–water adduct thus is basically different in structure from the chlorophyll oligomer. It is clear that there is no barrier to the repetition of the interaction that forms –Chl H_2O Chl–, and as a result very large $(Chl \cdot H_2O)_n$ adducts can be built up. That these entities are in fact large is indicated by ultracentrifugation, X-ray diffraction, and electron microscopy. Chlorophyll oligomers in aliphatic hydrocarbon solvents give no diffraction pattern. The $(Chl \cdot H_2O)_n$ species in aliphatic hydrocarbons gives a sharp X-ray diffraction line indicative of a 7.5 Å spacing, a distance very close to that estimated from models for the Mg–Mg distance between the parallel chlorophyll molecules separated by a water molecule (Fig. 11). Whereas chlorophyll oligomers sediment slowly in the ultracentrifuge, and, indeed, ultracentrifugation is very suitable for measuring oligomer size[85], $(Chl \cdot H_2O)_n$ is completely sedimented by ultracentrifugation at 130,000 x *g* for 30 min. Thus, $(Chl \cdot H_2O)_n$ in aliphatic hydrocarbon solvents occurs as entities of colloidal dimensions. From electron microscopy the smaller particles contain at least a hundred chlorophyll molecules, and the particles range in size up to entities that can be clearly resolved in the light microscope (Fig. 12). Solutions, dispersions, or films of $(Chl \cdot H_2O)_n$ are yellow-green, with an absorption maximum in the visible at about 740 nm. The $(Chl \cdot H_2O)_n$ species is thus the most red-shifted species so far observed. It is likely that the "crystalline" chlorophyll preparations of Jacobs *et al.*[74] and of Anderson and Calvin[86] are largely or entirely composed of $(Chl \cdot H_2O)_n$, and that e.s.r.[82], i.r.[75], visible absorption spectra[76], monolayers[87], and photoconductivity[88] of chlorophyll will require reinterpretation in terms of the $(Chl \cdot H_2O)_n$ species.

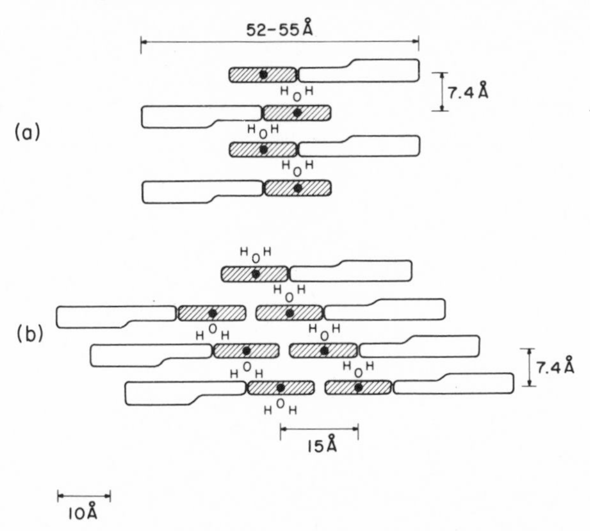

Fig. 12. The chlorophyll-water micelle in an aliphatic hydrocarbon solvent. The hydrogen bonds to the keto and ester C=O functions are not shown. The macrocycle is shaded, the phytyl chain is clear, and the Mg atom is a solid circle. Dimensions are from Courtauld models. (a) Side-view of staggered configuration; (b) side-view of step configuration[101].

3.5 Chlorophyll-pheophytin interaction[89]

The Mg-free chlorophyll derivative pheophytin *a* (Pheo) has been found to interact with Chl *a* in a surprising way in the presence of water. When dry Chl *a* and Pheo *a* are mixed in aliphatic hydrocarbon solvents or as a film, little spectral change results, either in the visible or the i.r. Water has no appreciable effect on the spectra of pure pheophytin *a*. However, when an anhydrous film of Chl *a* containing 10–60% Pheo *a* is exposed to water vapor at 55°C for 1 hour, a new absorption maximum appears at 712 ± 2 nm. When the Pheo *a* content of the film is greater than 10%, even prolonged hydration fails to form the $(Chl \cdot H_2O)_n$ species absorbing at 740 nm. Hydration of a Chl *a*–Pheo *a* solution in dodecane or n-octane produces a sharp absorption peak at 720 nm, without formation of the 740 nm $(Chl \cdot H_2O)_n$ species. Centrifugation at low speed quantitatively removes the 720 nm species. By spectral analysis, the ratio of Chl *a* to Pheo *a* is 2.00 ± 0.09. Because the aggregate is so easily sedimented at low centrifugal fields, the composition $(2Chl\ a{:}1\ Pheo\ a \cdot xH_2O)_n$ is assigned to the adduct. Infra-red spectra provide a clue to the structure of 720 nm species. In the carbonyl region, three principal absorptions are evident at 1735, 1675, and 1654 cm^{-1}. These receive reasonable interpretation as follows: 1735 cm^{-1}, free ester C=O; 1675 cm^{-1}, Chl keto C=O---Mg---OH_2; and 1654 cm^{-1}, Pheo keto C=O---H(OH)Mg. The 1654 cm^{-1} absorption is thus assigned to a hydrogen bonded pheophytin *a* keto C=O function. The chlorophyll-pheophytin-H_2O adduct thus can be considered to arise primarily from hydrogen-bonding interactions between chlorophyll dimer dihydrate and pheophytin. Two sites for hydrogen bonding are available in pheophytin, the keto C=O function in Ring V and the pyrrole nitrogen atoms, and thus, as both (H_2O Chl H_2O) and Pheo *a* are polyfunctional, large aggregates can form. Unlike the $(Chl \cdot H_2O)_n$ adduct, which can be represented as –Chl H_2O Chl H_2O Chl H_2O Chl–, the Chl *a*–Pheo *a*–H_2O adduct may be represented as –Chl H_2O pheo H_2O Chl Chl H_2O pheo H_2O Chl–. This structure differs from that of $(Chl \cdot H_2O)_n$ in that Chl–Chl interactions involving keto C=O---Mg still exist in the 720 nm species. Although both of these structures have been written in linear fashion, it should be recognized the presence of Chl–Chl interactions makes it impossible to arrange the macrocycle planes in a parallel fashion, as is the case for the $(Chl \cdot H_2O)_n$ species. In the $(2Chl \cdot Pheo \cdot xH_2O)_n$ species, the chlorophyll molecules may even be at right angles to each other, and the only parallel orientations in the adduct may be those of a chlorophyll and a pheophytin molecule, a circumstance that probably accounts for the smaller red shift observed in the $(2Chl \cdot Pheo \cdot xH_2O)_n$ species.

Absorption maxima at or near 695, 712, and 720 nm have been reported in green plants[90], generally under conditions where there is reason to suppose that large amounts of pheophytin are also present. The origin of these absorption maxima now appear clearer, and because pheophytin does not appear to occur as a normal component of healthy plants, there is reason to question whether these absorption maxima relate to chlorophyll species that have physiological significance in photosynthesis.

3.6 Equilibria between chlorophyll species[55]

The chlorophyll monomer, dimer, oligomer, and chlorophyll mono- and bi-functional species described above are related to each other by a series of equilibria. These equilibria provide a convenient framework to describe the effects of solvent, nucleophile, temperature, and concentration, all of which are important variables in laboratory systems. As of this writing, these equilibria can only be written in detail for chlorophyll *a*, but they also can be applied in a general way to the other chlorophylls. The chlorophyll–water interaction represents only a special case of the general nucleophile, L, interaction, but because of its remarkable nature, it is indicated explicitly.

The equilibria relating the various chlorophyll species can be described by the following equations:

$$2\,\text{Chl} \overset{K_1}{\rightleftharpoons} \text{Chl}_2 \qquad (1)$$

$$n(\text{Chl}_2) \overset{K_2}{\rightleftharpoons} (\text{Chl}_2)_n \qquad (2)$$

$$\text{Chl}_2 + \text{L} \overset{K_3}{\rightleftharpoons} \text{Chl}_2 \cdot \text{L} \qquad (3)$$

$$\text{Chl}_2 \cdot \text{L} + \text{L} \overset{K_4}{\rightleftharpoons} \text{Chl}_2 \cdot \text{L}_2 \qquad (4)$$

$$\text{Chl}_2 \cdot \text{L}_2 \overset{K_5}{\rightleftharpoons} 2\text{Chl} \cdot \text{L} \qquad (5)$$

$$\text{Chl}_2 + 2\text{L} \overset{K_6}{\rightleftharpoons} 2\text{Chl} \cdot \text{L} \qquad (6)$$

$$\text{Chl} \cdot \text{L} + \text{L} \overset{K_7}{\rightleftharpoons} \text{Chl} \cdot \text{L}_2 \qquad (7)$$

$$\text{Chl}_2 + \text{H}_2\text{O} \overset{K_8}{\rightleftharpoons} \text{Chl}_2 \cdot \text{H}_2\text{O} \qquad (8)$$

$$\text{Chl}_2 \cdot \text{H}_2\text{O} + \text{H}_2\text{O} \overset{K_9}{\rightleftharpoons} \text{Chl}_2 \cdot (\text{H}_2\text{O})_2 \qquad (9)$$

$$\text{Chl}_2 \cdot (\text{H}_2\text{O})_2 \overset{K_{10}}{\rightleftharpoons} 2\text{Chl} \cdot \text{H}_2\text{O} \qquad (10)$$

$$\text{Chl}_2 + 2\text{H}_2\text{O} \overset{K_{11}}{\rightleftharpoons} 2\text{Chl} \cdot \text{H}_2\text{O} \qquad (11)$$

$$n(\text{Chl} \cdot \text{H}_2\text{O}) \overset{K_{12}}{\rightleftharpoons} (\text{Chl} \cdot \text{H}_2\text{O})_n \qquad (12)$$

Equilibrium (1) describes dimer formation in non-polar solvents. At room temperature, in any non-polar solvent, for any concentration of chlorophyll, the equilibrium is displaced very far to the right. The equilibrium constant K_1, as judged from i.r. and visible absorption spectra (see Section 4), must be at least 10^6 and may even be considerably larger. This means that even in very dilute chlorophyll solutions, the concentration of monomeric chlorophyll is vanishingly small. The coordination interaction keto C=O---Mg must therefore be considered a rather strong one. The ready reversibility of equilibrium 6 or 11 is more a consequence of small differences in the coordination interaction between different ligands rather than an indication of a weak interaction.

Equilibrium (2) describes the continued aggregation of dimer to oligomer. Equilibrium constants for oligomerization are given in ref. 61. In CCl_4 or benzene, which are favorite solvents for chlorophyll laboratory investigations, this equilibrium is displaced to the left, and even in solutions of Chl *a* as concentrated as 0.1 *M*, the predominant form is still the dimer. In cycloaliphatic or aliphatic hydrocarbon solvents, however, the equilibrium constant for oligomerization is much larger, reflecting the poorer solvation characteristics of these solvents for the chlorophyll macrocycle. In these solvents, sizeable oligomers are present at chlorophyll concentrations larger than $10^{-3}\,M$.

Equilibria (3), (4) and (5), and (8), (9) and (10) have not been studied separately. They are included because titration data indicates differences in the stability of the chlorophyll dimer solvates in various solvents, and multiple equilibria will make it possible to describe ultimately the disaggregation process more precisely. The sums of these equilibria, K_6 and K_{11}, have been studied by i.r. and n.m.r. methods, and experimental values for these equilibrium constants are given in Table I. As indicated above, the rather small equilibrium constants are indicative of small energy differences between the chlorophyll–chlorophyll and the chlorophyll–ligand interactions. The agreement between equilibrium constants as derived by different experimental methods is seen to be quite good. The equilibrium constant for the formation of hexacoordinate Mg species, K_7, is much smaller than that for the formation of pentacoordinate Mg (K_6), and it is estimated from n.m.r. titration data that K_6 is about 100 K_7. The species $Chl \cdot L_2$ appears to be of importance primarily under forcing conditions, such as result from solution in a neat polar solvent.

The effect of water on a chlorophyll solution in CCl_4 or benzene is described by reaction (11). In these systems, the relative concentrations of Chl_2, $Chl_2 \cdot H_2O$, and $Chl \cdot H_2O$ will be determined primarily by chlorophyll concentration because of the limited solubility of water in CCl_4 ($\sim 8 \times 10^{-3} M$). In concentrated chlorophyll solutions in this solvent, the chlorophyll will occur mainly as Chl_2 and $Chl_2 \cdot H_2O$, and only a small amount of $Chl \cdot H_2O$ will be present. In dilute chlorophyll solutions in the range of 10^{-6}–10^{-4} M, such as are commonly used in spectrophotometry in the visible, water-saturated CCl_4 or benzene will have an enormous molar excess of water present, equilibrium (11) will be displaced far to the right, and the chlorophyll will be present as $Chl \cdot H_2O$. The disaggregation of Chl_2 dimers reported by some investigators to occur in very dilute solution as deduced from visible absorption spectra is not a reversal of equilibrium (1), but a consequence of the presence of adventitious water.

Equilibrium (12) is perhaps the most sensitive to solvent effects. This reaction describes the formation of the $(Chl \cdot H_2O)_n$ species that absorbs at 740 nm. In CCl_4 or benzene, this equilibrium is displaced far to the left and only in the most concentrated solutions can any evidence for the formation of the 740 nm species be found. In aliphatic hydrocarbon solvents, however, equilibrium (12) is strongly displaced to the right. Chlorophyll concentration is important here also; in a very dilute Chl *a* solution in hexadecane, saturated with water, only the monomeric species $Chl \cdot H_2O$, absorbing near 663 nm, can be seen, whereas in a 10^{-3} M solution, only the highly aggregated $(Chl \cdot H_2O)_n$ species absorbing near 740 nm is present.

An extensive literature on the spectroscopy of "aggregated" chlorophyll in the form of films exists, and it is therefore desirable to consider how the relationships described here for solutions apply in the special case of films. Films of chlorophyll may be thought of as limiting cases of solution in aliphatic hydrocarbon solvents, with the long aliphatic phytyl chains of the chlorophyll providing the aliphatic hydrocarbon environment. Chlorophyll–adduct interactions in solution result in adducts that assume configurations of minimum energy because molecular motions are not restricted. In films, molecular motions are much more constrained, and only those chlorophyll molecules that are already properly positioned can interact. Increasing the temperature increases the mobility of chlorophyll in films, thus permitting a closer approach to configurations of minimum energy. The keto C=O---Mg interaction requires a specific orientation of the two chlorophyll molecules. As a film forms from a solution in a non-polar solvent, only those chlorophyll molecules that are properly positioned, or which require only small motions to achieve the necessary orientation, will interact by additional coordination. Films are thus very sensitive to the exact conditions under which they are formed, and reflect the state of chlorophyll in the solvent from which the film is cast. Films prepared from CCl_4 solution, where chlorophyll is chiefly the dimer, have a smaller amount of keto C=O---Mg interactions, judged by i.r., than a film prepared from hexane, where Chl *a* occurs as an oligomer. Contrary to intuition a Chl *a* film cast from CCl_4 is less aggregated in terms of keto C=O---Mg interactions than is a solution of chlorophyll in dodecane. Recognizing these factors, the equilibria discussed in this section are useful in describing the behavior of chlorophyll films. For example, in a solution or dispersion of $(Chl \cdot H_2O)_n$, it is easy to remove the water and reform $(Chl_2)_n$, merely by bubbling with dry nitrogen gas. Reversal is possible because the chlorophyll molecules are free to reform keto C=O---Mg bonds, thus facilitating the displacement of water coordinated to Mg. In a hydrated film, however, only a portion of the chlorophyll molecules are in a position to replace $Mg \cdots OH_2$ by keto C=O---Mg interactions, and it is only this water that can be removed by pumping. At room temperature, even long protracted pumping will not remove all of the water from a film of hydrated chlorophyll. The usual co-distillation procedures for dehydrating chlorophyll solubilizes the chlorophyll and permits new coordination interactions to be established[72].

The situation with respect to the other chlorophylls is much more complicated, because of the additional electron donor function present in Chl *b* and BChl. It now appears that the fundamental unit for both Chl *b* and BChl is the trimer rather than the dimer, which of course affects the reactions written to describe the equilibria with ligands[61]. Both Chl *b* and BChl tend to oligomerize more extensively than does Chl *a*, and the trimer units are harder to disrupt by competing ligands. It can be readily seen that there are a very large number of ways in which a ligand such as water can be introduced into a Chl *b* or BChl oligomer. Although it appears that the same principles apply to these chlorophylls as hold for Chl *a*, additional information will be required before definite structures for Chl *b* and BChl oligomers and ligand adducts can be formulated.

References pp. 1063–1066

4.0 LONG WAVELENGTH FORMS OF CHLOROPHYLL[91]

4.1. Deconvolution of electronic transition spectra

Absorption spectroscopy in the visible is the tool that has been most widely used to study chlorophyll in the laboratory and to characterize the state of chlorophyll in the plant. Despite its wide use, however, absorption spectra in the visible [referred to here also as electronic transition (e.t.) spectra] are probably more difficult to record and interpret than those of any other spectroscopic technique. For one thing, visible absorption spectroscopy requires very dilute chlorophyll solutions because of the intense color. Chlorophyll in the plant, however, occurs in highly concentrated form. A cell has now been developed that makes it possible to record e.t. spectra of chlorophyll solutions as concentrated as 0.1 *M*, thus overlapping the concentration regions used in vapor phase osmometry and i.r. spectroscopy[92]. The characterization of the various chlorophyll species described in this chapter now makes it possible to correlate *in vitro* e.t. spectra with known chlorophyll species, which is essential for laboratory data to be applied properly to the interpretation of e.t. spectra in the plant.

Chlorophyll in the plant is known to have unusual light absorption properties as compared to chlorophyll dissolved in polar solvents in the laboratory. Chlorophyll in polar solvents absorbs in the red near 662 nm, whereas chlorophyll in the plant absorbs at longer wavelengths, the bulk of the chlorophyll absorbing near 680 nm[93]. It is also widely accepted that a small fraction of the chlorophyll that absorbs near 700 nm has special significance for the light conversion act in photosynthesis[46]. Despite the unusual spectral properties of chlorophyll which has led some investigators to conclude that multiple forms of chlorophyll exist in the plant, only one chlorophyll *a* with the usual structure has ever been isolated from the plant. The unusual spectral properties of Chl *a* in the plant have therefore been attributed to "aggregation"[48], to complexing with protein or other cellular components, or to some special environment[47], but none of these explanations have been very specific.

The solvent dependence of chlorophyll e.t. spectra in the laboratory have also long been recognized. A red shift is observed in the visible absorption spectra of Chl *a* solutions in non-polar solvents. This shift was originally interpreted as a reversal of the relative energies of the n, π^* and π,π^* singlet states by a change from a polar to a non-polar solvent[94,95]. The change in chlorophyll species from monomer to dimer or oligomer is an interpretation that supersedes the one based on the relative energies of excited states. When such an interpretation is advanced in a modern monograph to explain the spectral or fluorescence properties of chlorophyll[96,97], the explanation should at least be properly qualified. Chlorophyll films treated with water vapor or dioxane show even larger red-shifts. In these systems, "aggregation" again has been invoked to account for the unusual spectral properties.

In the past, the principal e.t. parameter of concern to investigations was the wavelength of the absorption maxima. Because of the complexity of the red absorption envelope, resort has now been had to computer deconvolution techniques to establish the nature of the chlorophyll absorption in the red in a more objective way. Computer deconvolution techniques have been used for plant material[43,98,99], but do not appear to have been applied previously to *in vitro* chlorophyll systems. An Argonne National Laboratory library program for deconvolution into Gaussian components was used and the deconvolutions were carried out on an IBM 360/75 computer. An initial error matrix is submitted, and the initial estimates are then improved by a variable metric minimization routine. These procedures make it possible to follow events over a large concentration range and to detect systematic trends in the e.t. spectra.

4.2 Electronic transition spectra of chlorophyll a dimers[100]

Chl *a* in CCl_4 solution has long been known to possess a shoulder on the red side of the red absorption envelope. The red absorption band can be resolved into Gaussian components with maxima at 628, 650, 662, 678, and 700 nm. The peak positions, half-band widths, and relative areas of the components are essentially concentration independent over the concentration range 10^{-6}–10^{-1} *M*. This is consistent with conclusions from vapor phase osmometry that Chl *a* in CCl_4 is largely composed of the dimer even in concentrated solution[72]. The visible absorption spectra and the molecular weight determinations are consistent with the idea that the state of chlorophyll in CCl_4 solution is more or less independent of concentration. Indeed, the e.t. spectra probably provide the best evidence for the conclusion that Chl *a* occurs in CCl_4 as the dimer in highly dilute solutions.

4.3 Electronic transition spectra of chlorophyll a oligomers[100]

The visible absorption spectra of $(Chl_2)_n$ oligomers in aliphatic hydrocarbon solvents are resolvable into Gaussian components at 628, 650, 662, 678 and 703 nm, with peak positions and half-band widths essentially identical with those observed in dimers in CCl_4 solution. However, in aliphatic hydrocarbon solvents, the relative areas of 662, 678 and 703 nm components vary with concentration, and thus with the size of the oligomer (Fig. 13). As the concentration increases and as the oligomer becomes larger, the area of the 662 nm component decreases and the area of the 678 nm component increases. In a 10^{-6} *M* solution of Chl *a* in dodecane, the ratio of the areas of the 662/678 nm components is near 1, whereas in a 0.1 *M* solution, the 662/678 nm ratio is about 0.2. The relative areas of these components are in fact a useful measure of oligomer size. Oligomer

References pp. 1063–1066

solutions more concentrated than 10^{-5} *M* always require an additional Gaussian component at 694–703 nm to fit the red end of the red band. Experiments with carefully controlled systems suggest that the enhanced absorption at the red edge of the red band is an intrinsic property of such systems. Whether the enhanced red-edge absorption is an artifact arising from light-scattering by the oligomers, or whether it is truly a low-lying electronic state important in the light conversion act in photosynthesis is still a matter of speculation.

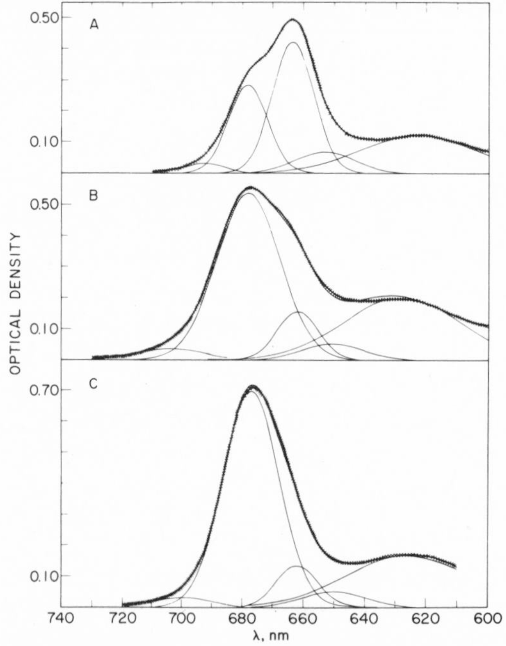

Fig. 13. Comparison of red band of chlorophyll *a* solutions in dodecane with that of an algal sonicate. A, 103 x 10^{-4} *M*; B, 7.73 x 10^{-2} *M*; C, *Tribonema* sonicate.

The physical significance of the Gaussian components is likewise an open question. From exciton theory, it might be expected that as the oligomers become larger, a red shift should occur in the e.t. spectra. Such is not observed, and even in the largest oligomers, the Gaussian components have the same absorption maxima peak wavelengths. What does change is the relative areas of the components. This suggests that the 662 nm component be assigned to chlorophyll acting as acceptor, and the 678 nm component to chlorophyll acting both as donor and acceptor. As the oligomer increases in size, the relative proportion of chlorophyll molecules that function only as acceptor decreases, and this could account for the changes in the relative areas of the Gaussian components. The long wavelength Gaussian components thus arise from the structural consequences of the coordination interactions.

4.4 *Electronic transition spectra of chlorophyll–bifunctional ligand adducts*

All chlorophyll adducts with bifunctional ligands are red-shifted, the chlorophyll–water adduct most strongly. From the available evidence, it is hard to avoid the conclusion that the red shift results from the juxtaposition of chlorophyll molecules, and that it is not the coordination interaction *per se* but the geometry imposed by the ligand on the chlorophyll molecules that is the important factor. Thus, the chlorophyll–pyrazine adduct:

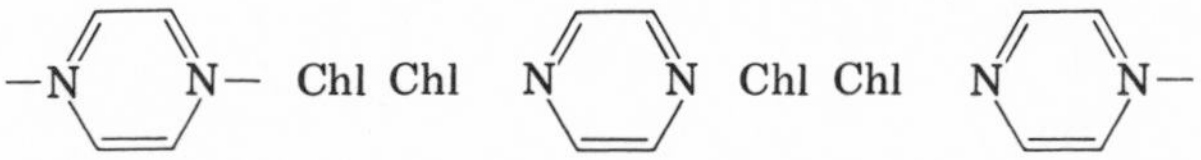

—N⟨pyrazine⟩N— Chl Chl N⟨pyrazine⟩N Chl Chl N⟨pyrazine⟩N—

is red shifted, but the species Chl · (pyrazine) and Chl · $(\text{pyrazine})_2$ formed with excess ligand in CCl_4 solution has the absorption spectrum of monomeric chlorophyll. It is thus in the relationship of the chlorophyll molecules in the adduct that the cause of the red shift must be sought. It should be pointed out again that the Chl–pyrazine adduct, although written in linear fashion here, cannot be linear, and the orientation of the chlorophyll molecules in the dimer units, and of the dimer units relative to each other may be very different in three dimensional space from the linear representation.

The chlorophyll–water species $(\text{Chl} \cdot H_2O)_n$, shows the largest red-shift. This species may be considered to arise from a chlorophyll oligomer absorbing at 678 nm, -Chl Chl Chl Chl-, by insertion of H_2O between all chlorophyll molecules, with the complete disruption of the keto C=O---Mg interaction characteristic of the oligomer. This species absorbs at 740 nm; the red shift is large probably because the water molecule that separates the chlorophyll is small so that the macrocycle planes are close to each other, and the staggered configuration of the parallel chlorophyll molecules leads to a large red shift. A 1:1 ratio of chlorophyll:H_2O is not the only adduct possible, however, and a series of chlorophyll–water adducts with different Chl:H_2O ratios and periodicities can be imagined:

A -Chl H_2O Chl H_2O Chl H_2O Chl H_2O Chl H_2O-
B -Chl Chl H_2O Chl H_2O Chl Chl H_2O Chl H_2O-
C -Chl Chl H_2O Chl Chl H_2O Chl Chl H_2O-
D -Chl $(\text{Chl})_n$ Chl H_2O Chl $(\text{Chl})_n$ Chl H_2O-

These species are expected to become less red shifted as the number of keto C=O---Mg interactions increase. The various long wavelength forms of chlorophyll in the plant could then be assigned to the structures shown above. Thus, A might correspond to the 740 nm form, B to the 720 nm form, C to a 710 nm species. The form absorbing at 700 nm can be considered to arise from D for values of $n \geqslant 1$. Where $n \gg 1$, the principal absorption band should be at 678 nm, which is characteristic of the oligomer, with a small contribution from -Chl H_2O Chl- absorbing at longer wavelengths. It should be clear that any bifunctional ligand, -X-, that positions the chlorophyll molecules *vis-à-vis* each other in the same way would have the same spectral consequences as H_2O. Although no substance other than water is known to function in this way, the possibility cannot be excluded that such a ligand, or perhaps a purely structural matrix, occurs in the cell.

The foregoing discussion is based on the idea that long wavelength forms of chlorophyll result from a cooperative interaction involving more than one chlorophyll molecule. In the laboratory, no matter what the coordinating ligand, monomeric chlorophyll species are always short wavelength forms. It is in interactions between chlorophyll molecules, whether joined directly or through the agency of intermediary bifunctional ligands, that the origin of the red shifted spectra must be sought.

4.5 Electronic transition spectra of Tribonema aequalis[100]

This is a fully competent photosynthetic alga that fixes CO_2 and evolves O_2. The only chlorophyll it possesses is chlorophyll *a*, and no other porphyrin, dihydroporphyrin, or phycocyanobilin is present as an auxiliary photosynthetic pigment. Sonicates of this organism have a red absorption envelope that can be deconvoluted into Gaussian components, that are highly similar in band positions, half-band widths, and relative areas to those of a 0.1 *M* Chl *a* solution in dodecane (Fig. 13). The bulk of the chlorophyll in the chloroplast thus appears from its e.t. spectrum to be very similar in its state of aggregation to a concentrated solution of Chl *a* in an aliphatic hydrocarbon solvent. Antenna chlorophyll must be similar, if not identical, in light absorption properties to the chlorophyll oligomer, $(Chl_2)_n$.

5.0 PHOTO-ACTIVITY OF CHLOROPHYLL SPECIES[101,102]

In 1956, Commoner *et al.*[103] discovered that electron spin resonance (e.s.r.) signals could be elicited from intact photosynthetic organisms, or chloroplast or bacterial chromatophore preparations by irradiation with light. The most prominent of the photo-e.s.r. signals (designated Signal I) is reversible, has the free-electron *g*-value 2.0025, a peak-to-peak linewidth (ΔH) of about 7 *g* (plants) or about 9.5 *g* (bacteria), a Gaussian line-shape, and no hyperfine structure. Much subsequent work, reviewed by Weaver[104], has established that Signal I arises by the photooxidation of special chlorophyll molecules in the photosynthetic reaction center, and that this is the chlorophyll responsible for the long wavelength absorption associated with the reaction center (P700, plants; P870, bacteria). The genesis of Signal I, then, has been assumed to be Chl^{+} or BChl^{+}. The e.s.r. signals produced by these *in vitro* species[105,106] are much broader than Signal I. It thus becomes of interest to see whether the chlorophyll species described in this chapter have any of the properties of reaction center chlorophyll.

5.1 *E.s.r. of monomeric chlorophyll*

The monomeric species Chl · L_1 and Chl · L_2 can be oxidized to the free radical $(Chl \cdot L)^{+\cdot}$ but only by chemical or electrochemical procedures. Values for $(^1H\text{-}Chl \cdot L)^{+\cdot}$, $(^2H\text{-}Chl \cdot L)^{+\cdot}$ are listed in Table II. Values obtained by chemical oxidation agree well with those recorded in $(Chl \cdot L)^{+\cdot}$ prepared by electrochemical oxidation. The nature of this species as a π-cation radical of Chl *a* was established unequivocally[105].

TABLE II

LINE WIDTHS (ΔH) OF *IN VITRO* CHLOROPHYLL e.s.r. SIGNALS[109]

System[a]	Solvent	λ_{max} (nm)	Oxidant	ΔH (g)
^{1}H-Chl *a*	b	663	$FeCl_3$ or I_2	9.3 ± 0.3[d]
^{2}H-Chl *a*	b	663	$FeCl_3$ or I_2	3.8 ± 0.2[d]
^{1}H-BChl	c	773	I_2	12.8 ± 0.5[d,e]
^{2}H-BChl	c	773	I_2	5.4 ± 0.2[d,e]
(Chl_2)	CCl_4	665, 678	I_2	9.0 ± 0.5
$(Chl_2)_n$	Film	665, 678	I_2	10.0 ± 0.5
$(Chl_2)_n$	$C_{16}H_{34}$	665, 678	I_2	9.0 ± 0.5
$(Chl_2)_n$	$C_{16}H_{34}$	665, 678	O_2	11 ± 1
$(Chl \cdot H_2O)_n$	Film	743	O_2	1.8 ± 0.2
$(Chl \cdot H_2O)_n$	Film	743	I_2	1.5 ± 0.3
$(Chl \cdot H_2O)_n$	Film	743	Light	1.8 ± 0.2
$(Chl \cdot H_2O)_n$	$C_{16}H_{34}$	743	Light	0.8 ± 0.2

[a] ^{1}H-, ordinary hydrogen; ^{2}H-, fully deuterated.
[b] Solvent: CH_3OH or $CH_3OH\text{-}CH_2Cl_2$. Concentration of Chl and oxidant, 10^{-4}–10^{-3} *M*.
[c] Solvent: CH_3OH-glycerol (1:1, v/v).
[d] *g*-Values, 2.00025 ± 0.0002. Spectra recorded at liquid N_2 temperatures.
[e] McElroy, Feher and Mauzerall[106].

5.2 *E.s.r. of chlorophyll dimer and oligomer*

Free radical signals can be obtained from these species only by chemical oxidation. Both $(Chl_2)^{+}$ and $(Chl_2)_n{}^{+}$ (Table II) have e.s.r. signals with linewidths of the order of 9–10 *g*, very similar in linewidth to the signals observed in the monomeric species (Chl · L). Even if the signal is generated in $(Chl_2)_n$ in solution, the signal is broad, meaning that the line is not broadened by restricted rotation.

5.3 *E.s.r. of* $(Chl \cdot H_2O)_n$

This is the only chlorophyll species that yields a reversible photo-e.s.r. signal (Fig. 14) sufficiently intense to be readily detected[101]. The signal from this species is unusually narrow (Table II). The nature and genesis of this e.p.r. signal has been discussed by Garcia-Morin *et al.*[102]. The effect of light is believed to cause hydrogen atom transfer in the water molecule between two chlorophylls, thus effecting charge separation. The extreme

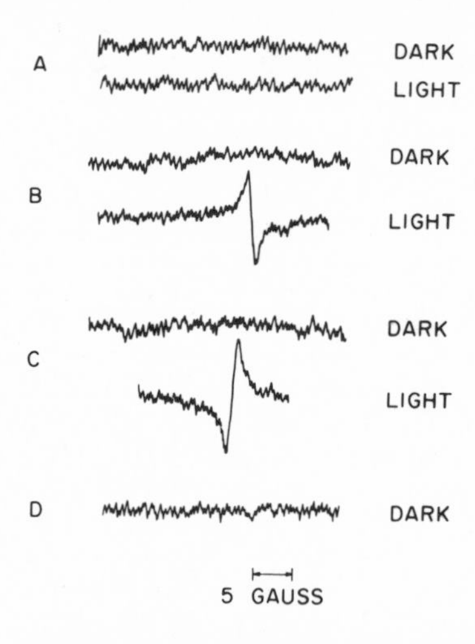

Fig. 14. E.s.r. spectra of chlorophyll oligomer and $(Chl \cdot H_2O)_n$ species in 0.001 *M* dodecane solution. Irradiated with red light (> 650 nm) at 25°C. A, $(Chl_2)_n$; B, $(Chl \cdot H_2O)_n$; C, $(^2H\text{-}Chl\ a \cdot H_2O)_n$; D, sample C after 1 h in the dark[101].

narrowness of the signal is attributed to the migration of the unpaired spin over the entire $(Chl \cdot H_2O)_n$ aggregate. When the unpaired spin is delocalized over a sufficiently large number of chlorophyll molecules at a sufficiently rapid rate, the e.s.r. signal collapses to a very narrow line. In $(Chl_2)^{\dagger}$ or $(Chl_2)_n{}^{\dagger}$ the spin is not delocalized to any appreciable extent, or, alternatively, spin delocalization is slow. In effect, the free radical produced in the oligomer is confined to the chlorophyll molecule in which it was formed; as a result, the e.s.r. signal is broad.

5.4 *E.s.r. of photosynthetic organisms*

Table III lists values for Signal I in a number of photosynthetic organisms, chloroplast preparations, and photosynthetic bacterial chromatophores. The e.s.r. parameters of Signal I are highly reproducible and are easy to replicate. It can be seen from Table III that the linewidth of Signal I is always significantly lower than the signal from (Chl · L$^{+\cdot}$, a difference much larger than can be accounted for by experimental error. All the chlorophyll species have e.s.r. signals either broader or narrower than Signal I, and absorb light at wavelengths considerably shorter or longer than does reaction center chlorophyll.

TABLE III

OBSERVED AND CALCULATED LINE WIDTHS (ΔH) OF e.s.r. SIGNAL I IN PLANTS AND BACTERIA[109]

Organisms[a]	*In vivo* ΔH, (g)		
	Observed	Calculated[b]	R[c]
^{1}H-*Synechococcus lividus*	7.1 ± 0.2[d]	6.6 ± 0.3	1.08 ± 0.06
^{2}H-*S. lividus*	2.95 ± 0.1[d]	2.7 ± 0.1	1.10 ± 0.05
^{1}H-*Chlorella vulgaris*	7.0 ± 0.2[d]	6.6 ± 0.3	1.06 ± 0.05
^{2}H-*C. vulgaris*	2.7 ± 0.1[d]	2.7 ± 0.1	1.00 ± 0.05
^{1}H-*Scenedesmus obliquus*	7.1 ± 0.2[d]	6.6 ± 0.8	1.08 ± 0.06
^{2}H-*S. obliquus*	2.7 ± 0.1[d]	2.7 ± 0.1	1.00 ± 0.05
^{1}H-HP700	7.0 ± 0.2[e]	6.6 ± 0.3	1.06 ± 0.05
^{1}H-*Rhodospirillum rubrum*	9.1 ± 0.5[f]	9.1 ± 0.4	1.01 ± 0.07
^{1}H-*R. rubrum*	9.2 ± 0.6[g]	9.1 ± 0.4	1.02 ± 0.08
^{1}H-*R. rubrum*	9.5 ± 0.5[h]	9.1 ± 0.4	1.05 ± 0.07
^{2}H-*R. rubrum*	4.0 ± 0.5[f]	3.8 ± 0.1	1.04 ± 0.14
^{2}H-*R. rubrum*	4.2 ± 0.3[h]	3.8 ± 0.1	1.10 ± 0.09
^{1}H-*Rhodopseudomonas spheroides*	9.6 ± 0.2[i]	9.1 ± 0.4	1.06 ± 0.05

[a] Whole cells except as otherwise indicated.
[b] Calculated from eqn. (1), using monomer chlorophyll values from Table II: ^{1}H-Chl *a*, 9.3 ± 0.3 g; ^{2}H-Chl *a*, 3.8 ± 0.2 g; ^{1}H-BChl, 12.8 ± 0.5 g; ^{2}H-BChl, 5.4 ± 0.2 g. All line shapes are Gaussian.
[c] Ratio of $\Delta H_{in\ vivo}/(\Delta H_{in\ vitro}/2)$; the *in vitro* values are cited in footnote b.
[d] Ref. 109.
[e] Measured on an active center preparation furnished by L. P. Vernon.
[f] Kohl *et al.*[107].
[g] P. A. Loach, personal communication. Chromatophore or active center preparation.
[h] McElroy, Feher and Mauzerall[106].
[i] Bolton, Clayton and Reed[108]. Chromatophore or reaction center preparations.

5.5 E.s.r. linewidths

In Section 5.3, the extreme narrowness of the e.s.r. signal from (Chl · $H_2O)_n$ is accounted for by spin migration that delocalizes the unpaired spin over a large number of chlorophyll molecules. The same process can be used to reconcile the *in vivo* and *in vitro* chlorophyll e.s.r. signals. From consideration of linewidths by the method of second moments, it can be shown[109] that the linewidth ΔH_N of an unpaired electron delocalized over N molecules is related to the linewidth of the signal from the monomer ΔH_M by the expression $\Delta H_N = 1/(N)^{\frac{1}{2}} H_M$. From this expression the linewidth in $(Chl \cdot H_2O)_n$ of about 1 *g* is accounted for by delocalization over about 100 chlorophyll molecules, as the linewidth ΔH_M of the monomer is around 10 *g*.

5.6 Origin of Signal I

If it is assumed that the narrowing of the *in vivo* e.s.r. signal is due to a process of spin delocalization, then the equation of Section 5.5 permits an estimate of the number of chlorophyll molecules over which the unpaired spin is delocalized. It can be seen that the *in vivo* Signal I listed in Table III are related to the linewidth of $(Chl \cdot L)^{+}$ very nearly by $1/(2)^{\frac{1}{2}}$ *i.e.*, $N = 2$ in the equation relating linewidth and delocalization. Thus, the characteristic linewidth of Signal I can be very well accounted for by the assumption that the signal arises in an entity $(Chl\ H_2O\ Chl)^{+}$. Linewidths calculated on the basis that the unpaired spin is delocalized over two properly positioned chlorophyll molecules accounts equally well for the e.s.r. Signal I observed in green algae, blue-green algae, photosynthetic bacteria, and active center preparations. Note that this unit is not a chlorophyll *a* dimer as described in this chapter.

5.7 Model of the photosynthetic unit

The bulk of the chlorophyll, the antenna chlorophyll, is considered to be the oligomer, $(Chl_2)_n$ with an absorption maximum near 680 nm. Charge separation occurs in an entity ($Chl\ H_2O\ Chl$) and the unpaired spin is confined to this entity. From exciton theory, it would be expected that the entity ($Chl\ H_2O\ Chl$) would absorb near 700 nm, based on a 740 nm absorption of $(Chl \cdot H_2O)_n$ and a 663 nm maximum for the monomer $(Chl \cdot L)$. The model thus appears to be compatible with both the e.s.r. and optical properties of reaction center chlorophyll. The antenna chlorophyll can be attached to the active center, as in the structures $[(Chl_2)_n\ Chl\ H_2O\ Chl\ (Chl_2)_n]$ or $[(Chl_2)_n\ Chl\ H_2O\ Chl]$ to form units that have the necessary optical and e.s.r. properties. The integrity of such a structure can be maintained only in a highly hydrophobic environment to which access of water is strictly regulated. The electron generated by light conducted to ($Chl\ H_2O\ Chl$) is removed via the quinones and other electron transport agents to the aqueous part of the chloroplast where the dark Calvin cycle reactions are carried out. Elsewhere, in the aqueous portion of the chloroplast, an electron is removed from water and restored to $(Chl\ H_2O\ Chl)^{+}$. Two chlorophyll molecules are available for electron transport so that removal and replacement of electrons can be simultaneous. The model of the photosynthetic unit described here deduced mainly from e.s.r., turns out to be very similar to one based on infra-red and visible absorption spectroscopy[91].

ACKNOWLEDGEMENT

This work was performed under the auspices of the U.S. Atomic Energy Commission.

REFERENCES

1 M. D. Kamen, *Primary Processes in Photosynthesis,* Academic Press, New York, 1963.
2 R. K. Clayton, *Molecular Physics in Photosynthesis,* Blaisdell Publishing Co., New York, 1965.
3 R. C. Dougherty, H. H. Strain, W. A. Svec, R. A. Uphaus and J. J. Katz, *J. Am. Chem. Soc.,* 92 (1970) 2826.
4 J. W. F. Wasley, W. T. Scott and A. S. Holt, *Can. J. Biochem.,* 48 (1970) 376.
5 H. H. Strain, B. T. Cope, Jr., G. N. McDonald, W. A. Svec, and J. J. Katz, *Phytochemistry,* 10 (1971) 1109.
6 A. S. Holt, *Can. J. Bot.,* 39 (1961) 327.
7 J. W. Purdie and A. S. Holt, *Can. J. Chem.,* 63 (1965) 3347.
8 A. S. Holt, J. W. Purdie and J. W. F. Wasley, *Can. J. Chem.,* 44 (1966) 88.
9 J. W. Mathewson, W. R. Richards and H. Rapoport, *J. Am. Chem. Soc.,* 85 (1963) 364.
10 H. H. Strain and W. A. Svec, in L. P. Vernon and G. R. Seely, *The Chlorophylls,* Academic Press, New York, 1966. Chapter 2, pp. 21-66.
11 J. S. Bunt; *Nature,* 203 (1964) 1261
12 Denise Y. C. Lynn Co and S. H. Schanderl, *J. Chromat.* 26 (1967) 442.
13 C. Sironval, M. R. Michel-Wolwertz and A. Madsen, *Biochim. Biophys. Acta,* 94 (1965) 344.
14 F. C. Pennington, H. H. Strain, W. A. Svec and J. J. Katz, *J. Am. Chem. Soc.,* 86 (1964) 1418.
15 F. C. Pennington, H. H. Strain, W. A. Svec and J. J. Katz, *J. Am. Chem. Soc.,* 89 (1967) 3875.
16 G. R. Seely, in L. P. Vernon and G. R. Seely, *The Chlorophylls,* Academic Press, New York, 1966, Chapter 3, pp. 91-93.
17 R. Willstätter and A. Stoll, *Investigations on Chlorophyll,* (translated by F. M. Shertz and A. R. Merz), Science Printing Press, Lancaster, Pennsylvania, 1928.
18 H. Fischer and A. Stern, *Die chemie des Pyrrols,* Vol. II, Part II, Akad. Verlag, Leipzig, 1940.
19 S. Aronoff, in W. Ruhland, *Handbuch der Pflanzenphysiologie,* Vol. 5, Part 1, Springer, Berlin, 1960, p. 234.
20 C. S. French, in W. Ruhland, *Handbuch der Pflanzenphysiologie,* Vol. 5, Part 1, Springer, Berlin, 1960, p. 252.
21 L. P. Vernon and G. R. Seely, (Eds.), *The Chlorophylls,* Academic Press, New York, 1966.
22 L. Bogorad, in L. P. Vernon and G. R. Seely, *The Chorophylls,* Academic Press, New York, 1966. Chapter 15, pp. 481-510.
23 I. Fleming, *J. Chem. Soc. (C),* (1968) 2765.
24 H. Brockmann, Jr., *Angew. Chem.,* 80 (1968) 233.
25 J. J. Katz, G. D. Norman, W. A. Svec and H. H. Strain, *J. Am. Chem. Soc.,* 90 (1968) 6841.
26 R. C. Dougherty, H. H. Strain and J. J. Katz, *J. Am. Chem. Soc.,* 87 (1965) 104.
27 G. Mackinney and M. A. Joslyn, *J. Am. Chem. Soc.,* 63 (1941) 2530.

28 L. J. Boucher and J. J. Katz, *J. Am. Chem. Soc.*, 89 (1967) 4703.
29 W. Hoppe, G. Will, J. Gassmann and H. Weichselgartner, *Z. Krist.*, 128 (1969) 18.
30 R. C. Pettersen, *Acta Cryst. Sect. B*, 25 (1969) 2527.
31 M. S. Fischer, D. H. Templeton, A. Zalkin and M. Calvin, *J. Am. Chem. Soc.*, 94 (1972) 3613.
32 R. Timkovich and A. Tulinsky, *J. Am. Chem. Soc.*, 91 (1969) 4430.
33 R. W. Parry, in J. C. Bailar, *The Chemistry of Coordination Compounds*, Reinhold, New York, 1956, p. 243.
34 G. L. Eichorn, *ibid.*, p. 740.
35 P. Hein, *Chemische Koordinationslehre*, S. Hirzel Verlag, Zurich, 1950, p. 380.
36 R. P. Linstead and A. R. Lowe, *J. Chem. Soc.*, (1934) 1022.
37 P. E. Wei, A. H. Corwin and R. Arellano, *J. Org. Chem.*, 27 (1962) 3344.
38 V. B. Evstigneev, V. A. Gavrilova and A. A. Krasnovskii, *Dokl.Akad. Nauk. SSSR.*, 70 (1950) 261.
39 J. J. Katz, G. L. Closs, F. C. Pennington, M. R. Thomas and H. H. Strain, *J. Am. Chem. Soc.*, 85 (1963) 3801.
40 G. L. Closs, J. J. Katz, F. C. Pennington, M. R. Thomas and H. H. Strain, *J. Am. Chem. Soc.*, 85 (1963) 3809.
41 A. F. H. Anderson and M. Calvin, *Arch. Biochem. Biophys.*, 107 (1964) 251.
42 J. J. Katz, *Dev. Appl. Spectrosc.*, 6 (1968) 201.
43 G. N. Cederstrand, E. Rabinowitch and Govindjee, *Biochim. Biophys. Acta*, 126 (1966) 1.
44 J. S. Brown and C. S. French, *Biophys. J.*, 1 (1961) 539.
45 J. B. Thomas, J. W. Kleinen and W. J. Arnolds, *Biochim. Biophys. Acta*, 102 (1965) 324.
46 B. Kok, *Plant Physiol.*, 34 (1959) 184.
47 H. Steffen and M. Calvin, *Biochem. Biophys. Res. Commun.*, 41 (1970) 282–286.
48 S. S. Brody and M. Brody, *Arch. Biochem. Biophys.*, 110 (1965) 583.
49 E. G. McRae and M. Kasha, *J. Chem. Phys.*, 28 (1958) 721.
50 R. S. Mulliken and W. B. Person, *Molecular Complexes*, Wiley Interscience, New York, 1969, p. 498.
51 R. Foster, *Organic Charge-Transfer Complexes*, Academic Press, New York, 1969, p. 470.
52 J. J. Katz, R. C. Dougherty and L. Boucher, in L. P. Vernon and G. R. Seely, *The Chlorophylls*, Academic Press, New York, 1966, Chapter 7, p. 186.
53 A. S. Holt and E. E. Jacobs, *Plant Physiol.*, 30 (1955) 553.
54 A. V. Karyakin and A. K. Chibisov, *Opt. Spectrosc. (USSR)*, (English translation) 13 (1962) 209.
55 K. Ballschmiter and J. J. Katz, *J. Am. Chem. Soc.*, 91 (1969) 2661.
56 L. J. Boucher, H. H. Strain and J. J. Katz, *J. Am. Chem. Soc.*, 88 (1966) 1341.
57 A. H. Siderov and A. H. Terenin, *Opt. Spectrosc. (USSR)*, 8 (1960) 254.
58 H. H. Inhoffen, J. W. Buchler and P. Jäger, *Fortsch. Chemie organ. Naturstoffe*, 26 (1968) 285.
59 R. J. Abraham, P. A. Burbidge, A. H. Jackson and T. W. Kenner, *Proc. Chem. Soc.*, (1963) 134.
60 C. Houssier and K. Sauer, *J. Am. Chem. Soc.*, 92 (1970) 779.
61 K. Ballschmiter, K. Truesdell and J. J. Katz, *Biochim. Biophys. Acta*, 184 (1969) 604.
62 S. Aronoff, *Arch. Biochem. Biophys.*, 98 (1962) 344.
63 R. Livingston, *Q. Rev.*, 14 (1960) 174.
64 J. J. Katz, H. H. Strain, D. L. Leussing and R. C. Dougherty, *J. Am. Chem. Soc.*, 90 (1968) 784.

65 C. B. Storm, A. H. Corwin, R. R. Arellano, M. Marz and R. Weintraub, *J. Am. Chem. Soc.*, 88 (1966) 2525; 92 (1970) 1423; J. R. Larry and Q. Van Winkle, *J. Phys. Chem.*, 73 (1969) 570.
66 R. J. Abraham, P. A. Burbidge, A. H. Jackson and D. B. McDonald *J. Chem. Soc. (B)*, (1966) 620.
67 B. B. Love and T. T. Bannister, *Biophys. J.*, 3 (1963) 99.
68 A. A. Krasnovskii and M. I. Bystrova, *Dokl. Akad. Nauk SSSR*, 174 (1967) 480.
69 G. Sherman and E. Fujimori, *Arch. Biochem. Biophys.*, 130 (1969) 624.
70 K. Ballschmiter, T. M. Cotton and J. J. Katz, to be published.
71 T. M. Sherman, *Nature*, 224 (1969) 1108.
72 K. Ballschmiter, T. M. Cotton, H. H. Strain and J. J. Katz, *Biochim. Biophys. Acta*, 180 (1969) 347.
73 R. Livingston and S. Weil, *Nature*, 170 (1956) 750.
74 E. E. Jacobs, A. E. Vetter and A. S. Holt, *Arch. Biochem. Biophys.*, 53 (1954) 228.
75 G. Sherman and S. Wang, *Photochem. Photobiol.*, 6 (1967) 239.
76 G. Sherman and H. Linchitz, *Nature*, 215 (1967) 511.
77 A. V. Karyakin, V. M. Kutyurin and A. K. Chebisov, *Dokl. Akad. Nauk. SSSR*, (English Translation) 140 (1961) 1321.
78 A. V. Karyakin and A. K. Chebisov, *Biofizika*, 8 (1963) 441.
79 A. N. Sidorov and A. N. Terenin, *Opt. Spectry. USSR*, (English translation) 8 (1960) 254.
80 A. F. H. Anderson and M. Calvin, *Nature*, 199 (1963) 241.
81 V. E. Kholmogorov and A. Terenin, *Naturwiss.*, 48 (1961) 158.
82 G. Sherman and E. Fujimori, *Nature*, 219 (1968) 375.
83 G. Sherman and E. Fujimori, *J. Phys. Chem.*, 72 (1968) 4345.
84 T. M. Cotton, K. Ballschmiter and J. J. Katz, *J. Chrom. Sci.*, 8 (1970) 546.
85 T. M. Cotton, R. Pedelty and J. J. Katz, to be published.
86 A. F. H. Anderson and M. Calvin, *Nature*, 194 (1962) 285.
87 F. F. Litvin and B. A. Gulyaev, *Dokl. Akad. Nauk SSSR*, 158 (1964) 460.
88 E. K. Putseiko in B. S. Neporent, *Elementary Photoprocesses in Molecules*, Consultants Bureau, New York, 1968, p. 289.
89 J. R. Norris, R. A. Uphaus, T. M. Cotton and J. J. Katz, *Biochim. Biophys. Acta*, 223 (1970) 446.
90 J. S. Brown, *Photochem. Photobiol.*, 2 (1963) 159.
91 K. Ballschmiter and J. J. Katz, *Nature*, 220 (1968) 1231.
92 R. A. Uphaus, T. M. Cotton, and J. J. Katz, *Rev. Sci. Instrum.*, 41 (1970) 1515.
93 W. L. Butler, in L. P. Vernon and G. R. Seely, *The Chlorophylls*, Academic Press, New York, 1966. Chap. 11, pp. 343–379.
94 J. Fernandez and R. L. Becker, *J. Chem. Phys.*, 31 (1959) 467.
95 J. Franck, J. L. Rosenberg and C. Weiss, Jr., in H. P. Kollmann and G. M. Spruch, *Luminescence of Organic and Inorganic Materials*, John Wiley, New York, 1962, p. 16 *et seq.*
96 G. K. Radda and G. H. Dodd, in E. J. Bowen, *Luminescence in Chemistry*, D. Van Nostrand, Princeton, N.J., 1968, p. 197–198.
97 C. A. Parker, *Photoluminescence of Solutions*, Elsevier, Amsterdam, 1968, p. 376.
98 C. S. French and L. Praeger, in H. Metzner, *Progress in Photosynthesis Research*, Vol. II, International Union of Biological Sciences, Tübinger, 1969, p. 555.
99 C. S. French *et al.*, in T. W. Goodwin, *Porphyrins and Related Substances*, Biochemical Society Symposia, 1968, p. 147.
100 T. M. Cotton, K. Ballschmiter, C. Foss and J. J. Katz, to be published.
101 J. J. Katz, K. Ballschmiter, M. Garcia-Morin, H. H. Strain, and R. A. Uphaus, *Proc. Natn. Acad. Sci. U.S.*, 60 (1968) 100.

102 M. Garcia-Morin, R. A. Uphaus, J. R. Norris and J. J. Katz, *J. Phys. Chem.*, 73 (1969) 1066.
103 B. Commoner, J. J. Heise and J. Townsend, *Proc. Natn. Acad. Sci. U.S.*, 42 (1956) 710.
104 E. C. Weaver, *A. Rev. Plant Physiol.* 19 (1968) 283.
105 D. C. Borg, J. Fajer, R. H. Felton and D. Dolphin, *Proc. Natn. Acad. Sci. U.S.* 67 (1970) 813.
106 J. D. McElroy, G. Feher and D. C. Mauzerall, *Biochim. Biophys. Acta*, 177 (1969) 180.
107 D. H. Kohl, J. Townsend, B. Commoner, H. L. Crespi, R. C. Dougherty and J. J. Katz, *Nature*, 206 (1965) 1105.
108 J. R. Bolton, R. C. Clayton and D. W. Reed, *Photochem. Photobiol.*, 9 (1969) 209.
109 J. R. Norris, R. A. Uphaus, H. L. Crespi and J. J. Katz, *Proc. Natn. Acad. Sci.*, 68 (1971) 625.

Chapter 30

CORRINOIDS*

H. A. O. HILL

Inorganic Chemistry Laboratory, South Parks Road, Oxford, Gt. Britain

INTRODUCTION

Since its isolation[1,2] in 1948, vitamin B_{12} and its derivatives have provided[3-18] a seemingly unending series of challenges and surprises. No sooner is one aspect of its multi-faceted chemistry "understood" than some new feature emerges. To the organic chemist, the initial task was to degrade the molecule in an attempt to determine its structure[3,4]. When this, in combination with invaluable X-ray diffraction studies[19-25], was successful, they accepted[17] the challenge of synthesizing it. This unfinished task has produced work remarkable for its elegance and perseverance.

The biological processes of synthesis and degradation have not remained uninvestigated[11]. One fascinating feature, at least to an inorganic chemist, emerged when a heretic corrinoid was isolated[26] which did not contain cobalt. Fortunately all coenzymes contain cobalt!

It could be said that the study of vitamin B_{12} chemistry is the quintessence of the interaction between inorganic chemistry and biochemistry: there has been much cross-fertilization but the confinement has had its difficult moments. Sometimes these have been due to a misconception of the role of the inorganic chemist but more often due to overstatements of the relevance of the properties of vitamin B_{12}, its derivatives, or other cobalt complexes, to the biochemical processes.

Not only inorganic chemists were surprised when, some ten years after the isolation of the vitamin, the light-sensitive coenzyme was isolated[27,28] from *Clostridium tetanomorphum* and shown[29] to be an organometallic derivative, having a bond between cobalt and the $C_{5'}$- of 5'-deoxyadenosine [Fig. 1 (i)]. The Co–$C_{5'}$-adenosyl derivative of the vitamin was later found[30] to be the principal corrinoid in the liver of man and other animals. Though the role of the coenzymes in enzymatic reactions has gradually been revealed, the dramatic therapeutic effects of the vitamin in the treatment of pernicious anaemia and other metabolic disorders are still far from understood[31,32].

---

* For abbreviations of "models" see Fig. 3, p. 1072.

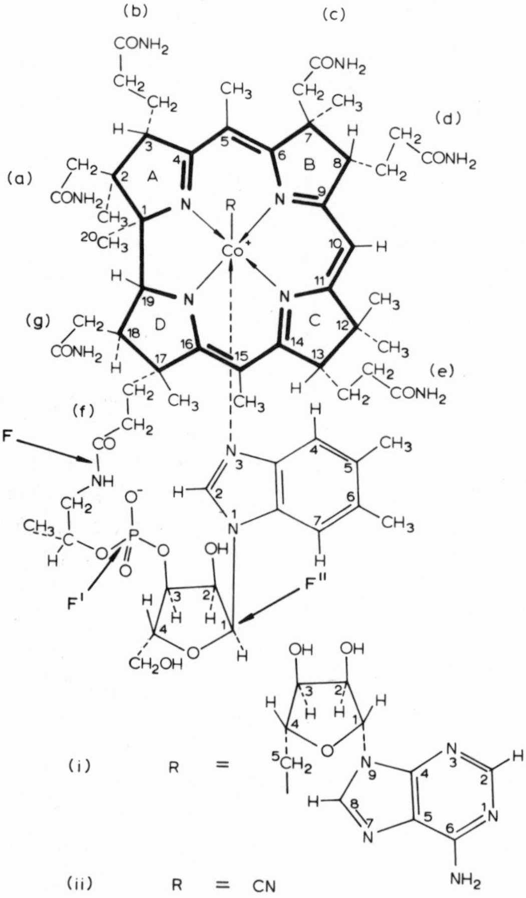

Fig. 1. Structure of (i) coenzyme B_{12}, 5′-deoxyadenosylcobalamin and (ii) vitamin B_{12}, cyanocobalamin.

NOMENCLATURE

The terminology used[33] in describing vitamin B_{12} and its derivatives is not without its idiosyncrasies. The structures of the vitamin and a coenzyme are shown in Fig. 1. The cobalt atom lies approximately in the plane of the four nitrogens of the corrin ligand (shown in bold). Compounds which contain this ligand are *corrinoids.* This macrocyclic mono-anionic ligand can be considered a reduced tetrapyrrole related to uroporphyrin III, but with the level of reduction "fixed" by the peripheral methyl groups. A striking difference from the porphyrins lies in the direct link between rings A and D. The conjugation therefore extends only over thirteen atoms, excluding the cobalt, and contains fourteen π-electrons.

The amide side-chains, which confer on the vitamin its solubility in protic and dipolar aprotic solvents, are arranged such that all the propionamide groups (*b*, *d*, *e*, and *f*) are on the same "side" of the corrin ring as the *α-5,6-dimethylbenzimidazole nucleotide* with the acetamide groups (*a*, *c*, and *g*) on the same side as the sixth and "upper" axial ligand. The hydrolysis of the amide side chains gives[4] various carboxylic acids. The hepta acid is *cobyrinic acid* and the mono-acid, cobyrinic acid *abcdeg*–hexamide, is *cobyric acid*, symbolized in Fig. 1 as fission at **F**. Hydrolysis at **F′** gives the important class of compounds, the *cobinamides*, which are amides of cobyric acid with 1-aminopropan-2-ol. Cobinamides, therefore, lack the nucleotide. Compounds lacking only the heterocyclic group, **F″** in Fig. 1, are *cobamides*, α-D-ribofuranose-3-phosphate esters of cobinamides. Nucleotides of this series can therefore be specified by the heterocyclic base; thus vitamin B_{12} is *α-(5,6-dimethylbenzimidazolyl) cobamide cyanide.* More conveniently derivatives of α-(5,6-dimethylbenzimidazolyl) cobamide are known as *cobalamins*, and the "upper" axial ligand is indicated as in hydroxocobalamin (Fig. 1, R = OH), also known as vitamin B_{12b}. Organic groups attached to the cobalt are similarly specified as in *methylcobalamin* (Fig. 1, R = CH_3) which is *α-(5,6-dimethylbenzimidazolyl)-Co-methylcobamide.* Coenzyme B_{12} may therefore be described as *5′-deoxyadenosylcobalamin* or *α-(5,6-dimethylbenzimidazolyl)-Co-5′-deoxyadenosylcobamide.* The coenzyme forms are photolabile giving a reduced cobalamin, B_{12r}, which is best described as *Co(II)cobalamin.* (The cobalamins, cobinamides, etc. are formally Co(III) complexes unless indicated otherwise.) Further reduction gives vitamin B_{12s}, *Co(I)cobalamin.* With these designations it should be possible to refer to most derivatives without ambiguity.

GENERAL CHEMICAL FEATURES

It will be helpful in providing a background to the detailed chemistry and biochemistry of the corrinoids to consider some properties[3-6,11,16] of a most useful vitamin B_{12} derivative, aquocobalamin, B_{12a}. Without prejudice, we begin at the central cobalt atom, since many reactions involve a change of ligand or oxidation state. Aquocobalamin is reduced by a wide variety of reagents including Sn(II) chloride, zinc in acetic acid, chromium(II) acetate(pH 5), various thiols, ascorbic acid, and also electrochemically and by catalytic hydrogenation to cobalt(II) cobalamin. It is further reduced by sodium borohydride, zinc in ammonium chloride, chromium acetate (pH 10), and electrochemically at –1.4 V *vs.* S.C.E. to cobalt(I) cobalamin. It takes part[34] in many equilibria, including loss of a proton to give hydroxocobalamin, pK_a = 7.8, exchange of water for a wide variety of ligands including cyanide to give the vitamin, azide, sulfite, iodide and selenocyanate. Displacement of the coordinated 5,6-dimethylbenzimidazole takes

References pp. 1122–1133

place[34] in strong acid with concomitant protonation of the base with pK_a of -2.4. This reaction to give the "base-off" form is much easier for other cobalamins, *e.g.* cyanocobalamin has a pK_a of +0.1. The 5,6-dimethylbenzimidazole can be displaced by cyanide to give dicyanocobalamin, a striking purple product, and having been displaced it can be alkylated[35].

Leaving cobalt and turning to the macrocyclic ligand, most reactions concern ring B and the C-10 position, though all the amide side-chains may be hydrolyzed, to give various derivatives of cobyrinic acid. Mild oxidation leads[36] to dehydrovitamin B_{12}, 8-amino-α-(5,6-dimethylbenzimidazolyl) cobamic acid *abdeg*-pentamide *c*-lactam, amide side-chain *c* now being attached at C-8. Treatment of cyanocobalamin with chloramine T gives[36] the corresponding lactone. The C-10 position is subject to electrophilic attack[37] and in *e.g.*, sulfito-cobalamin, substitution precedes lactone formation. Replacements by chlorine[37], bromine[37], deuterium[38] and the nitroso[37] group have been reported with the course of the reaction dependent on both axial ligands. The hydrolysis of the phosphate linkage in side chain *f* to give the important cobinamide derivatives is preferably performed[39,40] using a cerium(III) hydroxide suspension.

Returning to the cobalt atom, various ligands give derivatives which are photolabile. For example, cyanocobalamin gives[41] a cobalt(III) cobalamin but most other complexes containing a cobalt–carbon bond give cobalt(II) corrinoids. The preparation of the organo-cobalt corrinoids usually involves nucleophilic substitution or addition by cobalt(I) corrinoids and some examples[42–45] are shown in Fig. 2. The coenzyme, 5′-deoxyadenosylcobalamin, has been prepared[42,43,46] from B_{12s} (cobalt(I) cobalamin)

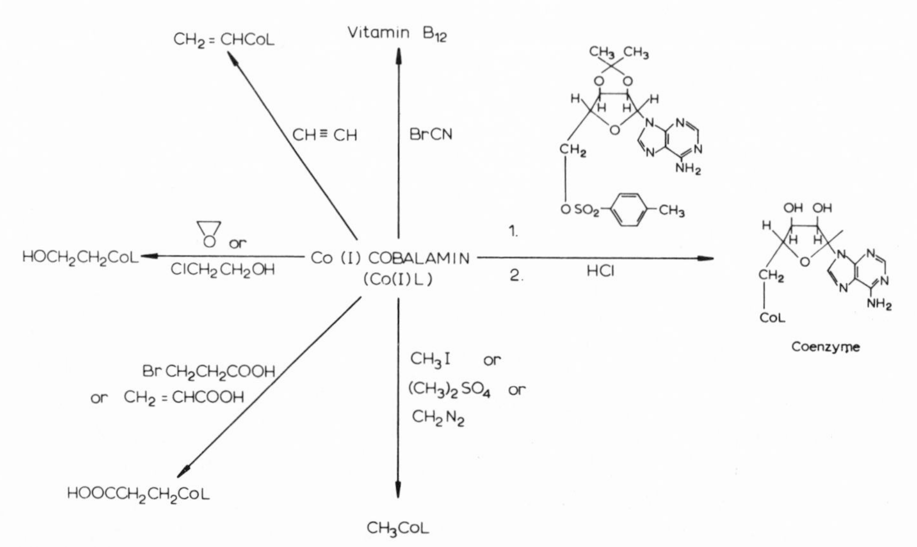

Fig. 2. Some reactions of vitamin B_{12s}, Co(I) cobalamin.

in like manner. The Co-alkyl, -alkenyl and -alkynyl derivatives are thermally quite stable in comparison with other cobalt-alkyls such as $CH_3Co(CO)_5$ or $[CH_3Co(CN)_5]^{3-}$. Co-alkyl groups have a marked effect on the properties of the corrinoids and they themselves take part in many interesting reactions, the enzymatic ones being the supreme examples.

Obviously the existence of alkyl derivatives of cobalt carbonyls and cyanides shows that the corrin ligand is not alone in "stabilizing" the cobalt–carbon bond. However, it has often been remarked that these complexes serve as poor "models" for the Co-alkylcorrinoids since their thermal stability is much lower. Though arguments based on undefined stability are fraught with danger, the characterization of coenzyme B_{12} as an organometallic derivative stimulated the search for "better models", complexes which contain a "stable" Co–carbon bond axial to a planar equatorial ligand. There has been some confusion about the role of such "models" which has probably arisen from an overstatement of their relevance. Since they will obviously not have all the properties identical to corrinoids, their role is not to duplicate the properties of the corrinoids but rather to isolate or represent particular features making the latter more amenable to experimental or theoretical investigation. The same comments apply to the study of cobalamins and cobinamides in as much as they serve as "models" for the enzyme-bound species. This is especially pertinent since there is no evidence that the coenzymes perform any biological function other than when bound to a protein. Some of the ligands which have been found to give Co–alkyl complexes are shown in Fig. 3. In addition aetioporphyrin[56,57] phthalocyanine[58] and its tetra-sulfonato derivative give cobalt–alkyl derivatives. All the ligands can provide four nitrogens, or two nitrogens and two oxygens, in a plane and they all are conjugated. Ligand VI is the only mono-anionic ligand, like corrin, and gives rise to water-soluble derivatives. The solubility of the other complexes in non-polar solvents makes it convenient to prepare the alkyl derivatives *via* the reaction of carbanions with cobalt(III) complexes as in the reaction[59] of the hepta-ethylester of cobyrinic acid with Grignard reagents.

COMPARATIVE CHEMISTRY OF COBALT(III) CORRINOIDS

In considering the chemistry of cobalt corrinoids, it is reasonable to bear in mind the questions: why the corrin ligand — why not a porphyrin, or, somewhat facetiously, dimethylgloxime? How does the corrin ligand affect the properties of the cobalt atom and the axial ligands and how, in turn, is it influenced by the axial ligands, particularly alkyl ligands? Are the features of the macrocyclic ligand concerned only in, for example, binding to the protein, or does it have structural or chemical features

References pp. 1122–1133

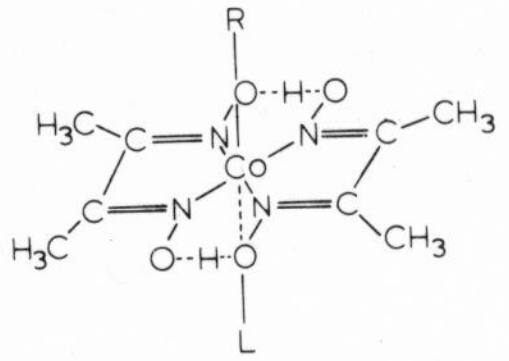

Fig. 3. Structures of some σ-bonded organocobalt complexes of the type RCo [chelate]L, where [chelate] is I, $(DMG)_2$[14] = bisdimethylglyoximato; II, SALEN[47,48] = *N,N'*-ethylenebis(salicylideneiminato); III, SALOPH[49] = *N,N'-O*-phenylenebis(salicyclideneiminato); IV, DIAPHEN[49] = *N,N'*-ethylenebis(α-methylsalicylideneiminato); V, BAE[50,51] = *N,N'*-ethylenebis(acetylacetoneiminato); VI, {(DO)(DOH)pn}[52,53] = diacetylmonoximeiminatodiacetylmonoximatoimino-propane 1,3; VII, CR[54,55] = 2,12-dimethyl-3,7,11,17-tetraazabicyclo [11.3.11]-peptadeca-1(17), 2,11,13,15-pentaene.

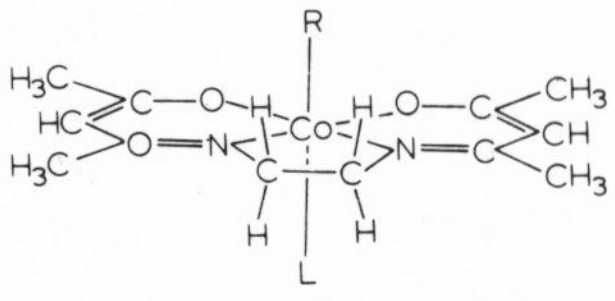

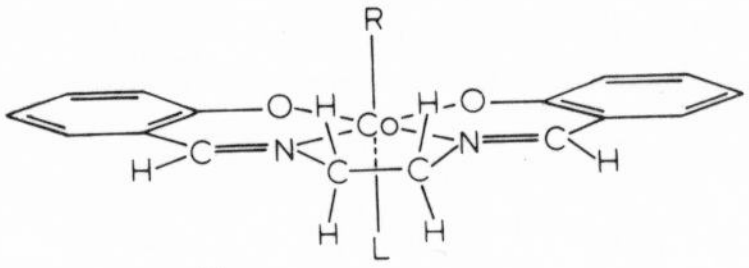

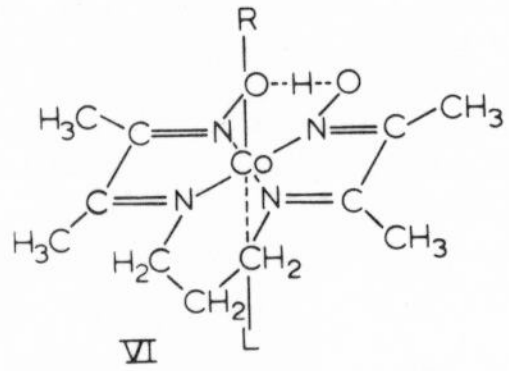

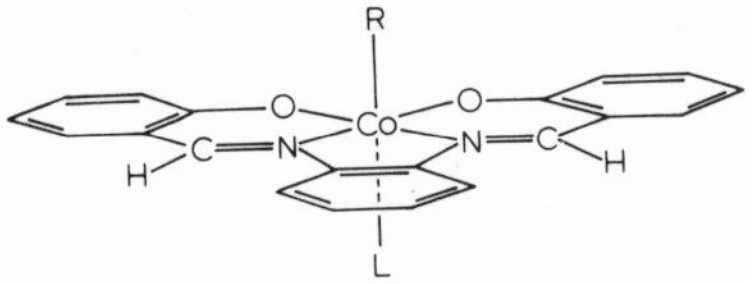

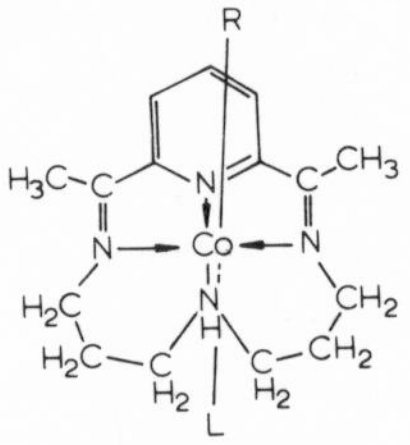

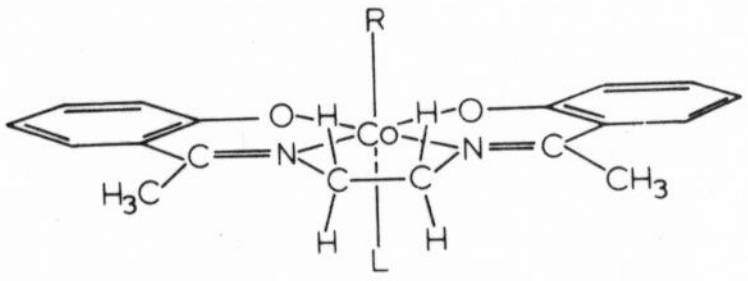

which are important, directly or indirectly, in influencing the course of the enzymatic reactions? In the previous section we dealt with what might be called the working chemistry of the corrinoids; let us now consider the corrinoids, as cobalt complexes, in greater detail.

Structural features

The structure of the corrin ring, as revealed by X-ray diffraction studies is quite similar in all derivatives studied to date. These include cyanocobalamin[19,23] in two crystal forms, the hexacarboxylic acid[22], (8-aminocobyrinic acid-*c*-lactam chloride cyanide), cobyric acid[10],

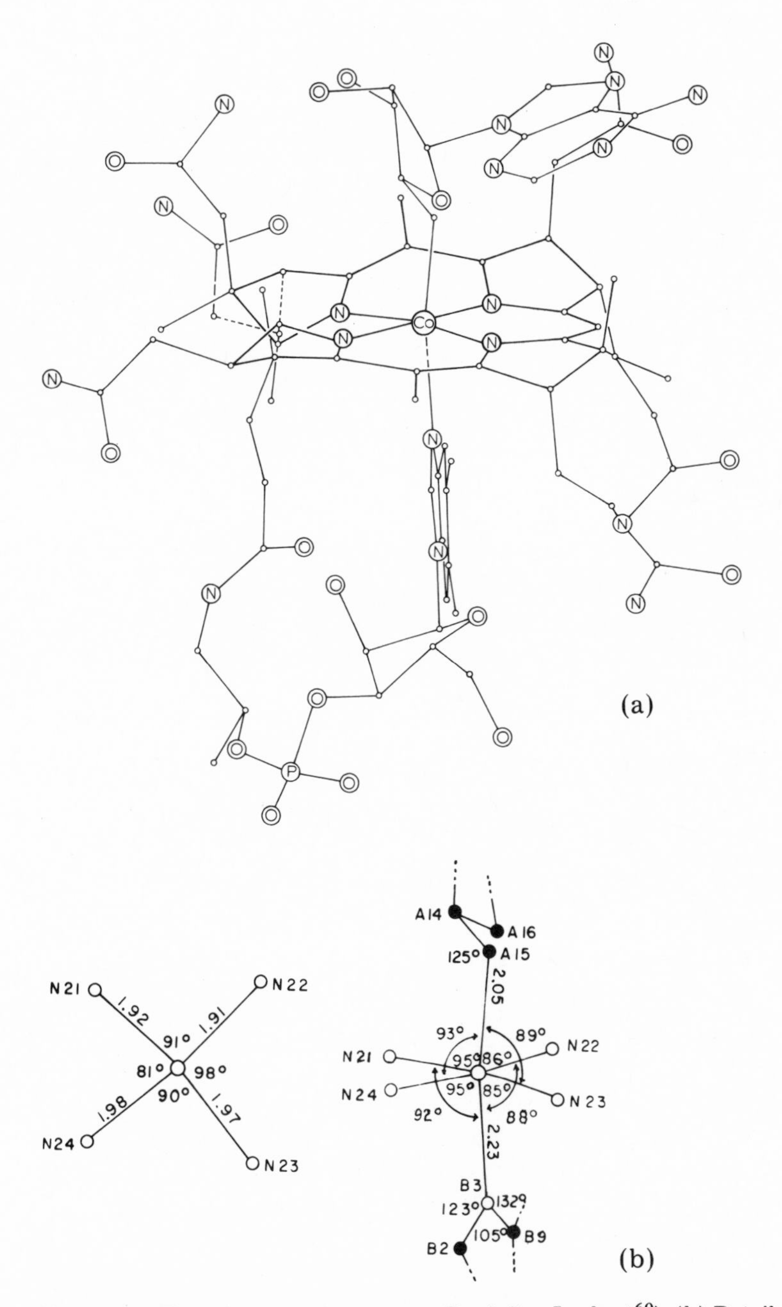

Fig. 4. (a) The structure of coenzyme B_{12} (after Lenhert[60]). (b) Details of ligand geometry around the central cobalt atom.

5′-deoxyadenosylcobalamin[29,60] (Fig. 4) and a synthetic nickel derivative[62], nickel(II),1,8,8,13,13-pentamethyl-5-cyano-*trans*-corrin, a *nirrin*. There is no evidence for any alteration in the conjugation. The main differences concern the overall planarity of the corrin and the puckering of the pyrroline rings. In cobyric acid, the side chains tend to distort the pyrroline rings from the nearly planar configuration found in the nirrin (Fig. 5) but

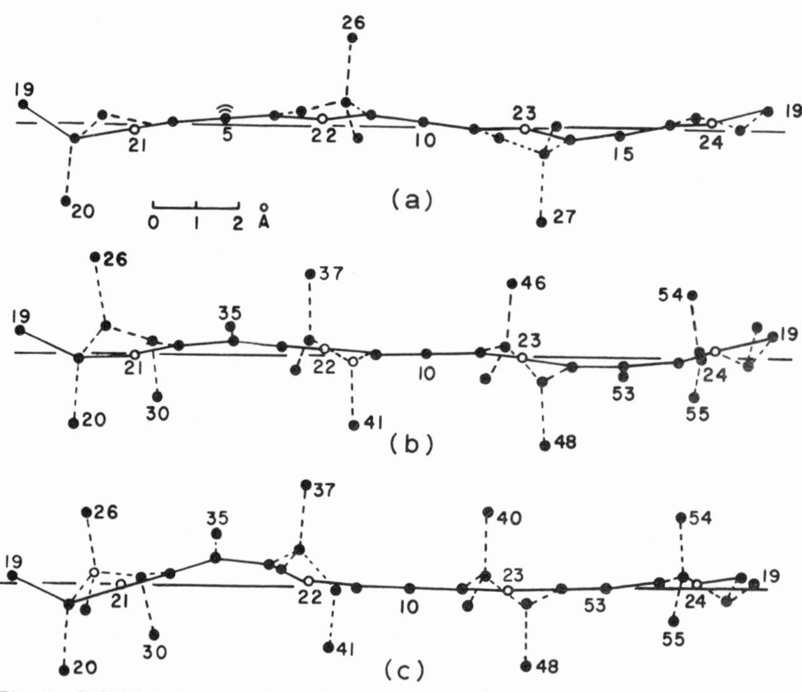

Fig. 5. Cylindrical projection of (a) nirrin (b) cobyric acid and (c) coenzyme B_{12}. The molecules are shown as they would be seen when viewed from the metal atom outwards. The atoms are projected radially on to a cylinder of 2.8 Å radius. The vertical displacement of each atom corresponds to the distance from the least squares plane of the four nitrogen atoms which are shown as open circles. (After Lenhert[60].)

the relationship of the atoms of the conjugated part of the macrocycle to the plane of the four nitrogens is closely similar, being slightly bow-shape. They differ from the cobalamins where the contact between C-5 and C-6 of the corrin nucleus and the 5,6-dimethylbenzimidazole C-4 hydrogen bends the macrocycle such that there is an angle of about 15° between the plane through N21, C4, C5, C6, N22, C9, and C10 and that through C10, C11, N23, C14, C16 and N24 (see Figs.1 and 4 for numbering). This ultimately affects the position of the 5′-deoxyadenosine in the structure of the coenzyme (Fig. 4) by making C26 and C37 better "stops" for rotation of the nucleotide about the Co–C bond. The representation of the structure in

Fig. 4 does not do justice to the compactness of the molecule. There is very little "room", most atoms being "in contact". The adenine "sits over" ring C and A7, A8, and A9 are close to the hydrogens of C46. As noted[63] previously, the Co–$C_{5'}$ bond is surrounded at about 6.5 Å by the two CH_2 groups of C26 and C37 and by the two methyl groups (C46 and 54) forming a square nearly 2 Å above the cobalt and perhaps designed to trap non-polar ligands and protect the Co–C bond from the approach of reagents. The Co–C–C bond angle is 125°, far removed from the normal tetrahedral angle. The important $C_{5'}$-hydrogens must be inequivalent and these and

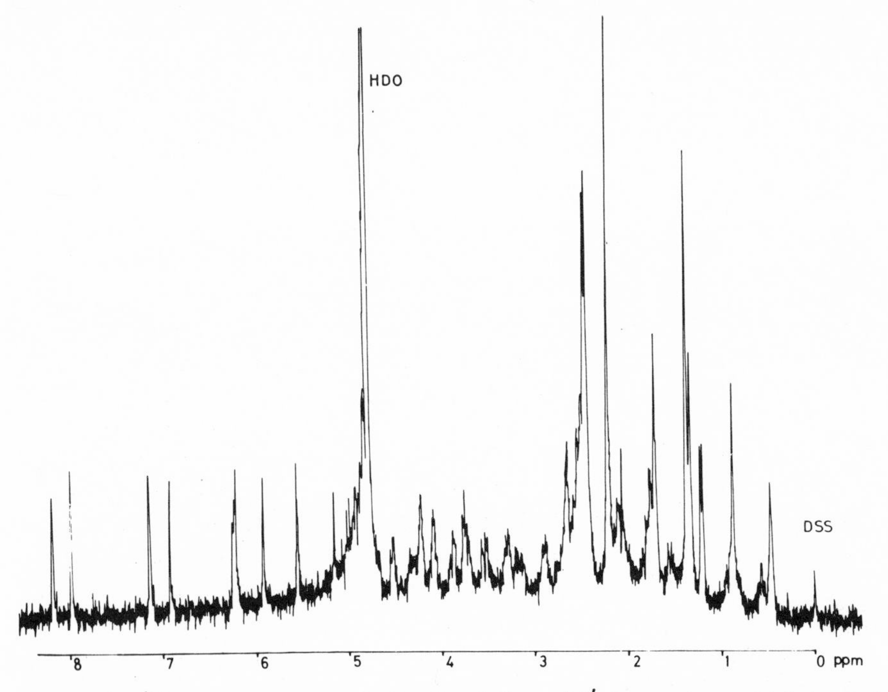

Fig. 6. The ^1H n.m.r. spectrum of a 7 m*M* solution of 5′-deoxyadenosylcobalamin in D_2O at 270 MHz.

other non-exchangeable protons give rise to resonances in the 270 MHz ^1H n.m.r. spectrum which has been assigned[64] (Fig. 6). The proximity of the 5,6-dimethylbenzimidazole to C-20 and other peripheral groups in cobalamins accounts for several high field resonances and this has proved useful in investigating base-on, base-off equilibria.

The cobalt is octahedrally coordinated in all these complexes though there is some distortion as shown in Fig. 4 for the coenzyme, the angle between N21–Co–N24 being caused by the direct link between rings A and D. The bond distances to the ligand atoms in these and some other cobalt complexes are given in Table I. Since the standard deviations of these values differ from one structure determination to another and some distances are determined by intermolecular interactions, one should be

TABLE I

COBALT–LIGAND BOND LENGTHS (Å)[a]

Cobalt complex	Planar ligands	Axial ligand atoms
Hexacarboxylic acid[22]	1.88(N)	2.41(Cl); 1.96(C)
Cobyric acid[10]	1.87(N)	2.02(O); 1.84(C)
Cyanocobalamin (wet)[19]	1.86(N)	1.97(N); 1.92(C)
Cyanocobalamin (dry)[23]	1.90(N)	2.07(N); 2.02(C)
5′-Deoxyadenosylcobalamin[29,60]	1.94(N)	2.23(N); 2.05(C)
$CH_3Co(BAE)$[65]	1.87(N); 1.87(O)	— ; 1.95(C)
$C_6H_5Co(BAE)H_2O$[66]	1.89(N); 1.91(O)	2.33(O); 1.93(C)
$CH_2{=}CHCo(BAE)H_2O$[67]		2.20(O); 1.89(C)
$CH_3OCOCH_2Co(DMG)_2$pyridine[68]	1.87(N)	2.04(N); 2.04(C)
$[CH_3Co\{(DO)(DOH)pn\}H_2O]ClO_4$[69]	1.90(N)	2.14(O); 1.99(C)
Co(II)(BAE)(C_6H_6)[70]	1.87(N); 1.85(O)	—
Co(II)SALEN[71]	1.89(N); 1.95(O)	—

[a] The bond lengths are mean values when there is more than one ligand atom of the same type; no estimate is given of the standard deviation.

careful in their interpretation. However the increase in the average cobalt–nitrogen (corrin) bond length in the coenzyme as compared with the vitamin may be significant but the increase in the cobalt–nitrogen (5,6-dimethylbenzimidazole) surely is. The bond lengths in the model complexes, even in the five-coordinate $CH_3Co(BAE)$[65], are quite similar, and most interestingly, differ but little from those in the analogous cobalt(II) complexes. The "planar" ligand in most of the complexes shows a slight distortion, due probably, as in the cobalamins, to the steric effects of the non-carbon ligand. In the five-coordinate complex the equatorial ligand is planar though the cobalt sits slightly out of the plane.

Some properties associated with cobalt and axial ligands

How do the properties of cobalt corrinoids compare with those of "typical" cobalt(III) complexes, inasmuch as the latter exist? Are there effects of the corrin ligand, or the alkyl ligands in particular, which are related to their role in enzymatic reactions? We will consider the Co-alkyl corrinoids as derivatives of cobalt(III) with a carbanionic ligand. It should be remembered that such a description says nothing about the charge distribution in the cobalt–carbon bond.

Both the vitamin and the coenzyme can therefore be described as low-spin, diamagnetic, d^6 cobalt(III) complexes. The energies of the orbitals containing the d-electrons will depend on the "strength" and symmetry of the ligand field. There have been[55,72–78] several more-or-less theoretical treatments of the corrinoids and similar complexes and though it is difficult to include the metal atom and axial ligands in the calculations with any rigor, it is obvious that the effects of changing one ligand will affect both the cobalt atom and the other ligands. The stability of the cobalt–carbon bond in Co–alkyl complexes has been discussed[78] in terms of the participation of the $4p_z$ orbital in bonding to the carbon-ligand. However the strong ligand field provided by the corrin ligand and "model" ligands may contribute to the kinetic stability of the cobalt–carbon bond; the availability of π*-orbitals in the ligand may "accommodate" the extra charge placed on the cobalt or the properties of the cobalt may be such that the cobalt–carbon bonds are only slightly polarized. All these points, inasmuch as they can be separated, are difficult to quantify. Rather let us compare the properties of the corrinoids with those of other cobalt complexes beginning with the effect of the ligands on those properties mainly associated with the cobalt atom.

TABLE II
SUBSTITUTION OF H_2O IN AQUOCOBALAMIN

Ligand	log $K(M^{-1})$	k_1 $(M^{-1}\ sec^{-1})$[f]	k_{-1} (sec^{-1})[f]	λ(nm)
CN^- [C][a]	> 12	1.3×10^3	10^{-9}	306.5
SO_3^{2-} [S][b]	7.3			364
$imid^-$ [N][c]	7.8 (1.3)			
OH^- [O][a]	6.2			357
histidine [N][c]	5.8 (3.6)			
cysteine [S][d]	6.0 (4.8)			371
N_3^- [N][b]	4.9	1.7×10^3	3×10^{-2}	358
CH_3NC [C][b]	4.8			360
imid [N][c]	4.6 (4.2)	27	6×10^{-4}	358
$S_2O_3^{2-}$ [S][e]	3.9			367
$NCSe^-$ [Se][e]	3.9			371
NCS^- [S][e]	3.1	7.1×10^3	1.8	357
NCO^- [N][e]	2.7	7.3×10^2	0.95	357
$C_6H_5O^-$ [O][d]	2.9 (-0.1)			355
I^- [I][e]	1.5			371
$CH_3CO_2^-$ [O][b]	0.7			352
Br^- [Br][e]	0.3			353
Cl^- [Cl][e]	0.1			352

[a] Ref. 34; [b] ref. 79; [c] ref. 80; [d] ref. 81; [e] ref. 82; [f] ref. 83.

The most common reaction of cobalamins is ligand replacement. Coordinated water in aquocobalamin is replaced rather quickly by other ligands, a fact which distinguishes this cobalt(III) complex from many others. It is difficult to find exactly analogous reactions but compare the data given in Table II with those for the reaction[84]:

$$Co(DMG)_2NO_2H_2O \xrightarrow{X} Co(DMG)_2NO_2X + H_2O$$

X	N_3^-	NCS^-	Cl^-	Br^-	HSO_3^-
$k(M^{-1}\ sec^{-1})(25°C)$	5.7×10^{-4}	5.8×10^{-4}	8×10^{-5}	1.6×10^{-4}	8.5×10^{-3}

or the rate of hydrolysis[83] of azidocobalamin (3×10^{-2} sec^{-1}) with the hydrolysis[85] of [*trans*-$Co(NH_3)_5N_3]^{2+}$, 2.1×10^{-9} sec^{-1} or[86] $[Co(CN)_5N_3]^{3-}$, 5.5×10^{-7} sec^{-7} sec^{-1}. Though the charge on the complex will affect the rate of replacement it has been concluded that the rates of analogous reactions in cobalt complexes probably increase in the order: $Co(DMG)_2 < Co(NH_3)_4 \ll$ Co(corrin) < Co(porphyrin).

The formation constants given in Table II are higher for small yet polarizable anions. The high formation constants with carbon ligands, CN^- and presumably alkyls, and sulfur ligands are notable since they are the ligands which have been implicated in the enzymatic reactions. The cobalt(III) in the corrinoids can be described[87] as a class "b" ion[88] (Chapter 2). Compare the formation constants for the halide complexes:

$\log_{10}K$	F^-	Cl^-	Br^-	I^-
[85] $[Co(CN)_5H_2O]^{2-}$	—	0	0	1.6
[82] Aquocobalamin	0	0.1	0.3	1.5
[89] $[Co(NH_3)_5H_2O]^{3+}$	—	0.1	-0.3	-0.7

which emphasize the lack of "typical cobalt(III)" properties.

A further difference lies in the reduction properties of cobalt(III) corrinoids and the model complexes compared to those of *e.g.*, cobalt(III) ammines. Whilst the latter are reduced[90] first to cobalt(II) complexes and then to the metal, the former are[51,91-98] reduced either in two one-electron steps; Co(III) → Co(II) → Co(I), or Co(III) → Co(I) with further reduction at very low potentials[51,91-98]. The half-wave potentials are given in Table III, and the ease of reduction is probably {(DO)(DOH)pn} > (DMG) > corrin > SALEN > BAE.

TABLE III

COMPARISON OF SOME PROPERTIES OF COBALT COMPLEXES

Ligand	^{59}Co n.m.r.[a]	$E_{\frac{1}{2}}$ Co(III) → Co(II)[b]	$E_{\frac{1}{2}}$ Co(II) → Co(I)[b]	Co–CH_3 n.m.r.[c]
BAE	-7.2	-0.61	-1.83	7.43
SALEN	-7.2	-0.39	-1.58	7.88
DMG	-3.6	-0.34	-1.20	9.18
{(DO)(DOH)pn}	-4.2			9.50
Corrin	-4.1			10.14

[a] Chemical shift (ppm $\times 10^{-3}$) relative to $K_3Co(CN)_6$ for Co-methyl derivatives except vinylcobalamin.
[b] In pyridine/THF *vs.* S.C.E.
[c] τ-values of Co-methyl derivatives with pyridine as sixth ligand except corrin (benzimidazole).

The ^{59}Co n.m.r. spectra[100] provide perhaps the most direct method of investigating the effect of the ligands on those properties associated with the cobalt. The spectra are broad, and only vinyl cobalamin has been examined to date. The ligand field strength of the planar ligands increases; BAE < SALEN < DMG ~ {(DO)(DOH)}pn ~ corrin, and methyl ligand has a much greater ligand field strength than iodide, which is not a surprise[55]. The many possible contributions to the proton chemical shift make it difficult to interpret that of coordinated methyl[101] but it is possible that the second-order Zeeman paramagnetism determines both ^{59}Co and ^{1}H chemical shifts. A large difference has been observed[102] between the ^{19}F chemical shift in trifluoromethylcobalamin and $CF_3Co(DMG)_2py$ which is surprising since other properties[14] suggest that they should be reasonably similar.

Cis- *and* trans-*effects*

Though further effects of the planar ligand on the properties of the cobalt will be discussed later when the properties of the Co(II) and Co(I) corrinoids are considered, we will turn now to the effect of the axial ligands on the rest of the molecule. It will be convenient to describe these as *cis*- or *trans*-effects, where the former refers to the effect of the axial ligand on the "planar" ligand or *vice versa* and the latter refers to the effect of one axial ligand on the other. These can be further considered in terms of the effects on thermodynamic processes, kinetics, or ground-state and excited-state effects, where the electronic state is referred to.

The best documented effect is the thermodynamic *trans-effect*[79,102-107] It has long been known that pK_a of 5,6-dimethylbenzimidazole in the cobalamins depends on the *trans-ligand* and early work[105,106] pointed out the much increased pK_a in the 5'-deoxyadenosylcobalamin. A more extensive series of alkyl ligands was reported[107] and a correlation was observed between the pK_a and the Hammett σ-constant for the substituent in a series of substituted ethyl and propyl ligands. The alkyl ligands are compared with others in four common reactions shown in Fig. 7; the results are plotted in Fig. 8. (The log K values of reaction are derived from the more commonly reported pK_a values by correcting for the protonation of the 5,6-dimethylbenzimidazole.) It will be seen that the alkyl ligands exert a marked *trans*-effect *irrespective* of the nature of the other axial-ligand. This means that not only are the formation constants low but there is little preference for any type of *trans*-ligand atom investigated to date, though it would be interesting to measure formation constants with sulfur containing ligands. This lack of enthusiasm for the *trans*-ligand is displayed most dramatically in the complexes RCoBAE[51], where a five-coordinate complex has been isolated and characterized[65]. With this ligand the five-coordinate complexes are easily isolated, and in the model complexes,

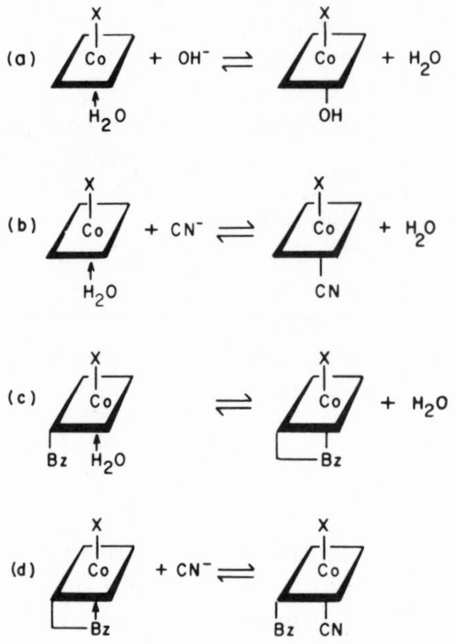

(a)
X
Co
H_2O
+ OH^- ⇌
X
Co
OH
+ H_2O
(b)
X
Co
H_2O
+ CN^- ⇌
X
Co
CN
+ H_2O
(c)
X
Co
Bz
H_2O
⇌
X
Co
Bz
+ H_2O
(d)
X
Co
Bz
+ CN^- ⇌
X
Co
Bz
CN

Fig. 7. Some reactions of cobinamides [(a) and (b)] and cobalamins [(c) and (d)].

CN^-
C_2H^-
$C_2H_3^-$
CH_3^-
$C_2H_5^-$
(b)
(c)
(d)
(a)
12
8
4
0
$Log_{10}K$
Coenzyme
Free CN^-
2130
2110
2090
Stretching frequency X-Co-CN (cm^{-1})

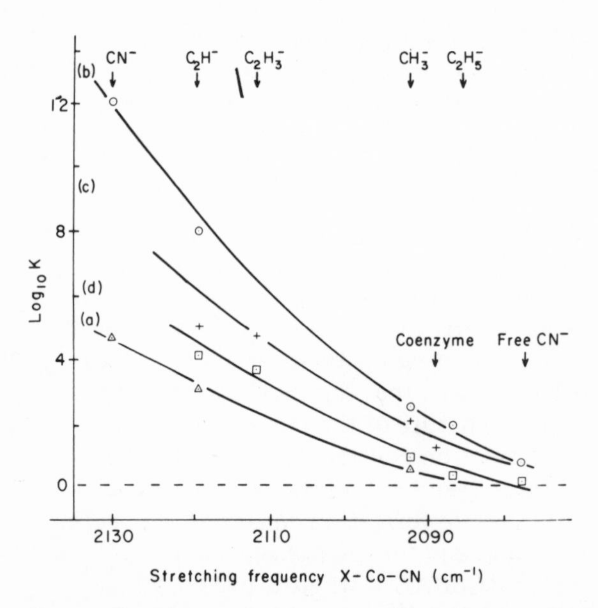

Fig. 8. Equilibrium constants for the reactions shown in Fig. 7 plotted against the cyanide stretching frequency in those complexes formed in reaction (b). (From Hill *et al.*[16].)

the ease of formation of five-coordinate complexes is: BAE > SALEN > DMG > {(DO)(DOH)pn}. Where does corrin fall in this series? The temperature-dependence[79,113] of the ^{1}H n.m.r. spectra of alkylcobinamides led to the suggestion that the equilibria involved the loss of water to give a five-coordinate species. No such equilibria were observed in cyanocobinamide or isopropylcobinamide, and it was proposed that these were the two end members of a series of increasing ease of formation of the five coordinate species; CN < CH_2=CH_2=CH^- < CH_3^- < $CH_3CH_2^-$ < 5′-deoxyadenosyl < $(CH_3)_2CH^-$ which is the same order as the *trans*-effect. The influence of the *trans*-ligand has been observed[109] in the equilibrium between the five and six-coordinate species in *p*-XC_6H_4Co(BAE) + L ⇌ *p*-XC_6H_4Co(BAE)L where the formation constants increase with increasing electron withdrawing power of X; in other words, as the cobalt becomes a better Lewis acid. Though the presence of five-coordinate species in solutions of alkyl cobinamides is not firmly established (it is notable that the equilibria are solvent dependent), the thermodynamic parameters derived from the temperature dependence of the spectra of *e.g.*, methylcobinamide; $\Delta H = 4.4 \pm 2.0$ kcal/mole; $\Delta S = 16 \pm 7$ e.u. support the loss of one weakly bound water. This could be lost, from[61] *e.g.*, a hydrogen bond between acetamide side chain, *g*, and the amino-propanol but it is difficult to see how this could affect the spectroscopic properties or, in turn, be affected by the change in axial ligand. Conformational changes in the corrin ring are possible but difficult to substantiate.

Similar temperature dependent spectra have been noted in alkylcobalamins including 5′-deoxyadenosylcobalamin and it was concluded that the latter in solution at pH 7, was 10% in the five coordinate form, *i.e.*, in which the 5,6-dimethylbenzimidazole is not coordinated and not replaced by solvent. Detailed ^{1}H n.m.r. studies[113] throw some doubt on this. Even so, it is safe to conclude that the 5,6-dimethylbenzimidazole is easily replaced, no ligand being strongly bound. Most importantly, the *rate* of replacement is fast, perhaps due to the low energy of intermediate five-coordinate species, if, as presumed, the reaction proceeds via a SN1 mechanism. The substitution reactions of $[RCo\{(DO)(DOH)pn\}H_2O]^+$ have been shown[114] to take place *via* a *lim* SN1 process.

$$[R(Co\{(DO)(DOH)pn\}H_2O]^+ \underset{k_2}{\overset{k_1}{\rightleftharpoons}} [R(CO\{(DOH)(DO)pn\}]^+ + H_2O$$

$$[RCo\{(DOH)(DOH)pn\}]^+ + \text{imidazole} \underset{k_4}{\overset{k_3}{\rightleftharpoons}} [RCo\{(DOH)(DOH)pn\}\text{imidazole}]^+$$

R	C_6H_5	CH_3^-	$C_6H_5CH_2^-$	$CH_3CH_2^-$	$CH_3CH_2CH_2^-$
k_1 sec^{-1}	3.6 ± 0.3	24.8 ± 5.6	226 ± 28	251 ± 23	620 ± 139
k_{obs} M^{-1} sec^{-1}	36	136	1900	1930	1720

The reactions are fast, show a marked *trans*-effect, and the cobalt shows a greater affinity for imidazole (and benzimidazole) than for water. The exchange rates for L in $CH_3Co(DMG)_2L$ decrease[115] in the following order: L = $CH_3CN > (CH_3)SO > (CH_3)_2S > P(OCH_3)_3 > P(C_6H_5)_3$ suggesting that, at least in the rate of dissociation, π-bonding effects may be important.

In the Co-alkylcobinamides and -cobalamins, the ligand *trans* to the alkyl group is easily removed, no matter which ligand atom is involved. This may be correlated with an increased bond length to the *trans*-ligand. Are there any other properties which reflect this marked electron donor characteristic of alkyl ligand? This should be mostly readily interpretable in (ground-state) properties and indeed, in a series of *m*- and *p*-$FC_6H_4Co\{(DO)(DOH)pn\}X$ complexes, it was found[116] that the chemical shift showed a similar dependence on the *trans*-ligand X as in a series[117] of $Pt(PR_3)RX$ complexes, reflecting both the σ and π-donor properties of the group X. In the corrinoids, the most direct evidence concerns[103] the C≡N stretching frequency in a series of cyanocobinamides. This depends markedly on the *trans*-ligand and correlates reasonably well for the equilibrium constants of the reactions. (Fig. 8). As the electron donating properties of the *trans*-ligand increase, the cyanide becomes more weakly bound and its stretching frequency approaches that of free cyanide. Obviously the cobalt atom relays charge to the axial ligand. Does it do so to the *cis*-ligand also? The chemical shift[38] of the C-10H, assuming it reflects changes in electron density[76,78] might act as a probe for this effect and indeed the chemical shift does correlate[101] with the CN stretching frequency. In a series of complexes, $XCo(DMG)_2P(C_6H_5)_3$ the chemical shift of the DMG methyl hydrogens was found[118] to correlate with the Hammett σ-constant of the ligand X. All these effects point to a common denominator which is probably the transmission of electron density *via* the σ-framework, and the π-bonding orbitals. It is no surprise therefore to find that the electronic absorption spectra of the corrinoids are also sensitive to the metal[17,119,120] and to the axial ligand, the excited state *cis*-effect. Since absorption spectroscopy is such a useful method of investigating corrinoids, let us examine the origin in a little more detail.

Electronic absorption spectra

The π-electron conjugation of the corrin extends over a bent chain of thirteen atoms, which carries a charge of −1, and the π-orbitals contain fourteen electrons. The spectrum of the metal-free corrin[26,119] so resembles that of metal derivatives, that there is no reason to suppose that any of the bands are due to *d–d transitions* but are rather π–π^* transitions modified both by the central atom and by the axial ligands[120] on the cobalt (Figs. 9 and 10).

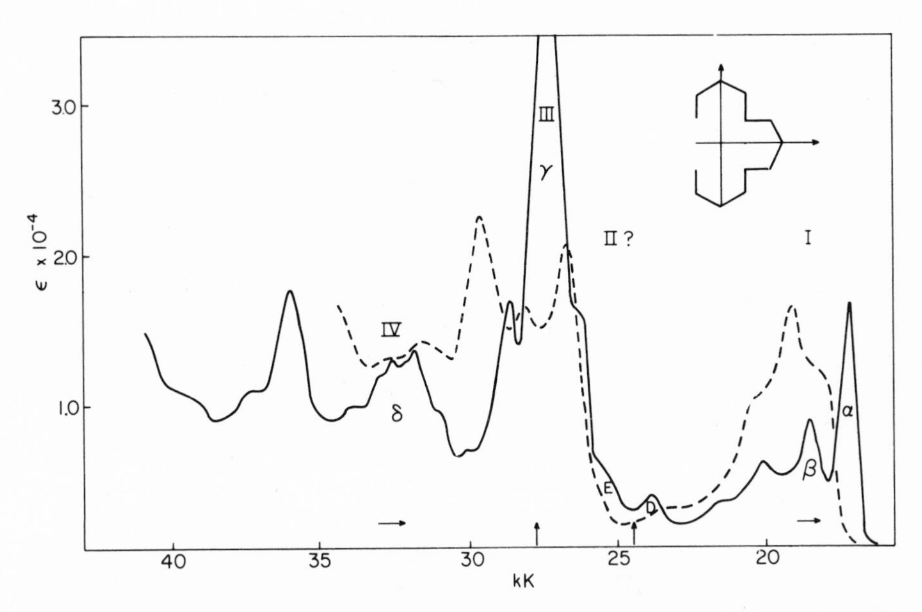

Fig. 9. The absorption spectra of dicyanocobinamide (—) and methylcobinamide (- - -) in ethanol at −180° C. The symbols, which are discussed in the text, refer only to the former spectrum.

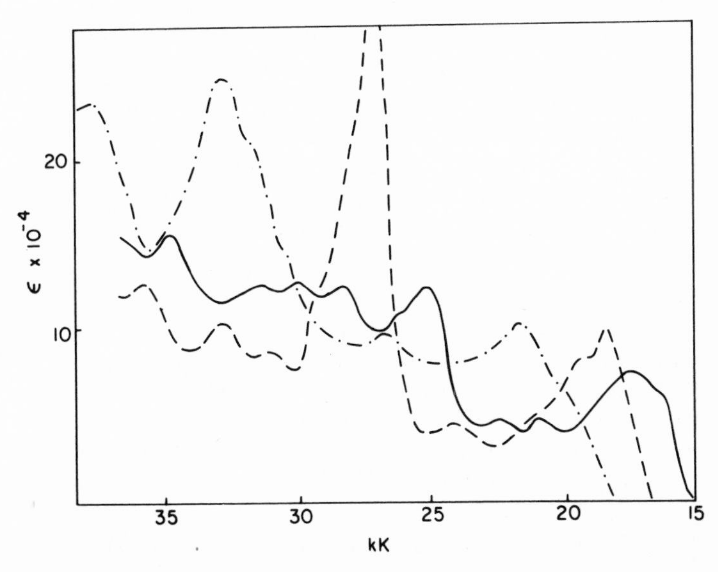

Fig. 10. The absorption spectra of cyanocobalamin (- - -), Co-methylcobinamide cyanide (—), and methylcobinamide (· · · ·) at room temperature.

References pp. 1122–1133

The α- and β-bands are vibrational components, 0–0 and 0–1 respectively of a transition from, in one electron terms, the highest-filled molecular orbital ψ_7 to the lowest unfilled ψ_8 of the corrin. The vibrational interval is related to the "aromatic" stretching frequency. Fluorescence studies[121] on the metal-free corrinoids show that the band is polarized as shown, which also follows from the theoretical treatments. The energy of the first transition decreases with increasing electron donation from the axial ligands in cobalt(III) corrinoids. The observed correlation[101] between the chemical shift of the C10-H and the energy of the β-band (the most readily identified band in most spectra) is expected inasmuch as chemical shift measures changes in electron density, since the lowest unfilled orbital has a node at C-10 and hence the transition also reflects changes in electron density. It is not surprising that substitution at C-10 has[37] a very marked effect on the energy of the α-band.

The γ-band region of the spectra of alkyl corrinoids and *e.g.*, cyanocobalamin are so dissimilar (Figs. 9 and 10) that at first it was thought that the chromophore was different. However, the spectrum of *e.g.*, ethynylcobalamin is "intermediate" in character and it was realized[82,122] that the γ-band region was very sensitive to the axial ligands and even more "anomalous" spectra[120] could be obtained *e.g.*, methylcobinamide or cobalt(II) cobalamin.

Theoretical treatments show[74] that indeed this region should be complex since, in the one electron model, assuming a symmetry of C_{2v} for the chromophore, transitions $\psi 7 \rightarrow \psi 9$ and $\psi 6 \rightarrow \psi 8$ are of the same symmetry and a model including electron interaction must be used. The degree of interaction between the excited states ${}^1\psi_7{}^9$ and ${}^1\psi_6{}^8$, to give transitions II and III, will depend very much on the charge on the nitrogens which presumably reflects the charge on the cobalt. As more charge is placed on the nitrogens the interaction between the ${}^1\psi_7{}^9$ and ${}^1\psi_6{}^8$ configurations will decrease until in its absence one might expect to have two bands of equal intensity. Perhaps this situation is present in methylcobalamin, in contrast to *e.g.*, cyanocobalamin or aquocobalamin in which we have one intense band, presumably the in phase linear combination of ${}^1\psi_7{}^9$ and ${}^1\psi_6{}^8$, III and one weak band, maybe bands D and E, II in Fig. 9. The δ band probably corresponds mainly to ${}^1\psi_7{}^9$, IV. All the bands show some dependence on the axial ligands and it is interesting that this dependence follows that of the CN stretching frequency, both presumably reflecting charge on cobalt. Other theoretical treatments[76,77] confirm the complexity of the transitions in this region, though there is still no generally accepted assignment of the bands, particularly D and E, and their relation to the γ-band.

The magnetic circular dichroism (c.d.) spectra[123] of a series of corrinoids, synthetic and natural, goes some way to aiding the assignments and relating the "anomalous" spectra of the nickel complexes to those of the other metal derivatives. This applies particularly to bands D and E which have both a negative magnetic c.d. in the synthetic corrinoids and a marked optical activity[120] in the natural corrinoids. These two techniques provide a challenge for the theoretician and will probably be useful in investigating binding to the apoenzyme, since often they are sensitive[120] to changes in the complexes which result in only minor effects on the absorption spectra. The temperature dependence[120] of the circular dichroism in the complexes may reflect various conformational equilibria.

Summarizing, one might say that the cobalt(III) corrinoids show high formation constants for polarizable ligands such as RS^- and R^-, that ligands *trans* to these are weakly bound and rapidly exchanged, that electron density changes are transmitted from one part of the molecule to another *via* the cobalt, and that there is some evidence that the corrin ring can accommodate both a high electron density on the cobalt and have considerable "flexibility".

Cobalt(II) Corrinoids

The additional electron in cobalt(II) corrinoids has an effect on the properties of cobalamins, not unlike that of a coordinated alkyl group[101,122]. Thus the pK_a for the removal and concomitant protonation of the 5,6-dimethylbenzimidazole to give the "base-off" form of Co(II)cobalamin, is ~2.5, identical to that of methylcobalamin. The absorption spectra of both the base-on and base-off forms are quite like that of the "high-temperature" forms of Co-alkyl-cobinamides with the important difference that the Co(II) corrinoids have additional absorption beyond 600 nm due possibly to *d–d* or π–π^* singlet–triplet transitions[113]. The base-off form and Co(II)cobinamide show[113] a similar temperature-dependence of the absorption spectra as the alkylcobinamides suggesting that they, too, take part in an equilibrium between five and six-coordinate forms. The Co(II) cobalamin may[122,124,125] also be five-coordinate having the nucleotide as the sole axial ligand. Other cobalt(II) complexes have similar five-coordinate forms[126] *e.g.*, $Co(CN)_5^{3-}$, though the formation of dimeric complexes is possible in some instances as in $[Co(II)(DMG)_2py]_2$[127]. Such dimeric complexes probably facilitate the disproportionation of Co(II) complexes to the Co(I) and Co(III) forms: $2Co(II)L \rightleftharpoons Co(I)L + Co(III)L$. For corrinoid complexes, the equilibrium probably lies well to the left but the possibility of disproportionation makes it difficult to be sure that reactions observed with the Co(II) complexes do not, in fact, involve the more reactive species.

All the ligands which give rise to Co-alkyl derivatives form low-spin cobalt(II) complexes. The single unpaired electron occupies an orbital well-described as d_{z^2} with some admixture of metal 4*s*. This description is compatible with the observed[125,129-133] electron paramagnetic resonance (e.p.r.) spectra which are[134-137] typical of low-spin d^7 complexes with significant tetragonal distortion. Though the spectra are observable at

References pp. 1122–1133

room temperature, more information can be derived from the powder or frozen solution spectra (Fig. 11), and the spin-Hamiltonian parameters extracted with reasonable accuracy. All complexes have $g_{\parallel} \sim 2.00$, the free-electron value; $g_{\perp}$ is > 2.0, but because of second-order effects this region of the spectrum is more difficult to interpret. The interaction of the unpaired electron with the cobalt nucleus ($I = 7/2$) gives rise to well-resolved hyperfine lines around $g_{\parallel}$ with $A_{\parallel Co} = 100 \times 10^{-4}$ cm^{-1} (Table IV). There is much less resolution in the $g_{\perp}$ region, and the derived values

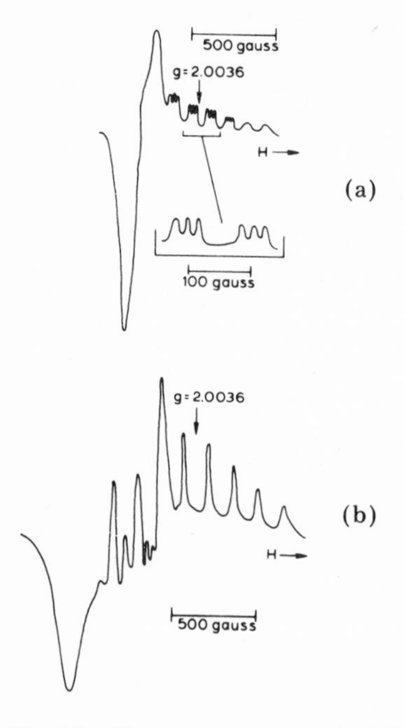

Fig. 11. The e.p.r. spectra of (a) Co(II) cobalamin, "base-on" and (b) Co(II) cobalamin "base-off", at X-band and 100K.

of $A_{\perp Co}$ are more uncertain. Additional superhyperfine structure is well-resolved on the $A_{\parallel Co}$ lines in all complexes containing a coordinated nucleotide. The resolution of the spectra is quite markedly affected by the medium, being best of all in[138] the enzyme where presumably the cobalt(II) corrinoid molecules have identical environments. Though the g values are relatively insensitive to the axial ligand, both $A_{\parallel Co}$ and $A_{\parallel N}$ depend[137] on the basicity of nitrogen ligands, though as $A_{\parallel Co}$ decreases, $A_{\parallel N}$ increases (Fig. 12). The absence of the nucleotide, as in Co(II) cobinamide, has a marked effect on the spectrum (Fig. 11), the superhyperfine splitting disappears, of course, and the $A_{\parallel Co}$ increases. The e.p.r. spectra of "model"

TABLE IV

ELECTRON PARAMAGNETIC RESONANCE SPECTRAL DATA FOR SOME CORRINOIDS

Corrinoid	$g_{\parallel}$	$g_{\perp}$	$A_{\parallel}^{Co}$(× 10^4 cm^{-1})	$A_{\perp}^{Co}$(×10^4 cm^{-1})	$A_{\parallel}^{N}$(× 10^4 cm^{-1})
Co(II) cobalamin (pH 7.0)	2.004	2.32	100	27	17.3
Co(II) cobalamin (pH 0.5)	2.005	2.59	134		
Co(II) cobalamin + cysteine (pH 1.0)	2.005	2.49	129		
Co(II) cobalamin + cysteine (pH 12)	2.001	2.30	97		
Co(II) cobinamide (pH 7.0)	2.005	2.58	133		
Co(II) cobinamide + 2-thioethanol (pH 6.0)	2.004	2.43	124		
Co(II) cobinamide + 2-thioethanol (pH 8)	2.001	2.275	93	23	
Co(II) cobinamide + dimethylsulfide (pH 6.5)	2.004	2.355	111		
Co(II) cobinamide + pyridine	2.004	2.320	103	27	17.5
Co(II) cobinamide + benzimidazole	2.004	2.320	102	27	17.2
Co(II) cobinamide + imidazole	2.005		106		17.3
Co(II) cobinamide + adenine	2.006		109		17.8

References pp. 1122–1133

complexes reveal[137] a number of differences. Most similar[136] are those of porphyrins but the $A_{\parallel Co}$ values are uniformly smaller for other complexes. The ability of the complexes to coordinate two axial nitrogenous ligands also varies, cobalt(II)corrinoids forming the bis-pyridine complex only at very high pyridine concentration, whereas cobalt(II)dimethylglyoximate takes on two pyridine ligands readily[125].

A reaction common to all complexes is the formation[139-144] as a more-or-less stable 1:1 cobalt oxygen complex which can be described as a cobalt(III) superoxide, though the e.p.r. spectra reveal (Fig. 13) con-

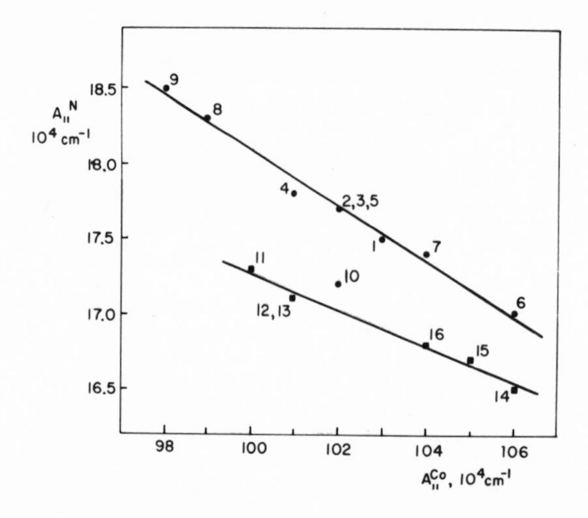

Fig. 12. Correlation of hyperfine coupling constants $A_{\parallel}{}^{N}$ and $A_{\parallel}{}^{Co}$ derived from the e.p.r. spectra of Co(II) cobinamide complexes with various bases; (1) pyridine, (2) 2-methylpyridine, (3) 4-methylpyridine, (4) 2,6-dimethylpyridine, (5) 4-ethylpyridine, (6) 4-cyanopyridine, (7) 4-chloropyridine, (8) 4-aminopyridine, (9) 4-hydroxypyridine, (10) benzimidazole; and Co(II) cobamides; (11) 5,6-dimethylbenzimidazolyl, (12) 5-hydroxybenzimidazolyl, (13) 5-methoxybenzimidazolyl, (14) adeninyl, (15) 5-methoxyadeninyl, and (16) 5-methylthioadeninyl.

siderable delocalization, the single unpaired electron being[144] "80%" on the oxygen and having a structure:

```
 .O
 /
O
|
CoL
```

The parameters derived from the complexes are similar, as expected if the above formulation is correct. These cobalt(III) superoxides can be further reduced[145] electrochemically and in most complexes, except the corrinoids, form[141] the μ-peroxycobalt(III) complexes. Though superoxy corrinoids

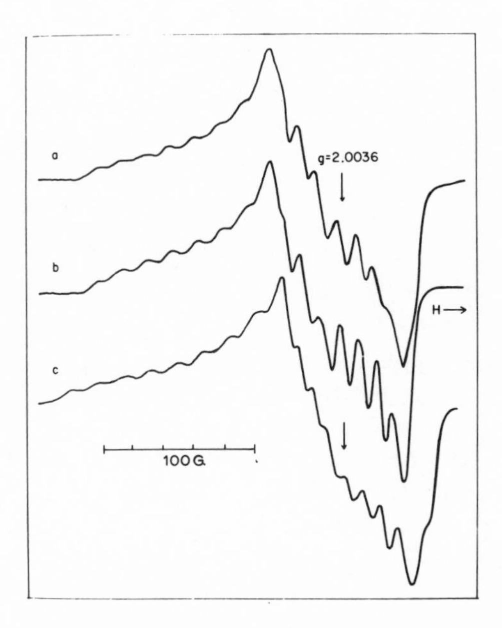

Fig. 13. E.p.r. spectra of some superoxocobalt(III) complexes: (a) cobalamin (b) bisdimethylglyoximato cobalt and (c) cobalt mesoporphyrin IX.

have probably no biological function, they may be intermediates in the oxidation of Co(II) corrinoids and may therefore be formed during enzymatic reactions studied under aerobic conditions. There is evidence[146] that they may be intermediates in the autoxidation[147] of thiols and in the reaction[146] of thiols with Co-alkylcorrinoids. These studies also showed[148] the existence of both a mercaptocobalt(II) cobalamin and the corresponding S-protonated derivatives; the two forms having distinct e.p.r. spectra (Table IV), *e.g.* R = $HOCH_2CH_2$ in (1).

$$\mathrm{LCo(II) + RSH \rightleftharpoons LCo(II)\overset{\overset{\textstyle H}{|}}{S}R} \quad (K = 11.5) \qquad (1a)$$

$$\mathrm{LCo(II) + RS^- \rightleftharpoons LCo(II)SR} \quad (K = 2500). \qquad (1b)$$

Co(II) cobalamin also forms a complex with thioethers, another example of the similarity to alkyl-Co complexes which coordinate thioethers. One can conclude that the charge density on the cobalt is similar in both the cobalt(II) complexes and the alkyl cobalt complex. This may have the corollary that the cobalt–carbon bond is not significantly polarized.

Cobalt(I) corrinoids

The two-electron reduction product of the cobalamins, vitamin B_{12s}, has attracted considerable attention and its novel properties prompted its invocation as an intermediate in a number of proposed mechanisms of the enzymatic reactions. Once called hydridocobalamin, it is now known[149] that in *aqueous solution* it exists mainly as the cobalt(I) unprotonated species, the reaction with the proton giving[92,97] Co(II)cobalamin and hydrogen, the intermediate hydride having only transitory existence:

$$Co(I) + H^+ \rightarrow Co(I)H^+ \equiv Co\text{-}H \rightarrow Co(II) + \tfrac{1}{2}H_2 \qquad (2)$$

The decomposition is[97] sensitive to anions present in solution and there is some evidence that the vitamin B_{12S} is formed[150] by "self-reduction" at high pH. However, some other cobalt(II) complexes such as $Co(II)(DMG)_2$ are reduced[14] by molecular hydrogen. Other complexes have been isolated[48,151] in the cobalt(I) form *e.g.*, Na^+[Co SALEN]$^-$ and there is evidence[48] for protonation to give a relatively stable hydride.

Cobalt(I) corrinoids show some properties "typical" of a low-spin d^8 complex. It is, of course, formally square-planar with the axial ligands weakly bound, if at all. Certainly the 5,6-dimethylbenzimidazole is[151] not coordinated in aqueous solutions of vitamin B_{12s} though π-accepting ligands such as isocyanides, phosphines may[152] coordinate to this and other cobalt(I) complexes[53].

The most striking property is that of a powerful nucleophile. Its reactions with haloalkanes have been extensively studied[152] and there is good evidence[153] that the reactions proceed *via* a classical S_N2 mechanism. Presumably those with unsaturated compounds can be described as nucleophilic additions, though there is some evidence[154], mainly spectral, that π-complexes are formed between cobalt(I) corrinoids and alkenes. The nucleophilicity of the various cobalt(I) complexes follows[53,152] the order of the Co(II) → Co(I) reduction potential BAE > SALEN > DMG ~ Corrin > [(DO)(DOH)pn]. That of Co(I)BAE is striking, since it reacts[51,53] with bromobenzene, uncatalyzed. The powerful nucleophilic properties of these Co(I) complexes are presumably due to a combination of a low potential and high polarizability of the d_{z^2} "lone-pair". It is possible, as has been suggested for Co(II) cobalamin, that the "flexibility" of the corrin ring allows the cobalt to "sit-out-of" the plane of the four nitrogens, thus allowing the lone-pair to be even more effectively directed. Indeed there seemed[152] to be little difference in the steric influence of the nucleophilic reactions between the Co(I) forms of corrin and dimethylglyoxime. The nucleophilicity of the cobalt (I) is also sensitive to coordinated ligands, at least in the dimethylglyoximates, π-accepting ligands such as phosphines, isocyanides and interestingly, sulfides, causing a reduction in the rates of substitution. In a most interesting series of experiments, Friedrich and his colleagues[155] have investigated the stereochemistry of the alkylcobalt corrinoids formed from the corresponding Co(I) derivatives. The major product is always the 'upper" isomer, *i.e.* with the alkyl group in the same position relative to the macrocycle as the 5′-deoxyadenosyl group in the coenzyme B_{12}, whereas with cyanoaquocorrinoids the isomers are present in approximately equal amounts[38,156,157].

Though there have been no reports of cobalt(II) corrinoids reacting with haloalkanes, $Co(CN)_5^{3-}$ and $Co(II)(DMG)_2L$[158] give alkyl–cobalt complexes. Disproportionation to give the corresponding Co(III) and Co(I) complexes is always a possibility and would allow nucleophilic substitution by the latter, but it has been proposed that the reactions proceed via a radical mechanism:

$$Co(II)(DMG)_2L + RX \rightarrow Co(III)(DMG)_2(L)X + R^{\cdot}$$

$$Co(II)(DMG)_2L + R^{\cdot} \rightarrow Co(III)(DMG)_2(L)R. \qquad (3)$$

If such a reaction were found in the corrinoids to proceed *via* this mechanism, it might have far-reaching implications for the enzymatic reactions.

REACTIONS OF COBALT CORRINOIDS

The detailed interpretation of the reactions of the corrinoids and the "model" complexes is made difficult by the significant number of intermediates which might be present in solution. Thus the disproportionation of cobalt(II) complexes or their dimerization, the formation of cobalt–oxygen complexes and reactions with reductive ligands may contribute to the disparity between results reported by different workers and defer decisions on the mechanisms. Reactions which appear, by the nature of reactants and products, to be related to those taking place in the enzymatic system, may proceed by a different mechanism. Nowhere is this more apparent than in reactions with sulfur-containing ligands.

Reactions with sulfur-containing ligands

Though cobalt(III) corrinoids are reduced by thiols in alkaline solution, at lower pH complexes are formed between aquocobalamin and cysteine[82,159,160], glutathione[161,162] and other thiols[163] in which the mercaptide acts as the ligand. Other complexes containing a Co–S bond are formed by the reaction of Co(I) corrinoids with arylsulfonyl chlorides[163,164], and sulfuryl chloride[164]. The product from the latter reaction, which is also obtained by reaction of aquocobalamin with sulfite, bears some resemblance to cobalt alkyl corrinoids. The absorption spectrum of sulfito cobinamide is[104] temperature-dependent like the alkyl cobinamide, both

References pp. 1122–1133

sulfitocobinamide and cobalamins are[159,165,166] photolabile and sulfitocobalamin has a pK_a intermediate between ethynyl- and vinyl-cobalamin. This comparison is interesting because, although the sulfitocobalamin can be considered a complex of SO_3^{2-} with a cobalt(III) corrinoid, the observed reaction[37]:

$$\underset{\text{Co L (10-Cl)}}{\overset{SO_3}{|}} \xrightarrow[\beta\text{-}C_{10}H_7NH_2]{CH_3I} \underset{\text{Co L (10-Cl)}}{\overset{CH_3}{|}} + \beta\text{-}C_{10}H_7NHSO_3 + HI \quad (4)$$

of C-10 chloro- or bromo-sulfitocobalamin with methyl iodide and β-naphthylamine under anaerobic conditions to give methylcobalamin and β-naphthylsulfamic acid, may be evidence of its ability to act as a non-anionic leaving group. Similarly, the mercapto Co(III) corrinoids are reported[161,163] to react with alkylating agents in the presence of a large excess of thiol to give alkyl cobalt complexes. With a stoichiometric ratio of methyl iodide to mercaptocobalt complex, alkylation of the thiol occurs[167]. The reaction of aquocobalamin with dithiols has been studied[15] and the following mechanism proposed:

$$\left(\begin{matrix}SH\\ SH\end{matrix}\right. + \text{Co(III) L} \underset{+H^+}{\overset{-H^+}{\rightleftharpoons}} \left(\begin{matrix}S\text{–}H\\ S\text{–Co(III) L}\end{matrix}\right. \longrightarrow \left(\begin{matrix}S\\ |\\ S\end{matrix}\right. \quad \text{Co(I) L} \quad (5)$$

The Co(I)cobalamin formed reacts with haloalkanes to give alkylcobalamins, or with aquocobalamin to give Co(II) cobalamin which is the observed product of the reaction in the absence of electrophils. There are two important features about this mechanism: it provides a feasible biochemical pathway for the formation of Co(I) cobalamin, and involves the reaction of the sulfur ligand in which it donates the electrons to the ligand. More straightforward are the reactions of sulfonium salts with Co(I) cobalamin *e.g.*, *S*-adenosylmethionine gives[45,168-170] methylcobalamin.

In alkaline solution the mercapto cobalt(III) complexes give the corresponding Co(II) complexes and disulfides. Indeed, cobalt(III) corrinoids catalyze[147] the autoxidation of thiols and recently it has been shown[146] that oxidation of both Co(II)HSR and Co(II)SR complex is involved.

It has been reported[167,171] that homocysteine reacts with methylcobalamin to give methionine, though a contradictory report has been published[172]. Previously it had been noted[173] that light irradiation increased the yield of methionine. At pH 10, dimethyl sulfide is formed[167] from methylcobalamin and CH_3S^-. Recently, however, evidence has been obtained[146] which shows that oxygen plays a key role in the reaction of methylcobalamin with various thiols in which either the thiol radicals or the products of the reduction of oxygen attack the cobalt-carbon bond. Certainly the reaction is more complex than at first sight and an analogy with the enzymatic methylation of homocysteine must be drawn with caution.

Reaction with carbon monoxide

The reaction of cobalt(III) complexes with carbon monoxide is equally complex. It had first been reported[124] that carbon monoxide reacts with aquocobalamin:

$$Co(III)L + CO + H_2O \rightarrow Co(II)L + CO_2 + 2H^+ \quad (6)$$

and there was evidence that Co(I) cobalamin was an intermediate. Recently some evidence has been presented[174] which suggests that the reaction is autocatalytic and that the initial Co(II) and Co(I) complexes are formed by "self-reduction" of aquocobalamin, involving intramolecular electron transfer between C-8 and the cobalt. Though there is as yet no evidence for complex formation between carbon monoxide and corrinoids in any of the three oxidation states, reactions of other cobalt(III) complexes which do not undergo "self-reduction" suggests that interaction of cobalt(III) complexes might be possible under certain circumstances. The reaction of various hydroxo complexes, *e.g.*, HOCo(III)(SALEN)H_2O in alcohol had previously[175] been shown to give the alkoxycarbonyl derivatives, ROC(O)Co(SALEN)H_2O. By analogy the hydroxycarbonyl complexes were proposed as intermediates in the reaction in aqueous tetrahydrofuran which gave[176] carbon dioxide, hydrogen and the cobalt(II) forms quantitatively. Most interestingly, the reaction with HOCo{(DO)-(DOH)pn}H_2O gave a crystalline derivative [Co{(DO)(DOH)pn}CO] which reacted with excess alkylating agents to give[177] *dialkylcobalt* complexes (Fig. 14). It is conceivable that a carbonyl cobalt(I) complex may be an intermediate in the reactions of the corrinoids; the possible implications of the dialkylcobalt complexes will be discussed later.

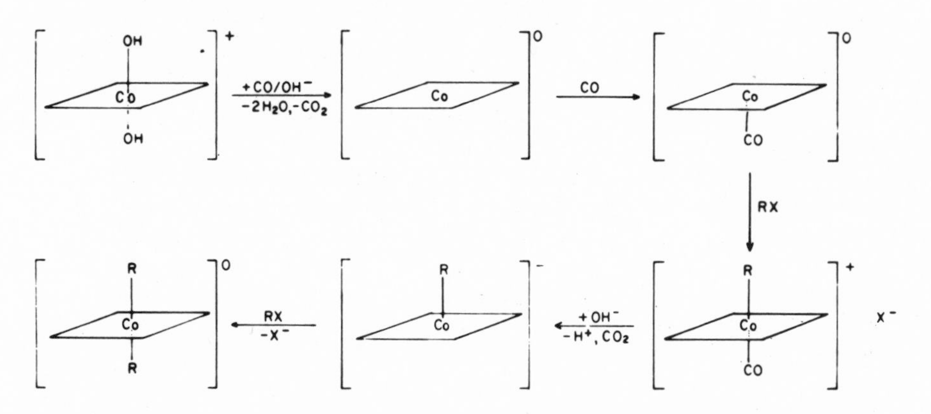

Fig. 14. Reduction of dihydroxy Co(III){(DO)(DOH)pn} by carbon monoxide.

REACTIONS OF ORGANOCOBALT COMPLEXES

Most reactions involve fission of the cobalt–carbon bond which can take place homolytically (a) or heterolytically (b) (c),

$$\mathrm{Co{-}R}\ \begin{cases} \xrightarrow{(a)} \mathrm{Co(II)} + \mathrm{R}^{\bullet} \\ \xrightarrow{(b)} \mathrm{Co(I)} + \mathrm{R}^{+} \\ \xrightarrow{(c)} \mathrm{Co(III)} + \mathrm{R}^{-} \end{cases} \tag{7}$$

Often, of course, fission is concerted with attack of reagent and the species, R, does not have a separate existence. In such cases, the mechanism of the reaction might be concluded from the corrinoid formed, though we have seen that such interpretations are fraught with danger. We will first consider a reaction which is very apparent to all who have handled organocobalt complexes; their instability to light.

Photolysis

The first step in the photolysis of an organocobalt corrinoid is[41,43,44,178–180] homolytic fission to give the corresponding Co(II) corrinoid and an organic radical, which has been detected[129,181] by e.p.r. spectroscopy in some instances. Since the photolysis of $CH_3Co(DMG)_2py$ in the presence of Co(II)cobalamin gives[182] methylcobalamin and interconversion[155] of the two ("upper" and "lower") isomers of alkylaquocorrinoids occurs on photolysis to give an equilibrium mixture, we may conclude that cobalt(II) corrinoids are good radical scavengers. Therefore one must consider that the reaction to give Co(II) and $R^{\bullet}$ is an equilibrium with the reverse reaction, *i.e.*, $\mathrm{Co{-}R} \rightleftharpoons \mathrm{Co(II)} + \mathrm{R}^{\bullet}$. The observed rate of photolysis and the products, will therefore depend on the pathways available for the consumption of both Co(II) and $R^{\bullet}$. Thus the marked dependence[41,183] of both rate and products on oxygen, may be due to the elimination of the reverse reaction by the formation of $ROO^{\bullet}$, CoO_2 or indeed CoOOR species. Similarly the dependence[184] of the rate of photolysis of methylcobalamin on the presence of alcohols in aqueous solution, isopropanol > n-propanol ≃ ethanol > methanol > t-butanol, could be related to the ease of abstraction of the α-hydrogen. Homocysteine and *p*-benzoquinone also increased the rate and we can conclude that the rate of photolysis is increased in the presence of

[Scheme (8): 5'-deoxyadenosyl–Co L (structure with OH OH, O, CH₂, N, NH₂) $\xrightarrow{h\nu}$ Co(II) L + cyclized adenosine (CH₂, O, N, NH₂)]   (8)

compounds which are readily attacked by radicals. In the case of coenzyme B_{12}, which is very photolabile, a pathway is available in the form of an intramolecular cyclization to give[178] 5′,8-cycloadenosine (reaction (8)). In air, adenosine 5′-aldehyde is formed[178] in addition to the cyclic derivative.

With the overall rate of photolysis largely dependent on the fate of the radical or the cobalt(II) complex it is not surprising that the products of any one ligand *e.g.*, methyl, depend on the conditions of the experiment. Thus, the anaerobic photolysis[182,185] of methylcobalamin gives methane, ethane and Co(II)cobalamin, and it is not clear whether the ethane is formed by radical coupling[182] or methyl abstraction[185] from the corrin ring. Reaction of the photolytically generated species with the corrin has been observed[186] in the photolysis of dichloromethylcobalamin, though there is evidence[181] that a chlorocarbene is formed in this case. Aerobic photolysis of methylcobalamin gives formaldehyde and methanol; anaerobic photolysis[175] of $CH_3Co(SALEN)H_2O$ in methane gives formaldehyde formed from the *solvent*, whereas photolysis of $CH_3Co(DMG)_2$pyridine in benzene gives[187] toluene. Reduction of the radical species has been observed[186] during photolysis under hydrogen.

In view of the sensitivity of the observed rate of photolysis to the experimental conditions, conclusions regarding the influence of axial ligands on the rates of photolysis must be made with care. If, as in coenzyme B_{12} there is a "built-in" reaction for the consumption of the radical, the rates will be fast. In the absence of such a mechanism, the rate seems to depend[105] on the donor properties of the carbon ligand, *e.g.*, $CF_3Co- > CF_2ClCo- > CF_2HCo$-cobalamin[102] though there are discrepancies, and on the coordinated nucleotide[188]. An attempt has been made[55] in the complexes RCo(CR)L to relate the photolability to the assigned electronic transitions.

Thermolysis

The thermal stability of organocobaltcorrinoids is greater than *e.g.*, methyltetracarbonylcobalt[189] $CH_3Co(CO)_4$. The observed stability may not necessarily reflect any great difference in the "Co–C" bond energy of the carbonyl and the corrinoid, since the stability is probably due to kinetic factors, and the mechanism of decomposition of the alkylcobalt carbonyl may be different from that of the alkyl corrinoid.

The thermolysis of alkylcobalt corrinoids is complicated by reactions of the side chains in the molecules though the interconversion[155] of the "upper" and "lower" isomer of alkylaquocorrinoids proceeds to equilibrium at 95°. However, the thermal decomposition of "model" cobalt complexes[175,187] gives the corresponding cobalt(II) species and alkanes, with small amounts of alkanes. Both in the model complexes and in the corrinoids, it has been noted that "bulky" groups reduce the thermal stability. Interestingly, though isopropyl-cobinamide is easily prepared[104] isopropylcobalamin is much less stable. It is not known whether the greater instability of the cobalamins is due to increased steric interactions between axial ligand and corrin ring, or an example of the *trans*-effect, or both.

Reductive cleavage

Though the reduction of cyanocobalamin to give Co(II)cobalamin and hydrogen cyanide[190] or methylamine[191,192] was observed some time ago, little attention has been given to the reduction of organocobalt corrinoids. Alkylcobalt corrinoids are reduced catalytically to give the corresponding alkanes[106], and sodium borohydride reduces[193] certain alkyl cobalamins, particularly where the alkyl group is bulky. The electrochemical reduction of alkylcobalt corrinoids[165,194] has been investigated and though the half-wave potential is moderately sensitive to the alkyl ligand, it is particularly sensitive to the fifth ligand. Thus, whereas the alkylcobalamins show a two-electron reduction $E_{\frac{1}{2}}$ (*vs.* S.C.E.) of ~ -1.37, alkylcobinamides have two one-electron waves, $E_{\frac{1}{2}}$ - 1.27, -1.44 for methylcobinamide. It would be most interesting to know the products of the electrochemical reduction but the controlled potential reduction of alkylcobalt corrinoids at a mercury electrode is complicated[195] by the formation of alkylmercury compounds. If the ligand *trans* to the alkyl ligand can alter both the reduction potential and the mode of electron transfer to the cobalt–carbon bond, the presence or absence of the fifth ligand could be extremely[138,196] relevant to the biochemical reactions.

Nucleophilic displacement of carbon ligands

We have seen that ligand exchange is a common feature of corrinoids with non-carbon ligands. Though the exchange of one carbon ligand by another has been observed[57] with alkylcobalt porphyrins and other cobalt complexes the work with alkylcobalt corrinoids has been limited mainly to displacement by cyanide. The reaction of coenzyme B_{12} with cyanide gives[197] adenine and the cyanohydrin of 2,3-dihydroxy-pentenal. Interestingly, one of the products from the controlled potential reduction of the coenzyme was found[195] to be adenine, suggesting that both Co(I) cobalamin and 5′-deoxyadenosine carbanion are formed in the reduction step.

Though n-alkylcobalamines are not susceptible to nucleophilic displacement by CN^-, s-butyl, cyclohexyl[196], β-trimethylaminoethyl[198] chlorodifluoro- and trifluoro-methylcobalamin[102] are attacked by alkaline cyanide. Presumably the halomethyl groups are easily displaced because of considerable polarity, $Co^{\delta+}$–$C^{\delta-}$, in the cobalt–carbon bond, whereas the two s-alkyl ligands may be "weakly" bound because of steric interactions. It is presumed that the CN^- attacks from the same side as the leaving group[197] though there is evidence that the coordination of cyanide *trans* to the alkyl group facilitates displacement. These reactions assume the carbon ligand leaves as an anion. Interestingly, the reported[14] product of the decomposition of $CH_3Co(DMG)_2H_2O$ is CH_3CN. The decomposition[105] of β-substituted ethylcobalamins was shown[199] later not to proceed *via* nucleophilic attack on the cobalt. Rather for β-cyanoethylcobalamin, the reaction involved the loss of a proton to give acrylonitrile and Co(I)-cobalamin, the reaction being reversible:

$$\underset{\displaystyle \mathrm{CoI}}{\mathrm{CH_2}}\cdot\mathrm{CH_2}\cdot\mathrm{CN} \underset{+\mathrm{H}^+}{\overset{-\mathrm{H}^+}{\rightleftharpoons}} \mathrm{Co(I)L} + \mathrm{CH_2{=}CHCN} \qquad (9)$$

The other reactions reported proceed[105] presumably by a similar E2 mechanism.

Acylcobalt corrinoid complexes are also attacked by[95] nucleophiles *e.g.*

$$\mathrm{O{=}C(CH_3)(CoL)} \leftarrow \mathrm{NH_2OH} \xrightarrow{-\mathrm{H}^+} \mathrm{Co(I)L} + \mathrm{CH_3CONHOH} \qquad (10)$$

In the presence of methyl iodide, methylcobalamin is formed. Acetyl-cobalamin is decomposed[200] by alkali to acetate and Co(I)cobalamin:

$$\mathrm{COOH{-}CH_2{-}CoL} \xrightarrow{\mathrm{OH}^-} \mathrm{CH_3COO^-} + \mathrm{Co(I)L} \qquad (11)$$

Interestingly, acetyl cobinamide is decomposed much faster.

A most important reaction[167,201] involves the decomposition of β-hydroxylethylcobalt complexes by alkali to give acetaldehyde (equation 12). This reaction is particularly fast for dimethylglyoximate and related complexes but much slower for cobalamin, cobinamides, BAE, and does not take place for the porphyrin[57] complex. The decomposition of the cobinamide is much faster than the cobalamin (cf. the decomposition of acetyl compound). The lability to base decreases with decreasing Co(II)/Co(I) reduction potential of the chelate though steric factors have been invoked[201] to account for the decreased reactivity of the corrinoids and porphyrin complexes. A 1,2-hydride shift was proposed mainly on the basis that β-alkoxyethyl cobalt complexes are stable to alkali:

$$\mathrm{H{-}C(OH){=}CH_2} + \mathrm{Co(I)L} \xleftarrow{?} \mathrm{H{-}C(OH)(H){-}C(H)(H){-}CoL} \underset{+\mathrm{H}^+}{\overset{-\mathrm{H}^+}{\rightleftharpoons}} \mathrm{H{-}C(O^-)(H){-}C(H)(H){-}CoL} \longrightarrow \mathrm{CH_3CHO} + \mathrm{Co(I)L} \qquad (12)$$

Though the relationship between 5′-deoxyadenosyl cobalt complexes[201-203] and β-alkoxyethyl cobalt complexes has been noted[167] it was found later[201] that the 5′-deoxyadenosylcobalt(III)bis-dimethylglyoximates and coenzyme B_{12} are unstable at high pH to give 4′,5′-dehydro-5′-deoxyadenosine

$$\text{HO}^- + \text{5'-deoxyadenosyl-CH}_2\text{-Co L} \longrightarrow \text{Co(I) L} + \text{CH}_2\text{=(4',5'-dehydro-5'-deoxyadenosine)} \qquad (13)$$

presumably by proton abstraction at $C_{4'}$.

Thus if the coordinated alkyl group contains a suitable substituent, the Co–C bond may be broken fairly readily. Whether such reactions are relevant to the enzymatic mechanism remains to be seen.

REACTION WITH ELECTROPHILES

Acid decomposition

Mild acid hydrolysis of coenzyme B_{12} gives adenine, erythro-2,3-dehydroxy-1-penten-5-al and aquocobalamin[204]. Other coenzyme forms behave[205] similarly. Protonation of the ligand precedes decomposition and substituted alkyl ligands which are protonated in acid also decompose[107], *e.g.* 2-hydroxyethyl, 2-methoxyethyl, 2-aminoethylcobalamin[11]:

$$\underset{\text{Co L}}{\overset{CH_2CH_2OH}{|}} \xrightarrow{H^+} \text{Co(III)L} + CH_2{=}CH_2 + H_2O \qquad (14)$$

Under the conditions of the experiment the cobalamins are in the "base-off" form.

Reaction with iodine

An interesting reaction[165] which has been little investigated is that of iodine with alkylcobalamins. Coenzyme B_{12} gives 5′-iodo-5′-deoxyadenosine and iodocobalamin and experiments with iodine monochloride suggest that the reaction is with the iodine cation. 5′-Deoxyadenosylcobinamide reacted at approximately the same rate as the coenzyme but both were much faster than ethyl- or methylcobalamin. The reaction of the Co–C bond with such reagents is worthy of further investigation.

Reaction with metal ions

A simple reaction[206,207] which turns out to have implications beyond B_{12} chemistry, is the reaction of alkylcorrinoids with mercury(II) to give alkyl mercury derivatives and cobalt(III) corrinoids. Hg(II) is a better electrophile towards coordinated methyl in methylcobalamin than I_2 or H^+. As seen in Table V, methylcobalamin is attacked most readily, other

TABLE V

COMPARISON OF COBALAMINS AND COBINAMIDES

Axial ligand	Property	Cobalamin	Cobinamide
CH_3-	+ $Hg(OOCCH_3)_2$ k mole^{-1} sec^{-1}	3.7×10^{2}	1.2×10^{-1}
CH_2=CH-	+ $Hg(OOCCH_3)_2$ k mole^{-1} sec^{-1}	7.05×10^{-1}	1.75×10^{-1}
CH_3CH_2	+ $Hg(OOCCH_3)_2$ k mole^{-1} sec^{-1}	2×10^{-1}	10^{-5}
$^{-}OOCCH_2$-	OH^-	slow	fast
CH_3	$E_{\frac{1}{2}}$ *vs.* S.C.E.	-1.39	-1.17
$CH_3CH_2CH_2$	$E_{\frac{1}{2}}$ *vs.* S.C.E.	-1.37	-1.24
Co(I)	+ CH_3CH_2Br k mole^{-1} sec^{-1}	31	31

n-alkyl ligands less so. However, isopropylcobinamide reacts, presumably due to the rather weak Co–C band. Most interestingly in contrast to those reactions which give Co(I) corrinoids, the coordination of 5,6-methylbenzimidazole, *increases* the rate of the reaction. Presumably increased electron donation to the cobalt facilitates the loss of the alkyl group as a carbanion as in the reactions of other organometallic complexes[208-210]:

$$\underset{\text{CoL}}{\overset{\text{CH}_3}{|}} \xrightarrow{Hg^{2+}} \text{Co(III)L} + [HgCH_3]^+ \qquad (15)$$

The reaction with the cobalamins is complex since mercury(II), like Cu^{2+}, Zn^{2+} and Co^{2+}, coordinates to the 5,6-dimethylbenzyimidazole. The data given in Table V allow for this equilibrium. Although silver(I) promotes the removal of coordinated cyanide, there is only evidence for π-complexes with both ethynyl- and vinyl-cobalamins. Since methyl corrinoids are found in many bacteria, the reaction, both enzymatically and non-enzymatically, to give highly toxic methyl mercury derivatives, considerably increases the hazards from mercury contamination[206,211].

CORRINOID—DEPENDENT ENZYMATIC REACTIONS

It would be disappointing if the enzymatic reactions which require coenzyme B_{12} were trivial. Nothing could be further from the truth.

TABLE VI

SOME FEATURES OF COENZYME B_{12}-DEPENDENT ENZYMES

Reaction	Monovalent cation requirements	^{3}H transfer		Added thiol required	Sulfhydryl groups	
		From substrate to coenzyme	From coenzyme to product		On same subunit as cobamide	On dissimilar protein from cobamide binding site
Methylmalonyl CoA mutase		+	+		+	
Glutamate mutase		+	+	+		+
α-Methylene-glutarate mutase			+	+	+	
Dioldehydrase	K^+ NH_4^+, Tl^+, Rb^+	+	+		+	
Glycerol dehydrase	K^+, NH_4^+					+
Ethanoldeaminase	K^+, NH_4^+, Rb^+	+	+		+	
L-β-lysine mutase	K^+, Rb^+	+	+	+		+
D-α-lysine mutase	K^+, Rb^+, NH_4^+	+	+	+		+

Indeed the isomerizations catalyzed by coenzyme B_{12} have few analogies in organic chemistry, and the astonishment when they were first announced has been noted[212] on at least one occasion. The reactions can be considered in two categories; those requiring a co-5′-deoxyadenosyl cobamide and involved in hydrogen transfer, and those implicated in methyl transfer. The reactions in the former group are summarized, with one exception and the formal relationship between them is indicated in Fig. 15 and some experimental features are given in Table VI. It would be fair to say that all involve

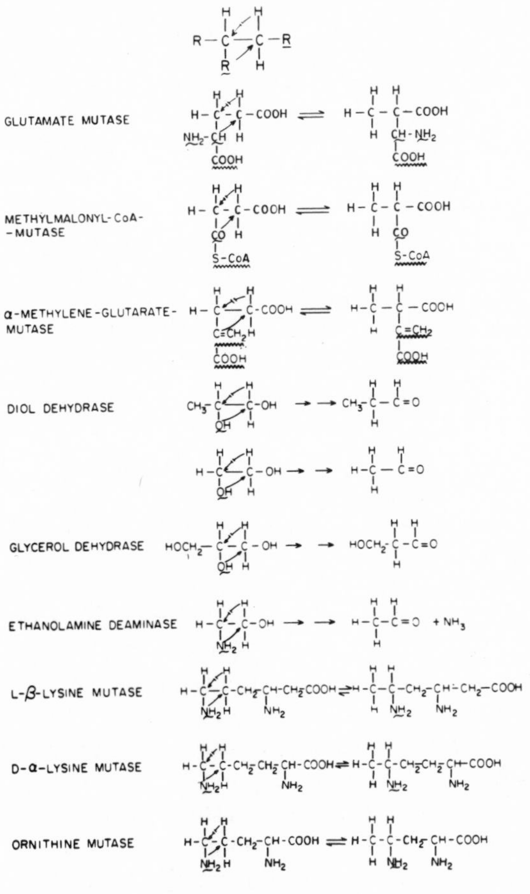

GLUTAMATE MUTASE

METHYLMALONYL-CoA-MUTASE

α-METHYLENE-GLUTARATE-MUTASE

DIOL DEHYDRASE

GLYCEROL DEHYDRASE

ETHANOLAMINE DEAMINASE

L-β-LYSINE MUTASE

D-α-LYSINE MUTASE

ORNITHINE MUTASE

Fig. 15.

a 1,2-hydrogen shift with an associated movement of another, larger group in the "opposite" direction. Ribonucleotide reductase, from *Lactobacillus leichmannii*, though requiring coenzyme B_{12} and involving hydrogen transfer, differs from the rest in that the hydrogen donor and acceptor are different molecules. The methyl transfer reactions shown in Fig. 16 all involve an enzyme-bound Co-corrinoid.

Most enzyme systems studied have been derived from bacterial sources and though it was once said[213] that "the reactions (requiring corrinoids) are vital only to unimportant species and unimportant to vital ones," this has become less apt in the intervening years. Indeed, bacteria and other micro-organisms hold a unique place in B_{12} chemistry since only

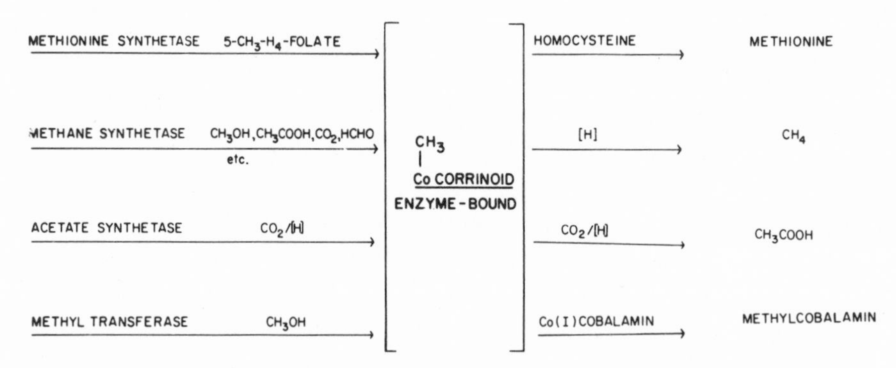

Fig. 16. Corrinoid-dependent enzymatic reactions involving one-carbon transfers.

they synthesize the corrinoids. "We" are dependent on B_{12} supplied by the diet and do not utilize any of the vitamin synthesized by intestinal bacteria. The dietary requirement is extremely low[32] some 2.5 μg/day, so the amount stored in the normal adult, ~5000 μg, mostly in the liver, constitutes a large storage pool since chemical evidence of deficiency is not observed until the total body content approaches 500 μg. Inefficient absorption of the vitamin is the most common cause of vitamin B_{12} deficiency and is associated with the absence of sufficient amounts of a gastric secretion called the "intrinsic factor"[3], a glycoprotein which has a high binding affinity for vitamin B_{12}. Both vitamin B_{12} and folic acid are essential[15,31] to DNA synthesis and a defect in this synthesis is probably responsible for megaloblastic anaemias, of which pernicious anaemia is a most common form. There are many horrendous symptoms associated with the disease, and if due to vitamin B_{12} deficiency, degenerative changes in the nervous system also occur.

The excessive excretion[214,215] of methylmalonate in the urine also accompanies an inherited metabolic disorder, methylmalonic acidaemia. Though, as we shall see, the enzymatic reactions discovered so far are still incompletely understood, there are signs that they may indeed be relevant to "vital species."

Biosynthesis of 5′-deoxyadenosyl coenzymes

The biosynthesis of the corrin ring has been adequately considered in previous reviews[3,4,11]. The formation of the Co-5′-deoxyadenosyl bond is more pertinent to the chemistry discussed in this chapter.

The enzyme systems from[216] *Propionibacterium shermanii* and[217-220] *Clostridium tetanomorphum* require ATP, a divalent metal ion and a variety of reducing agents including thiols[221], reduced flavins[222,223], or a reduced ferredoxin[222]. It has been shown for *C. tetanomorphum* that two separate reduction systems, both requiring DPNH-dependent flavoproteins, are necessary, a B_{12a} (aquocobalamin) reductase utilizing FAD, and a B_{12r} reductase utilizing FAD or FMN. Dithioerythritol can replace DPNH in the latter system. B_{12s} was not detected as the product of the second reduction since the reaction was coupled to the adenosylating system[221] which requires only a divalent metal ion and releases a tripolyphosphate[220]. Presumably Co(I)cobalamin, which need be formed in only small amounts, acts as a nucleophile displacing tripolyphosphate from $C_{5'}$ of deoxyadenosine:

$$\text{Co(I)L} + \text{ATP} \rightarrow 5'\text{-deoxyadenosyl CoL} + \text{O}-\underset{\displaystyle \text{O}}{\overset{\displaystyle \text{O}}{\overset{|}{\underset{|}{\text{P}}}}}-\text{O}-\underset{\displaystyle \text{O}}{\overset{\displaystyle \text{O}}{\overset{|}{\underset{|}{\text{P}}}}}-\text{O}-\underset{\displaystyle \text{O}}{\overset{\displaystyle \text{O}}{\overset{|}{\underset{|}{\text{P}}}}}-\text{O} \quad (16)$$

The relationship between the three systems, whether they form a structural and functional complex, the role of thiol and the activation of ATP in the adenosylating step[223] are not yet understood.

Methylmalonyl-CoA mutase

The interconversion of methylmalonyl-CoA (a) and succinyl-CoA (b) (Fig. 17) is a necessary step in the metabolism, in animal tissues, of propionate and its formation by propionic acid bacteria[224]. Following the observation[226] that the mutase activity of rat liver is markedly reduced in animals fed a vitamin B_{12}-deficient diet, coenzyme B_{12} was implicated in both animal[225] and bacterial[226] metabolism. The enzyme has been purified from propionic acid bacteria[227,228] and from sheep liver[229]. The mammalian enzyme is larger (mol. wt. 165,000) than that from *Propionibacterium shermanii* (mol. wt. 56,000); is active with either coenzyme B_{12} or its benzimidazole analog, whereas the bacterial enzyme, in addition, is active with α-(adenyl)-cobamide coenzyme; binds two coenzyme molecules (one subunit) with the bacterial enzyme binding one, and though the K_m values are comparable ($\sim 2.0 \times 10^{-8}$), the coenzyme is more readily removed from the bacterial enzyme. A more striking difference between the two enzymes lies in their sensitivity to "sulfhydryl" reagents. The bacterial enzyme is not significantly affected by mercurials *etc.*, whereas the mammalian apoenzyme is markedly inhibited though the bound coenzyme affords almost complete protection. The properties of the coenzyme bound to the mammalian enzyme are quite different from that free in solution; it is stable to light and cyanide which may indicate that the cobalt–carbon bond is significantly altered on binding.

Fig. 17. Isomerization of *R*-2-D_1-methylmalonyl-CoA (a) to succinyl-CoA (b) by methylmalonyl-CoA mutase.

The stereochemical course of the reaction has been thoroughly studied. The reaction has been shown[228,230-236] to be intramolecular in the sense of not involving solvent or other non-enzyme bound species. The reaction of methyl-d_3-malonyl-CoA gave[237] all the deuterium in the product. The reaction of *R*-2-D_1-methylmalonyl-Co-A (a) with the coenzyme was shown[238,239] to lead to (+)-*S*-succinic acid (Fig. 17). The isomerization occurs with retention of configuration at C-2, the hydrogen atom shifted from the methyl group of methylmalonyl-CoA takes the place of the -CO–S–CoA group. The hydrogen transfer[240-242] occurs via the $C_{5'}$ position of the bound coenzyme, hydrogen being removed in the rate-determining step and becoming one of three equivalent hydrogens before being returned to regenerate methylmalonyl CoA or yield succinyl CoA.

Glutamate mutase and α-methyleneglutarate mutase

The study[27,243] of glutamic acid fermentation by *Clostridium tetanomorphum* by Barker and his colleagues[218,243-245] led to the isolation of the coenzyme form of α-(adenyl)-cobamide. The enzyme catalyzes[246,247] the reversible conversion of L-glutamate to L-*threo*-β-methylaspartate with the equilibrium favoring the accumulation of glutamate (K = 10.7 at 25°C). The enzyme has been separated into[246-250] two fractions. The larger component E (mol. wt. 128,000) binds glutamate and β-methylaspartate and, in the absence of the smaller component S, binds 1 mole of coenzyme B_{12}/mole. In the presence of excess S it binds 2 moles, with K_m decreasing as the mole ratio of S:E increases. Component S (mol. wt. 17,000) is inhibited by sulfhydryl reagents and contains six to seven-SH groups, two of which are exposed. A considerable number of coenzyme forms are active[251,252] and though the main one in *C. tetanomorphum* is adenylcobamide coenzyme, its K_m is greater (binding weaker) than benzimidazole or a 5(6)-monosubstituted benzimidazole. There is little exchange between free and bound coenzyme. The loss of base to give a cobamide coenzyme, or nucleotide to give a cobinamide coenzyme, results in complete loss of activity, as does modification of the 5′-deoxyadenosyl group[205]. The Co-methyl- and Co-carboxymethylcobalamins are inhibitory.

The reaction can be considered to involve transfer of a glycine moiety and a hydrogen atom; free glycine, ammonium ion, acrylate and α-ketoglutarate are[246,247,249,253] not intermediates and no hydrogen is incorporated from the solvent. Though there are many formal analogies with the methylmalonyl CoA mutase reaction, the most striking difference is that, whereas the isomerization of R-methylmalonyl-CoA to succinyl-CoA occurs with

(a) 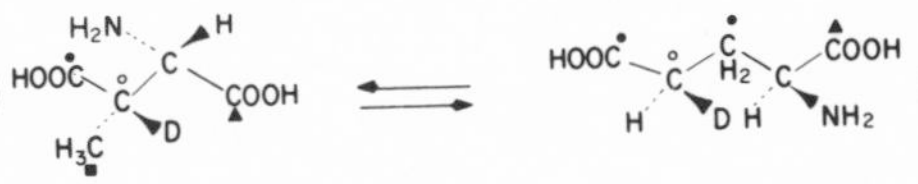 HOOC–C(H₃C)(D)–C(H₂N)(H)–COOH ⇌ HOOC–C(H)(D)–CH₂–C(H)(NH₂)–COOH (b)

Fig. 18. Isomerization of 2*S*-3*S*-3-methylaspartate (a) to *S*-4-D_1-glutamate (b) by glutamate mutase.

retention of configuration at C-2, the isomerization of 2*S*-3*S*-3-methylaspartate (a) to glutamate (b) involves inversion of configuration at C-3 of methylaspartate (Fig. 18). Consequently one can say that the hydrogen migrating from the methyl group of methylaspartate attacks the C-3 of methylaspartate from a direction opposite to the departing –CH(NH_2)COOH group. Hydrogen transfer proceeds *via* bound 5′-deoxyadenosylcobalamin with C'_5 of deoxyadenosyl group acting as hydrogen carrier.

α-Methyleneglutarate mutase, isolated[18,254,255] from *Clostridium barkeri*, has a molecular weight of ~200,000, and catalyzes the reversible formation of α-methylene-β-methylsuccinate. Tritium is transferred from 5′-^{3}H-5′-deoxyadenosylcobalamin to both substrate and product.

Dioldehydrase

It was shown[256] sometime ago that some species of *Aerobacter* and *Escherichia* are able to form large amounts of 1,3-propanediol from glycerol. Cell-free extracts from *Aerobacter aerogenes* ATCC 8724, grown anaerobically on glycerol, convert[257] ethylene glycol to acetaldehyde and 1,2-propanediol to propionaldehyde irreversibly (Fig. 19). Though the cells convert glycerol to β-hydroxypropionaldehyde, the cell-free extracts do not. However, the enzyme, glycerol dehydrase, isolated[258-262] from another strain of *A. aerogenes* (No. 572 PZH) carries out all three conversions and both it and the 1,2-propanediol dehydrase[263] have been purified.

The latter enzyme has been thoroughly studied. It requires a monovalent cation with K^+, NH_4^+, Tl^+ and Rb^+ effective and α-(5,6-dimethylbenzimidazolyl)-, α-(benzimidazolyl)- and α-(adenyl)-Co-5′-deoxyadenosyl cobamides give active holoenzymes. The vitamin itself, cyanocobalamin and hydroxocobalamin are inhibitors and both vitamins and coenzymes are tightly bound. The apoenzyme but not the holoenzyme is[264] inhibited by sulfhydryl reagents; it is inactivated by coenzyme in the absence of substrate, and there is a report that the anaerobic illumination of the holoenzyme under the same conditions, causes[265] a subsequent two-fold increase in dioldehydrase activity.

The activity of the coenzyme is sensitive to changes in many parts of the molecule. Modification of the Co-C_5'-deoxyadenosyl group affects

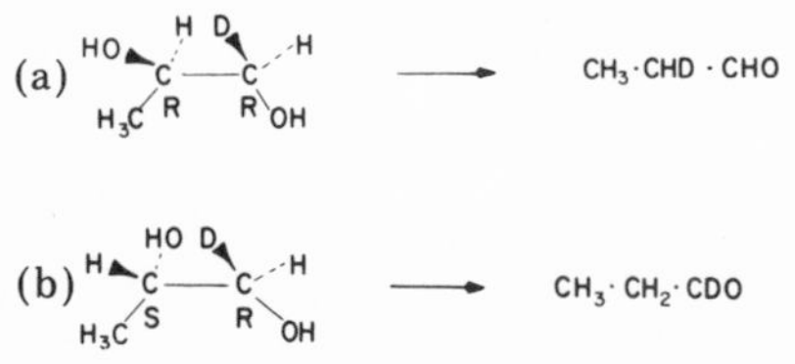

(a) HO, H, H_3C, R — C — C — D, H, R, OH ⟶ $CH_3 \cdot CHD \cdot CHO$

(b) H, HO, H_3C, S — C — C — D, H, R, OH ⟶ $CH_3 \cdot CH_2 \cdot CDO$

Fig. 19. Propanedioldehydrase-catalyzed reaction of 1-*R*-1-D-2-*R*-propane-1, 2-diol (a) and 1-*R*-1-D-2-*S*-propane-1,2-diol (b).

activity, 2′,5′-dideoxyadenosylcobalamin being[205] a less effective coenzyme though prior photolysis of the enzyme-bound coenzyme increases[266] activity to a level comparable with that of coenzyme B_{12}. 5′-Deoxyadenosylcobinamide has reduced activity as has 5′-deoxy-N_6-methyladenosylcobalamin[180]. Even substitution of chlorine at C-10 in cobalamin coenzyme, though it does not affect binding, markedly reduces the catalytic activity[267].

The glycerol dehydrase, purified from *A. aerogenes*, forms[262] a stable inactive complex with hydroxocobalamin with a molecular weight of 188,000. The hydroxocobalamin apoenzyme complex is comparatively stable to dissociation but Na^+, low pH, and EDTA promote[260,262] the dissociation of the apoenzyme into subunits. Hydroxocobalamin also protects the apoenzyme against attack by sulfhydryl reagents and increases its thermal stability. In the presence of Mg^{2+} and SO_3^{2-}, the hydroxocobalamin can be replaced by 5′-deoxyadenosylcobalamin to give active holoenzyme. Strangely, the holoenzyme is unstable giving the hydroxocobalamin complex, though cobinamides do not cause[268,269] this loss of activity. An enzyme isolated[261] from the same strain by a different procedure can use ethylene-glycol, propane-1,2-diol or glycerol as substrates though the propanediol is preferred. The enzyme has a molecular weight of $\simeq$ 178,000, binds[270] one molecule of 5′-deoxyadenosylcobalamin with an equivalent weight of 145,000 ± 10,000. The enzyme is also activated by monovalent cations, and, in the absence of substrate, slowly loses activity, though butane-1,2-diol affords some protection. It may eventually appear that there are more similarities than differences between glycerol dehydrase and propane-1,2-diol-dehydrase.

The fate of the reagent atoms in the propane-1,2-diol reaction has been thoroughly investigated. No hydrogen is incorporated into substrates or products from solvent and the reaction does[264,271,272] not involve enzyme-free intermediates. Convincing evidence for hydrogen migration between bound substrate and enzyme-bound coenzyme and thence to

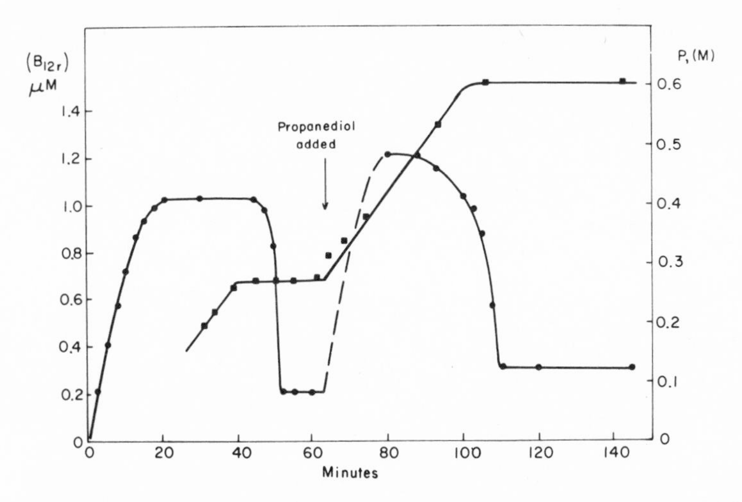

Fig. 20. The formation of Co(II) cobalamin B_{12r} (●) and propionaldehyde P (■) from propane-1,2-diol and coenzyme B_{12} in the presence of dioldehydrase under anaerobic conditions.

product has been described[240,273,274]. The stereochemical features of the reaction have been[271,272,275-277] elegantly investigated. Thus, 1-*R*-1-D-2-*R*-propane-1,2-diol (a) gives propionaldehyde with the deuterium in the C-2 position; conversely 1-*R*-1-D-2-*S*-propane-1,2-diol (b) gave propionaldehyde with the deuterium in the C-1 position (Fig. 19). Therefore, it can be concluded that hydrogen migration from C-1 to C-2 proceeds with inversion of configuration at C-2. Propane-1,2-diol, labelled with ^{18}O, showed that the hydrogen migration from C-1 to C-2 is accompanied by an oxygen shift from C-2 to C-1 followed by an enzyme catalyzed dehydration of the intermediate propane-1,1-diol. The mechanism of the glycerol reaction is[278] similar.

It had been suggested[279,280] that the Co–CH_2 bond could act as the hydrogen recipient. The holoenzyme is reduced by[281,282] glycolaldehyde to give 5′-deoxyadenosine, which is suggestive. It was shown that hydrogen transfer takes place *via* the $C_{5'}$-deoxyadenosylcobalamin. In doing so, the two $C_{5'}$-hydrogens become equivalent. Only the $C_{5'}$-hydrogens of deoxyadenosylcobalamin appear to be involved[283] in hydrogen transfer since there is no evidence for the involvement of $C_{4'}$ hydrogen in the reaction. Even the carbocyclic analog of 5′-deoxyadenosylcobalamins, *i.e.*, with the ribose oxygen replaced by a CH_2 group, has some activity[284]. The spectrum of the holoenzyme, during the enzymatic reaction, resembles those of Co(II) corrinoids. Though, in itself, this is not evidence for the participation of the Co(II) corrinoids in the catalytic step, a more-than-casual connection between the formation of Co(II) cobalamin and the enzymatic

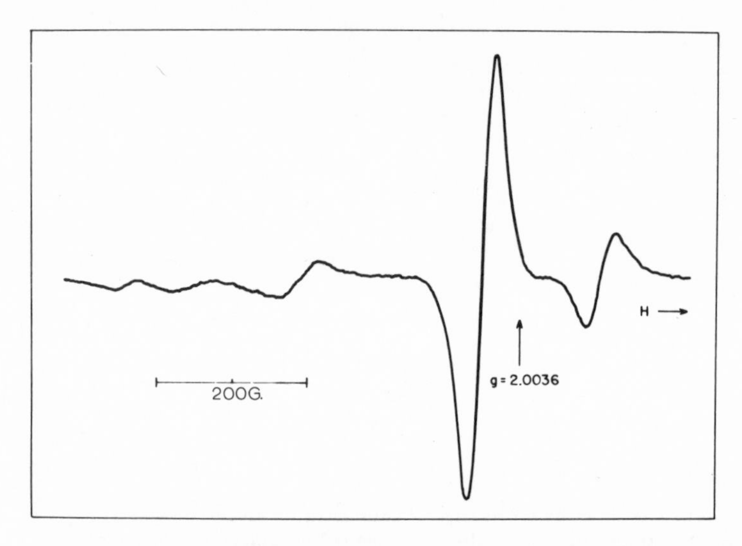

Fig. 21. The e.p.r. spectrum of coenzyme B_{12}, dioldehydrase and propane-1,2 diol in frozen aqueous solution at 100K.

reaction has been observed[270]. As shown in Fig. 20, when the substrate is consumed the absorption band associated with Co(II) cobalamin disappears; when more substrate is added, it reappears. Under the same anaerobic conditions, an e.p.r. spectrum is observed (Fig. 21) which has features, principally two resonances near $g = 2$, which are not present in the spectra of Co(II) corrinoids but which have been reported[285] in the spectrum of a preparation of ribonucleotide reductase. Such a spectrum may be due to the interaction of organic radicals with the Co(II) cobalamin.

Ethanolamine deaminase

This enzyme, elaborated by a clostridium which can utilize choline[286] or ethanolamine[287,288] as substrates for growth, requires a cobamide coenzyme. It has been purified[289,290], has a molecular weight of 520,000 with eight to ten subunits and requires K^+, NH_4^+, Rb^+, with Na^+ and Li^+ inhibitory. α-(Adenyl)-5′-deoxyadenosylcobamide has a K_m value of

7.7×10^{-6}, with the analogs, α-(5,6-dimethylbenzimidazolyl)-cobamide coenzyme, 1.5×10^{-6}, α-(benzimidazolyl)-cobamide coenzyme, 1.9×10^{-7} somewhat different. The V_{max} values are equal. 2′,5′-Dideoxyadenosylcobalamin is also effective with a lower V_{max} but B_{12}, B_{12a} and methylcobalamin are inhibitory[291]. The enzyme is inactivated by coenzyme in the absence of oxygen. It is also deactivated by mercurials. The enzyme binds two coenzyme molecules, each associated with a separate and independent active site[292].

Like dioldehydrase, the reaction involves the transfer of hydrogen to the bound coenzyme, probably in the rate-determining step[290,293,294] and thence to the carbon from which the amino group has moved to form presumably 1-aminoethanol from which ammonia is lost to give the product acetaldehyde. There is some evidence[294] that the enzyme can distinguish between the carbinol hydrogens and, as expected, the carbinol oxygen remains in the molecule. There is also evidence[295] for a reversible homolytic dissociation of the cobalt–carbon bond since species containing unpaired electrons have been detected[296,297] during catalysis. 5′-Deoxyadenosine has been detected as one of the products of this dissociation in the presence of substrate.

L-*β-Lysine mutase*

The anaerobic decomposition of lysine by various clostridia to give carboxylic acids and ammonia

$$\overset{6}{NH_2}\overset{5}{CH_2}\overset{4}{CH_2}\overset{3}{CH_2}\overset{2}{CH_2}CH(NH_2)\overset{1}{COOH} + H_2O + PO_4^{3-} + ADP \rightarrow$$

$$\overset{6}{CH_3}\overset{5}{CH_2}\overset{4}{CH_2}\overset{3}{COOH} + \overset{2}{CH_3}\overset{1}{COOH} + 2NH_3 + ATP \qquad (17)$$

requires a cobamide coenzyme as a catalyst[298-300]. The first intermediate in the process is the amino group transfer from carbon-2 to carbon-3 to give β-lysine (3,6-diaminohexanoate) which does not require[301] a cobamide. Free ammonia is not an intermediate in this reaction nor is it in the next, cobamide-dependent and reversible step to give[302] 3,5-diaminohexanoate.

The β-lysine mutase from *C. sticklandii* has been partly purified[303] and fractionated into two proteins, one of which contains a tightly bound cobinamide α-(adenyl)-5′-deoxyadenosylcobamide (mol. wt. 160,000) whilst the other is a sulfhydryl protein (mol. wt. 60,000). Both proteins and a monovalent cation, K^+, Rb^+ are necessary for activity and the incompletely purified system requires ATP, a thiol, FAD, Mg^{2+} and

References pp. 1122–1133

pyruvate, though all except the latter may simply be involved in the synthesis of a 5′-deoxyadenosylcobamide. The cobamide coenzyme functions as a hydrogen carrier[304] for the hydrogen which migrates from C-5 of β-lysine to C-6. Two other cobamide-dependent enzymes have been isolated from *C. sticklandii*; D-α-lysine mutase[305] which catalyzes the formation of 2,5-diaminohexanoate and the ornithine mutase[306,307] which gives 2,4-diaminopentanoate. These enzymes are quite similar to β-lysine mutase, consisting[308] of a cobamide-binding protein (mol. wt. 160,000) and a sulfhydryl protein (mol. wt. 60,000). D-α-Lysine mutase requires a monovalent cation (K^+, Rb^+), and a divalent metal ion ATP and pyridoxal phosphate for activity, the latter perhaps to react with the 6-amino group, making it more able to migrate to C-5. It has been suggested that pyruvate may have a similar function in L-β-lysine mutase. ATP increases[308] the affinity of D-α-lysine mutase for the substrate. Tritium is transferred[18,309] both from substrate to 5′-deoxyadenosylcobalamin and from hence to both substrate and product.

Ribonucleotide reductase

The implication of a vitamin B_{12} derivative in the conversion (Fig. 22) of ribonucleotides to 2′-deoxyribonucleotides in *Lactobacillus leichmannii*

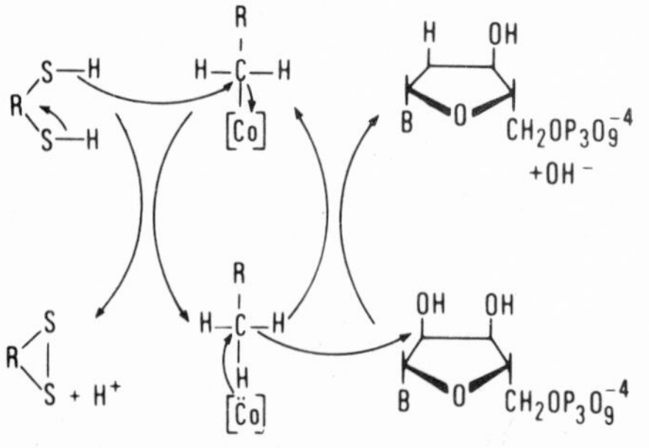

Fig. 22. Proposed mechanism for ribonucleotide reductase.

was suspected[31] sometime prior to the preparation of a cell-free system[310-315] with reductase activity and the purification[316-318] of the enzyme. Ribonucleoside triphosphates GTP, ITP, ATP, CTP and UTP (in order of decreasing effectiveness)[311,317,319-321] are the preferred substrates for this enzyme; that from *E. coli*, has ribonucleoside diphosphates as substrates, does not require a cobamide coenzyme[31]. The purified reductase has[315] a molecular weight of 115,000 though another preparation[317] had a lower molecular weight (70,000). The requirements for cytosine triphosphate reduction are α-(5,6-dimethylbenzimidozolyl) 5′-deoxyadenosylcobamide, ATP, a dithiol reductant and a relative requirement for Mg^{2+} (or Mn^{2+} or Ca^{2+}). The dithiol must be capable of intramolecular cyclization on oxidation, *e.g.*, dihydrolipoate, dithioerythritol or dithiothreitol. *E. coli* thioredoxin and thioredoxin reductase can serve as hydrogen donor[320] though presumably that in *L. leichmannii* serves *in vivo*[322]. The oxidation of dehydrolipoate is stoichiometrically related to the reduction of the nucleotide[323] and the enzyme after treatment with dehydrolipoate or the thioredoxin system is inactivated with iodoacetamide, one disulfide bond being opened[316] on reduction. 5′-Deoxyadenosylcobalamin is partly replaceable by α-(benzimidazolyl)-5′-deoxyadenosylcobamide but not by methylcobalamin, cyanocobalamin or hydroxocobalamin[313,318].

A most interesting feature of the system is that the 2′-deoxyriboside triphosphates act as[317,318] positive allosteric effectors, the reduction of each triphosphate being maximally stimulated by a different 2′-deoxyribonucleoside triphosphate; dATP for CTP; dCTP for UTP, dGTP for ATP and dTTP for GTP. Both Mg^{2+} and the effector molecules change[138] the physical state of the molecule.

Unlike the dioldehydrase reaction, reduction is accompanied by introduction into the product of hydrogen from the solvent and most importantly only into the C-2′ position[324,325] with retention of configuration[326] which is similar to the cobamide-independent[327] reduction in *E. coli.*

Again, in contrast to the dioldehydrase reaction, tritium from 5′-deoxyadenosylcobalamin-5′-$^3H_2$ is not transferred to product but to water[328,329] even in the absence of substrate but always requiring the presence of the effector[330]. In the other reactions requiring a Co-5′-deoxyadenosylcobamide, the hydrogen which is transferred does not exchange with solvent, whereas in the reductase reaction, the sulfur-bound hydrogens exchange readily with solvent hydrogens. The reaction scheme shown in Fig. 22 has been proposed[331]. The dithiol reduces the Co-5′-deoxyadenosylcobalamin to give 5′-deoxyadenosine as intermediate, the reaction being reversible and the rates faster than the subsequent transfer of the hydrogen from the 5′-deoxyadenosine cobamide complex to the ribonucleotide, hence no label originating in the coenzyme is found in the deoxyribonucleotide. Though cobalt(II) cobalamin has been detected[138,332] in the holoenzyme in the presence of dihydrolipoate and a nucleoside triphosphate, it was concluded that it was formed by oxidative degradation of cobalt(I) cobalamin, since the rate of formation was much slower than the rate of hydrogen exchange with solvent or reduction of the ribonucleotide. However, the generation of the species giving rise to the e.p.r. spectrum similar to that found in the diol dehydrase, is formed at a rate comparable with that of ribonucleotide reduction[285].

References pp. 1122–1133

Methionine synthetase (N^5-methyltetrahydrofolate-homocysteine cobalamin transmethylase)

The biosynthesis of methionine involves[333] the transfer of a methyl group from N^5-methyltetrahydrofolate (5-CH_3-H_4-folate) to homocysteine:

$$\text{5-}CH_3\text{-}H_4\text{-folate} + HSCH_2CH_2CH(NH_2)COOH \xrightarrow{\text{SAM}} H_4\text{-folate} + CH_3SCH_2CH_2CH(NH_2)COOH \qquad (18)$$

It so requires a corrinoid and has as cofactors *S*-adenosylmethionine (SAM), a reduced flavin and 1,4-dithiothreitol (DTT). It has been partially purified from *E. coli*[334–340], *Streptoccus faecalis*[342], *Salmonella typhimurium*[343], mammalian[170, 336, 344] and avian liver[336]. The enzyme, purified[338] from *E. coli*, has a molecular weight of 140,000, and methylcobalamin has been isolated from both the bacterial[345] and mammalian[344] enzymes. Methylcobalamin had long been realized[345] to be an intermediate in the methylation reaction though a thiol-cobalamin was possibly present in the inactive form of the enzyme[335, 346]. The roles of the two methylating agents, 5-CH_3-H_4-folate and SAM, have gradually been unravelled. Only catalytic amounts of SAM are required to sustain the transmethylase reaction[168,346] and indeed an enzyme which is independent of SAM has been described[347-349] though the dependence can be restored by reaction with homocysteine in the absence of 5-CH_3-H_4-folate or by a "demethylating" protein[350].

Inhibition studies using n-propyl iodide[351] and elegant tracer experiments[347,352] have shown that there is probably a sequential transfer of methyl groups to bound cobalamin (Fig. 23). In the presence of reduced flavin and DTT, SAM gives a methylcobalamin, which reacts with homocysteine to give methionine and an active reduced cobalamin. The latter demethylates 5-CH_3-H_4-folate and the reversible cyclic process continues unless the active intermediate, perhaps Co(I) cobalamin, is blocked by propylation (which can be reversed by photolysis) or inactivated by oxidation. SAM and the reductants are required to compensate for oxidative inactivation during catalysis.

Though the evidence to hand is certainly consistent with this mechanism, there are many details still unclear. Neither SAM nor 5-CH_3-H_4-folate react with hydroxocobalamin in the presence of the flavin reducing system but in the absence of the apoenzyme[338]. The reduced enzyme does react with SAM, however, to give holoenzyme containing methylcobalamin and the rate of the reaction is increased by 5-CH_3-H_4-folate. If the reaction is analogous to that of SAM with Co(I) cobalamin, some as yet unknown property of the protein must be responsible for the conversion of the corrinoid, probably Co(II) cobalamin, to a form which can demethylate SAM. The role of the thiol *e.g.* DTT in the reduction is not understood[348,353,354] and though it could give Co(I) cobalamin if mechanism (5) on p. 1092, is applicable, it is also possible its function is to reduce -S-S-

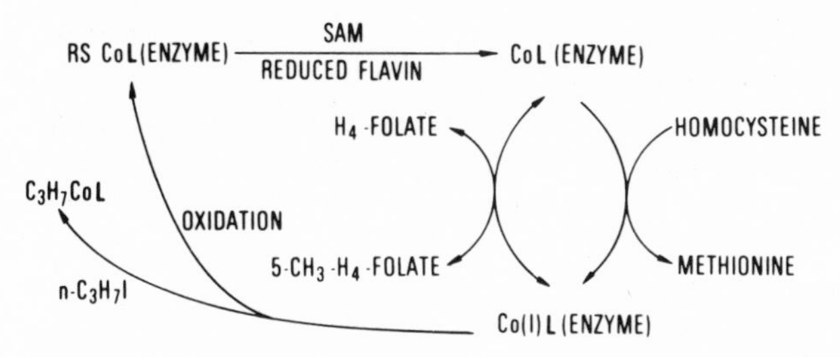

Fig. 23. Proposed mechanism for methionine synthetase.

bonds in the protein or in some way to make the cobalt more accessible to the reductant and/or methylating agent.

The non-enzymatic reaction of methylcobalamin with thiols is complex, so it is not obvious that the subsequent methyl transfer to homocysteine will generate a Co(I) intermediate though, of course, the protein may alter the capacity of the methyl group to "leave" as CH_3^+. If the products of methyl transfer are methionine and a Co(II) cobalamin, further reduction would be necessary to give the supposedly active Co(I) corrinoid.

It is this next step in the reaction which is the most puzzling feature, since 5-CH_3-H_4-folate is a relatively poor methyl donor. There are no examples[167,177] of *in vitro* methylation of corrinoids by *N*-methyl donors but the possible modification of the corrinoid or 5-CH_3-H_4-folate by the protein means that such observations are not necessarily relevant. However, the identity of the cobalamin which can demethylate 5-CH_3-H_4-folate is still not known. Though it appears to be derived from a Co-methyl complex formed by the SAM methylation reaction this is not absolutely necessary for the methylation of homocystein by 5-CH_3-H_4-folate, since the product of the reaction of methylcobalamin with apoenzyme is alone capable[348] of methyl transfer. SAM and the reducing system are required to achieve a steady-state transfer. One possibility is that the protein can stabilize protonated Co(I) cobalamin *i.e.*, hydridocobalamin, which otherwise has only a transient existence in aqueous solution; before decomposing to Co(II) cobalamin and hydrogen. Perhaps hydridocobalamin is formed only in the locus of the methyl by the transfer reaction shown below:

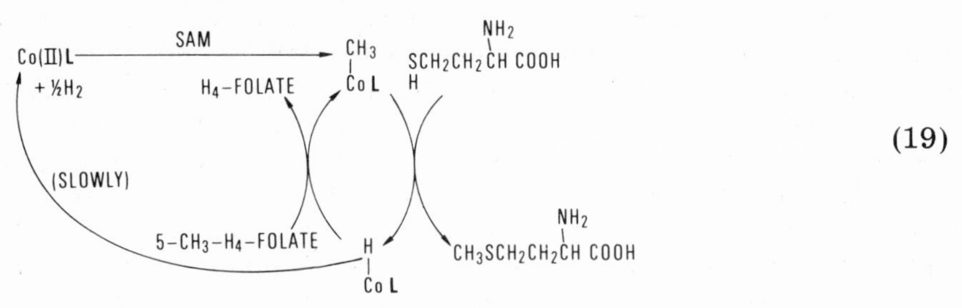

(19)

and not by other reductants. Therefore at least one methyl transfer would be necessary before a cobalamin was formed which could demethylate 5-CH_3-H_4-folate. The concentration of hydridocobalamin would gradually diminish and consequently remethylation would be required. The transferase activity of the methylcobalamin enzyme complex[348] apparently disposes of the intriguing possibilities[15] of the involvement of dimethyl cobalamins, unless further reduction of methylcobalamin by the protein is

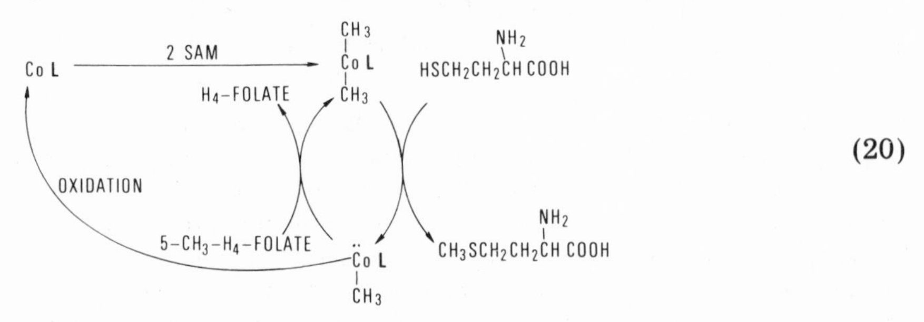

(20)

possible. Otherwise (20) may be feasible, where the *reduced* methyl corrinoid would fulfil the role of the demethylating agent for 5-CH_3-H_4-folate and would obviously have to be formed by some prior methylation.

Methane synthetase and methyl transferase

Some bacteria form methane during the anaerobic decomposition of certain organic compounds including methanol and acetate, the methyl group being preserved intact during the reduction[355,356]. Cell-free extracts of *Methanosarcina barkeri*[357] and *Methanobacillus omelianskii*[358] have been shown to form methane from carbon dioxide, formate, formaldehyde, methanol, pyruvate, ^5N-methyltetrahydrofolate, 5,10-methylenetetrahydrofolate, and Co-methylcorrinoids, and in addition *M. omelianskii* converts[359,360] the C-3 of serine to methane. A variety of Co-methyl derivatives are converted to methane, including methylcobalamin[358,361] methylcobinamide[362] and α-(5-hydroxybenzimidazolyl)-methylcobamide. In the presence of catalytic amounts of Co(II) cobalamin, some Co-methyl DMG complexes[363,364] acted as methyl donors, though the methyl group is first transferred to cobalamin, an interesting observation in itself[102]. Ethyl- and propyl-cobalamins are[365] inhibitors though difluoromethylcobalamin is a substrate giving methane by further reduction of the intermediate Co-monofluoromethylcobalamin. Interestingly reduction of a fluorocarbene intermediate was also proposed[102]. Both *M. omelianskii*[83] and *M. Barkeri* preparations gave at least two protein fractions[366-368] and from a red one (5-hydroxybenzimid-azolyl)-aquocobamide was isolated[252,365,369,370]. A Co-methyl derivative is probably present as an enzyme-bound intermediate and indeed photolysis of a corrinoid-containing protein yielded a product whose e.p.r. spectrum was much like that of Co(II) cobalamin. However, the spectrum of the active protein resembled[371] that of a B_{12}–thiol complex, which, of course, would also be photosensitive.

The normal overall reaction in the presence of a reductase such as pyruvate dehydrogenase, is the formation of methane. In the absence of a reducing system, the methyl group can be transferred to Co(I) cobalamin and the same red protein is required[368] for transferase activity. The mechanisms of reduction or transfer are not known. Comparison with the *in vitro* reductive cleavage, of Co-methyl corrinoids suggests that both one- and two-electron reductions are possible, though very recent evidence[372] suggests that the methyl group is transferred to a small molecule, coenzyme M and then reduction to methine takes place. This important observation demonstrates only too well the difficulty of relating *in vitro* or model experiments to the mechanism of biochemical reactions.

Acetate synthetase

The conversion of carbon dioxide to acetate by *Clostridium thermoaceticum* requires a corrinoid[373]. Cell-free extracts catalyze the synthesis of Co-methyl corrinoids from carbon dioxide and their conversion to acetate and hence both carbon atoms of acetate derive[374] from carbon dioxide. There is good evidence that the key intermediate[375,376] is α-(5-methoxybenzimidazolyl)-Co-methylcobamide. Two protein fractions have been obtained[377,378] from the cell-free extracts and in combination with coenzyme A, pyruvate and a ferredoxin, convert methylcobalamin to acetate, and one of them converts carboxymethylcobalamin to acetate. The available evidence suggests that one mole of carbon dioxide gives a methylated derivative of H_4-folate which is able to transfer this methyl group to a reduced cobamide. The methyl cobamide could be carboxylated to give perhaps a carboxymethyl cobamide[375] which would then be reduced to give acetate:

$$\underset{\text{CoL}}{\overset{\text{CH}_3}{|}} \xrightarrow{\text{CO}_2} \underset{\text{CoL}}{\overset{\text{CH}_2\text{COOH}}{|}} \xrightarrow{[\text{H}]} \text{CoL} + \text{CH}_3\text{COOH} \qquad (21)$$

In this case only two of the hydrogens of the methyl group would appear in the acetate unless the hydrogen lost in the carboxylation was retained by protein and used in the reduction of the carboxymethyl complex.

Another possibility[280] is that heterolytic fission of the cobalt–carbon bond gives a Co(III) . . . CH_3^- intermediate which then reacts to give $CH_3CO \cdot O \cdot CoL$, which yields acetate. Recent evidence[379] provides some support for this mechanism. It is also possible that prior reduction of the Co-methyl intermediate makes it a better CH_3^--donor; here to CO_2, and in methane synthetase, to H^+.

MECHANISM OF REACTIONS REQUIRING COENZYME B_{12}

Though the mechanisms of the enzymatic reactions requiring 5′-deoxyadenosyl corrinoids will probably differ in detail, there are enough features in common, particularly concerning hydrogen transfer (see Table VI), to justify considering them together. Any discussion of possible mechanisms tends to concentrate on the properties of the coenzymes but in the absence of more detailed information on the role of the apoenzymes, it is an understandable failing. Indeed the speculative content of any mechanistic discussion probably varies inversely with our knowledge of the role of the protein.

Most of the apoenzymes bind cobalamins irrespective of the sixth ligand, with the amide side chains probably playing the major role in binding. The chemistry of alkylcobalamins and cobinamides suggests that the fifth ligand is readily, both thermodynamically and kinetically, replaced. Therefore, it is possible that a group from the protein displaces, for example, the nucleotide, though the protein, by binding tightly to chain *f*, could result in the absence of any fifth ligand. The latter situation would be tolerated by Co-alkyl derivatives, but if the *trans*-cobalt–carbon bond broke subsequently, the presence or absence of a fifth ligand would significantly affect the properties of the cobalt. If cobalt(I), the effect would be slight, but a "four" coordinate cobalt(II) and particularly cobalt(III) corrinoid, would have properties much removed from those of the corresponding complexes in solution where the coordination number was higher. Thus the Lewis acidity of a Co(II) or Co(III) corrinoid, which had "no" axial ligands would be much greater than if an axial nitrogen ligand was present. Consequently, in the absence of detailed knowledge of the nature and number of the axial ligands in enzyme-bound coenzymes, one must be very careful about extrapolation from the properties of the corrinoids in solution.

There is good evidence that the protein certainly modifies the properties of the Co–C bond, presumably by binding the 5′-deoxyadenosyl group. The altered sensitivity to light and cyanide of the coenzymes when bound to apoenzymes may indicate that the Co–C bond is broken but such evidence is ambiguous since the protein could "trap" the 5′-deoxyadenosine radical and the cobalt–carbon bond subsequently re-form. More telling evidence is the instability of the holoenzyme in the absence of substrate, which suggests that a reactive species is formed which attacks the protein. This could either be the 5′-deoxyadenosyl radical, or, since the role of oxygen has been noted in this decomposition, superoxide or peroxide formed by reaction with Co(II) or Co(I) intermediates.

This raises the first important point in the discussion of the mechanism. Assuming the cobalt–carbon bond breaks, is the fission heterolytic or homolytic? Most mechanisms proposed previously have assumed that heterolysis took place, but, in at least three of the enzymatic reactions, cobalt(II) corrinoids and radical intermediates have been detected. It had been noted earlier that the environment of the cobalt–carbon bond was non-polar. If this is reflected in a low local dielectric, the homolytic[380] pathway might be favored. Concomitantly, the properties of the cobalt(II) corrinoid, which has often been described as "unreactive," may be significantly altered in the low dielectric as, of course, would be a cobalt(I) or cobalt(III) intermediate. Present evidence certainly does not exclude a homolytic pathway but, of course, identification of the cobalt(II) form does not necessarily mean that it is the effective catalyst, since a low concentration of another oxidation state would remain undetected.

The reaction of the coenzyme with the apoenzyme need only generate a low concentration, which might be increased by substrate, of the complex with the cobalt–carbon bond no longer intact. Indeed, as mentioned above, it is better to consider the fission products in equilibrium with the coenzyme (Fig. 24). In propanediol dehydrase, it has been shown that the fission is not promoted by base-catalyzed cleavage to give 4′,5′-dehydrodeoxyadenosine, nor does the ribose oxygen participate in the reaction. Consequently we will consider only the reactions of the species produced by the cleavage reactions shown in Fig. 24. Further electron transfer could give equilibria between all three species.

The next step in the reaction involves hydrogen transfer as a hydride (a), hydrogen atom (b) or proton (c) to give respectively 5′-deoxyadenosine and (d), (e) or (f). 5′-Deoxyadenosine has been detected in the *reaction* of

References pp. 1122–1133

propanedioldehydrase with glyoxal, which is suggestive. The residence time of the hydrogen transferred to $C_{5'}$ must be sufficiently long that it becomes one of three equivalent hydrogens, presumably by rotation about the $C_{4'}$-$C_{5'}$ bond. Is this initial transfer of hydrogen followed by formation of a bond

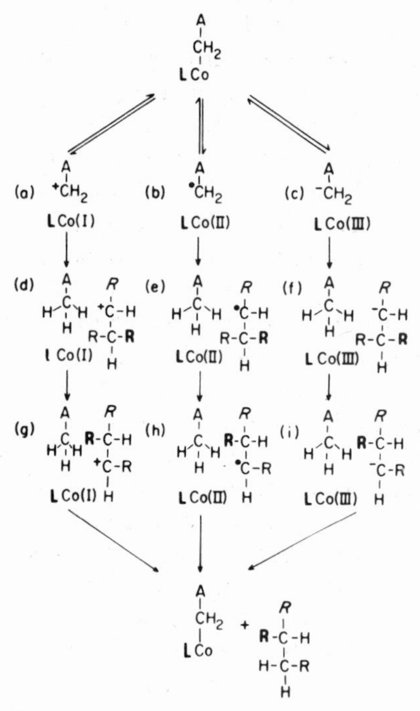

Fig. 24. Possible mechanisms for enzymatic reactions requiring coenzyme B_{12}.

between cobalt and the hydrogen-deficient carbon? This has often been suggested, with the subsequent 1,2-shift of R, taking place on the cobalt perhaps via an intermediate π-complex:

$$\mathrm{A{-}CH_3} \;+\; \underset{\mathrm{LCo}}{\underset{\vdots}{(R)(H)C}}{-}C(R)(R)(H) \quad \text{(a)} \longrightarrow \underset{\mathrm{LCo^{*}}}{\mathrm{A{-}CH_3}} \;\; R{-}C(R)(H){-}{}^{*}C(R)(H) \qquad (22)$$

with the number of electrons on cobalt and carbon unspecified (and denoted by the asterisk), though the total number of electrons involved is two. Some reactions of Co-alkyl corrinoids in solution have been assumed to provide support for this reaction. Indeed, it is possible that one role of the cobalt could be to provide a good "leaving group" and in addition one of *variable charge* in the sense that cobalt, acting as a leaving group as

Co(III), Co(II) or Co(I), could thereby accommodate the shift preferences of the substrate. It is also possible that the rearrangement of any of the species in Fig. 24 could proceed without the intervention of (a) in eqn. (22). No experiment has been reported which discriminates between these two possibilities. In either case, the ability of R to undergo a 1,2-shift will be important, and it is possible that some of the co-factors required, *e.g.* pyridoxal phosphate in β-lysine mutase, are required to alter the ability of R in this migration. Not all reactions need proceed via the same mechanism (eqn. (22) or Fig. 24).

The next intermediate will be (g), (h) or (i), formed from (d), (e) and (f) respectively. In (g), the Co(I) corrinoid could act as a nucleophile displacing hydride from $C_{5'}$ of deoxyadenosine to reform the coenzyme and give "product".

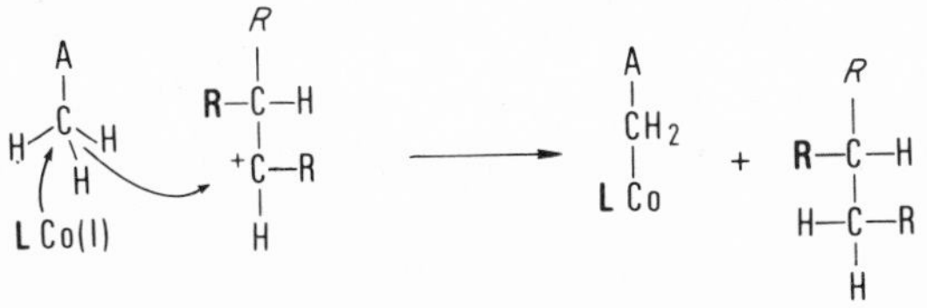

$$\text{A–CH}_2\text{(H)}\cdots\text{LCo(I)} + \text{R–}\overset{R}{\text{C}}\text{H–}{}^{+}\text{CH–R} \longrightarrow \text{A–CH}_2\text{–LCo} + \text{R–}\overset{R}{\text{C}}\text{H–CH}_2\text{–R} \qquad (23)$$

The nucleophilicity could be affected by the coordination of a good donor to the cobalt. The close proximity of the cobalt to the $C_{5'}$ carbon and of the carbonium ion to the hydrogen could make this mechanism feasible. The presumed lack of reactivity of the $C_{5'}$ methyl has made this mechanism and any of those derived from (g), (h) or (i) seem unreasonable. However, it is probably misleading to consider the $C_{5'}$ methyl as an *isolated* methyl group. The proteins may ensure that it is forced close to the cobalt and in that case it is better to consider the cobalt–hydrogen–$C_{5'}$ system as a unit. If that is the case, the properties of the hydrogens may bear little relation to those of conventional methyl groups. Such alterations in the properties of functional groups in proteins is a well-known phenomenon. Even in inorganic chemistry, proximity of carbon–hydrogen bonds to metals may cause "unusual reactions", witness the transformation of metal triarylphosphines to give hydridoaryl metal phosphines.

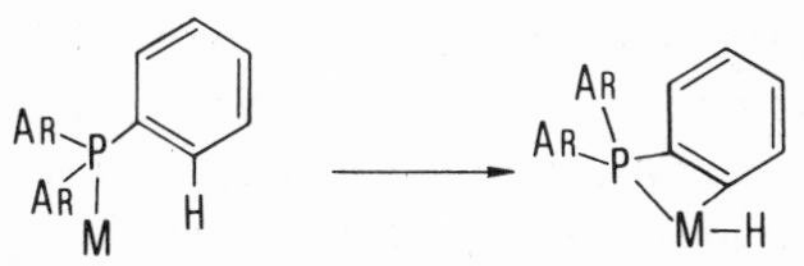

$$\text{Ar}_2\text{P(C}_6\text{H}_5\text{)–M} \longrightarrow \text{Ar}_2\text{P(C}_6\text{H}_4\text{)M–H} \qquad (24)$$

There is a possible analogy here with eqn. (23) though, in the latter, carbon would be the hydride acceptor. To emphasize the interaction between the cobalt and the neighboring hydrogen, it may be more appropriate to consider the three-centered system; C. . .H. . .Co. The transfer of hydrogen to

References pp. 1122–1133

"substrate" carbon then proceeds with the concomitant shift of two, one and zero electrons in (g), (h) and (i) and the simultaneous formation of the Co-5′-deoxyadenosyl bond.

This discussion has tried to emphasize the possible alteration of the properties of both the 5′-deoxyadenosyl moiety and the cobalt corrinoid by the protein. Lack of information has restricted the role of the apoenzyme to a chemically "passive" function. This is undoubtedly not so. The function of sulfhydryl groups, whether on a separate protein, or not; the role of cofactors, of monovalent cations, and particularly the mechanism by which the apoenzyme modifies the Co–C bond, remain unknown. As more detailed kinetic experiments are reported and the state of the corrinoids, during the catalytic step, detected, so a distinction might be made between the various mechanisms described here. To end on a cautionary note; though a knowledge of the chemistry of the corrinoids is indispensable to an understanding of their biochemical role, it would be naive to expect their properties in solution to be replicated in the enzyme. It is difficult to believe that there are no more surprises in store.

ACKNOWLEDGEMENTS

I wish to thank Drs. M. A. Foster, B. A. Hill, F. S. Kennedy, R. J. P. Williams, Profs. J. D. Brodie, H. P. C. Hogenkamp, B. L. Vallee and J. M. Wood for their generous help.

APPENDIX

Since the time that the manuscript of this review was completed, interest in the many and various aspects of the chemistry and biochemistry of the corrinoids has continued undiminished. Indeed, interest has even been aroused[382] in political circles in connection with the role of methylcobalt corrinoids in mercury pollution. However in calmer areas, the search for intermediates in enzymatic reactions and novel properties of the cobalt–carbon bond, has been persued with ever-increasing vigour.

A recent book describes[383] the inorganic chemistry of corrinoids in considerable detail and an entertaining review has been published[384] which emphasizes the role of spin-labelled coenzyme analogues in the study of B_{12}-dependent enzymatic reactions. An alkylcobalt derivative has been prepared[385] from a high-spin cobalt(II) complex and the work of Costa and his colleagues has been reviewed[386]. Full details of the structure of cobyric acid[387] and cyanocobalamin-5-phosphate[388] have been published. Neovitamin B_{12} has been shown[389] to result from epimerization at C_{13}. Structural studies on some model complexes have been reviewed[390].

The importance of n.m.r. spectroscopy increases with increasing sophistication in instrumentation. Attempts have been made[391,392] to deduce the structures of cobalamins and cobinamides in solution from ^{1}H n.m.r. spectra. The initial results are encouraging and work employing shift reagents may allow[393] a meaningful comparison between structure in solution and in the crystalline state.

The rates of substitution reactions in cobalamins[394] and bis-dimethylglyoximatocobalt(III) complexes[395,396] have been determined. Electrochemical investigations[397] have shown that, though the one-electron reduction products of alkylcobalt complexes decompose fairly rapidly, those from aryl cobalt complexes are relatively stable. Recent calculations, which include explicitly the eigen-functions of the cobalt atom, show[398] that the electronic absorption spectra are very sensitive to the charge placed on the cobalt atom, and that the so-called anomalous spectra of, for example, methylcobinamide can be understood.

It was concluded[399] that hydridocobalamin formed using zinc dust in glacial acetic acid as reductant since the products formed by reaction of the reduced cobalamin with substituted alkenes differ from those formed by reaction in water. Cobalt(I)cobalamin is probably the effective reductant in the catalysis by cobalamins of the borohydride reduction of dihalo compounds[400] and nitro and nitroso compounds[401].

Quantum yields have been reported[402] for the photolysis of methylcobalamin in water. In dry *N,N*-dimethylformamide, coenzyme B_{12} is completely stable to cyanide[403], suggesting that prior protonation of the adenosine is a prerequisite for fission of the cobalt–carbon bond. Further studies of the solvent dependence would be useful since the environment of the coenzyme when bound to protein may differ considerably from that in water.

In a very interesting reaction methanolysis[404] of 2-acetoxyethyl-(pyridine)bis-dimethylglyoximatocobalt(III) takes place perhaps *via*[405] a π-complex. Alkyl transfer to mercury[406–409], cobalt[410] and the reaction of alkylcobalt corrinoids with other metal ions[411] have emphasized the mobility of the coordinated methyl group under various conditions. A useful collection of articles on preparative and analytical features of corrinoids has been published[412].

There have been [413,414] two interesting papers on the mammalian biochemistry of B_{12}. ^{1}H n.m.r. spectral investigations of methylmalonyl CoA mutase suggest[415] that the 5,6-dimethylbenzimidazole is not involved in binding of coenzyme to the protein. Investigations of glutamate mutase show[416] that hydrogen which is transferred to coenzyme becomes one of three equivalent hydrogens. The details of the preparation of α-methyleneglutarate mutase have been published[417].

There have been two important papers on ethanolamine deaminase (ethanolamine ammonia lyase) which show that the rate-determining step[418]

References pp. 1122–1133

is hydrogen transfer from coenzyme to form product and the formation[419] of 5′-deoxyadenosine during reaction with acetaldehyde and the ammonium ion. Ribonucleotide reductase has been purified[420] (molecular weight 76,000) and though it has no quaternary structure, it is subject to regulatory control. There is[421] further evidence for the formation of Co(II)-cobalamin under certain conditions. In an interesting paper[422] on the kinetic aspects of the methionine synthetase it was concluded that the data are consistent with a ping-pong mechanism, in which SAM and $5CH_3$–H_4-folate react with distinct enzyme forms, each of which gives rise to an enzyme-bound methylcobalamin. There have been[423,424] two papers on the regulatory aspects of methione synthesis and an interesting paper on binding of cobalamins to the intrinsic factor[425]. So a selection from the year's output shows that corrinoids continue to fascinate scientists from a wide range of disciplines.

REFERENCES

1 E. L. Rickes, N. G. Brink, F. R. Koniuszy, T. R. Wood and K. Folkers, *Science*, 107 (1948) 396.
2 E. Lester Smith and L. F. Parker, *Biochem. J.*, 43 (1948) viii.
3 E. Lester Smith, *Vitamin B_{12}* 3rd Edn., Methuen, London, 1965.
4 R. Bonnett, *Chem. Rev.*, 63 (1963) 573.
5 K. Bernhauer, O. Müller and F. Wagner, *Angew. Chem. (Int. Edn.)*, 3 (1964) 200.
6 K. Bernhauer, O. Müller and F. Wagner, *Adv. Enzymol.*, 26 (1964) 233.
7 D. Perlman, *Ann. N.Y. Acad. Sci.*, 112 (1964) 547.
8 A. F. Wagner and K. Folkers, *Vitamins and Coenzymes*, J. Wiley, New York, 1964, p. 194.
9 L. Jaenicke, *A. Rev. Biochem.*, 33 (1964) 287.
10 D. C. Hodgkin, Discussion on Corrin, *Proc. R. Soc. (A)*, 288 (1965) 294 *et seq.*
11 F. Wagner, *A. Rev. Biochem.*, 35 (1966) 405.
12 T. C. Stadtman, *A. Rev. Microbiol.*, 21 (1967) 121.
13 H. P. C. Hogenkamp, *A. Rev. Biochem.*, 37 (1968) 668.
14 G. N. Schrauzer, *Accounts Chem. Res.*, 1 (1968) 97.
15 F. M. Huennekens, in T. P. Singer, *Biological Oxidations*, Interscience, New York, 1968, p. 439.
16 H. A. O. Hill, J. M. Pratt and R. J. P. Williams, *Chem. Brit.*, 5 (1969) 156.
17 A. Eschenmoser, *Q. Rev.*, 24 (1970) 366.
18 T. C. Stadtman, *Science*, 171 (1971) 859.
19 C. Brink-Shoemaker, D. W. J. Cruickshank, D. C. Hodgkin, J. Kamper and D. Pilling, *Proc. R. Soc. (A)*, 278 (1964) 1.
20 D. C. Hodgkin, J. Pickworth, J. H. Robertson, K. N. Trueblood, R. J. Prosen and J. G. White, *Nature*, 176 (1955) 325.
21 D. C. Hodgkin, J. Kamper, J. Lindsey, *et. al.*, *Proc. R. Soc. (A)*, 242 (1957) 228.
22 D. C. Hodgkin, J. Pickworth, J. H. Robertson, R. J. Prosen, R. A. Sparkes and K. N. Trueblood, *Proc. R. Soc. (A)*, 251 (1959) 306.

23 D. C. Hodgkin, J. Lindsey, M. MacKay and K. N. Trueblood, *Proc. R. Soc. (A)*, 266 (1962) 475.
24 D. C. Hodgkin, J. Lindsey, R. A. Sparkes, K. N. Trueblood, and J. G. White, *Proc. R. Soc. (A)*, 266 (1962) 494.
25 J. G. White, *Proc. R. Soc, (A)*, 266 (1962) 440.
26 J. I. Toohey, *Proc. Natn. Acad. Sci. U.S.*, 54 (1965) 394.
27 H. A. Barker, H. Weissbach and R. D. Smyth, *Proc. Natn. Acad. Sci. U.S.*, 44 (1958) 1093.
28 H. Weissbach, J. I. Toohey and H. A. Barker, *Proc. Natn. Acad. Sci. U.S.*, 45 (1959) 521.
29 P. G. Lenhert and D. C. Hodgkin, *Nature*, 192 (1961) 937.
30 J. I. Toohey and H. A. Barker, *J. Biol. Chem.*, 236 (1961) 560.
31 W. S. Beck, *Vitamins and Hormones*, 20 (1968) 413.
32 M. M. Wintrobe and G. R. Lee, in M. M. Wintrobe *et. al.*, *Harrison's Principles of Internal Medicine*, 6th Edn., McGraw-Hill, New York, 1970.
33 IUPAC-IUB Commission on Biochemical Nomenclature, *J. Biol. Chem.*, 241 (1966) 2987.
34 G. C. Hayward, H. A. O. Hill, J. M. Pratt, N. J. Vanston and R. J. P. Williams, *J. Chem. Soc.*, (1965) 6485.
35 W. Friedrich and K. Bernhauer, *Chem. Ber.*, 89 (1956) 2030.
36 R. Bonnett, J. R. Canon, V. M. Clark, A. W. Johnson, L. F. Parker, E. L. Smith and A. R. Todd, *J. Chem. Soc.*, (1957) 1158.
37 F. Wagner, *Proc. R. Soc. (A)*, 288 (1965) 344.
38 H. A. O. Hill, B. E. Mann, J. M. Pratt, and R. J. P. Williams, *J. Chem. Soc. (A)*, (1968) 564.
39 W. Friedrich and K. Bernhauer, *Z. Naturforsch.*, 9b (1954) 685.
40 W. Friedrich and K. Bernhauer, *Chem. Ber.*, 89 (1956) 2507.
41 J. M. Pratt, *J. Chem. Soc.*, (1964) 5154.
42 E. Lester Smith, L. Mervyn, A. W. Johnson and N. Shaw, *Nature*, 194 (1962) 1175.
43 A. W. Johnson, L. Mervyn, N. Shaw and E. Lester Smith, *J. Chem. Soc.*, (1963) 4146.
44 O. Müller and G. Müller, *Biochem. Z.*, 336 (1962) 299.
45 O. Müller and G. Müller, *Biochem. Z.*, 337 (1963) 179.
46 K. Bernhauer, O. Müller and G. Müller, *Biochem. Z.*, 336 (1962) 102.
47 G. Costa, G. Mestroni and L. Stefani, *J. Organometal. Chem.*, 7 (1967) 493.
48 G. Costa, G. Mestroni and G. Pellizer, *J. Organometal. Chem.*, 11 (1968) 333.
49 A. Bigotto, G. Costa, G. Mestroni, *et. al.*, *Inorg. Chim. Acta Rev.*, in the press.
50 G. Costa, G. Mestroni, G. Tauzher and L. Stefani, *J. Organometal. Chem.*, 6 (1966) 181.
51 G. Costa and G. Mestroni, *J. Organometal. Chem.*, 11 (1968) 325.
52 G. Costa and G. Mestroni, *Tetrahedron Lett.*, (1967) 4005.
53 G. Costa, G. Mestroni and E. L. Savorgnani, *Inorg. Chim. Acta*, 3 (1969) 323.
54 E. Ochiai and D. H. Busch, *Chem. Commun.*, (1968) 905.
55 E. Ochiai, L. M. Long, C. R. Sperati and D. H. Busch, *J. Am. Chem. Soc.*, 91 (1969) 3201.
56 D. A. Clarke, R. Grigg and A. W. Johnson, *Chem. Commun.*, (1966), 208.
57 D. A. Clarke, D. C. Dolphin, R. Grigg, A. W. Johnson and H. A. Pinnock, *J. Chem. Soc. (C)*, (1968) 881.
58 P. Day, H. A. O. Hill and M. G. Price, *J. Chem. Soc. (A)*, (1968) 90.
59 D. Rietz, *Doctoral Thesis*, Techn. Hochschule, Stuttgart, 1964.
60 P. G. Lenhert, *Proc. R. Soc. (A)*, 303 (1968) 45.
61 J. D. Dunitz and E. F. Meyer, *Proc. R. Soc. (A)*, 288 (1965) 324.

62 F. M. Moore, B. T. M. Willis and D. C. Hodgkin, *Nature*, 214 (1967) 130.
63 D. C. Hodgkin, in O. Bastiansen, *The Law of Mass Action*, Det Norske Videnskaps-Aka Demi I, Oslo, Universitetsforlaget, 1964.
64 S. A. Cockle, H. A. O. Hill, B. E. Mann and R. J. P. Williams, *Biochim. Biophys. Acta*, 215 (1970) 415; H. A. O. Hill, J. W. Kendall, A. M. Turner and R. J. P. Williams, to be published.
65 S. Brückner, M. Calligaris, G. Nardin and L. Randaccio, *Inorg. Chim. Acta*, 3 (1968) 308.
66 S. Brückner, M. Calligaris, G. Nardin and L. Randaccio, *Chem. Commun.*, (1970) 152.
67 S. Brückner, M. Calligaris, G. Nardin and L. Randaccio, *Inorg. Chim. Acta*, 2 (1969)
68 P. G. Lenhert, *Chem. Commun.*, (1967) 980.
69 S. Brückner, M. Calligaris, G. Nardin and L. Randaccio, *Inorg. Chim. Acta*, 3 (1969) 278.
70 S. Brückner, M. Calligaris, G. Nardin and L. Randaccio, *Inorg. Chim. Acta*, 2 (1968) 386.
71 S. Brückner, M. Calligaris, G. Nardin and L. Randaccio, *Acta. Cryst.*, B25 (1969) 1671.
72 G. De Alti, V. Galasso, A. Bigotto and G. Costa, *Inorg. Chim. Acta*, 3 (1969) 523.
73 H. Kuhn, *Fortschr. Chem. Org. Naturstoffe*, 17 (1959) 404; *Proc. R. Soc. (A)*, 288 (1965) 348.
74 P. Day, *Theor. Chim. Acta*, 7 (1967) 328; *Coordination Chem. Rev.*, 2 (1967) 109.
75 G. N. Schrauzer, *Naturwissenschaften*, 53 (1966) 459.
76 P. O'. D. Offenhartz, B. H. Offenhartz and M. M. Fung, *J. Am. Chem. Soc.*, 92 (1970) 2966.
77 L. L. Ingraham and J. P. Fox, *Ann. N.Y. Acad. Sci.*, 153 (1969) 728.
78 H. Johansen and L. L. Ingraham, *J. Theor. Biol.*, 23 (1969) 191.
79 R. A. Firth, H. A. O. Hill, J. M. Pratt, R. G. Thorp and R. J. P. Williams, *J. Chem. Soc. (A)*, (1969) 381.
80 G. I. H. Hanania and D. H. Irvine, *J. Chem. Soc.*, (1964) 5694.
81 H. A. O. Hill, J. M. Pratt, R. G. Thorp, B. Ward and R. J. P. Williams, *Biochem. J.*, 120 (1970) 263.
82 J. M. Pratt and R. G. Thorp, *J. Chem. Soc. (A)*, (1966) 187.
83 W. C. Randall and R. A. Alberty, *Biochemistry*, 6 (1967) 1520.
84 J. Halpern and D. N. Hague, *Inorg. Chem.*, 6 (1967) 2059.
85 J. Halpern, R. A. Palmer and L. M. Blakley, *J. Am. Chem. Soc.*, 88 (1967) 2877.
86 A. Haim, R. J. Grassi, and W. K. Wilmarth, *Adv. Chem. Ser.*, 49 (1965) 31.
87 J. M. Pratt and R. G. Thorp, *Adv. Inorg. Radiochem.*, 12 (1970) 375.
88 S. Ahrland, J. Chatt and N. R. Davies, *Q. Rev.*, 12 (1958) 265.
89 R. G. Yalman, *Inorg. Chem.*, 1 (1962) 16.
90 J. Heyrovsky, *Principles of Polarography*, Academic Press, New York, 1966, p. 535.
91 B. Jaselskis and H. Diehl, *J. Am. Chem. Soc.*, 76 (1954) 4345.
92 S. L. Tackett, J. W. Collat and J. C. Abbott, *Biochemistry*, 2 (1963) 919.
93 H. A. O. Hill, J. M. Pratt and R. J. P. Williams, *Chem. Ind.*, (1964) 197.
94 O. Müller and G. Müller, *Biochem. Z.*, 335 (1962), 340.
95 K. Bernhauer and E. Irion, *Biochem. Z.*, 339 (1964) 530.
96 G. N. Schrauzer, K. J. Windgassen and J. Kohnle, *Chem. Ber.*, 98 (1965) 3324.
97 P. K. Das, H. A. O. Hill, J. M. Pratt and R. J. P. Williams, *J. Chem. Soc. (A)*, (1968) 1261.

98 G. Costa, G. Tauzher and A. Puxeddu, *Inorg. Chim. Acta*, 3 (1969) 41.
99 G. Costa, G. Mestroni, A. Puxeddu and E. Reisenhofer, *J. Chem. Soc. (A)*, (1970) 2870.
100 H. A. O. Hill, K. G. Morallee, G. Costa, G. Pellizer and A. Loewenstein, in C. Franconi (Ed.), *Magnetic Resonance in Biological Research*, Gordon and Breach, New York, 1971.
101 H. A. O. Hill, J. M. Pratt and R. J. P. Williams, *Discuss. Faraday Soc.*, 47 (1969) 165.
102 M. W. Penley, D. G. Brown and J. M. Wood, *Biochemistry*, 9 (1970) 4302.
103 R. A. Firth, H. A. O. Hill, J. M. Pratt, R. G. Thorp and R. J. P. Williams, *J. Chem. Soc. (A)*, (1968) 2428.
104 R. A. Firth, H. A. O. Hill, B. E. Mann, J. M. Pratt, R. G. Thorp and R. J. P. Williams, *J. Chem. Soc. (A)*, (1968) 2419.
105 E. Lester Smith, L. Mervyn, R. W. Muggleton, A. W. Johnson and N. Shaw, *Ann. N.Y. Acad. Sci.*, 112 (1964) 565.
106 D. Dolphin, A. W. Johnson and G. Rodrigo, *Ann. N.Y. Acad. Sci.*, 112 (1964) 590.
107 H. P. C. Hogenkamp, J. E. Rush and C. A. Swenson, *J. Biol. Chem.*, 240 (1965) 3641.
108 H. A. O. Hill, K. G. Morallee, G. Pellizer, G. Mestroni and G. Costa, *J. Organometal. Chem.*, 11 (1968) 167.
109 H. A. O. Hill, K. G. Morallee and G. Pellizer, *J. Chem. Soc. (A)*, (1969) 2096.
110 A. V. Ablov and N. M. Samus, *Russ. J. Inorg. Chem.*, 4 (1959) 790.
111 A. V. Ablov and N. M. Samus, *Russ. J. Inorg. Chem.*, 5 (1960) 410.
112 A. V. Ablov, and G. P. Syrtsova, *Russ. J. Inorg. Chem.*, 10 (1965) 1079.
113 S. A. Cockle, *D. Phil. Thesis*, Oxford University, 1970.
114 G. Costa, G. Mestroni, G. Tauzher, D. M. Goodall, M. Green and H. A. O. Hill, *Chem. Commun.* (1970) 34.
115 L. M. Ludwick and T. L. Brown, *J. Am. Chem. Soc.*, 91 (1969) 5188.
116 H. A. O. Hill, K. G. Morallee and G. Pellizer, *J. Am. Chem. Soc.*, 94 (1972) 277.
117 G. Parshall, *J. Am. Chem. Soc.*, 88 (1966) 704.
118 H. A. O. Hill and K. G. Morallee, *J. Chem. Soc. (A)*, (1969) 554.
119 V. B. Koppenhagen and J. J. Pfiffner, *J. Biol. Chem.*, 245 (1970) 5865.
120 R. A. Firth, H. A. O. Hill, J. M. Pratt, R. J. P. Williams and W. R. Jackson, *Biochemistry*, 6 (1968) 2178.
121 A. J. Thomson, *J. Am. Chem. Soc.*, 91 (1968) 2780.
122 R. A. Firth, H. A. O. Hill, B. E. Mann, J. M. Pratt and R. G. Thorp, *Chem. Commun.*, (1967) 1013.
123 B. Briat and C. Djerassi, *Bull. Soc. Chim. Fr.*, (1969) 135.
124 J. H. Bayston, and M. E. Winfield, *J. Catalysis*, 9 (1967) 217.
125 G. M. Schrauzer and L. P. Lee, *J. Am. Chem. Soc.*, 90 (1968) 6541.
126 J. M. Pratt and R. P. Silverman, *J. Chem. Soc. (A)*, (1967) 1286; 1291.
127 G. N. Schrauzer and R. J. Windgassen, *Chem. Ber.*, 99 (1966) 602.
128 R. Yamada, S. Shimizu and S. Fukui, *Biochemistry*, 7 (1968) 1713.
129 H. P. C. Hogenkamp, H. A. Barker and H. S. Mason, *Arch. Biochem. Biophys.*, 100 (1963) 353.
130 H. A. O. Hill, J. M. Pratt and R. J. P. Williams, *Proc. R. Soc. (A)*, 288 (1965) 352.
131 R. Yamada, S. Shimizu and S. Fukui, *Arch. Biochem. Biophys.*, 117 (1966) 675.
132 S. A. Cockle, H. A. O. Hill, J. M. Pratt and R. J. P. Williams, *Biochim. Biophys. Acta*, 177 (1969) 686.
133 J. H. Bayston, F. D. Looney, J. R. Pilbrow and M. E. Winfield, *Biochemistry*, 9 (1970) 2164.

134 J. M. Assour, *J. Chem. Phys.*, 43 (1965) 2477.
135 J. M. Assour, and W. K. Kahn, *J. Am. Chem. Soc.*, 87 (1965) 207.
136 F. A. Walker, *J. Am. Chem. Soc.*, 92 (1970) 4235.
137 S. A. Cockle, H. A. O. Hill and R. J. P. Williams, to be published.
138 J. A. Hamilton, R. Yamada, R. L. Blakley, H. P. C. Hogenkamp. F. D. Looney and M. E. Winfield, *Biochemistry*, 10 (1971) 347.
139 J. H. Bayston, N. K. King, F. D. Looney and M. E. Winfield, *J. Am. Chem. Soc.*, 91 (1969) 2775.
140 S. A. Cockle, H. A. O. Hill and R. J. P. Williams, *Inorg. Nucl. Chem. Lett.*, 6 (1970) 131.
141 G. N. Schrauzer and L. P. Lee, *J. Am. Chem. Soc.*, 92 (1970) 1551.
142 A. L. Crumbliss and F. Basolo, *J. Am. Chem. Soc.*, 92 (1970) 55.
143 D. Diemente, B. M. Hoffman and F. Basolo, *Chem. Commun.*, (1970) 467.
144 B. M. Hoffman, D. L. Dimente and F. Basolo, *J. Am. Chem. Soc.*, 92 (1970) 61.
145 G. Costa, A. Puxeddu and L. N. Stefani, *Inorg. Nucl. Chem. Lett.*, 6 (1970) 191.
146 H. A. O. Hill, S. Ridsdale and R. J. P. Williams, *J. Chem. Soc. (Dalton)*, (1972) 297.
147 J. L. Peel, *Biochem. J.*, 88 (1963) 296.
148 S. A. Cockle, H. A. O. Hill, S. Ridsdale and R. J. P. Williams, to be published.
149 P. K. Das, H. A. O. Hill, J. M. Pratt and R. J. P. Williams, *Biochim. Biophys. Acta*, 141 (1967) 644.
150 R. H. Yamada, T. Kato, S. Shimizu, and S. Fukui, *Biochim. Biophys. Acta*, 117 (1966) 13.
151 F. Calderazzo and C. Floriani, *Chem. Commun.*, (1967) 139.
152 G. N. Schrauzer and E. Deutsch, *J. Am. Chem. Soc.*, 91 (1969) 3341.
153 F. R. Jensen, V. Madan and D. H. Buchanan, *J. Am. Chem. Soc.*, 92 (1970) 1414.
154 G. N. Schrauzer, J. H. Weber, and T. M. Beckham, *J. Am. Chem. Soc.*, 92 (1970) 7078.
155 W. Freidrich and J. P. Nordmeyer, *Z. Naturforsch.*, 23b (1968) 1119; 24b (1969) 588; W. Freidrich and R. Messerschmidt, *Z. Naturforsch.*, 24b (1969) 465.
156 W. Friedrich, *Z. Naturforsch.*, 21b (1966) 138; W. Freidrich and M. Moskophidio, *Z. Naturforsch.*, 23b (1968) 804.
157 R. A. Firth, H. A. O. Hill, J. M. Pratt and R. G. Thorp, *Anal. Biochem.*, 23 (1968) 429; *J. Chem. Soc. A*, (1968) 453.
158 P. W. Schneider, P. F. Phelan and J. Halpern, *J. Am. Chem. Soc.*, 91 (1969) 77.
159 J. A. Hill, J. M. Pratt and R. J. Williams, *J. Theor. Biol.* 3 (1962) 423.
160 N. Alder, T. Medwick, and T. J. Poznanski, *J. Am. Chem. Soc.*, 88 (1966) 5018.
161 F. Wagner and K. Bernhauer, *Ann. N.Y. Acad. Sci.*, 112 (1964) 580.
162 J. W. Dubnoff, *Biochem. Biophys. Res. Commun.*, 16 (1964) 484.
163 D. Dolphin and A. W. Johnson, *J. Chem. Soc.*, (1965) 2174.
164 K. Bernhauer and O. Wagner, *Biochem. Z.*, 337 (1963) 366.
165 K. Bernhauer, P. Renz and F. Wagner, *Biochem. Z.*, 335 (1962) 443.
166 D. H. Dolphin and A. W. Johnson, *Proc. Chem. Soc.*, (1963) 311.
167 G. N. Schrauzer and R. J. Windgassen, *J. Am. Chem. Soc.*, 89 (1967) 143.
168 M. A. Foster, M. J. Dilworth and D. D. Woods, *Nature*, 201 (1964) 39.
169 W. Freidrich, and E. König, *Biochem. Z.*, 336 (1962) 444.
170 S. S. Kerwar, J. H. Mangum, K. G. Scrimgedur, J. D. Brodie and F. M. Huennekens, *Arch. Biochem. Biophys*, 116 (1966) 305.
171 J. R. Guest, S. Friedman, M. J. Dilworth and D. D. Woods, *Ann. N.Y. Acad. Sci.*, 112 (1964) 774.
172 H. Weissbach, B. G. Redfield and H. Dickerman, *J. Biol. Chem.*, 239 (1964) 1942.

173 A. W. Johnson, N. Shaw and F. Wagner, *Biochim. Biophys. Acta,* 72 (1963) 107.
174 G. N. Schrauzer and L. P. Lee, *Arch. Biochem. Biophys.,* 138 (1970) 16.
175 G. Costa, G. Mestroni and G. Pellizer, *J. Organometal. Chem.,* 15 (1968) 187.
176 G. Costa, G. Mestroni, G. Pellizer, G. Tauzher and T. Licari, *Inorg. Nucl. Chem. Lett.,* 5 (1969) 515.
177 G. Costa, G. Mestroni, T. Licari and G. Mestroni, *Inorg. Nucl. Chem. Lett.,* 5 (1969) 561.
178 H. P. C. Hogenkamp, *Ann. N.Y. Acad. Sci.,* 112 (1964) 552.
179 R. O. Brady and H. A. Barker, *Biochem. Biophys. Res. Commun.,* 4 (1961) 373.
180 B. Zagalak and J. Pawelkiewicz, *Acta Biochim. Polon.,* 12 (1965) 103.
181 F. S. Kennedy, T. Buchman and J. M. Wood, *Biochim. Biophys. Acta,* 177 (1969) 661.
182 G. N. Schrauzer, J. Sibert and R. J. Windgassen, *J. Am. Chem. Soc.,* 90 (1968) 668.
183 D. H. Dolphin, A. W. Johnson and R. Rodrigo, *J. Chem. Soc.,* (1964) 3186.
184 R. Yamada, S. Shimizu and S. Fukui, *Biochim. Biophys. Acta,* 124 (1966) 195.
185 H. P. C. Hogenkamp, *Biochemistry,* 5 (1966) 417.
186 J. M. Wood, F. S. Kennedy and R. S. Wolfe, *Biochemistry,* 7 (1968) 1707.
187 G. N. Schrauzer and R. J. Windgassen, *J. Am. Chem. Soc.,* 88 (1966) 3738.
188 W. H. Pailes and H. P. C. Hogenkamp, *Biochemistry,* 7 (1968) 4160.
189 W. Hieber, O. Vohler and G. Braun, *Z. Naturforsch.,* 13b (1958) 192.
190 G. E. Boxer and J. C. Rickards, *Arch. Biochem. Biophys.,* 30 (1951) 379.
191 J. L. Ellingboe, J. I. Morrison and H. Diehl, *Iowa State College J. Sci.,* 30 (1955) 263.
192 H. Schmid, A. Ebnöther and P. Karrer, *Helv. Chim. Acta,* 36 (1953) 65.
193 H. A. O. Hill, M. O'Riordan, F. Williams and R. J. P. Williams, to be published.
194 H. P. C. Hogenkamp and S. Holmes, *Biochemistry,* 9 (1970) 1886.
195 C. Bottom, P. K. Das, H. A. O. Hill, J. M. Pratt and R. J. P. Williams, unpublished.
196 J. D. Brodie, *Proc. Natn. Acad. Sci. U.S.,* 62 (1969) 461.
197 A. W. Johnson and N. Shaw, *J. Chem. Soc.,* (1962) 4608.
198 R. Yamada, T. Umetani, S. Shimizu and S. Fukui, *J. Vitaminol.,* 14 (1968) 316.
199 R. Barnett, H. P. C. Hogenkamp, and R. H. Abeles, *J. Biol. Chem.,* 241 (1966) 1483.
200 S. Fukui, S. Shimizu, R. Yamada and T. Umetani, *Vitamins,* 40 (1969) 113.
201 G. N. Schrauzer and J. W. Sibert, *J. Am. Chem. Soc.,* 92 (1970) 1022.
202 I. N. Rudakova, B. N. Shevcenko, T. A. Posnelova and A. M. Urkevitch, *Zh. Obshch. Chim.,* 37 (1967) 1748.
203 A. M. Urkevitch, I. N. Rudakova, and H. A. Preodrasheenskij, *Chim. Prirod. Svedinenij,* (1967) 48.
204 H. P. C. Hogenkamp and H. A. Barker, *J. Biol. Chem.,* 236 (1961) 3097.
205 H. P. C. Hogenkamp and T. G. Oikawa, *J. Biol. Chem.,* 239 (1964) 1911.
206 J. M. Wood, F. S. Kennedy and C. G. Rosen, *Nature,* 220 (1968) 173.
207 H. A. O. Hill, S. Ridsdale, J. M. Pratt, F. R. Williams and R. J. P. Williams, *Chem. Commun.,* (1970) 341.
208 S. Jensen and A. Jernlov, *Nordforsk,* 14 (1968) 3.
209 M. H. Abraham, and T. R. Spalding, *J. Chem. Soc. (A),* (1969) 784.
210 M. H. Abraham, and G. F. Johnson, *J. Chem. Soc. (A),* (1970) 188.
211 J. M. Wood, in R. Metcalf and N. J. Pitts, *Advances in Environmental Science,* Vol. II., Interscience, New York, 1970.
212 G. Popjak, in F. D. Boyer, *The Enzymes,* 3rd Edn., Academic Press, New York, 1970, Vol. 2, p. 206.
213 Ref. 3, p. 168.

214 G. Morrow, III, L. A. Barness, G. J. Cardinale, R. H. Abeles and J. G. Flaks, *Proc. Natn. Acad. Sci. U.S.*, 63 (1969) 191.
215 L. E. Rosenberg, A. C. Lilljequist and Y. E. Hsia, *Science*, 162 (1968) 805.
216 R. O. Brady, E. G. Castanera and H. A. Barker, *J. Biol. Chem.*, 237 (1962) 2325.
217 H. Weissbach, R. Redfield and A. Peterkofsky, *J. Biol. Chem.*, 236 (1961) PC 41.
218 H. Weissbach, J. N. Ladd, B. E. Volcani, R. D. Smyth and H. A. Barker, *J. Biol. Chem.*, 235 (1960) 1462.
219 A. Peterkofsky and H. Weissbach, *J. Biol. Chem.*, 238 (1963) 1491.
220 A. Peterkofsky and H. Weissbach, *Ann. N.Y. Acad. Sci.*, 112 (1964) 622.
221 E. Vitols, G. Walker and F. M. Huennekens, *Biochem. Biophys. Res. Commun.*, 15 (1964) 372.
222 H. Weissbach, N. Brot and W. Loverberg, *J. Biol. Chem.*, 241 (1966) 317.
223 G. A. Walker, S. Murphy and F. M. Huennekens, *Arch. Biochem. Biophys.*, 134 (1969) 95.
224 Y. Kaziro and S. Ochoa, *Adv. Enzymol.*, 26 (1964) 283.
225 S. Gurnani, S. P. Mistry and B. C. Johnson, *Biochim. Biophys. Acta*, 38 (1960) 187; J. R. Stern and D. L. Friedman, *Biochem. Biophys. Res. Commun.*, 2 (1960) 82.
226 E. R. Stadtman, P. Overath, H. Eggerer and F. Lynen, *Biochem. Biophys. Res. Commun.*, 2 (1960) 1; R. Stjernholm and H. G. Wood, *Proc. Natn. Acad. Sci. U.S.*, 47 (1961) 303.
227 H. G. Wood, R. W. Kellermeyer, R. Stjernholm and S. H. G. Allen, *Ann. N. Y. Acad. Sci.*, 112 (1964) 661; R. W. Kellermeyer, S. H. G. Allen, R. Stjernholm and H. G. Wood, *J. Biol. Chem.*, 239 (1964) 2562.
228 P. Overath, E. R. Stadtman, G. M. Kellerman and F. Lynen, *Biochem. Z.*, 336 (1962) 77.
229 J. J. B. Cannata, A. Focesi, R. Mazumder, R. C. Warner and S. Ochoa, *J. Biol. Chem.*, 240 (1965) 3249.
230 H. A. Barker, R. D. Smyth, R. M. Wilson and H. Weissbach, *J. Biol. Chem.*, 234 (1959) 320.
231 R. Mazumder, T. Sasakawa, Y. Kaziro and S. Ochoa, *J. Biol. Chem.*, 236 (1961) PC 53.
232 H. Eggerer, E. R. Stadtman, P. Overath and F. Lynen, *Biochem. Z.*, 333 (1960) 1.
233 R. W. Swick, *Proc. Natn. Acad. Sci. U.S.*, 48 (1962) 288.
234 R. W. Kellermeyer and H. G. Wood, *Biochemistry*, 1 (1962) 1124.
235 E. F. Phares, M. V. Long and S. F. Carson, *Biochem. Biophys. Res. Commun.*, 8 (1962) 142.
236 E. F. Phares, M. V. Long and S. F. Carson, *Ann. N.Y. Acad. Sci.*, 112 (1964) 680.
237 J. D. Enfle, J. M. Clark and B. C. Johnson, *Ann. N.Y. Acad. Sci.*, 112 (1964) 684.
238 M. Sprecher and D. B. Sprinson, *Ann. N.Y. Acad. Sci.*, 112 (1964) 665.
239 M. Sprecher, R. L. Switzer and D. B. Sprinson, *J. Biol. Chem.*, 241 (1966) 864.
240 G. T. Cardinale and R. H. Abeles, *Biochim. Biophys. Acta*, 132 (1967) 517.
241 J. Retey and D. Arigoni, *Experientia*, 22 (1966) 783.
242 W. W. Miller and J. H. Richards, *J. Am. Chem. Soc.*, 91 (1969) 1498.
243 H. A. Barker, R. D. Smyth, H. Weissbach *et al.*, *J. Biol. Chem.*, 235 (1960) 181.
244 H. P. C. Hogenkamp, J. N. Ladd and H. A. Barker, *J. Biol. Chem.*, 237 (1962) 1950.
245 H. P. C. Hogenkamp, *J. Biol. Chem.*, 238 (1963) 477.
246 H. A. Barker, F. Suzuki, A. A. Iodice and V. Rooze, *Ann. N.Y. Acad. Sci.*, 112 (1964) 644.

247 H. A. Barker, V. Rooze, F. Suzuki and A. A. Iodice, *J. Biol. Chem.*, 239 (1964) 3260.
248 F. Suzuki, V. Rooze and H. A. Barker, *Federation Proc.*, 22 (1963) 231.
249 F. Suzuki and H. A. Barker, *J. Biol. Chem.*, 241 (1966) 878.
250 R. L. Switzer and H. A. Barker, *J. Biol. Chem.*, 242 (1967) 2658.
251 J. I. Toohey, D. Perlman and H. A. Barker, *J. Biol. Chem.*, 236 (1961) 2119.
252 A. G. Lezius and H. A. Barker, *Biochemistry*, 4 (1965) 510.
253 A. A. Iodice and H. A. Barker, *J. Biol. Chem.*, 238 (1963) 2094.
254 H. F. Kung, S. Cederbaum, L. Tsai, and T. C. Stadtman, *Proc. Natn. Acad. Sci. U.S.*, 65 (1970) 978.
255 H. F. Kung and T. C. Stadtman, *J. Biol. Chem.*, 246 (1971) 3378.
256 M. N. Michelson and C. H. Werkman, *Enzymologia*, 8 (1940) 252.
257 R. H. Abeles and H. A. Lee, *J. Biol. Chem.*, 236 (1961) 2347.
258 J. Pawelkiewicz, B. Zagalak, *Acta Biochim. Polon.*, 12 (1965) 207.
259 Z. Schneider, K. Pech and J. Pawelkiewicz, *Bull. Acad. Polon. Sci.*, 14 (1966) 7.
260 Z. Schneider and J. Pawelkiewicz, *Acta Biochim. Polon.*, 13 (1966) 311.
261 S. P. Davies, M. A. Foster, H. A. O. Hill and R. J. P. Williams, to be published.
262 Z. Schneider, E. G. Larsen, G. Jacobson, B. C. Johnson and J. Pawelkiewicz, *J. Biol. Chem.*, 245 (1970) 3388.
263 H. A. Lee and R. H. Abeles, *J. Biol. Chem.*, 248 (1963) 2367.
264 R. H. Abeles and H. A. Lee, *Ann. N.Y. Acad. Sci.*, 112 (1964) 695.
265 T. Yamane, S. Shimizu and S. Fukui, *Biochim. Biophys. Acta*, 110 (1965) 616.
266 T. Yamane, S. Shimizu and S. Fukui, *J. Vitaminol. (Kyoto)*, 12 (1966) 10.
267 Y. Tamoa, Y. Morikawa, S. Shimizu and S. Fukui, *Biochem. Biophys. Res. Commun.*, 28 (1967) 692.
268 A. Stroiniski, Z. Schneider and J. Pawelkiewicz, *Bull. Acad. Polon. Sci., Ser. Sci. Biol.*, 15 (1967) 727.
269 J. Pawelkiewicz and Z. Schneider, *Bull. Acad. Polon. Sci., Ser. Sci. Biol.*, 15 (1967) 65.
270 S. A. Cockle, H. A. O. Hill, R. J. P. Williams, S. P. Davies and M. A. Foster, *J. Am. Chem. Soc.*, 94 (1972) 275.
271 R. H. Abeles and H. A. Lee, *Brookhaven Symp. Biol.*, 15 (1962) 310.
272 B. Zagalak, P. A. Frey, G. L. Karabatsos and R. H. Abeles, *J. Biol. Chem.*, 241 (1966) 3028.
273 P. A. Frey, S. S. Kewar and R. H. Abeles, *Biochem. Biophys. Res. Commun.*, 29 (1967) 873.
274 P. A. Frey, M. K. Essenberg and R. H. Abeles, *J. Biol. Chem.*, 242 (1967) 5369.
275 J. Retey, A. Umani-Ronchi and D. Arigoni, *Experientia*, 22 (1966) 72.
276 J. Retey, A. Umani-Ronchi, J. Seibl and D. Arigoni, *Experientia*, 22 (1966) 502.
277 G. J. Karabatsos, J. S. Fleming, N. Hsi and R. H. Abeles, *J. Am. Chem. Soc.*, 88 (1966) 849.
278 B. Zagalak, *Bull. Acad. Pol. Sci.*, 16 (1968) 67.
279 H. A. Barker, *Federation Proc.*, 20 (1961) 956.
280 L. L. Ingraham, *Ann. N.Y. Acad. Sci.*, 112 (1964) 713.
281 O. W. Wagner, H. A. Lee, P. A. Frey and R. H. Abeles, *J. Biol. Chem.*, 241 (1966) 1751.
282 R. H. Abeles and B. Zagalak, *J. Biol. Chem.*, 241 (1966) 1245.
283 P. A. Frey, M. K. Essenberg, S. S. Kerwar and R. H. Abeles, *J. Am. Chem. Soc.*, 92 (1970) 4488.
284 S. S. Kewar, T. A. Smith and R. H. Abeles, *J. Biol. Chem.*, 245 (1970) 1169.
285 J. A. Hamilton and R. L. Blakley, *Biochim. Biophys. Acta*, 184 (1969) 224.
286 H. R. Hayward, *Bact. Proc.*, (1961) 68.

287 C. Bradbeer, *J. Biol. Chem.*, 240 (1965) 4675.
288 C. Bradbeer, *J. Biol. Chem.*, 240 (1965) 4669.
289 B. H. Kaplan and E. R. Stadtman, *J. Biol. Chem.*, 213 (1968) 1787.
290 B. H. Kaplan and E. R. Stadtman, *J. Biol. Chem.*, 243 (1968) 1794.
291 B. M. Babior, *J. Biol. Chem.*, 244 (1969) 2917.
292 B. M. Babior and T. K. Li, *Biochemistry*, 8 (1969) 154.
293 B. M. Babior, *Biochim. Biophys. Acta*, 167 (1968) 456.
294 B. M. Babior, *J. Biol. Chem.*, 244 (1969) 449.
295 B. M. Babior, *J. Biol. Chem.*, 245 (1970) 6125.
296 B. M. Babior and D. C. Gould, *Biochem. Biophys. Res. Commun.*, 34 (1969) 441.
297 P. Y. Law, D. G. Brown, E. L. Lien, B. M. Babior and J. M. Wood, *Biochemistry*, 10 (1971) 3428.
298 T. C. Stadtman, *J. Biol. Chem.*, 235 (1962) PC 2409.
299 T. C. Stadtman, *J. Biol. Chem.*, 238 (1963) 2766.
300 T. C. Stadtman, *Ann. N.Y. Acad. Sci.*, 112 (1964) 728.
301 R. N. Costilow, O. M. Rochovansky and H. A. Barker, *J. Biol. Chem.*, 241 (1966) 1573.
302 L. Tsai and T. C. Stadtman, *Arch. Biochem. Biophys.*, 125 (1968) 210.
303 T. C. Stadtman and P. Renz, *Arch. Biochem. Biophys.*, 125 (1968) 226.
304 J. Retey, F. Kunz, T. C. Stadtman and D. Arigoni, *Experientia*, 25 (1969) 801.
305 T. C. Stadtman and L. Tsai, *Biochem. Biophys. Res. Commun.*, 28 (1967) 920.
306 Y. Tsuda and H. C. Friedmann, *Federation Proc.*, 29 (1970) 597.
307 J. K. Dyer and R. N. Costilov, *J. Bact.*, 101 (1970) 77.
308 C. G. D. Morley and T. C. Stadtman, *Biochemistry*, 9 (1970) 4890.
309 C. G. D. Morley and T. C. Stadtman, unpublished.
310 R. L. Blakley, *Biochem. Biophys. Res. Commun.*, 16 (1964) 391.
311 R. L. Blakley, R. K. Ghambeer, P. F. Nixon and E. Vituls, *Biochem. Biophys. Res. Commun.*, 20 (1965) 439.
312 R. L. Blakley, *Federation Proc.*, 25 (1966) 1633.
313 W. S. Beck and J. Hardy, *Proc. Natn. Acad. Sci. U.S.*, 54 (1965) 286.
314 R. Abrams and S. Duraiswami, *Biochem. Biophys. Res. Commun.*, 18 (1965) 409.
315 R. L. Blakley, *J. Biol. Chem.*, 240 (1965) 2173.
316 E. Vitols, H. P. C. Hogenkamp, C. Brownson, R. L. Blakley and J. Connellan, *Biochem. J.*, 104 (1967) 58.
317 E. Vitols, C. Brownson, W. Gardiner and R. L. Blakley, *J. Biol. Chem.*, 242 (1967) 3035.
318 M. Goulian and W. S. Beck, *J. Biol. Chem.*, 241 (1966) 4233.
319 R. Abrams, *J. Biol. Chem.*, 240 (1965) PC 3697.
320 W. S. Beck, M. Goulian, A. Larsson and P, Reichard, *J. Biol. Chem.*, 241 (1966) 2177.
321 R. L. Blakley, *J. Biol. Chem.*, 241 (1966) 176.
322 M. D. Orr and E. Vitols, *Biochem. Biophys. Res. Commun.*, 25 (1966) 109.
323 E. Vitols and R. L. Blakley, *Biochem. Biophys. Res. Commun.*, 21 (1965) 466.
324 R. L. Blakley, R. K. Ghambeer, T. J. Batterham and C. Brownson, *Biochem. Biophys. Res. Commun.*, 24 (1966) 418.
325 M. M. Gottesman and W. S. Beck, *Biochem. Biophys. Res. Commun.*, 24 (1966) 353.
326 T. J. Batterham, R. K. Ghambeer, R. L. Blakley, C. Brownson, *Biochemistry*, 6 (1967) 1203.
327 L. J. Durham, A. Larsson and P. Reichard, *Eur. J. Biochem.*, 1 (1967) 92.

328 W. S. Beck, R. H. Abeles and W. G. Robinson, *Biochem. Biophys. Res. Commun.*, 25 (1966) 421.
329 R. H. Abeles and W. S. Beck, *J. Biol. Chem.*, 242 (1967) 3589.
330 H. P. C. Hogenkamp, R. K. Ghambeer, C. Brownson, R. L. Blakley, *Biochem. J.*, 103 (1967) 5.
331 H. P. C. Hogenkamp, R. K. Ghambeer, C. Brownson, R. L. Blakley and E. Vitols, *J. Biol. Chem.*, 243 (1968) 799.
332 J. A. Hamilton, R. L. Blakley, F. D. Looney and M. E. Winfield, *Biochim. Biophys. Acta*, 177 (1969) 374.
333 R. L. Blakley, *The Biochemistry of Folic Acid and Related Pteridines*, North-Holland, Amsterdam, 1969.
334 J. R. Guest, C. W. Helleiner, M. J. Cross and D. D. Woods, *Biochem. J.*, 76 (1960) 396.
335 S. Takeyama and J. M. Buchanan, *J. Biochem. (Tokyo)*, 49 (1961) 578.
336 H. Weissbach, A. Peterkofsky, B. G. Redfield and H. Dickerman, *J. Biol. Chem.*, 238 (1963) 3318.
337 R. T. Taylor and H. Weissbach, *Biochem. Biophys. Res. Commun.*, 27 (1967) 398.
338 R. T. Taylor and H. Weissbach, *J. Biol. Chem.*, 242 (1967) 1502, 1509.
339 R. T. Taylor and H. Weissbach, *J. Biol. Chem.*, 242 (1967) 1517.
340 R. T. Taylor and H. Weissbach, *Arch. Biochem. Biophys.*, 119 (1967) 572.
341 J. F. Morningstar and R. L. Kisliuk, *J. Gen. Microbiol.*, 39 (1965) 43.
342 L. Jaenicke, *Biochem. J.*, 99 (1966) 21.
343 S. E. Cauthen, M. A. Foster and D. D. Woods, *Biochem. J.*, 98 (1966) 630.
344 G. T. Burke, J. H. Mangum and J. D. Brodie, *Biochemistry*, 9 (1970) 4297.
345 R. E. Ertel, N. Brot, R. T. Taylor and H. Weissbach, *Arch. Biochem. Biophys.*, 125 (1968) 353.
345 J. R. Guest, S. Friedman, D. D. Woods and E. L. Smith, *Nature*, 195 (1962) 340.
346 S. Rosenthal and J. M. Buchanan, *Acta Chem. Scand. (Suppl.)*, 17 (1963) 288.
347 R. T. Taylor and H. Weissbach, *Arch. Biochem. Biophys.*, 29 (1969) 728.
348 R. T. Taylor and M. C. Hanna, *Arch. Biochem. Biophys.*, 139 (1970) 149.
349 H. Rüdiger and L. Jaenicke, *Eur. J. Biochem.*, 10 (1969) 557.
350 H. Rüdiger and L. Jaenicke, *Eur. J. Biochem.*, 16 (1970) 92.
351 R. T. Taylor and H. Weissbach, *Arch. Biochem. Biophys.*, 123 (1968) 109.
352 R. T. Taylor and H. Weissbach, *Arch. Biochem. Biophys.*, 129 (1969) 745.
353 J. Galivan and F. M. Huennekens, *Biochem. Biophys. Res. Commun.*, 38 (1970) 46.
354 J. Galivan, S. Murphy and D. Jacobsen, *Federation Proc.*, 29 (1970) 334.
355 M. J. Pine and H. A. Barker, *J. Bact.*, 71 (1956) 644.
356 M. J. Pine and W. Vishniac, *J. Bact.*, 73 (1957) 736.
357 B. A. Blaylock and T. C. Stadtman, *Biochem. Biophys. Res. Commun.*, 11 (1963) 34.
358 M. J. Wolin, E. A. Wolin and R. S. Wolfe, *Biochem. Biophys. Res. Commun.*, 12 (1963) 464.
359 J. M. Wood, A. M. Allam, W. J. Bill and R. S. Wolfe, *J. Biol. Chem.*, 240 (1965) 4564.
360 J. M. Wood and R. S. Wolfe, *Biochem. Biophys. Res. Commun.*, 19 (1965) 306.
361 B. A. Blaylock and T. C. Stadtman, *Ann. N.Y. Acad. Sci.*, 112 (1964) 799.
362 J. M. Wood and R. S. Wolfe, *Biochemistry*, 5 (1966) 3598.
363 B. C. McBride, J. M. Wood, J. W. Sibert and G. N. Schrauzer, *J. Am. Chem. Soc.*, 90 (1968) 5276.
364 J. W. Sibert and G. N. Schrauzer, *J. Am. Chem. Soc.*, 92 (1970) 1421.

365 J. M. Wood and R. S. Wolfe, *Biochem. Biophys. Res. Commun.*, 22 (1966) 119.
366 B. A. Blaylock and T. C. Stadtman, *Biochem. Biophys. Res. Commun.*, 17 (1964) 475.
367 B. A. Blaylock and T. C. Stadtman, *Arch. Biochem. Biophys.*, 116 (1966) 138.
368 B. A. Blaylock, *Arch. Biochem. Biophys.*, 124 (1968) 314.
369 T. C. Stadtman and B. A. Blaylock, *Federation Proc.*, 25 (1966) 1657.
370 J. M. Wood and R. S. Wolfe, *J. Bact.*, 92 (1966) 696.
371 J. M. Wolin, E. A. Wolin and R. S. Wolfe, *Biochem. Biophys. Res. Commun.*, 15 (1964) 420.
372 B. C. McBride and R. S. Wolfe, *Biochemistry*, 10 (1971) 2317.
373 J. M. Poston and E. R. Stadtman, *Ann. N.Y. Acad. Sci.*, 112 (1964) 804.
374 K. Kuratomi, J. M. Poston and E. R. Stadtman, *Biochem. Biophys. Res. Commun.*, 23 (1966) 691.
375 L. Ljungdahl, E. Irion and H. G. Wood, *Biochemistry*, 4 (1965) 2771.
376 L. Ljungdahl, E. Irion and H. G. Wood, *Federation Proc.*, 25 (1966) 1642.
377 L. Ljungdahl, D. Glatzle, J. Goodyear and H. G. Wood, *Abstr. Am. Soc. Microbiol.*, (1967) 128.
378 J. M. Poston and E. R. Stadtman, *Biochem. Biophys. Res. Commun.*, 26 (1967) 550.
379 D. J. Parker, H. G. Wood, L. G. Ljungdahl and R. K. Ghambeer, *Federation Proc.*, 30 (1971).
380 H. Eggerer, P. Overath, F. Lynen and E. R. Stadtman, *J. Am. Chem. Soc.*, 82 (1960) 2643.
381 G. Parshall, *Accounts Chem. Res.*, 3 (1970) 139.
382 See *Chem. Eng. News*, 22, July 5, 1971.
383 J. M. Pratt, *Inorganic Chemistry of Vitamin* B_{12}, Academic Press, London, 1972.
384 J. M. Wood and D. G. Brown, *Struct. Bonding*, 11 (1972) 47.
385 W. M. Coleman and L. T. Taylor, *J. Am. Chem. Soc.*, 93 (1971) 5446.
386 A. Bigotto, G. Costa, G. Mestroni, G. Pellizer, A. Puxeddu, E. Reisenhofer, L. Stefani and G. Tauzher, *Inorg. Chim. Acta Rev.*, 4 (1970) 41.
387 K. Venkatesan, D. Dale, D. C. Hodgkin, C. E. Nockolds, F. H. Moore and B. H. O'Connor, *Proc. R. Soc. A*, 323 (1971) 455.
388 S. W. Hawkinson, C. L. Coulter and M. L. Greaves, *Proc. R. Soc. A*, 318 (1970) 143.
389 R. Bonnett, J. M. Godfrey, V. B. Math, E. Edmond, H. Evans and O. J. R. Hodder, *Nature*, 229 (1971) 473.
390 M. Calligaris, G. Nardin and L. Randaccio, *Coord. Chem. Rev.*, 7 (1972) 385.
391 J. D. Brodie and M. Poe, *Biochemistry*, 10 (1971) 914; 11 (1972) 2534.
392 D. Doddrell and A. Allerhand, *Chem. Commun.*, (1971) 728.
393 H. A. O. Hill, J. W. Kendall, A. M. Turner and R. J. P. Williams, to be published.
394 D. Thusius, *J. Am. Chem. Soc.*, 93 (1971) 2629.
395 T. L. Brown, L. M. Ludwick and R. S. Stewart, *J. Am. Chem. Soc.*, 94 (1972) 384.
396 K. L. Brown and R. G. Kallen, *J. Am. Chem. Soc.*, 94 (1972) 1894.
397 G. Costa, A. Puxeddu and E. Reisenhofer, *J. Chem. Soc. Dalton*, (1972) 1519.
398 J. W. Kendall and N. Sanders, unpublished observations.
399 G. N. Schrauzer and R. J. Holland, *J. Am. Chem. Soc.*, 93 (1971) 4060.
400 D. Bieniek, P. N. Moza, W. Klein and F. Korte, *Tetrahedron Lett.*, (1970) 4055.
401 A. E. Brearley, H. Gott, H. A. O. Hill, M. O'Riordan, J. M. Pratt and R. J. P. Williams, *J. Chem. Soc. A*, (1971) 612.
402 J. M. Pratt and B. R. D. Whitear, *J. Chem. Soc. A*, (1971) 252.
403 H. A. O. Hill, J. W. Kendall and R. J. P. Williams, unpublished work.

404 B. T. Golding, H. L. Holland, U. Horn and S. Sakrikar, *Angew. Chem. Int. Ed. Engl.*, 9 (1970) 959.
405 R. B. Silverman, D. Dolphin and B. M. Babior, *J. Am. Chem. Soc.*, 94 (1972) 4028; see also R. M. Acheson, *Accounts Chem. Res.*, 4 (1971) 177.
406 G. N. Schrauzer, J. H. Weber, T. M. Beckham, and R. K. Y. Ho, *Tetrahedron Lett.*, 3 (1971) 275.
407 A. Adin and J. H. Espenson, *Chem. Commun.*, 13 (1971) 653.
408 F. R. Jensen, V. Madan and D. H. Buchanan, *J. Am. Chem. Soc.*, 93 (1971) 5283.
409 L. Bertilsson and H. Y. Neujahr, *Biochemistry*, 10 (1971) 2805.
410 G. Costa, G. Mestroni and C. Cocevar, *Chem. Commun.*, (1971) 706.
411 G. Agnes, H. A. O. Hill, J. M. Pratt, S. C. Ridsdale, F. S. Kennedy and R. J. P. Williams, *Biochim. Biophys. Acta*, 242 (1971) 207; G. Agnes, S. Bendle, H. A. O. Hill, F. R. Williams and R. J. P. Williams, *Chem. Commun.*, (1971) 850.
412 *Vitamins and Coenzymes Part C*, in D. B. McCormick and L. D. Wright (Eds.), *Methods in Enzymology*, Vol. 18, Academic Press, New York, 1971.
413 M. J. Mahoney and L. E. Rosenberg, *Am. J. Med.*, 48 (1970) 584.
414 R. Silber and C. F. Moldow, *Am. J. Med.*, 48 (1970) 549.
415 J. D. Brodie, H. A. O. Hill, J. W. Kendall, A. M. Turner, A. D. Woodhams and R. J. P. Williams, unpublished observations.
416 R. G. Eager, B. G. Baltimore, M. M. Herbst, H. A. Barker and J. H. Richards, *Biochemistry*, 11 (1972) 253.
417 H. F. Kung and T. C. Stadtman, *J. Biol. Chem.*, 246 (1971) 3378.
418 D. A. Weisblat and B. M. Babior, *J. Biol. Chem.*, 246 (1971) 6064.
419 T. J. Carty, B. M. Babior and R. H. Abeles, *J. Biol. Chem.*, 246 (1971) 6313.
420 D. Panagon, M. D. Orr, J. R. Dunstone and R. L. Blakley, *Biochemistry*, 11 (1972) 2378.
421 R. Yamada, Y. Tamao and R. L. Blakley, *Biochemistry*, 10 (1971) 3959.
422 G. T. Burke, J. H. Mangum and J. D. Brodie, *Biochemistry*, 10 (1971) 3079.
423 H.-F. Kung, C. Spears, R. C. Greene and H. Weissbach, *Arch. Biochem. Biophys.*, 150 (1972) 23.
424 J. Dawes and M. A. Foster, *Biochim. Biophys. Acta*, 237 (1971) 455.
425 E. Hippe and H. Olsen, *Biochim. Biophys. Acta*, 243 (1971) 83.

PART VII

METAL INTERACTIONS WITH OTHER PROSTHETIC GROUPS

Chapter 31

VITAMIN B_6 COMPLEXES

R. H. HOLM

Department of Chemistry, Massachusetts Institute of Chemistry, Cambridge, Massachusetts 02139, U.S.A.

I.. INTRODUCTION

The term vitamin B_6 as it is now employed refers to a family of tetrasubstituted pyridine derivatives composed of pyridoxol (I), pyridoxal (PL, II), pyridoxamine (PM, III), pyridoxal phosphate (PLP, IV), and pyridoxamine phosphate (V). Vitamin B_6 was originally sought as that component of the vitamin B complex responsible for the arrest and cure of dermatitis in rats. It was first obtained in crystalline form in 1938 and a year later its structure was determined to be **I** by chemical degradation and direct synthesis. Subsequent studies of the growth-promoting activity of pyridoxol for lactic acid bacteria led to the recognition of the existence in natural substances of pyridoxol-like compounds which possessed greater growth promoting activity than **I**. These compounds were identified in 1944 as pyridoxal and pyridoxamine. The biocatalytically active or coenzyme form of vitamin B_6 is now well established to be pyridoxal phosphate (codecarboxylase). It functions in non-oxidative enzymatic transformations of amino acids and catalyzes a remarkably diverse series of reactions in amino acid metabolism including decarboxylation, racemization, transamination, β-elimination, and γ-elimination. Accounts of the isolation and synthesis of vitamin B_6 components and of the nutritional and therapeutic roles of the vitamin in the metabolism of animals and humans are available elsewhere[1-4].

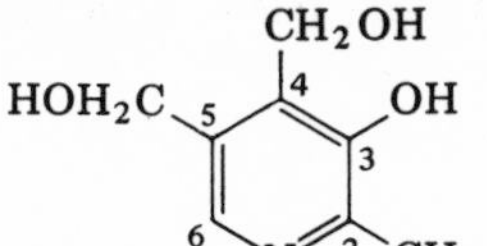

(I) 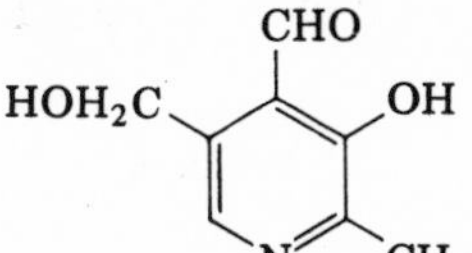 (II) 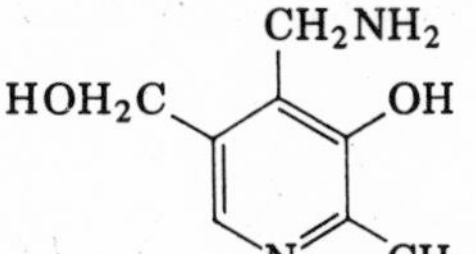 (III)

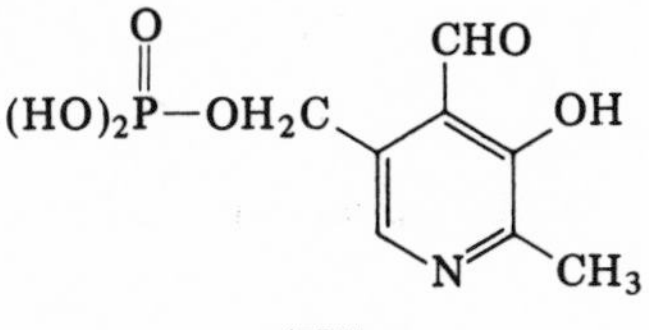

(IV) 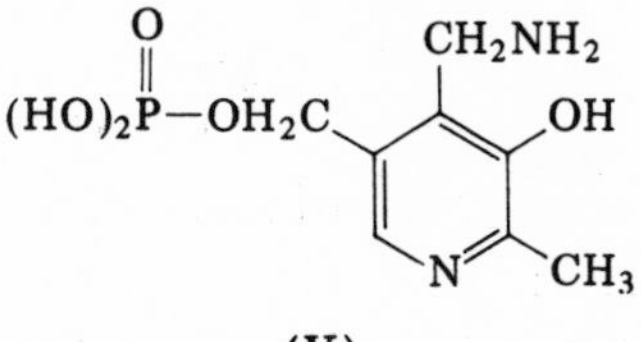 (V)

The most thoroughly studied enzymatic transformation catalyzed by pyridoxal phosphate is the transamination reaction (1) between an α-amino acid and an α-keto acid. This reaction:

$$\underset{\displaystyle NH_2}{R{-}\underset{|}{C}HCOOH} + R'{-}\overset{\displaystyle O}{\overset{\|}{C}}{-}COOH \rightleftharpoons \underset{\displaystyle NH_2}{R'{-}\underset{|}{C}H{-}COOH} + R{-}\overset{\displaystyle O}{\overset{\|}{C}}{-}COOH \qquad (1)$$

was first reproduced in a non-enzymatic system in 1945 by Snell[5], who found that when pyridoxal and glutamic acid or pyridoxamine and α-ketoglutaric acid were heated in aqueous solutions near neutral pH transamination, reaction (2), occurred. The reaction was observed to be rever-

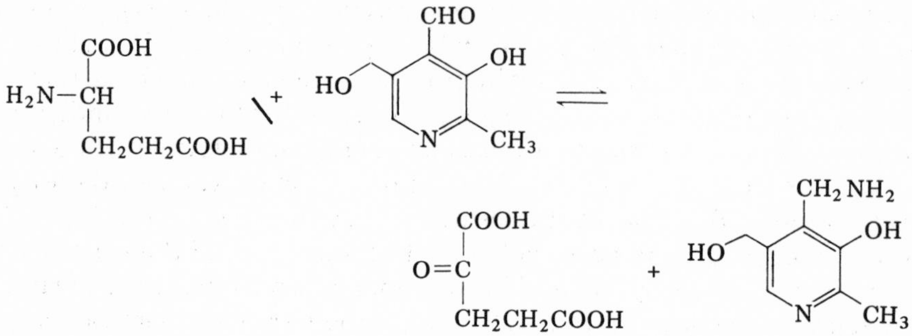

$$H_2N{-}CH(COOH)CH_2CH_2COOH + \text{pyridoxal (CHO, OH, } CH_3\text{, HO}CH_2\text{)} \rightleftharpoons O{=}C(COOH)CH_2CH_2COOH + \text{pyridoxamine (}CH_2NH_2\text{, OH, } CH_3\text{, HO}CH_2\text{)} \qquad (2)$$

sible and to require a large excess of amino acid to effect completion of keto acid formation. The first indication that metal ions might be involved in enzymatic or non-enzymatic pyridoxal-catalyzed reactions was found in 1952 when Metzler and Snell[6] determined that the rate of non-enzymatic transamination between pyridoxal or pyridoxal phosphate and amino acids in aqueous solution was significantly enhanced by certain metal ions through the intermediacy of pyridoxylidene (aldimine) and pyridoxylimino (ketimine) Schiff base complexes* postulated to have the basic structural

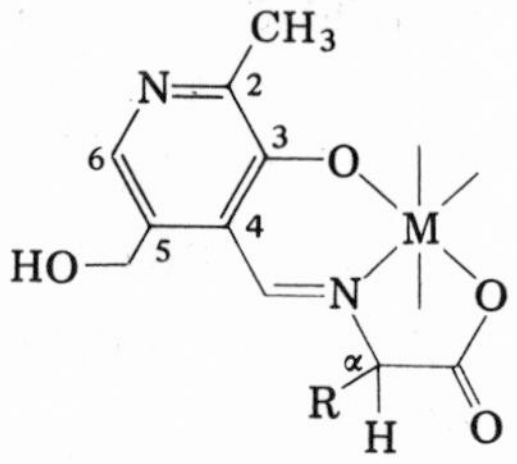

(VI) M(pyr-aa)

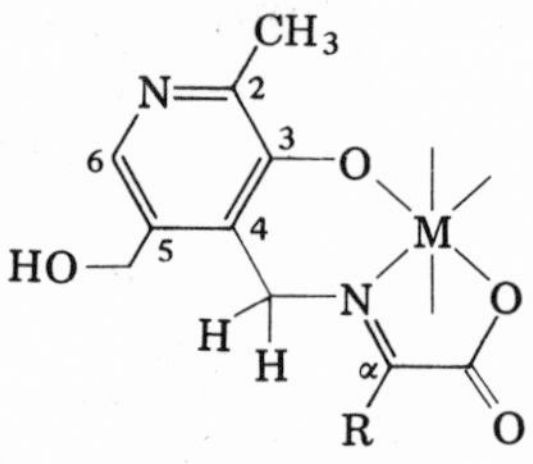

(VII) M(pym-ka)

* A tabulation of ligand and other abbreviations is given at the end of the chapter.

features shown in **VI** and **VII**, respectively. The similarity of the six-membered chelate ring in **VI** to that present in the familiar and extensive group of metal salicylaldimine chelates, derived from the Schiff bases formed by salicylaldehyde and primary amines, was quickly recognized, and soon thereafter a number of pyridoxylidene complexes were isolated[7-9].

The structures and functions of complexes such as **VI** and **VII** and other Schiff base metal chelates in PL-catalyzed transformations of amino acids are dealt with in some detail in subsequent sections. As will be seen, it is the role of these species in catalysis which is chiefly responsible for the past and present significance of "vitamin B_6 complexes." However, it is noted at this point that the presence of metal ions in model systems is not obligatory to the accomplishment of these transformations, although it generally increases the rate of reaction and the extent of product formation at equilibrium. Further, there is no convincing evidence that metal ions are active in the functioning of pyridoxal phosphate-dependent enzymes, and in one enzyme, glutamic–aspartic transminase, direct assay has revealed less than 0.4 mole of metal ion of any kind per mole of active site[10]. Despite the lack of proof that these enzymes function as metalloenzymes, a high degree of interest continues to be maintained in systems which effect amino acid synthesis or transformations in the presence of pyridoxal or pyridoxamine, their phosphorylated derivatives, or their analogs (*vide infra*), and metal ions. The latter may simulate certain of the features of enzymatic active sites in the sense of facilitating Schiff base formation and of labilizing bonds formed by the α-carbon in the condensed amino acid portion of complexes of type **VI**. The formation of aldimine or ketimine Schiff bases is an essential feature of all currently accepted mechanisms of vitamin B_6 catalysis, many of which also require lability of the α-hydrogen in those reactions proceeding to products through an intermediate aldimine. Recent symposia on pyridoxal catalysis[11,12] have included a number of contributions detailing studies of metal ion-containing systems, which are commonly referred to as model systems for enzymatic reactions in spite of the lack of evidence for participation of metal ions in these reactions.

This chapter is primarily concerned with the structures, formation and reactions of metal complexes formed from the vitamin B_6 compounds **II–V** or their analogs and amino acids or α-keto acids. The role of these complexes in transformations of amino acids in model systems is considered, with particular emphasis placed upon transamination. Relevant information on the chemistry of other Schiff base complexes, particularly the salicylaldimines, is also included. This report serves to update, although not exhaustively, existing summaries or reviews[1-3,13-17] of reaction mechanisms in model systems. Excluded from consideration are model systems not containing metal ions, and the enzymatic reactions themselves, discussions of which may be found elsewhere[1-4,11-16,18,19].

References pp. 1165–1167

II. TYPES AND STRUCTURES OF COMPLEXES

In any consideration of reactions catalyzed or otherwise effected by vitamin B_6 compounds in the presence of metal ions, it is desirable at the outset to specify the types and structures of complexes involved. Despite the central importance of aldimine and ketimine complexes such as **VI** and **VII**, relatively little effort has been directed toward the fundamental matters of their isolation, structural determination, and investigation of their electronic properties. Attention is directed at this point to complexes which have been isolated and the structural and electronic information derived therefrom.

Relevant complexes are those formed from the condensation products of pyridoxal and pyridoxamine, or their analogs, with amino acids and α-keto acids, respectively. Because of the tridentate nature of the Schiff base ligand, complexes with 2:1 and 1:1 ligand: metal stoichiometry can be formed, and may be protonated at the pyridine nitrogens. The 1:1 complexes have the general structures **VI–IX** in which the vacant coordination positions are occupied wholly or in part by water molecules when isolated from aqueous solution. The latter two structures represent aldimine complexes formed from amino acids and two isomeric hydroxypyridinealdehydes[20], both of which function as transamination catalysts. Also of some interest are the salicylaldimine complexes **X**, whose structural and electronic properties are similar in several respects to those of **VI–IX**.

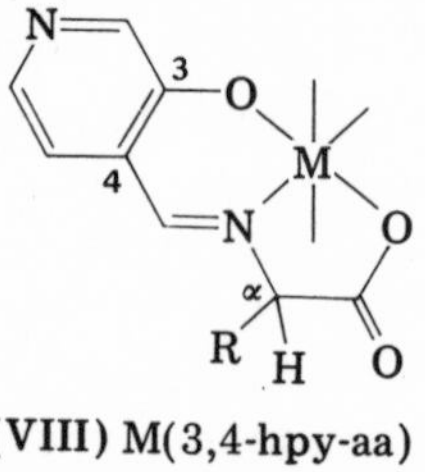

(VIII) M(3,4-hpy-aa) 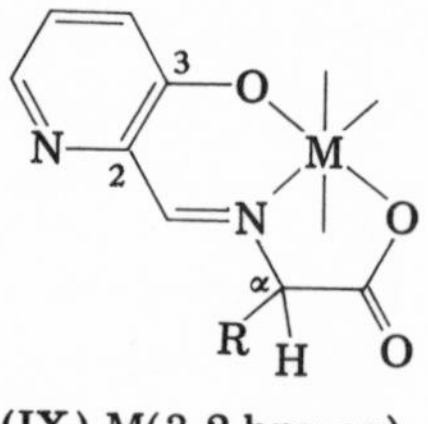   (IX) M(3,2-hpy-aa) 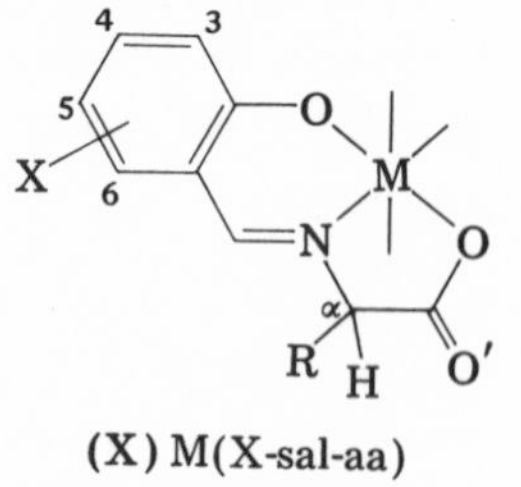   (X) M(X-sal-aa)

However, salicylaldehyde does not possess the catalytic properties of pyridoxal or the simpler hydroxypyridinealdehydes in amino acid transformations. A number of complexes of types **VI**[7-9,21-23], **VIII**[24], **IX**[23,24], and **X**[23,25-29], including some with the pyridine nitrogen protonated[21,23,24], have been isolated with metal ions such as Mn(II), Cu(II), and Zn(II). Verification of the chelate ring structures in **VI** and **VIII–X** has recently been afforded by X-ray determination of [Cu(sal-gly)(H_2O)] · $\frac{1}{2}H_2O$[30], Cu(pyr-DL-val)[31], and [Cu(Ppyr-DL-Phala)(H_2O)][32]*. In each case an approximate square pyramidal coordination sphere exists, in which three

* The structure of PLP-oxime (metal-free) has also been reported[32a].

positions in the basal plane are occupied by the aldimine ligand. In $[Cu(sal\text{-}gly)(H_2O)] \cdot \frac{1}{2}H_2O$ the remaining basal and the apical position are filled by a water molecule and the carboxylate oxygen (**X**, O′) of an adjoining molecule. The structures of the Cu(II) pyridoxylidene complexes are shown in Fig. 1. The fourth basal position is occupied by a pyridine nitrogen or a water molecule and the apical position by a hydroxymethyl (A) or phosphate (B) oxygen atom. It is not known whether or not five-coordination is retained in solution, but it is highly likely that the Cu–pyridine bond in crystalline Cu(pyr-DL-val) does not persist in aqueous or other polar media.

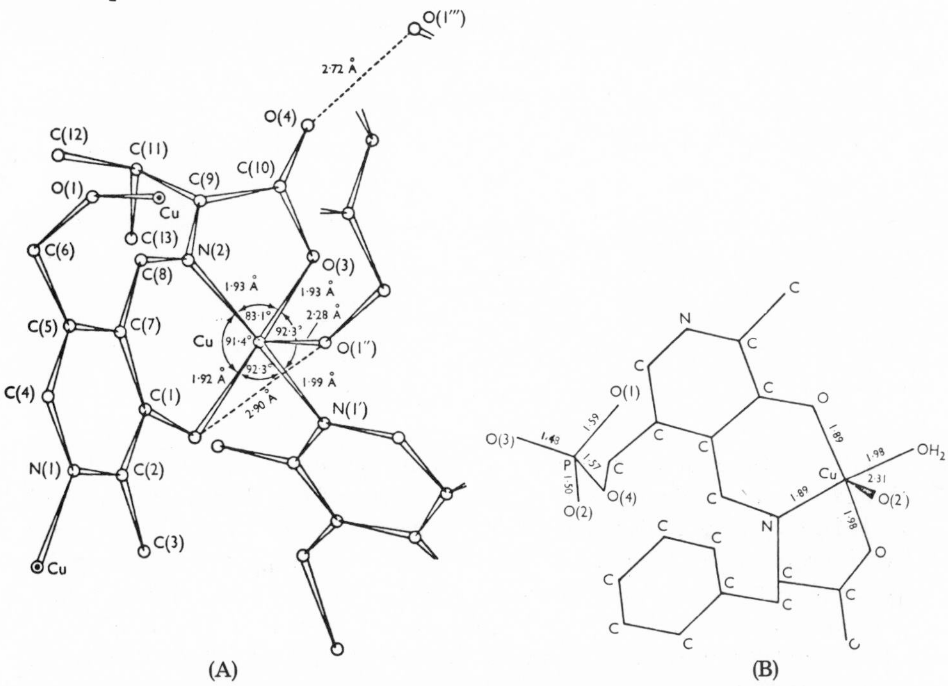

Fig. 1. Structures of Cu(pyr-DL-val)[31] (A) and $[Cu(Ppyr\text{-}DL\text{-}Phala)(H_2O)]$[32] (B) illustrating the square pyramidal coordination around the metal ion.

Complexes of the 2:1 type are somewhat more interesting from a structural point of view. Bis-tridentate octahedral chelate complexes in which each ligand possesses three inequivalent donor atoms have six potential geometric forms. However, the presence of a trigonal nitrogen as the central donor atom renders the tridentate ligand relatively rigid, such that chelation on an octahedral edge rather than a face is strongly favored. The resultant structure has the two azomethine nitrogens *trans* and the two phenolate and carboxylate oxygens mutually *cis*. It is dissymmetric and the two absolute configurations are represented by Δ and Λ. The presence of an asymmetric center in the condensed amino acid generates the following diastereoisomeric species, illustrated by the bis-pyridoxylidene complexes **XI, XII,** and **XIII:**

$\Delta(LL) \equiv \Lambda(DD)$
$\Delta(DD) \equiv \Lambda(LL)$
$\Delta(DL) \equiv \Lambda(LD)$

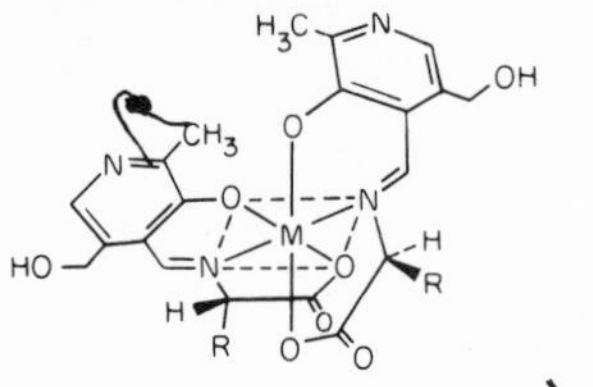

(XI) Δ-M(pyr-L-aa)$_2$

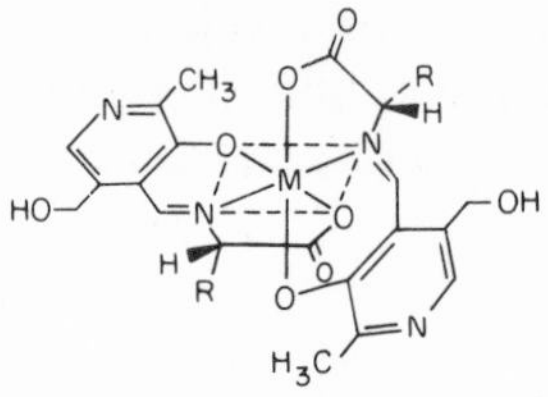

(XII) Λ-M(pyr-L-aa)$_2$

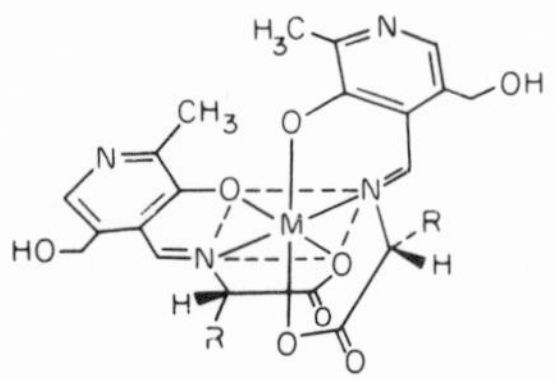

(XIII) Δ-M(pyr-DL-aa)$_2$

The designations of the indicated absolute configurations as Δ and Λ are arbitrary. Six-coordinate bischelate complexes of metal ions such as Mn(II), Fe(II,III), Co(III), Ni(II), and Zn(II) derived from pyridoxal[8,22,24], hydroxy-pyridinealdehydes[24], salicylaldehyde[34], and amino acids have been isolated. X-ray studies of [Mn(Hpyr-DL-val)$_2$] · 4H_2O[35] and Zn(Hpyr-val)$_2$[36] have established the geometrical isomer as that shown in **XI–XIII** and thereby invalidate an improbable earlier claim[22] that the phenolic oxygens are not coordinated.

Comparatively few attempts have been made to isolate the ketimine tautomers (*e.g.*, **VII**) of the aldimine species **VI** and **VIII–XIII**. Consequently, there is no definite structural information on such complexes. However, the basic formulation **VII** depicting the ketimine ligand as a planar tridentate is highly probable by analogy with known aldimine structures and is assumed throughout. Although the existence of both aldimine and ketimine complexes in solution is now well documented (*vide infra*), there exist only several reports of presumed isolation of ketimine species[27,33]. Reaction of a nickel(II) salt with pyridoxal and potassium alaninate and with pyridoxamine and potassium pyruvate is claimed to yield the tautomers bis(*N*-pyridoxylidenealaninato)Ni(II) and bis(α-pyridoxyliminopropionato)Ni(II)[33], respectively, which are presumably diprotonated. The tautomeric pair *N*-

salicylidenealaninato-Cu(II) and α-*o*-hydroxybenzyliminopropionate-Cu(II) has also been claimed[27]. The existence of this tautomeric pair has recently been confirmed and the corresponding pair of Zn(II) complexes prepared[37]. Judging from the reaction of other complexes considered in subsequent sections, rather facile conversion in solution to the aldimine tautomer★ is expected and base-catalyzed ketimine → aldimine conversion has been observed for several Cu(II) systems[37].

Little information on the fundamental electronic properties of either 1:1 or 2:1 complexes of any type is available. Magnetic moments of hydrated 1:1 Cu(II) complexes of types **VI, IX** and **X** [1.7–1.9 Bohr magnetons (BM)] are consistent with the expected spin-doublet ground state. Those of variously prepared anhydrous modifications of Cu(sal–gly) (1.3–1.6 BM[40]) possibly indicate some degree of antiferromagnetic spin exchange, which might arise from a dimeric or trimeric bridged structure postulated for other tridentate Schiff base complexes of Cu(II) with comparable room temperature moments[41]. Measurements of the temperature dependence of the susceptibility which might establish the nature of magnetic interaction have not been reported. Salts of $[Co(sal\text{–}aa)_2]^-$ are diamagnetic and the moment of $[Fe(sal\text{–}gly)_3]^{-1}$ (5.99 BM) is that expected for a high-spin octahedral d^5 complex. Magnetic measurements on $[Fe(sal\text{–}gly)_2]^{-2}$ are claimed to be consistent with a distribution over spin-singlet and quintet states with the latter populated to the extent of ~90% at 25°C[42]. Unfortunately, no data were presented to support this contention, which is deserving of further investigation in view of the current interest in spin state equilibria[43]. Optical rotatory dispersion (o.r.d.) and circular dichroism (c.d.) spectra of a considerable number of Cu(II) complexes of types **VI, IX,** and **X** (X = H, 4-, 6-NO_2) have been obtained[23,44,45]. The most conspicuous feature is the presence of a negative Cotton effect or negative c.d. band associated with an absorption at ~360–400 nm in complexes derived from L-amino acids. Finally, the proton magnetic resonance (p.m.r.) spectra of Zn(II) and Al(III) pyridoxylidene complexes in aqueous solution have been

---

★ Other attempts to prepare tautomeric pairs of complexes have not succeeded owing to the apparently greater stability of one form. Contrary to earlier reports[38], reaction of Cu(II) with the Schiff bases derived from pyruvic acid and glycine and from glyoxylic acid and alanine give one and the same species, *viz.*, **XIV** (L = H_2O)[39]. Similar reactions with palladium(II) chloride yield only the anionic complex **XIV** (L = Cl^-)[39].

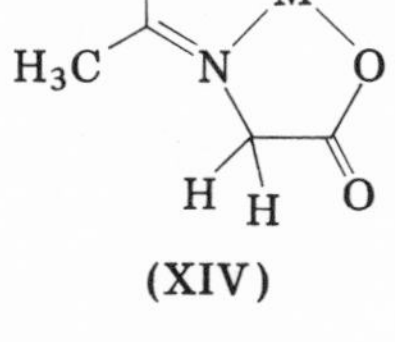

O O L M H_3C N O H H O

(XIV)

determined in the last several years[46-51]. P.m.r. spectra have proven quite valuable for inferring solution structures of non-labile complexes and examining the details of transamination and other reactions. They are dealt with in some detail in Section IVA.

III. MECHANISM OF FORMATION

The large majority of reactions performed with metal ion-containing model systems have been transformations of amino acids involving pyridoxal as reactant or catalyst, in order that the latter might simulate the action of the coenzyme PLP. Consequently, the first key stage in the reaction sequence of such systems is the formation of aldimine complexes, which react further to generate products in subsequent steps. The reactions of Schiff base complexes themselves are examined in the next section. Here the available information concerning the mechanism of formation of these complexes is considered.

Condensation of a primary amine with a carbonyl group coordinated to, or otherwise in the presence of, a metal ion to form a Schiff base complex is one of the oldest and most significant synthetic reactions in coordination chemistry[52]. Despite this fact the kinetics and probable mechanisms of formation of such complexes have remained undefined

TABLE I

SELECTED FORMATION CONSTANTS[a] AND KINETIC PARAMETERS FOR *N*-SALICYLIDENE- AND *N*-PYRIDOXYLIDENE-GLYCINATO METAL(II) COMPLEXES IN AQUEOUS SOLUTION AT 25°C

(a) M^{2+} + pyr-gly^{2-} ⇌ M(pyr-gly)
(b) M^{2+} + sal-gly^{2-} ⇌ M(sal-gly)
(c) M^{2+} + sal^{-} + gly^{-} ⇌ M(sal-gly) (XVII)
(d) M^{2+} + sal^{-} + gly^{-} ⇌ M(sal)(gly) (XV)

M^{2+}	β_{11}[b](a)	β_{11}[b](b)	β_{11}[c](c)	β^{*}_{11}[c](d)	k_{P1}[c](sec^{-1})
H^{+}	10.86	11.21	—	—	—
Mg^{2+}	—	—	4.77	3.08	0.015
Mn^{2+}	—	6.78	7.29	4.66	0.015
Co^{2+}	—	—	—	6.94	$< 10^{-4}$
Ni^{2+}	9.02	10.27	—	—	$< 10^{-6}$
Cu^{2+}	14.1-14.7	15.67	—	—	$< 2 \times 10^{-7}$
Zn^{2+}	7.15	9.17	9.65	7.36	0.0026
Cd^{2+}	—	—	7.31	5.20	0.062
Pb^{2+}	—	—	8.86	6.58	1.9

[a] β = formation constant, log values given; [b] data from ref. 42;
[c] data from ref. 55.

until recently[53-57]. Considerably greater effort has been expended in the establishment of solute structures and equilibrium compositions of aqueous and methanol solutions containing, in particular, pyridoxylidene and salicylidene complexes derived from divalent metal ions and simple amino acids such as glycine, alanine, and valine[21,22,42,47,58-61]. Formation constants for 1:1 and 2:1 pyr–aa and sal–aa complexes have been determined and some comparative data for 1:1 species are given in Table I.

The most detailed kinetic studies have dealt with the formation of M(sal–gly) complexes in aqueous solution as followed by pH-stat and spectrophotometric techniques[54,55]. In the range pH 5-9 these systems exhibit two types of kinetic behavior: a metal-independent path only (Cu^{2+}, Ni^{2+}, Co^{2+}), or simultaneous metal-independent and metal-dependent paths (Mg^{2+}, Mn^{2+}, Zn^{2+}, Cd^{2+}, Pb^{2+}). The former paths are identical for all metal ion systems and conform to the rate law $d(SB)/dt = (k_a a_H^2 + k_b a_H + k_c)(sal^-)(gly^-)$, which has been interpreted in terms of rate-limiting attack of glycinate ion on salicylaldehyde anion (k_c) and on neutral salicylaldehyde (k_b), with the latter process also exhibiting an acid-catalyzed path (k_a). In these cases the metal ion acts in effect as a stabilizing trap for the intermediate carbinolamine and final product, but does not kinetically catalyze the reaction. Metal-dependent reactions follow the rate law (3) where $k_{P1} = k_0/\beta_{11}^*$, $k_{PH} = k_1/\beta_{11}^*$, and $k_{P2} = k_2/\beta_{12}^*$. β_{11}^* is the formation constant

$$\begin{aligned} d(SB)_{M^{2+}}/dt &= [k_0 + k_1 a_H + k_2 (gly^-)](M^{2+})(sal^-)(gly^-) \\ &= (k_{P1} + k_{PH} a_H)[M(sal)(gly)] + k_{P2}[M(sal)(gly)_2^-] \end{aligned} \qquad (3)$$

for the ternary (uncondensed) complex M(sal)(gly) (reaction (d), Table I) and β_{12}^* is defined analogously. Barring improbable ternary collisions, the form of the rate law indicates a rapid pre-equilibrium (4) among reactants to form labile M(sal)(gly) **(XV)**, in which the ligands are independently coordinated, followed by the rate-determining step to form the product M(sal–gly) **(XVII)**. The data in Table I indicate that for M(sal)(gly) $\rightleftharpoons$ M(sal–gly) $K_{eq} = k_{12}k_{23}/k_{21}k_{32} = \beta_{11}/\beta_{11}^* \sim 100$. The slow step most consistent with the kinetic data is proposed to be formation of the carbinolamine **(XVI)**, which is succeeded by rapid dehydration to **XVII**[55]. If this scheme is correct k_{12} is equal to the parameter k_{P1}, which provides an index of the kinetic catalytic effectiveness of metal ions in Schiff base formation. Values of this parameter (Table I) lead to the order Pb $\gg$ Cd $>$ Mn $\sim$ Mg $>$ Zn $\gg$ Co, Ni, Cu (very small). The most intriguing feature of this order is that the metal ions effective as catalysts are those with no partly filled *d*-orbitals. Complexes of these ions are inherently more labile than those of catalytically inactive Co^{2+}, Ni^{2+}, and Cu^{2+}, and further, do not necessarily conform to particular geometries, which in ions with incomplete *d*-subshells are stabilized by ligand field effects. Consequently, the rate-limiting conversion **XV** → **XVI** is facilitated. Schiff base formation by this mechanism is consistent with the usual concept of the kinetic template effect, in which the metal ion binds reactants in fixed spatial positions suitable for specific multistep reactions. Recently, the more incisively descriptive term "promnastic" ("matchmaking") has been coined to convey the role of a kinetic catalytic agent "which forms a ternary complex with two reactants but which imposes a minimum steric requirement upon them"[55].

$$\text{salicylaldehyde anion} + H_2NCH_2CO_2^- + M^{2+} \rightleftharpoons \text{(XV)} \quad (4)$$

$$\text{(XV)} \underset{k_{21}}{\overset{k_{12}}{\rightleftharpoons}} \text{(XVI)} \underset{k_{23}}{\overset{k_{32}}{\rightleftharpoons}} \text{(XVII)}$$

(XV) (XVI) (XVII)

The generality of the above mechanism of Schiff base complex formation with respect to other *o*-hydroxyarylcarbonyl and amino acid reactants, including the specific matter of mechanistic similarity in the formation of pyr–aa complexes, remains to be established. A recent study[56] of the aqueous system PLP:glutamate:Cu^{2+} has shown that at pH ~ 4 Schiff base complex formation proceeds by a metal ion-independent path, which is kinetically analogous to that yielding Cu(sal-gly). Finally, the results of a spectrophotometric investigation[57] of the aqueous system containing a 1:1:1 mole ratio of 5-sulfosalicylaldehyde (Ssal), glycine, and Cu^{2+}, to which an equivalent of ethylamine was added, has shown that a kinetic template or promnastic effect is not likely to be operative in the presence of unbound amine. The first detectable Schiff base species is the thermodynamically less stable complex $[Cu(Ssal\text{-}Et)(gly)]^-$, which subsequently converts to $[Cu(Ssal\text{-}gly)]^-$ in a thermodynamic template reaction.

IV. REACTIONS IN MODEL SYSTEMS

The preceding remarks and the contents of this section serve to show that the principal significance of vitamin B_6 complexes rests in the formation and reactions of Schiff base species of types **VI** and **VII** and their ring-protonated forms. Complexes of the group of vitamin B_6 compounds themselves (**I–V**) do not play an important role in catalytic reactions of α-amino acids. In model systems they may be formed in rapid reactions prior to the slower processes which result in Schiff base formation. Solution spectral features and formation constants of complexes of pyridoxamine[58,62–64] and pyridoxal[42,59,60] with ions such as Zn^{2+}, Cu^{2+}, and Ni^{2+} have been reported, but none of these complexes has been isolated.

The classic pioneering work of Snell, Metzler, and their co-workers in the early 1950s demonstrated that many reactions of pyridoxal phosphate enzymes could be duplicated in model systems containing pyridoxal or some other suitable *o*-hydroxyarylcarbonyl compound and various metal ions as catalysts[3]. These non-enzymatic reactions in general proceed readily in slightly acidic, neutral, or slightly basic aqueous solutions at 100°C and require ~30 min to several hours for equilibrium or near-equilibrium to be reached. Investigations of model systems resulted in the independent formulation by Braunstein and Shemyakin[65] and Metzler, Ikawa, and Snell[66] in 1953–1954 of a series of reaction pathways and mechanisms which accounted for the catalytical role of pyridoxal and metal ions in model systems, and, by implication, the role of pyridoxal phosphate in enzymatic reactions. Reactions of α-amino acids in model systems are listed in Table II, which closely follows the division of reaction types given by Snell[67] but is less complete. A common feature of these reactions involving pyridoxal catalysis is the formation of a 1:1 or 2:1 aldimine complex of types **VI** or **XVIII**, depending upon conditions, followed by breaking of

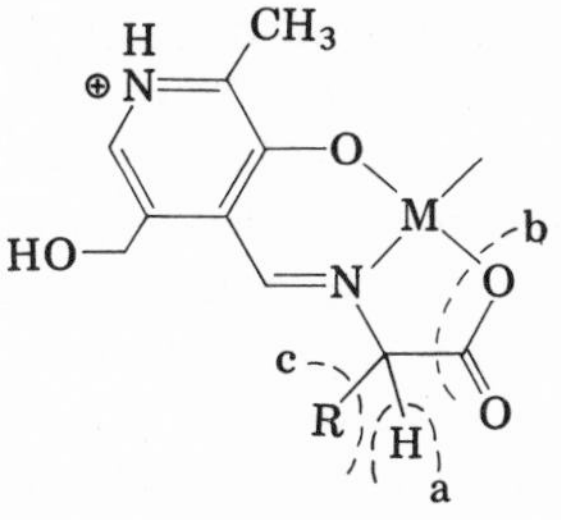

(XVIII)

the three bonds a, b, c in the condensed amino acid portion. The first three reaction types in Table II derive from the breaking of these bonds, the notation of which is that of Snell. Similar tabulations and discussions of reactions catalyzed *in vivo* by pyridoxal phosphate enzymes are available in reviews by Snell[3], Braunstein[13], and Fasella[69]. In the discussions which follow, the reactions listed in Table II are considered in turn. Emphasis is placed on the function of Schiff base complexes in the various transformations, for which mechanisms consistent with the available information are presented.

TABLE II

REACTIONS OF α-AMINO ACIDS CATALYZED BY PYRIDOXAL AND METAL IONS IN MODEL SYSTEMS

A. Reactions resulting from labilization of an α-hydrogen (XVIII, bond a)
1. transamination: $RCH(NH_2)COOH + R'COCOOH \rightleftharpoons RCOCOOH + R'CH(NH_2)COOH$
2. racemization: L-$RCH(NH_2)COOH \rightleftharpoons$ D-$RCH(NH_2)COOH$
3. β-elimination: dehydration of α-amino-β-hydroxyacids
$HOCH_2CH(NH_2)COOH \rightarrow CH_3COCOOH + NH_3$
desulfhydration of cysteine
$HSCH_2CH(NH_2)COOH \rightarrow CH_3COCOOH + H_2S + NH_3$
4. tryptophan synthesis from serine and indole
5. β-proton exchange: $R_2CHCH(NH_2)COOH \rightleftharpoons R_2CDCD(NH_2)COOH$
6. γ-elimination: desulfhydration of homocysteine
$HSCH_2CH_2CH(NH_2)COOH \rightarrow CH_3CH_2COCOOH + H_2S + NH_3$
7. Synthesis of α-amino-β-hydroxyacids from glycine and an aldehyde

B. Reactions resulting from labilization of the carboxyl group (XVIII, bond b)
decarboxylation of amino acids
$RCH(NH_2)COOH \rightarrow RCH_2NH_2 + CO_2$

C. Reactions resulting from labilization of an R group (XVIII, bond c)
degradation of α-amino-β-hydroxyacids to aldehydes and glycine
$RCH(OH)CH(NH_2)COOH \rightleftharpoons RCHO + H_2NCH_2COOH$

D. Oxidative deamination
$RCH(NH_2)COOH + O_2 + H_2O \rightarrow RCOCOOH + NH_3 + H_2O_2$

Before considering the reactions in model systems, several simplifying features in the presentation of mechanisms in Figs. 2 and 5–8 are noted. Because many studies of these systems were performed in dilute acid solutions in order to simulate physiological conditions, the corresponding pH values were usually less than pK_a values of pyridinium protons in the complexes. These have been determined for a number of species in aqueous solutions and are summarized in Table III. Complexes are usually give in their protonated forms and, for ease in formulation, as 1:1 species; those of 2:1 stoichiometry function in a similar way as reactive intermediates. Overall charges on complexes are unspecified with the understanding that the metal ion M in species such as **XVIII** may be di- or tri-valent. Pyridoxal and pyridoxamine are shown in their predominant forms[70] in weakly acidic and neutral solution in Fig. 2; in other figures the appropriate forms are represented by PL or PM. Finally, free Schiff bases, which under certain conditions of concentration, pH, and temperature may exist in appreciable equilibrium quantities with their complexes, are omitted without loss of essential mechanistic detail.

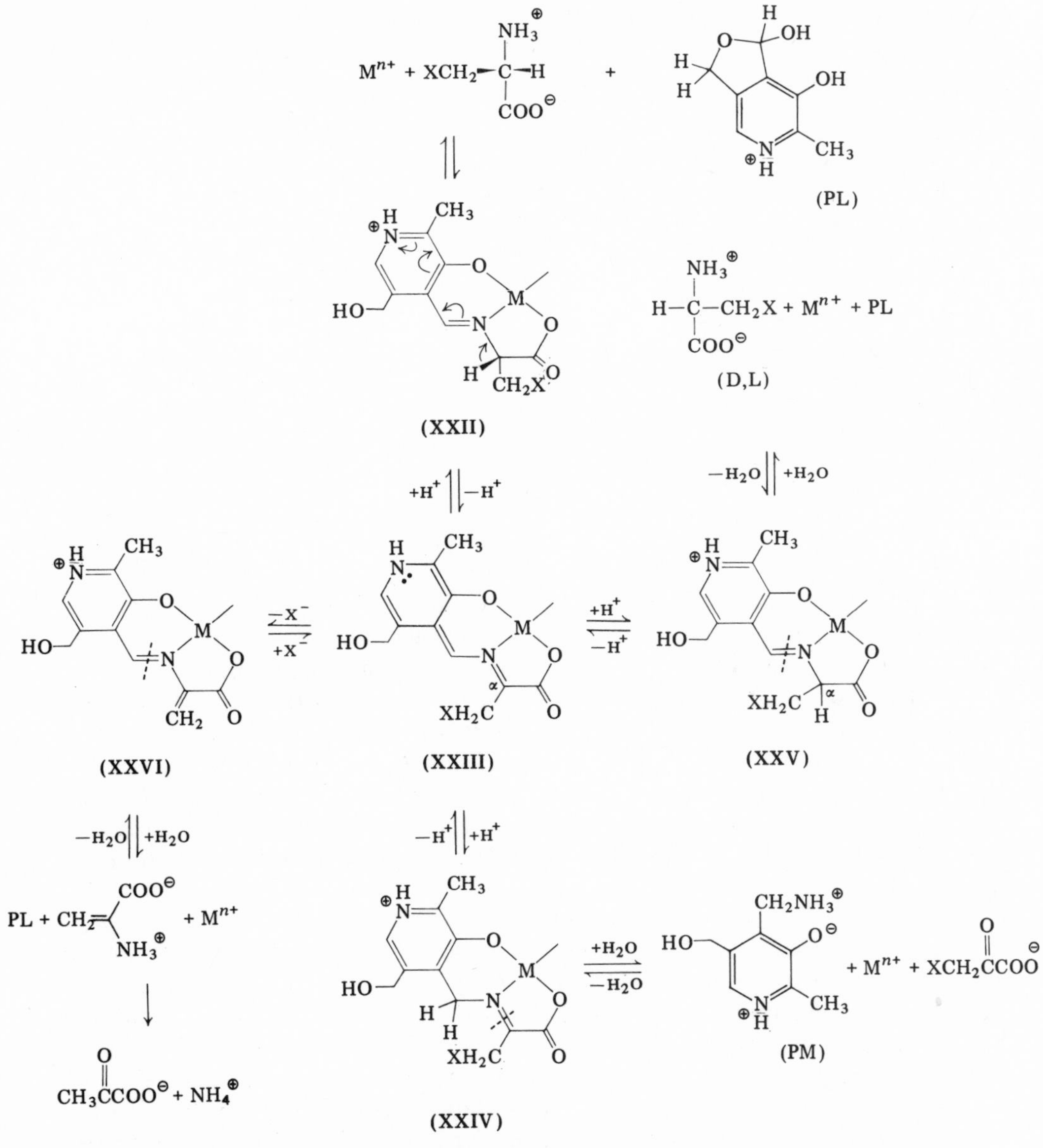

Fig. 2. Mechanisms for pyridoxal-catalyzed transamination, racemization, and β-elimination (X = OH, SH, *etc.*) of L-amino acids.

References pp. 1165–1167

TABLE III

pK_a VALUES FOR PYRIDINIUM PROTONS OF METAL COMPLEXES (XVIII) IN AQUEOUS SOLUTION

Complex	pK_a [a]	Reference
$Al(Hpyr\text{-}ala)^{2+}$	~5.2	47
$Al(Hpym\text{-}pv)^{2+}$ [b]	~6.3	48
$Ni(Hpyr\text{-}gly)^{+}$	7.18	59
$Ni(Hpyr\text{-}val)^{+}$	6.7	21
$Ni(Hpyr\text{-}val)_2$	~7.3, ~8.1	22
$Cu(Hpyr\text{-}gly)^{+}$	6.05	21
$Cu(Hpyr\text{-}val)^{+}$	5.6	21
$Zn(Hpyr\text{-}gly)^{+}$	7.34	59
$Zn(Hpyr\text{-}ala)^{+}$	~6.6	47
$Zn(Hpyr\text{-}val)^{+}$	6.5	21
$Zn(Hpym\text{-}pv)_2$ [b]	~7.9	48

[a] All values not strictly comparable, cf. ref. 59.
[b] Ketimine complex.
pK_a data for pyridinium protons of PLP Schiff base complexes[61]:
$Zn(HPLP\text{-}aa)^-$, 7.45 (gly), 7.40 (ala); $Zn[(H)(PLP\text{-}aa)_2]^{5-}$, 8.32 (gly); 8.60 (ala); $Zn(H_2PLP\text{-}aa)_2^{4-}$, 6.94 (gly), 7.16 (ala).

A. Reactions resulting from labilization of α-hydrogen

Of the reactions occurring as a consequence of breaking bond *a*, transamination has been investigated in the greatest detail. A result of prime importance has been the demonstration that the minimum structural and electronic properties of pyridoxal requisite to its catalytic activity in this reaction[6,66,71] and, apparently, in the other reactions listed in Table II, are found in 3-hydroxypyridine-2- and -4-aldehyde. Like pyridoxal, these molecules contain hydroxyl and formyl groups *ortho* to each other, an obvious requirement for the formation of chelate rings as in **VI, VIII, IX,** and **XVIII**, and a conjugative electron-withdrawing unit ($-N=$ or $-\overset{\oplus H}{N}=$) *ortho* or *para* to the azomethine function. Schiff base complexes derived from these aldehydes possess the common electronic features of (i) a planar conjugated double bond system extending from the pyridine to the azomethine nitrogen, and (ii) weakening of the bonds a, b, and c by virtue of the inductive electron-withdrawing property of the HC=N–M grouping. In the case of loss of α-H, for example, the importance of feature (i) is evident in the resonance stabilization **XIX** ⇌ **XX** ⇌ **XXI** of the intermediate, which is responsible for the kinetic catalytic effectiveness of pyridoxal and its analogs by decreasing the activation energy necessary for formation of the intermediate. In contrast to these compounds, salicylaldehyde, which does not possess a low energy resonance form analogous to **XXI**, fails to catalyze the conversion of glutamate to α-ketoglutarate[3] and forms 1:1 Cu(II) complexes which racemize relatively slowly[23] (*vide infra*). Resonance structures similar to **XIX–XXI** are accessible to unprotonated complexes, which are active in both transamination and racemization. These reactions are competitive processes with the former usually prevalent at pH values near or below the pK_a values of the complexes. In the presentation of reaction mechanisms involving the breaking of bonds a, b, and c the intermediates are depicted in the quinoid form with the understanding that such a form implies others in which negative charge is delocalized over the α- and azomethine carbons.

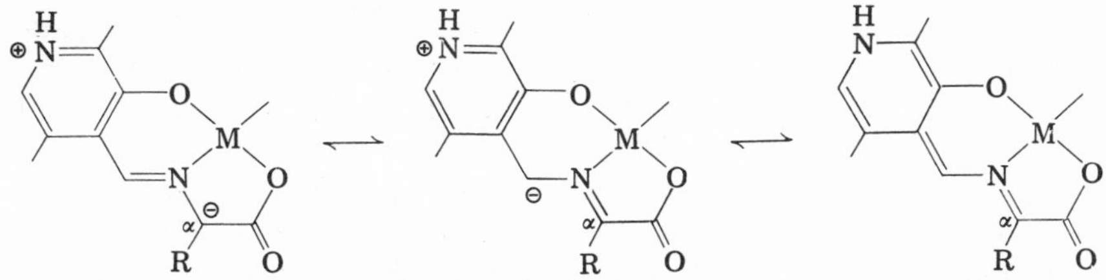

(XIX) (XX) (XXI)

(1) Transamination

The overall transamination reaction (1) results from the coupling of the sequential reactions (5) and (6):

$$RCH(NH_2)COOH + \text{pyridoxal} \rightleftharpoons RCOCOOH + \text{pyridoxamine} \quad (5)$$

$$\text{pyridoxamine} + R'COCOOH \rightleftharpoons R'CH(NH_2)COOH + \text{pyridoxal} \quad (6)$$

$$RCH(NH_2)COOH + R'COCOOH \rightleftharpoons R'CH(NH_2)COOH + RCOCOOH \quad (1)$$

which form the basis for the generally accepted mechanism for transamination of amino acids catalyzed by pyridoxal and metal ions. This mechanism is summarized in Fig. 2 for neutral or weakly acidic solutions. The initially formed pyridoxylidene chelate **XXII** undergoes α-proton loss (bond a rupture) to yield the resonance-stabilized intermediate **XXIII**, which combines with a proton at the azomethine carbon to afford the ketimine species (**XXIV**). Hydrolysis of the imine bond yields pyridoxamine and the metal ion, free or complexed, and the ketoacid, thus completing reaction (5). Reversal of this process involving formation of **XXIV** from the initially present keto acid in the first step results in reaction (6) and accomplishment of the overall reaction (1). The component reactions are reversible and the same equilibrium position for each is attained in systems with R = R′. While the overall reaction can be achieved in model systems, it has proven experimentally simpler to investigate the mechanistic course of the reaction by examination of the separate reactions, each of which is a transamination process itself.

Indication that metal ions are implicated in these reactions came originally from observation of enhanced rates of transamination in their presence[6,72], and the detection of optical selectivity in transamination between D- or L-alanine or phenylalanine and α-ketoglutarate catalyzed by pyridoxal and Cu(II)[73]. The comparative activities of metal ions in catalyzing reaction (2) starting from pyridoxamine and glutaric acid have been investigated. Under conditions where reaction rates are proportional to metal ion concentrations and protonated 2:1 complexes are presumably involved, the following rate order was established[72]: Ga(III) > Cu(II) > Al(III) > Fe(II) > Fe(III) $\gtrsim$ Zn(II) > In(III) $\gtrsim$ Ni(II) > Co(II) > Sc(III). This series represents rate enhancements at 100°C of ~7–35 compared to the metal-free case. Slight catalytic activity was observed for the series Cd(II) > Cr(II) > Mn(II) > Mg(II). A detailed interpretation of these catalytic orders is difficult. However, for divalent metal ions their relative positions correlate with formation constants for the reaction $M^{2+} + PL^- + val^- = M(pyr\text{–}val)$[21] except for inversion of Zn(II) and Ni(II). Data for the corresponding reaction with glutamate are lacking. Optical selectivity was observed for the overall transamination reaction between alanine or phenylalanine (present in excess) and α-ketoglutarate in aqueous solution with pH ~5[73]. It follows from the mechanism in Fig. 2 that formation of optically enriched glutamic acid must arise as a consequence of stereoselectivity in the ketimine (**XXIV**) → aldimine (**XXII**) conversion. This in turn requires a dissymmetric intermediate, which is most simply formulated as one containing the optically active amino acid coordinated to Cu(II) for a time exceeding that required for tautomeric conversion. Possible species involved in the conversion are **XXVII** and **XXVIII**; hydrolysis of the latter affords L-glutamic acid.

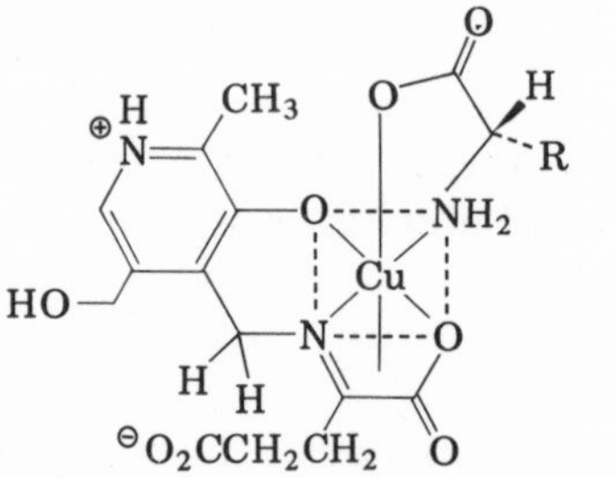

(**XXVII**) [Cu(Hpym–glt) (*L*-aa)]$^-$

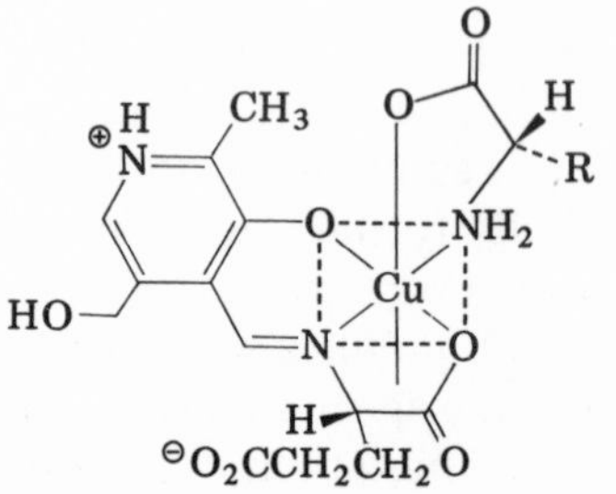

(**XXVIII**) [Cu(Hpyr-*L*-glu) (*L*-aa)]$^-$

This compound was formed with 2–3% optical enrichment with the L-form predominant if L-alanine was present. On this basis the LL diastereoisomer (**XXVIII**) is more stable than the DL form. Although stereoselectivity of product formation is small and the structures of species such as **XXVII** and **XXVIII** entirely speculative, it is difficult to reconcile any selectivity without metal complex formation.

Subsequent spectrophotometric and proton magnetic resonance studies of the individual reactions (5) and (6) have provided abundant evidence for the formation of Schiff base complexes prior to the hydrolytic steps. The recent p.m.r. investigations[46–51] have proven most useful in detecting complex formation and measuring total aldimine and ketimine formation as a function of pH, providing semi-quantitative pK_a values for complexes, and yielding structural information for complexes which do not undergo fast ligand exchange on the p.m.r. time-scale. For the presumably typical systems PL:alanine and PM:pyruvate containing Zn(II)

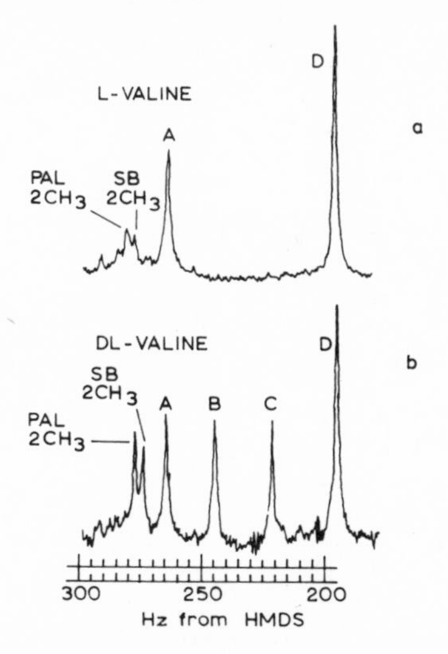

Fig. 3. 100 MHz p.m.r. spectra[50] of D_2O solutions initially containing 0.10 *M* valine, 0.10 *M* pyridoxal, and 0.050 *M* Al(III) in the 2-CH_3 region illustrating formation of diastereoisomeric complexes: (a) L-valine, pD 8.9; (b) DL-valine, pD 9.7 (HMDS = hexamethyldisiloxane external reference, PAL = pyridoxal, SB = free Schiff base).

or Al(III) it has been found that the amount of free and complexed Schiff base exceeds that in the corresponding metal-free system at any pH less than 12, and that formation of 2:1 complexes is favored as the pH is increased and maximizes at pH ~9–11. Ligand exchange is fast in Zn(II) systems but slow in those containing Al(III), thereby permitting observation of signals of the individual complexes. This effect, together with the favorable juxtaposition of the pyridoxylidene 2-CH_3 substituent with respect to the magnetically anisotropic pyridine ring and azomethine group, has permitted detection of the various Al(III) complexes formed at high pH. As illustrated by Fig. 3, L-valine produces two 2-CH_3 signals (A and D) which arise from the Δ(LL) (**XI**) and Λ(LL) (**XII**) diastereoisomers of Al(pyr–L-val)$_2^-$. The use of DL-valine gives in addition of A and D the equally intense signals B and C, which are assigned to the two inequivalent ring methyl substituents of the isomer Δ(DL) ≡ Λ(LD). If the diastereo-isomers are differentiated energetically by 2-CH_3-R(iPr) steric interactions as proposed[50,51], signals A and D are due to Λ(LL) and Δ(LL), respectively. On the basis of signal intensities the latter is more stable than Λ(LL) by

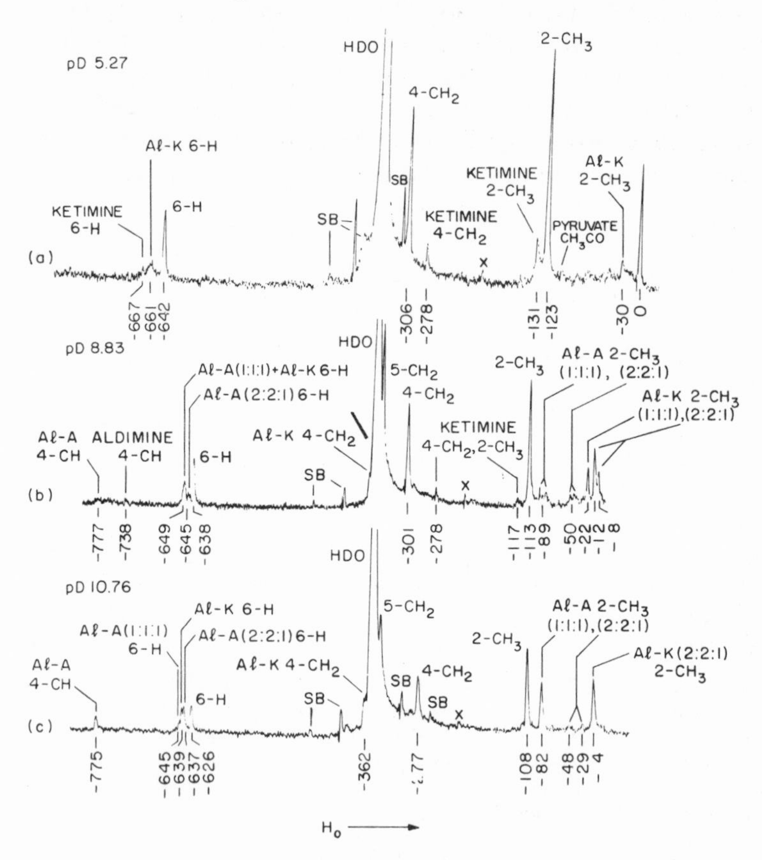

Fig. 4. P.m.r. spectra (100 MHz) of D_2O solutions initially 0.1 *M* in both pyridoxamine and pyruvate, 0.05 *M* in Al(III)[48]: (a) fresh solution showing the formation of free ketimine and the 1:1 Al complex; (b), (c) spectra revealing transamination of Al-ketimine complexes to yield 1:1 and 2:1 Al-aldimine species. The three solutions are not at equilibrium. Unprefixed signals refer to free pyridoxamine: Al-A, aluminum aldimine, Al-K, aluminum ketimine complex; SB, side band; X, impurity.

~0.9 kcal/mole and more stable than Λ(LD) ≡ Δ(DL) by ~0.5 kcal/mole. The correctness of this interpretation, which requires slow racemization of the complexes on the p.m.r. time-scale, is supported by corresponding spectral features of complexes formed by six other asymmetric amino acids[50,51]. Al(pyr–gly)$_2^-$ gives a single 2-CH_3 signal. This interpretation is to be preferred to an earlier assignment of the spectra of Al(pyr–DL-ala)$_2^-$, which assumed rapid racemization and the existence of an uncoordinated carboxyl group[47].

The key step in the transamination reaction mechanism is the aldimine → ketimine conversion **XXII** → **XXIV**, which at the time of its proposal[65,66] had not been observed as a separate process uncomplicated by side reaction or subsequent hydrolysis steps. Other than the detection of uncomplexed transamination products in model systems, early evidence of this reaction was obtained from spectral studies of the system PM:pyruvate:Ni(II)[74], and separation of ketimine and aldimine complexes from the system PL:glycine:Al(III)[75]. More recent work has clearly established this conversion as part of the transamination reaction sequence. It is more easily approached from the ketimine side due to the greater stability and more rapid rate of formation of the aldimine tautomers. The p.m.r. spectra in Fig. 4 clearly reveal the conversion of Al(III)-ketimine complexes[48] to the corresponding aldimine species, the signals of which had been identified in a complementary study of the PL:alanine:Al(III) system[47]. Similar observations have been made for the corresponding Zn(II) system[46-48], except that the averaged chemical chifts due to fast ligand exchange preclude detection of the tautomerization of individual species. An important study has demonstrated that the unprotonated complexes **XXIX** formed from pyridoxamine and α-ketoisovalerate convert to the aldimine form **XXX** by a first-order process in methanol solution at ambient temperature[62,76]. The catalytic order of metal ions, Cu(II) > Zn(II) > Ni(II), is the same as that found for PM-glutarate transamination[72] and results in an approximately 1000-fold rate enhancement over the metal-free case. The presence of base accelerates the reaction. The intermediate is proposed[76] to be the unprotonated analog of **XIX–XXI** which preferentially recombines with a proton at the α-carbon to form the thermodynamically more stable aldimine chelate.

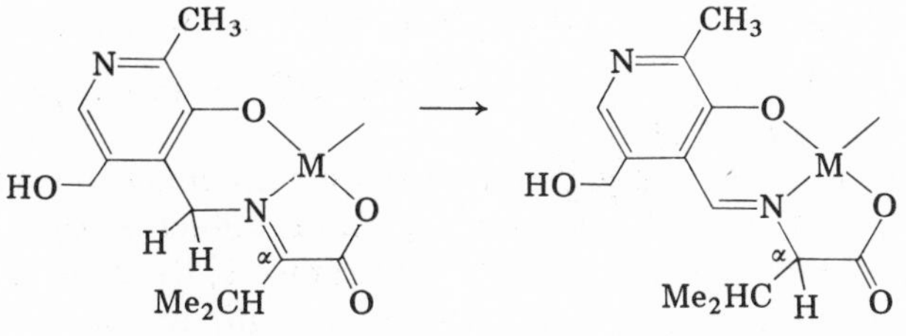

(XXIX) → (XXX)

(2) Racemization

An early study of this reaction in model systems containing pyridoxal and amino acids such as L-alanine, L-phenylalanine, and L-valine showed that its rate was enhanced by Al(III), Fe(III), and Cu(II)[77]. The pH-rate profile for transamination and racemization in the system PL:L-alanine:Cu(II) or Al(III)[77] at 100°C indicates that transamination is the predominant reaction up to pH ~9 and is optimized at pH values equal to or near pK_a values of pyridinium protons of the complexes. The pH maxima for racemization correspond to values at which the pyridine nitrogen is unprotonated. The two reactions differ in the site of proton recombination with the intermediate, the electronic properties of which apparently favor electrophilic attack at the azomethine carbon in pyridinium species. The sequence **XXII** ⇌ **XXIII** ⇌ **XXV** in Fig. 2, followed by hydrolysis, represents a possible mechanism for racemization involving pyridinium species, the key feature of which is protonation of the trigonal α-carbon of the intermediate **XXIII**. However, racemization can also take place by reversal of transamination without formation of the keto acid and this pathway, rather than the preceding reaction sequence, is considered operative in dilute acid solution[77].

Racemization has also been investigated under conditions where the nature of the complexes involved and the course of the reaction are more certain. Base-catalyzed racemization rates of the following four preformed aldimine complexes in 95% ethanol at 50°C were found to increase in the order Cu(sal-L-val) ≪ Cu(pyr-L-val) < Cu(3,2-hpy-L-val) < Cu(4-NO_2sal-L-val)[23]. The reactions are first-order in complex and occur without detectable transamination. A probable mechanism involves removal of the α-proton in a slow step followed by rapid protonation of the α-carbon. The qualitative order of various aldehyde compounds as transamination catalysts in acid solution[3,66] (salicylaldehyde ≪ 4-nitrosalicylaldehyde ~ 3-hydroxypyridine-2-aldehyde ~ pyridoxal) is essentially the same as that of base-catalyzed racemization rates of complexes derived from them. This relationship supports the contention that appreciable reaction rates for transamination in acid solution and racemization in basic solution derive from resonance stabilization of the appropriate intermediates, for which an electron-withdrawing group (–N=, –$\overset{+}{N}$H=, –NO_2) *ortho* or *para* to the azomethine portion is required.

Further evidence that racemization of aldimine complexes occurs under basic conditions (pD ~ 10) is afforded by p.m.r. detection of H–D exchange at the α-carbon in D_2O solutions[51]. As examples, the alanine methyl doublet in Al(pyr-L-ala)$_2^-$ ★ is replaced by a singlet and in several Al(pyr-L-aa)$_2^-$ species the 2-CH_3 signals B and C (cf. Fig. 3) increase in intensity with time. These observations are consistent with racemization accompanying exchange and producing the DL and, eventually, DD diastereoisomers from the initial LL species.

★ An earlier report[47] which claimed no detectable exchange of this species under slightly different conditions is apparently incorrect due to unfortunate overlap of methyl signals of the free and complexed Schiff base.

(3) β-Elimination. α-Amino acids possessing an electronegative substituent undergo a β-elimination reaction more rapidly than transamination when heated with pyridoxal and metal salts[8,66,72,78,79]. Examples are the dehydration of serine to pyruvate and ammonia and the desulfhydration of cysteine to pyruvate, ammonia, and hydrogen sulfide; these reactions are given in Table II. The proposed mechanism is summarized in Fig. 2 for the first of these reactions (X = OH). The intermediate **XXIII** loses hydroxide to generate the unsaturated species **XXVI**, the hydrolysis of which yields α-aminoacrylic acid. This compound spontaneously hydrolyzes to pyruvate. β-Elimination proceeds in acidic, neutral, or alkaline solution and is most effectively catalyzed by those metal ions which are also the better transamination catalysts[72].

(4) Tryptophan synthesis

The overall β-elimination reaction illustrated in Fig. 2 is not reversible due to the decomposition of α-aminoacrylic acid. However, if the species

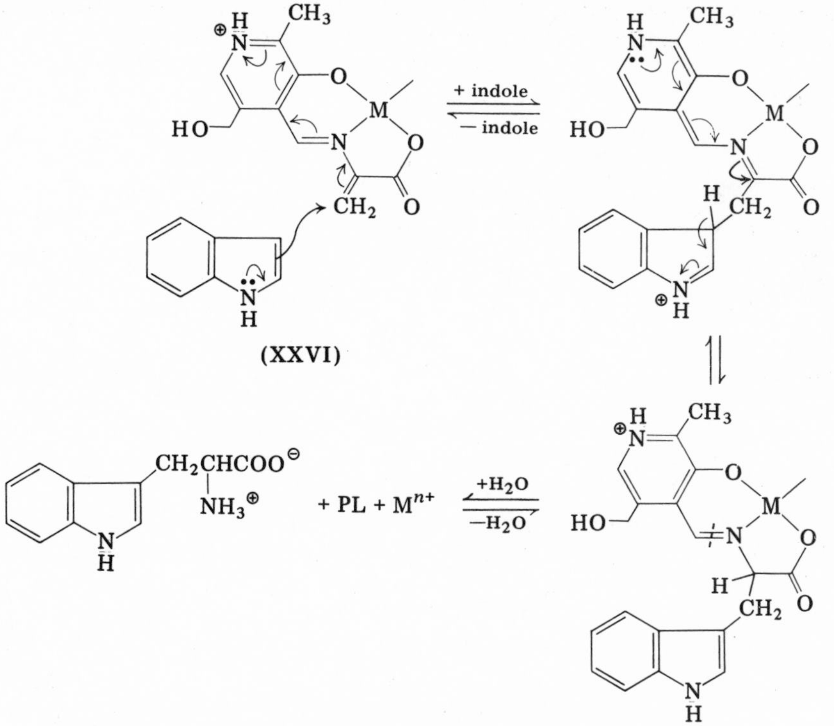

Fig. 5. Synthesis of tryptophan from serine and indole.

References pp. 1165–1167

XXVI could be steadily generated in the presence of a suitably reactive molecule Y, the reaction might be reversed and the β-substituted amino acid $YCH_2CH(NH_2)COOH$ produced. This has actually been accomplished in the synthesis of tryptophan from serine and indole in the presence of pyridoxal and Al(III) at pH 5–6[66]. The very small yields (< 1%) are due in part to the competing dehydration of serine *via* **XXVI**. A possible reaction mechanism is illustrated in Fig. 5.

(5) β-Proton-exchange

This is the newest reaction discovered to be catalyzed by pyridoxal or pyridoxamine and metal ions. P.m.r. spectra of D_2O solutions containing pyridoxal, Al(III) or Zn(II), and valine or α-aminobutyric acid at pD 5–7 have clearly revealed through changes in signal intensities and spin multiplets that α- *and* β-proton exchange occurs upon heating[49]. In a complementary system composed of α-ketobutyric acid, pyridoxamine, and either metal ion facile exchange was observed in neutral solution without heating[49]. Similar observations were also made on systems containing pyruvate, and it was found that the exchange rate of pyruvidene protons qualitatively increased with increasing pD[48]. The exchange reactions have been rationalized in terms of the mechanisms[49] shown in Fig. 6. The initially formed pyrid-

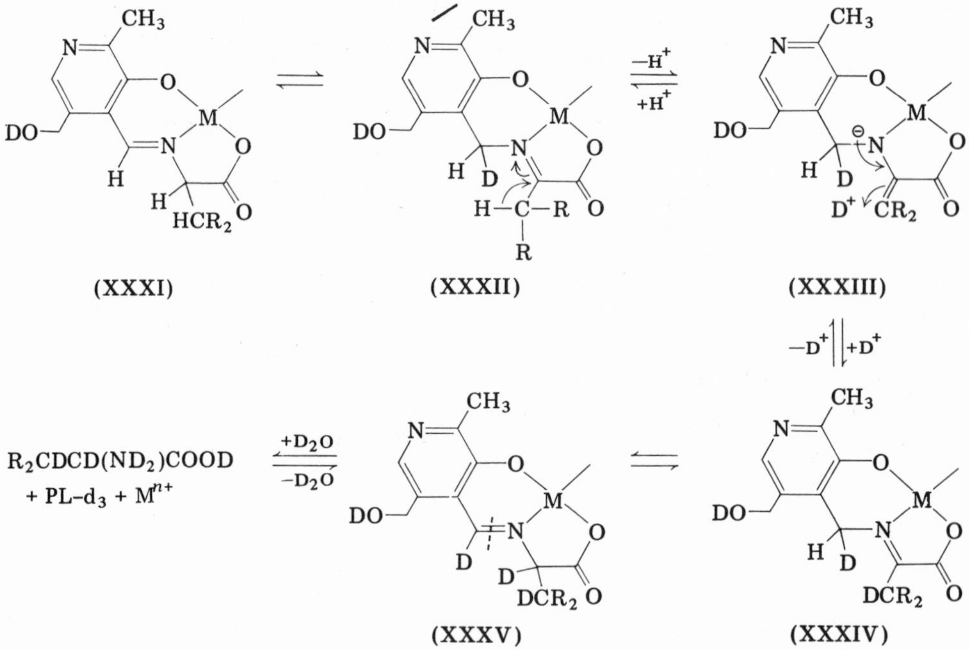

(XXXI) ⇌ (XXXII) ⇌ (−H⁺ / +H⁺) (XXXIII) ⇌ (−D⁺ / +D⁺) (XXXIV) ⇌ (XXXV) ⇌ (+D₂O / −D₂O) $R_2CDCD(ND_2)COOD + PL\text{-}d_3 + M^{n+}$

Fig. 6. Mechanism for α,β-deuteration of α-amino acids (R = CH_3; R = CH_3, H).

oxylidene complex (**XXXI**) tautomerizes on heating to the ketimine (**XXXII**), which undergoes proton loss and deuteron attack (**XXXIII** ⇌ **XXXIV**), and conversion to the deuterated aldimine (**XXXV**). Hydrolysis of this complex yields the α,β-deuterated amino acid. The proposed intervention of ketimine species is consistent with their formation under neutral or weakly acidic conditions in transamination. Alternatively, the ketimine complex may be formed from pyridoxamine and an α-keto acid and β-deuteration allowed to proceed. Because this process appears to be more rapid than tautomerization[48,49], the fully deuterated keto acid can be obtained before appreciable transamination has occurred. Demonstration of α,β H–D exchange could lead to a useful synthesis of preferentially β-deuterated amino acids[49]. After catalytic α,β-deuteration in weakly acidic or neutral solution, the amino acid could be separated or the D_2O removed, H_2O added, and the acid subjected to α-protonation under alkaline conditions (pH 9–10) where racemization has been shown to occur[77]. The latter reaction does not require, and presumably does not involve, an intermediate such as **XXXIII**, which is necessary for β-proton exchange. Consequently, the resultant amino acid should be mainly β-deuterated and α-protonated.

(6) γ-Elimination

In contrast to β-elimination this reaction does not proceed rapidly in model systems and may be obscured by faster reactions such as transamination[3,66]. The desulfhydration of homocysteine to α-ketobutyrate, ammonia, and hydrogen sulfide is one of several γ-eliminations catalyzed by PLP enzymes. A possible pathway for this reaction[16,66] and a revised mechanism for γ-eliminations[49] are given elsewhere.

(7) Synthesis of α-amino-β-hydroxyacids

These compounds are formed by the reverse of a reaction resulting in loss of the R group of **XVIII**, and their synthesis in model systems is discussed in section C.

B. Reactions resulting from labilization of the carboxyl group

Decarboxylation of α-amino acids has been accomplished in model systems containing pyridoxal[15] and does not require the presence of an α-hydrogen. The reaction is inhibited but not completely repressed by metal ions. This effect may be due to the necessity of breaking the metal–carboxylate oxygen bond in **XVIII** upon carbon dioxide release and to completing reactions which are catalyzed by metal ions[3]. A possible mechanism is shown in Fig. 7. This formulation may not be complete inasmuch as decarboxylation of α-aminoisobutyrate in the presence of pyridoxal

References pp. 1165–1167

yielded, in addition to carbon dioxide, isopropylamine, and pyridoxal, acetone and pyridoxamine[80]. In this system a proton presumably replaces the metal ion and acts as a catalyst. When metal complexes are involved, protonation of the intermediate (**XXXVI**) might lead to the aldimine complex (**XXXVII**) and a ketimine species, the hydrolysis of which would yield RCOR′ and pyridoxamine.

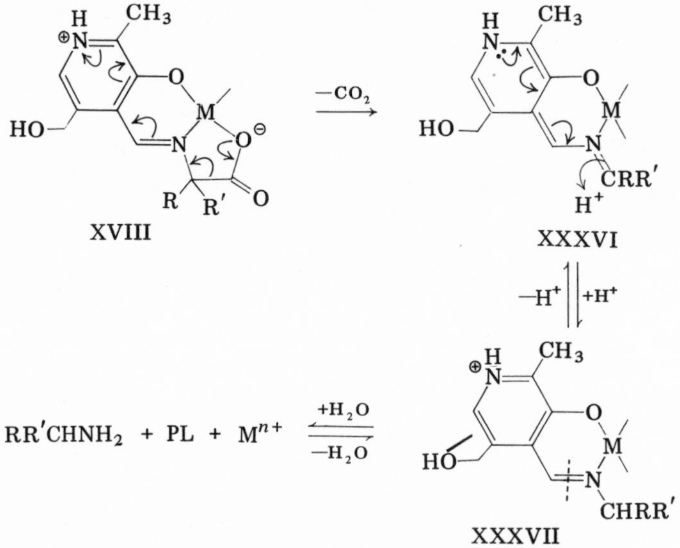

XVIII $\xrightarrow{-CO_2}$ XXXVI $\underset{+H^+}{\overset{-H^+}{\rightleftharpoons}}$ XXXVII

$RR'CHNH_2 + PL + M^{n+} \underset{-H_2O}{\overset{+H_2O}{\rightleftharpoons}}$ XXXVII

Fig. 7. Mechanism for decarboxylation of α-amino acids.

C. Reactions resulting from labilization of an R group

A more thorough assay of the products formed in the catalytic dehydration of serine to pyruvate[78] led to the detection of glycine and the discovery of degradation reactions of β-hydroxy-α-amino acids[8,81]. This transformation of serine, which is one of dealdolization, is given by reaction (7) and, in the mechanism referred to below, occurs as a conse-

$$HOCH_2CH(NH_2)COOH \rightleftharpoons HCHO + H_2NCH_2COOH \qquad (7)$$

quence of bond c rupture in **XVIII.** Detailed studies of this type of reaction have been carried out [8,72,81,82] with a number of β-hydroxy-α-amino acids. Of these threonine and allothreonine are particularly favorable for they undergo the degradation reaction (8) to acetaldehyde and glycine more rapidly than β-elimination to α-ketobutyrate. Reaction (8) is catalyzed

$$CH_3CH(OH)CH(NH_2)COOH \rightleftharpoons CH_3CHO + H_2NCH_2COOH \qquad (8)$$

by pyridoxal and metal ions such as Cu(II), Fe(II,III), Zn(II), and Al(III)[72] over the pH range 4–10, and has been established to be reversible with the equilibrium position favoring the cleavage products. pH optima vary with the metal ion and reaction conditions[8,72]. Reaction (7) is likewise reversible but is partly obscured by β-elimination and transamination. The proposed reaction sequence, illustrated in Fig. 8, proceeds by elimination of an aldehyde from the initial complex **(XXXVIII)** followed by protonation of the intermediate **(XXXIX)** and formation of the *N*-pyridoxylideneglycinato chelate **(XL)**; hydrolysis of this species generates glycine.

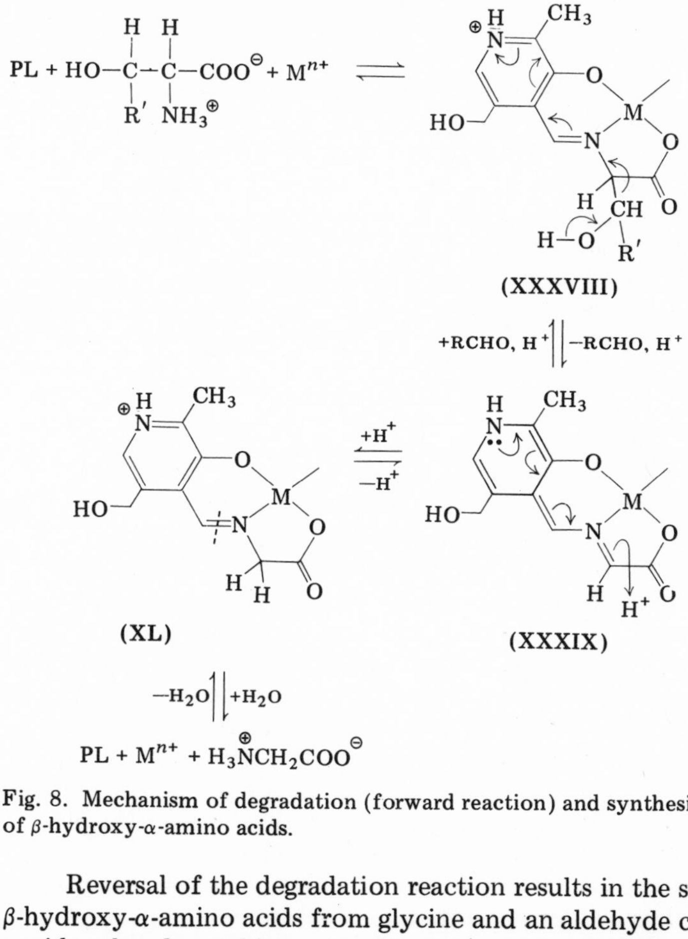

$$PL + HO{-}\underset{R'}{\overset{H}{C}}{-}\underset{NH_3^{\oplus}}{\overset{H}{C}}{-}COO^{\ominus} + M^{n+} \rightleftharpoons \text{(XXXVIII)}$$

(XXXVIII)

$$\text{(XXXVIII)} \underset{+RCHO,\ H^+}{\overset{-RCHO,\ H^+}{\rightleftharpoons}} \text{(XXXIX)}$$

(XXXIX)

$$\text{(XXXIX)} \underset{-H^+}{\overset{+H^+}{\rightleftharpoons}} \text{(XL)}$$

(XL)

$$\text{(XL)} \underset{-H_2O}{\overset{+H_2O}{\rightleftharpoons}} PL + M^{n+} + H_3\overset{\oplus}{N}CH_2COO^{\ominus}$$

Fig. 8. Mechanism of degradation (forward reaction) and synthesis (reverse reaction) of β-hydroxy-α-amino acids.

Reversal of the degradation reaction results in the synthesis of β-hydroxy-α-amino acids from glycine and an aldehyde catalyzed by pyridoxal and metal ions. At pH values which minimize β-elimination formation of serine and pyruvate from formaldehyde and glycine has been observed[8]. Other aldehydes react similarly, and, in addition, it has been found possible to produce in low yield threonyl or seryl peptides from glycyl peptides under catalytic conditions[8]. These reactions require loss of an α-proton of **XL** and thus resemble the synthesis of higher amino acids such as valine, leucine, and aspartic acid from Cu(sal–gly)[83] (**XLI**). This process is summarized by reaction (9),

$$\text{(XLI)} + \text{RBr} \xrightarrow[\text{DMF, } 65^{\circ}\text{C}]{\text{KOH}} \text{(XLII)} \qquad (9)$$

(XLI) (XLII)

which presumably proceeds by base-assisted deprotonation of **XLI** to an anionic intermediate which is an effective nucleophile towards the alkyl bromide. Hydrolytic decomposition of the product complex (**XLII**) affords the new amino acid in yields of about 60%[83]. Finally, it is noted that activation of the α-C–H bonds required for amino acid synthesis following the path in Fig. 8 can be achieved without formation of a Schiff base complex such as **XL**. In reactions of preparative value bis(glycinato)Cu(II) readily reacts with a variety of aldehydes under basic conditions to yield β-hydroxy-α-amino acids[84].

D. Oxidative deamination

This reaction, whose general form is given in Table II, was first observed in 1954 when it was found that certain amino acids yielded α-keto acids and ammonia when heated with metal ions and pyridoxal under catalytic conditions at 100°C[85]. The reaction was investigated at pH 4 and 9.6, under which conditions pyridoxamine was also observed to be oxidatively deaminated to pyridoxal and ammonia in the presence of metal ions. Because pyridoxal also catalyzes transamination of amino acids under these conditions, part of the deamination could occur as a consequence of this reaction succeeded by oxidative deamination of the pyridoxamine formed. However, it was believed that a separate reaction pathway was operative inasmuch as transamination is slow at pH 9.6 and glycine and valine, which transaminate relatively slowly even at the pH optimum of 4–5, were found to react readily. No specific reaction mechanism was proposed at that time other than the suggestion that the process proceeded through decomposition of a pyridoxylidene complex.

Further experimentation has revealed that catalytic systems containing Mn(II) and pyridoxal effect ready oxidative deamination of glycine, alanine, and leucine at pH 9 to the corresponding α-keto acids and ammonia[86,87]. Hydrogen peroxide is apparently the other reaction product but this has not been clearly established. Pyridoxal can be replaced with 3-hydroxypyridine-4-aldehyde but not with salicylaldehyde. The reaction is not general for all amino acids; α-methylalanine, *N*-methylalanine, and β-alanine are not oxidized under conditions where alanine itself is rapidly oxidized. The mechanism of the reaction is not yet established. A possible mechanism is given in Fig. 11, Chapter 20. Unpublished results cited by Martell[17] indicate that the catalytic order of metal ions is Mn(II) > Co(II) > Cu(II) ≫ Ni(II), a result which emphasizes the significance of relatively stable oxidation levels above the divalent state in catalytic activity. Although the overall oxidative deamination reaction in model systems bears a resemblance to reaction (10) catalyzed by several PLP-dependent amine oxidases which require a metal ion as an additional cofactor, several differences are apparent. The substrates oxidized enzymatically are simple amines whereas in model systems amino acids are more readily oxidized, and Cu(II) appears to be the metal ion involved in enzymatic catalysis. Speculation as to the origin of these differences is given by Hamilton[87]. Finally, oxidative deamination has been classified in Table II as a reaction distinct from those involving rupture of bonds a, b, or c because it is the only amino acid transformation requiring an oxidizing agent and change in oxidation state of the catalytic metal ion. This classification is obviously arbitrary since according to the proposed mechanisms[17,86,87] loss of α-hydrogen is an integral part of the process as well.

$$R_2CHNH_2 + O_2 + H_2O \rightarrow R_2CO + NH_3 + H_2O_2 \qquad (10)$$

The results summarized in this chapter serve to demonstrate the enormous versatility of model systems containing pyridoxal or pyridoxamine and metal ions in catalyzing reactions which are accomplished by enzymes requiring pyridoxal phosphate as cofactor. Indeed, with the possible exception of the cobaloximes[88], there is no other group of metal complexes which has been as successful as M(pyr–aa) and M(pym–ka) species in reproducing the reactions of metal-free or metal-containing biological molecules. The remarkable work of Snell and his associates commencing in 1952[6] must be considered as one of the principal starting points in the development of "inorganic biochemistry", the status and accomplishments of which are the subjects of these volumes. Despite the fact that PLP-dependent enzymes do not appear to require metal ions in order to effect the transformations considered here, investigations of model systems containing metal ions are likely to continue at a significant pace in the next few years for at least several reasons. First, there is no other group of metal complexes which manifest as great a diversity of reactions of coordinated ligands, whether of biological significance or not, as do pyridoxylidene and pyridoxylimino chelates. Discovery and elucidation of reactions of coordinated ligands is clearly one of the rapidly developing and significant research activities in coordination chemistry[89] and catalysis[17,90]. Second, there remains much to be understood about the basic structural and electronic features of the complexes themselves. In particular, more effort should be devoted to precise structural determinations, definition of ligand field electronic features of complexes with incomplete *d*-subshells, and assessment of charge distribution and other ground state electronic properties of ring-protonated and unprotonated complexes and intermediates derived thereform, presumably by semi-empirical molecular orbital methods, in order to provide a clearer understanding of reactivities. A useful start on the last problem has been made by Perault *et al.*[91], who have performed molecular orbital calculations of metal-free Schiff bases and intermediates involved in reactions of types A, B, and C (Table II). Finally, further detailed kinetic studies of component steps in the various overall transformations will continue to be of much value.

ABBREVIATIONS

glu	free or condensed glutamate
glt	free or condensed glutarate
gly	free or condensed glycine
3,2-hpy–aa	*N*-(3-hydroxopyridyl-2-methylene)amino acidato dianion
3,4-hpy–aa	*N*-(3-hydroxopyridyl-4-methylene)amino acidato dianion
Hpyr–aa	ring-protonated *N*-pyridoxylideneamino acidato anion
Hpym–ka	ring-protonated α-pyridoxyliminoketoacidato anion
M	di- or tri-valent metal ion
Phala	free or condensed phenylalanine
PL	pyridoxal
PLP	pyridoxal phosphate
PM	pyridoxamine
Ppyr	pyridoxylidene phosphate
pv	free or condensed pyruvate
pyr–aa	*N*-pyridoxylideneamino acidato dianion
pym–ka	α-pyridoxyliminoketoacidato dianion
X-sal–aa	ring-substituted *N*-salicylideneamino acidato dianion (X = H not explicitly stated)
SB	Schiff base (metal-free)
Ssal	5-sulfosalicylaldehyde anion
val	free or condensed valine
Δ,Λ	absolute configurations of bis(tridentate) complexes (**XI, XII, XIII**)

REFERENCES

1 A. F. Wagner and K. Folkers, *Vitamins and Coenzymes*, Interscience, New York, 1964, chapter IX.
2 F. A. Robinson, *The Vitamin Cofactors of Enzyme Systems*, Pergamon Press, Oxford, 1966, chapter V.
3 E. E. Snell, *Vitamins and Hormones*, 16 (1958) 77.
4 *Cf.* collected papers in *Vitamins and Hormones*, 22 (1968) 361–884.
5 E. E. Snell, *J. Am. Chem. Soc.*, 67 (1945) 194.
6 D. E. Metzler and E. E. Snell, *J. Am. Chem. Soc.*, 74 (1952) 979.
7 J. Baddiley, *Nature*, 170 (1952) 711.
8 D. E. Metzler, J. B. Longenecker and E. E. Snell, *J. Am. Chem. Soc.*, 76 (1954) 639.
9 H. N. Christensen and S. Collins, *J. Biol. Chem.*, 220 (1956) 279.
10 P. Fasella, G. G. Hammes and B. L. Vallee, *Biochim. Biophys. Acta*, 65 (1962) 142.
11 E. E. Snell, P. M. Fasella, A. E. Braunstein and A. Rossi-Fanelli, (Eds.), *Chemical and Biological Aspects of Pyridoxal Catalysis*, Macmillan, New York, 1963.
12 E. E. Snell, A. E. Braunstein, E. S. Severin, and Yu. M. Torchinsky, (Eds.), *Pyridoxal Catalysis: Enzymes and Model Systems*, Interscience, New York, 1968.
13 A. E. Braunstein, in P. D. Boyer, H. Lardy and K. Myrbäck, *The Enzymes*, Vol. 2, 2nd edn., Academic Press, New York, 1960, chapter 6.
14 B. M. Guirard and E. E. Snell, in M. Florkin and E. H. Stotz, *Comprehensive Biochemistry*, Elsevier, New York, 1964, chapter V.
15 T. C. Bruice and S. J. Benkovic, *Bioorganic Mechanisms*, Vol. II, W. A. Benjamin, New York, 1966, chapter 8.
16 H. R. Mahler and E. H. Cordes, *Biological Chemistry*, Harper and Row, New York, 1966, chapter 8.
17 A. E. Martell, *Pure Appl. Chem.*, 17 (1968) 129.
18 H. C. Dunathan, L. Davis, P. G. Kury and M. Kaplan, *Biochemistry*, 7 (1968) 4532.
19 J. E. Ayling, H. C. Dunathan and E. E. Snell, *Biochemistry*, 7 (1968) 4537.
20 D. Heinert and A. E. Martell, *Tetrahedron*, 3 (1958) 49.
21 L. Davis, F. Roddy and D. E. Metzler, *J. Am. Chem. Soc.*, 83 (1961) 127.
22 H. N. Christensen, *J. Am. Chem. Soc.*, 79 (1957) 4073.
23 G. N. Weinstein, M. J. O'Connor and R. H. Holm, *Inorg. Chem.*, 9 (1970) 2104.
24 D. Heinert and A. E. Martell, *J. Am. Chem. Soc.*, 85 (1963) 1334.
25 R. P. Houghton and D. J. Pointer, *J. Chem. Soc.*, (1964) 3302.
26 A. Nakahara, *Bull. Chem. Soc. Japan*, 32 (1959) 1195.
27 Y. Nakao, S. Sasaki, K. Sakurai and A. Nakahara, *Bull. Chem. Soc. Japan*, 40 (1967) 241.
28 Y. Nakao, K. Sakurai and A. Nakahara, *Bull. Chem. Soc. Japan*, 40 (1967) 1536.
29 M. J. O'Connor, R. E. Ernst, J. E. Schoenborn and R. H. Holm, *J. Am. Chem. Soc.*, 90 (1968) 1744.
30 T. Ueki, T. Ashida, Y. Sasada and M. Kakudo, *Acta Cryst.*, 22 (1967) 870.
31 J. F. Cutfield, D. Hall and T. N. Waters, *Chem. Commun.*, (1967) 785.
32 G. A. Bentley, J. M. Waters and T. N. Waters, *Chem. Commun.*, (1968) 988.
32a A. N. Barrett and R. A. Palmer, *Acta Cryst.*, B25 (1969) 688.
33 B. E. C. Banks, A. A. Diamantis and C. A. Vernon, *J. Chem. Soc.*, (1961) 4235.
34 R. C. Burrows and J. C. Bailar, Jr., *J. Am. Chem. Soc.*, 88 (1966) 4150; *J. Inorg. Nucl. Chem.*, 29 (1967) 709.
35 E. Willstadter, T. A. Hamor and J. L. Hoard, *J. Am. Chem. Soc.*, 85 (1963) 1205.
36 I. Branden, unpublished work; quoted by H. C. Freeman, *Adv. Protein Chem.*, 22 (1967) 257.

37 G. N. Weinstein and R. H. Holm, *Inorg. Chem.*, in press.
38 Y. Nakao, K. Sakurai and A. Nakahara, *Bull. Chem. Soc. Japan*, 38 (1965) 687; 39 (1966) 1471.
39 H. Yoneda, Y. Morimoto, Y. Nakao and A. Nakahara, *Bull. Chem. Soc. Japan*, 41 (1968) 255.
40 M. Kishita, A. Nakahara and M. Kubo, *Aust. J. Chem.*, 17 (1964) 810.
41 W. E. Hatfield and F. L. Bunger, *Inorg. Chem.*, 5 (1966) 1161; 8 (1969) 1194.
42 D. L. Leussing and K. S. Bai, *Anal. Chem.*, 40 (1968) 575.
43 R. L. Martin and A. H. White, *Transition Metal Chem.*, 4 (1968) 113.
44 Yu. M. Torchinskii and L. G. Koreneva, *Dokl. Biochem.*, 155 (1964) 110; *Biochemistry*, 30 (1965) 31.
45 Yu. M. Torchinskii, *Biochemistry*, 31 (1966) 909.
46 O. A. Gansow and R. H. Holm, *J. Am. Chem. Soc.*, 90 (1968) 5735.
47 O. A. Gansow and R. H. Holm, *J. Am. Chem. Soc.*, 91 (1969) 573.
48 O. A. Gansow and R. H. Holm, *J. Am. Chem. Soc.*, 91 (1969) 5984.
49 E. H. Abbott and A. E. Martell, *Chem. Commun.*, (1968) 1501; *J. Am. Chem. Soc.*, 91 (1969) 6931.
50 E. H. Abbott and A. E. Martell, *J. Am. Chem. Soc.*, 91 (1969) 6866.
51 E. H. Abbott and A. E. Martell, *J. Am. Chem. Soc.*, 92 (1970) 5845.
52 R. H. Holm, G. W. Everett, Jr. and A. Chakravorty, *Progr. Inorg. Chem.*, 7 (1966) 83.
53 L. J. Nunez and G. L. Eichhorn, *J. Am. Chem. Soc.*, 84 (1962) 901.
54 K. S. Bai and D. L. Leussing, *J. Am. Chem. Soc.*, 89 (1967) 6126.
55 D. Hopgood and D. L. Leussing, *J. Am. Chem. Soc.*, 91 (1969) 3740.
56 M. E. Farago and T. Matthews, *J. Chem. Soc. (A)*, (1970) 609.
57 C. V. McDonnell, Jr., M. S. Michailidis and R. B. Martin, *J. Phys. Chem.*, 74 (1970) 26.
58 Y. Matsushima and A. E. Martell, *J. Am. Chem. Soc.*, 89 (1967) 1322.
59 D. L. Leussing and N. Huq, *Anal. Chem.*, 38 (1966) 1388.
60 Y. Matsushima, *Chem. Pharm. Bull. Tokyo*, 16 (1968) 2143.
61 W. L. Felty, C. G. Ekstrom and D. L. Leussing, *J. Am. Chem. Soc.*, 92 (1970) 3006.
62 A. E. Martell and Y. Matsushima, in E. E. Snell *et al.*, *Chemical and Biological Aspects of Pyridoxal Catalysis;* Macmillan, New York, 1963 pp. 33–52.
63 V. R. Williams and J. B. Neilands, *Arch. Biochim. Biophys.*, 53 (1954) 56.
64 R. L. Gustafson and A. E. Martell, *Arch. Biochim. Biophys.*, 68 (1957) 485.
65 A. E. Braunstein and M. M. Shemyakin, *Biokhimiya*, 18 (1953) 393.
66 D. E. Metzler, M. Ikawa and E. E. Snell, *J. Am. Chem. Soc.*, 76 (1954) 648.
67 E. E. Snell, in ref. 11, pp. 1–12.
68 H. N. Christensen, *J. Am. Chem. Soc.*, 80 (1958) 2305.
69 P. Fasella, *A. Rev. Biochem.*, 36 (1967) 185.
70 O. A. Gansow and R. H. Holm, *Tetrahedron*, 24 (1968) 4477, and references therein.
71 M. Ikawa and E. E. Snell, *J. Am. Chem. Soc.*, 76 (1954) 633.
72 J. B. Longenecker and E. E. Snell, *J. Am. Chem. Soc.*, 79 (1957) 142.
73 J. B. Longenecker and E. E. Snell, *Proc. Natn. Acad. Sci. U.S.*, 42 (1956) 221.
74 G. L. Eichhorn and J. W. Dawes, *J. Am. Chem. Soc.*, 76 (1954) 5663.
75 P. Fasella, H. Lis, N. Siliprandi and C. Baglioni, *Biochim. Biophys. Acta*, 23 (1957) 417.
76 Y. Matsushima and A. E. Martell, *J. Am. Chem. Soc.*, 89 (1967) 1331.
77 J. Olivard, D. E. Metzler and E. E. Snell, *J. Biol. Chem.*, 199 (1952) 669.
78 D. E. Metzler and E. E. Snell, *J. Biol. Chem.*, 198 (1952) 353.

79 J. B. Longenecker and E. E. Snell, *J. Biol. Chem.*, 225 (1957) 409.
80 G. D. Kalyankar and E. E. Snell, *Biochemistry*, 1 (1962) 594.
81 D. E. Metzler, J. B. Longenecker and E. E. Snell, *J. Am. Chem. Soc.*, 75 (1953) 2786.
82 J. B. Longenecker and E. E. Snell, *J. Biol. Chem.*, 225 (1957) 409.
83 A. Nakahara, S. Nishikawa and J. Mitani, *Bull. Chem. Soc. Japan*, 40 (1967) 2212.
84 T. T. Otani and M. Winitz, *Arch. Biochim. Biophys.*, 90 (1960) 254; 102 (1963) 464; A. Mix, *Z. Physiol. Chem.*, 327 (1961) 41.
85 M. Ikawa and E. E. Snell, *J. Am. Chem. Soc.*, 76 (1954) 4900.
86 G. A. Hamilton and A. Revesz, *J. Am. Chem. Soc.*, 88 (1966) 2069.
87 G. A. Hamilton, in ref. 12, pp. 375–390, and references cited therein.
88 G. N. Schrauzer, *Accounts Chem. Res.*, 1 (1968) 97.
89 D. H. Busch, *Helv. Chim. Acta*, Alfred Werner Commemoration Volume, (1967) 174.
90 E. Ochiai, *Coordination Chem. Rev.*, 3 (1968) 49.
91 A.-M. Perault, B. Pullman and C. Valdemoro, *Biochim. Biophys. Acta*, 46 (1961) 555.

Chapter 32

THE STRUCTURE AND REACTIVITY OF FLAVIN-METAL COMPLEXES

P. HEMMERICH AND J. LAUTERWEIN

Fachbereich Biologie, Universität Konstanz, 775 Konstanz, Germany

INTRODUCTION

The study of the metal complexes of flavins constitutes a prerequisite for an understanding of the structure and mechanism of action of metallo-flavoproteins. The first question to ask in this context is whether in those enzymes there is any direct contact between metal and flavin. The "model

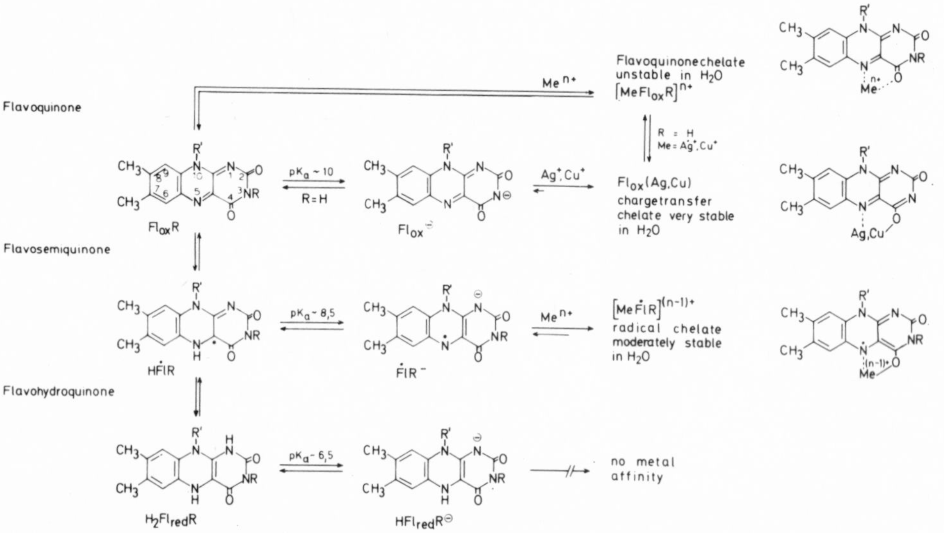

Scheme 1. Structure and formation of flavin metal chelates. Charge transfer chelates are also formed with Mo^{V}, Ti^{III} and Fe^{II} as centers, but these chelates are less stable. The "charge-transfer chelates" arise from "flavoquinone chelates" by loss of the proton from N(3). At the same time, they can arise from "radical chelates" by loss of one electron. With all "non-donating" metal ions, the charge transfer chelates are unstable toward solvolysis.

studies" described in the present chapter have been undertaken in order to establish the structures of all possible kinds of flavin–metal complexes, their range of stability and their physical properties. The formulae of compounds of the flavin redox system are depicted in scheme 1. The heterocycle can coordinate metal ions in several positions; other portions of the flavin may also interact, *e.g.* riboflavin (scheme 1: R = H, R′ = ribityl) may complex through the hydroxyl groups, an interaction which becomes important at high pH and low ionic strength. The complexity of the systems and the low stability of some metal–flavin complexes are the reasons why the early studies in the field result in erroneous conclusions. Potentiometric titrations[1] were apparently complicated by metal hydrolysis[2]. Mixing stoichiometric quantities of flavin, metal and alkali[3] produces a precipitate which has an irreproducible composition, which does not yield data characteristic of "flavoquinone metal chelates" and which is not soluble without dissociation in any solvent. Infrared (i.r.) spectroscopy in the solid state does not distinguish coordination compounds from simple salts (Fig. 1) as it was supposed[4,5].

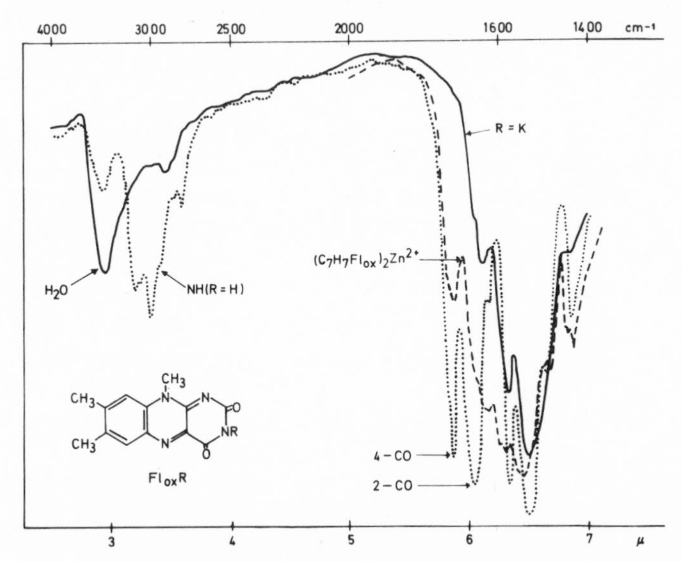

Fig. 1. Infra-red spectra (KBr) of lumiflavins in the neutral flavoquinone state $Fl_{ox}H$ (. . .), as potassium salt $Fl_{ox}K$ (——)[5] and as bis(3-benzyl-lumiflavoquinone)-Zn-perchlorate (- - -)[13]. Some free flavoquinone may have been liberated by interaction with water present in the KBr. Chelation and salt formation can not be distinguished by this method.

From these corrected observations, it can be concluded that, in spite of its bidentate profile, flavoquinone does not show any affinity for common *d*-metal ions under aqueous conditions. One exception has been known, however, since the early times of research in the flavin field, *viz.*, the red silver salt of riboflavin, which had been utilized for the specific precipitation of flavin from complex mixtures[6]. Subsequent studies of the silver compound[7,8] revealed an obviously bidentate complex $[Fl_{ox}Ag]^0$ (cf. scheme 1) of extremely high thermodynamic stability ($\log K^{Ag}_{AgFl} \sim 10$).

For comparison, the upper limit for the stability of other flavoquinone *d*-metal chelates, as estimated from their "non-existence" under aqueous conditions, is more than 6 orders of magnitude lower[2], *e.g.* for Cu^{II} $\log K^{Cu}_{FlCu} < 4$. The apparent "specific affinity" of Ag^I and Cu^I for flavoquinone[8,9] was found to be related to the extremely "soft"[10] (or "class b"[11]) character of these ions (Chapter 2). This was later also demonstrated for Mo^V and Fe^{II} [12,13]. We have, therefore, termed the $Fl_{ox}Me^{(n-1)+}$ class of flavin metal complexes as *charge-transfer chelates*, since their stability as well as color are largely determined by the high extent of d,π-back donation from the metal towards the electron-deficient ligand. This charge-transfer is marked not only in the excited state but also in the ground state of the chelate.

From these data it was obvious that, in order to characterize flavoquinone complexes with the metal ions of hard or moderately soft character, *non-aqueous conditions* had to be applied. A prerequisite for this was the synthesis of "lipophilic" flavin derivatives. This can be achieved in two ways: starting from riboflavin (scheme 1, R = H, R′ = ribityl), acylation of the polyhydroxylic side chain leads to suitable model compounds whereas, starting from lumiflavin (scheme 1, R = H, R′ = CH_3), introduction of a large hydrophobic residue in position 3 would fulfill this purpose (*e.g.* benzyl or ethoxycarbonylmethyl). The advantage of the riboflavin derivative lies in the fact, that in this case the choice exists for having N(3) either protonated or alkylated.

Such *flavoquinone chelates* $(MeFl_{ox}R)^{n+}$ have indeed been found to be stable[9,13] in non-aqueous polar solvents. In some cases, well shaped crystals $Me(Fl_{ox}R)_2(ClO_4)_2 \cdot 2H_2O$ have been isolated, which are at present under crystallographic examination[14].

Scheme 1 is formulated for 1:1-chelates only, but can be extended to 1:2- and 1:3-chelates. Chelates involving neutral flavoquinone $Fl_{ox}R$ as ligand and chelates involving flavoquinone anion Fl_{ox}^- as ligand, have to be distinguished sharply. For the type $(MeFl_{ox}R)^{n+}$, the nature of R (alkyl or H) is unimportant. If R = H, type $(MeFl_{ox})^{(n-1)+}$ becomes competitive. This type, however, is realized only in the case of "soft" metal ions such as Cu^+ and Ag^+, whose affinities towards monodentate bases such as OH^-, OR^-, or NH_3 are low, and then only 1:1-chelates are formed. For most of the other complexes $(MeFl_{ox}H)^{n+}$, neutralization results in immediate hydrolysis, since equivalent amounts of water are always present, unless water-free soluble *d*-metal salts are used throughout, a condition which is hard to achieve for most heavy metal ions. If, however, an aprotic medium were provided by *e.g.*, working with acetonitrile solvates of Me^{2+}, addition of base would lead to slow irreversible dimerization[15] of flavoquinone.

In addition to the very strong "charge-transfer" chelates Ag^IFl_{ox} and Cu^IFl_{ox}, which are not hydrolyzed by alkali, and to the weak flavoquinone chelates $Me^{II}(Fl_{ox}R)_m^{2+}$, which are hydrolyzed by water, the existence of a third class of flavin–metal complexes could be suspected, namely those derived from "reduced" ligand. As scheme 1 demonstrates, there is only one reduced flavin species which exhibits a bidentate profile with electron pairs available for chelation at N(5) *and* O(4), *i.e.* the radical anion $\dot{\mathrm{F}}\mathrm{lR}^-$. This indeed proved to be the flavin species of highest general metal affinity, yielding chelates $[\mathrm{Me^{II}}\dot{\mathrm{F}}\mathrm{lR}]^+$, which are stable under aqueous conditions at neutral or alkaline pH for all valence-stable bivalent metal ions.

These three types of flavin–metal complexes, *viz.* flavoquinone chelates, flavin charge-transfer chelates and flavosemiquinone chelates, will be dealt with separately in the following sections.

CHARGE-TRANSFER CHELATES

Removal of a proton from N(3) of $\mathrm{Fl_{ox}H}$ results, as mentioned above, in hydrolysis of any flavoquinone metal complex, except for those of the soft metal ions[8,9] $\mathrm{Ag^I}$, $\mathrm{Cu^I}$ and $\mathrm{Fe^{II}}$[12,13], whose affinity for water and the hard base $\mathrm{OH^-}$ is comparatively low. The change from $\mathrm{(Ag, Cu)Fl_{ox}H^+}$ towards $\mathrm{(Ag, Cu)Fl_{ox}}$ increases, however, the back donation of *d*-electrons from the metal toward the π-electron pool of the ligand. This changes the

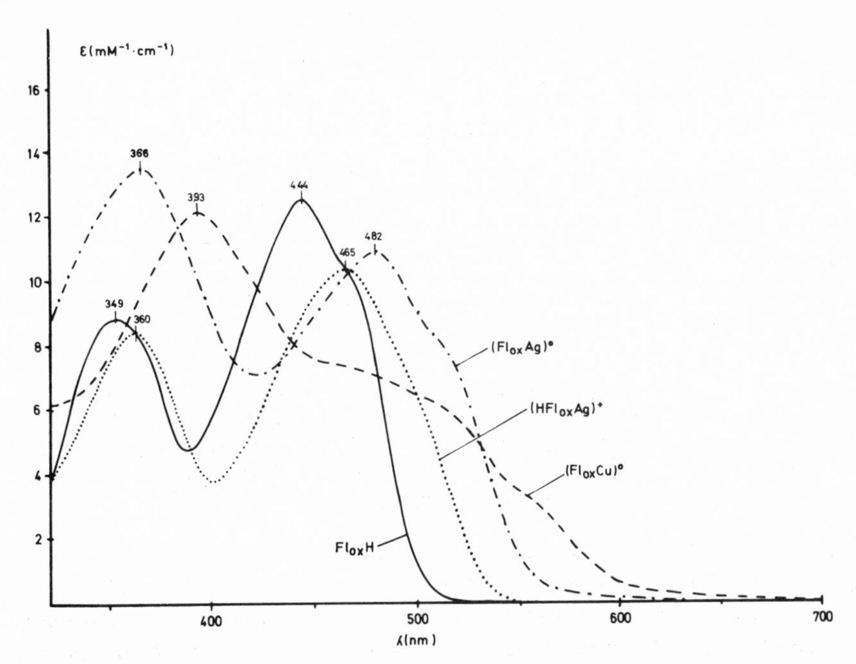

Fig. 2. Charge transfer chelates: Comparison of optical spectra[13]. (——): 5×10^{-4} *M* in tetraacetyl riboflavin ($\mathrm{Fl_{ox}H}$) in acetone–water 7:3, (. . .): the same in the presence of 10^{-2} *M* $\mathrm{AgClO_4}$, the measured pH (glass electrode) is 4.5; (–·–·–·): the same system brought to pH 7.6 with NaOH; (– – – –): 7×10^{-4} *M* riboflavin in argon-flushed dimethyl formamide, 10^{-2} *M* in triethylamine and 2×10^{-3} *M* in $\mathrm{Cu(CH_3CN)_4ClO_4}$[5]. Admission of air removes the $\mathrm{Cu^I}$-chelate only slowly in the absence of water (cf. Fig. 4).

spectra from those of flavoquinone chelates (Fig. 6) towards those of flavin "charge-transfer" chelates (Fig. 2) according to

$$[(Ag^{I},Cu^{I},Fe^{II})Fl_{ox}^{-} \longleftrightarrow (Ag^{II},Cu^{II},Fe^{III})\dot{F}l^{2-}]$$

This electron transfer nevertheless does not give any information on the spin state *i.e.* it does not by itself involve a spin transition from *e.g.* singlet to triplet. It demonstrates, however, the intermediate position of "charge-

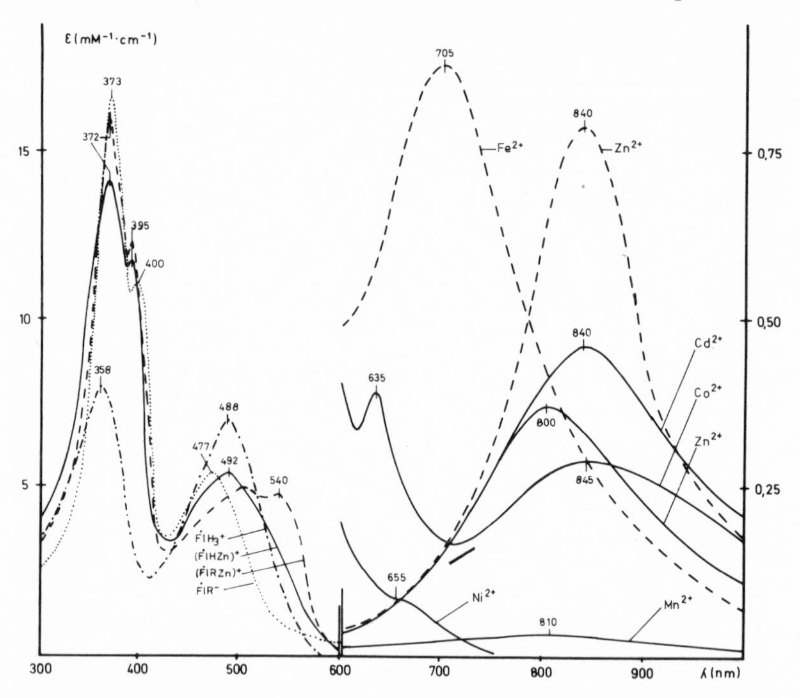

Fig. 3. Optical and near infra-red spectra of flavin radical chelates[26]. In the optical range, the spectrum is practically independent of the nature of the metal, but dependent on whether N(3) is alkylated or not; therefore, only the spectra for $\dot{F}lHZn^{+}$ (——) and $\dot{F}lRZn^{+}$ (- - -) are shown, along with the spectra of the free radical cation (-·-·-·) and the free radical anion (. . . .)[22] for comparison. In the near i.r. region, spectra for different metals are shown. (——) in dimethyl formamide; (- - - -) in pyridine.

transfer chelate" (Fig. 2) spectra between "radical chelate" spectra (Fig. 3) and "flavoquinone chelate" spectra (see below and Fig. 6). The effective magnetic moment of the silver chelate has been found to be zero[16]. However, one would suspect a slight contribution from a triplet state, which might be in equilibrium with the singlet, since in contrast to the situation with $AgFl_{ox}H^{+}$, it is impossible to evaluate $AgFl_{ox}$ by nuclear magnetic resonance (n.m.r.) spectroscopy[13] because of paramagnetic broadening. Clearly, a very small amount of a paramagnetic species, which might be undetectable by susceptibility measurements, could suffice for this.

It is a characteristic feature of charge-transfer chelates that only 1:1 - species exist, even under conditions of ligand excess (cf. Fig. 4). Evidence for the chelate structures as given in scheme 1 can be provided by very different methods:

(a) For the "charge-transfer chelate" $Fl_{ox}Cu$, evidence for the bidentate character of flavoquinone can be obtained[5] from the fact that the reaction:

$$Cu(CH_3CN)_2{}^+ + Fl_{ox}H \rightleftarrows Fl_{ox}Cu + 2CH_3CN + H^+ \quad (2)$$

indeed depends on the square of the CH_3CN concentration. This appears to exclude any monodentate complex $Fl_{ox}CuCH_3CN$ (Fig. 4).

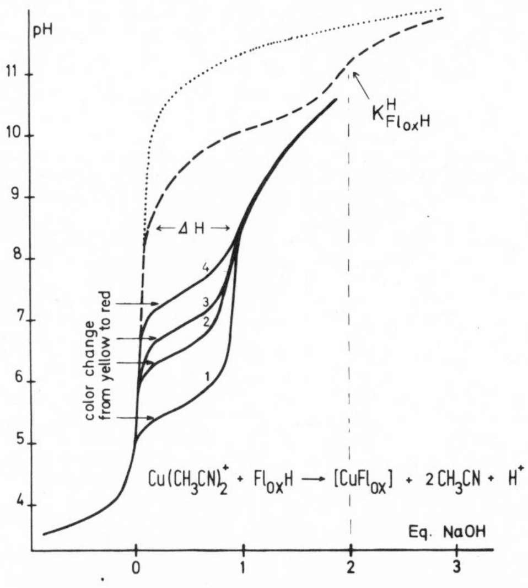

Fig. 4. pH titration in 50% ethanol-water of a system containing 10^{-3} M $HClO_4$, 0.1 M $NaClO_4$ and 2 x 10^{-3} M riboflavin[5]; (. . . .): blank titration with $HClO_4$ alone; (- - -): titration of ligand only. The pK of about 10 of the uncomplexed ligand is reflected in the difference between the dashed and dotted lines. Curves 1–4: pH titration as before, but under argon and in the presence of 10^{-3} M $Cu(CH_3CN)_4ClO_4$. The ligand/metal ratio is 2:1, but in spite of this, the 1:1-chelate is formed exclusively. The metal is added from a $10^{-1}M$ solution in acetonitrile, and the total concentration of acetonitrile is adjusted to 0.25/0.75/1.15/2.25 M for curves 1–4 respectively. The dependence of chelation on acetonitrile concentration indicates liberation of two moles of acetonitrile from $Cu(CH_3CN)^{2+}$ per proton liberated during titration, *i.e.* per chelated flavin ligand. This is evidence for flavin acting as strictly bidentate ligand. Admission of air to the reddish-brown chelate solution, leads to formation of $Cu(OH)_2$ and free flavin.

(b) For the "flavoquinone chelates" it can be shown by n.m.r.[13] (see below and Fig. 5), that the chelating metal ions Zn^{2+} or Cd^{2+} cause different diamagnetic shifts with CH(9) and CH(6) of the flavin nucleus, which indicates metal binding at the neighboring position N(5). The weakly complexing ions Mg^{2+} and Mn^{2+}, however, prefer the formation of monodentate complexes with the flavoquinone ligand. This is demonstrated by the fact that the diamagnetic shifts of CH(6) and CH(9) are equal for Mg^{2+} (Fig. 5) as well as by the difference in absorption spectra (Fig. 6) between "chelates" $ZnFl_{ox}R^{2+}$ and monodentate complexes $MgFl_{ox}R^{2+}$. In agreement with this, any substitution of flavoquinone at N(5), *e.g.* by alkylation[17,18] produces a large bathochromic shift, while the same substitution in the "pyrimidine"

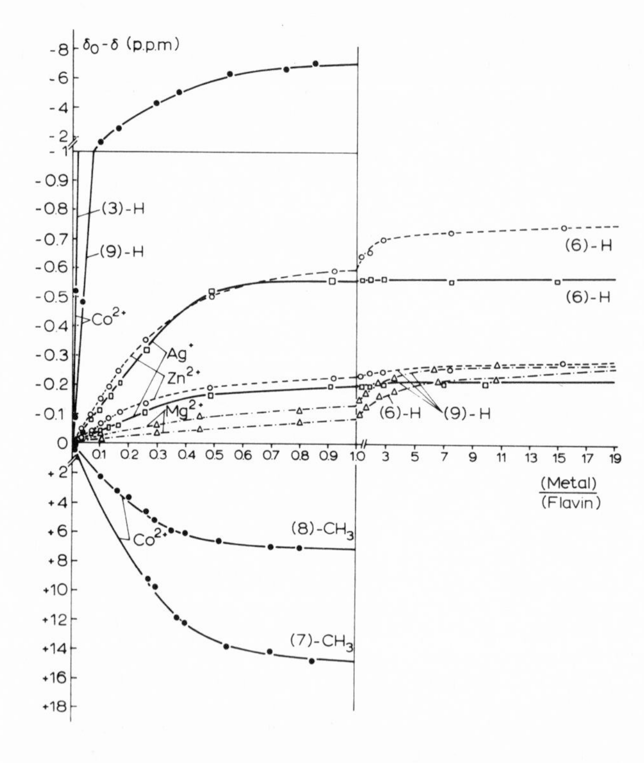

Fig. 5. Diamagnetic and paramagnetic shifts (ppm) as a function of metal to flavin ratio (tetraacetyl riboflavin 3 x 10^{-2} *M* in d_6-acetone)[13]. The initial slopes of the curves indicate 3:1 complexes for Co^{II}, 2:1 for Ag^{I} and 1:1 for Mg^{II}. Negative values correspond to shifts toward low field; positive values to shifts toward high field.

positions 1–3 produces a slight hypsochromic shift (*e.g.* protonation or alkylation at N(1)[17]) or no shift at all (*e.g.* deprotonation or metallation of N(3)).

(c) For the "radical chelates" it can be shown by electron spin resonance (e.s.r.) spectroscopy (Table I) that the metal must be bonded to a ligand position of high spin density, and, furthermore, it can be shown that the only position of the flavosemiquinone anion, which combines high spin and high charge density, is N(5)[19].

(d) For all three types of chelates the Me–N(5) bond is documented by the fact that replacement of C(6)H by C(6)–CH_3 ("isoflavins") abolishes flavin–metal affinity entirely.

Charge transfer chelates have been found with the following metals in decreasing order of stability: $Cu^I > Ag^I \gg Ti^{III} \sim Mo^V \sim Fe^{II}$. Since stability depends on the water affinity of the metal ion in the given state of oxidation, only Cu^I (Fig. 4) and Ag^I form chelates which are very stable in dilute aqueous solution. The optical absorption of these chelates (Fig. 2) is characterized by

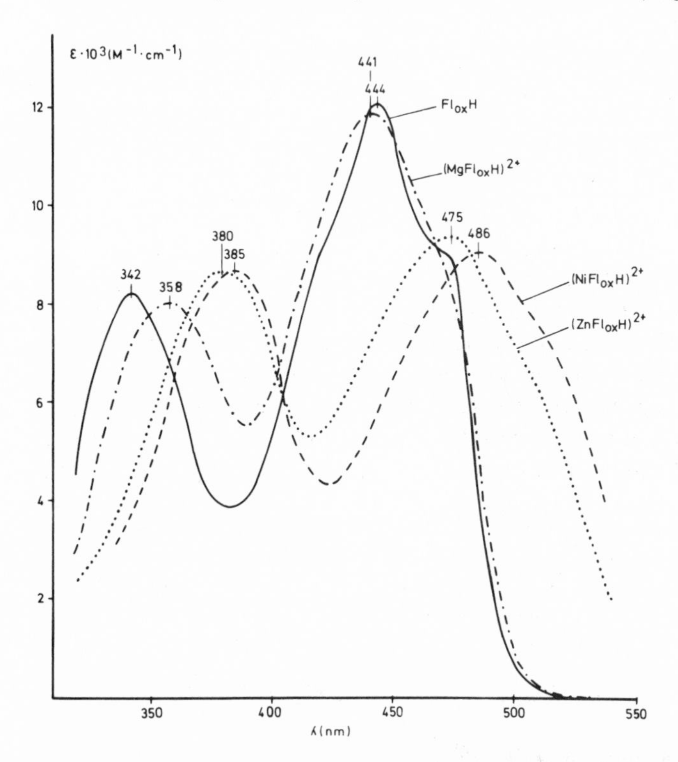

Fig. 6. Optical spectra of tetraacetyl riboflavin 8 x 10^{-4} M in acetone containing 0.5 M benzyltrimethyl ammonium perchlorate for constant ionic strength (——) and in the presence of Zn^{2+} (. . .); Ni^{2+} (- - -) both 5 x 10^{-2} M and Mg^{2+} 10^{-1} M (-· · -·). The red shift of the first band is due to chelation, *i.e.* formation of a Me–N(5) bond. Mg^{2+} causes a shift of the second band only, which indicates polarization of C=O alone presumably in position 2[13].

TABLE I

ELECTRON SPIN-NUCLEAR SPIN COUPLING CONSTANTS

Coupling constant	Radical chelate[a] (G)	Flavoquinone chelate[b] (G)
A_5^N	7.7 ± 0.3	
A_6^H	3.5 ± 0.5	-?
$A_7^H(CCH_3)$	small	-0.0027
$A_8^H(CCH_3)$	3.9 ± 0.2	+0.0002
A_9^H	small	+0.0142
A_{10}^N	3.1 ± 0.5	
$A_{10}^H(NCH_3)$	3.1 ± 0.2	-0.0030
A_{4a}^C	~ 3[30]	
A_3^H	small	-0.0016
$A^{111}Cd$	16.7 ± 0.10	
$A^{113}Cd$	17.6 ± 0.10	
$A^{67}Zn$	2.6 ± 0.05	

[a] Measured for Zn^{2+} or Cd^{2+} flavosemiquinone chelates by e.s.r.[19]
[b] Measured for flavoquinone–Fe^{II} chelate by n.m.r.[13]

a strong π–π^* transition in the range 370–400 nm and long wave end-absorption stretching out into the near infra-red. This end-absorption does not exhibit any well resolved bands, in contrast to the optical spectra of flavosemiquinone chelates (cf. Fig. 3). The intensity of the long wavelength tail of the spectrum seems to reflect the degree of back donation, which is, however, by no means directly related to the complex stability [eqn. (3), step 1]. The back donation as illustrated in the molecular orbital picture of Fig. 7 seems to parallel the ease of oxidation [eqn. (3), step 2]: while $AgFl_{ox}$

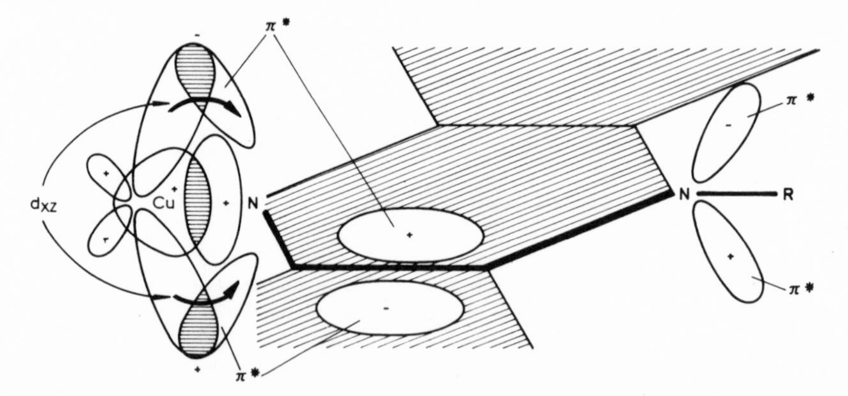

Fig. 7. Molecular orbital scheme of metal–flavin back donation.

is stable against O_2 in water, $CuFl_{ox}$ is stable only under perfectly anhydrous conditions, *e.g.* in dimethyl formamide; the addition of water causes immediate autoxidation. The other charge transfer chelates (Ti^{III}, Mo^{V}, Fe^{II}) remain intact for some time only in the absence of O_2 *and* H_2O. The overall reaction is then a "flavin-catalyzed autoxidation of metal" according to:

$$Me_{red} + Fl_{ox}H \overset{1}{\rightleftharpoons} [Me_{red}Fl_{ox} \longleftrightarrow Me_{ox}\dot{F}l] + H^+ \overset{2}{\rightleftharpoons}$$

$$Me_{ox} + \dot{F}lH^- \overset{3}{\rightleftharpoons} Me_{ox} + \tfrac{1}{2}(Fl_{red}H_2^- + Fl_{ox}^-) \xrightarrow{O_2} Fl_{ox}^- + H_2O_2 \quad (3)$$

FLAVOQUINONE METAL CHELATES

After all attempts to observe flavoquinone metal interaction in dilute aqueous solution had failed, except with the back-donating ions, we tried

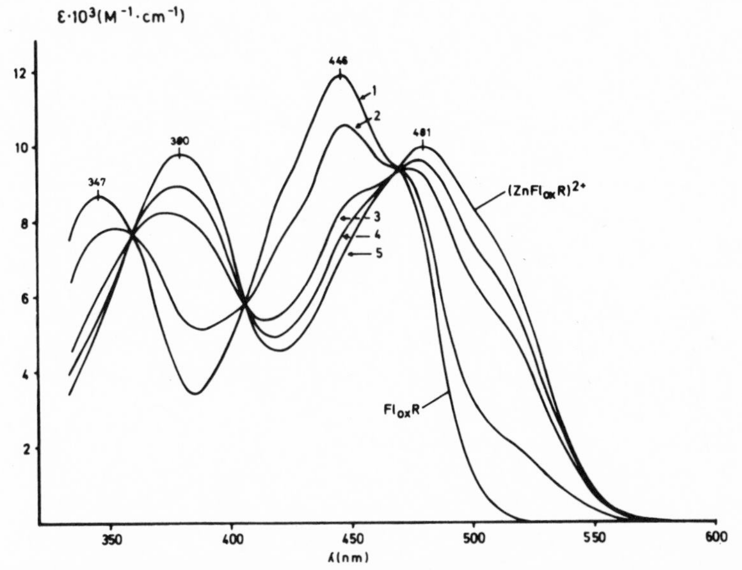

Fig. 8. 3-Ethyltetraacetyl riboflavin 8×10^{-4} *M* in acetone containing 0.5 *M* benzyltrimethyl ammonium perchlorate to induce constant ionic strength, titrated with $Zn(ClO_4)_2 \cdot 6H_2O$ dissolved in acetone. Curves 1–5 correspond to 0, 10^{-3}, 5×10^{-3}, 10^{-2}, and 5×10^{-2} *M* Zn^{2+} [13].

aprotic polar media[11,13] and found suitable conditions for chelate formation in acetone containing water at a less than tenfold molar excess over flavin. The presence of water in such quantities does not interfere with metal–flavin chelation and makes it feasible to work with metal perchlorate hydrates. Typical absorption spectra of chelates $MeFl_{ox}R^{2+}$ are presented in Figs. 6 and 8.

Under the conditions of Figs. 6 and 8, the metal is present in excess over flavin. If the flavin concentration exceeds that of the metal, chelates $Me(Fl_{ox}R)_m{}^{n+}$ are preferentially formed in solution (Fig. 5). No significant spectral differences exist for m = 1,2,3. This is apparent from the isosbestic course of chelate formation (Fig. 8). Chelates of intermediate composition $Me(Fl_{ox}R)_2(ClO_4)_2 \cdot 2H_2O$ are obtained in crystalline form from acetone solutions of suitable composition.

R in the formula $Fl_{ox}R$, scheme 1, may be hydrogen as well as alkyl, and addition of base, *e.g.* triethylamine, leads to hydrolysis irrespective of whether R is H or alkyl. Chelates of type $Me(Fl_{ox}^{-})_m{}^{n-m}$ are, therefore, very unstable, even in acetone, except for those of Ag^{I}, Cu^{I} and Fe^{II} (cf. "charge-transfer chelates").

As mentioned above, no chelation takes place with the less polarizable S-ions — Mg^{2+} and Mn^{2+} — as demonstrated from n.m.r. (Fig. 5) and optical spectra (Fig. 6).

In Fig. 5 the diamagnetic and paramagnetic shifts for flavin protons are plotted against the metal/ligand ratio. Under these conditions, the magnitude of the shift of flavin protons is at first linear with metal-ion concentration, then reaches a plateau indicating complete complexation.

The stoichiometry of the complexes is reflected in the initial slopes of the curves, which can be extrapolated to intersect the plateau line, a process which is reliable when complex stability is relatively high. From such evaluation, the stoichiometry metal:ligand can be estimated as 1:3 for Zn^{II}, Cd^{II}, Fe^{II}, Ni^{II} and 1:2 for Ag^{I}, Cu^{I}, Cu^{II}. The stability can be estimated from the percentage of chelate formed (= percentage of plateau value reached at 1:3- or 1:2-metal/ligand ratio). The sequence of stabilities ($\log K^{Me}_{FIRMe}$) is $Ni^{2+} > Co^{2+} > Ag^{+} > Zn^{2+} > Cd^{2+} > Fe^{2+} \gg Mn^{2+}$, Mg^{2+}.

Furthermore, Fig. 5 depicts bidentate *vs.* monodentate structures as seen from the difference between C(6)H- and C(9)H-shifts, which is negligible for Mg^{2+}.

The much larger shifts and linewidth broadening observed with the paramagnetic complexes (Cu^{2+}, Co^{2+}, Ni^{2+}, Fe^{2+}) are mostly due to electronic paramagnetism as is evident from the temperature dependence. Shifts of comparable magnitude up to 15 ppm are observed for protons located all around the flavin ring, mainly down-field for methine protons and upfield for methyl protons, which suggests contact interaction with antiparallel or β-polarized π-electron spin. The shifts are roughly proportional to the effective magnetic moment of the metal ions. In particular, C(7)–CH_3 is more strongly coupled than C(8)–CH_3, while the reverse is true for flavosemiquinone chelates (Table I). Simple π-delocalization of an unpaired spin either from the metal ion to the ligand or from the ligand to the metal ion cannot be considered to be the mechanism for this interaction in these high-spin complexes since this should correspond to parallel or α-polarization of spin and a coupling strength similar to that observed with flavin radical chelates (see below). Thus the metal–flavin interaction is essentially of σ-type, regardless of the nature of the divalent transition metal ion, and the contact shifts are well explained in sign and magnitude by σ–π spin polarization.

Furthermore, the exceptional dependence of paramagnetic line-broadening on metal/ligand ratio as observed with Co^{2+} (Fig. 9) is interesting.

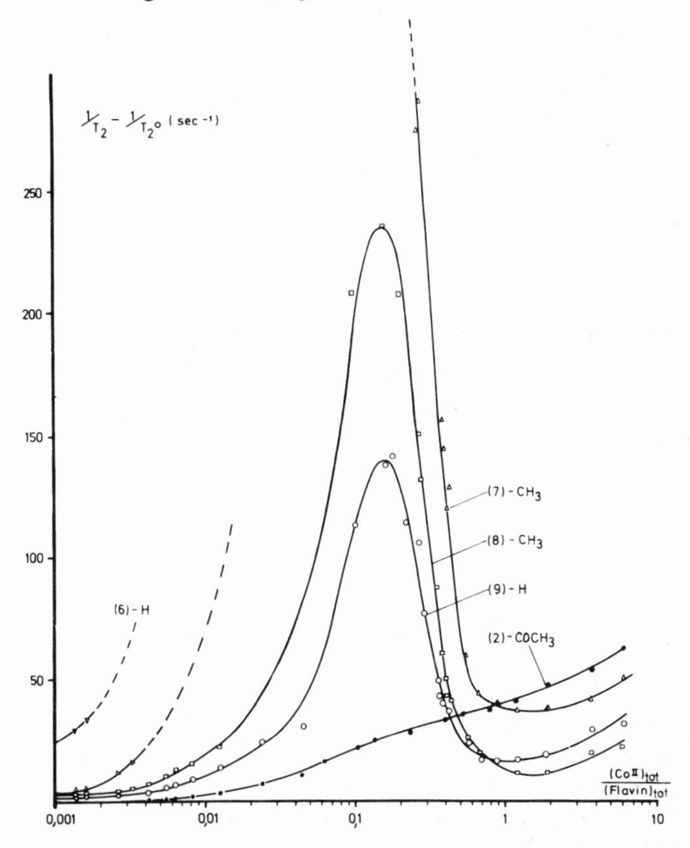

Fig. 9. Proton relaxation rates of tetraacetyl riboflavin as a function of increasing amounts of Co^{2+}[13]. T_2 = measured relaxation time; T_2° = relaxation time of the uncomplexed flavin. The broadening reaches a maximum below a Co/Fl ratio of about 1:3 in agreement with the formation of a 1:3 chelate.

Proton relaxation is governed by the isotropic spin exchange interaction which is proportional to the amount of chelate formed and to the square of the hyperfine interaction constant and is a function of the correlation time τ_e for isotropic hyperfine interaction, where

$$\frac{1}{\tau_e} = \frac{1}{\tau_s} + \frac{1}{\tau_h} \qquad (4)$$

With most metal ions, the electron spin relaxation time τ_s dominates throughout. In the case of Co^{II}, however, only at low metal/ligand ratio, where formation of a 1:3-chelate is found, τ_s is dominating ($\tau_s < \tau_h$). With increasing metal concentration, the line width goes through a maximum and line narrowing by chemical exchange rate $1/\tau_h$ becomes dominant ($\tau_h < \tau_s$).

FLAVOSEMIQUINONE METAL CHELATES

Flavosemiquinone is, as mentioned, the flavin species of highest general metal affinity for two conceivable reasons: (a) the proton affinity of $\dot{F}lH^-$ — in contrast to Fl_{ox}^- — is low enough so as to exclude competition of OH^- with ligand complexation; (b) in contrast to $\dot{F}l_{ox}^-$, where the negative charge is highly localized at position 2–4, $\dot{F}lH^-$ exhibits a considerable charge density at N(5), as documented by the fact that N(5) is the site of protonation in the neutral-radical $H\dot{F}lR$ and not N(1)[20] (cf. scheme 1 and Table I).

The disadvantage of $\dot{F}lH^-$, as compared to the oxidized or fully reduced flavin anions, is its thermodynamic instability.

Flavin represents a thermodynamically reversible redox system. At half reduction, a radical dismutation equilibrium (eqn. 5) is established, which is sensitive to pH[21,22].

$$\begin{array}{ccccc} Fl_{ox}R + Fl_{red}RH_2 & \underset{A}{\rightleftarrows} & 2H\dot{F}lR & & \\ \downarrow\uparrow C & & \downarrow\uparrow D & & \\ Fl_{ox}R + Fl_{red}RH^- + H^+ & \underset{B}{\rightleftarrows} & 2\dot{F}lR^- + 2H^+ & \underset{E}{\overset{Me^{2+}}{\rightleftarrows}} & 2[FlRMe]^+ + 2H^+ \end{array} \tag{5}$$

The parameters which control the thermodynamic stability of the flavosemiquinone or radical species, *i.e.* equilibria A,B of eqn. (5), are the acid dissociation constants of all the flavin species involved, as defined by equilibria C, D. Figure 10 shows a graphic representation of this pH-dependence of the radical stability.

When the acid dissociation and dismutation equations are combined for the neutral range of pH, namely, with respect to neutral flavoquinone $Fl_{ox}R$, neutral radical $H\dot{F}lR$, radical anion $\dot{F}lR^-$, neutral hydroquinone $Fl_{red}RH_2$ and its anion $Fl_{red}RH^-$ as the flavin species involved, eqn. (6)[21] results for the range of:

$$6 < pH \begin{cases} < 11 \text{ if } R = \text{Alkyl} \\ < 9 \text{ if } R = H \end{cases}$$

$$k_{tot}(H) = k_{neutr} \cdot (1 + \dot{K}/[H])^2/(1 + K_{red}/[H]) \tag{6}$$

where K_{red} and $\dot{K}$ are the acid dissociation constants of flavohydroquinone and flavosemiquinone, respectively, and k_{tot} is the pH-dependent "total radical formation constant":

$$k_{tot}(H) = \frac{[\dot{F}l(tot)]^2}{[Fl_{ox}(tot)]\,[Fl_{red}(tot)]} \tag{7}$$

whereas k_{neutr} is the pH-independent formation constant of the neutral radical species alone:

$$k_{neutr} = \frac{[H\dot{F}lR]^2}{[Fl_{ox}R]\,[Fl_{red}RH_2]} \tag{8}$$

Presentation of k_{tot} as function of pH indicates a minimum of radical

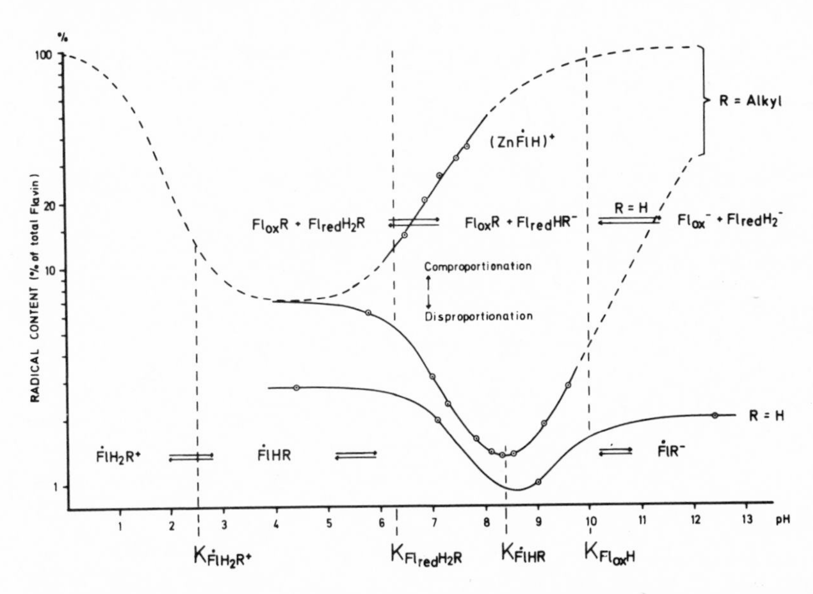

Fig. 10. pH dependence of flavosemiquinone disproportionation for N(3)-alkylated and N(3)-unsubstituted flavin[11,31]. The full lines indicate calculated values; the dashed lines are estimated values; ⊙ indicates an experimentally determined value; the acid dissociation constants of the flavin species concerned are indicated below; they are the same within ± 0.1 for R = H or alkyl. In the presence of metal, the minimum of radical formation at pH 8.4 is absent (upper curve).

content (= a maximum of disproportionation in the neutral range of pH) for $p\dot{K} \sim 8.5$ and $pK_{red} \sim 6.5$ (values somewhat depending on the kind of flavin used). Whereas K_{red} can be determined easily by pH titration (cf. Fig. 11), direct measurement of $\dot{K}$ calls for a method for selective determination of radical concentration. This has been done by e.s.r.[22]. At the same time, the early potentiometric data of Schwarzenbach and Michaelis[23] have been refined by Ingraham[24]. After this treatment, they showed perfect agreement with the direct determination by e.s.r.

It follows from eqn. (3) that the thermodynamic stability of flavoquinone reaches a minimum only for $\dot{K} < K_{red}$. Because of the selective metal affinity of the radical anion $\dot{F}lR^-$, $\dot{K}$ is shifted in the presence of heavy metal ions according to eqn. (5E). The proton balance of this equation shows that one half proton per metal ion is liberated in addition to the amount liberated in the absence of metal (eqn. 9):

$$Fl_{ox}R + Fl_{red}RH^- + 2Me^{2+} \xrightleftharpoons{pH > 5} 2[\dot{F}lRMe]^+ + H^+ \quad (9)$$

which has been verified by experiment[21] as shown in Fig. 11. Consequently,

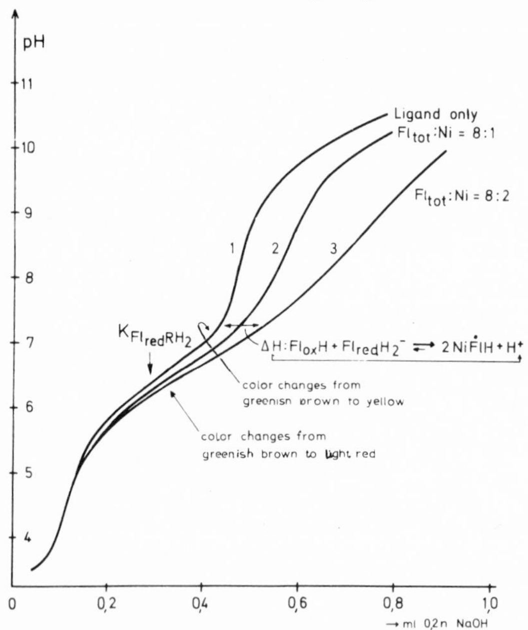

Fig. 11. pH titration of $4 \times 10^{-3}\,M$ riboflavin in 50% ethanol–water half reduced 0.1 M in $NaClO_4$ and $10^{-3}\,M$ in $HClO_4$ at 50°C[21]. Curve 1: ligand alone. The pK seen is that of flavohydroquinone. Since disproportionation is practically complete under these conditions, the amount of pK = 6.4-acid is equal to one half of the total flavin. Curves 2 and 3 show titration of the same system in the presence of increasing amounts of Ni^{2+}. ΔH indicates the additional protons liberated through formation of a flavin radical chelate.

the radical content *increases* steadily at pH 5 according to the upper curve of Fig. 10, when metal ions are present. This increase can be demonstrated even more easily by e.s.r.[19,25] or by spectrophotometry[26].

Whereas the optical spectrum of "free" neutral flavosemiquinone $H\dot{F}lR$[27] cannot be completely evaluated, since the small amount of free flavosemiquinone absorbing at 565 nm (Fig. 12) is hidden under the excess of flavoquinone present in the disproportionated half-reduced flavin system, pure spectra can be obtained either with flavoenzymes, where the protein prevents disproportionation of the radical[20] (Fig. 12), or chemically by alkylation of flavin in position 5[18,28,29]. In fact, stabilization of the neutral radical by the protein, *i.e.* by H-bridging of N(5)[20], and stabilization by alkylation of N(5), shift the spectrum of the blue flavosemiquinone about 15 nm towards longer wavelengths compared to the free flavosemiquinone as shown in Fig. 12. When the acidic function N(5)H is dissociated, one obtains, with color change from blue to red, the flavosemiquinone anion $\dot{F}lR^-$ (the residual substituent R is either alkyl or H in position 3). This

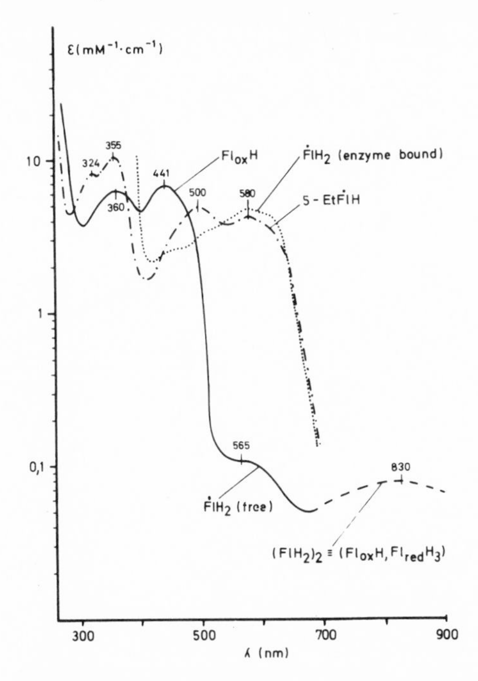

Fig. 12. Logarithmic plot of the optical spectrum of a half-reduced neutral FMN solution (——)[27], showing the equilibrium amount of free radical present (λ = 565 nm) in the nearly totally disproportionated system. The dashed part of the spectrum indicates flavoquinhydrone transfer complex present in the solution. For comparison, spectra are shown of a pure blue flavosemiquinone *viz.* chemically stabilized *i.e.* neutral 3-methyl-5-ethyl lumiflavin semiquinone (-··-·)[29] and a blue neutral flavosemiquinone stabilized by protein (phytoflavinsemiquinone) (. . . .)[34].

anion can not be observed optically under aqueous conditions but it can be stabilized, as discussed above for the neutral radical, either in flavoproteins[20] or (if R = alkyl) under half reduced aprotic conditions[22]. The spectrum is shown in Fig. 3. Chelation of the anion $\dot{F}lR^-$ with d-metal ions causes a shift of the first band (477 → 492 for R = H and → 540 nm for R = alkyl) (Fig. 3), but the general shape of the $\dot{F}lR^-$-spectrum remains very much the same. In addition, however, there is a new band of lower intensity in the near infra-red which is metal-specific and reflects even the solvation, *i.e.* the type of ternary complexes flavin–metal–solvent, very strongly, as outlined in Fig. 3 on the right. This characteristic band could not be assigned to a well-defined transition up to now[26].

Figure 13 shows the e.s.r. spectrum of a Cd-flavosemiquinone[19] with Cd-isotopes of different nuclear spin[19]. The e.s.r.-active isotope ^{113}Cd ($I = \frac{1}{2}$) gives a spectrum composed of the ligand hyperfine structure plus an additional splitting of 17.55 G from the metal.

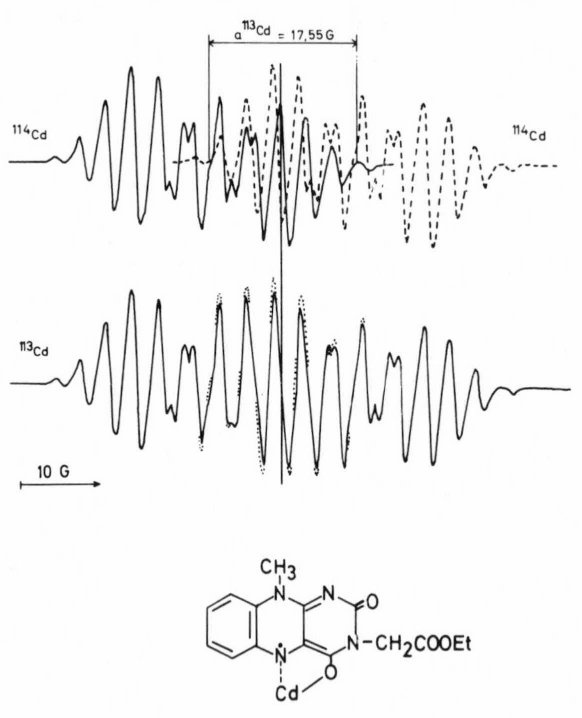

Fig. 13. E.s.r. spectrum of isotopically labelled flavin radical chelate[19]. The lower trace gives the spectrum as measured for ^{113}Cd, $I = \frac{1}{2}$ (——) and the deviations (. . .) from a ^{114}Cd, $I = 0$, spectrum to which an additional splitting of 17.55 G has been added graphically. The upper trace shows the ^{114}Cd components superimposed in a distance of 17.55 G.

The radical hyperfine coupling constants are given in Table I as far as measurable. Unfortunately, it is not possible to arrive at a complete picture of spin distribution as the bridge atoms between the rings do not readily lend themselves to isotopic substitution except for C(4a)[30]. Anyhow, about 50% of the unpaired electron is located in the C(4a)–N(5) region[30]. A further general distinction can be made, however: the spin density of the positions 1–4 is always negligibly small. It can, therefore, be concluded that at least one-electron transfer with flavins — if not all types of redox-reactions — is limited to the quinoxaline part of the flavin system, and this means it is limited to (4a)C–N(5).

The hyperfine coupling of the diamagnetic metal to the radical ligand seems to be stronger than might be explained by spin polarization and calls for π-delocalization of electrons from the metal to the ligand, at least for the filled shell ions Zn^{2+} and Cd^{2+}, although not of sufficient strength to make the electron distribution of the ligand dependent on the nature of the metal. Hence, whereas flavoquinone metal chelates have only very small metal–ligand hyperfine coupling (the spin is localized nearly entirely on the metal), radical chelates have their spin localized mainly on the ligand, although marked delocalization toward the metal does occur.

"Charge-transfer chelates" occupy an intermediate position in this respect. To date a precise evaluation of the spin distribution has not been possible with these latter compounds. They do not exhibit e.s.r. signals, either because they are diamagnetic or because of spin–spin interaction in a complex of total spin $S > \frac{1}{2}$, and n.m.r. spectroscopy is encumbered by the presence of paramagnetism. This paramagnetism may be due to thermal changes of multiplicity or rapid dissociation equilibria A and/or B (eqn. 10):

$$\mathrm{Me_{ox}} + \dot{\mathrm{Fl}} \overset{A}{\rightleftarrows} [\mathrm{Me_{ox}}\dot{\mathrm{Fl}} \longleftrightarrow \mathrm{Me_{red}Fl_{ox}}] \overset{B}{\rightleftarrows} \mathrm{Me_{red}} + \mathrm{Fl_{ox}} \qquad (10)$$

CONCLUSIONS ABOUT THE BIOCHEMICAL FUNCTION OF FLAVINS

The role of flavin in biochemical oxidoreduction is mainly twofold: CH-activation and electron pair splitting. Flavin-dependent substrate dehydrogenation is generally agreed to be a two-electron transfer with regard to the fact that flavin *radicals* are not found to be essential intermediates in pure dehydrogenase reactions (for review, see ref. 31). However, among the two possible modes of heterolytic CH-activation only one has been considered to date, *viz.* transfer of a H^- unit towards flavin with release of a carbonium ion, while the other possibility — transfer of a carbanion unit towards flavin with release of a proton — has been largely neglected. Arguments in favor of the latter mechanism have been collected in a recent review[31]. In the present context of flavin–metal interaction, a solution to the question of hydride- *vs.* group-transfer is important insofar as it would decide whether the electron transfer from reduced flavin toward ubiquinone, hemoproteins or iron–sulfur proteins, starts from a flavin substrate complex Fl–S or from "uncomplexed" reduced flavin $Fl_{red}H_2$ (scheme 2). In contrast to the two-electron step of CH-activation (which is the low potential- or input-site in biological oxidation and the high potential- or output-site in photosynthesis[31]) flavin reacts with metal in one-electron steps (which is the output or the high potential site of the flavin range in the respiratory chain and the input- or low potential-site in

References pp. 1189–1190

the photosynthesis chain). This mode of reaction implies that the stability of the flavin radical is enhanced by the protein[20].

In Tables II and III we have compiled the main enzyme systems in which electron transfer from flavin to metal is known to occur. For our purposes, it is of minor importance whether a metallo-flavoprotein complex can be separated to yield intact flavoprotein and metalloprotein fragments. This has been achieved thus far only for the systems contained in Table III. It is more important to understand what the implications are of the apparent

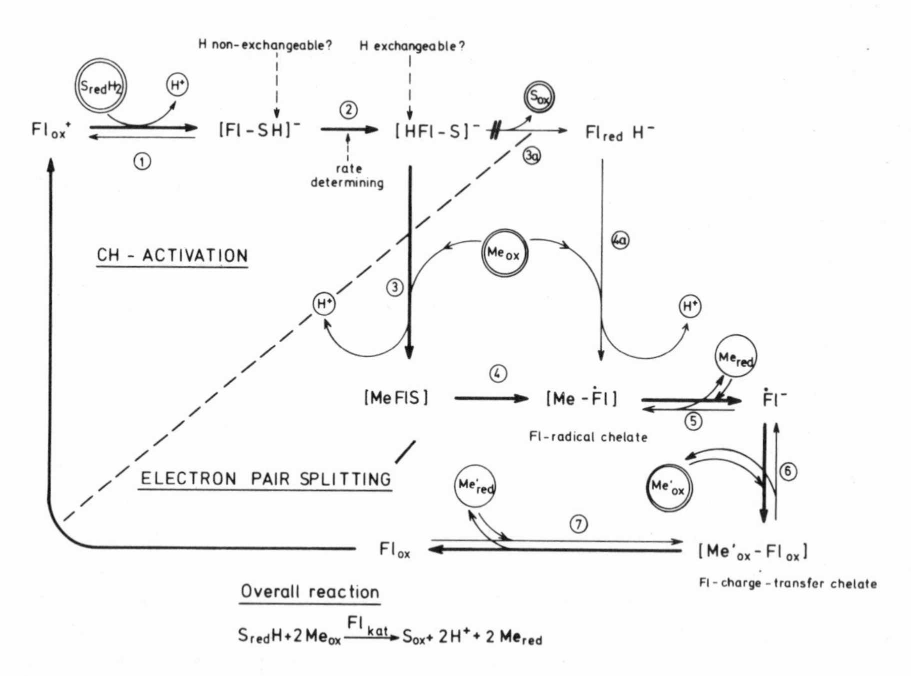

Scheme 2. Reaction scheme proposed for succinate dehydrogenase (flavin-iron-sulfur-protein), based on kinetic data[33] and the results obtained from "chemical" interaction of flavin-metal interaction, including flavin-substrate and flavin-metal complexes, as presented above. The bolt arrows depict the main pathway of electron flow and the catalytic cycle of flavin.

splitting of electron pairs: if there are two consecutive steps of reoxidation, one would expect a rather large difference in potential between the first step and the second step. This assumption is verified by the findings of Massey *et al.*[32] who for the first time measured the one-electron potentials between reduced flavin and radical on the one hand, and radical and oxidized flavin on the other, and showed that they differ by 260 mV. From this, the kinetics of reoxidation should be expected to be rapid in the first step and slower in the second one. In fact, however, one observes the reverse: the

TABLE II

METAL FLAVOPROTEINS

Enzyme	Flavins per mole protein	Metals	Substrate red./ox.	% Radical (max)	Reference
Xanthine oxidase (Milk)	2 (FAD)	Mo, SFe[a]	-CHO/-COOH -CH=N-/CONH-	3–10	35,36
Aldehyde oxidase (Liver)	2 (FAD)	Mo, SFe	-CHO/-COOH	24	37
Succinate dehydrogenase (Heart mitochondria)	1 (FAD)	SFe	$HOOCCH_2CH_2COO$/ HOOCCH=CHCOOH	20–30	33
Glycerol phosphate dehydrogenase (Mitochondria)	1 (FAD)	Fe	$CH_2OHCHOHCH_2$-P/ $OHCCHOHCH_2O$-P		
Dihydroorotic dehydrogenase (Zymobact.)	2 (FMN)+ 2 (FAD)	SFe	dihydro orotic acid orotic acid	32–35	38
Choline dehydrogenase (Liver mitochondria)	1 (FAD	Fe	$(CH_3)_3\overset{+}{N}CH_2CH_2OH$/ $(CH_3)_3NCH_2CHO$		
NADH dehydrogenase (Heart mitochondria)	1 (FMN)	SFe	NADH/NAD		
Sulfate reductase (Bact. Desulfovibrio)	1 (FAD)	SFe	AMP-SO_3H/AMP + SO_3^{2-}		
Nitrate reductase (Bacteria)	? (FAD)	Mo	NO_3^- NO_2^- NADPH/NADP		
Sulfite reductase (*Salmonella*)	4 (FMN) 4 (FAD)	SFe, hFe[b]	SO_3^{2-}/S^{2-}	25–37	39

[a] SFe = iron-sulfur protein
[b] hFe = heme-iron
References are given only for the measured radical concentration values; for other enzymes, see ref. 31

first step is slow, whereas the second one is too rapid to be detectable in the kinetic course. This is demonstrated, for instance, by the fact that the reoxidation of reduced succinate dehydrogenase[33] does not depend on the square of the ferricyanide concentration as would be expected if there were two steps of about equal rate. In contrast, it is tempting to assume that the relatively low rate of the first step is due to the fact that the metal has to react with the enzyme–substrate complex, expelling the oxidized substrate at about the same time, whereas in the second step, the reaction is supported by "substrate-free" radical Fl^- (cf. scheme 2).

We believe that it is a meaningful result that none of the known metal flavoproteins ever shows — under any conditions tried so far — more than 50%

radical formation per total flavin, in contrast to metal-free flavoproteins[31]. We conclude, therefore, that the "missing" percentage of flavin is bound to the metal intramolecularly as supported by the results from the "chemical" model systems discussed in this chapter. Analytically, flavin-iron binding would not be easily detectable, as it is not at all detected by e.s.r. spectroscopy and barely by spectrophotometry. Nevertheless, it appears worthwhile to search in flavoprotein systems for the strong π–π^* transition at 370 nm (Fig. 3) and the "near i.r." absorption of flavin radical chelates, as well as for paramagnetic ^{1}H-n.m.r. shifts.

In scheme 2 we visualize in the first step the formation of an enzyme substrate complex which rearranges in the rate-limiting step to a tautomeric enzyme–product complex[33]. Now it seems that this latter complex cannot dissociate directly (step 3a) to release the oxidized substrate to form normal reduced flavin. Apparently this hydrolytic step is more easily, if not exclusively, achieved after removal of one electron by step 3, yielding a ternary metal–flavin–substrate complex which decays extremely rapidly with release of the oxidized substrate ("or product") and formation of the binary metal flavin complex of the type termed "flavin radical chelate" in the present review (step 4).

TABLE III

FLAVOPROTEINS HAVING IRON PROTEINS AS NATURAL ACCEPTORS

Enzyme (Source)	Flavins per mole proteins	Donor	Acceptor
Pyruvate dehydrogenase (*E. coli*)	4 (FAD)	$CH_3COCOOH$	hFe
NADH-cyt. b_5 reductase (Liver micros.)	1 (FAD)	NADH	hFe
NADPH-cyt. *c* reductase (Liver micros.)	2? (FAD)	NADPH	hFe
Lactate dehydrogenase (Yeast)	4 (FAD)	$CH_3CHOHCOOH$	hFe
Ferredoxin-NADP reductase (Spinach)	1 (FAD)	NADP, NADPH	SFe

This chelate is in equilibrium with metal free flavosemiquinone (step 5). If a second oxidized acceptor site is available, this flavosemiquinone now reacts to yield another chelate of the "charge-transfer" type, which again dissociates to yield a second molecule of reduced acceptor and oxidized flavin (steps 6–7).

This tentative scheme is not to be taken as established reality. It is meant to demonstrate the manifold complications encountered in the study of the biological action of flavins, if one tries to reconcile the available enzymological data with the structure and reactivity of suitable chemical models.

The biological importance of flavin metal contact still remains to be established. Measurement of "near -i.r." electronic spectra and paramagnetic shift of ^{1}H-n.m.r. should be applicable not only to coenzymes and coenzyme models, but also to the holoproteins.

ACKNOWLEDGEMENTS

Thanks are due to Dr. H. Beinert for critical revision and to Miss A. Castillo for help in preparing the manuscript. The studies of our group on metal flavin interactions have been supported by the Swiss National Foundation for the Advancement of Scientific Research 1960-67 and by the German "Forschungsgemeinschaft" 1968-1970. During all these years, we have been indebted for intense cooperation, to Dr. Anders Ehrenberg, University and Nobel Institute, Stockholm, Sweden, to Dr. H. Beinert, University of Wisconsin and Dr. V. Massey, University of Michigan, as well as to Dr. J.-M. Lhoste, Institut Radium, Paris, France, Dr. J. van Voorst, University of Amsterdam and Dr. C. Veeger, Agricultural University, Wageningen, The Netherlands.

REFERENCES

1 A. Albert, *Biochem. J.*, 47 (1950) 27; 54 (1953) 646.
2 P. Hemmerich and S. Fallab, *Helv. Chim. Acta*, 41 (1958) 498.
3 W. O. Foye and W. E. Lange, *J. Am. Chem. Soc.*, 76 (1954) 2199.
4 J. T. Spence and E. R. Peterson, *J. Inorg. Nucl. Chem.*, 24 (1962) 601.
5 P. Hemmerich, unpublished results.
6 R. Kuhn, T. György and T. Wagner-Jauregg, *Ber. Dtsch. Chem. Ges.*, 66 (1933) 576.
7 I. F. Baarda and D. F. Metzler, *Biochim. Biophys. Acta*, 50 (1961) 463.
8 P. Bamberg and P. Hemmerich, *Helv. Chim. Acta*, 44 (1961) 1001.
9 P. Hemmerich, F. Müller and A. Ehrenberg, in T. E. King, H. S. Mason and M. Morrison, *Oxidases and Related Redox Systems*, Vol. I, J. Wiley and Sons, New York, 1965, p. 157.
10 S. Ahrland, J. Chatt and N. R. Davies, *Q. Rev.*, 12 (1958) 265.
11 R. G. Pearson, *J. Am. Chem. Soc.*, 85 (1963) 3533.
12 P. Hemmerich and J. Spence, in E. C. Slater, *Flavins and Flavoproteins*, Elsevier, Amsterdam, 1966, p. 82.
13 J. Lauterwein, P. Hemmerich and J.-M. Lhoste, to be published.
14 P. Kierkegaard, to be published.
15 P. Hemmerich, B. Prijs and H. Erlenmeyer, *Helv. Chim. Acta*, 42 (1959) 2164.
16 L. E. G. Eriksson, *Biochim. Biophys. Acta*, 208 (1970) 528.
17 K. H. Dudley, A. Ehrenberg, P. Hemmerich and F. Müller, *Helv. Chim. Acta*, 47 (1964) 1354.
18 W. H. Walker, P. Hemmerich and V. Massey, *Eur. J. Biochem.*, 13 (1970) 258.
19 F. Müller, L. E. G. Eriksson and A. Ehrenberg, *Eur. J. Biochem.*, 12 (1970) 93.
20 F. Müller, P. Hemmerich, A. Ehrenberg, G. Palmer and V. Massey, *Eur. J. Biochem.*, 14 (1970) 185.

21 P. Hemmerich, *Helv. Chim. Acta,* 47 (1964) 464.
22 A. Ehrenberg, F. Müller and P. Hemmerich, *Eur. J. Biochem.*, 2 (1967) 286.
23 L. Michaelis and G. Schwarzenbach, *J. Biol. Chem.*, 123 (1938) 527.
24 R. D. Draper and L. L. Ingraham, *Arch. Biochem. Biophys.*, 125 (1968) 802.
25 P. Hemmerich, D. V. Dervartanian, C. Veeger and J. D. W. van Voorst, *Biochem. Biophys. Acta,* 77 (1963) 504.
26 F. Müller, P. Hemmerich and A. Ehrenberg, *Eur. J. Biochem.*, 5 (1968) 158.
27 H. Beinert, *J. Am. Chem. Soc.*, 78 (1956) 5323.
28 P. Hemmerich, S. Ghisla, U. Hartmann and F. Müller, in H. Kamin, *Flavins and Flavoproteins,* University Park Press, Baltimore, 1970.
29 M. Brüstlein, P. Hemmerich, F. Müller and W. Walker, *Eur. J. Biochem.*, 25 (1972) 573.
30 W. H. Walker, A. Ehrenberg and J.-M. Lhoste, *Biochim. Biophys. Acta*, 215 (1970) 166.
31 P. Hemmerich, G. Nagelschneider and C. Veeger, *FEBS Lett.*, 8 (1970) 69.
32 S. G. Mayhew, G. P. Foust and V. Massey, *J. Biol. Chem.*, 244 (1969) 803.
33 D. V. Dervartanian, C. Veeger, W. H. Orme-Johnson and H. Beinert, *Biochim. Biophys. Acta,* 191 (1969) 22. W. P. Zeijlemaker, D. V. Dervartanian, C. Veeger and E. C. Slater, *Biochim. Biophys. Acta,* 178 (1969) 213.
34 H. Bothe, H. Sund and P. Hemmerich, in H. Kamin, *Flavins and Flavoproteins,* University Park Press, Baltimore, 1970.
35 H. Beinert and W. H. Orme-Johnson, in H. Kamin, *Flavins and Flavoproteins,* University Park Press, Baltimore, 1970.
36 G. Palmer and V. Massey, *J. Biol. Chem.*, 244 (1969) 2614.
37 K. V. Rajagopola, P. Handler, G. Palmer and H. Beinert, *J. Biol. Chem.*, 243 (1968) 3784.
38 V. Alleman, P. Handler, G. Palmer and H. Beinert, *J. Biol. Chem.*, 243 (1968) 2560, 2569.
39 L. M. Siegel and H. Kamin, in K. Yagi, *Flavins and Flavoproteins,* University of Tokyo Press (University Park Press) Nagoya (Baltimore) 1968, p. 15.

Chapter 33

COMPLEXES OF NUCLEOSIDES AND NUCLEOTIDES

G. L. EICHHORN

Laboratory of Molecular Aging, Gerontology Research Center, National Institutes of Health, National Institute of Child Health and Human Development, Baltimore City Hospitals, Baltimore, Maryland 21224, U.S.A.

INTRODUCTION

The genetic information that is used to build the many varieties of protein molecules from their amino acid constituents is contained in nucleic acid molecules, which are macromolecules built from nucleotides by phosphodiester bond formation, just as proteins are macromolecules built from amino acids by peptide bond formation. The sequence of nucleotides within the nucleic acid molecule in fact constitutes the genetic code. The biological functioning of the nucleic acids involves the participation of metal ions. Moreover, the nucleotides as monomers are of great importance in many metabolic processes, most of which are also mediated by metal ions (Chapters 17 and 18). It is therefore of fundamental importance to understand the nature of the metal complexes of nucleotides, both for their intrinsic importance and for their involvement in the reactions of the nucleic acids. A number of reviews have recently appeared on this subject[1-4].

As has been indicated in Chapter 2, a nucleotide consists of a heterocyclic base, *i.e.*, a 1-ring pyrimidine or a 2-ring purine, attached to ribose, which in turn is attached to phosphate. When the phosphate is removed, the remaining base–ribose combination is called a nucleoside. The four most common bases derived from ribonucleic acid (RNA) are the purines adenine and guanine and the pyrimidines cytosine and uracil; thymine, or 5-methyl-uracil, occurs in deoxyribonucleic acid (DNA) in place of uracil. The structures of these bases, as well as hypoxanthine (deaminoguanine) are as follows:

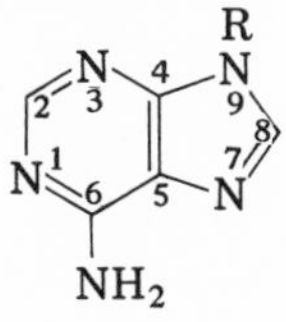

Adenine 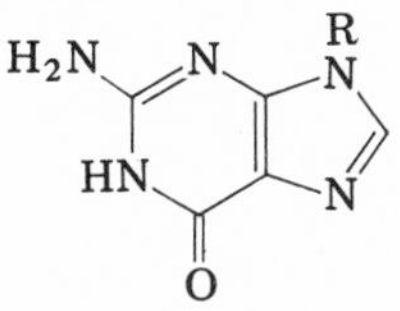 | Guanine 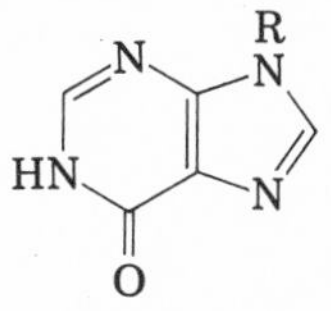 | Hypoxanthine

References pp. 1207–1209

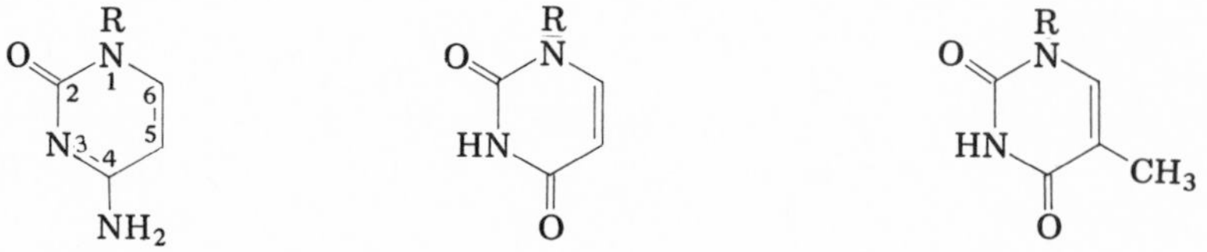

Cytosine Uracil Thymine

R represents the attachment of ribose in a nucleoside or ribose phosphate in a nucleotide. The numbering of adenine and cytosine is representative of the purines and pyrimidines. The numbering of the ribose ring is as follows:

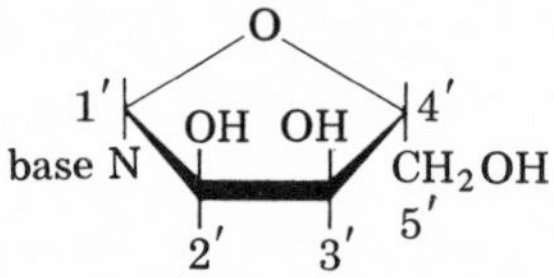

Since DNA contains no 2′-hydroxyl group, its monomeric constituents are 2′-deoxynucleotides, which become 2′-deoxynucleosides with the phosphate removed. The presence or absence of the 2′-OH group is sometimes indicated by the designation ribonucleosides and deoxynucleosides.

The names of the major bases and the corresponding nucleosides and nucleotides are as follows:

Base	Nucleoside	Nucleotide
adenine	adenosine	adenylic acid, adenosine monophosphate (AMP)
guanine	guanosine	guanylic acid, guanosine monophosphate (GMP)
cytosine	cytidine	cytidylic acid, cytidine monophosphate (CMP)
uracil	uridine	uridylic acid, uridine monophosphate (UMP)
thymine	thymidine	thymidylic acid, thymidine monophosphate (TMP)
hypoxanthine	inosine	inosinic acid, inosine monophosphate (IMP)

The di- and tri- phosphates of the nucleosides, and their metallic complexes also have substantial biological importance (Chapters 17 and 18).

POTENTIAL METAL BINDING SITES

It is obvious that the heterocyclic bases contain oxygen and nitrogen electron donors for potential coordination of metal atoms. In addition, the nucleosides contain ribose hydroxyl groups and the nucleotides have phosphate groups as well available for coordination.

The binding of metal ions to the purines or pyrimidines unattached to ribose is of marginal biological interest, since the metal can bind to the N-9 of purines and the N-1 of pyrimidines by proton displacement, although these atoms are not available for electron donation in the nucleosides and nucleotides. The major questions to be resolved are the binding sites of metals on the bases in nucleosides and nucleotides, the extent to which the phosphate competes with these base sites for the metal ions, and the possibilities of metals binding to base sites and phosphates simultaneously.

STABILITIES OF METAL COMPLEXES OF BASES, NUCLEOSIDES AND NUCLEOTIDES

Before discussing the structural characteristics of the metal complexes, we shall consider the thermodynamics of the reaction of metals with the various sites. Of the three types of sites, base, ribose and phosphate, the ribose appears to be generally the weakest donor group. The phosphate group is the strongest coordinating group for transition metals and alkaline earth metal ions so that stability constants of the nucleotides generally reflect the stability of bonds to the phosphate. This is immediately apparent from the similarity of the stabilities of complexes of methyl triphosphate and ATP, indicating virtually no contribution of the base to the stability of the complexes[5].

Stability constants of the adenine nucleosides and nucleotides, as well as the di- and tri-phosphates, have been collected in Table I[★]. Examination of the table underscores the lack of stability of nucleoside complexes, and reveals that stability increases with increasing phosphate content[7]. The Irving–Williams series (Chapter 2) is followed for metals of the first transition series, and the usual decrease in stability with increasing atomic number is observed for the alkaline earth series. Thus, these complexes observe the usual trends in stability.

Very few studies are available comparing the stabilities of complexes of the different nucleosides. Fiskin and Beer[10] found the following values for copper complexes by a pH-stat technique: adenosine, $\log K = 0.70$, cytidine, $\log K = 1.59$, guanosine, $\log K = 2.15$, whereas uridine formed a complex with copper too weak to detect. Ropars and Viovy[11] placed the deoxynucleosides in a similar order, based upon electron paramagnetic resonance (e.p.r.) measurements. The nucleoside bases can frequently be placed in the following order of relative stabilities of their metal complexes: G > A, C > U, T. Such an order was established by the work of Frieden and Alles[12] on the relative ability of various nucleosides and nucleotides to

★A much more extensive compilation of thermodynamic data has been made by Phillips[3]. Table I lists comparisons of data obtained by the same investigators, rather than averages. In view of the differences in ionic strength, techniques, *etc.*, averages could be misleading.

TABLE I

STABILITIES OF 1:1 COMPLEXES OF ADENOSINE NUCLEOSIDES AND NUCLEOTIDES (UNPROTONATED SPECIES)

	log K_1						
	Adenosine	5′-AMP		5′-ADP		5′-ATP	
	(Ref. 5)	(Ref. 7)	(Ref. 8)	(Ref. 7)	(Ref. 8)	(Ref. 7)	(Ref. 9)
Mn	-0.82	2.2	2.4	3.5	3.5		4.8
Co	-0.30		2.6		4.2		4.7
Ni	-0.17		2.8		4.5		5.0
Cu	+0.84		3.2		5.9		6.1
Zn	-0.28		2.7		4.3		4.9
Mg		1.7	2.0	3.0	3.2	3.4	4.2
Ca		1.4	1.8	2.8	2.9	3.3	4.0
Sr		1.3	1.8	2.5		3.0	3.5
Ba			1.7				3.3
	(Ref. 6)						
Hg	4.3						

inhibit the copper activated enzyme ascorbic acid oxidase. This order, although frequently followed, is by no means universal. Thus Fiskin and Beer[10] found log K values for the lead complexes to be, for adenosine -0.5, for cytidine, 1, and for guanosine, 0.5. The lead complex for cytidine is thus the most stable. The difference in the order of stabilities for copper and lead is made possible by the variety of sites to which metal ions can bind on these base molecules.

Comparison of the stability of the copper complex of guanosine (log K = 2.15) with that of deoxyguanosine (log K = 2.9)[11] indicate merely a similarity in order of magnitude; the results from diverse techniques cannot be taken as evidence for a difference due to the presence or absence of a 2′-hydroxyl group. Frieden and Alles[12] had concluded from their ascorbic acid oxidase inhibition studies that the deoxynucleotides form more stable complexes than the ribonucleotides.

The extraordinary stability of the mercury complex of adenosine reflects the much greater affinity of the bases for metals such as mercury and silver than for the first group transition metals. Binding of these metals to nucleotide bases is accompanied by drastic changes in the ultra-violet absorption spectrum of the bases[13-17,42].

A number of studies have been carried out on the affinity of metal ions for the heterocyclic bases unattached to ribose or phosphate. Even though such studies cannot always be related to the chemistry of nucleotides and nucleic acids, since the bases contain one potential coordinating group that is absent in the ribosides, some of these studies have proved useful in obtaining information on the binding sites of metals. The following constants for log K_1 for Cu(II) binding have been obtained by Reinert and Weiss[18]:

purine	1.9
adenine	2.7
7-methyl adenine	2.7
9-methyl adenine	1.7

Since methylating adenine at the 7-position has no effect on stability, but substitution at the 9-position greatly diminishes it, it is concluded that copper preferentially binds at N-9 in adenine. We shall see that this thermodynamic reasoning is in accord with crystallographic findings.

STRUCTURES OF METAL COMPLEXES OF NUCLEOSIDES AND NUCLEOTIDES

Complexes of the bases

Despite the conclusion from the stability studies that the nucleotide bases are not necessarily adequate models for the nucleosides and nucleotides in complexing ability, it is instructive to consider some structural studies on the metal complexes of these bases, since crystallographic studies have been carried out on them, and no crystallographic studies are as yet available on nucleoside and nucleotide complexes.

Carrabine and Sundaralingam[19] have demonstrated that Cu(II) binds N-9 in guanine to form a complex (with $CuCl_2$) having the structure depicted in Fig. 1. Thus the N-9 position is implicated as the favored binding site of

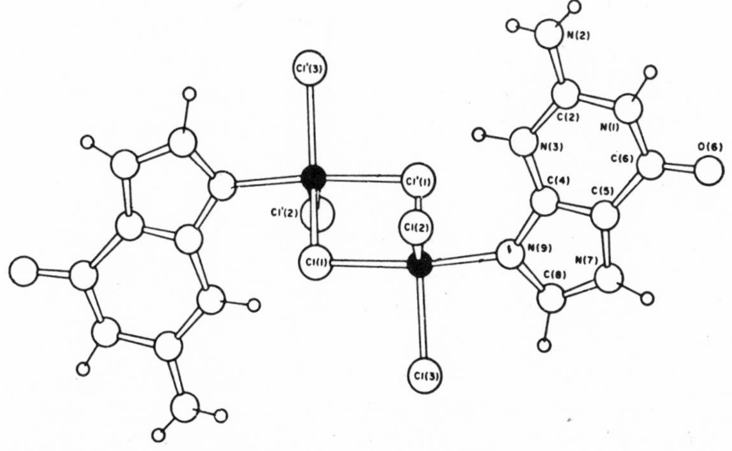

Fig. 1. Structure of complex formed from guanine and $CuCl_2$ (from ref. 19).

Cu(II) on guanine as well as in adenine. Although N-9 binding is not possible in nucleosides and nucleotides, this finding is nevertheless of interest for complexes of these substances, since it raises the question as to why N-9 binding is preferred over chelation to N-7 and O-6. If such chelation were possible, presumably the copper would chelate in preference to binding N-9. Since such chelation does not occur with the guanine base, there is some question whether chelation occurs even when the N-9 position is blocked. We shall consider this matter again later.

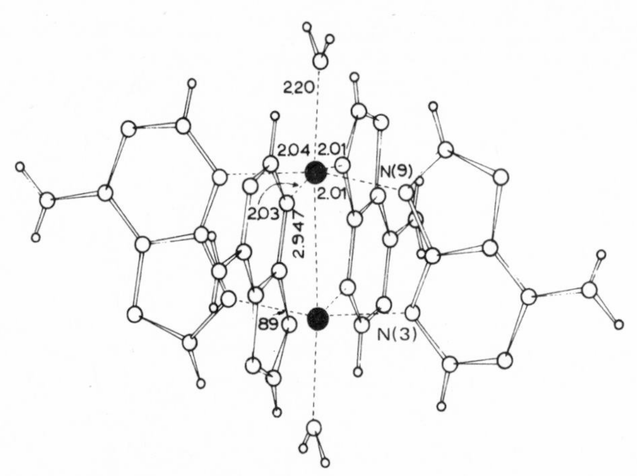

Fig. 2. Structure of Cu(II) adenine complex (from Ref. 20).

Sletten[20] has determined the structure of the Cu(II) complex of adenine, and found each adenine bound to two copper atoms, at N-3 as well as at N-9, the copper atoms presumably forming a copper–copper bond which serves as an axis around which four adenines are held. The structure is shown in Fig. 2. The N-9 of the purine ring is again implicated, as predicted by the thermodynamic data. Structures involving N-3 and N-9 binding had been previously postulated on the basis of color relationships among isolated complexes[21].

Although these structures of copper adenine and guanine complexes cannot apply to adenosine and guanosine, the structure determined for the copper complex of cytosine[22] does not have such a drawback, since N-3 is not blocked by riboside bond formation. Copper is bound to N-3, and this atom is available for bonding in cytidine and cytidylic acid, so that N-3 can be expected to be the binding site of copper in these substances also.

The only other crystallographic structure determination that has been carried out on these complexes is that of mercury(II) uracil; surprisingly, mercury is bound to the oxygen atoms on position 4[23]. This finding is also of interest in nucleic acid chemistry, since this group is, of course, also not involved in riboside formation.

The crystallographic evidence thus indicates that Cu binds to N-9 in guanine, N-9 and N-3 in adenine, and N-3 in cytosine, while Hg binds to O-4 in uracil. We shall now consider the structural evidence that has been accumulated for the complexes of nucleosides and nucleotides.

Adenine nucleoside and nucleotide complexes

Probably because of the great interest in ATP, the complexes of adenine nucleosides and nucleotides have been much more thoroughly investigated than those of the others. A very successful technique for the elucidation of the structures of these complexes has been nuclear magnetic resonance (n.m.r.). Two rather different techniques have been applied, depending on the nature of the metal. With paramagnetic metal ions the broadening of the lines of the ligand spectra has been used as an indication of proximity of the metal ion to the atom whose characteristic line has been broadened. With diamagnetic ions line shifts have been similarly applied.

The first application of these techniques was by Cohn and Hughes[24] who studied the Mg(II), Ca(II), and Zn(II) complexes of ADP and ATP by line shifts and the Cu(II) and Mn(II) complexes of ADP, ATP, as well as AMP, by line broadening. They found no shifts in the proton magnetic resonance (p.m.r.) spectra of ATP solutions with magnesium and calcium, but zinc ions caused a shift of the H-8, but not the H-2 line. This result was interpreted as indicating binding of zinc to N-7 (near H-8), but no binding of Mg and Ca to base at all. All these diamagnetic ions displaced the β- and γ-phosphate lines of ATP, thus indicating that these metals bind the terminal two phosphate positions, presumably through chelation, but not the position adjacent to the ribose. In ADP only two phosphate positions (α and β) are available, and lines corresponding to these groups are shifted by magnesium(II); Ca and Zn could not be determined because of insufficient solubility. Line broadening studies suggested that Cu forms complexes similar to those of zinc, broadening of H-8 and β- and γ-phosphates of ATP being interpreted as resulting from Cu binding to N-7 and the two terminal phosphates. Similarly, Cu apparently binds N-7 and the α-, β-phosphate grouping of ADP and N-7 and phosphate of AMP. Mn(II) also was found to bind N-7 in ADP and ATP, but unlike the situation with Cu, all three phosphate positions are affected; as with Cu, both phosphate positions in ADP are involved.

The difference between Cu(II) and Mn(II) in binding two and three phosphate positions, respectively, in ATP was confirmed by Sternlicht *et al.*[25], who showed that Ca(II) and Ni(II) behaved like Mn(II), and that therefore Cu(II) was unique in this series in binding only to the two terminal phosphates.

It can be concluded from these investigations that Mg(II) and Ca(II) bind only to phosphate, whereas Mn(II), Ni(II), Cu(II) and Zn(II) bind to both phosphate and base sites.

Many other techniques have confirmed the ability of metal ions to bind both phosphate and base in adenine nucleotides. Two further questions to which we shall address ourselves are: (1) where on the base does the binding occur? and (2) what type of interplay results from the binding to base and phosphate sites?

Since many groups have confirmed the binding of metal ions to N-7 (and other heterocyclic nitrogens, see below) of adenine in nucleosides and nucleotides, a remaining problem concerning the binding site on the base is whether the amino group is also coordinated in addition to the N-7, to form a chelate ring. It has been tempting for many investigators to speculate on the formation of such a ring and to devise structural formulations to depict this speculation. However, Schneider *et al.*[5] first suggested that the amino group is probably not involved. On a theoretical basis, they felt that the would-be free electron pair on the nitrogen is required to stabilize an electron deficient ring system, and such electronic interaction requires coplanarity of the amino group with the ring. For the amino group to participate in chelation it would have to be reoriented in a plane perpendicular to the ring, and such a reorientation seems implausible in view of the aforementioned electron deficiency.

Evidence that the amino group is not involved in chelation with copper was obtained by Eichhorn *et al.*[26] for the coordination of copper by deoxyadenosine in dimethylsulfoxide (DMSO), using p.m.r. broadening (the amino hydrogens exchange with D_2O, requiring the use of a solvent such as DMSO to enable detection of the NH_2 group.) It was found that the NH_2 protons were not broadened by copper and that therefore chelation apparently is ruled out (the possibility that binding in DMSO is different from that in water must, however, be kept in mind). NH_2 shifts in adenosine due to complexation by zinc(II) were indeed interpreted by Wang and Li[27] as indicative of NH_2-binding, but similar shifts with cytidine were interpreted as resulting from electronic perturbation of the amino group by zinc binding elsewhere, and perhaps the NH_2 shift in adenosine can be attributed to the same phenomenon as the similar shift in cytidine. Happe and Morales[28] discovered that zinc(II) shifts ^{15}N resonance peaks at ATP for N-7, N-9 and the amino group, a circumstance which can be explained, again, on the basis of a perturbation of NH_2 without its being directly bound to the metal. Reinert and Weiss[18] showed that the ethylation of the amino group of adenine results in an increase in stability, thus indicating a failure of the amino group to participate in chelation. These considerations, along with the fact that the crystallographic study of copper–adenine complex reveals binding to the N-3, N-9 side of adenine[19], indicate that N-7, without amino group chelation, appears to be the favored site of binding of metal ions to the adenine base in nucleosides and nucleotides. Evidence for amino group binding has been obtained, but is not conclusive[27,29].

The second question concerns the relationship between phosphate and base binding. Various investigators[3,4,30] have speculated on the possibility that metal ions form a large chelate ring by simultaneous binding of phosphate and base, as follows:

O
OH HO
CH$_2$
N N
O
N N
Cu—O—P—O
NH$_2$
O

These speculations are generally based on experiments showing that the same copper atom is bound to both phosphate and a base site. For example, it was shown that the rate of hydrolysis of ATP is enhanced 53 times by copper ions, compared to a tenfold rate for methyl triphosphate[31]. It is concluded that the copper that catalyzes the hydrolysis of the P–O–P bond by electron withdrawal must also be bound to the base. Another experiment[32] revealed that the relaxation times of the proton and ^{31}P resonance spectra are the same in metal complexes of ATP and therefore led to the postulation of the chelate structure.

However, the evidence for the binding of an individual copper atom simultaneously to base and phosphate does not, of course, require binding to these two moieties on the same ligand molecule. Sternlicht *et al.*[25] have indeed found that either Mn(II) or Ni(II), when mixed with AMP and ATP in a 1:1:1 system, revealed a broadening of the H-8 proton magnetic resonance peak of both AMP and ATP, even though the binding constant of Mn(II) ATP, is, *e.g.*, 300 times that of Mn(II) AMP. This evidence, in conjunction with the discovery of the simultaneous binding to base and phosphate[25,32] led to the postulation of a metal bound to N-7 of one molecule and phosphate of another molecule.

The manner in which base and phosphate binding of metal ions to adenine nucleotides are related was further elucidated by Berger and Eichhorn[34] through a study of p.m.r. broadening of AMP isomers by Cu(II) ions. It was found that copper affected the p.m.r. of 3′- and 5′-AMP by broadening H-8 preferably to H-2, in line with previous studies (with 5′-AMP). However, 2′-AMP, as well as 2′3′-cyclic and 3′5′-cyclic AMP, when reacted with copper, had its H-2 and H-8 protons broadened equally. Presumably Cu(II) binds primarily to N-7 only in 3′- and 5′-AMP, but not in the other isomers. The dichotomy between 3′- and 5′-AMP and the other isomers in

References pp. 1207–1209

their p.m.r. behavior with Cu(II) can be explained by combining the discovery that copper in AMP can bind N-7 in one molecule and phosphate in another, and the tendency of purines to stack through π-interactions. By use of CPK models it can be shown that a binuclear 2:2 Cu:AMP complex, in which each copper is bound to N-7 on one molecule and phosphate of the other, can be formed with maximum π-interaction between the purine rings, when the phosphates are either 3′- or 5′- (Fig. 3). A similar binuclear complex constructed from 2′-AMP would produce much less extensive

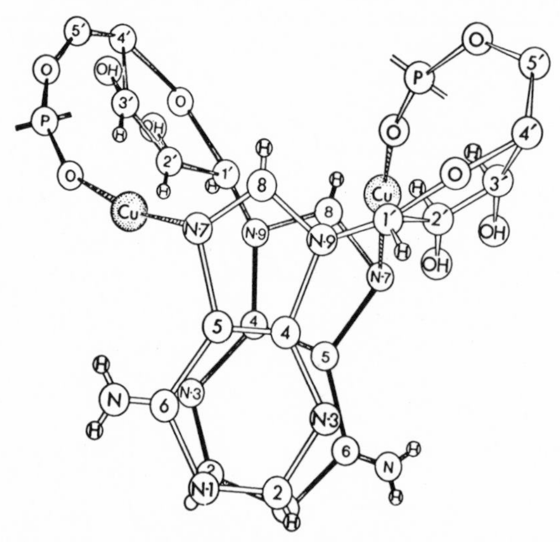

Fig. 3. Structure postulated for the Cu(II)–complex of 3′-AMP and 5′-AMP (but not 2′-AMP) (from ref. 48).

overlapping of purines and therefore less π-interaction; such a complex could not be formed at all from the cyclic phosphates. Further evidence for the structure of Fig. 3 for the Cu complexes of 3′- and 5′-AMP comes from the fact that the 2′- and 3′-protons on ribose are broadened, while the 1′-proton is unaffected, in accordance with the proximities of these protons to Cu in Fig. 3. 2′-AMP may have a chelate structure, involving N-3 and phosphate, in accord with the broadening of H-2 as well as H-1′, and the lack of broadening of H-3′ and H-4′.

In order to assure the validity of n.m.r. broadening studies in the assignment of metal binding sites in these molecules, two potential objections must be overcome. The first arises from the finding by Carrabine and Sundaralingam[19] that the attachment of copper(II) to one position on a purine has a profound effect on bond lengths elsewhere in the purine. It follows that electronic perturbations occur throughout the ring system, and the question therefore arises whether these electronic perturbations can induce broadening of proton resonances at loci that are distant from the metal binding site. To test this possibility, the p.m.r. spectrum of the copper complex of tubercidin:

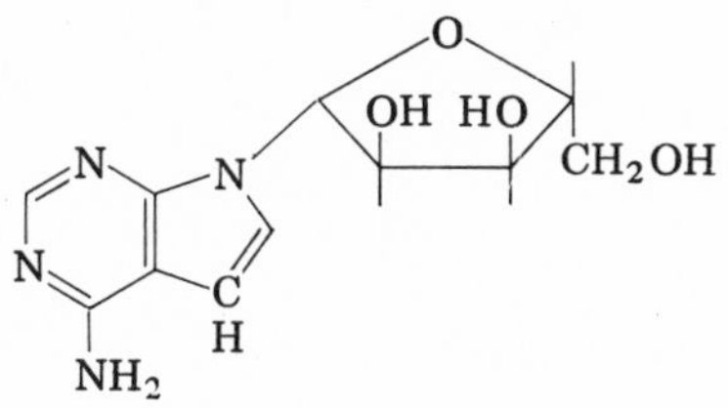

was determined[34]. Tubercidin is 7-deaza-adenosine; broadening of H-8 would indicate a lack of reliability of the n.m.r. line broadening technique, since there is no electron donor group at the 7-position. Indeed, no broadening occurred at H-8, but H-2 was extensively broadened. Thus, the electronic perturbations in the molecule do not appear to invalidate the p.m.r broadening technique.

Another source of difficulty is the possibility that copper ions bound to phosphate could approach the purine in such a manner as to induce broadening without any attachment of Cu(II) to base. Such a possibility exists for 5′-AMP, which, in its characteristic *anti*-conformation, has the phosphate approaching the H-8 position very closely. H-8 broadening in the copper complex of 5′-AMP could result from perturbation by the phosphate-bound copper. However, in 3′-AMP the phosphate group cannot be made to come anywhere near the H-8, and since copper ions affect 3′- and 5′-AMP identically, the preferential broadening must be associated with copper binding the base at N-7.

Chan and Nelson[35] have used H-8 broadening by Mn(II) complexes of oligonucleotides to establish proximity of Mn(II) bound phosphate to the H-8. The lower tendency of Mn(II), compared with Cu(II), to bind heterocyclic nitrogen appears to justify this technique.

Sternlicht *et al.*[32] calculated distances of Co(II), Mn(II), and Ni(II) ions from the protons of ATP. Although these investigators did not conclude that there are multiple binding sites, and assumed binding to N-7, the average distances that they calculated, 3.5 Å from H-8, 4.8 Å from H-2, and 5.35 Å from H-1′ are in better agreement with some metal ions binding on N-7 and other metals on N-1 or N-3. The finding by Berger and Eichhorn[34] that H-2 in tubercidin is strongly broadened by Cu(II), and that H-2 and H-8 are equally broadened in adenosine, 2′-AMP, and 2′3′- and 3′5-cyclic AMP can best be understood by assuming binding to multiple sites on the purine ring. It is generally found that metal ions will select one specific site on a molecule containing a variety of potential electron donors, because that site can lead to a more stable complex than the other sites. If the

References pp. 1207–1209

purine ring in adenine nucleotides does not lead to such specific interaction, it is probably because the three nitrogen atoms have similar electronegativities[36], and thus do not markedly differ from each other in their tendency to bind metal ions. However, as we have seen, the presence of phosphate groups in locations favorable for simultaneous binding of metal, along with the base site, can greatly enhance the ability of one electron donor site on the molecule to produce a complex involving that site and phosphate.

Although n.m.r. techniques have been particularly successful in elucidating the structures of these complexes, both base binding and phosphate binding have been confirmed by other methods, *e.g.* infra-red[29,37-39] and Raman spectroscopy[39a] and potentiometric titration[40,41].

Although the N-7 of adenine appears to be frequently useful as electron donor in many of the complexes discussed, we have already noted the multiplicity of binding sites on the adenine ring and the potential for the formation of different complex types with different metal ions. The difference between copper and mercury in binding ATP was dramatically demonstrated by Schneider *et al.*[5], who showed that, whereas Cu(II) preferentially broadens H-8 in 5′-AMP, Hg(II) preferentially broadens H-2, thus indicating that mercury binding involves the 6-membered instead of the 5-membered ring.

Mercury(II) has much greater affinity for the adenine base than copper(II) and the other first transition metals as the stability constants indicate (Table I). Log β_{102} of the 2:1 mercury complex of adenosine is 8.5, in comparison with 0.84 for copper. Unlike the first transition metal ions, which affect the intensity of absorption with little displacement of the location of the adenosine absorption peak, mercury drastically shifts the absorption spectrum of adenosine[13-17,42]. Spectrophotometric and titration studies[16,42] showing sensitivity of the mercury(II) system to pH suggest that mercury in adenosine is bound to an amino group by displacement of one of the amino protons; this conclusion is borne out by the failure of mercury to bind when the amino group is blocked by reaction with formaldehyde[16]. Simpson[42] has demonstrated that, in addition to the amino group, the N-1 also serves as electron donor toward mercury, in line with the finding[5] that Hg(II) preferentially affects H-2 of 5′-ATP.

Silver(I) resembles mercury(II) in shifting the absorbance peak of adenine[14] and adenosine[17] at alkaline pH, and titration indicates that adenosine binds silver also on the amino group by proton displacement[17]. Gillen *et al.*[43] present good evidence for the involvement of the amino group by the demonstration that methylation of the amino group on adenine strongly decreases the affinity for silver. They suggest a structure for the silver adenine complex in which two silver ions bridge two adenine molecules, with each silver bound to the amino group of one adenine and N-1 of the other, in line with Simpson's[42] binding sites.

We have seen that the adenine nucleotides can produce a variety of complexes with metal ions, depending on the position of the phosphate, and also depending on the nature of the complexing ion. We have concentrated on binding to base sites and phosphate, ignoring the ribose portion of the molecule that is generally less involved in complex formation. Metal binding to ribose is, however, of some significance, as noted in Chapter 34, since metal ions appear capable of distinguishing between ribonucleotides and deoxyribonucleotides in enzymatic reactions. Indeed, at high pH copper(II) ions do replace the ribose hydroxyl protons in 5′-AMP[44]. Such a conclusion has been deduced from effects in the titration curves of alkaline solutions of Cu(II)–adenosine, which are not observed when the

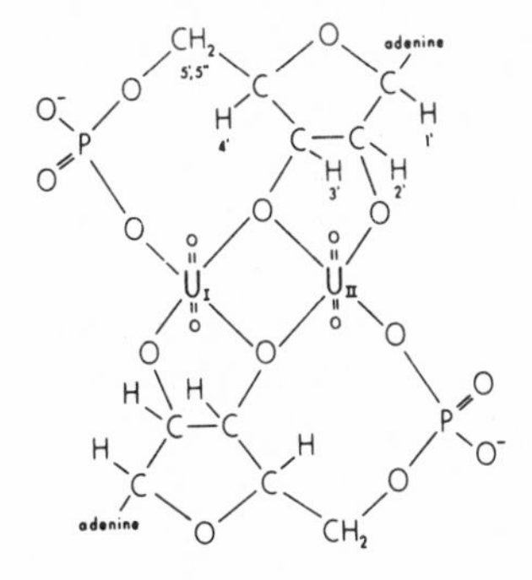

Fig. 4. Structure postulated for the uranyl complex of 5′-AMP (from ref. 46).

adenosine is replaced by 9-methyl adenine or deoxyadenosine; it is therefore assumed that copper binds to the 2′ and 3′ oxygens on the ribose. The affinity of Cu(II) in Cu(II) acetate for the ribose is enhanced in DMSO, where differences in the visible spectra of the Cu(II) complexes of adenosine and deoxyadenosine point to the ribose hydroxyls as the preferred binding site on the nucleoside[45].

The tendency to bind ribose appears to be particularly strong with uranyl ions, whose reaction with adenine nucleotides has been investigated by Agarwal and Feldman[46]. They have discovered, from n.m.r. studies, that uranyl ions bind to phosphate and ribose in 5′-AMP, rather than to phosphate and base. Their structure for the uranyl complex of 5′-AMP[23] is given in Fig. 4. Later studies indicated OH^- ions also bound to U[65].

Metal complexes of other nucleosides and nucleotides.

The adenine nucleosides and nucleotides, including the reaction of these molecules with metal ions, have received much more attention than the nucleosides and nucleotides of guanine, inosine, cytosine, uracil and thymine, but studies on the structures of the metal complexes of the other nucleosides and nucleotides have recently intensified.

References pp. 1207–1209

The structure of guanosine and inosine resembles that of adenosine in that all of these nucleosides are ribosides of substituted purines. Guanosine and inosine, like adenosine, contain a potential chelating group, in these cases N-7 and O-6, in addition to electron donor sites elsewhere in the molecules.

The N-7 position in 5′-dGMP as well as in deoxyadenosine has been implicated as the binding site for Cu(II) by the broadening of H-8 in the n.m.r. spectra of both compounds in the presence of copper[26]. With guanosine the line broadening studies were carried out in DMSO, where the H-1 and NH_2 protons are observed, and it can be seen that they are not broadened[26]. N-7 binding is also postulated for Cu–guanosine by Tu and Friederich[47] on the basis of conductometric and spectrophotometric titration of Cu with guanosine, GMP, inosine, IMP, and theophylline, indicating a reaction of one mole Cu(II) per mole ligand, but no reaction with caffeine. Since theophylline has methylated N-1 and N-3, but a free N-7 position, whereas caffeine has all three of these nitrogen atoms methylated, the N-7 group appears implicated. Tu and Friederich[47] also carried out infra-red studies on these complexes and interpreted a diminution of the carbonyl stretching vibration band as indicative of binding to O-6, so that it was concluded that Cu(II) probably chelates N-7 and O-6, although a binuclear 2:2 complex involving each copper binding to the oxygen of one ring and N-7 of the other was considered possible.

P.m.r. line broadening of 5′-IMP by Cu(II) brings about broadening of both H-2 and the H-8[48]. This finding is in line with the possibility of a chelate ring in Cu(II)–IMP, and by analogy also in Cu(II)–GMP, since the amino group in the 2-position in guanine eliminates the H-2 in GMP, so that binding on the 6-membered ring cannot be detected. The p.m.r. results with Cu(II)–IMP can also be explained by copper ions binding to alternate sites on the 5- and 6-membered rings, *e.g.* on N-7 and on N-1.

A chelate ring involving N-7 and O-6 has been postulated by other investigators[11,50,51]; it is even more tempting to envisage a chelate for guanosine than for adenosine, since guanosine forms more stable complexes (at least with copper) than the other nucleosides, and the N-7, O-6 chelation resembles the stable chelates of 8-hydroxyquinoline. On the other hand, the best evidence for the involvement of O-6 is the infra-red perturbation[47]. Against the chelate hypothesis must be considered the observation that in guanine base Cu(II) prefers binding to N-9, rather than to chelate O-6 and N-7, and, if the latter were possible, it would seem that it would be preferred. Although there is evidence for chelation, the matter cannot be considered as settled.

Tu and Reinosa[49] have used evidence similar to that for Cu–guanosine to postulate N-7, O-6 chelation of silver(I) with guanosine. Silver has a profound effect on the ultra-violet spectrum of guanosine (much more than adenosine)[14,17], and affects its titration curve by removing a proton (that corresponding to N-1) at low pH that is ordinarily titrated at high pH. This evidence has been used[17] as evidence for Ag(I) binding to the N-1, O-6 grouping. (It is not possible to presume that competition with the N-1 proton implies binding to N-1, since metal binding could involve tautomerization, with the metal bound to O-6 instead of N-1.)

Mercury binding to guanosine has been somewhat more thoroughly investigated. Simpson's[42] spectrophotometric thermodynamic studies indicate that Hg binds to N-1 at high pH and to the NH_2 group at low pH. Eichhorn and Clark[16] have used a combination of experiments to assign the N-1, O-6 grouping as the Hg binding site at high pH; this site had also been suggested by Yamane and Davidson[15]. Cheng *et al.*[52] have carried out a spectrophotometric investigation demonstrating the formation of a novel absorption peak by complexation of Hg(II) with 5′-GMP. This effect is remarkable since it requires guanine as the base, and phosphate in the 5′-position; it is not exhibited in the absence of the amino group or with 2′,3′-GMP. Consequently a chelate involving Hg(II) bound to the amino group and the phosphate group was postulated. This specific u.v. effect of mercury with 5′-AMP is reminiscent of the specific n.m.r. effect of Cu(II) with 3′- and 5′-AMP[34] and can perhaps also be explained by a 2:2 complex such as that shown in Fig. 3.

The structure of complexes of cytidine nucleosides and nucleotides, at least with Cu(II) can be guessed from the crystallographic examination of Cu–cytosine[22], showing binding at N-3. There is no reason to suspect that the placement of a ribose or ribose phosphate moiety at N-1 should alter the preference of Cu(II) for this site. This expectation is confirmed by the results of the effect of Cu(II) in selectively broadening H-5, rather than H-6 in cytidine, 5′-CMP and 5′-dCMP[26]; the amino group protons are not affected.

Simpson[42] has identified H-3 also as the binding site of cytidine for mercury at low pH and the amino group at high pH. Eichhorn and Clark[16] have carried out potentiometric and spectrophotometric studies that indicate amino group binding. Similar studies with Ag–cytidine[17] have also implicated the amino group, deprotonated, as the binding site for that metal at alkaline pH.

We have previously noted that, of all the nucleosides, uridine and thymidine form the least stable complexes with metal ions[10,47]. No binding of Cu to 5′-TMP and thymidine could be detected by n.m.r. broadening[26], and no line shift occurs through reaction of zinc(II) with uridine[27]. In contrast to TMP and thymidine, Cu(II) ions do affect the n.m.r. spectra of 5′-UMP and uridine by broadening H-5[48], indicating coordination at N-3. In DMSO, Cu(II) ions in Cu(II) acetate readily bind to the hydroxyl groups of ribose[45], as indeed they also do at high pH in aqueous solution[18].

The crystallographic demonstration that mercury(II) binds to O-4 in uracil[23] suggests the probability that mercury binds to this site also in uridine and UMP, since, again, it is not likely that the placement of a riboside or ribose phosphate group at N-1 will diminish the affinity of O-4 for the metal ion. Simpson[42] had assumed mercury binding to N-3 on the basis of spectrophotometric proton competition studies, but tautomerization phenomena can account for these results even if the Hg is on the oxygen. The titration of the N-3 hydrogen, that ordinarily begins at pH 8, commences below pH 4 in the presence of mercury; this phenomenon has been interpreted as indicating binding of mercury to N-3 or O-4[16]. Silver binding was assigned to the same site, using similar techniques[17]. The tendency of "soft" mercury to prefer "soft" nitrogen binding sites to "hard" oxygen binding sites probably has no relevance to pyrimidine (or purine) rings since the electrons of both oxygen and nitrogen atoms are intimately associated with the heterocyclic ring, so that a comparison with these donor atoms in, *e.g.*, water and ammonia is not applicable.

* * *

It has occurred to a number of investigators that the ability of metal ions to react preferentially with some of the nucleoside bases, rather than others, can be very useful in the study of nucleotide sequence in polynucleotides. The preference of Cu(II) for guanine and its lack of affinity for uracil provides such a selective phenomenon. Beer *et al.* have found that the mercury(II) complex of *p*-hydroquinone selectively binds uridine and guanosine[53] and that OsO_4 reacts preferentially with dTMP, only slightly with dCMP, and practically not at all with dAMP and dGMP[54]; this reaction involves an oxidative degradation of the heterocycles.

We have previously noted the drastic shifts in the absorption spectra of the nucleotides in the presence of mercury(II). Cheng[55], noting that the optical rotatory dispersion of nucleotides is differentially affected by Hg, measured the optical rotational dispersion curves of numerous dinucleoside phosphates, and noted that they can be identified by their optical rotatory characteristics.

The principal importance of these metal binding studies is of course to provide an understanding of the participation of metal ions in the biological functions of nucleotides (Chapters 17 and 18) and nucleic acids (Chapter 34).

ACKNOWLEDGEMENT

I am grateful to Drs. Nathan A. Berger, James J. Butzow, Patricia Clark, Josef Pitha, Joseph Rifkind, and Yong A. Shin for helpful comments.

NOTE ADDED IN PROOF

Copper(II) acetate exists in non-aqueous solution as a dimer containing a copper–copper bond with a length of approximately 2.7 Å, which is also the approximate distance between the 2′ and 3′ hydroxyl groups of nucleosides; the copper acetate dimer consequently reacts with all ribonucleosides, which form a bridge between the two copper atoms, thus distinguishing ribonucleosides from deoxynucleosides, which cannot form such a bridge[56]. On the basis of nuclear magnetic resonance studies, Swift *et al.*[57,58] have postulated that the metal ion in ATP complexes can coordinate to the phosphate moiety and simultaneously to a water molecule which in turn is hydrogen-bonded to N-7. Such a structure had been previously proposed by Brintzinger on the basis of the simultaneous binding of metal to phosphate and base[6], but was not published at the time partly because of criticism from this author. Heller *et al.*[59] have indicated that Mn(II) binds negligibly to the bases of ATP, as compared to phosphate, whereas Anderson *et al.* have shown that Mn(II) binds preferably to guanine in comparison with other nucleoside bases[60]. Berger and Eichhorn[61] have shown that Cu(II) can bind to both N-7 and N-1 of inosine, the former preferably at low pH, the latter at high pH. Kan and Li[62] have calculated stability constants in DMSO for the mercury complexes of cytidine, adenosine and guanosine as 33.9, 7.2 and 5.9, respectively. These may be compared with constants reported earlier in water for cytidine, adenosine and thymidine as $\log K_1$ = 5.45, 4.25 and 10.6, respectively[6]. The relative base affinity of mercury(II) thus appears to be thymidine > cytidine > adenosine > guanosine. This sequence is the opposite of that obtained with copper(II). Weser and Dönnicke[63] have determined stability constants of the silver complexes of adenosine, AMP, ADP, ATP, DNA and RNA as $\log K$ = 3.9, 4.2, 4.2, 4.3 and 4.4, respectively, indicating the lack of importance of the phosphate group in silver binding.

REFERENCES

1 G. L. Eichhorn, *Adv. Chem.*, 63 (1966) 378.
2 G. L. Eichhorn, J. J. Butzow, P. Clark and Y. A. Shin, in J. Maniloff, J. R. Coleman and M. W. Miller, *Effects of Metals on Cells, Subcellular Elements, and Macromolecules*, C. C. Thomas, Springfield, Illinois, 1970, p. 77.
3 R. Phillips, *Chem. Rev.*, 66 (1966) 501.
4 U. Weser, *Structure and Bonding*, 5 (1968) 41.
5 P. W. Schneider, H. Brintzinger and H. Erlenmeyer, *Helv. Chim. Acta*, 47 (1964) 992.
6 H. T. Writh and H. Davidson, *J. Am. Chem. Soc.*, 86 (1964) 4325.
7 R. M. Smith and R. A. Alberty, *J. Am. Chem. Soc.*, 78 (1956) 2376.
8 M. M. Taqui Khan and A. E. Martell, *J. Am. Chem. Soc.*, 84 (1962) 3037.
9 M. M. Taqui Khan and A. E. Martell, *J. Am. Chem. Soc.*, 86 (1964) 4325.

10 M. Fiskin and M. Beer, *Biochemistry*, 4 (1965) 1289.
11 C. Ropars and R. Viovy, *J. Chim. Phys.*, (1965) 408.
12 E. Frieden and J. Alles, *J. Biol. Chem.*, 230 (1958) 797.
13 C. A. Thomas, Jr., *J. Am. Chem. Soc.*, 76 (1954) 6052.
14 T. Yamane and N. Davidson, *Biochim. Biophys. Acta*, 55 (1962) 609.
15 T. Yamane and N. Davidson, *J. Am. Chem. Soc.*, 83 (1961) 2599.
16 G. L. Eichhorn and P. Clark, *J. Am. Chem. Soc.*, 85 (1963) 4020.
17 G. L. Eichhorn, J. J. Butzow, P. Clark and E. Tarien, *Biopolymers*, 5 (1967) 283.
18 H. Reinert and R. Weiss, *Z. Physiol. Chem.*, 350 (1969) 1310.
19 J. A. Carrabine and M. Sundarilangam, *J. Am. Chem. Soc.*, 92 (1970) 369.
20 E. Sletten, *Chem. Commun.*, (1967) 1119.
21 R. Weiss and H. Venner, *Z. Physiol. Chem.*, 333 (1963) 169.
22 J. A. Carrabine and M. Sundaralangam, *Chem. Commun.*, (1968) 746.
23 M. Sundaralangam and J. A. Carrabine, *Acta Cryst.*, A25 (1969) 5179.
24 M. Cohn and T. R. Hughes, Jr., *J. Biol. Chem.*, 237 (1962) 176.
25 H. Sternlicht, R. G. Shulman and E. W. Anderson, *J. Chem. Phys.*, 43 (1965) 3123.
26 G. L. Eichhorn, P. Clark and E. D. Becker, *Biochemistry*, 5 (1966) 245.
27 S. M. Wang and N. C. Li, *J. Am. Chem. Soc.*, 90 (1968) 5069.
28 J. A. Hoppe and M. Morales, *J. Am. Chem. Soc.*, 88 (1966) 2077.
29 J. Brigando and D. Colaitis, *Bull. Soc. Chim. France*, (1969) 3445.
30 H. Sigel, *Helv. Chim. Acta*, 50 (1967) 582.
31 P. W. Schneider and H. Brintzinger, *Helv. Chim. Acta*, 47 (1964) 1717.
32 H. Sternlicht, R. G. Shulman and E. W. Anderson, *J. Chem. Phys.*, 43 (1965) 3133.
33 H. Sternlicht, D. E. Jones and K. Kustin, *J. Am. Chem. Soc.*, 90 (1968) 7110.
34 N. A. Berger and G. L. Eichhorn, *Biochemistry*, 10 (1971) 1847.
35 S. I. Chan and J. A. Nelson, *J. Am. Chem. Soc.*, 91 (1969) 168.
36 H. Pullman, *Ann. N.Y. Acad. Sci.*, 158 (1969) 65.
37 H. Brintzinger, *Biochim. Biophys. Acta*, 77 (1963) 343.
38 H. Fritzsche and C. Zimmer, *Eur. J. Biochem.*, 5 (1968) 42.
39 K. A. Hartman, Jr., *Biochim. Biophys. Acta*, 138 (1967) 192.
39a L. Rimai, M. E. Heyde and E. B. Caren, *Biochem. Biophys. Res. Commun.*, 38 (1970) 231.
40 A. Albert, *Biochem. J.*, 54 (1953) 646.
41 T. R. Harkins and H. Freiser, *J. Am. Chem. Soc.*, 80 (1958) 1132.
42 R. B. Simpson, *J. Am. Chem. Soc.*, 86 (1964) 2059.
43 K. Gillen, R. Jensen and N. Davidson, *J. Am. Chem. Soc.*, 86 (1964) 2792.
44 H. Reinert and R. Weiss, *Z. Physiol. Chem.*, 350 (1969) 1321.
45 N. A. Berger, unpublished observations.
46 R. P. Agarwal and I. Feldman, *J. Am. Chem. Soc.*, 90 (1968) 6635.
47 A. T. Tu and C. G. Friederich, *Biochemistry*, 7 (1968) 4367.
48 N. A. Berger and G. L. Eichhorn, *Biochemistry*, 10 (1971) 1857.
49 A. T. Tu and J. A. Reinosa, *Biochemistry*, 5 (1966) 3375.
50 L. E. Minchenkova and V. I. Ivanov, *Biopolymers*, 5 (1967) 615.
51 L. N. Drozdov-Tikhomirov and L. I. Kikoin, *Biofizika*, 12 (1967) 407.
52 P. Y. Cheng, D. S. Honbo and J. Rozsnyai, *Biochemistry*, 8 (1969) 4470.
53 A. M. Fiskin and M. Beer, *Biochim. Biophys. Acta*, 108 (1965) 159.
54 M. Beer, S. Stern, D. Carmalt and K. M. Mohlenreich, *Biochemistry*, 5 (1966) 2283.
55 P. Y. Cheng, *Biochem. Biophys. Res. Commun.*, 33 (1968) 746.
56 N. A. Berger and G. L. Eichhorn, *Nature*, (1972) in press.
57 T. A. Glassman, C. Cooper, L. W. Harrison and T. J. Swift, *Biochemistry*, 10 (1971) 843.

58 G. P. P. Kuntz, T. A. Glassman, C. Cooper and T. J. Swift, *Biochemistry*, 11 (1972) 538.
59 M. J. Heller, A. J. Jones and A. T. Tu, *Biochemistry*, 9 (1970) 4981.
60 J. A. Anderson, G. P. P. Kuntz, H. H. Evans and T. J. Swift, *Biochemistry*, 10 (1971) 4368.
61 N. A. Berger and G. L. Eichhorn, *J. Am. Chem. Soc.*, 93 (1971) 7062.
62 L. S. Kan and N. C. Li, *J. Am. Chem. Soc.*, 92 (1970) 4823.
63 U. Weser and M. Dönnicke, *Z. Naturforsch.*, 25 (1970) 592.
64 R. M. Izatt, J. J. Christensen and J. H. Rytting, *Chem. Rev.*, 71 (1971) 439.
65 I. Feldman and K. E. Rich, *J. Am. Chem. Soc.*, 92 (1970) 4559.

Chapter 34

COMPLEXES OF POLYNUCLEOTIDES AND NUCLEIC ACIDS

G. L. EICHHORN

Laboratory of Molecular Aging, Gerontology Research Center, National Institutes of Health, National Institute of Child Health and Human Development, Baltimore City Hospitals, Baltimore, Maryland 21224, U.S.A.

INTRODUCTION

There are a variety of naturally occurring polynucleotides, or *nucleic acids*, and all of them are involved in some manner in the transfer of genetic information. The nucleic acids are conveniently classified into two types, ribonucleic acid (RNA) and deoxyribonucleic acid (DNA),

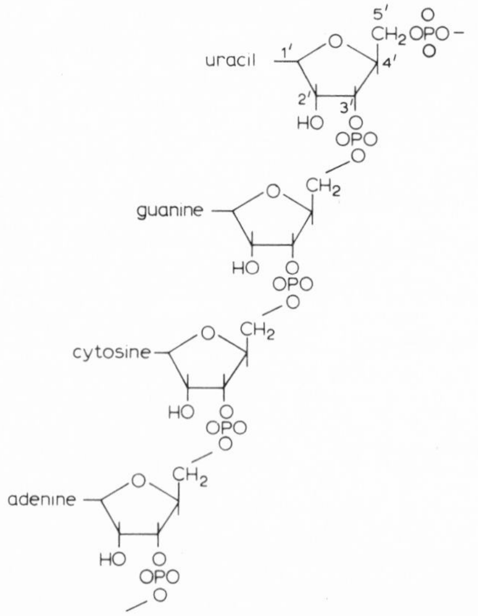

Fig. 1. Primary structure of RNA. The structure of DNA is similar, without 2′-hydroxyl and with thymine substituting uracil.

depending upon the presence or absence of a 2′-hydroxyl group, or, to put it differently, upon the constitution of the macromolecules from ribonucleotides or deoxyribonucleotides, respectively. Both DNA and RNA are constructed from monomers by phosphodiester linkages of the 3′-hydroxyl of one nucleotide to the 5′-hydroxyl group of the adjacent nucleotides, as shown in Fig. 1 for RNA.

DNA lacks the 2′-hydroxyl groups shown in this Figure, and uracil is substituted by thymine. The sequence of bases along the DNA chain constitutes the genetic code. The DNA molecule consists of two intertwined helices, with each heterocyclic base hydrogen-bonded to a complementary base on the other chain, so that each guanine is bound to a cytosine, and each adenine to a thymine, as shown in Fig. 2. Thus each strand of the double helix contains a form of the genetic code, and the unwinding into separate strands, with the concomitant formation of complementary strands for each of the original strands, will lead to the production of two new DNA double helices, each of which is a replicate of the original. This replication process requires the presence of metal ions, and can also be thwarted by metal ions, thus illustrating the significance of metal complexation in the biochemistry of the nucleic acids.

Not only the replication of DNA, but virtually every step in the utilization of the genetic code for the eventual production of the proteins

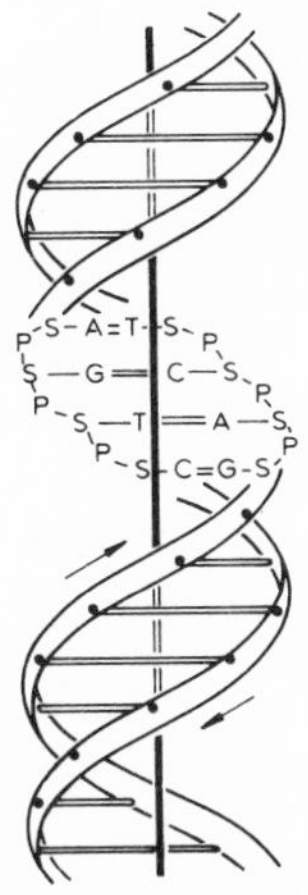

Fig. 2. Schematic structure of DNA. P = phosphate, S = sugar (deoxyribose), A = adenine, T = thymine, G = guanine, C = cytosine (from ref. 138).

specified by the code is governed in some way by the presence of metal ions. A very schematic representation of the processes involved is shown in Fig. 3. The DNA, in the nucleus of the cell, serves as a template for the formation of messenger RNA (mRNA), which then carries the genetic message into the cytoplasm. This transcription of DNA into mRNA is also influenced by metal ions. The genetic code in mRNA is contained in trinucleotide sequences, called codons; each codon "codes" for a specific amino acid. The codons are "recognized" by anticodons, trinucleotide sequences in transfer RNA (tRNA) that contain the bases that are complementary to those in the codon. Binding of tRNA anticodon to mRNA codon occurs on the surface of ribosomes, whose structure is determined by metal ion constituents. While one part of the tRNA contains the anticodon, another part binds the specific amino acid for which the particular tRNA containing that anticodon is constructed. To illustrate, a codon for phenylalanine is the trinucleotide CUU; thus the phenylalanyl tRNA contains GAA at the anticodon site, and specifically binds phenylalanine, and, barring error, no other amino acid. In this way the nucleotide code of the mRNA is translated into the proper sequential placement of amino acids; after a sequential amino acid is positioned, peptide linkage with the preceding amino acid occurs. The interpretation of the nucleotide code by the proper sequential placement of amino acids is called translation.

We shall now see how each of the three fundamental processes, replication, transcription, and translation is influenced by metal complex formation.

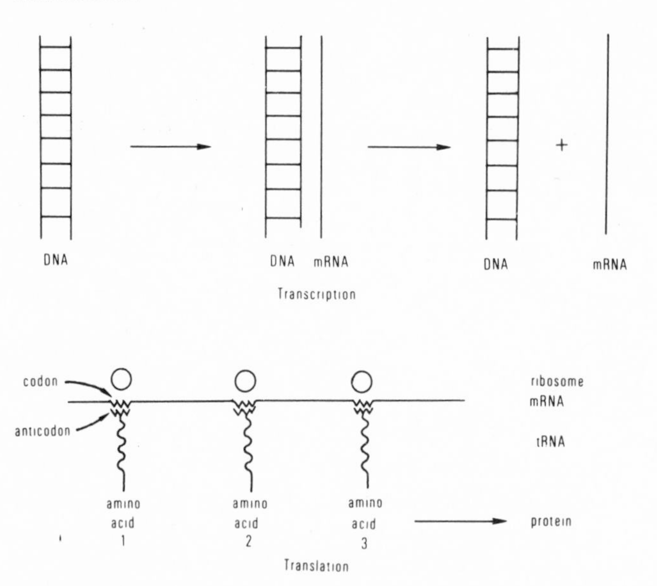

Fig. 3. Schematic representation of the utilization of the DNA code in protein synthesis.

METAL IONS AND DNA REPLICATION

The mechanism of the replication of the DNA molecule in living cells is not at all well understood. Evidence has recently accumulated that the thoroughly investigated process catalyzed by the enzyme DNA polymerase I from *E. coli* may not account for the principal mechanism of biological DNA replication, but may perhaps be involved in the repair of damaged DNA molecules. Nevertheless, the DNA polymerase reaction is a reaction capable of synthesizing new DNA molecules with the nucleotide sequence of the existing DNA, and therefore does constitute a biological synthesis of DNA[1]. The DNA polymerase reaction requires, in addition to DNA template, the four deoxynucleotide triphosphates, dATP, dCTP, dGTP and dTTP, and divalent metal ions, generally magnesium(II). The reaction involves the cleavage of pyrophosphate from the triphosphates and the formation of phosphodiester linkages between adjacent deoxynucleotides.

Since this polymerization involves the cleavage and the formation of phosphate bonds, the metal requirement is indeed to be anticipated from the routine involvement of metal ions in such processes (see Chapters 17 and 18). There are some very curious aspects of the metal participation in this reaction that indicate that other functions as well may be ascribed to the metal ions.

The most interesting of these aspects is the fact that, whereas manganese(II) can replace magnesium ions in the reaction, the selectivity of the reaction is modified by such a substitution. Magnesium ions permit the incorporation of only deoxyribonucleotides into DNA. Manganese ions, on the other hand, permit the incorporation of both deoxynucleotides and ribonucleotides into the "DNA"[2]. When formed in the presence of manganese, the DNA thus contains both the deoxynucleotides that it should contain and the ribonucleotides that it should not contain. Thus magnesium ions lead to the formation of the desired product by screening unwanted monomers, and manganese ions can lead to errors in the synthesis of DNA by an inability to carry out such screening.

The fact that this polymerization reaction occurs with both magnesium and manganese, but with an essential selectivity in the former case that is lacking in the latter, implies that in the presence of magnesium ions differentiation between nucleotides that do or do not contain a 2′-hydroxyl group is possible. In the presence of manganese ions the presence or absence of the 2′-hydroxyl group is not so definitively detected. The mechanism by which a metal ion can produce such selectivity, or perhaps can prevent it, is not known, but we are reminded of the studies discussed in Chapter 33 that reveal metal ion binding to ribose hydroxyls in nucleotides and not in deoxynucleotides. It is hoped that further studies of this nature may help to clear up the role of the metal ion in the polymerization reaction.

The function of the Mg^{2+} ions in this polymerization reaction appears to involve binding the deoxynucleoside triphosphates to the enzyme[3]. In addition to this requirement for Mg^{2+}, DNA polymerase apparently contains two tightly bound atoms of zinc(II) per molecule of enzyme[4]. The zinc ions seem to have a function different from that of Mg(II) ions in that they apparently aid in the binding of DNA, rather than nucleotides, to the enzyme[4]. It is interesting to note in this connection that Weser *et al.*[5] have localized the nucleus, which is the site of replication, as a major site of zinc activity and that they have found[6] that the concentration of zinc is highest in the nuclei. Mg^{2+} ions, necessary for binding the triphosphates, are not required for the binding of DNA[7]. One atom of manganese(II)

References pp. 1240–1243

can be tightly bound to the enzyme, while another 3 atoms are less tightly bound and another 6 atoms are very weakly bound[4].

Another deoxynucleotide polymerizing enzyme is the terminal deoxynucleotidyl transferase of Bollum and coworkers which is capable of adding deoxynucleotides to the end of a polydeoxynucleotide chain, and indeed of synthesizing polydeoxynucleotides from small initiator molecules such as trimers[8]. This enzyme also involves the cleavage of phosphate from deoxynucleoside triphosphates and requires metal ions; Mg^{2+}, Mn^{2+}, and Co^{2+} all have been used in this reaction. A different type of selectivity is now encountered. Magnesium ions are most effective for incorporation of purine nucleotides, but cobalt ions are most effective for pyrimidine nucleotides. This phenomenon also cannot be readily explained by our present knowledge of monomer complexes. (Possibly complexing to the bases serves to inhibit the reaction, and base complexation, of which Mg(II) is not capable, is in fact stronger with the purines than with the pyrimidines.)

The terminal deoxynucleotidyl transferase is also a metal enzyme[9] and recent studies have indicated that the metal binds the enzyme to the growing end of the polydeoxynucleotide chain, and it has been suggested that the metal may be involved in the translational motion of the enzyme *vis-à-vis* the growing chain.

In a system, isolated from chick embryo, for the incorporation of *ribo*nucleotides into DNA (involving deoxygenation to deoxynucleotides), it was found[10] that only a narrow range of magnesium concentration is effective; it was postulated that variations in magnesium ion concentration could indeed exercise control over DNA synthesis.

METAL IONS AND TRANSCRIPTION

The copying of the DNA code in polyribonucleotide form, as mRNA, is brought about through the RNA polymerase enzymes which have been isolated from various bacterial sources, *e.g. E. coli* and *M. lysodeicticus* as well as from mammalian cells. These polymerases require DNA as template, ribonucleoside triphosphates (generally ATP, CTP, GTP and UTP), and, again metal ions[11-19]; pyrophosphate is produced in this reaction also. Several metal ions, Mg^{2+}, Mn^{2+}, and Co^{2+} have been shown to be active. As with DNA polymerase, the involvement of these metals can be anticipated because of the phosphate bond cleavage and formation, but again this function does not appear solely responsible for the metal requirement.

With the *M. lysodeicticus* enzyme, magnesium again showed itself to be highly selective for substrate, only this time the presence of magnesium ions insured the incorporation of ribonucleotides instead of deoxynucleotides[12]. Thus, with this RNA polymerase, magnesium has the opposite selective effect as with *E. coli* DNA polymerase in the selection of the nucleotide with or without a 2′-hydroxyl group. It may be noted that magnesium ions cause incorporation of the *proper* nucleotide in both DNA and RNA synthesis.

Manganese, while quantitatively promoting more rapid reaction[13,14] can bring about the *error* incorporation of deoxynucleotides alongside ribonucleosides[12]. Manganese produces other non-physiological variations of RNA synthesis[19a]. It brings about the synthesis of homopolymers, with DNA template, using only one nucleoside triphosphate[20-22], the slow synthesis of poly rA and poly rU, with no template[23], and the copying of poly rA and poly rU[24]. It even causes the polymerization of 2′-O-methylcytidylic acid forming a polymer similar to poly rC[25]. None of these reactions are observed with magnesium.

The reason for the anomalous behavior is not apparent; perhaps because manganese, in contrast with magnesium, has varied coordinating tendencies, while Mg is confined to binding phosphate, the manganese exhibits less discrimination and the RNA polymerization consequently loses specificity. The manner in which any of the metals influence the polymerization is not known, although it is known that metal is required both in the initiation of the reaction and in the propagation of polymerization[12]. Cations in general stimulate the initiation of the reaction, possibly by affecting the organization of the subunits of which the enzyme is composed[26]. Fox *et al.*[27] have shown that metal ions are not required to achieve binding of enzyme to the DNA. (The substitution of Mn(II) for Mg(II) also lowers the specificity of an enzyme tRNA nucleotidyl transferase, which is required in the synthesis of the CCA terminal trinucleotide present in all tRNA molecules; with Mn(II) atypical terminal sequences are produced, through incorporation of U instead of C. CMP is incorporated more rapidly with Mg, UMP more rapidly with Mn[19a].)

In addition to Mg^{2+} and Mn^{2+}, Co^{2+} has also been used as an activator for RNA polymerase[19,28], while other metal ions have been found to be ineffective[29,30].

Whereas most of the work on RNA polymerase and therefore also the effect of metal ions has been carried out using bacterial enzymes, there has been some work on mammalian RNA polymerase. RNA polymerase activity has been found in different fractions of liver when Mg or Mn was used for activation[31], leading to the speculation that at least two polymerases exist that are differently affected by these metal ions. The magnesium activated fraction is inhibited by Cu(II), Cd(II) and Co(II) (but not Zn(II)), whereas Hg(II), Cd(II) and Cu(II), but not Zn(II) and Cd(II) are effective inhibitors of the manganese activated fraction[32].

It may be noted in this connection that copper deficient rats show a decline in their protein synthesizing ability that can be ascribed to a decrease in mRNA formation[33]. Iron deficiency fails to produce a similar decrease.

References pp. 1240–1243

METAL IONS AND TRANSLATION

The final step in the interpretation of the DNA code to synthesize protein is recognition by amino acid specific tRNA molecules of the codons on mRNA on the surface of the ribosome. Later in this chapter we shall see how the translation phenomenon can be very much influenced by the effect of metal ions on the conformation of two of the constituents in the process, tRNA and ribosome.

A very striking illustration of the importance of metal ions in the translation process is the demonstration that the concentration of divalent metal ions present can determine which of a number of amino acids will be incorporated into protein by a given codon. Generally, magnesium ions are

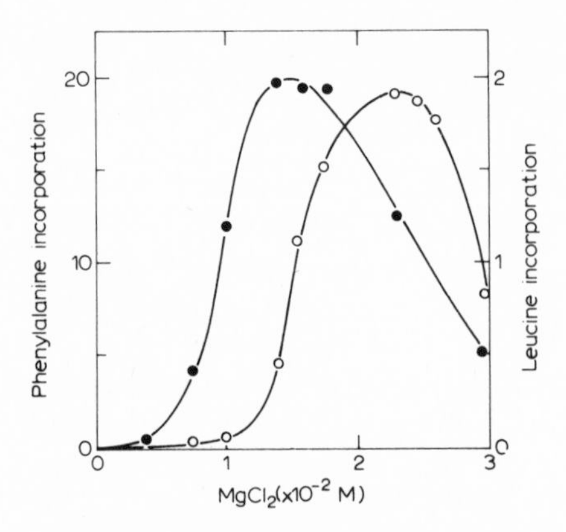

Fig. 4. The effect of Mg(II) concentration on the incorporation of phenylalanine and leucine into protein, using poly rU as template (from ref. 36).

required for *in vitro* protein synthesis[34]. When poly U is substituted for mRNA in such an *in vitro* system, only UUU codons, which code for phenylalanine, are available, and the protein synthesized is indeed poly-phenylalanine[34]. Although phenylalanine incorporation is the primary reaction, some leucine incorporation is also induced by poly U[35], the relative amounts of phenylalanine and leucine incorporation depending upon the magnesium concentration[36], as shown in Fig. 4. The figure reveals that at 1×10^{-2} *M* Mg^{2+} only phenylalanine, and virtually no leucine, is incorporated, but that at higher Mg^{2+} concentrations, as the phenylalanine incorporation diminishes, leucine incorporation increases. Both UUU and UUC are codons for phenylalanine; UUG and UUA are codons for leucine. It is apparent that at high metal concentration the recognition of the third nucleotide in the codon trimer is lost, permitting the "mistaken" incorporation of leucine.

A similar miscoding phenomenon is observed with the use of copolymer poly (U-G)[37], which contains alternating G and U. In the presence of 1×10^{-2} M Mg^{2+}, the polypeptide contains valine and cysteine, as indicated by the code letters GUG and UGU for these amino acids, respectively. At 2×10^{-2} M Mg^{2+}, however, valine is extensively substituted by tyrosine, which has a UAU codon. Serine and arginine (codons UCU and CGU, respectively) are also substituted. Crick[38] had postulated that the "degeneracy" of the code is due to ambivalent reading of the third letter in the codon; the magnesium induced "errors" seem to indicate that the first and second letters, as well as the third letter, of the codon can be misread at high metal ion concentration.

Although Mg^{2+} is frequently used for *in vitro* protein synthesis schemes, Ca^{2+} and Mn^{2+} have been substituted in the poly U directed synthesis of phenylalanine[39]; Mn^{2+} gives poor though significant incorporation, but Ca^{2+} actually works better than Mg^{2+}.

* * *

We have seen that the presence of metal ions is of great importance in the normal processes of replication, transcription and translation, and that metal ions in the wrong amounts or of the wrong kinds can cause these processes to go wrong. The mechanisms by which metal ions produce these effects are poorly understood. It can be assumed that they involve the ability of metal ions to interact with so many electron donor sites on these molecules: the phosphates, the bases, as well as the ribose hydroxyls. We have observed, in Chapter 33, that different metal ions have different affinities for these sites. When metal ions bind to the various sites in polynucleotides, they exert rather dramatic effects upon them by bringing about ligand reactions that greatly change the structures of the macromolecules. Before considering the ligand reactions of the macromolecules, we shall first summarize the knowledge available on the stability and binding sites of metal–polynucleotide complexes.

STABILITY OF METAL COMPLEXES OF POLYNUCLEOTIDES

Several investigators have determined a stability constant for the Cu(II) complex of DNA of the order of 10^4 [40-43], although higher[44] values have been reported. Assuming 10^4 as the order of magnitude of the complex, the stability of Cu(II) DNA falls between that of Cu(II) AMP and Cu(II) ADP (Chapter 33). It is reasonable that a cooperative effect in the binding of Cu(II) to the polymer should increase the stability over that of a nucleoside monophosphate. Several efforts have been made[44,41] to measure binding to the bases separately and to compare such binding to the total stability of the DNA complex, including phosphate and base binding.

The result is that base binding alone produces a much lower constant than the combination of base and phosphate binding, with phosphate binding consequently constituting the considerably stronger of the two forces. Binding of Cu(II) to native and denatured DNA[43,42] reveals that denatured DNA binds Cu(II) more strongly, reflecting the added base-binding capacity of DNA in the denatured form. Cd(II) appears to bind DNA somewhat more weakly than Cu(II)[43]. (A higher stability constant of 2×10^5 has been reported, curiously, for Mg(II) DNA[45] than for Cu(II) DNA.) A constant for Mn(II) DNA of similar magnitude as that for Cu(II) has been observed[46]. It is best to restrict this discussion to orders of magnitude, since it is apparent that the different techniques for the determination of the constants, the different conditions under which they were obtained, as well as the different preparations of the DNA itself, have led to the variability of the results. Bach and Miller[43] have studied polarographically the extreme ionic strength dependence for these stability constants.

The stabilities of the Mn(II) complexes of tRNA, as well as of poly rA, poly rC and poly rI are similar to those of DNA[46], but that of Mn(II) poly rU is very much lower. Similar stabilities (4×10^4 and 3×10^4, respectively)[47] have been reported for the Mg(II) complexes of poly rA and poly rU.

BINDING SITES OF METALS ON POLYNUCLEOTIDES

Just as the stabilities of the metal complexes of polynucleotides are similar to those of mononucleotides, the binding sites, insofar as they have been determined, are also alike.

Recent nuclear magnetic resonance (n.m.r.) studies have indicated that Cu(II) ions bind N-7 in poly rA just as in 5′-AMP[48], N-7 in poly rI as well as in 5′-IMP[49], N-3 in poly rC[49] as in 5′-CMP[50] and N-3 in poly rU as in 5′-UMP[49].

There has been some interest in the nature of the binding of Cu(II) to DNA because of the Cu(II) induced unwinding and rewinding of DNA (see below). Infra-red evidence[51,52] and electron spin resonance (e.s.r.) studies[40] have suggested binding primarily to the GC rich regions of DNA; visible absorption spectra indicate primarily binding to guanine[53]. A great deal of evidence has been accumulated to indicate that Cu(II) binds to the DNA bases primarily in the unwound state[50,54,55]; Cu(II) ions, under conditions that do not give rise to any effect on the spectrum of DNA at room temperature, perturb the ultra-violet spectrum of the heterocycles at elevated temperature. The ability of Cu(II) ions to promote the unwinding of DNA at elevated temperature, but not at low temperature (see below), is in line with these observations. It has, however, been shown, that Cu(II) ions will react with DNA bases even at room temperature, if enough time is allowed for the reaction to occur or if the DNA is subjected to gel filtration on Sephadex[41].

We noted in Chapter 33 that Ag(I) and Hg(II) have very different base affinities from those of Cu(II); this is true for polynucleotides as well as with the monomers. Ag(I), like Cu(II), prefers the GC regions of DNA[56], whereas Hg(II) selectively binds the AT regions[57], due apparently to its great affinity for thymine[58]. It is of interest that thymine, which is least strongly bound to Cu(II)[50,59] is most strongly bound to Hg(II). We shall consider the binding sites of Hg(II) and Ag(I) to DNA more thoroughly in a later section of this chapter.

A comparison of the reactions of Ag(I) with poly rA, poly rC, poly rI and poly rU reveals that Ag ions deprotonate poly rI and poly rU beginning at pH 5; they have no such effect on poly rA and poly rC[110].

N.m.r. studies have been interpreted to indicate phosphate binding of Mn(II), and Co(II) in RNA[60] and Mn(II), Co(II), Ni(II) and Fe(II) in DNA[64], whereas Fe(III) was similarly construed to be possibly bound to the bases. The evidence that Mg(II) is bound tobases[61-63,65] is not convincing; in line with the effect of Mg(II) in the stabilization of the DNA double helix, Mg(II) ions seem to be confined to phosphate sites in DNA as well as in other polynucleotides[45,66].

It is evident that many metal ions bind both phosphate and base groups in polynucleotides, as well as in the monomers. Although little is known from direct thermodynamic measurements about the relative affinity of the various metal ions for phosphate and base sites, the effect of these ions on the macromolecules, as we shall see, permits a qualitative assessment of these affinities.

EFFECT OF METAL IONS ON DNA

Stabilization

The attempted solution of native double-stranded DNA in distilled water results in the unwinding of the double helical structure; an electrolyte concentration of at least 2×10^{-3} M is required to prevent the unwinding of the double helix[67]. The ability of metal ions to stabilize the DNA double helix was first discovered by Shack *et al.*[68] and elucidated by Dove and Davidson[69], who found that T_m, the "melting temperature", or the temperature at which DNA unwinds into single strands, increases linearly with the logarithm of ionic strength (Fig. 5). Divalent ions (Mg^{2+} and Co^{2+}) are effective at much lower concentrations than univalent ions.

The stabilization of the DNA double helix by metal ions can be readily explained. In the double helix the negatively charged phosphate groups on adjacent nucleotides repel each other; unwinding of the double helix increases the distance between the phosphates, thus decreasing the repulsion. If the charges on the phosphate are neutralized by counter ions, then of course the driving force for the unwinding has been removed. Divalent ions, of course, should be much more effective than univalent ions.

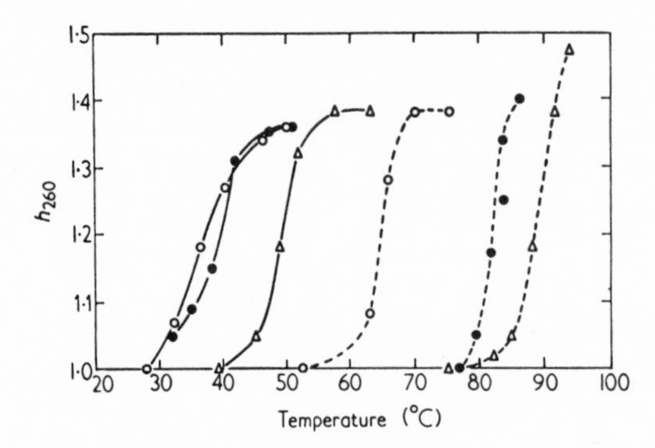

Fig. 5. Effect of ionic strength on the heating curve of *B. magaterium* DNA at pH 7. Ionic strengths, by $NaClO_4$ addition, are as follows:

	μ
○—○	3.1×10^{-4}
●—●	4.1×10^{-4}
△—△	1.3×10^{-3}
○--○	1.0×10^{-2}
●--●	0.10
△--△	0.50

The ordinate represents the absorbance at the ambient temperature/the absorbance at room temperature. The double helix has a low absorbance caused by π-interaction of the stacked bases, the unwound single coils have high absorbance. T_m is the midpoint of the transition between double helix and single coils (from ref. 69).

Unwinding and rewinding

Although the DNA double helix is stabilized by some metal ions, this phenomenon is not a universal characteristic of metal ions; in fact, different metal ions have very different influences on the conformation of DNA. This fact is dramatically illustrated by the opposite effects of Mg^{2+} and Cu^{2+} ions on the melting curves of DNA (Fig. 6)[70]. Curve A represents the transition from the low absorbance double helix to the more highly absorbing random coil, *i.e.*, the unwinding of the double helix, in the absence of divalent metal ions. The decrease in absorbance on cooling results from hydrogen bonding within the single strands, forming "hairpin loops" (Fig. 7). The randomly distributed single chains are not in a position in which the complementary bases are able to reunite at the lower temperature favorable for the reformation of the ordered structure. Mg^{2+} ions increase the melting temperature (see legend to Fig. 5), as expected for a metal ion that stabilizes the double helix; more energy is required to destroy the helix in the presence of the stabilizing magnesium. Cu^{2+} ions, on the other hand, decrease the melting temperature. Such a

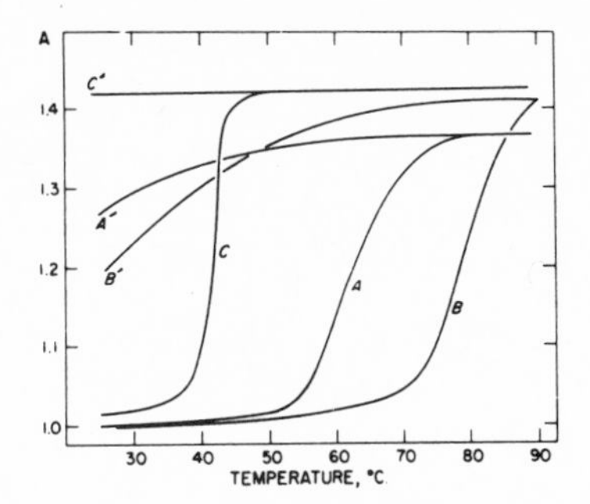

Fig. 6. The opposite effects of Mg^{2+} and Cu^{2+} on the heating and cooling curves of 5×10^{-5} M(P) DNA in 5×10^{-3} M $NaNO_3$. (A) Heating curve with no divalent metal, (B) with Mg^{2+}, (C) with Cu^{2+}. (A′), (B′) and (C′) are cooling curves.

decrease indicates that less energy is required to "melt" the DNA double helix, and that the Cu^{2+} ions therefore destabilize the DNA. The reason for the difference in behavior of the two ions with respect to DNA can be readily explained by differences in binding tendency. Mg^{2+} binds to phosphate, stabilizing the ordered structure by the counter ion effect. Cu^{2+} ions bind not only to phosphate, but to base as well, and binding to base competes with the hydrogen-bonding of the double helix and therefore

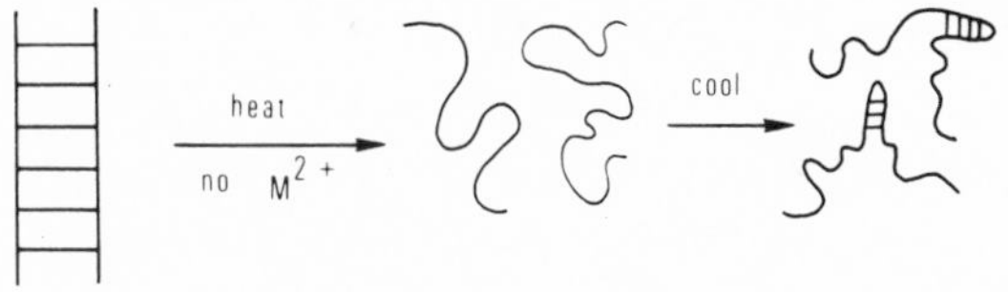

heat

no M^{2+}

cool

Fig. 7. Diagram for the effect of heating and then cooling double stranded DNA, in the absence of divalent metals.

helps to destroy it. The opposite effects of Mg^{2+} and Cu^{2+} can thus be readily explained by binding of Mg^{2+} to phosphate and of Cu^{2+} to base.

The cooling of a solution of Cu(II) DNA is not accompanied by a decrease in absorbance that is generally characteristic of cooling of DNA solutions and explained by the formation of hairpins. It is apparent that copper ions continue to be bound to DNA bases at the lowered temperature, thus preventing the hydrogen bonding required for the hairpins.

References pp. 1240–1243

Cu(II) ions help to bring about unwinding of DNA only at low ionic strength. Higher electrolyte concentrations than those used in the experiment in Fig. 6 prevent the copper from causing the drastic decrease in the melting temperature of DNA shown in Fig. 6c[71]. The high electrolyte concentration stabilizes the double helix to such an extent that copper ions are unable to counteract the stabilization. Essentially, the reaction of copper with DNA involves a competition between hydrogen bonding (double helix) and copper binding.

It is perhaps not surprising that such a competitive effect can be used to obtain a reversible unwinding and rewinding of DNA[54,55,71,72]. If a Cu(II) DNA solution is heated as in Fig. 6a until the DNA is unwound, the solution is cooled, and solid electrolyte is then added, the double helix is

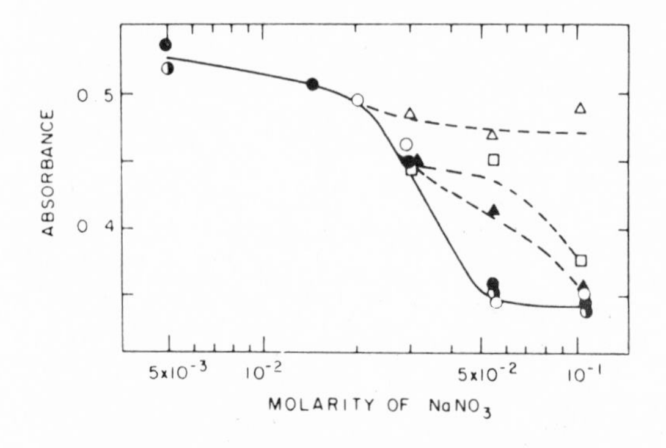

Fig. 8. Reversal of Cu(II) unwinding of DNA with excess electrolyte. The initial samples, containing 5×10^{-5} M(P)DNA, 10^{-4} M $Cu(NO_3)_2$, and 5×10^{-3} M $NaNO_3$, were heated to 50°C at a rate of 1°C per min and allowed to cool to room temperature. Solid sodium nitrate was then added to give the concentrations indicated on the abscissa. The absorbance was measured after: △ 5 min; □ 1 h; ▲ 3 h; ○ 5 h; ◑ ● 24 h. The initial absorbance was 0.355 (from ref. 71).

regenerated as indicated not only by the attainment of the original low absorbance, but by many other criteria, including density gradient sedimentation[71] and biological transforming activity[54] (see Fig. 10).

Since heating DNA in the absence of Cu^{2+} ions results in the dissociation of the single strands in the solution in such a way that the complementary bases are unable to find each other again when conditions are favorable for reassociation to the double helix, the ability of the bases to recognize each other and to reform native DNA must be attributed to copper ions in the denatured state forming cross-links between the DNA strands. Only a few such cross-links between the bases are required to maintain the single strands in close proximity, so that, under thermodynamically favorable conditions, *e.g.* high electrolyte concentration, the double helix can be regenerated.

The equilibrium between DNA denatured in the presence of Cu^{2+} and the double helix is dependent on the concentration of added electrolyte in such a way that an S-shaped curve describes the transition from single-stranded to double-stranded DNA (Fig. 8). Intermediate electrolyte concentrations lead to mixtures of single and double-stranded species. The midpoint in the electrolyte-dependent transition can be named the freezing concentration (C_f)★, analogous to the midpoint in the transition of double to single stranded DNA on heating, which is called the melting temperature (T_m). This curve again emphasizes the opposite effects exerted by different metal ions, in this case Na^+ and Cu^{2+}, on the structure of DNA, and

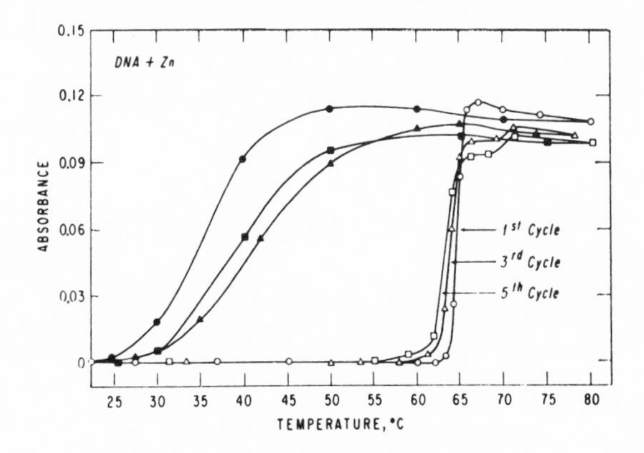

Fig. 9. Reversible unwinding and rewinding of DNA in the presence of Zn(II) by heating and cooling. Solution contains 5 x 10^{-5} M(P)DNA, 2 x 10^{-4} M zinc nitrate, and 5 x 10^{-3} M $NaNO_3$. (○) First heating, (●) first cooling, (△) third heating, (▲) third cooling, (□) fifth heating, and (■) fifth cooling. The second and fourth cycles were omitted for clarity (from ref. 73a).

demonstrates the extent to which the chemistry of DNA is dependent on the concentrations and the nature of the metal ions with which it is associated.

The rewinding of the two DNA strands by Na^+ ions is particularly remarkable because of this demonstration of diverse effects of different metal ions simultaneously acting on DNA, but the rewinding can also take place by removal of the Cu^{2+} ions, either by a complexing agent, *e.g.* EDTA, or by dialysis[54].

Cu^{2+} ions are not unique in their ability to bring about the reversible unwinding and rewinding of DNA. When DNA is heated with zinc, rewinding occurs simply by cooling, as is indicated by the reversal of the absorbance increase induced by heating[73a] (Fig. 9). It can be seen that repeated heating

★ The midpoint has also been termed I_f, for ionic freezing[73].

and cooling of zinc(II) DNA solutions results always in unwinding on heating and rewinding on cooling. The reactions of zinc with DNA can therefore be used to force the equilibrium between single and double-stranded DNA in either direction simply by temperature manipulation.

It is of some interest to consider why copper and zinc ions, although both making possible the rewinding of unwound DNA, do so under such different conditions. Let us consider the equilibrium expressed by the diagram of Fig. 10. The DNA double helix is unwound in the presence of either Cu^{2+} or Zn^{2+}, and both metal ions presumably form crosslinks between the bases of the single strands, since in both cases the double helix can be regenerated. With zinc(II), all that is required for this regeneration is cooling; with Cu(II), cooling is not enough; it is also necessary to increase the electrolyte concentration. The difference can be understood in terms of the relative affinity of these metal ions for the nucleotide bases. Zinc binds less strongly to the bases, and can therefore be readily displaced simply by

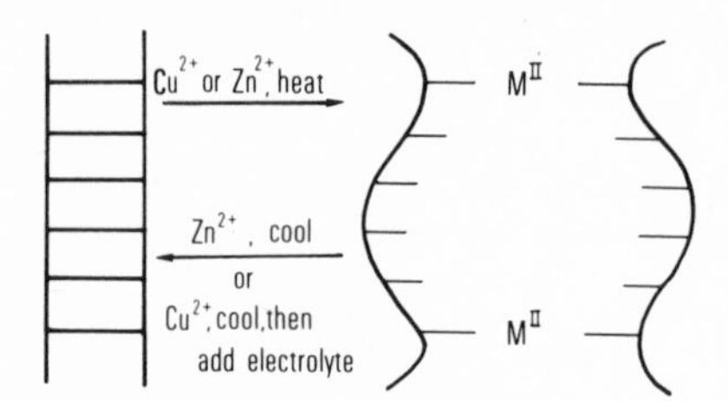

Fig. 10. Diagrammatic scheme for the reversible unwinding and rewinding of DNA in the presence of metal ions.

cooling the solutions; copper binds more strongly, is therefore not displaced by cooling the solution, but can be removed by the more drastic procedure of increasing the concentration of electrolyte.

The difference in the effect of various metal ions on the structure of DNA, as exemplified by the difference between magnesium(II) and copper(II) shown in Fig. 6, is in fact not a dichotomy that places metals into two categories, those that bind phosphate and those that bind bases. Many metal ions bind to both phosphate and base sites, and their effect on DNA depends on their relative affinity for the two types of binding sites. This phenomenon is illustrated in Fig. 11, which compares some transition series metals and magnesium in their effect on the T_m of DNA as a function of metal ion concentration[74]. It can be seen that Mg^{2+} ions continue to augment the T_m of DNA with every increase in Mg^{2+} ion concentration. Such an effect is in line with the known tendency of Mg^{2+} to bind to phosphate and the lack of evidence for Mg^{2+} base binding. Co^{2+} and Ni^{2+} resemble Mg^{2+} in this respect, though these metals effect less of an increase in T_m than Mg^{2+} with increasing concentration. Mn^{2+}, Zn^{2+}, Cd^{2+} and Cu^{2+} all give increasing T_m with initial metal increments, but after reaching a maximum T_m, higher metal concentrations decrease T_m. Evidently all of these metals stabilize DNA at low concentrations by phosphate binding, and this stabilization is counteracted by H-bond-breaking base binding at high metal ion concentrations. This figure can then be taken as an index of the relative affinity of phosphate to DNA base of these metal ions, and places them in the order Mg(II) > Co(II) > Ni(II) > Mn(II) > Zn(II) > Cd(II) > Cu(II).

This order is readily correlated with the effect of these metals on the rewinding of DNA that has been unwound in their presence[74]. In the

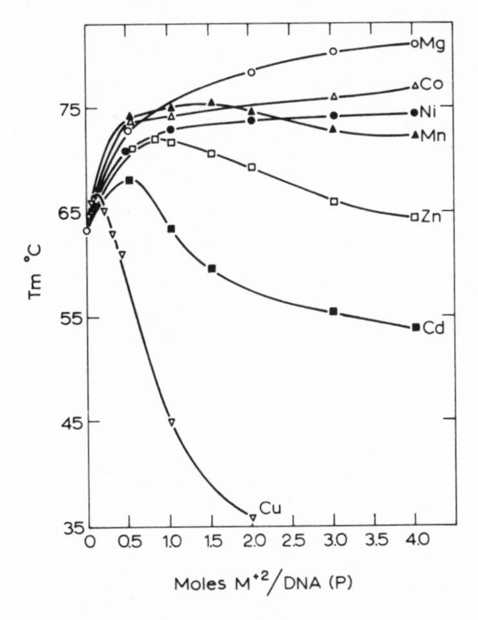

Fig. 11. Variations of T_m of solutions of DNA as a function of divalent metal ion concentration (from ref. 74).

presence of Mg^{2+} there is no rewinding, in the presence of Co^{2+} and Ni^{2+} a small portion of the DNA is rewound, with Mn^{2+} most of the DNA is rewound on cooling and with Zn^{2+} all of the DNA is rewound on cooling. With Cd^{2+} and Cu^{2+} there is no rewinding at all on cooling, but Cd^{2+} gives instantaneous rewinding on addition of 0.1 *M* electrolyte and Cu^{2+} gives rewinding after 5 hours. Thus there is a continuous spectrum of renaturation behavior from Mg^{2+} to Cu^{2+}, and all of this behavior can be explained by an increasing relative affinity for base *vs.* phosphate in this series.

Both replication and transcription require the unwinding of double stranded DNA and its subsequent rewinding. It is tempting to speculate that metal ions, which are necessary for both of these processes, may be partly responsible for the unwinding and rewinding in the biological system, but there is not yet any direct evidence that such is the case.

References pp. 1240–1243

Reaction of Hg(II) and Ag(I) with DNA

We have noted that Hg(II) and Ag(I) ions are very effective in binding the nucleotide bases and that there is no evidence that these metal ions bind phosphate at all (Chapter 33). It could be expected therefore that these ions would be particularly effective in the reversible unwinding and rewinding of DNA. Mercury and silver ions bind the bases so strongly, however, that, instead of unwinding the double strands, these ions in stoichiometric quantity interpolate between the strands so that a very rigid Hg(II) DNA or Ag(I) DNA complex is produced. Every other base pair is apparently linked by metal ions. The mercury and silver ions can be removed only by competing complexing agents (Cl^-, Br^-, CN^-) regenerating native DNA[57]. Thus Hg(II) and Ag(I) exhibit much stronger adhesion to the DNA bases than any of the metals discussed in the preceding section, and Hg and Ag can therefore be placed at the right of the list of metals in decreasing order of phosphate/base binding ability, as follows: Mg(II) > Co(II) > Ni(II) > Mn(II) > Zn(II) > Cd(II) > Cu(II) > Ag(I) > Hg(II).

The reversible reaction of Hg(II) or Ag(I) with DNA cannot be considered as unwinding and rewinding, because the Hg and Ag complexes are not unwound. These complexes can be considered as double helices with a diameter enhanced by the diameter of the metal ions over that of the uncomplexed DNA double helix. The reversible reaction is of great interest because of the selective base binding exhibited by these metal ions and the application of this selectivity to the separation of DNA molecules of varying base content[75].

After the reaction of DNA with mercury had been discovered by Katz[75a], Thomas[76] determined the drastic effect of Hg(II) on the ultraviolet spectrum of DNA, suggesting that mercury was indeed binding to the bases. A comparison of the wavelength shift produced by mercury (11 nm)[57] and copper ions (4 nm)[74] on DNA reveals the very much stronger perturbation of the electronic structure of the heterocycles by the former. The rotatory dispersion of DNA is also much more affected by Hg than by Cu or other metal ions[76a]. Yamane and Davidson[57] noted that the affinity of DNA for mercury increases with AT content, the strongest complex being formed between mercury and the poly dAdT complex[57]. Katz[58] uncovered evidence for the preference of mercury for thymidine sites and assumed the following structure for the Hg(II) poly dAdT complex:

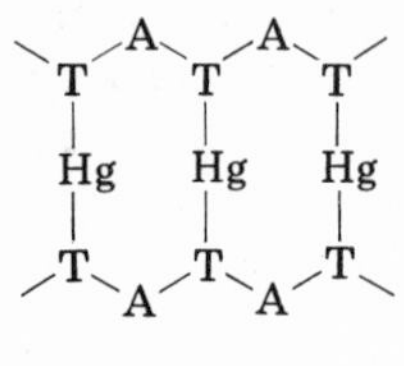

```
 \   A   A   /
  T/  \T/  \T
  |    |    |
  Hg   Hg   Hg
  |    |    |
 /T\  /T\  /T\
    A    A
```

He postulated that some slippage occurs in the association of the single strands to enable Hg(II) ions to be associated exclusively with thymidine.

Just as Hg(II) reacts more strongly with DNA containing a high fraction of AT, so Ag(I) reacts more strongly with DNA that is rich in GC[56]. Davidson *et al.*[56,77] and Daune *et al.*[78] thoroughly investigated the consecutive formation of two types of complexes between DNA and silver. The first complex occurs with no release of protons, and is assumed to involve binding to N-3 or N-7 of guanine, or else π-interaction with guanine and the base adjacent to it[56]. The formation of the second complex which occurs simultaneously at high pH, and consecutively at lower pH, is accompanied by a loss of proton and therefore attributed to binding N-3 of pyrimidines and N-1 of purines. Binding strength increases

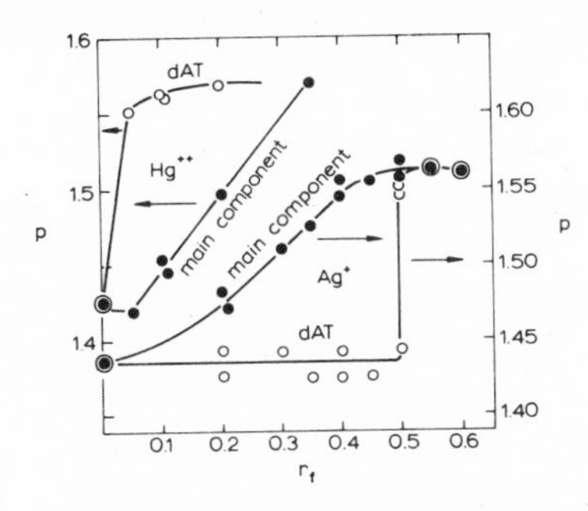

Fig. 12. The opposite effects of Hg(II) and Ag(I) on the densities of the main component and the dAdT from crab DNA (dAT in Figure). Buoyant densities of dAdT and main component of whole crab DNA as a function of r_f, the ratio of added metal ion (silver or mercury) to total DNA phosphorus. Ultracentrifuge runs are at 44,770 rev/min in Cs_2SO_4, ρ = 1.50. The curves in the upper left and the ordinate scale on the left are for Hg^{2+} experiments, the curves in the lower right and the ordinate scale on the right are for Ag^+ experiments (from ref. 74).

with increasing GC content of the DNA[56]. The reaction of DNA with Ag is just as reversible as the reaction of DNA with Hg[56].

The high affinity of mercury for AT rich DNA and of silver for GC rich DNA is highlighted in the separation of the two DNA components from crabs[75]. One of these components is a conventional DNA containing about 60% AT and 40% GC; the other component contains almost exclusively AT, so that it can be considered poly dAdT. Fig. 12 reveals what happens to the buoyant density of these two DNA components in a Cs_2SO_4 density gradient upon ultracentrifugation, in the presence of Hg in one instance, and Ag in the other. Small increments of mercury drastically increase the density of the poly dAdT component before the density of the main component is affected at all. With silver, on the other hand, the main DNA component registers a gradual increase in density until 0.5 mole of Ag have been added per DNA nucleotide, before the density of the poly dAdT is affected[75]. These experiments show dramatically that Hg and Ag preferentially affect different portions of the DNA molecule. The differential binding capacity of DNA of varying GC content for Hg and Ag has been applied by Davidson *et al.* in the separation of the dAdT component of crab DNA from the main component[56,75,79]. The potential of these complexing reactions for DNA separations is very great, as shown by its use in the separation of the two halves of DNA molecules[80] and of heterogeneous human DNA fractions[81].

The crosslinking of DNA strands by Hg(II) ions was confirmed[82,83] by studying the reaction of CH_3HgOH with DNA. Since methyl groups occupy one of the two linear coordination positions of mercury in this molecule, only one binding site, instead of two, is available for attachment to the nucleotide bases, after loss of hydroxyl ions; consequently, the mercury in CH_3Hg^+ cannot hold two strands together. It is reasonable to expect, therefore, that the reaction of native DNA with CH_3HgOH, followed by removal of the methyl mercury ions, will not regenerate native DNA. It is, in fact, observed, that methyl mercury ions react with DNA in a manner similar to Hg(II), prefering AT-rich DNA, and binding first to N-3 of thymine and then to N-1 of guanine. Removal of the methyl mercury does not, however, regenerate the double helix. The failure of methyl mercury ions to make possible the reformation of double helix, while reacting in a similar way as Hg(II) with the single strands, strongly suggests that the Hg(II) ions form cross-links between the DNA strands.

It is worth pointing out that the crosslinking function of the metal ions in the regeneration of native DNA from its complex with Hg(II), Cu(II) or Zn(II) is reminiscent of the ability of metal ions to condense small ligands in the manner governed by the stereochemistry of the metal ion, sometimes called the template effect in coordination chemistry.* The metal ions organize the single strands in a conformation that makes possible a reformation of double helix (Fig. 10) that is not possible in the absence of metal ions (Fig. 7).

EFFECT OF METAL IONS ON RNA

Stabilization

It is a comparatively easy task to study the effect of metal ions on the ordered structure of DNA, since ordered DNA is a double helix, and the denaturation of ordered DNA brings about the formation of randomly coiled single strands. Studies of transitions between ordered and disordered

---

* Not to be confused with the template activity of DNA in arranging nucleotide monomers during replication and transcription.

forms of RNA are not quite so clear-cut, because RNA is generally not double stranded and "order" is produced by hydrogen-bonding between complementary bases of the same strand (as in Fig. 7). The ordered conformations of RNA are nevertheless stabilized by metal ions in the same way as the double helix of DNA, as shown by the increase in T_m that divalent metal ions induce in RNA preparations[84-88]. The extraction of a large variety of metal ions with RNA preparations has been reported[89,90] some time ago, but the significance of this finding is not understood.

The structure of tRNA is now better understood than that of any other naturally occurring RNA molecules. All tRNA molecules that have been studied have been shown to contain several stretches of hydrogen bonded regions. It is thus quite apparent that the ordered structure of such a molecule can be stabilized by metal ions in the same way as the DNA double helix. Lindahl *et al.*[91] have demonstrated that "denatured" leucyl- tRNA, which has lost its ability to bind leucine, completely regains such activity through the addition of Mg^{2+}; other divalent ions, *e.g.* Co^{2+}, Ca^{2+} and Mn^{2+}, partly reactivate the tRNA. tRNA so reactivated by metal ions in turn can be made to lose its activity by treatment with EDTA. The renatured tRNA is not identical with the native form, since the latter cannot be deactivated by EDTA. Nevertheless, it is evident that metal ions are essential in converting an inactive form of tRNA to an active form, and the evidence indicates that the metal ions are bound to specific sites on the tRNA. Cohn *et al.*[92] have confirmed the existence of strong binding sites (6–7 per molecule) for Mn^{2+}, using n.m.r. water proton enhancement studies, while at the same time revealing the presence of weak Mn^{2+} binding sites to the extent of $\frac{1}{2}$ mole/mole of nucleotide phosphate. It is not yet clear exactly how the metal ions produce the active tRNA form, but perhaps metal ions are needed both for stabilizing specific sites and also for a generalized counterion effect stabilization of the ordered structure.

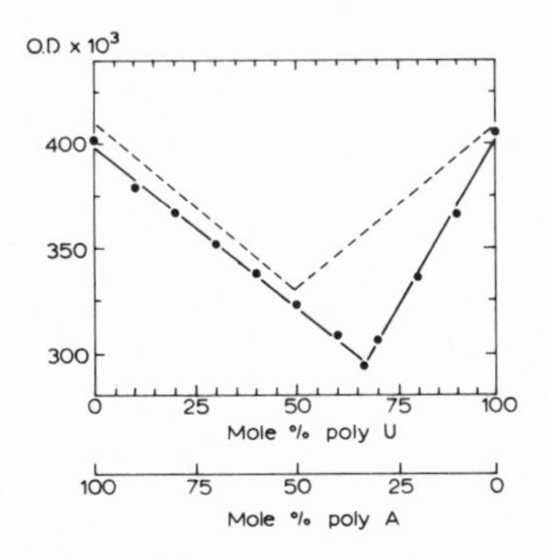

Fig. 13. Continuous variation studies demonstrating the formation of poly rA · poly rU in the presence of 0.1 *M* NaCl (dashed line) and of poly rA · poly rU after the addition of 1.2 x 10^{-2} *M* $MgCl_2$ (solid line) (from ref. 93).

References pp. 1240–1243

The ability of metal ions to influence the structure of polynucleotides is very much in evidence in the interaction of different polynucleotide strands to form multiple stranded helices. Felsenfeld and Rich[93] furnished a striking demonstration of the importance of metal ions in determining what kind of multiple helix is produced. In the presence of 0.1 M NaCl poly rA and poly rU combine to form a double stranded helix of poly rA · poly rU. When 1.2×10^{-2} M Mg^{2+} is added to such a solution the double stranded helix combines with a second strand of poly rU to form the triple stranded helix, poly rA · 2 poly rU. These reactions are summarized in Fig. 13, which illustrates the continuous variation studies for the formation of the double- and triple-stranded helices.

Stabilization and destabilization of different conformations by different metal ions

Recent studies by Shin and Eichhorn[94] have demonstrated not only how sensitive polyribonucleotides are to the presence of metal ions, but also how different metal ions can stabilize different forms of the polynucleotides under otherwise identical conditions. Consider, for example, the effect of a variety of metal ions on the optical rotatory dispersion (o.r.d.) of polyriboadenylic acid (poly rA), as shown in Fig. 14. At pH 6 this molecule, in the absence of divalent metal ions, is stable in a double helical configuration. The double helix is only partly converted to single helix on addition of zinc(II), but the presence of nickel(II) produces an o.r.d. curve characteristic of a single stranded helix, while the addition of copper(II) leads to the formation of randomly coiled single strands.

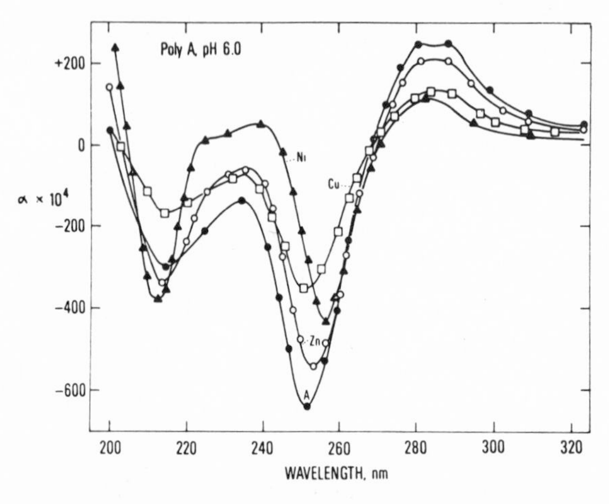

Fig. 14. Optical rotatory dispersion curves of poly rA in the absence of divalent metal ions and in the presence of Zn(II), Ni(II), and Cu(II).

Another striking illustration[94] of the differing effects of metal ions on polyribonucleotides is provided by the various degrees in which different metal ions influence the helix → random coil equilibrium of poly rA at pH 7, where a single helix is the generally stable form. Fig. 15 is a plot of the decrease in optical rotation at 257 nm as a function of temperature; the fall in α indicates the decreasing amount of helix present. Various points on this curve indicate the corresponding decrease in helicity produced at room temperature by the addition of a number of metal ions. It can be seen that these metal ions at room temperature can bring the helix–coil equilibrium to positions corresponding to those at various temperatures when no metal ions are present.

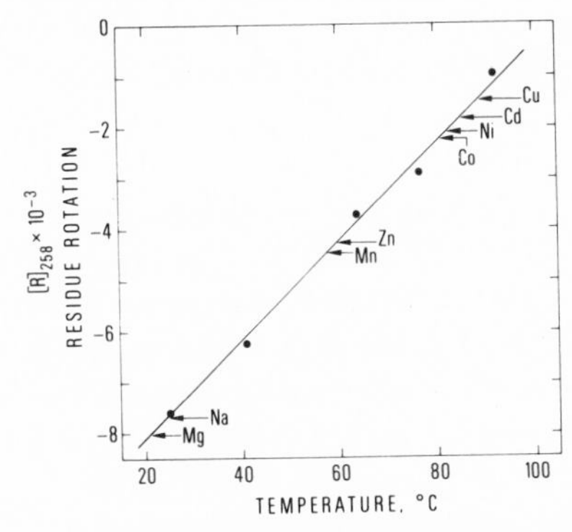

Fig. 15. Helix → coil transition of poly rA by heat or metal ions. The straight line shows the decrease in rotation, corresponding to unwinding of the helix, as a function of temperature. The effect of various metal ions at room temperature is indicated by the arrows. The Na arrow represents the rotation at room temperature in the absence of divalent ions (from ref. 94).

Destabilization of ordered structure of polyribonucleotides by metal ions

RNA and polyribonucleotides, like DNA, are very differently affected by various metal ions, as we have just noted. The phenomena described by Figs. 14 and 15 are evidence that the ordered structures of polyribonucleotides, like that of DNA, can be destabilized by metal ions.

We have seen (Fig. 15) that Mg^{2+} ions are capable of stabilizing ordered conformations of polynucleotides, just as Mg^{2+} ions stabilize the DNA double helix. Cu^{2+} ions destabilize native DNA, and we have already seen that they also destabilize double helical poly rA.

Copper ions have a destabilizing influence upon multiple helical structures of polynucleotides in general. Eichhorn and Tarien[95] have demonstrated that poly rA · poly rU will not form in the presence of excess Cu(II) per mole of nucleotide phosphate. Moreover, if poly rA ·

poly rU is allowed to form in the absence of Cu(II), the subsequent addition of Cu(II) causes the unwinding of the double helix. Similarly, the poly rI · poly rC double helix is prevented from forming, when Cu(II) is present in the mixture of polynucleotides, and the double helix is unwound by adding Cu(II) to preformed poly rI · poly rC. Moreover, just as the unwinding of DNA by Cu(II) is reversible, the poly rI · poly rC unwound by Cu(II) may be rewound by the addition of electrolyte. RNA and DNA then are very similar in their susceptibility to either the stabilization or the destabilization of ordered structures by metal ions.

Depolymerization of RNA and polyribonucleotides by metal ions

RNA and DNA do not by any means exhibit similar behavior in all of their reactions with metal ions. There is one reaction that proceeds exclusively with RNA and virtually does not affect DNA — the depolymerization of the polynucleotide structure by the cleavage of phosphodiester bonds[95a].

Metal hydroxides have been used to degrade RNA, leading to the formation of a variety of end products, including nucleosides and nucleotides, and have indeed been employed for the synthesis of these substances[96-99]. $Cd(OH)_2$ and $Zn(OH)_2$ produce nucleotides, $Pb(OH)_2$ yields nucleosides, while $Bi(OH)_3$ and $Al(OH)_3$ lead predominantly to dinucleoside phosphate, dinucleotides, *etc.* The hydrolysis of RNA takes place in the presence of a large variety of metal ions, such as the rare earths[100,101], thorium[102], as well as many divalent ions[103-106].

The rate of degradation of polynucleotides varies markedly with the nature of the metal ions, as is shown when the formation of acid soluble fragments is compared for the metal ions of the first transition series[106]. The rate of degradation with Zn(II) ions is very much greater than with the other metal ions.

A comparison of the rates of degradation of RNA, poly rA, poly rC, poly rU, and poly rI by zinc ions reveals that the rates vary considerably, and particularly the degradation of poly rI is very much more sluggish than that of the other polynucleotides. It is thus apparent that the cleavage of the phosphodiester linkages of the polynucleotide chain is influenced by the nature of the bases in the chain, and that metal ions can distinguish between phosphodiester bonds in different homopolyribonucleotides.

The question arises whether metal ions can also differentiate between phosphodiester bonds in a heteropolynucleotide, like RNA, in which different bases are attached to the same ribose phosphate backbone. Recent experiments have demonstrated indeed that metal ions favor cleavage of the phosphodiester bonds adjacent to uracil, and under some conditions, cytosine, while tending not to attack those bonds adjacent to guanine[107].

The reaction conditions have been varied in such a way as to achieve optimal differentiation between the RNA bases; at pH 7, 64°C, and a zinc/phosphate ratio of 20:1, the following cleavage pattern emerges:

	Cleavage sites (%)	RNA composition
G	9	28
A	23	25
C	30	19
U	40	28

It is apparent that the number of bonds adjacent to guanine that are cleaved is considerably smaller than expected from the amount of guanine originally in the RNA and that the number of bonds adjacent to uracil and cytosine that are cleaved is considerably greater than the composition of the RNA would lead one to anticipate. This ability of zinc ions to recognize the bases bound to the linkages to be cleaved is potentially useful, in conjunction with enzymatic cleavage, for the determination of the sequence of bases in a polynucleotide chain. The metal ions mimic enzymes in exhibiting specificity toward this cleavage reaction. Other metals may exhibit specificities different from zinc; Pb(II), for example, degrades the homopolynucleotides in the order poly A > poly U > poly I > poly C[108]. Lead, in fact, degrades RNA faster than any other metal ion studied[108].

The fact that the degradation reaction occurs with RNA, but not with DNA, has led to the assumption that the 2′-hydroxyl group, present in RNA but not in DNA, must be involved. The logical intermediate proposed by a number of investigators is one in which the metal ion is chelated to the phosphate group and the 2′-hydroxyl group[100,101,103]. The cleavage reaction would proceed from this intermediate, when the positively charged zinc ions withdraw electrons from the phosphate group, weakening the phosphodiester linkage and then breaking it.

It has been established that, when RNA is degraded by techniques other than the use of metal ions, *i.e.*, enzymatic, acid, or alkaline hydrolysis, a 2′,3′-cyclic phosphate intermediate is generally produced. When the products from the zinc degradation of poly rA are analyzed, it is found that cyclic phosphate products are indeed produced by zinc degradation also, challenging the chelate intermediate hypothesis[109].

The formation of a cyclic phosphate intermediate would be explained by the following mechanism (Fig. 16): zinc ions binding to the phosphate would place a positive dipole upon the phosphorus, which would then attract the negative dipole on the 2′-hydroxyl group, producing the cyclic phosphate. The resulting structure with five bonds to the phosphorus would constitute an unstable arrangement that would be relieved by cleavage of the phosphodiester linkage.

To confirm the cyclic phosphate intermediate for the zinc degradation reaction, the kinetics of degradation of a number of trinucleotides were followed[109], as indicated for tri-inosinic acid (IpIpI) in Fig. 17. It can be seen that the formation of inosine-3′-phosphate (I3′p) is S-shaped, exhibiting

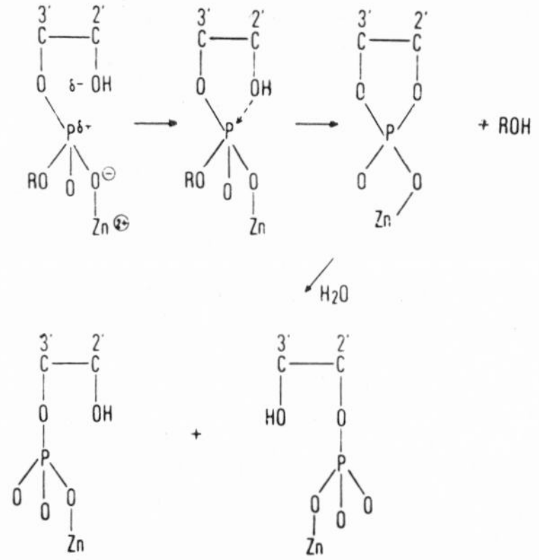

Fig. 16. Mechanism of cleavage of phosphodiester bonds in RNA by Zn^{2+} (from ref. 109).

an initial lag followed by an accelerated increase. The formation of inosine 3′,5′-cyclic phosphate (I > p) precedes that of the open chain phosphate,

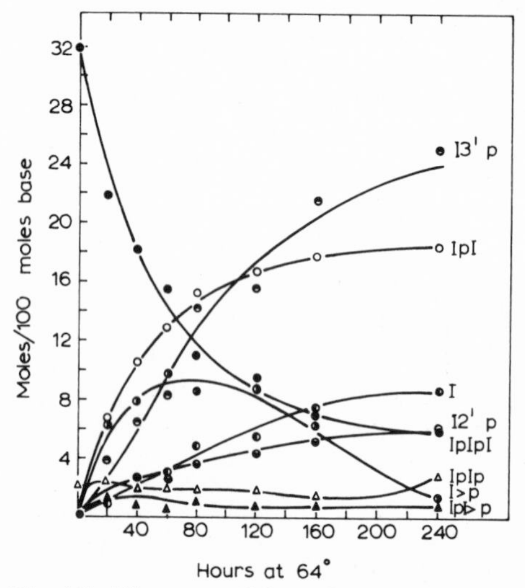

Fig. 17. The rates of degradation of the trimer and formation of products in the degradation of tri-inosinic acid (IpIpI) by Zn(II) ions. Compare the curve for I3′p and I > p (from ref. 109).

passes through a maximum, and eventually the cyclic phosphate concentration reaches zero, while the open chain phosphate (Ip) has levelled off. Since the degradation of IpIpI is typical of the degradation of all the trinucleotides, it can be concluded that the intermediate production of the cyclic phosphate, and not chelation involving the 2′-hydroxyl group, is a general characteristic of the degradation by zinc ions of the phosphodiester linkage.

The depolymerization of RNA by metal ions is thus mechanistically similar to the enzymatic degradation of RNA. Another resemblance between zinc and the enzyme in the degradation reaction is the inhibition of the reaction by other metal ions. Enzymes are characteristically inhibited by metal ions. The zinc depolymerization of ribopolynucleotides is dramatically inhibited by base-binding metal ions, *e.g.* silver[110]. In fact, the degradation of poly I and poly U is virtually prevented in the presence of silver. A possible explanation of the inhibitory effect of silver ions on the zinc degradation is that, while zinc ions binding to phosphate tend to weaken the phosphodiester links, the whole polynucleotide structure is strengthened by the attachment of silver ions to the bases. The analogy to the inhibition of enzymatic degradation by metal ions may be coincidental, since the inhibition of enzymes frequently results from the denaturation of the protein by metal ions.

EFFECT OF METAL IONS ON THE ENZYMATIC DEGRADATION OF RNA AND DNA

While metal ions alone are effective only for the depolymerization of RNA, there are enzymes that specifically depolymerize either RNA (ribonucleases, or RNAses,) or DNA (deoxyribonucleases, or DNAses). It is of some interest to compare the effect of metal ions on these two types of enzymes.

The most thoroughly investigated DNA degrading enzyme is DNAse I isolated from bovine pancreas. This enzyme requires metal ions for its activity. It is activated by a large number of metal ions[111-115]; Mg(II) and Mn(II) are the most frequently used activators, although Co(II) is a much better activator than either of these[116]. The mechanism of the degradation of DNA by DNAse I can be quite different in the presence of different metal ions[117]. With Mg(II), there is an initial lag in the hydrolysis of the phosphodiester bonds, which is attributed to a "double hit" mechanism, in which only one chain of the double helix is cleaved at a time, thus requiring a chance hit at the same location on the opposite chain to effect cleavage of the molecule. In the presence of Mn(II), Ca(II) or Co(II) there is no lag in cleavage, and it is apparent that these metal ions bring about simultaneous cleavage of the two chains.

Just as Mg(II) and Mn(II) differ in their effects on the mechanism of action of DNAse I, they also differ in their selectivity for phosphodiester bonds. In a manner reminiscent of the effects of these metals on DNA and

References pp. 1240–1243

RNA polymerase, Mn(II) ions render the reaction less selective than do Mg(II) ions. DNAse I degradation of poly dG · poly dC, will occur with Mn(II), but not with Mg(II)[118]. The ability of Mn(II), in contrast to Mg(II), to facilitate cleavage of both chains simultaneously at the same site could be due to the lower selectivity of cleavage sites by Mn ions[117].

Although the Mg(II) dependent DNAse activity exhibits some selectivity dependent on the nucleotide composition, the reaction is relatively non-specific in comparison with the metal degradation of polyribonucleotides. It has been possible to enhance specificity of the DNAse reaction very markedly by using Cu, instead of Mg, as the activating metal ion. The following Table[119] reveals that Cu(II) ions in conjunction with DNAse elicit a specificity toward cleavage in DNA similar to the specificity in the non-enzymatic cleavage of RNA by zinc ions. The cleavage at G is dramatically reduced, while cleavage at T is dramatically increased. Possibly metal ions binding to the DNA bases inhibit the hydrolysis of the bonds adjacent to the bound bases. Since Cu(II) ions bind most readily to G and least readily to T, these data can be rationalized in this way.

	% Composition in DNA	Cleavage sites with Mg(II) DNAse	Cleavage sites with Cu(II) DNAse
G	22	24	8
A	28	20	16
C	22	18	23
T	28	37	53

Aposhian *et al.*[120] have studied an interesting DNA degrading enzyme (SP3 DNAse) that works from the 5′ terminal and splits off one dinucleotide after another. This enzyme generally requires both Mg(II) and Mn(II), although reactions of some small oligomers will occur in the presence of Mg(II) only.

The cleavage of RNA by the widely studied (bovine pancreatic) enzyme ribonuclease does not require metal ions for its activity. Nevertheless, metal ions at appropriate concentrations are effective activators of ribonuclease[116,121–124], as shown in Fig. 18. It can be seen that all of the metals activate at low metal concentration, but finally, above an optimum concentration inhibit the enzyme. The metals that generally coordinate most strongly reach optimum activating ability at low concentration, while the weaker coordinating metal ions activate maximally at higher metal concentrations. When the metal ions are placed in the order of metal concentrations for maximum activation, the order is Mn $>$ Co $>$ Ni $>$ Cu $<$ Zn; this is the opposite of the Irving-Williams stability series

(Chapter 2). The shapes of the curves can be explained by the fact that the more strongly binding metal ions coordinate most strongly not only to the activating site, but also to the inhibiting site. The more strongly binding metals then inhibit at the lower concentrations, thus accounting for the correlation with the Irving–Williams series. The activating sites are probably DNA phosphate groups, whereas the inhibiting sites may be on the enzyme, whose binding characteristics with metal ions have been thoroughly investigated (Chapter 7).

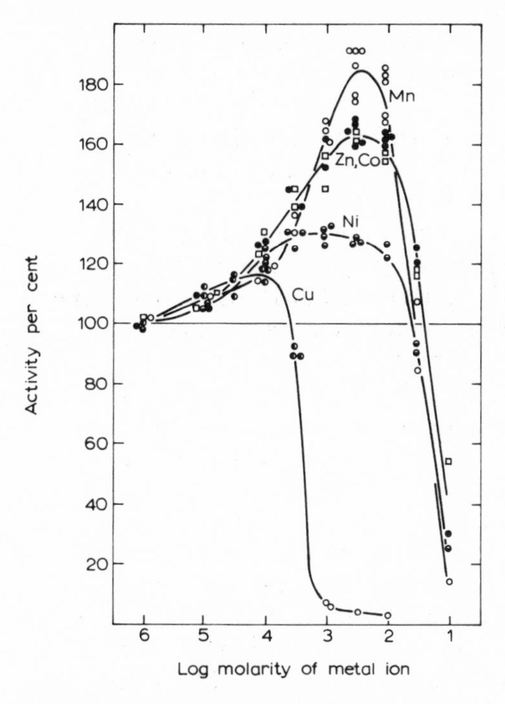

Fig. 18. The effect of transition metals on the activity of ribonuclease (from ref. 116).

METAL IONS AND NUCLEOPROTEINS

We have already noted that the DNA double helix is unstable in the absence of positively charged counter ions. We have seen that the addition of phosphate-binding metal ions stabilizes the double helix against unwinding to random coils. The stabilization of the double helix in the cell is accomplished by binding to proteins which contain large amounts of the positively charged lysine and arginine amino acids. The combination of the nucleic acid with protein is called nucleoprotein.

The association of the protein with DNA blocks transcription of messenger RNA. This circumstance gives rise to a popular notion about the mechanism of cellular differentiation. This mechanism rationalizes the paradox that DNA molecules carry all of the genetic information of an organism throughout its life span, but that different portions of this information are expressed during different portions of this life span. A possible explanation of the "turning on" and "turning off" of different portions of DNA at the proper moments is that the protein part of the nucleoprotein blocks the expression of the DNA code and that the removal of the protein from any portion of DNA removes the block, thus permitting transcription of that segment of DNA.

Since both metal ions and histones (positively charged nuclear proteins) can bind to the phosphate groups of DNA, it is not surprising that metal ions can be instrumental in the removal of protein from DNA, by competition for the phosphate groups. The stepwise addition of electrolyte to nucleoprotein indeed results in a stepwise removal of histones from the DNA molecule[125].

Different cations have different specific effects in their ability to remove a polylysine derivative from poly A · poly U and poly I · poly C[126]. The following metal ions, Mg(II), Ca(II) and Mn(II), as well as some alkali metal ions, preferentially remove the polypeptide from poly A · poly U, indicating a stronger affinity of these metal ions for this polynucleotide structure over that of poly I · poly C. On the other hand, some organic cations such as lysine discriminate little between the two types of polymers, whereas arginine and histidine exhibit a preference for poly I · poly C.

Although metal ions can compete with positively charged proteins for binding to DNA and thus destabilize nucleoproteins containing large amounts of lysine and arginine, divalent ions are required for the tight packing of other nucleoproteins[127], such as ribosomes[128,129], which consist of ribosomal RNA (rRNA) bound to protein. We have previously noted the importance of ribosomes in the recognition of amino-acid specific tRNA molecules by mRNA. These ribosomes consist of particles which, in *E. coli*, are known as 30S and 50S ribosomes, designated by their ultracentrifugal sedimentation behavior. If 10^{-4} *M* Mg(II) is present, the 30S and 50S ribosomes are not bound together, but at concentrations of Mg(II) higher than 10^{-4} *M*, these particles associate to a larger ribosome with a sedimentation of 70S and at 10^{-2} *M* Mg(II) the association also produces some 100S ribosomes[129]. It is believed that the 30S ribosome is required in the recognition of tRNA by mRNA and that the 50S ribosome is needed for the formation of peptide linkages (see Fig. 3). The two processes are, of course, related and probably require the association of the two ribosomal components. Heating the 70S ribosome causes reversal of the association process without loss of Mg(II)[130].

The association of metals with ribosomes occurs almost exclusively through the rRNA[131]. Removal of metal ions not only causes dissociation of 70S particles into 30S and 50S components, but also brings about the unwinding of the smaller fragments[128,132-134]. Readdition of metal ions (*e.g.* Mg^{2+}, Zn^{2+} and Ni^{2+}) rewinds the ribosomal particles in a manner reminiscent of the reversible unwinding and rewinding of DNA[134].

Not only are divalent metal ions required for the association of the ribosomal components with each other; they are also essential for the binding of tRNA to the ribosome[129,135-137]. Thus the importance of metal ions in protein synthesis is involved in many aspects of the synthetic process.

ACKNOWLEDGEMENT

I am grateful to Drs. Nathan A. Berger, James J. Butzow, Patricia Clark, Josef Pitha, Joseph Rifkind, and Yong A. Shin for helpful comments.

NOTE ADDED IN PROOF

Scrutton *et al.*[139] have demonstrated that RNA polymerase from *E. coli* contains two tightly bound zinc atoms per molecule of enzyme. Salser *et al.*[140] have utilized the ability of DNA polymerase I in the presence of Mn^{2+} to insert ribonucleotides into DNA in an ingenious effort to develop a technique for the determination of DNA base sequences. The idea is to furnish one base as a ribonucleoside triphosphate and all other bases as deoxynucleoside triphosphates. The selected base is then incorporated into DNA as a ribonucleotide and can be selectively cleaved by alkali[140], or, as might be suggested here, by metal ions. Sander and Ts'o[141] have determined stability constants for complexes of Mg^{2+} with various polynucleotides; the results indicate remarkable stability for the complex with tRNA. Schreiber and Daune[142] have postulated two types of binding sites for the reaction of Cu(II) with DNA, and Liebe and Stuehr[143] have reexamined the increase of T_m at low Cu(II) concentration, attributed to stabilization by phosphate binding, and have noted that such an increase does not occur when equilibrium Cu(II) concentrations are plotted. Eichhorn *et al.*[144] have carried out a series of studies that demonstrate that high concentrations of divalent metal ions can induce the mispairing of nucleotide bases, thus constituting a source of error in the propagation of genetic information.

References pp. 1240–1243

REFERENCES

1 C. C. Richardson, C. L. Schildkraut, H. V. Aposhian and A. Kornberg, *J. Biol. Chem.*, 239 (1964) 222.
2 P. Berg, H. Fancher and M. Chamberlain, in H. Vogel, *Symposium on Informational Macromolecules*, Academic Press, New York, 1963, p. 467.
3 P. T. Englund, J. A. Huberman, T. M. Jovin and A. Kornberg, *J. Biol. Chem.*, 244 (1969) 3038.
4 J. P. Slater, A. S. Mildvan and L. A. Loeb, personal communication.
5 U. Weser, *FEBS Lett*, 7 (1970) 356.
6 U. Weser and E. Bischoff, *Eur. J. Biochem.*, (1970) in press.
7 P. T. Englund, R. B. Kelly and A. Kornberg, *J. Biol. Chem.*, 244 (1969) 3045.
8 K. I. Kato, J. M. Goncalves, G. E. Houts and F. J. Bollum, *J. Biol. Chem.*, 242 (1967) 2780.
9 L. M. S. Chang and F. J. Bollum, *Federation Proc.*, 29 (1970) 406.
10 H. K. Miller, N. Valanju and M. E. Balis, *Biochemistry*, 4 (1965) 1295.
11 E. Fuchs, R. L. Millette, W. Zillig and G. Walter, *Eur. J. Biochem.*, 3 (1967) 183.
12 T. L. Steck, M. J. Caicuts and R. G. Wilson, *J. Biol. Chem.*, 243 (1968) 2769.
13 C. F. Fox, W. S. Robinson, R. Haselkorn and S. B. Weiss, *J. Biol. Chem.*, 239 (1964) 186.
14 T. Nakamoto, C. F. Fox and S. B. Weiss, *J. Biol. Chem.*, 239 (1964) 167.
15 B. B. Biswas and R. Abrams, *Biochim. Biophys. Acta* 55 (1962) 827.
16 G. Richter, *Biochim. Biophys. Acta*, 72 (1963) 342.
17 R. L. Hancock, R. F. Zells, M. Shaw and H. G. Williams-Ashman, *Biochem. Biophys. Res. Commun.*, 55 (1962) 257.
18 R. J. Mans and G. D. Novelli, *Biochim. Biophys. Acta*, 91 (1964) 186.
19 P. Ballard and H. G. Williams-Ashman, *Nature*, 203 (1964) 150.
19a H. G. Klemperer and G. R. Haynes, *Biochem. J.*, 104 (1967) 537.
20 C. F. Fox and S. B. Weiss, *J. Biol. Chem.*, 239 (1964) 175.
21 M. Chamberlain and P. Berg, *Proc. Natn. Acad. Sci. U.S.*, 48 (1962) 81.
22 A. Stevens, *J. Biol. Chem.*, 239 (1964) 204.
23 D. A. Smith, R. L. Ratliff, D. L. Williams and A. M. Martinez, *J. Biol. Chem.*, 242 (1967) 590.
24 I. Haruna and S. Spiegelman, *Proc. Natn. Acad. Sci. U.S.*, 54 (1965) 1189.
25 B. Zmudzka, C. Janion and D. Shugar, *Biochem. Biophys. Res. Commun.*, 6 (1969) 895.
26 A. G. So, E. W. Davie, R. Epstein and A. Tissieres, *Proc. Natn. Acad. Sci. U.S.*, 58 (1967) 1739.
27 C. F. Fox, R. I. Gumport and S. B. Weiss, *J. Biol. Chem.*, 240 (1965) 2101.
28 C. C. Widnell and J. R. Tata, *Biochem. J.*, 92 (1964) 313.
29 J. J. Furth, M. Rosenberg and P. L. Ho, *J. Cell. Physiol.*, 69 (1967) 209.
30 P. O. Ballard and H. G. Williams-Ashman, *J. Biol. Chem.*, 240 (1966) 1602.
31 S. Liao, D. Sagher, A. H. Lin and S. Fang, *Nature*, 223 (1969) 297.
32 F. Novello and F. Stirpe, *Biochem. J.*, 111 (1969) 115.
33 A. Halbreich and E. Weissenberg, *Proc. Israel J. Chem.*, 5 (1967) 118p.
34 M. W. Nirenberg and J. H. Matthaei, *Proc. Natn. Acad. Sci. U.S.*, 52 (1961) 1494.
35 M. W. Nirenberg and O. W. Jones, Jr., in H. J. Vogel, V. Bryson and J. O. Lampen, *Informational Macromolecules*, Academic Press, New York, 1963, p. 451.
36 W. Szer and S. Ochoa, *J. Mol. Biol.*, 8 (1964) 823.
37 S. Nishimura, F. Harada and M. Hirabayashi, *J. Mol. Biol.*, 40 (1969) 173.
38 F. H. C. Crick, *J. Mol. Biol.*, 19 (1966) 548.
39 J. Gordon and F. Lipmann, *J. Mol. Biol.*, 23 (1967) 23.

40 C. Ropars and R. Viovy, *J. Chim. Phys.*, (1965) 408.
41 S. E. Bryan and E. Frieden, *Biochemistry*, 6 (1967) 2728.
42 V. K. Srivastava, *Ind. J. Biochem.*, 6 (1969) 93.
43 D. Bach and I. R. Miller, *Biopolymers*, 5 (1967) 161.
44 K. B. Yatsimirskii, E. E. Kriss and T. I. Akhrameeva, *Dokl. Akad. Nauk. U.S.S.R.*, 168 (1966) 840.
45 J. Shack and B. S. Bynum, *Nature*, 184 (1959) 635.
46 J. Eisinger, I. Fawaz-Estrup and R. G. Shulman, *J. Chem. Phys.*, 42 (1965) 43.
47 A. M. Willemsen, *Thesis*, Catholic University of Nijmegen, Netherlands, 1969, p. 93.
48 N. A. Berger and G. L. Eichhorn, *Biochemistry*, 10 (1971) 1847.
49 N. A. Berger and G. L. Eichhorn, *Biochemistry*, 10 (1971) 1857.
50 G. L. Eichhorn, P. Clark and E. D. Becker, *Biochemistry*, 5 (1966) 245.
51 H. Fritzsche and C. Zimmer, *Eur. J. Biochem.*, 5 (1968) 42.
52 H. Fritzsche, *Studia Biophysica*, (1968) 31.
53 C. Zimmer and H. Venner, *Studia Biophysica*, (1967) 207.
54 S. Hiai, *J. Mol. Biol.*, 11 (1965) 672.
55 J. H. Coates, D. O. Jordan and V. K. Srivastava, *Biochem. Biophys. Res. Commun.*, 20 (1965) 611.
56 R. H. Jensen and N. Davidson, *Biopolymers*, 4 (1966) 17.
57 T. Yamane and N. Davidson, *J. Am. Chem. Soc.*, 83 (1961) 2599.
58 S. Katz, *Biochim. Biophys. Acta*, 68 (1963) 240.
59 M. Fiskin and M. Beer, *Biochemistry*, 4 (1965) 1289.
60 R. G. Shulman, H. Sternlicht and B. J. Wyluda, *J. Chem. Phys.*, 43 (1965) 3116.
61 G. Zubay and P. Doty, *Biochim. Biophys. Acta*, 29 (1958) 47.
62 G. Zubay, *Biochim. Biophys. Acta*, 32 (1959) 233.
63 M. M. Fishman, J. Isaac, S. Schwartz and S. Stein, *Biochem. Biophys. Res. Commun.*, 29 (1967) 378.
64 J. Eisinger, R. G. Shulman and B. M. Szymanski, *J. Chem. Phys.*, 36 (1962) 1721.
65 L. E. Perlgut and K. Amemiya, *Arch. Biochem. Biophys.*, 132 (1969) 370.
66 G. Felsenfeld and S. Huang, *Biochim. Biophys. Acta*, 34 (1959) 234.
67 R. Thomas, *Trans. Faraday Soc.*, 50 (1954) 304.
68 J. Shack, R. J. Jenkins and J. M. Thompsett, *J. Biol. Chem.*, 203 (1953) 373.
69 W. F. Dove and N. Davidson, *J. Mol. Biol.*, 5 (1962) 467.
70 G. L. Eichhorn, *Nature*, 194 (1962) 474.
71 G. L. Eichhorn and P. Clark, *Proc. Natn. Acad. Sci. U.S.*, 53 (1965) 586.
72 H. Venner and C. Zimmer, *Biopolymers*, 4 (1966) 321.
73 B. Baranowska, T. Baranowski and M. Laskowski, Sr., *Eur. J. Biochem.*, 4 (1968) 345.
73a Y. A. Shin and G. L. Eichhorn, *Biochemistry*, 7 (1968) 1026.
74 G. L. Eichhorn and Y. A. Shin, *J. Am. Chem. Soc.*, 90 (1968) 7323.
75 N. Davidson, J. Widholm, U. S. Nandi, R. Jensen, B. M. Olivera and J. C. Wang, *Proc. Natn. Acad. Sci. U.S.*, 53 (1965) 111.
75a S. Katz, *J. Am. Chem. Soc.*, 74 (1952) 2238.
76 C. A. Thomas, *J. Am. Chem. Soc.*, 76 (1954) 6052.
76a P. Y. Cheng, *Biochim. Biophys. Acta*, 102 (1965) 314.
77 T. Yamane and N. Davidson, *Biochim. Biophys. Acta*, 55 (1962) 609.
78 M. Duane, C. A. Dekker and H. K. Schachman, *Biopolymers*, 4 (1966) 51.
79 U. S. Nandi, J. C. Wang, N. Davidson, *Biochemistry*, 4 (1965) 1687.
80 J. C. Wang, U. S. Nandi, O. S. Hogness and N. Davidson, *Biochemistry*, 4 (1965) 1697.
81 G. Corneo, E. Ginelli and E. Polli, *J. Mol. Biol.*, 48 (1970) 319.

82 D. W. Gruenwedel and N. Davidson, *J. Mol. Biol.*, 21 (1966) 129.
83 D. W. Gruenwedel and N. Davidson, *Biopolymers*, 5 (1967) 847.
84 K. Fuwa, W. E. C. Wacker, R. Druyan, A. F. Bartholomay and B. L. Vallee, *Proc. Natn. Acad. Sci. U.S.*, 46 (1960) 1298.
85 A. Tissieres, *J. Mol. Biol.*, 1 (1959) 365.
86 D. Giacomoni and S. Spiegelman, *Science*, 138 (1962) 1328.
87 H. R. Mahler, G. Dutton and H. Mehrota, *Biochim. Biophys. Acta*, 68 (1963) 199.
88 S. Nishimura and G. D. Novelli, *Biochem. Biophys. Res. Commun.*, 11 (1963) 161.
89 W. E. C. Wacker and B. L. Vallee, *J. Biol. Chem.*, 234 (1959) 3257.
90 W. E. C. Wacker, M. P. Gordon and J. W. Huff, *Biochemistry*, 2 (1963) 717.
91 T. Lindahl, A. Adams and J. R. Fresco, *Proc. Natn. Acad. Sci. U.S.*, 55 (1966) 941.
92 M. Cohn, A. Danchin and M. Grunberg-Manago, *J. Mol. Biol.*, 39 (1969) 199.
93 G. Felsenfeld and A. Rich, *Biochim. Biophys. Acta*, 26 (1957) 457.
94 Y. A. Shin, J. Mitteim and G. L. Eichhorn, *Bioinorganic Chemistry*, 1 (1972) 149.
95 G. L. Eichhorn and E. Tarien, *Biopolymers*, 5 (1967) 273.
96 K. Dimroth, L. Jaenicke and D. Heinzel, *Ann. Chem.*, 566 (1950) 206.
97 K. Dimroth, H. Witzel, W. Hulsen and H. Mirbach, *Ann. Chem.*, 620 (1959) 94.
98 K. Dimroth and H. Witzel, *Ann. Chem.*, 566 (1959) 109.
99 H. Witzel, *Ann. Chem.*, 566 (1959) 122.
100 E. Bamann, H. Trapmann and F. Fischler, *Biochem. Z.*, 328 (1954) 89.
101 G. L. Eichhorn and J. J. Butzow, *Biopolymers*, 3 (1965) 79.
102 H. Trapmann and M. Devani, *Z. Physiol. Chem.*, 340 (1965) 81.
103 S. Matsushita and F. Ibuki, *Mem. Res. Inst. Food Sci. Kyoto Univ.*, 22 (1960) 32, 38.
104 J. W. Huff, K. S. Sastry, M. P. Gordon and W. E. C. Wacker, *Biochemistry*, 3 (1964) 501.
105 P. Berg, personal communication.
106 J. J. Butzow and G. L. Eichhorn, *Biopolymers*, 3 (1965) 95.
107 G. L. Eichhorn, E. Tarien and J. J. Butzow, *Biochemistry*, 10 (1971) 2014.
108 W. R. Farkas, *Biochim. Biophys. Acta*, 155 (1968) 401.
109 J. J. Butzow and G. L. Eichhorn, *Biochemistry*, 10 (1971) 2019.
110 G. L. Eichhorn, J. J. Butzow, P. Clark and E. Tarien, *Biopolymers*, 5 (1967) 283.
111 M. Laskowski, in P. D. Boyer, H. Lardy and K. Myrbäck. *The Enzymes*, Vol. 5, Academic Press, New York, 1961, p. 123.
112 T. Miyaji and J. P. Greenstein, *Arch. Biochem. Biophys.*, 32 (1951) 414.
113 J. S. Wiberg, *Arch. Biochem. Biophys.*, 73 (1958) 337.
114 J. Erkama and P. Suntarinen, *Acta Chem. Scand.*, 13 (1959) 323.
115 J. Shack and B. S. Bynum, *J. Biol. Chem.*, 239 (1964) 3843.
116 G. L. Eichhorn, P. Clark and E. Tarien, *J. Biol. Chem.*, 244 (1969) 937.
117 E. Melgar and D. A. Goldthwait, *J. Biol. Chem.*, 243 (1968) 4409.
118 G. E. Becking and H. O. Hurst, *Can. J. Biochem. Physiol.*, 41 (1963) 1433.
119 P. Clark and G. L. Eichhorn, unpublished observations.
120 H. V. Aposhian, N. Friedman, M. Nishahara, E. P. Hehner and A. L. Nussbaum, *J. Mol. Biol.*, 49 (1970) 367.
121 H. S. Kaplan and L. A. Heppel, *J. Biol. Chem.*, 222 (1956) 907.
122 S. R. Dickman, J. P. Aroskar and R. B. Kropf, *Biochim. Biophys. Acta*, 21 (1956) 539.
123 C. E. Carter and J. P. Greenstein, *J. Natn. Cancer Inst.*, 7 (1946) 29.
124 C. Ceriotti, *Nature*, 163 (1949) 874.

125 E. O. Akinrimisi, J. Bonner and P. O. P. Ts'o, *J. Mol. Biol.*, 11 (1965) 128.
126 S. A. Latt and H. A. Sober, *Biochemistry*, 6 (1967) 3307.
127 D. Kabat, *Biochemistry*, 6 (1967) 3443.
128 L. P. Gavrilova, D. A. Ivanov and A. S. Spirin, *J. Mol. Biol.*, 16 (1966) 473.
129 A. Goldberg, *J. Mol. Biol.*, 15 (1966) 663.
130 M. L. Petermann and A. Paulovec, *Biochemistry*, 6 (1967) 2950.
131 P. O. P. Ts'o, J. Bonner and J. Vinograd, *Biochim. Biophys. Acta*, 30 (1958) 570.
132 D. L. Weller and J. Horwitz, *Biochim. Biophys. Acta*, 87 (1964) 361.
133 R. F. Gesteland, *J. Mol. Biol.*, 18 (1966) 356.
134 M. Tal, *Biochim. Biophys. Acta*, 169 (1968) 564.
135 J. S. Anderson, J. E. Dahlberg, M. S. Bretscher, M. Revel and B. F. C. Clark, *Nature*, 216 (1967) 1072.
136 T. Ohta, S. Sarkar and R. E. Thack, *Proc. Natn. Acad. Sci. U.S.*, 58 (1967) 1638.
137 J. M. Ravel, *Proc. Natn. Acad. Sci. U.S.*, 57 (1967) 1811.
138 R. M. Steiner and R. F. Beers, Jr., *Polynucleotides*, Elsevier, Amsterdam, 1961, p. 187.
139 M. C. Scrutton, C. W. Wu and D. A. Goldthwait, *Proc. Natn. Acad. Sci. U.S.*, 68 (1971) 2497.
140 W. Salser, K. Fry, C. Brunk and R. Poon, *Proc. Natn. Acad. Sci. U.S.*, 69 (1972) 238.
141 C. Sander and P. O. P. Ts'o, *J. Mol. Biol.*, 55 (1971) 1.
142 J. P. Schreiber and M. Daune, *Biopolymers*, 8 (1969) 139.
143 D. C. Liebe and J. E. Stuehr, *Biopolymers*, 11 (1972) 145.
144 G. L. Eichhorn, J. Pitha, E. Tarien and C. Richardson, unpublished observations.

Index

Acetate synthetase, 1115
Acetazolamide, 519, 520
Acetazolamide, binding to carbonic anhydrase, 512
Acetyl-coenzyme A synthetase, 386, 423
Acetylene, reduction by nitrogenase and model, 766
Acidity of ligand, increase through coordination, 362
Aconitase, 419
Actinonin, 198
Actins, 205, 214
Activated complex, 620
Activation of small molecules via coordination, 374
Active transport of iron, 177
Adair equation, 875
Adenine, 1191
Adenine copper complex, 1196
Adenine nucleoside and nucleotide complexes, 1197
Adenosine nucleosides, complex stabilities, 1194
Adenosine nucleotides, complex stabilities, 1194
Adenosine triphosphate, see ATP
Adenylate kinase, 383, 587, 590, 593, 600
Adrenal iron-sulfur protein, 723
Adrenodoxin, 712, 719
Aerobactin, 169, 179, 191, 192
Aetioporphyrin, 1071
Alamethicin, 213
Albomycin, 177, 178, 179, 188, 189
Albomycin δ_1, 187
Albomycin δ_2, 187
Albomycin ϵ, 187
Alcohol dehydrogenase, 395
Aldehyde oxidase, 713, 1187
Aldolases, 415
Alkali ions, 26
Alkali metal chelators, 203
Alkali peptide complexes, 132
Alkaline earth complexes of X-537A, 217
Alkaline earth ions, 26, 68
Alkaline earth peptide complexes, 132
Alkaline phosphatase, 549, 560, 577
Alkaline phosphatase, cadmium, 394
Alkanes, binding to myoglobin, 857
Allosteric linkage, in myoglobin, 856
Aluminium, in ferrichrome, 172
Aluminium, in transamination, 1155
Aluminium mycobactin P, 194
Amino acids, 91, 121
Amino acids, oxidation by metal ions, 679
Amino acid ester, hydrolysis, 23
Amino acid oxidases, 680
Aminoacyl-tRNA synthetases, 424
Amino groups, terminal, as donors, 123
Amino nitrogen donor, 83
Ammonia lyases, 420
Amylases, 422
Antibiotics, ionophores as, 221
Anticodon, 1211
Antiprism, square, 15
Apo-carbonic anhydrase, 512
Apo-carbonic anhydrase, rate of recombination of zinc(II), 506
Apo-carboxypeptidase A, 456
Apoceruloplasmin, 312
Apoenzyme preparation, 394
Apoferritin, 255, 256
Apomyoglobin copper complex, 241
Aquo complex, 63
Arginase, 383, 422
Arginine kinase, 383, 593, 597
Arginine side chains, metal binding, 230
Aromatic systems, reaction with metal ions, 377
Ascorbate oxidase, see Ascorbic acid oxydase
Ascorbic acid, oxidation, 675, 676
Ascorbic acid oxidase, 679, 680, 690, 704, 1194
Aspergillic acid, 171, 197
ATP, 552
ATP, in creatine kinase reaction, 593
ATP, hydrolysis of, by metal ions, 569, 603
ATP complexes, in nitrogenase reaction, 756
ATPase, 207
Autoxidation in hemoglobin and myoglobin, 851
Azofer, 745, 749
Azofermo, 745, 749
Azoferredoxin, 749
Azurin, 690, 692, 695

Bacteriochlorophyll, 1022, 1023, 1036
Beauvaricin, 208
Benzylamine oxidase, 690
Binuclear iron, 725
Bioenergetics, phosphates and, 549
Biotin carboxylases, 425
Biotin enzymes, 401
Biuret cadmium complex, 131
Biuret mercury complex, 131
"Blue" copper oxidases, 689
Bohr effect, 322, 858, 862, 874
Bridge, ligand, in metal enzyme, 382
Bridge, metal, in metal enzyme, 383
Bridge, molybdenum-iron in nitrogenase, 773
Bridging structures, oxo and dihydroxo, 108

Cadmium, in phosphoglucomutase, 414
Cadmium alkaline phosphatase, 394, 561, 577
Cadmium biuret, 131
Cadmium carbonic anhydrase, 504, 505, 512, 531
Cadmium carboxypeptidase A, 456, 471
Cadmium flavoquinone, 1173
Cadmium flavosemiquinone, 1184
Cadmium phosphatase, 563
Cadmium transferrin, 290
Calcium, and adenylate kinase, 600
Calcium, in aminoacyl-tRNA synthetases, 424
Calcium, in amylases, 422
Calcium, in creatine kinase, 593
Calcium, in dinitrogen fixation, 776
Calcium, effect on glutamate dehydrogenase, 398
Calcium, effect on hemocyanin oxygenation, 350
Calcium, in glutamine synthetase, 426
Calcium, in hydrolases, 422
Calcium, in kinases, 582, 588
Calcium, in phospholipases, 422
Calcium, and protein synthesis, 1217
Calcium ADP, 595
Calcium ATP, 586
Calcium glycylglycylglycine, 130, 131
Calcium nucleotides, 1197
Cancer chemotherapy, 181
Carbonic anhydrase, 384, 488
Carbonic anhydrase, absorption spectra, 499, 508
Carbonic anhydrase, absorption spectra and circular dichroism, 521
Carbonic anhydrase, amino acid composition, 492, 494
Carbonic anhydrase, amino acid sequences, 498
Carbonic anhydrase, binding of carbon dioxide, azide, 513
Carbonic anhydrase, binding of sulfonamide, 512
Carbonic anhydrase, coordinated water, 517
Carbonic anhydrase, enzymatic activity and stability, 504
Carbonic anhydrase, e.s.r. spectrum, 530
Carbonic anhydrase, fluorescence studies, 525
Carbonic anhydrase, hydrogen ion equilibria, 497
Carbonic anhydrase, inhibitor binding, 511
Carbonic anhydrase, interactions with sulfonamides containing chromophores, 523
Carbonic anhydrase, kinetics, 538
Carbonic anhydrase, magnetic susceptibility, 531
Carbonic anhydrase, mechanism of action, 536
Carbonic anhydrase, nuclear magnetic resonance, 526
Carbonic anhydrase, optical rotatory properties, 500
Carbonic anhydrase, phosphorescence studies, 525
Carbonic anhydrase, physico-chemical properties, 491, 494
Carbonic anhydrase, reaction with cyanide and sulfide, 518
Carbonic anhydrase, removal of zinc from, 503
Carbonic anhydrase, species and isozyme variants, 489
Carbonic anhydrase, spin-labelled, 528
Carbonic anhydrase, sulfonamide interaction, 511
Carbonic anhydrase, X-ray structure, 531
Carbonic anhydrase, zinc in, 503
Carbonic anhydrase, Zn binding site, 535
Carbonic anhydrase cobalt complex, circular dichroism, 510
Carbonic anhydrase cobalt complex, spectrum with iodoacetate, 515
Carbonic anhydrase complexes, absorption spectra and optical activity, 507

Carbonic anhydrase metal complexes, stability of, 505
Carbonic anhydrase sulfonamide complexes, 518
Carbonyl-magnesium interaction, 1028, 1030
Carboxylases, 401
Carbosylate donor groups, 83
Carboxylation, 364
Carboxyl groups, as donors, 126
Carboxypeptidase, 384, 438, 873, 877
Carboxypeptidase, zinc in, 159
Carboxypeptidase A, 94, 438
Carboxypeptidase A, active center, 447
Carboxypeptidase A, activity as a function of pH, 464
Carboxypeptidase A, binding of substrates and inhibitors, 462
Carboxypeptidase A, enzyme-substrate interactions, 455
Carboxypeptidase A, function of metal, 470
Carboxypeptidase A, kinetics, 457
Carboxypeptidase A, mechanism of action, 473
Carboxypeptidase A, metal ligands, 468
Carboxypeptidase A, specificity, 475
Carboxypeptidase A, stability and kinetic constants of metal complexes, 471
Carboxypeptidase A, stereo drawing, 444
Carboxypeptidase A, zinc-ligand bond angles, 450
Carboxypeptidase B, 478
Carboxypeptidase G, 480
Carnosine copper complex, 143
Catalase, 686, 799, 988, 1015
Catalase action, 654
Catalase mechanism, 1016
Catalase models, 684, 1018
Catechols, oxidation, 678
Catoptromers, 41
Cerium, in phosphate ester hydrolysis, 570
Ceruloplasmin, 293, 306, 690, 692, 695, 697, 698, 704, 706
Ceruloplasmin, amino acid composition, 307, 308
Ceruloplasmin, biosynthesis, 316
Ceruloplasmin, carbohydrate composition, 314
Ceruloplasmin, copper dissociation, 312
Ceruloplasmin, copper, incorporation of, 316
Ceruloplasmin, copper exchange, 311
Ceruloplasmin, electrophoretic properties, 309
Ceruloplasmin, heterogeneity, 313
Ceruloplasmin, optical properties, 309
Ceruloplasmin, oxidase action, 312
Ceruloplasmin, physical properties, 309
Cesium complex of X-537A, 217
Cesium nonactin, 207
CFSE, see Crystal field stabilization energy
Charge transfer, 349
Charge transfer chelates, 1170
Charge transfer spectra, 108
Chelate, 79
Chelate complex, 122
Chelate rings, 39
Chelation, 6
Chelation by histidine, 146
Chlorobium chlorophylls, 1022
Chlorophyll, 1022
Chlorophyll, bifunctional ligand adducts, 1057
Chlorophyll, as electron donor-acceptor, 1027
Chlorophyll, electronic transition spectra, 1054
Chlorophyll, electron paramagnetic resonance, 1059
Chlorophyll, magnesium in, 1025
Chlorophyll, photoactivity, 1058
Chlorophyll, spectroscopy, 1026
Chlorophyll a, 1022, 1023
Chlorophyll b, 1022, 1023, 1036
Chlorophyll c_1, 1022, 1023
Chlorophyll c_2, 1022, 1023
Chlorophyll d, 1022
Chlorophyll-chlorophyll interactions, 1028
Chlorophyll-chlorophyll oligomers, 1037
Chlorophyll-ligand interactions, 1042, 1043
Chlorophyll-pheophytin, 1049
Chlorophyll species, equilibria, 1050
Chlorophyll-water interactions, 1046
Chlorothiazide, 519
Choline dehydrogenase, 1187
Chromium, in carboxypeptidase A, 441
Chromium, stereochemistry, 19
Chromium mycobactin, 172
Chromium mycobactin P, 194
Chromium transferrin, 286, 290, 297, 301
Cis isomers, 41
Citramalate hydrolase, 420
Citrate lyase, 418

Citrate manganese complex, 420
Citrulline, conversion of copper complex to ornithine, 372
Class (a) hard acids, bases, 67
Class (a) metal ions, 70
Class (b) metal ions, 70, 153, 156
Cobalamins, 1069
Cobaloximes, 1163
Cobalt, and adenylate kinase, 600
Cobalt, in alcohol dehydrogenase, 395
Cobalt, in B_{12}, 1067
Cobalt, in B_{12} mechanisms, 1116
Cobalt, bisacetatobis(ethylenethiourea) Co(II), model of cobalt carbonic anhydrase, 507
Cobalt, in carboxypeptidase A, 441, 442
Cobalt, cause of accumulation of siderochromes, 177
Cobalt, in corrinoids, 1067
Cobalt, in creatine kinase, 593
Cobalt, in deoxynucleotidyl transferase, 1214
Cobalt, in dinitrogen fixation, 776
Cobalt, with DNAse I, 1235
Cobalt, in kinases, 582
Cobalt, in mercury pollution, 1120
Cobalt, oxidation-reduction, 631
Cobalt, oxygen carrier, 670, 671
Cobalt, redox rates, 627
Cobalt, and RNA polymerase, 1214
Cobalt, stereochemistry, 21
Cobalt aldolase, 416
Cobalt alkaline phosphatase, 561, 577
Cobalt carbonic anhydrase, 504, 505, 508, 509, 512, 513, 526, 531
Cobalt carboxypeptidase A, 442, 471
Cobalt chlorophyll, 1025
Cobalt(I) cobalamin, 1113
Cobalt(II) cobalamin, 1108, 1111
Cobalt complex, in hydrolysis of methyl phosphate, 575, 576
Cobalt complex, reaction with H_2 and N_2, 767
Cobalt(I) corrinoids, 1090
Cobalt(II) corrinoids, 1085
Cobalt dimethylglyoxime, 1072
Cobalt dinitrogen complex, 777, 784
Cobalt glycine, 127
Cobalt glycylglycine, 674
Cobalt histidine, 89, 127, 147, 148
Cobalt imidazole complex, 146
Cobalt leucine aminopeptidase, 481
Cobalt nitrile complex, 763
Cobalt oxygen carriers, 674
Cobalt peptide complexes, 134, 142
Cobalt phosphatase, 562
Cobalt porphyrin, 1096, 1097
Cobalt protoporphyrin, 841
Cobalt pyruvate carboxylase, 385
Cobalt pyruvate kinase, 385
Cobalt riboflavin, tetraacetyl, 1179
Cobalt Schiff base complex, 1142
Cobalt transferrin, 286, 290
Cobamides, 1069
Cobinamides, 1069
Coboglobin, 841, 842
Cobyric acid, 1069
Cobyrinic acid, 1069
Codecarboxylase, 1137
Codon, 1211
Coenzyme B_{12}, 1067
Coenzyme B_{12}, mechanism of reactions, 1116
Coenzyme B_{12}, structure, 1073
Conalbumin, 280, 282
Conalbumin copper complex, 287
Conalbumin and lactoferrin, biologic functions, 302
Concanavalin A, 389
Configuration, 4
Conformational isomerism, 55, 56
Conformations of ligand, 4
Coordination number, 4, 5
Coordination number, high, 39
Copper, in ascorbic acid oxidase, 679
Copper, in ATP hydrolysis, 569
Copper, binding peptide nitrogen, 236
Copper, "blue", in oxidases, 691
Copper, bound to peptide N, 241
Copper, in catalase models, 684
Copper, catalysis of ascorbic acid oxidation, 676
Copper, in ceruloplasmin, 310
Copper, coordination preferences, 232
Copper, in cytochrome oxidase, 647, 957, 970, 971, 981
Copper deficiency, 306
Copper, in dinitrogen fixation, 776
Copper, in dopamine-β-hydroxylase, 666
Copper, e.p.r. non-detectable, in oxidases, 700
Copper, in hemocyanin, 116, 344
Copper, in hemoglobin, 843
Copper, in hydrolysis of methylphosphonofluoridate, 571
Copper, and mRNA formation, 1215

Copper, in oxidase enzymes, 680
Copper, in oxidation of benzoic acid, 682
Copper, in phosphate ester hydrolysis, 570
Copper, reaction with cystine, 158
Copper, reaction with poly (rA), 1230
Copper, in ribulose diphosphate carboxylase, 409
Copper, stereochemistry, 24
Copper, in tyrosinase, 681
Copper, unwinding DNA reversibly, 1220
Copper adenine, 1196
Copper alkaline phosphatase, 561
Copper amino acid complex, 91
Copper biuret, 132
Copper carbonic anhydrase, 504, 505, 508, 512
Copper carboxypeptidase A, 456, 457, 471
Copper carnosine, 143
Copper chelates, models of catalase, 684
Copper chlorophyll, 1025
Copper complexes of nucleoside bases, 1195
Copper complexes of protein functional groups, stability, 231
Copper conalbumin, 287
Copper-copper pair in oxidases, 702, 705
Copper cytidine, 1205
Copper cytosine, 1196
Copper DNA, 1217, 1218
Copper flavins, 1168, 1175
Copper glycine, condensation with acetaldehyde, 373
Copper glycylglycylglycine, 132, 135
Copper glycylglycylglycylglycine, 237
Copper guanine, 1195
Copper guanosine, 1204
Copper histamine, 151
Copper histidine, 150
Copper imidazole, 144, 230, 238
Copper inosine phosphate, 1204
Copper ions, generation of free radicals, 683
Copper myoglobin, 143, 238, 239
Copper nucleosides, 1193
Copper nucleotides, 1197
Copper oxidases, 689
Copper oxidases, catalytic activity, 703
Copper oxidases, forms of copper in, 691
Copper oxytocin, 236
Copper peptide complexes, 134, 242
Copper porphyrin, 845
Copper(II) pyridoxylidene, 1141
Copper ribonuclease, 245
Copper riboside, 1207
Copper Schiff base complexes, 1140
Copper serum albumin, 242
Copper transferrin, 297, 301
Copper tubercidin, 1201
Copper uridine, 1205
Copper uroporphyrin, 802
Copper vasopressin, 236
Coprogen, 191
Corrin, 1068
Corrins, axial ligands, 1076
Corrins, electronic absorption spectra, 1082
Corrinoids, 1067, 1068
Corrinoids, cis- and trans-effects, 1079
Corrinoids, cobalt, acid decomposition, 1098
Corrinoids, cobalt, nucleophilic displacement of carbon ligands, 1096
Corrinoids, cobalt, photolysis, 1094
Corrinoids, cobalt, reaction with carbon monoxide, 1093
Corrinoids, cobalt, reaction with iodine, 1098
Corrinoids, cobalt, reaction with metal ions, 1099
Corrinoids, cobalt, reaction with sulfur ligands, 1091
Corrinoids, cobalt, reductive cleavage, 1096
Corrinoids, cobalt, thermolysis, 1095
Corrinoid-dependent enzymatic reactions, 1099
Corrinoid models, 1072
Creatine kinase, 383, 389, 583, 587, 590, 591, 593
Creatine kinase, inhibition of, 589
Crown polyethers, 205, 211
Crystal field stabilization energy, 30, 31, 33
Cyanocobalamin, 1070
Cysteine, 92
Cysteine, reaction with copper, 158
Cysteine cobalt complex, 155
Cysteine complexes, 154
Cysteine iron complex, 155, 710, 733, 738, 944
Cysteine molybdenum complex, 156, 753
Cysteine palladium complex, 155
Cytidine copper complex, 1205
Cytidine mercury complex, 1205
3′-Cytidine monophosphate, ternary complex with RNase and copper and zinc, 244

Cytidine silver complex, 1205
Cytochromes, in bacteroids, 776
Cytochromes, circular dichroism spectra, 938
Cytochrome, reaction with oxygen, 647
Cytochrome a, 798, 799, 956, 961, 962, 970
Cytochrome a_3, 799, 956, 961, 962, 970
Cytochrome b, 798, 799, 902, 933
Cytochrome b_2, 939
Cytochrome b_5, 934
Cytochrome b_5, amino acid sequence, 936
Cytochrome b-555, 937
Cytochrome b-562, 937
Cytochrome b-562, amino acid sequence, 939
Cytochrome c, 798, 800, 902, 903, 925
Cytochrome c, amino acid composition, 906
Cytochrome c, binding to cytochrome oxidase, 915
Cytochrome c, electron transfer, 643, 648, 922
Cytochrome c from prokaryotes, 926, 927
Cytochrome c, heme crevice, 913
Cytochrome c, heme group, 630
Cytochrome c, ligand groups, 907
Cytochrome c, mammalian type, 903, 904
Cytochrome c peroxidase, 646, 990
Cytochrome c, propionate linkages, 913
Cytochrome c, protein—prosthetic group interaction, 916
Cytochrome c, redox rate constant, 627
Cytochrome c′, 932
Cytochrome c″, 932
Cytochrome c_1, 902, 925
Cytochrome c_2, 929
Cytochrome c_3, 931
Cytochrome c^{III}, electron exchange, 630, 631
Cytochrome d, 956
Cytochrome f, 926
Cytochrome P_{450}, 799, 942
Cytochrome oxidase, 690, 798, 828, 955
Cytochrome oxidase, composition, 957, 959
Cytochrome oxidase, e.p.r. studies on copper, 974
Cytochrome oxidase, molecular weight, 957
Cytochrome oxidase, oxidation-reduction potentials, 976
Cytochrome oxidase, reaction with cytochrome c, 647, 978
Cytochrome oxidase, reaction with inhibitors, 981
Cytochrome oxidase, reaction with oxygen, 979
Cytodeuteroheme dimethyl ester, 802
Cytodeuteroporphyrin, 965
Cytosine, 1191
Cytosine copper complex, 1196

Danomycin, 197
Decarboxylases, 401, 409
Decarboxylases, oxidative, pyridine nucleotide-dependent, 399
Decarboxylation, 364
Deferriferrioxamine B, 181, 276
Deferrischizokinen, 169
Deferrisiderochrome, 170
Deferroferroverdin, 199
Dehydrogenase, 394, 1185
Dehydrogenase, catalytic role of metal, 396
Dentate, 213, 218
5′-Deoxyadenosyl cobalt bond, biosynthesis, 1103
Deoxygenation of hemoglobin, conformational changes, 864
Deoxyhemerythrin, 331
Deoxyhemerythrin, structure, 339
Deoxyhemoglobin, 836
Deoxymyoglobin, 836
Deoxyrhodotorulic acid, 190
Deoxyribonucleic acid, see DNA
Desferal, 181
Desferrioxamine, see Deferriferrioxamine B
Deuteroheme, 803
Diamine oxidase, 690, 696, 705
Dianemycin, 204, 217, 223
Diastereoisomers, 41
Dibenzo-18-crown-6, 218
Dicyclohexyl-18-crown-6, 211
Dihydroorotate dehydrogenase, 713, 1187
Dihydroxybenzoylglycine, 185
Dihydroxybenzoyl-L-Lysine, 184
Dimerum acid, 191
Dimethylglyoxime cobalt complex, 1072
Dinactin, 205
Dinitrogen, ruthenium complex, 760
Dinitrogen, transition metal complexes, 781
Dinitrogen complexes, 777
Dinitrogen complexes, phosphine-stabilized, 783

Dinitrogen fixation, 745
Dinitrogen fixation, biological, 748
Dinitrogen fixation, commercial, 777
Dinitrogen fixation, inorganic, 769
Dinitrogen fixation, mechanisms, 768
Dinitrogen reduction, by metal ions, 778
Dioldehydrase, 1100, 1105
Diphosphoglyceric acid, 867
Diphosphoglyceric acid, effect on hemoglobin, 858
Disulphide as donor, 158
DNA, 1191, 1210
DNA, effect of metal ions on, 1219
DNA, melting, and metal ions, 1225
DNA, reaction with Hg(II) and Ag(I), 1226
DNA, unwinding and rewinding, 1220
DNA copper complex, 1217
DNA polymerase, 1213
DNA polymerase, manganese and, 1239
DNAse I, activation by metals, 1235
Dodecahedron, 15
Donor atoms, 4, 71
Dopamine-β-hydroxylase, 666, 690, 695, 696

Eight coordination, 15
Ekkrinosiderophilin, 281
Electron exchange reactions, 626
Electron paramagnetic resonance, 388, 591, 692, 716
Enantiomers, 41
Endonucleases, 557
Enniatin, 204, 205, 208, 209, 214
Enniatin B, 210, 218
Enolase, 418, 593
Entatic state, 385
Enterobactin, 169, 171, 173, 174, 175, 177, 179, 183
Enterochelin, 200
Enzymatic phosphate transfer, metal ion requirements, 559
Enzyme, active site, 361
Enzymes, metallo-metal activated, 381
Enzymes, schemes for binding metal and ligand, 382
Enzyme bridge complex, 386
Enzyme-substrate complex, carboxypeptidase A, 448
Ethanolamine deaminase, 1108
Ethanoldeaminase, 1100
Ethoxzolamide, 519, 522
Ethylenediamine, 74
Ethylenediaminetetraacetate, 75
Etioporphyrin I, 804
Exonucleases, 557

Fenton's reagent, 659
Ferredoxin, 155, 710, 1115
Ferredoxin, chloroplast, 712
Ferredoxins, in nitrogenase reaction, 754
Ferredoxins, reductants in oxygenation, 664
Ferredoxin models, 739, 753
Ferribactin, 178, 187, 197
Ferrichrome, 169, 171, 172, 175, 177, 186, 187, 188
Ferrichrome A, 176, 186, 187, 188, 189
Ferrichrome C, 187, 188, 189, 190
Ferrichrysin, 187, 188
Ferricrocin, 187
Ferrihemoglobin, 836
Ferrimycin, 179, 195
Ferrimycin A_1, 196
Ferrimyoglobin, 836
Ferrioxamine, 179, 195
Ferrioxamine A_1, B, D_1, D_2, E, G, 196
Ferrireductase enzyme, 276
Ferrirhodin, 187, 188
Ferrirubin, 187, 188
Ferritin, 179, 253
Ferritin, amino acid composition, 257
Ferritin, arrangement of core in, 268
Ferritin, biosynthesis, 270
Ferritin, effect of iron component on properties and structure, 269
Ferritin, formation from apoferritin, 274
Ferritin, iron core analogues, 263, 264
Ferritin, iron cores, Mössbauer spectra, 266
Ferritin, iron mobilization from, 276
Ferritin, magnetic susceptibility, 264
Ferritin, Mössbauer spectra, 265
Ferritin, optical rotatory dispersion, 270
Ferritin, preparation, 254
Ferritin, quarternary structure, 257
Ferritin, reconstitution from apoferritin and iron, 272
Ferritin, removal of iron, 256
Ferritin cores, atomic structure, 265
Ferritin cores, electron microscopy, 260
Ferritin cores, structure, 258
Ferritin cores, X-ray scattering, 260
Ferrochelatase, 178
Ferroxidase, 312, 313
Ferroverdin, 199
Five coordination, 12
Five-membered rings, 56

Flavins, biochemical function, 1185
Flavins, charge-transfer chelates, 1171
Flavins, reductants in oxygenation, 664
Flavin complexes, 1168, 1176
Flavin complexes, electron spin-nuclear spin coupling constants, 1176
Flavodoxins, in nitrogenase reaction, 754
Flavohydroquinone, 1168
Flavoproteins, 1168, 1183
Flavoproteins, metal, 1187
Flavoquinone, 1168
Flavoquinone chelates, 1170
Flavoquinone metal chelates, 1177
Flavosemiquinone, 1168
Flavosemiquinone metal chelates, 1180
Formyltetrahydrofolate synthetase, 425
Four coordination, 9
Free radical addition of oxygen, 667
Fusarinine, 194, 195
Fusarinine A, 195
Fusarinine B, 195
Fusarinine C, 195
Fusigen, 195

Galactose oxidase, 680, 690, 696
Gallium mycobactin P, 194
Gallium transferrin, 290, 297
Genetic code, 1211
Glutamate dehydrogenase, 398
Glutamate mutase, 1100, 1104, 1121
Glutamate zinc complex, 129
Glutamine synthetase, 386, 426
Glycerol dehydrase, 1100, 1105
Glycerol phosphate dehydrogenase, 1187
Glycine, 74
Glycine cobalt complex, 127
Glycine copper complex, reaction with acetaldehyde, 373
Glycine neodymium complex, 129
Glycine platinum complexes, 126
Glycine silver complex, 125, 128
Glycine zinc complex, 124
Glycylglycine, 94
Glycylglycine cobalt complex, 670, 674
Glycylglycine zinc complex, 124
Glycylglycylglycine calcium complex, 130, 131
Glycylglycylglycine copper complex, 132
Glycylglycylglycine zinc complex, 129
Glycylglycylglycylglycine copper complex, 237
Gramicidin, 211, 214, 221
Grisein, 178, 187
Grisorixin, 216, 218
Guanine, 1191
Guanine copper complex, 1195
Guanosine copper complex, 1204
Guanosine mercury complex, 1205
Guanosine silver complex, 1204

Hadacidin, 171, 198
Haemo-, see Hemo-
Halide bridging, 640
Hard acids, 67
"Hard" base-"hard" acid, 34
Hard bases, 67
Hard metal ions, 69
Hematins, 818
Hematoheme, 803, 804
Hematoporphyrin, 905
Heme, 72, 797
Heme derivatives, 801
Heme a, 798, 963, 964
Heme a, derivatives, 802
Heme a, structure, 798
Heme b, 798, 803
Heme b, structure, 799
Heme c, 798, 800
Heme c, structure, 801
Heme s, 803
Heme carbonyls, C—O stretch frequencies, 807
Heme-heme interaction, 836, 874, 883, 887
Heme synthetases, 427
Hemerythrin, 113, 320
Hemerythrin, absorption spectrum, 114
Hemerythrin, absorption spectrum and circular dichroism, 328
Hemerythrin, distribution and preparation, 321
Hemerythrin, electron spin resonance, 335
Hemerythrin, interconversions between chemical forms, 322
Hemerythrin, ligands to iron, 324
Hemerythrin, magnetic susceptibility, 330
Hemerythrin, model compounds, 326
Hemerythrin, Mössbauer spectra, 330, 332
Hemerythrin, oxidation and reactions with small ligands, 323
Hemerythrin, oxygen equilibrium, 322
Hemerythrin, primary sequence, 325
Hemerythrin, structure of the active site, 335
Hemerythrin, subunits, 324
Hemin a, oxo linkage, 827
Hemin dimers, 817
Hemin hydroxides, 818

Hemins, 797, 799
Hemins, μ-oxo-bridged, 826
Hemins, with one axial ligand, 818
Hemins, with two axial ligands, 824
Hemochrome, 827, 846, 908, 963
Hemocyanin, 116, 344
Hemocyanin, absorption spectra, 347
Hemocyanin, ageing and regeneration, 351
Hemocyanins, catalase activity, 352
Hemocyanin, circular dichroism, 348
Hemocyanin, copper binding sites, 353
Hemocyanin, ligand binding, 350
Hemocyanins, molecular weight, 345
Hemocyanin, oxygen binding, 349
Hemocyanin, quarternary structure, 346
Hemocyanin, removal of copper from, 352
Hemoglobin, 799, 832
Hemoglobin, abnormal, 848, 872, 877
Hemoglobin, autoxidation, 851
Hemoglobin, CO binding, 853
Hemoglobin, conformational changes, 860
Hemoglobin, $\alpha_1\beta_1$ contact, 863, 865
Hemoglobin, $\alpha_1\beta_2$ contact, 865, 866
Hemoglobin, cyanide binding, 853
Hemoglobin, dissociation, 880, 888
Hemoglobin, distal side of the heme, 872
Hemoglobin, electron transfer from iron to oxygen, 850
Hemoglobin, environment of the heme, 851
Hemoglobin, heme-heme interaction, 883
Hemoglobin, iron in, 845
Hemoglobin, iron out-of-plane, 824
Hemoglobin, iron-oxygen bond, 848
Hemoglobin, kinetics of ligand reactions, 877
Hemoglobin, ligand binding, 858
Hemoglobin, ligand binding equilibrium, 874
Hemoglobin, liganded form, 860
Hemoglobin, orientation of oxygen relative to the heme, 849
Hemoglobin, oxygen binding, 673
Hemoglobin, porphyrin, planarity of, 846
Hemoglobin, porphyrin interactions, 843
Hemoglobin, proximal side of the heme, 873
Hemoglobin, reversible oxygenation, 851
Hemoglobin, role of the fifth ligand, 846
Hemoglobin, spin state of iron, 839
Hemoglobin, subunit contacts, 887
Hemoglobin, X-ray crystallography, 860
Hemoglobin M, abnormal, 884
Hemopexin, 282
Hemoprotein hydroxides, 112
Hemosiderin, 253, 276
Hexokinase, 583, 593, 601
High-spin complex, 103
Hill equation, 875
Histamine copper complex, 151
Histidinato nickel(II) isomers, stereoselectivity of ester hydrolysis, 366
Histidine, 92, 931, 935
Histidine, as ligand, 847
Histidine cobalt complex, 89, 127, 148, 670
Histidine complexes, 147
Histidine copper complex, 150
Histidine deaminase, 421
Histidine iron complex, 907, 931
Histidine molybdenum complex, 753
Histidine nickel complex, 89
Histidine platinum and palladium complexes, 149
Histidine zinc complex, 448, 532
Histohematin, 955
Hydration of CO_2, by carbonic anhydrase, 489
Hydridocobalamin, 1090
Hydrogen transfer, 1101
Hydrolases, 421
Hydro-lyases, 418
Hydrolysis, 23, 364
Hydrolysis, of amino acid esters, metal catalysis, 365
Hydrolysis, catalysis by metal ions, 378
Hydrolysis, equilibrium, 639
Hydrolysis, metal catalysis of, 367
Hydrolysis, of peptide or ester bonds, by cobalt complexes, 472
Hydrolysis, of peptides by metal ions, 367
Hydrolysis, of phosphate derivatives, standard free energy of, 550
Hydrolytic processes, 365
Hydroxamate, 175
Hydroxamates iron complexes, 186
Hydroxide bridging, 639
Hydroxides, metal, 78
Hydroxocobalamin, 1069
Hydroxylation, 662, 719, 720
Hydroxylation reactions, 659
Hydroxyl groups as ligands, 152
Hydroxy-3-nitrobenzene-sulfonamide, 519
Hypoxanthine, 1191

Imidazole, binding in transferrin, 297
Imidazole, as ligand, 143, 847
Imidazole cobalt complex, 146
Imidazole copper complex, 144, 146, 230, 238
Imidazole zinc complex, 144, 232, 235, 238
Indium transferrin, 291
Inner shell reorganization, 623
Inosine phosphate copper complex, 1204
Inositol oxygenase, 680
Insertion reactions, 654, 656
Insertion reactions, via coordination, 375
Insulin, reaction with zinc, 233
Interaction energy, in electron transfer, 618
Iodoacetate, reaction with carbonic anhydrase, 515
Ionic character, 27
Ionophores, 203
Ionophores, dynamics of complex formation, 218
Ionophores, as tools for perturbing subcellular organelles, 222
Iridium, oxygen carrier, 672, 847
Iridium acetylene complex, 765
Iridium dinitrogen complex, 783
Iron, 935
Iron, in aconitase, 419
Iron, binding to transferrin, 287
Iron, binding to transferrin conalbumin, 287
Iron, in carboxypeptidase A, 441
Iron, in catalase models, 685, 1018
Iron, catalysis of ascorbic acid oxidation, 676
Iron, in citramalate hydrolase and tartrate hydrolase, 420
Iron, in cytochromes, 902
Iron, in cytochrome oxidase, 955, 968
Iron, in dinitrogen reduction, 780
Iron, in ferritin, 253
Iron, in hemerythrin, 320
Iron, in hemoglobin, 832
Iron, insertion into heme, 427
Iron, ligands to hemerythrin, 324
Iron, ligand field in hemoglobin and myoglobin, 837
Iron, in mixed function oxidases, 665
Iron, in model peroxidase systems, 660
Iron, in myoglobin, 832
Iron, in nitrogenase, 745, 748
Iron, in oxygenases, 664
Iron, peroxidase activity, 1015
Iron, in phosphate ester hydrolysis, 570
Iron, in pyrocatechase, 657
Iron, redox rates, 627
Iron, removing pathological deposits of, 181
Iron(II), requirement in hemoglobin and myoglobin, 841
Iron, stereochemistry, 20
Iron, storage, 253
Iron, in succinate dehydrogenase, 1186
Iron, transport by transferrin, 300
Iron, transport compounds, 167
Iron aerobactin, 192
Iron aldolase, 416
Iron amino acid complex, 91
Iron-binding proteins, 280
Iron(III) binuclear complexes, 108, 336
Iron(III) binuclear complexes, structural and magnetic properties, 327
Iron chlorophyll, 1025
Iron(II) complexes, 103
Iron(III) complexes, 104
Iron complexes, electronic structures, 102
Iron cysteine, 155, 710, 733, 738, 944
Iron dinitrogen complex, 777, 784
Iron enterobactin, 183, 184
Iron flavins, 1168, 1175
Iron histidine, 907
Iron hydroxamates, 169
Iron methionine, 909
Iron mycobactin, 192
Iron oxide hydrate, 263
Iron oxide hydrate in ferritin, 268
Iron phenolates, 169
Iron phosphatase, 558
Iron porphyrin, 797, 845
Iron(III) porphyrin dimers, 111
Iron proteins, high potential, 731
Iron protein, in nitrogenase, 752
Iron pyrimine, 199
Iron rhodotorulate, 189, 190
Iron Schiff base complex, 1142
Iron-sulfur proteins, 72, 710, 752, 753, 1185
Iron-sulfur proteins, characteristics, 712
Iron-sulfur proteins, 1-Fe-S and 2-Fe-S, 715
Iron-sulfur proteins, 2-Fe-S, 719
Iron-sulfur proteins, 4-Fe-S, 731
Iron-sulfur proteins, 8-Fe-S, 734
Iron-sulfur proteins, Mössbauer spectra, 728, 729
Iron-sulfur proteins, in nitrogenase reaction, 754

Irving-Williams order, 73, 74, 87, 94, 96, 97
Isocitrate dehydrogenase, 400
Isomerases, 411
Isomerism, configurational, 40, 41
Isomerism, conformational, 40, 55
Isomerism, geometrical, 41
Isomers, ligand, 46
Isomers, linkage, 46, 640
Isomers, optical, 47
Isomers, structural, 46
Itoic acid, 172, 173, 176, 179, 185

Keto-acid lyases, 417
Kinases, 554, 582
Kinases, classification, 593
Kinases, kinetic studies, 587
Kinases, metal-substrate equilibria, 584

Laccase, 690, 691, 692, 694, 695, 697, 698, 700–706
Lactoferrin, 281, 282, 288
Lactoperoxidase, 990
Lanthanum, in phosphate ester hydrolysis, 570
Lanthanides, 68
Lead, in phosphate ester hydrolysis, 570
Lead, in RNA degradation, 1233
Lead carboxypeptidase A, 442
Lead nucleosides, 1194
Leghemoglobin, 775
Leucine aminopeptidase, 480
Lewis acid, 34, 76
Lewis base, 71, 76
Ligands, 4
Ligand, size of, 35
Ligand basicity, 76
Ligand bridge complex, 382
Liganded form of hemoglobin, 860
Ligand field theory, 102
Ligand isomers, 46
Ligand properties, 33
Ligand reactions, metal induced, 361
Linkage isomers, 46, 640
Low-spin complex, 103
Lumiflavin, 1170
D-α-Lysine mutase, 1100
L-β-Lysine mutase, 1100, 1109

Macrocyclic effect, 84
Macrocyclic rings, 203
Magnesium, in aminoacyl-tRNA synthetases, 424
Magnesium, in ATP hydrolysis, 569
Magnesium, in chlorophyll, 1023, 1025, 1042
Magnesium, in creatine kinase, 593
Magnesium, in deoxynucleotidyl transferase, 1214
Magnesium, with DNA polymerase, 1213
Magnesium, DNA stabilization, 1219
Magnesium, with DNAse I, 1235
Magnesium, effect on glutamate dehydrogenase, 398
Magnesium, enolase reaction, 419
Magnesium, in glutamine synthetase, 426
Magnesium, in kinases, 582, 588
Magnesium, in phosphate hydrolysis, 550
Magnesium, in phosphoglucomutase, 413, 559
Magnesium, and polynucleotides, 1239
Magnesium, and protein synthesis, 1216
Magnesium, and pyruvate kinase, 598
Magnesium, in ribosomes, 1238
Magnesium, and RNA polymerase, 1214
Magnesium ADP, 595
Magnesium ATP, 586
Magnesium flavin, 1174
Magnesium leucine aminopeptidase, 481
Magnesium nucleotides, 1197
Magnesium porphyrins, 802
Magnesium riboflavin, 1175
Magnesium tetraphenylporphyrin, 1026
Magnetic resonance, 590
Malonate complexes, 81
Manganese, and adenylate kinase, 600
Manganese, in aminoacyl-tRNA synthetases, 424
Manganese, in carboxypeptidase A, 441
Manganese, in catalase models, 685
Manganese, in creatine kinase, 593
Manganese, with DNA polymerase, 1213, 1239
Manganese, with DNAse I, 1235
Manganese, effect on glutamate dehydrogenase, 398
Manganese, enolase reaction, 419
Manganese, in glutamine synthetase, 426
Manganese, in hemoglobin, 843
Manganese, in kinases, 582, 588
Manganese, nuclear magnetic relaxation rate, 390
Manganese, in O—O cleavage, 658
Manganese, oxidation by peroxidase, 1014
Manganese, in phosphoglucomutase, 414
Manganese, in pyruvate carboxylase, 402
Manganese, reaction with pyruvate kinase, 383

Manganese, and RNA polymerase, 1214
Manganese, stereochemistry, 19
Manganese, in xylose isomerase, 412
Manganese activated enzymes, 592
Manganese ADP, 587, 595
Manganese aldolase, 416
Manganese alkaline phosphatase, 561
Manganese ATP, 586, 587
Manganese ATP and ADP complexes, 594
Manganese carbonic anhydrase, 504, 505, 512
Manganese carboxypeptidase A, 457, 470, 471
Manganese citrate, 420
Manganese citrate lyase complex, 417
Manganese complexes, nmr enhancements, 601
Manganese concanavalin A complex, 389
Manganese creatine kinase complex, 389
Manganese flavin, 1174
Manganese leucine aminopeptidase, 481
Manganese nucleotides, 1197
Manganese phosphatase, 563
Manganese polynucleotide complexes, 1218
Manganese porphyrins, 802, 890
Manganese Schiff base complexes, 1140, 1142
Manganese transferrin, 286, 290, 301
Masking of ligand reactions, 363
Masking of ligand reactivity, 370
Membranes, biological, 218
Mercaptoethylamine, 92
Mercury, in DNA separations, 1227
Mercury, pollution, 1120
Mercury, reaction with cobalt corrinoids, 1099
Mercury adenosine, 1194
Mercury alkaline phosphatase, 561
Mercury biuret, 131
Mercury carbonic anhydrase, 504, 505, 512, 531
Mercury carboxypeptidase A, 442, 456, 457, 471
Mercury cysteine, 154
Mercury cytidine, 1205
Mercury DNA, 1219, 1226
Mercury guanosine, 1205
Mercury nucleotides, 1207
Mercury uracil, 1196
Mercury uridine, 1205
Mesoheme, 803
Mesoporphyrin dimethyl ester, zinc complex, 85
Messenger RNA, 1211
Metabolism, enzymes of carbon dioxide, 401
Metal-activated enzymes, 381
Metal bridge complex, 383, 399, 417, 418, 421
Metal enzymes, experimental approaches to the study of, 386
Metal induced ligand reactions, 361
Metallo-enzymes, 381
Metallo-enzymes, criteria for, 395
Metal-metal replacements, in metal enzymes, 393
Methane synthetase, 1114
Methemerythrin, 321, 323, 331
Methemerythrin, structure, 336
Methemoglobin, 836
Methionine complexes, 157, 158
Methionine iron complex, 909
Methionine platinum complex, 156
Methionine synthetase, 1112, 1122
Methylaspartase, 382
Methylcobalamin, 1069, 1112
α-Methylene-glutamate mutase, 1100, 1104
Methylmalonyl CoA mutase, 1100, 1103, 1121
Methylpheophorbide a, 1023
Methyl transfer, 1101
Methyl transferase, 1114
Metmyoglobin, 836
MFO enzymes, 664
Mixed-function oxidase, 942
Mixed (or ternary) complexes, 85
Model compounds, 159
Molecular orbital energy level diagram, 628
Molybdenum, in dinitrogen reduction, 780
Molybdenum, in nitrogenase, 745, 748
Molybdenum, in reduction of ethylene, 765
Molybdenum, stereochemistry, 25
Molybdenum cysteine, 156, 753
Molybdenum dinitrogen complex, 785
Molybdenum flavins, 1168, 1175
Molybdenum histidine, 753
Molybdenum-iron protein, in nitrogenase, 750
Molybdenum isonitrile complex, 764
Molybdoferredoxin, 749
Molybdothiol, acetylene reduction, 766
Molybdothiol-borohydride, 773, 774, 780
Molybdothiol-borohydride model, 759, 775

Monactin, 205
Monamycins, 214
Monazomycin, 213
Monensin, 216, 217, 218
Mono-oxygenases, 664
Mutases, 413
Mycelianamide, 197, 198
Mycobactin, 169, 171, 175, 179, 192, 193
Mycobactin M, 194
Mycobactin P, 194
Mycobactin chromium complex, 172
Myeloperoxidase, 990
Myoglobin, 95, 799, 832
Myoglobin, autoxidation, 851
Myoglobin, azide binding, 854
Myoglobin, CO binding, 853
Myoglobin, coordination of fluoride, 389
Myoglobin, cyanide binding, 853
Myoglobin, electron transfer from iron to oxygen, 850
Myoglobin, environment of the heme, 851
Myoglobin, iron in, 845
Myoglobin, iron out-of-plane, 824
Myoglobin, iron-oxygen bond, 848
Myoglobin, ligand binding, 853
Myoglobin, ligand induced conformational changes, 855
Myoglobin, orientation of oxygen relative to the heme, 849
Myoglobin, porphyrin, planarity of, 846
Myoglobin, porphyrin interactions, 843
Myoglobin, reversible oxygenation, 851
Myoglobin, role of the fifth ligand, 846
Myoglobin, role of protein residues in vicinity of heme, 853
Myoglobin, spin state of iron, 839
Myoglobin, total protein conformation, 854
Myoglobin copper complex, 143, 238, 239
Myoglobin zinc complex, 238, 239
Myohematin, 955

NADH dehydrogenase, 724, 1187
Neodymium glycine, 129
Nickel, in carboxypeptidase A, 441
Nickel, effect on ribosomes, 1239
Nickel, in hemoglobin, 843
Nickel, in phosphoglucomutase, 414
Nickel, and pyruvate kinase, 598
Nickel, reaction with poly (rA), 1230
Nickel, stereochemistry, 23
Nickel, in transamination, 1155
Nickel aldolase, 416
Nickel alkaline phosphatase, 561
Nickel amino acid complex, 91
Nickel biuret, 132
Nickel carbonic anhydrase, 504, 505, 508, 512
Nickel carboxypeptidase A, 471
Nickel chlorophyll, 1025
Nickel complexes, 152
Nickel corrinoids, 1084
Nickel dinitrogen complex, 784
Nickel histidine, 89, 147
Nickel isonitrile complex, 764
Nickel leucine aminopeptidase, 481
Nickel nucleotides, 1197
Nickel peptide complexes, 134, 141
Nickel porphyrin, 802, 806, 824, 843, 845
Nickel riboflavin, 1175, 1182
Nickel Schiff base complex, 1142
Nickel transferrin, 290
Nigericin, 204, 216, 217, 218, 222
Nine coordination, 15
Nitrate reductase, 1187
Nitrogen complexes, see Dinitrogen
Nitrogen fixation, see Dinitrogen fixation
Nitrogenase, 734, 745, 748
Nitrogenase, acetylene reduction, 766
Nitrogenase, ATP-dependent H_2 evolution, 759
Nitrogenase, ATP reaction, 756
Nitrogenase, chemical and physical characteristics, 749
Nitrogenase, cleavage of NO bond, 760
Nitrogenase, cleavage of triple bonds, 759, 761
Nitrogenase, diazene intermediate, 769
Nitrogenase, dihydrogen reactions, 766
Nitrogenase, inhibitors, 767
Nitrogenase, models, 753
Nitrogenase, nitride intermediate, 769
Nitrogenase, reaction, 748, 754
Nitrogenase, reduction of triple bonds, 764
Nitrogenase, substrates, 747, 758
Nitrogenase-catalyzed reductions, 758
Nonactin, 204, 205, 206, 207, 218
"Non-blue" copper oxidases, 689
Norleucine silver complex, 125
Nuclear magnetic relaxation rates of ligand nuclei, paramagnetic effects on, 390
Nuclear magnetic resonance, 591
Nuclear magnetic resonance, enhancement phenomenon, 391
Nucleic acids, complexes of, 95, 1210

Nucleophilic attack via coordination, 363
Nucleoproteins, metal ions and, 1237
Nucleoside complexes, 1191, 1193, 1194
Nucleoside diphosphates, in creatine kinase, 594
Nucleoside diphosphate kinase, 603
Nucleotide bases, mispairing induced by metal ions, 1239
Nucleotide complexes, 95, 425, 585, 586, 1191
Nucleotide-di and -triphosphates, 72

Octaethylporphyrin, 804
Octahedral complexes, 47
Octahedral geometry, 14
Olation, 111
Operator gene, 167
Optical isomerism, 6
Optical isomers, 47
Optically active ligands, 89
Orbitals, overlap of, 619
Ornithine, from copper complex of citrulline, 372
Osmium tetroxide, reaction with dTMP, 1206
Ovotransferrin, 280
Oxalate complexes, 81
Oxalic acid oxidase, 680
Oxaloacetate decarboxylation, 384, 403
Oxidases, mixed function, 828
Oxidase, "non-blue" copper in, 695
Oxidase reaction, roles of the different copper atoms in, 704
Oxidation-reduction, bridging groups, 639
Oxidation-reduction, conformation- and substitution-controlled, 643
Oxidation-reduction, in coordination compounds, 611
Oxidation-reduction, cross-reactions, 633
Oxidation-reduction, one-electron and two-electron transfers, 637
Oxidation-reduction, solvent reorganization, 624
Oxidative addition, 645
Oxo and dihydroxo bridging structures, 108
Oxygen, complexes of Co(II) and polyamines, 670
Oxygen, complexes of glycylglycine and histidine, 670
Oxygen, insertion, 656
Oxygen, metal ion catalysis of reactions of, 654
Oxygen, reactions of, 655
Oxygen, reduction, 662
Oxygen, reduction products, 656
Oxygen, storage, 322
Oxygen carriers, reactions, 674
Oxygen carriers, synthetic, 669
Oxygen-carrying proteins, comparison of properties, 320
Oxygenases, 664, 689
Oxygenases, models, 663
Oxygenation of hemoglobin and myoglobin, 851, 853
Oxyhemes, 827
Oxyhemerythrin, 113, 323, 332
Oxyhemerythrin, absorption spectrum, 115
Oxyhemerythrin, magnetic moment data, 115
Oxyhemerythrin, structure, 338
Oxyhemoglobin, 836
Oxymyoglobin, 836
Oxytocin copper complex, 236

Pacifarin, 181, 184
Palladium histidine complex, 149
Palladium isonitrile complex, 764
Palladium methionine, 157
Palladium peptide complexes, 134, 142
Palladium porphyrin, 845
Penicillamine, 93
Pentose isomerases, 411
PEP-carboxykinase, 593
Peptides, 23, 94, 121
Peptide, as donor, 229
Peptides, liganded through N, 133
Peptides, liganded through O, 131, 470
Peptide bond hydrolysis, 384
Peptide complexes, alkali and alkaline earth, 132
Peptide groups, as donors, 130
Peptide hydrolysis, 23
Peptide nitrogen, copper binding, 241
Peptide nitrogen, copper complex, 236
Peroxidase, 661, 799, 988
Peroxidase, in benzidine blood test, 1005
Peroxidase, cytochrome-c, 646
Peroxidase, Horseradish, 990
Peroxidase, Horseradish, intermediates, 992
Peroxidase, Horseradish, properties, 990
Peroxidase, models, 684
Peroxidase, model systems, 660
Peroxidase, oxidation of amines, 993

Peroxidase, oxidation of phenols, 1009
Peroxidase, oxidations involving inorganic materials, 1014
Peroxidase, transalkylation by, 1004
Peroxidases, transiodination by, 1001
Peroxidase action, mechanism, 1012
Peroxides, organo, 668
Perturbation, 616
pH dependence of metal binding, 230
Phenolate iron complexes, 183
Phenolate iron hydroxamates, 169
Pheophytin a, 1023, 1030, 1037, 1049
Phosphatase, 557
Phosphatase, alkaline, 558, 560
Phosphatase, alkaline, mechanism of catalysis, 562
Phosphatase, alkaline, model of active site, 566
Phosphatase, alkaline, phosphate binding, 563
Phosphatase, alkaline, spectra, 564
Phosphatase, alkaline, strength of metal binding, 565
Phosphate, methyl, hydrolysis by cobalt complexes, 575
Phosphate binding in nucleotides, 1199
Phosphate transfer, activation by metal ions, 549
Phosphate transfer, mechanisms, 553
Phosphodiesterases, 557
Phosphoenolpyruvate carboxykinase, 599
Phosphoenolpyruvate carboxylating enzymes, 407
Phosphofructokinase, 602
Phosphoglucomutase, 413, 414, 558, 559
Phosphoglycerate kinase, 593, 602
Phospholipases, 422
Phosphomannose isomerase, 412
Phosphomonesterases, 557
Phosphorolysis, 552
Phosphorylation, 552
Phosphorylation, nonenzymatic model systems, 568
Phosphoryl transfer, with kinases, 582
Phosphotransferases, 557
Photo-activity, chlorophyll, 1058
Photochemical reactions of complexes, 377
Photosynthesis, 1022, 1185
Photosynthetic reaction center, 1058
Photosynthetic unit, model, 1062
Phthalocyanine, 1071
Phyllochlorin, 1025
Plastocyanin, 690, 693, 694, 704
Platinum, in reduction of alkynes, 765
Platinum carboxypeptidase A, 442
Platinum complexes, anti-tumor activity, XI
Platinum complex, reduction of diazonium salt by, 775
Platinum glycine, 126
Platinum histidine complexes, 149
Platinum methionine, 156, 157
Platinum peptide complexes, 134, 142
Polarizability, 71
Polynucleotides, binding sites of metals on, 1218
Polynucleotides, effect of metals on conformation of, 1230
Polynucleotides, stability of metal complexes, 1217
Polynucleotide complexes, 1210
Polynucleotide phosphorylase, 554
Polyribonucleotides, depolymerization by metal ions, 1232
Polyribonucleotides, destabilization of ordered structure by metal ions, 1231
Porphyrins, 85, 96
Porphyrins, absorption, 809, 810, 811, 813
Porphyrins, basicities absorption, 806
Porphyrin, displacement of iron from plane of, 818, 845, 861
Porphyrins, iron, 797
Porphyrins, iron, absorption, 816
Porphyrins, iron, Mössbauer spectra, 808
Porphyrins, iron, n.m.r. chemical shifts, 819
Porphyrin, iron(III) dimers, 111
Porphyrins, iron zero-field splitting, 809
Porphyrins, manganese, 890
Porphyrin, meso-, dimethyl ester, Zn(II) complex, 97
Porphyrin, metal complexes, stability, 97
Porphyrins, nickel, absorption, 812
Porphyrins, nickel, ΔH and ΔS, 807
Porphyrin, planarity of, 846
Porphyrins, synthetic, 804
Porphyrin c, 904, 905
Porphyrin IX, deutero-, dimethyl ester, nickel complex, 98
Porphyrin structure, effects on properties, 805
Potassium monamycins, 214
Potassium nonactin, 207
Potassium transport, 204
Potassium valinomycin, 210
Potential energy surfaces, 612
Precursor complex, 620, 622

1,2-Propanediol dehydrase, 1105
Prosthetic groups, metal interactions with, 1136
Protease, B. subtilis, 482
Protein complexes, factors affecting metal ion-ligand affinity, 229
Protein complexes of metals, 227
Protein functional groups, stability of copper and zinc complexes, 231
Protein side chains as donors, 229
Protochlorophyll, 1022
Protochlorophyllide, 1023
Protohematin IX, 991
Protohemin, 799
Protohemin IX, 990
Protoporphyrin IX, 799, 934, 1015
Pteridines, 663, 665
Pteridines, reductants in oxygenation, 664
Pulcherriminic acid, 197, 198
Purine, 1191
Putidaredoxin, 712, 720, 722, 724
Pyridine nucleotide-dependent oxidative decarboxylases, 399
Pyridoxal, 1137, see also Vitamin B_6
Pyridoxal, reactions catalyzed by, 1148
Pyridoxal phosphate, 1110, 1137
Pyridoxamine, 1137
Pyridoxamine phosphate, 1137
Pyridoxine decarboxylase, 410
Pyridoxol, 1137
Pyridoxylidene, 1138
Pyrimidine, 1191
Pyrimine iron complex, 199
Pyrocatechase, 657
Pyrocatechol, oxidation, 678
Pyrochlorophyll, 1023
Pyrochlorophyll a, 1023
Pyrophosphate, 68
Pyruvate carboxylase, 402, 427, 593
Pyruvate carboxylase, cobalt complex, 385
Pyruvate decarboxylase, 410
Pyruvate kinase, 383, 407, 576, 583, 590, 591, 593, 597
Pyruvate kinase, cobalt complex, 385

Radii (Å), crystal, 27
Radii (Å), hydrated, 27
Rate constants, for electron exchange reactions, 627, 629
Rate constants, redox, 635
Redox properties of ligands, 368
Redox reactions, 611
Relaxation methods, 592
Replication of DNA, metal ions and, 1212
Replication of genetic code, 1211
Repressor gene, 167
Resonance, 615
Retro-rhodotorulic acids, 189
Rhodotorulic acid, 176, 178, 189, 190
Riboflavin, 1170
Riboflavin, silver, 1169
Ribonuclease, effect of metal ions on, 1236
Ribonuclease, metal ion interactions, 243
Ribonucleic acid, see RNA
Ribonucleosides, reaction with copper, 1207
Ribonucleotide reductase, 1101, 1110, 1122
Ribose binding in nucleosides, 1203
Ribosomes, 1211
Ribosomes, effect of magnesium on, 1238
Ribotide reductase, 181
Ribulose-1,5-diphosphate carboxylase, 409
RNA, 1210
RNA, depolymerization by metal ions, 1232
RNA, effect of metal ions on, 1228
RNA polymerase, 1214
RNA polymerase, zinc in, 1239
Rubredoxin, 712, 715
Rubredoxin, X-ray crystallography, 718
Ruthenium, redox rates, 627, 628
Ruthenium dinitrogen complex, 760, 777, 781, 784
Ruthenium nitrile complex, 763
Ruthenium N_2O complex, 760, 761

Sake colorant, 187
Sake colorant A, 188
Salicylaldimine complexes, 1140
Schiff base complexes, 1139
Schizokinen, 172, 186, 192
Selenium, substitution in ferredoxin, 720, 726
Serine complexes, 152
Serine zinc complex, 565
Serum albumin, bovine, 95
Serum albumin, copper complex, 242
Serum albumin, reaction with zinc ions, 232
Serum transferrin, 281
Seven coordination, 14
Sideramines, 168
Siderochrome dependent auxotrophic micro-organisms, 180

Siderochromes, 167, 168
Siderochromes, stability constants of iron complexes, 176
Sideromycins, 168
Siderophilin, 280, see also Transferrin
Silver, 154
Silver, in carboxypeptidase A, 442
Silver, in DNA separations, 1227
Silver, inhibition of RNA degradation by, 1235
Silver actins and polyethers, 216
Silver carboxypeptidase A, 442
Silver cysteine, 154
Silver cytidine, 1205
Silver DNA, 1219, 1226
Silver flavins, 1168, 1175
Silver glycine, 125, 128
Silver guanosine, 1204
Silver methionine, 157
Silver monensin complex, 216
Silver nigericin, 216
Silver norleucine, 125
Silver nucleotides, 1207
Silver riboflavin, 1169
Silver uridine, 1206
Six coordination, 13
Six-membered rings, 59
Sodium antamanide, 213
Sodium monensin complex, 216
Sodium nonactin, 207
"Soft" base-"soft" acid, 34
Soft metal ions, 69
Spin-label, 389
Square planar complexes, 49
Square planar geometry, 11
Square pyramid, 13
Stability, order of, 73
Stability constant, 63
Stellacyanin, 690, 691, 694, 695, 704
Stereochemistry, ligand induced, 36
Stereoisomerism, 40
Stereoselectivity, 89, 148, 366
Structural isomers, 46
Succinate complexes, 81
Succinate dehydrogenase, 1187
Succinimycin, 197
Sulfanilamide, 518, 521
Sulfate reductase, 1187
Sulfhydryl donors, 153
Sulfite reductase, 1187
Sulfonamide, binding to carbonic anhydrase, 512
Sulfur, in iron-sulfur proteins, 710
Synthetase, 422, 423
Tartrate dehydrase, 420
Taste acuity, copper dependence, 306
Taste acuity, metal and, X
Testodoxin, 712, 720
Tetragonal distortion, 14
Tetrahedral complexes, 48
Tetrahedral geometry, 9
Tetrahydrofolate synthetase, 383
Tetraphenylporphin, 805
Thallium actins and polyethers, 216
Thallium pyruvate kinase, 599
Thermolysin, 482
Thiamine pyrophosphate-dependent enzymes, 410
Thioglycollate complexes, 155
Thorium, in phosphate ester hydrolysis, 570
Three coordination, 8
Thymine, 1191
Titanium, in dinitrogen reduction, 778
Titanium, stereochemistry, 17
Titanium flavins, 1168, 1175
Transamination, 364, 1138, 1151, 1155
Transcarboxylation, 404
Transcription, of genetic code, 1211
Transcription, metal ions and, 1214
Trans effect, 846
Transferases, 415
Transferrins, 280
Transferrins, amino acid composition, 283
Transferrin, binding constants of iron to, 289
Transferrin, biologic functions, 300
Transferrins, carbohydrate content, 284
Transferrin, chemical modification, 297
Transferrins, hydrodynamic properties, 286
Transferrins, immunochemistry, 285
Transferrin, as an iron buffer, 302
Transferrin, iron uptake, 306
Transferrin, kinetics of iron binding, 292
Transferrin, magnetic resonance spectroscopy, 295
Transferrin, metal-binding sites, 299
Transferrins, molecular weight, 285
Transferrins, optical properties, 286
Transferrin, optical rotatory dispersion and circular dichroism, 294
Transferrin, optical studies, 293
Transferrin, preparation, 281
Transferrin, role of bicarbonate, 291
Transferrin, serum, 281
Transferrin, transfer of iron to ferritin, 274

Transferrin chromium complex, 286, 290, 297, 301
Transferrin copper complex, 297, 301
Transferrin gallium complex, 290, 297
Transferrin metal complexes, 290
Transferrin-reticulocyte interaction, 300
Transfer RNA, 1211
Transfer RNA, aminoacylation, 424
Transfer RNA, stabilization by metal ions, 1229
Trans isomers, 41
Transition metal ions, 29
Translation, of genetic code, 1212
Translation, metal ions and, 1216
Trigonal bipyramid, 13
Trigonal distortion, 14
Trinactin, 205
Tryptophanpyrrolase, 657
Tubercidin copper complex, 1201
Tungsten, stereochemistry, 25
Tunneling, 617
Two coordination, 8
Tyrocidin, 211
Tyrosinase, 681, 690, 700
Tyrosine, binding site in transferrin, 293
Tyrosine, in carboxypeptidase A, active site, 465
Tyrosine, metal binding in transferrin, 299
Tyrosyl group, copper interaction, 153

Uracil, 1191
Uracil mercury complex, 1196
Uricase, 680, 690
Uridine copper complex, 1205
Uridine diphosphate glucose, 554
Uridine mercury complex, 1205
Uridine silver complex, 1206
Uroporphyrin, 1068
L-Valine-N-monoacetato copper(II), 90
Valinomycin, 203, 204, 205, 209, 214, 218, 222
Valinomycin potassium complex, 210
Vanadium, in phosphate ester hydrolysis, 570
Vanadium, stereochemistry, 18
Vanadium, substitution in nitrogenase, 750
Vanadium nitrogenase, 752, 757, 763, 768, 773
Vanadyl deoxophylloerythroetioporphyrin, 1025
Vanadyl porphyrins, 802
Vaska's compound, 340
Vasopressin copper complex, 236
Verdoperoxidase, 990
Vicinal effects, 50
Vitamin B_6 complexes, 1137
Vitamin B_6 complexes, decarboxylation, 1159
Vitamin B_6 complexes, degradation reactions, 1160
Vitamin B_6 complexes, β-elimination, 1157
Vitamin B_6 complexes, γ-elimination, 1159
Vitamin B_6 complexes, mechanism of formation, 1144
Vitamin B_6 complexes, oxidative deamination, 1162
Vitamin B_6 complexes, β-proton exchange, 1158
Vitamin B_6 complexes, racemization, 1155
Vitamin B_6 complex, reactions in model systems, 1147
Vitamin B_6 complexes, reactions resulting from labilization of α-hydrogen, 1150
Vitamin B_6 complexes, tryptophan synthesis, 1157
Vitamin B_6 complexes, types and structures, 1140
Vitamin B_{12}, 22, 1067, see also Corrinoids
Vitamin B_{12a}, 1069
Vitamin B_{12r}, 1069
Vitamin B_{12s}, 1069, 1070, 1090

Wilson's disease, 306, 313

X-206, 217, 223
X-537 A, 216, 217, 218
Xanthine dehydrogenase, 713
Xanthine oxidase, 713, 1187
Xenon, binding to myoglobin, 857
Xylose isomerase, 412

Zinc, as activator of isocitrate dehydrogenase, 400
Zinc, in alcohol dehydrogenase, 395
Zinc, in alkaline phosphatase, 560
Zinc, binding site in carbonic anhydrase, 535
Zinc, in carbonic anhydrase, 488, 503
Zinc, in carboxypeptidase A, 94, 159, 438, 441, 442, 471
Zinc, coordination preferences, 232
Zinc, in DNA polymerase, 1213

Zinc, effect on myoglobin ligand binding, 857
Zinc, effect on ribosomes, 1239
Zinc biuret, 132
Zinc carbonic anhydrase, 531
Zinc chlorophyll, 1025
Zinc chlorophyll a, 1037
Zinc complexes, bond-angles, 161
Zinc complexes, of protein functional groups, stability, 231
Zinc complexes, structural parameters, 160
Zinc cysteine, 154
Zinc flavoquinone, 1173
Zinc, in hemoglobin, 843
Zinc, in kinases, 582
Zinc, in nuclei, 1213
Zinc, in phosphoglucomutase, 414
Zinc, in phosphomannose isomerase, 412
Zinc, and pyruvate kinase, 598
Zinc, reaction with poly (rA), 1230
Zinc, in RNA degradation, 1232
Zinc, in RNA polymerase, 1239
Zinc, in thermolysin, 482
Zinc, in transamination, 1155
Zinc, unwinding DNA reversibly, 1223
Zinc aldolase, 416
Zinc alkaline phosphatase, 561, 577
Zinc amino acid complex, 91
Zinc glutamate, 129
Zinc glycine, 124
Zinc glycylglycine, 124
Zinc glycylglycylglycine, 129
Zinc histidine, 147, 448, 532
Zinc imidazole, 144, 232, 235, 238
Zinc insulin complex, 233
Zinc ions, reaction with serum albumin, 232
Zinc leucine aminopeptidase, 481
Zinc ligands, in carboxypeptidase A, 448, 450
Zinc ligand bond angles, 450
Zinc ligand bond lengths, 161
Zinc mesoporphyrin dimethyl ester, 85
Zinc metallo-enzymes, 394
Zinc myoglobin, 238, 239
Zinc nucleotides, 1197
Zinc peptide bond, 470
Zinc porphyrin, 802, 843
Zinc riboflavin, 1175
Zinc Schiff base complex, 1140, 1142
Zinc serine, 565
Zinc serine complexes, 152
Zinc stereochemistry, 25
Zinc transferrin, 290, 301